Switching, Protection and Distribution
in Low-Voltage Networks

Coordination:

Georg Haberl

Editing:

Georg Schöllhorn

Translation:

Jan Dommisse

Authors:

Petra Belzner
Erich Boehl
Jan Dommisse
Otmar Fleig
Gudrun Frank
Richard Freitag
Dieter Fuchs
Siegbert Gern
Jorge Guillen
Egbert Handwerker
Manfred Hartz
Johann Hauer
Werner Hörmann
Hans Joseph Huesmann
Holger Jansen
Günther Kachelrieß
Hermann Klärner
André Kling
Anton Kohling
Peter Kreppel
Josef Lobnig
Andreas Meindl
Lothar Mickel
Harald Middendorf
Reinhard Müller
Roland Nagel
Manfred Schäfer
Reinhold Schroll
Reinhard Steger
Ludmilla Stupp
Werner Sturm
Hiltrud Theiss
Friedrich Witibschlager
Johann Wolf
Gert Zauscher

Switching, Protection and Distribution in Low-Voltage Networks

Handbook
with selection criteria
and planning guidelines
for switchgear,
switchboards and
distribution systems

2nd revised edition, 1994

Publicis MCD Verlag

Die Deutsche Bibliothek – CIP-Einheitsaufnahme

Switching, protection and distribution in low-voltage networks:
handbook with selection criteria and planning guidlines for switchgear, switchboards and distribution systems / [ed.: Siemens-Aktiengesellschaft, Berlin and Munich. Authors: Petra Belzner ... Ed. Georg Schöllhorn. Transl.: Jan Dommisse]. – 2., rev. ed. – München : Publicis-MCD-Verl., 1994
    Einheitssacht.: Schalten, Schützen, Verteilen in
    Niederspannungsnetzen ⟨engl.⟩
    Früher u.d.T.: Schmelcher, Theodor: Low voltage handbook
    ISBN 3-89578-000-6
NE: Belzner, Petra; Schöllhorn, Georg [Hrsg.]; Siemens-
    Aktiengesellschaft ⟨Berlin; München⟩; EST

Registered trademarks or tradenames are not indicated as such in this handbook. This, however, does not imply that no proprietary rights are attached to any such names or words.

The following product and range names used in this book are registered trade names of Siemens AG:

ASIST, AUSTER, BERO, DIAZED, DULUX, ELLE, GEAFOL, KOALA, KUBS, N SYSTEM, NEOZED, OPTO-BERO, OPTOMAT, PROTODUR, SICONTROL, SIENOPYR, SIKUFEST, SIFLA, SIGUT, SIKOSTART, SIKUS, SILIZED, SIMATIC, SIMICONT, SIMOCODE, SIPRO, SITAS, SIVACON, SIVOLT, SONAR-BERO, SONPROG, STAB-VERTEILUNG, VIDEOMAT, ZSS.

Published jointly by
Publicis MCD Werbeagentur GmbH, Munich (Germany)
VCH Publishers, Inc., New York, NY (USA)

Library of Congress Card No. applied for

ISBN 3-89578-000-6

2nd rev. and enl. ed. – 1994

Editor: Siemens Aktiengesellschaft, Berlin and Munich
Publisher: Publicis MCD Verlag, Erlangen

© 1984 by Siemens Aktiengesellschaft, Berlin and Munich
© 1994 by Publicis MCD Werbeagentur GmbH, Munich
This publication and all parts thereof are protected by copyright. All rigts reserved. Any use of it outside the strict provisions of the copyrigt law without the consent of the publisher is forbidden and will incur penalties. This applies particularly to reproduction, translation, microfilming or other processing, and to storage or processing in electronic systems. It also applies to the use of individual illustrations and extracts from the text.

Printed in Germany

Preface

This handbook is published to assist the user of low-voltage switchgear, switchboards and distribution systems in the planning of installations as well as in the specification, selection, operation and maintenance of equipment. It not only deals with the broader fundamental and theoretical principles, but also provides quick and precise answers to specialized questions in this field of electrical engineering. The particular problems related to project planning receive the same attention as those arising during construction, installation, operation and maintenance.

A host of selection guidelines as well as planning and circuit diagram examples help the reader to achieve both technically and economically sound solutions for his specific application. The book provides the experienced reader with several new ideas for future projects, but also describes possibilities to optimize existing installations and systems.

To enhance the practical value of this handbook, reference is made throughout the text to the relevant Siemens products and catalogues as well as to pertinent national and international specifications, regulations, standards and IEC publications. Where considered useful, extracts are quoted. However, since specifications and standards are constantly being revised, updated and extended, these extracts should be cross-checked with the originals before they are applied.

Particular attention is given to factors of ever increasing importance such as the interaction between electromechanical and electronic devices, economical installation and mounting methods, simple operation and sensible maintenance.

Should there be any aspects which are not covered or questions which remain unanswered by this book, or which arise from further technical developments in the field of low-voltage switchgear, our technical specialists located in branches and offices around the world are at your disposal and will be pleased to assist.

We continue to welcome any suggestions which may enhance or improve future additions of this handbook. Please address your comments and suggestions to:

Siemens AG
Department ASI 2 V 28
P.O. Box 3240
D-91050 Erlangen
Federal Republic of Germany

Erlangen, September 1994

Siemens Aktiengesellschaft

Translator's Note:

The English edition has been updated and extended to include additional information regarding a number national specifications, standards and locally used methods or techniques. Owing to differences in the English language and its usage in various countries, and particularly to the use of different technical terms to describe similar devices, quantities, processes, etc. Section 9.11 has been extended to include a number of synonymous and regionally used terms. The English edition, therefor, does not correspond exactly with the original German publication.

Contents

1	**Specifications for low-voltage devices and switchgear assemblies**	15
1.1	**Nomenclature of the standards authorities**	15
1.2	**Low-voltage switchgear and devices**	17
1.2.1	Summary of specifications and approvals	17
1.2.2	Regulations and approvals in European countries	17
1.2.3	Standards and approvals in the U.S.A., Canada and Australia	22
1.3	**Type-tested and partially type-tested switchgear assemblies (TTA and PTTA)**	26
1.3.1	Summary of specifications and standards	26
1.3.1.1	Construction specifications	26
1.3.1.2	Generic specifications	31
1.3.1.3	Equipping specifications	31
1.3.1.4	Erection specifications	32
1.3.1.5	Relevant DIN standards for switchgear assemblies	33
1.4	**Protection of persons and material assets**	33
1.4.1	Protection against direct contact	33
1.4.1.1	Protection against electrical shock to DIN VDE 0100	33
1.4.1.2	Protection against electrical shock, positioning of operating elements in the vicinity of parts which are dangerous to touch	34
1.4.1.3	Measures to be taken in the modification or extension of switchgear assemblies already in service	38
1.4.1.4	"Safety separation" with low-voltage switchgear	38
1.4.2	Protection against indirect contact	40
1.4.2.1	Comments related to the protective measures specific to the network configuration	42
1.4.3	Protection by total insulation	53
1.4.4	Erection and connection of switch and distribution boards	54
1.4.5	Clearances and creepage distances	56
1.5	**Equipping specifications and relevant standards**	57
1.5.1	Main switches	58
1.5.2	Emergency-off apparatus	58
1.5.3	Safety disconnecting switches for mechanical maintenance (service/repair switches)	60
1.5.4	The electrical equipping of industrial machinery scheduled for export to countries outside Germany	60
1.5.5	Colours for pushbuttons, illuminated pushbuttons and indicator lights	62
1.5.6	IP degrees of protection against contact with live parts, ingress of foreign bodies and ingress of liquid	63
1.5.6.1	IP degrees of protection to DIN and IEC	63
1.5.6.2	Degrees of protection according to other national specifications	64
1.6	**Operating and ambient conditions**	65
1.6.1	Normal conditions	65
1.6.1.1	Ambient temperature	65
1.6.1.2	Altitude	65
1.6.1.3	Environmental influences	66
1.6.1.4	Parameters and pollution levels of the climate classes to IEC 721 and the Siemens standard SN 29070, part 1	68
1.6.1.5	Protection against corrosion, type of finish for technical products	68
1.6.1.6	Decontaminability	69
1.6.1.7	Climatic operating conditions for electronic equipment	70
1.6.2	Abnormal conditions for transport, storage and operation	72
1.6.2.1	Arduous operating conditions	72
1.6.2.2	Influence of α, β and γ radiation	72
1.6.2.3	Stressing due to vibrations and impact	73
1.6.2.4	Measures against the effects of induced vibration	73

1.6.2.5	Resistance to termites	74
1.6.2.6	Environmental acceptability	74
1.6.2.7	Suppression of radio interference	74

2 Network data and duty types . . . 76

2.1	**Network data**	**76**
2.1.1	Nominal voltages and frequencies	76
2.1.2	Short-circuit current	79
2.1.3	Types of short-circuits	81
2.1.3.1	Contribution to the short-circuit current by connected motors	83
2.1.4	Effects of short-circuit currents	84
2.1.5	Diagrams for determining resistance values and short-circuit currents, computer program "KUBS" for short-circuit calculations and product selection	85
2.1.6	Influence of the transformers and conductors on the short-circuit current	96
2.2	**Duty types**	**98**
2.2.1	Continuous operation duty	98
2.2.2	Short-time operation duty	98
2.2.3	Intermittent periodic duty	99
2.2.4	Operation at inconsistent loading	102

3 Selection criteria for low-voltage switchgear in main circuits 104

3.1	**Network and operating conditions**	**104**
3.1.1	Rated voltage and rated frequency	104
3.1.2	Rated short-circuit strength and rated switching capacity	104
3.1.3	Rated currents	106
3.2	**Switching tasks and conditions**	**106**
3.2.1	Switching tasks	106
3.2.1.1	Disconnection	106
3.2.1.2	Off-load switching	107
3.2.1.3	On-load switching	107
3.2.1.4	Motor switching	107
3.2.1.5	Power switching, switching of short-circuits	107
3.2.2	Switching of components in an electrical installation	110
3.2.2.1	Starting of low-voltage motors	111
3.2.2.2	Switching of high-voltage motors	114
3.2.2.3	Switching of capacitors	114
3.2.2.4	Switching of electric heating equipment	115
3.2.2.5	Switching of lamps in lighting installations	115
3.2.2.6	Switching of low-voltage transformers	117
3.3	**Switching frequency and service life**	**117**
3.3.1	Permissible switching frequency	117
3.3.2	Mechanical service life	118
3.3.3	Electrical service life	118
3.3.4	Selection in terms of utilization categories	118
3.3.4.1	Selection of contactors	118
3.3.4.2	Selection of load switches, disconnectors, on-load disconnectors and combination fuse switch units	121
3.4	**Protection against overcurrent and excessive temperature rise**	**122**
3.4.1	General functions	122
3.4.1.1	Protection against overload	122
3.4.1.2	Protection against the efects of short-circuits	123
3.4.1.3	Protection against excessive temperature rise	123
3.4.2	Specifications	123
3.4.2.1	Specifications for overload protection devices	123
3.4.2.2	Service factors in terms of UL and CSA specifications	124
3.4.2.3	Specifications for overcurrent protection devices	125
3.4.2.4	Specifications for temperature dependent protection devices	125
3.4.3	Protection devices	125
3.4.3.1	Fuses	125
3.4.3.2	Circuit-breakers	128
3.4.3.3	Line-protection (miniature) circuit-breakers	130
3.4.3.4	Overload relays	131
3.4.3.5	Thermistor motor protection devices	138
3.4.3.6	Instantaneous electromagnetic over-current relays	139
3.4.4	Switchgear combinations	139
3.4.4.1	Swichgear combinations with fuses	139
3.4.4.2	Switchgear combinations without fuses	141
3.4.4.3	Switchgear combinations with thermistor motor protection devices	143
3.4.4.4	Protection properties of switchgear combinations	144
3.4.4.5	Comparison between the protection properties of switchgear combinations	146

3.4.4.6	Selection of circuit-breakers for power distribution systems with or without fuses	148
3.4.5	Protection of plant components . . .	150
3.4.5.1	Protection of three-phase induction motors	150
3.4.5.2	Protection of conductors and cables outside the switchgear combinations .	159
3.4.5.3	Protection of transformers	160
3.4.5.4	Protection of capacitors	163
3.4.6	Discriminative protection (selectivity) .	163
3.4.6.1	Discrimination in radial networks . .	164
3.4.6.2	Use of discrimination tables	172
3.4.6.3	Discrimination in meshed networks .	174
3.5	**Protection against overvoltage**	175
3.5.1	Overvoltage transients associated with vacuum switchgear	175
3.5.2	High voltage vacuum contactors, type 3TL6, for the switching of three-phase inductive motors with slip-rings or short-circuited rotors over 1 kV up to 12 kV	177
3.5.3	Vacuum contactors, type 3TF6, for the switching of three-phase induction motors with slip-ring or short-circuited rotors up to 1000 V	182
3.6	**Leakage-current and earth-fault protection**	183
3.6.1	Construction and method of operation	183
3.6.2	Leakage-current circuit-breakers for a.c. and pulsing d.c. leakage currents	184
3.6.3	Discriminative leakage-current circuit-breakers	184
3.6.4	Product range of Siemens leakage-current circuit-breakers and tripping devices	184
3.7	**Application of low-voltage switchgear in main circuits**	185
3.7.1	Parallel and series connection of current path assemblies	185
3.7.2	Application of four-pole switchgear .	186
3.7.3	Influence of network frequency and harmonic currents on the operation of switching devices	187
3.7.3.1	Thermal load carrying capacity of the current path assemblies and conductors in dependence of the network frequency	187
3.7.3.2	Switching capacity at network frequencies other than 50 Hz	188
3.7.3.3	Contact service life	189
3.7.3.4	Tripping response of releases and relays	189
3.7.3.5	The effect of harmonic currents on the tripping response of overload releases and relays	191
3.7.3.6	Electrical operating mechanisms for switchgear	191
3.7.4	Use of a.c. switchgear in d.c. networks	192
3.7.4.1	Load carrying capacity of the current path assemblies	192
3.7.4.2	Contact service life	193
3.7.4.3	Direct-current switching capacity . .	193
3.7.5	Use of a.c. contactors in networks with square-wave voltages	194
3.7.6	Switchgear for the switching of three-phase capacitors	195
3.7.6.1	Switching of capacitors with circuit-breakers	195
3.7.6.2	Switching of capacitors with contactors	195
3.7.6.3	Switching on of single capacitors . .	195
3.7.6.4	Switching of capacitor banks	195
3.7.6.5	Switching of capacitors with lower power ratings	196
3.7.7	Selection of 3TF and 3TB contactors in terms of contact service life and utilization category	197
3.7.8	Selection of 3TF contactors for short-time and intermittent periodic duty	201
3.7.9	Selection of contactors for three-phase pole-changing induction motors	204
3.7.10	Selection of 3TF and 3TK contactors as well as circuit-breakers for the switching of lamps	206
3.7.11	Switching of three-phase transformers up to 1000 V with 3TF contactors	210
3.7.12	Starting of three-phase induction motors with slip-ring rotors	210
3.7.12.1	3PA3 Oil-cooled starters	211
3.7.12.2	Starter and start-control switches . .	219
3.7.13	Starting of three-phase induction motors with stator-resistance starters	219
3.7.14	Direct-on-line starting of three-phase induction motors with type 3TW motor starters	223

3.7.15	A.C. semiconductor motor controllers and starters, SIKOSTART 3RW22 . 223		5	**Installation, operation and maintenance of low-voltage switchgear** . 273	
			5.1	**Installation** 273	
			5.1.1	Mounting aids 273	
			5.1.2	Mounting position 274	
4	**Selection criteria for low-voltage switchgear in auxiliary circuits** . . 233		5.1.3	Clearance for switching-arc gases . . 275	
			5.2	**Termination** 275	
			5.2.1	SIGUT termination technique 275	
4.1	**Operating voltages in auxiliary circuits** 233		5.2.2	Tab connectors 276	
			5.2.3	Box terminals 277	
4.1.1	Contact reliability in the case of low-voltages 233		5.3	**Operation** 277	
			5.3.1	Manual operation 277	
4.1.2	Voltage instability in auxiliary circuits 233		5.3.2	Powered operation 279	
4.2	**Operating conditions** 234				
4.2.1	Utilization categories to DIN VDE and IEC 234		5.4	**Measures to facilitate the checking/ replacement of parts and maintenance work** 279	
4.2.2	Special considerations for the selection and the use of low-voltage switchgear in Canada and the USA . 234		5.5	**Checking the condition of contact pieces in 3TF contactors; assessment criteria** 280	
4.2.3	Short-circuit protection in auxiliary circuits 236				
4.2.4	Short-circuit and overload protection of control transformers 237		**6**	**Transducing sensors and signal processing systems** 282	
4.2.5	Thermistor protection in control transformers 237		6.1	**Selection criteria for BERO proximity switches** 283	
4.3	**Operating conditions for low-voltage switchgear in auxiliary circuits** 237		6.1.1	Inductive and capacitive proximity switches for operating distances up to 65 mm 283	
4.3.1	Prevention of operational down-time in contactor control systems 237		6.1.2	Opto-BERO photoelectric proximity switches 286	
4.3.2	Long control conductors - problems and solutions 242		6.1.3	Sonar-BERO ultrasonic proximity switches 287	
4.3.3	Limiting of overvoltage spikes caused by the switching off of contactors (overvoltage suppression) 248		6.2	**Electronically compatible control and signalling by low-voltage switchgear** . 291	
4.3.3.1	Causes of overvoltages 248		6.2.1	Reliable operation by electronic output stages 291	
4.3.3.2	Overvoltage suppression with RC elements 249		6.2.2	Adaptation of the operating voltage tolerances 292	
4.3.3.3	Overvoltage suppression with diodes 253		6.2.3	Overvoltage suppression 292	
4.3.3.4	Overvoltage suppression with varistors 253		6.2.4	Contact reliability 293	
4.3.4	The use of contactor relays in safety circuits 254		6.3	**Assessment criteria for electromechanical and electronic controls** . . 297	
4.3.5	Selection criteria for low-voltage control transformers 257		**7**	**Type-tested switchgear assemblies (TTA)** 299	
4.3.5.1	Operating conditions 257				
4.3.5.2	Duty types 259		7.1	**General** 299	
4.3.5.3	Transformer types 260		7.1.1	Versions and designs 299	
4.3.6	Application and selection of position switches, type 3SE 266		7.1.2	Types of construction 302	
4.3.6.1	Position switches with safety function 271		7.1.3	Selection criteria 306	

7.2	Switchboards in standardized design	308
7.2.1	Introduction	308
7.2.2	Standard switchboards Type 8PU	313
7.2.3	Transformer load centre (S) substations for up to 24 kV and 1250 kVA	320
7.3	Distribution board systems	328
7.3.1	8HS sheet-steel enclosed distribution boards	328
7.3.2	8HU sheet-steel enclosures	329
7.3.3	8HP insulated distribution board system	329
7.3.4	8PL insulated busbar trunking system (L-system)	330
7.3.5	8L. mounting and wiring systems for control circuits	331
7.3.6	8MF cubicle system for switch, distribution and control boards	332
7.4	Guidelines for project planning of low-voltage switch, distribution and control boards or systems	333
7.4.1	General	333
7.4.2	8PU low-voltage switchboards	338
7.4.3	8HS, 8HP and 8HU distribution systems	340
7.4.4	8L. mounting systems for control circuits	340
7.4.5	Domestic and utility distribution boards	340
7.4.5.1	8GB Small distribution boards	343
7.4.5.2	8GD and 8GA STAB wall-mounting/ SIKUS floor-standing distribution board systems	344
7.4.5.3	8GA SIKUS free-standing distribution board cubicles	346
7.4.5.4	SIPRO universal system meter cabinets, meter/distribution cubicles, distribution cabinets, free-standing distribution boards	348
7.4.5.5	8MB, 8MM and 8GR cable, distribution and meter cubicles for outdoor use	350
7.4.6	Air-conditioning of installations, switchboards and cubicles	351
7.4.6.1	General	351
7.4.6.2	8ME78 heat exchangers	354
7.4.6.3	8MR11 filtered fans	355
7.4.6.4	8MR17 refrigeration units	356
7.4.6.5	8MR21 heating units	357
7.4.6.6	Temperature rise inside insulation material and sheet steel enclosures	358
7.4.7	Degrees of protection, climatic and other ambient conditions	360
7.4.8	Power factor correction in networks with or without harmonics	363
7.4.8.1	Basic principles	363
7.4.8.2	Types of power factor correction (reactive power compensation)	365
7.4.8.3	Power factor correction of three-phase induction motors and transformers	367
7.4.8.4	Project planning	369
7.4.8.5	Voltage rise caused by capacitors	372
7.4.8.6	Compensation in networks with harmonics	373
7.4.8.7	The use of audio-frequency remote control systems	377
7.4.8.8	Range of products for power factor correction	380
7.5	Charging units for stationary standby battery installations	382
7.6	Current transformers	385
7.6.1	Basic designs	385
7.6.2	Current transformers for specific applications	386
7.6.2.1	Interposing current transformers	386
7.6.2.2	Summation current transformers	386
7.6.2.3	Thread-through (or pin-wound) current transformers	388
7.6.2.4	Cable or busbar current transformers (split-core c.t.'s)	388
7.6.2.5	Cast resin current transformers	388
7.6.2.6	Explosion-protected current transformers and current transformers for mining applications	388
7.6.2.7	Current transformers for protection purposes	389
7.6.2.8	Current transformers for power factor correction controllers	389
7.6.3	Accuracy classes of current transformers	390
7.6.4	Secondary currents of current transformers	391
7.6.5	Rated output power and overcurrent factor of current transformers	391
7.6.6	Voltages across the secondary terminals of a current transformer	392
7.6.7	Selection criteria for current transformers	394
7.6.8	Power consumption and losses in current transformer secondary circuits	394

8 Fundamental circuit diagrams . . 397

8.1 General information 397
8.1.1 Terminal designations 397
8.1.2 Graphic symbols according to DIN, ANSI, BS and IEC 400
8.1.3 Designation of equipment, conductors and general functions 408
8.1.4 Circuit diagrams 410
8.1.4.1 Types of circuit diagrams 410
8.1.4.2 Making use of the fundamental circuit diagrams or modifying them 412
8.1.5 Switching via contactors 412
8.1.5.1 Contactors with drop-out delay unit for fluttering command signals . . . 412
8.1.5.2 Extended (early-make / late-break) auxiliary contacts in contactors (mainly for d.c. excitation of the coils) 414
8.1.5.3 Drop-out delay units for contactors . 414
8.1.5.4 Contactor safety combinations . . . 414

8.2 Direct switching of three-phase induction motors 415
8.2.1 Switching on and off of three-phase induction motors 415
8.2.2 Switch-over of a three-phase induction motor from one supply network to another 416
8.2.3 Automatic sequential starting of three-phase induction motors 418
8.2.4 Reversing the direction of rotation of three-phase induction motors (reversing starters) 420
8.2.5 Switching of pole-changing three-phase induction motors 422
8.2.5.1 Pole-changing three-phase induction motor with one winding (Dahlander connection), two speeds, one direction of rotation 422
8.2.5.2 Pole-changing three-phase induction motor with one winding (Dahlander connection), two speeds, two directions of rotation 424
8.2.5.3 Pole-changing three-phase induction motor with two separate windings, two speeds, one direction of rotation 426
8.2.5.4 Pole-changing three-phase induction motor with two separate windings, two speeds, two directions of rotation 428
8.2.5.5 Pole-changing three-phase induction motor with three speeds, one direction of rotation, one winding in a Dahlander connection, and one separate winding for low speed . . . 430
8.2.5.6 Pole-changing three-phase induction motor with three speeds, one direction of rotation, one winding in a Dahlander connection, one separate winding for the intermediate speed . . 432
8.2.5.7 Pole-changing three-phase induction motor with three speeds, one direction of rotation, one winding in a Dahlander connection, one separate winding for high speed 434
8.2.5.8 Pole-changing three-phase induction motor with three speeds, two directions of rotation, one winding in a Dahlander connection, one separate winding for low speed . . . 436
8.2.5.9 Pole-changing three-phase induction motor with four speeds, one direction of rotation and two separate windings 440

8.3 Starting of three-phase induction motors 444
8.3.1 Star-delta starting of three-phase induction motors with star contactor, delta contactor and line contactor 444
8.3.2 Closed transition star-delta starting of three-phase induction motors . . . 446
8.3.3 Four stage star-delta starting 448
8.3.4 Star-delta starting of three-phase induction motors in two directions of rotation 450
8.3.5 Star-delta starting of three-phase induction motors with power factor correction 452
8.3.6 Automatic starting of three-phase squirrel-cage motors via a single-pole stator-resistor (KUSA connection) using a time relay 454
8.3.7 Automatic starting of three-phase induction motors via a three-pole stator-resistor using a time relay . . . 456
8.3.8 Automatic starting of three-phase slip-ring motors 458
8.3.9 Closed transition autotransformer starting of three-phase squirrel-cage motors (Korndörfer connection) . . . 460

8.4	**Circuits with thermistor motor protection** 462	**8.9**	**Interface units** 487
8.4.1	Thermistor motor protection with positive temperature coefficient (PTC) temperature sensors 462	**8.10**	**Auxiliary circuits incorporating time relays** 488
8.4.1.1	Thermistor motor protection for a pole-changing three-phase induction motor with two separate windings and two speeds 462	8.10.1	Star-delta starting of three-phase induction motors with star contactor, delta contactor, line contactor and time relay 488
8.4.1.2	PTC thermistor motor protection for alarm and switch-off of a three-phase induction motor with six sensors via a circuit-breaker for motor protection equipped with overload and short-circuit releases 464	8.10.2	Functions of the motor-driven time relay 7PR4140 490
		8.11	**Switching of an electrical heating system using a thermostat and contactor combination** 491
8.4.1.3	Thermistor motor protection for the switching off of six three-phase induction motors via contactors . . . 466	**8.12**	**Stand-by power supply installations** . . 492
		8.12.1	Three-pole change-over from network supply to stand-by supply using contactors (generator operation) . . . 493
8.4.2	Thermistor motor protection with negative temperature coefficient (NTC) temperature sensors 468	8.12.2	Change-over from network supply to stand-by supply with four-pole disconnection of the distribution system via two three-pole contactors 494
8.5	**Circuits with monitors** 470		
8.5.1	Circuits with speed monitors 470		
8.5.1.1	Direct-on-line starting of three-phase induction motors. Stopping by reverse-current braking (plug braking) 470	**8.13**	**Project planning and engineering aids** 496
8.5.1.2	Direct-on-line starting of three-phase induction motors. Stopping by reverse-current braking (plug braking). Circuit with contactor relay 472	**9**	**Appendix** 499
		9.1	**Fundamental equations, characteristic quantities and units of electricity** . . . 499
8.5.1.3	Direct reversal of three-phase induction motors with reverse-current or plug braking in both directions of rotation 474	9.1.1	Fundamental equations of electrical engineering 499
		9.1.2	Characteristic quantities and units of electricity in accordance with DIN VDE and IEC 500
8.5.2	Circuits with conveyor belt monitors . 476		
8.5.3	Circuits of contactor control systems with pressure monitors 478	9.1.3	Differences in the IEC 157-1 and IEC 947-2 publications 502
8.6	**Circuits with position switches** . . . 479	9.1.4	Equation symbols and SI units International System of Units (SI) . . 503
8.6.1	Reverser circuits with position switches (e.g. gate control) 479	9.1.5	Conversion of international, British and American units 507
8.6.2	Position switches with indicator lights 480		
8.7	**Terminal blocks** 482	**9.2**	**Enclosures for electrical equipment to American and Canadian standards** . . 512
8.7.1	Circuits with isolating terminal blocks for current transformers 482		
8.7.2	Circuit-breaker terminals for auxiliary circuits 484	**9.3**	**Climatic values, influence of temperature and thermal conduction** . 514
		9.3.1	Climatic values 514
8.8	**Circuits with leakage-current (residual-current) protective devices** . . 486	9.3.2	Effects of temperature and thermal conduction 514

9.4	**Current carrying capacity and overcurrent protection of insulated wires, cables and busbars** 516		9.7.3	Characteristic curves and tripping behaviour of circuit-breakers 539		
9.4.1	Coordination of protection devices . . 517		9.7.4	Current-limiting diagrams of fuses . . 541		
9.4.1.1	Overload protection 517		9.7.5	Discrimination (selectivity) between fuses and circuit-breakers 543		
9.4.1.2	Short-circuit protection 518					
9.4.2	Current carrying capacity 521		**9.8**	**Short-circuit currents** 546		
9.4.3	Load ratings of insulated conductors at ambient temperatures of 30° to 70 °C and the assignment of cable protection fuses in accordance with USA and Canadian standards 525		9.8.1	Limiting effect of conductors and cables on short-circuit currents . . . 546		
			9.8.2	Dynamic forces created by short-circuit currents 547		
9.4.4	Thermal ratings of busbars and device terminals 527		**9.9**	**Number of switching operations of switching devices subjected to different operating periods per day** 548		
9.4.5	Resistance of copper and aluminium conductors 529					
9.5	**Rated currents of three-phase induction motors** 530		**9.10**	**International network voltages and frequencies** 549		
9.6	**Three-phase power transformers** . . . 532		**9.11**	**EC guidelines for low voltage equipment** 555		
9.6.1	Graphic symbols and vector groups of three-phase power transformers . . 535					
9.7	**Tripping behaviour of line protection and switchgear protection devices** . . . 536		**9.12**	**Glossary - Brief explanations of some technical terms** 556		
9.7.1	Time-current tripping characteristics of circuit-breakers, miniature circuit-breakers and overload relays 536		**9.13**	**Addresses of important specification, standards and testing bodies** 643		
9.7.2	Pre-arcing time-current characteristics of fuses (operating classes gL/gG and aM) 536		**Index** 645			

1 Specifications for low-voltage devices and switchgear assemblies

1.1 Nomenclature of the standards authorities

Specifications and standards in the Federal Republic of Germany

The specifications, regulations and standards for low-voltage switchgear and switchgear assemblies applicable in the Federal Republic of Germany are issued by

▷ DIN (Deutsches Institut für Normung e.V.), and

▷ VDE (Verband Deutscher Elektrotechniker e.V.).

Documents produced by the electrotechnical standardization committees which contain stipulations related to safety, are not only entered as a German norm in the DIN standards, but are also adopted, as DIN standards with an additional identification code to indicate compliance with VDE standards, into the VDE set of specifications. Such documents are all labelled with DIN VDE ... code numbers.

International, regional, and national specifications and standards.

Table 1.1 lists the technical associations and testing authorities responsible for international and regional specifications and standards. Table 1.2 lists a number of national standards authorities.

The European committee for electrotechnical standardization (CENELEC) is an amalgamation of the national standards institutions and electrotechnical committees. The European norms (EN) and harmonizing documents (HD) adopted by CENELEC are, as far as possible, aligned with the international norms and standards of the IEC. The results of the work carried out by the European standards committees must be adopted by all the member countries and apply directly as national standards and norms.

All German specifications published after 1.6.93 are labelled as follows:
  Specifications derived from IEC with DIN IEC...
  Specifications adopted from CENELEC with DIN EN...
In addition, both DIN IEC and DIN EN publications are still labelled as VDE specifications, so that a correlation remains.
Harmonizing documents are only labelled DIN VDE...

Table 1.1 International and regional authorities for specifications and standards.

Abbreviation	Meaning
CEE	International **C**ommission on Rules for the approval of **E**lectrical **E**quipment: (partially used by the Scandinavian countries as a basis for low-voltage switchgear with rated currents up to 63 A)
CEN	Comité **É**uropéen de **N**ormalisation: European committee for standardization
CENELEC	Comité **É**uropéen de **N**ormalisation **Élec**trotechnique: European committee for electrotechnical standardization (general secretariat in Brussels)
IEC	**I**nternational **E**lectrotechnical **C**ommission: (French: CEI) All the major industrialized countries are involved in the work of the International Electrotechnical Commission. The resulting IEC recommendations are to some extent either adopted directly into national specifications and standards, or the national specifications and standards are adapted to harmonized with these recommendations.
ISO	**I**nternational **S**tandards **O**rganization

1 Specifications for low-voltage devices and switchgear assemblies

Table 1.2 National authorities for standards and specifications

Abbreviation	Meaning
ANSI	**A**merican **N**ational **S**tandards **I**nstitute: Publishes specifications and standards in virtually all fields (not only electrical). For low-voltage switchgear, the ANSI has adopted the American NEMA and UL specifications to a large extent
AS	**A**ustralian **S**tandards (already partially adapted to IEC)
BS	**B**ritish **S**tandard (already partially adapted to IEC)
CEI	**C**omitato **E**lettrotecnico **I**taliano: Italian electrotechnical committee
CEMA	**C**anadian **E**lectrical **M**anufacturers **A**ssociation
CSA	**C**anadian **S**tandards **A**ssociation: Responsible for publishing standards and granting approvals
DEMKO	**D**anmarks **E**lektriske **M**ateriel**ko**ntrol: Danish board of control for electrotechnical products. Responsible for publishing standards and granting approvals
EEMAC	**E**lectrical and **E**lectronic **M**anufactures **A**ssociation **C**anada
IEEE	**I**nstitute of **E**lectrical and **E**lectronics **E**ngineers: Institute of engineers in the USA
IS	**I**ndian **S**tandard (already partially adapted to IEC)
JIS	**J**apanese **I**ndustrial **S**tandard
KEMA	**K**euring van **E**lektrotechnische **Ma**terialen: Dutch testing authority, which also e.g. performs CSA approval tests for European manufacturers
NBN	**N**ormes de l'Institut **B**elge de **N**ormalisation: Standards issued by the Belgian standards institute (already partially adapted to IEC)
NEMA	**N**ational **E**lectrical **M**anufacturers **A**ssociation (USA)
NEMKO	**N**orges **E**lektriske **M**ateriell**ko**ntroll: Norwegian controls authority for electrotechnical products, responsible for publishing standards and granting approvals
NEN	**N**ederlandse **N**orm: Dutch standard
ÖVE	**Ö**sterreichischer **V**erband für **E**lektrotechnik: Austrian association for electrotechnology. Conforms to a large extent with the DIN VDE and IEC specifications
RER	**R**afmagnseftrilit **ri**kisins: Finnish testing authority for electrotechnical products
SABS	**S**outh **A**frican **B**ureau of **S**tandards
SASO	**S**audi **A**rabien **S**tandard **O**rganisation
SEMKO	**S**venska **E**lektriska **M**ateriel**ko**ntrollanstalten: Swedish controls authority for electrotechnical products, responsible for publishing standards and granting approvals
SEN	**S**venska **E**lektrotekniska **N**ormer: Swedish electrotechnical standards
SETI	Finnish electrotechnical testing authority.
SEV	Swiss electrotechnical association
UL	**U**nderwriters' **L**aboratories Inc.: Testing authority of the national fire insurance in the USA. Among other activities, it carries out testing on electrotechnical products, and issues the corresponding specifications and regulations.
UTE	**U**nion **T**echnique de l'**É**lectricité: French electrotechnical association

1.2 Low-voltage switchgear and devices

1.2.1 Summary of specifications and approvals

Low-voltage switchgear and switchgear combinations manufactured by Siemens AG are developed constructed and tested in accordance with the relevant DIN VDE specifications as well as relevant DIN standards, CENELEC harmonizing documents, European Norms (EN) and IEC recommendations (Table 1.4). They also fulfil the safety regulations of the EC-guidelines for low-voltage equipment of 19.2.1973 (also refer to Section 9.10, page 527).

In the Federal Republic of Germany, the manufacturers of low-voltage switchgear themselves carry the responsibility for adherence to DIN VDE, EN and IEC recommendations. The quality of the products is ensured by adherence to the international standards for quality assurance systems DIN ISO 9000 to 9004 (alternatively EN 29 000 to EN 29 004).

Further specifications for low-voltage switchgear assemblies (TTA and PTTA) are described in section 1.3.1, page 26.

Table 1.5 and 1.6 respectively show summaries of important specifications for transformers and low-voltage fuses.

Siemens low-voltage switchgear can therefore be installed and used without problem in many countries around the world which have either aligned their national regulations with IEC recommendations, or which recognize the DIN VDE specifications and IEC documents as valid electrotechnical standards.

Owing to the ever increasing degree of international standardization, specifications and standards are continuously updated and modified. In cases of doubt, it is therefore always advisable to consult the currently valid editions of the standards and specifications in question.

The regulations of some European countries (e.g. the Nordic countries and Switzerland), the U.S.A. and Canada as well as the regulations for circuit-breakers in Australia, differ considerably from the IEC recommendations. In these countries some items of low-voltage switchgear may only be sold if they carry the approval of the relevant testing authority.

Use of low-voltage devices in marine installations

For the use of low-voltage switchgear or other low-voltage devices in marine applications, the specifications of the marine classification societies must also

Table 1.3 Marine classification societies

Abbreviation	Meaning
ABS	**A**merican **B**ureau of **S**hipping: Country of origin USA
BV	**B**ureau **V**eritas: Country of origin France
DNV	**D**et **N**orske **V**eritas: Country of origin Norway
GL	**G**ermanischer **L**loyd: Country of origin Federal Republic of Germany
LRS	**L**loyd's **R**egister of **S**hipping: Country of origin Great Britain
PRS	**P**olski **R**ejestre **S**tatkow: Country of origin Poland
RINA	**R**egistro **I**taliano **Na**vala: Country of origin Italy
USSR	**USSR**-Register of Shipping: Country of origin USSR. The name has recently been changed to "Maritime Register of Shipping" (MRS)
ZC	Chinese Classification Society: Country of origin China

be taken into account (refer Table 1.3). To some extent, these specifications require a special testing of the devices. Once successfully tested, the devices are awarded the relevant mark of approval.

The relevant sections of the Siemens catalogues indicate which items of low-voltage switchgear are approved by the various marine classification societies. Items which comply with the particular specifications without the need for special type-testing, are also listed. (For addresses of the marine classification societies, refer to Section 9.12, page 576)

1.2.2 Regulations and approvals in European countries

The Swiss regulations and those in the Nordic countries are based, to some extent, on CEE-Publications. Here it is taken into consideration that low-voltage switchgear, particularly such with lower current ratings, is not only used in industrial applications, but may also be installed in a domestic environment or e.g. in domestic appliances. In such instances low-voltage switchgear may not only be installed and operated by trained personnel, but also by laymen. It should therefore guarantee a particularly high degree of safety protection.

1 Specifications for low-voltage devices and switchgear assemblies

Table 1.4 Specifications for low-voltage switchgear and switchgear assemblies

Currently valid specifications (1991/92)		Relation to	
DIN VDE 0660	Contents	**IEC** Publication	**CENELEC** Harmonizing Documents (HD), alt. **European Norm (EN)**

Generic specifications

Supplement 1	Switchgear, index of the standards in the DIN VDE 0660 series	–	–
Supplement 2	Switchgear, quoted and further standards in the DIN VDE 0660 series	–	–
Part 12	Protective-conductor terminals	–	–
Part 14	Supplementary specifications for railways	–	–
Part 99	Connectable conductor cross-sections	–	–
Part 100	General rules; Definitions and requirements	947-1	EN 60947-1

Individual specifications

Part 101	Low-voltage switchgear Circuit-breakers	947-2 [3]	EN 60947-2
Part 102	Low-voltage switchgear Contactors	947-4-1 [3]	EN 60947-4-1
Part 104 [2]	Low-voltage motor-starters a.c. motor-starters up to 1000 V	947-4-1 [3]	EN 60947-4-1
Part 106	a.c. motor-starters, star-delta starters	947-4-1	EN 60947-4-1
Part 107 [1]	Switches, disconnectors, switch-disconnectors and fuse combination units	947-3 [3]	422 (prov. EN 60947-3)
Part 108	Circuit-breakers, supplementary requirements for d.c. circuit-breakers over 1200 to 3000 V	–	–
Part 109 [1]	Semiconductor contactors	158-2	419.2
Part 114, D	Automatic transfer switching equipment	947-6	–
Part 200 [4]	Control circuit devices and switching elements, general requirements	947-5-1 [3] Chapter 1	EN 60947-5-1
Part 201 [4]	Control circuit devices and switching elements, supplementary specifications for pushbuttons and similar auxiliary switches	947-5-1 [3]	EN 60947-5-1
Part 202 [4]	Control circuit devices and switching elements, supplementary specifications for rotary switches	947-4-1 [3]	EN 60947-5-1
Part 203 [4]	Control circuit devices and switching elements, supplementary specifications for contactor relays	947-5-1 [3]	EN 60947-5-1
Part 204 [4]	Control circuit devices and switching elements, supplementary specifications for automatic auxiliary switches with pilot function	947-5-1 [3]	EN 60947-5-1
Part 205 [4]	Control circuit devices and switching elements, supplementary specifications for indicator lights	947-5-1 [3]	EN 60947-5-1
Part 206 [4]	Control circuit devices and switching elements, supplementary specification for position switches with positive opening operation in safety applications	947-5-1 Chapter 3	EN 60947-5-1
Part 207 [4]	Control circuit devices and switching elements, supplementary specifications for EMERGENCY-OFF command devices	–	–

D Draft only

[1] Current standard being updated, draft has been issued. By the end of 1992, the drafts were due to be issued as "white papers". As from this publication date, the previously valid specifications no longer apply.
[2] To be superceded by the draft DIN VDE 0660 Part 102, 10.87 (Low-voltage switchgear, contactors and motor starters)
[3] Correlation with the respective drafts of the DIN VDE specifications
[4] Draft DIN VDE 0660 part 200, 07.92 is intended as replacement for DIN VDE 0660 part 200 to part 207.

Table 1.4 *(continued)*

Currently valid specifications (1991/92)		Relation to	
DIN VDE	Contents	**IEC** Publication	**CENELEC** Harmonizing Documents (HD), alt. **European Norm (EN)**
0660 Part 208 [1]	Control circuit devices and switching elements, supplementary specifications for inductive proximity switches	947-5-2	–
0660 Part 209	Supplementary specifications for proximity-type position switches in safety applications	–	–
0660 Part 301	Low-voltage motor-starters Rotor resistance starters	947-4-1 [3]	EN 60947-4-1
0660 Part 302	Thermal machine protection for rotating electrical machines; Temperature sensors and tripping devices	34-11-2, Section 1	–
0660 Part 303	Thermal machine protection for rotating electrical machines; PTC thermistor temperature sensors and tripping devices	34-11-2, Section 2	–
0660 Part 500	Low-voltage switchgear assemblies	439-1	EN 60439, part 1
0660 Part 501	Particular requirements for building-site distribution boards	439-4	EN 60439, part 4
0660 Part 502, D	Particular requirements for busbar distribution systems	439-2	EN 60439, part 2
0660 Part 503 [1]	Supplementary specifications for cable distribution cubicles	–	–
0660 Part 504	Supplementary specifications for type-tested switchgear assemblies intended for operation by laymen	439-3	EN 60439
0660 Part 506	Switchgear assemblies, switchboards cable channels; Requirements, tests	–	–
0660 Part 507	Low-voltage switchgear assemblies; Method for determining heat rise in partially type-tested switchgear assemblies (PTTA) by extrapolation	890	528 S1
0435 Part 201	Rules for electrical relays in heavy current installations, switching relays	255-1-00	–
0435 Part 2021	Relays with specified time delays (time relays) Requirements, tests	–	–
0609 [1]	Clamping points of screw-type terminals for the connection or joining of copper conductors up to 240 mm²	–	–
0611 Part 1	Low-voltage switchgear and controlgear-ancillary equipment – Terminal blocks for copper conductors	947-7-1 [3]	EN 60947-7-1
0611 Part 3	Terminal blocks for protective conductors up to 120 mm²	–	–
0611 Part 4	Multi-storey distribution terminal blocks up to 6 mm²	–	–
0611 Part 20	Terminal blocks up to AC 1000 V and DC 1200 V, tests for inflammability, and flame propagation	–	–
0641 Part 11	Miniature circuit-breakers for domestic use and similar applications	898	EN 60898
0641 Part 2	Miniature circuit-breakers up to 63 A rated current, and up to DC 440 V	–	–
0641 Part 3	Miniature circuit-breakers up to 63 A rated current, up to AC 415 V and up to DC 440 V	–	–

D Draft only

[1] Current standard being updated, draft has been issued By the end of 1992, the drafts were due to be issued as "white papers". As from this publication date, the previously valid specifications no longer apply.
[2] To be superceded by the draft DIN VDE 0660 Part 102, 07.92 (Low-voltage switchgear, contactors and motor starters)
[3] Correlation with the respective drafts of the DIN VDE specifications

1 Specifications for low-voltage devices and switchgear assemblies

Table 1.5   Specifications for current and voltage transformers (CT's and VT's) and auxiliary power supply transformers

Currently valid specifications (1991/92)		Relation to	
DIN VDE	Contents	**IEC** Publication	**CENELEC** Harmonizing Documents (HD), alt. **European Norm (EN)**
0414	Specifications for instrument transformers	185/186	prov. HD 553
0532 [1] various Parts	Specifications for transformers and reactor coils	76	398
0550 Part 1 [3]	Specifications for small transformers, General specifications	989 [2]	—
0550 Part 3 [3]	Particular specifications for isolating and control transformers as well as power supply transformers and safety isolating transformers over 1000 V	—	—
0551 Part	Isolating transformers and safety transformers, requirements	742	EN60742
0552	Specifications for variable ratio transformers with moving contact perpendicular to coil windings	—	—

[1] Draft has been issued. "White paper" due by the end of 1992.
[2] Correlation with the respective drafts of the DIN VDE specifications.
[3] DIN VDE 0550 parts 1 and 3 are still partially valid until 10.92. Superceded by DIN VDE 0551 part 1.

Table 1.6   Specifications for low-voltage fuses

Currently valid specifications (1991/92)		Relation to	
DIN VDE 0636	Contents	**IEC** Publication	**CENELEC** Harmonizing Documents (HD), alt. **European Norm (EN)**
Part 1 [1]	General requirements	269, 241, 291	EN 60269-1 [2]
Part 21 [1]	LV HRC system; Cable and conductor protection up to 1250 A and ∼500 V, ⎓400 V as well as ∼660 V	269-2	—
Part 22 [1]	LV HRC system; HRC installation protection fuses up to 1250 A and ∼1000 V aM, gTr, gB	269-2	—
Part 23	LV HRC system; Semiconductor protection fuses up to 1600 A and up to 3000 V	269-2	—
Part 31 [1]	D system; Cable and conductor protection up to 100 A and 500 V, alternatively 63 A and ∼660 V, ⎓600 V	269-3 269-3A	—
Part 33	D system; Semiconductor protection fuses up to 100 A and 500 V	269-4	—
Part 41 [1]	D0 system; Cable and conductor protection up to 100 A and ∼380 V, ⎓250 V	—	—

[1] Current standard being updated. Draft has been issue. "White paper" due by the end of 1992.
[2] Correlation with the respective drafts of the DIN VDE specifications.

Table 1.7 Approvals for low-voltage switchgear in Europe, obligatory approval and marking

Country	Denmark	Finland	Iceland	Norway	Sweden	Switzerland
Authorized testing body (appointed by government, privately and legally recognized)	DEMKO (Danmarks Elektriske Materielkontrol)	SETI (Suomi Elektrical Inspectorate)	RER (Rafmagnseftirlit ríkisins)	NEMKO (Norges Elektriske Materiellkontroll)	SEMKO (Svenska Elektriska Materielkontrollanstalten)	SEV (Schweizerischer Elektrotechnischer Verein)
Test mark for approved equipment	Ⓓ	Ⓕⓘ	ῗŜ	Ⓝ	Ⓢ	Ⓢ̂
Obligatory marking and general guidelines on approval obligations	Obligatory approval and registration for specific devices for retail sale	No obligatory approval or marking requirements for devices which are only installed in industry	No obligatory approval or marking requirements for devices which are only installed in industry	Obligatory marking and approval for certain devices intended for retail sale	No obligatory marking requirements; Obligatory approval only for use in domestic installations and appliances	Obligatory marking and registration for devices intended for retail sale [3]

Obligatory approval of:

	Denmark	Finland	Iceland	Norway	Sweden	Switzerland
Fuse combination units, switchable in-line fuse switch disconnectors, load disconnectors with fuses	–	up to 800 A	up to 125 A	up to 250 A	–	all
HRC switchgear protection fuses	–	up to 800 A	up to 125 A	up to 250 A	–	all
Switch disconnectors	–	up to 800 A	up to 63 A	up to 32 A	–	– [4]
Circuit-breakers for motor protection	all	up to 25 A	up to 63 A	up to 32 A	up to 25 A, 500 V [1]	all
Circuit-breakers	–	up to 1000 A as main switch	up to 63 A	up to 32 A	–	all
Miniature circuit-breakers	up to 63 A	up to 63 A	up to 63 A	up to 63 A	up to 63 A	all
Residual current protection devices	up to 63 A	up to 63 A	up to 63 A	–	up to 63 A, 500 V [1]	all
Pushbuttons, control switches, indicator lights	–	all (position switches only as safety position switches)	all	–	up to 32 A, 660 V [1]	– [4]
Position switches				all	–	–
Proximity switches, BERO				–		
Thermistor motor protection	–	up to 63 A	up to 32 A	up to 32 A	up to 25 A, 500 V [1]	all
Monitors	–	–	up to 32 A	–	–	– [4]
Control switches for auxiliary circuits	–	up to 63 A	up to 32 A	up to 32 A	up to 32 A, 660 V [1]	– [4]
Main switches		up to 1000 A	up to 63 A	up to 63 A		
Control switches for main circuits		up to 63 A	up to 63 A	up to 32 A		
Time relays	–	all	all	–	up to 32 A, 660 V [1]	– [4]
Contactor relays	all	all	all	all	up to 32 A, 660 V [1]	– [4]
Contactors	up to 32 A, including housings if used	up to 63 A	up to 32 A	up to 32 A		
Contactors with overload relays, overload relays for separate mounting	up to 32 A [2]	up to 25 A	up to 32 A	up to 32 A	up to 25 A, 500 V [1]	all
Contactor-based reversers, contactor-based star-delta starters	up to 32 A [2]	up to 63 A [2]	up to 32 A [2]	–	–	all [2]
Electronic motor starters	–	–	up to 32 A	–	–	– [4]
Terminal blocks	–	up to 800 A	up to 63 A, 16 mm²	up to 400 A	up to 32 A, 750 V [1]	– [4]
Transformers	–	up to 10 kVA	up to 1,5 kVA	–	–	all safety and transformers

[1] Only when used in domestic installations, domestic appliances and similar equipment
[2] Approval of the individual components which are used in the assemblies. Approval obligatory for complete devices with own order numbers
[3] >42 V to 1000 V ≈/1500 V   alt. ≤42V & >2A
[4] Proof of conformity with or without mark approval

The obligatory testing and authorization, the registration in approval records and, in certain instances the marking of the equipment with a specified mark of approval, confirm to the user that the equipment complies in full with the national regulations.

A summary of obligatory testing, registration and marking requirements i.e. obligatory approvals for low-voltage switchgear in western Europe can be found in Table 1.7 on page 21.

In the east European countries Czechoslovakia, Hungary and Poland obligatory approvals exist for all electrical devices operated at higher than 50 V. Obligatory markings do not exist. The use of a mark of conformity is possible. Furthermore, a test for the suitability of the terminals to accept aluminium conductors is required in the Czechoslovakia. A reciprocal recognition of the receptive approvals by the approval authorities in the Czechoslovakia, Hungary and Poland is planned for 1992. Also refer to Section 9.13, page 643.

1.2.3 Standards and approvals in the U.S.A., Canada and Australia

To a large extent the ANSI, NEMA and UL standards in the U.S.A., the CSA standards in Canada and the AS standards for circuit-breakers in Australia are identical to one another. They do, however, still differ from the VDE specifications and the IEC recommendations.

As far as it is economically justifiable, the UL and CSA specifications as well as the AS specifications for circuit-breakers, are also taken into consideration in the design and construction of Siemens low-voltage switchgear. In some instances it may be necessary to base the selection of the switchgear on lower ratings. Only in isolated cases must a Siemens low-voltage device be supplied as a special model in order to meet the requirements of the UL, CSA or AS specifications.

Obligatory approvals apply for all low-voltage switchgear in Canada and for circuit-breakers in Australia. In the majority of the states in the U.S.A. it is prescribed by law, that only switchgear which has been tested and approved by an authorized testing laboratory e.g. UL, the best known neutral testing authority, may be used in an installation. Therefore, it is imperative that the end user decides whether the switchgear to be installed requires the UL approval, or not.

A summary of the differing approval procedures for CSA and UL is given in the Tables 1.8 to 1.10. The marking of low-voltage switchgear with the specified test mark can be executed only after successful testing and authorization by CSA or UL, respectively. The manufacture of low-voltage switchgear bearing the CSA or UL test mark is inspected on a regular basis to ensure that the standards are adhered to.

Based on the local conventions of switchboard design and construction, and on different safety considerations in the respective countries, CSA and UL distinguish between circuit-breakers and other industrial switchgear. The differences apply to the relevant specifications, regular inspections (production control procedure), regulations concerning test marks (special label) for circuit-breakers as well as the permissible applications. For example, an item of industrial switchgear must, in accordance with the NEC (**N**ational **E**lectrical **C**ode), have some form of short circuit protection e.g. fuses or circuit-breaker on the supply side. Switches that are intended to provide short-circuit protection must possess the necessary switching capability (stated in kA) and must have passed the required switching capacity tests. They must also have longer creepage paths and clearance distances than other industrial switchgear.

Delegates from the UL, CSA and IEC standards committees are presently working on harmonizing the various specifications.

Specifications and Approvals in USA, Canada, Australia

Table 1.8
Approval of low-voltage switchgear as "Industrial Control Equipment" to CSA and UL specifications (the term "Industrial Control Equipment" refers to control devices and in particular to all devices intended for the switching and overload protection of motors)

Country	Canada	USA	
Authorized testing body (appointed by government, privately and legally recognized)	CSA (Canadian Standards Association), represented in Europe by KEMA (Arnheim)	UL (Underwriter's Laboratories)	
Approval procedures	Approvals are generally obligatory: "Certified Industrial Control Equipment"	UL approval, to a large extent legally recognized, differentiates between two types:	
		"*Listed* Industrial Control Equipment"	"*Recognized* Component Industrial Control Equipment"
Applicable specifications	CSA Standard C22.2 No 0: "General Requirements-Canadian Electrical Code, Part II" CSA Stardard C22.2 No 14: "Industrial Control Equipment"	UL 508: "Electric Industrial Control Equipment" (unrestricted) UL 486: "Wire Connectors and Soldering Lugs"	UL 508: "Electric Industrial Control Equipment" (with some restrictions)
Scope of the approvals	Devices approved for "field wiring", i.e. – devices intended for installation in control systems which are fully wired at the factory or in a workshop, – or which are sold as individual items in Canada	Devices approved for "field wiring" (this includes "factory wiring"), i.e. – devices intended for installation in control systems which are fully wired at the factory or in a workshop, – or which are sold as individual items in the USA	Devices approved as components for "factory wiring", i.e. – devices intended for installation in control systems which are fully wired at the factory or in a workshop by specially trained personnel, and which are specifically selected for the application by qualified personnel
Marking	Devices are marked on the rating plate with the CSA monogram Ⓢ	Devices are marked on the rating plate with the Ⓤ "Listing Mark"	Devices may be marked with the ᴙᴜ sign on the rating plate (The approval is valid even if the device does not display the mark)
Registration (certification records, guide cards, reports)	Registration in the "List of certified Electrical Equipment" (CSA issues "Certification Records" with guide and file numbers	Registration in (green) "Electrical Construction Materials Directory" (UL issues white "guide cards" with guide and file numbers	Registration in (yellow) "Recognized Component Directory" (UL issues yellow "guide cards" with guide and file numbers
Regular verification (factory inspection)	CSA inspectors carry out a regular "Re-examination Service" at the place of manufacture (visual inspection, in some cases several times a year)	UL inspectors carry out a regular "Type R Follow-up Service" at the place of manufacture (visual inspection, 4 times per year)	

Table 1.9
Approval of "Circuit-Breakers" to CSA and UL specifications (circuit-breakers: particularly suited to short-circuit protection, and with a switching capacity of at least 10 kA)

Country	Canada	USA
Authorized testing body (appointed by government, privately and legally recognized)	CSA (Canadian Standards Association), represented in Europe by KEMA (Arnheim, Holland)	UL (Underwriter's Laboratories)
Approval procedures	Approvals are generally obligatory "Certified Circuit-Breakers"	UL approval, to a large extent legally recognized "Listed Molded-Case Circuit-Breakers"
Applicable specifications	CSA-Standard C22.2 No. 0: "General Requirements – Canadian Electrical Code, Part II" CSA-Standard C22.2 No. 5: "Molded Case Circuit Breakers" CSA C22.2 No. 65: "Construction and Test of Wire Connectors"	UL 489: "Molded-Case Circuit Breakers and Circuit Breaker Enclosures" UL 486: "Wire Connectors and Soldering Lugs"
Scope of the approvals	Devices approved for "field wiring", i.e. devices intended for installation in control systems which are fully wired at the factory or in a workshop, or which are sold as individual items in Canada	Devices approved for "field wiring" (this includes "factory wiring"), i.e. devices intended for installation in control systems which are fully wired at the factory or in a workshop, or which are sold as individual items in the USA
Marking with special label	In addition to the rating plate, the devices are marked with a special label. This label displays the CSA monogram, the CSA guide card number and a running registration number	In addition to the rating plate, the devices are marked with a special label. This label displays the UL sign and a running registration number. The label is issued by UL
Registration (certification records, guide cards, reports)	Registration in the "List of certified Electrical Equipment" (CSA issues "Certification Records" with guide and file numbers	Registration in (green) "Electrical Construction Materials Directory" (UL issues white "guide cards" with guide and file numbers
Regular verification (factory inspection)	CSA inspectors carry out a regular "Label Service" at the place of manufacture (annual, fixed and specified test procedure similar to original acceptance tests)	UL inspectors carry out a regular "Type L follow-up Service" at the place of manufacture (regular monitoring of calibrations, and up to 4 times per year, fixed and specified test procedure similar to original acceptance tests)

Table 1.10 Approval of control transformers to CSA and UL specifications

Country	Canada	USA
Authorized testing body (appointed by government, privately and legally recognized)	CSA (Canadian Standards Association), represented in Europe by KEMA (Arnhiem, Holland)	UL (Underwriter's Laboratories)
Approval procedures	Approvals are generally obligatory	UL approval, to a large extent legally recognized
Applicable specifications	CSA-Standard C22.2 No. 0: "General Requirements – Canadian Electrical Code, Part II" CSA-Standard C22.2 No 66: "Specialty Transformers"	US-Standard UL 506: "Specialty Transformers"
Scope of the approvals	Devices approved for "field wiring", i.e. devices intended for installation in control systems which are fully wired at the factory or in a workshop, or which are sold as individual items in Canada	Devices approved for "field wiring" (this includes "factory wiring"), i.e. devices intended for installation in control systems which are fully wired at the factory or in a workshop, or which are sold as individual items in the USA
Registration (certification records, guide cards, reports)	Registration in the "List of certified Electrical Equipment" (CSA issues "Certification Records" with guide and file numbers	Registration in (green) "Electrical Construction Materials Directory" (UL issues white "guide cards" with guide and file numbers
Regular verification (factory inspection)	CSA inspectors carry out a regular "Label Service" at the place of manufacture (annual, fixed and specified test procedure similar to original acceptance tests)	UL inspectors carry out a regular "Type L Follow-up Service" at the place of manufacture (regular monitoring of calibrations and, up to 4 times per year, fixed and specified test procedure similar to original acceptance tests)

1 Specifications for low-voltage devices and switchgear assemblies

1.3 Type-tested and partially type-tested switchgear assemblies (TTA and PTTA)

For the planning of factory-built switchgear assemblies the knowledge and understanding of the pertinent specifications are essential. In Germany, for example, the specific requirements related to the application of the switchgear, and to the proof of adherence to recognized engineering practice, are stipulated in associated laws or regulations, e.g.

▷ in the Power Industries Act (2nd regulation),
▷ in the Equipment Safety Act, and
▷ in the Accident Prevention Regulations of the various trade associations (of particular interest are the regulations for "Electrical Installations and Equipment" (VBG 4).

These contain clear obligations relating to the documentation of the adherence to technical specifications (e.g. in catalogues, on rating plates and labels or in other associated records).

Even before the planning stage, the applicable specifications and standards must be determined. This is necessary, since it is the only way in which the demands to be met by the assembly in its intended application and point of installation can be described exactly.

1.3.1 Summary of specifications and standards

The general requirements, regulations, test specifications, etc. for

▷ the construction,
▷ equipping and
▷ erection

of low-voltage switchgear installations and distribution systems, are contained in the relevant national and international specifications and standards. In future, there will also be comprehensive regional standards (European Standards).

In order to assist the project planner in becoming acquainted with the more device-orientated system of DIN VDE specifications, it is recommended that these be divided into various categories.

▷ Construction specifications,
▷ Generic specifications,
▷ Equipping specifications and
▷ Erection specifications.

In addition, there are several DIN standards relating specifically to switchgear assemblies which must to be taken into account.

1.3.1.1 Construction specifications

Construction specifications include regulations and requirements for the manufacture of switchgear and switchgear assemblies, i.e. for switchgear installations and distribution systems including their relevant test specifications.

The construction specifications for individual types and construction forms of switchgear assemblies and switchboards are given in the Siemens NV catalogues.

The most important construction specifications are:

DIN VDE 0160	Specifications for the equipment in power installations with electronic devices,
DIN VDE 0660 parts 101 to 303	Specifications for low-voltage switchgear,
EN 60439 part 1	Low-voltage switcgear and controlgear assemblies, Part 1
DIN VDE 0660 part 500	Switchgear, low-voltage switchgear assemblies
IEC Publication 439-1	"Low-voltage switchgear and controlgear assemblies" Requirements for type tested and partially type tested assemblies.
BS 5486-1 (Great Britain)	Identical to EN 60439 part 1, 02.90 ("Low-voltage switchgear and controlgear assemblies"),
NEMA-ICS 2 (USA)	"Standards for industrial control devices, controllers and assemblies"
NF/UTE C63-410 (France)	Low-voltage switchgear; factory built low-voltage switch and controlgear assemblies" ("Ensembles d'appareillage à basse tension – Première partie: Règles pour les ensembles de série et les ensembles dérivés de série")

TGL 200-0645/02 (East Germany)[1]	Electrotechnical installations. Factory assembled units for rated voltage up to AC 1000 V or DC 1500 V"

The DIN VDE 0100 is no longer grouped with the construction specifications since § 30b has been transferred to DIN VDE 0660 part 500.

For the sake of completeness, reference must also be made to the following national construction specifications for factory built assemblies:

DIN VDE 0603	Small distribution boards and meter boxes with up to 250 V to earth
DIN VDE 0660 part 501, 02.92[1]	Switchgear; low voltage switchgear assemblies; Section 4: Particular reqquirements for building site distribution boards (IEC 439-4:1990),
DIN VDE 0660 part 504, 04.92[2]	Switchgear; low voltage switchgear assemblies; Section 3: Particular requirements for switchgear assemblies to which laymen have access; domestic distribution boards (IEC 439-3:1990, modified)

These three specifications will shortly be covered by DIN VDE 0660 part 500. For necessary deviations, corresponding new parts will be issued.

The switchboard and distribution systems featured in the Siemens NV catalogs comply with the specifications

– EN 60439 part 1/DIN VDE 0660 part 500, 04.91,
– IEC publication 439-1, 1984 edition,

and, in part, they also fulfill the requirements of DIN VDE 0660 part 504.

They are offered as type tested switchgear assemblies (TTA) and are delivered ready for connection. Type tested sub-assemblies, modules, cabinets, housings as well as individual items of switchgear are also available. These can be assembled in workshops or at the place of installation to form fully type tested (TTA) or partially type tested switchgear assemblies (PTTA).

As far as is economically justifiable, important specifications of other countries are also taken into account in the construction and manufacture of the standard factory-built assemblies and their modules (e.g. clearances to NEMA standards).

The harmonization of the IEC publication 439-1 with the DIN VDE specifications has not yet been fully completed. The texts of these two specifications are therefore not congruent, i.e. compliance with the one does not automatically mean that the conditions of the other are fulfilled, even if the deviations are insignificant (refer e.g. to appendix of EN 60439 part 1, and DIN VDE 0660 part 500).

The distinction between TTA's and PTTA's which is made in both EN 60439 part 1/DIN VDE 0660 part 500 and IEC 439-1, should not lead to the classification of the two types into different quality levels. Both version are to be valued equally.

The difference between TTA's and PTTA's lies essentially in the nature and the extent of the tests to be carried out (refer also to Table 1.13, page 30, for main points of difference). In the case of PTTA's, the tests to be carried out are somewhat "simpler". For example, the determination of the heat rise may also be done by calculation instead of by actual measurement.

Hot spots are however not identified by a heat rise calculation. As a result, the responsibility for the determination of the wiring cross-sections cannot be taken by the manufacturer alone. The selection of the wiring cross-sections must be based on DIN VDE 0660 part 500 A5 (presently available as draft[3]). Also refer to the Tables 1.11 and 1.12, pages 28 and 29.

Note

The wiring selection given in the tables may also be applied to TTA's. A comparison of the cross-sections in mm^2 with the corresponding AWG, kcmil (MCM) and SWG sizes is given in Table 9.10 of Section 9.1.5 (page 511).

---

[1] German version of EN 60439, 1991. In the case of products which complied with the preceeding standard DIN VDE 0612 of May 1974 before 1 December 1991, the expired standard may be applied untill 1 December 1996.
[2] German Version of EN 60439-3, 1990
[3] Owing to the current Harmonizing process, a white paper procedure is not possible. It is however recommended that this draft be taken into consideration.

Table 1.11
Current carrying capacity, overload and short-circuit protection of insulated cables in PTTA's at an ambient temperature of **35 °C** around the cable (e.g. cabling on open cable tray), permissible operating temperature 70 °C.
The values given in the table may be converted in terms of DIN VDE 0100 part 430 for short-time and intermittent periodic loading

Column 1	2	3	4	5	6	7	8	9	10	11	12	13
Conductor cross-section	Current carrying capacity and setting of overload relay (a-release)	Overload protection with fuse, miniature circuit-breaker	Short-circuit protection with fuse, miniature circuit-breaker	Short-circuit protection with circuit-breaker (n-release)[2]	Current carrying capacity and setting of overload relay (a-release)	Overload protection with fuse, miniature circuit-breaker	Short-circuit protection with fuse, miniature circuit-breaker	Short-circuit protection with circuit-breaker (n-release)[2]	Current carrying capacity and setting of overload relay (a-release)	Overload protection with fuse, miniature circuit-breaker	Short-circuit protection with fuse, miniature circuit-breaker	Short-circuit protection with circuit-breaker (n-release)[2]
mm²	A	A	A	A	A	A	A	A	A	A	A	A
0.5	6	6	20	90	8	6	20	120	8	6	20	120
0.75	8	6	20	120	8	6	20	120	8	6	20	120
1	8	6	20	120	8	6	20	120	8	6	20	120
1.5	12	10	25	180	12	10	25	180	12	10	25	180
2.5	17	16	32	255	20	20	35	300	20	20	35	300
4	22	20	35	330	25	25	40	375	25	25	40	375
6	28	25	40	420	32	32	50	480	32	32	50	480
10	38	35	63	570	48	40	80	720	50	50	100	750
16	52	50	100	780	64	63	125	960	65	63	125	975
25					85	80	160	1275	85	80	160	1275
35					104	100	200	1560	115	100	200	1725
50					130	125	250	1950	150	125	250	2250
70					161	160	315	2415	175	160	315	2625
95					192	160	315	2880	225	200	400	3375
120					226	200	400	3390	250	250	500	3750
150					275	250	500	4125	275	250	500	4125
185					295	250	500	4425	350	315	630	5250
240					347	315	630	5205	400	400	800	6000
300					400	400	800	6000	460	400	800	6900

[1] In this case, a random laying (or "bunching") of the conductors is possible. The given values refer to 6 cores, all simultaneously loaded with 100%, within a multi-core bundle
[2] Instantaneous electromagnetic overcurrent release

For "free wiring", e.g. 8LW, the values given in columns 2 to 5 apply. Multi-core plastic-sheathed cable, or cable within the PTTA (e.g. for wiring to maintenance socket outlet or cabinet interior light fitting) may be selected from columns 6 to 9 of the table.

The given cross-sections are adequately protected against overload, if the rated or set current of the protection device does not exceed the values given in the columns 2 or 3, 6 or 7 and 10 or 11 respectively. For short-circuit protection, the fuses or cable protec-

Table 1.12
Current carrying capacity, overload and short-circuit protection of insulated cables in PTTA's at an ambient temperature of **55 °C** around the cable (e.g. cabling on open cable tray), permissible operating temperature 70 °C.
The values given in the table may be converted in terms of DIN VDE 0100 part 430 for short-time and intermittent periodic loading

Column 1	2	3	4	5	6	7	8	9	10	11	12	13
Conductor cross-section	Current carrying capacity and setting of overload relay (a-release)	Overload protection with fuse, miniature circuit-breaker	Short-circuit protection with fuse, miniature circuit-breaker	Short-circuit protection with circuit-breaker (n-release)[2]	Current carrying capacity and setting of overload relay (a-release)	Overload protection with fuse, miniature circuit-breaker	Short-circuit protection with fuse, miniature circuit-breaker	Short-circuit protection with circuit-breaker (n-release)[2]	Current carrying capacity and setting of overload relay (a-release)	Overload protection with fuse, miniature circuit-breaker	Short-circuit protection with fuse, miniature circuit-breaker	Short-circuit protection with circuit-breaker (n-release)[2]
mm²	A	A	A	A	A	A	A	A	A	A	A	A
0.5	4	4	16	60	5	4	16	75	6	6	20	90
0.75	5	4	16	75	6	6	20	90	6	6	20	90
1	6	6	20	90	6	6	20	90	6	6	20	90
1.5	8	6	20	120	8	6	20	120	8	6	20	120
2.5	11	10	25	165	12	10	25	180	12	10	25	180
4	14	10	25	210	18	16	32	270	20	20	35	300
6	18	16	32	270	23	20	35	345	25	25	40	375
10	25	25	40	375	31	25	40	465	32	32	50	480
16	34	32	50	510	42	40	80	630	50	50	100	750
25					55	50	100	825	65	63	125	975
35					67	63	125	1005	85	80	160	1275
50					85	80	160	1275	115	100	200	1725
70					105	100	200	1575	149	125	250	2235
95					125	125	250	1875	175	160	315	2625
120					147	125	250	2205	210	200	400	3150
150					167	160	315	2505	239	200	400	3585
185					191	160	315	2865	273	250	500	4095
240					225	200	400	3375	322	315	630	4830
300					260	250	500	3900	371	315	630	5565

[1] In this case, a random laying (or "bunching") of the conductors is possible. The given values refer to 6 cores, all simultaneously loaded with 100%, within a multi-core bundle
[2] Instantaneous electromagnetic overcurrent release

tion circuit-breakers may be selected from up to 3 steps higher. Here the values in columns 4, 8 and 12 apply.
Power circuit-breakers may be adjusted to 15 times the value from column 2, 6 or 12 respectively. The corresponding calculated values are shown in columns 5, 9 and 13 respectively.

Table 1.13 Main differences between TTA (type-tested assemblies) and PTTA (partially type-tested assemblies)

TTA	PTTA
A rated diversity factor of between 0.6 and 1 may be assumed.	Unless otherwise agreed, a rated diversity factor of 1 must be used.
Construction parts may be used as an additional or as the only protective conductor.	Type-tested arrangement or separate protective conductor required. Cross-sections in accordance with DIN VDE 0660 Part 500 A5 (Draft).
Temperature-rise must be verified by means of tests.	Temperature-rise may be verified either by means of testing or by calculation using DIN VDE 0660 part 507.
Short-circuit withstand capability must be verified by means of tests.	Short-circuit withstand capability can be verified by tests or by extrapolation of type-tested arrangements.
Verification of the dielectric strength by means of high voltage tests (type and routine tests).	Verification of the dielectric strength or verification of the insulation resistance.

Note

Since only short-circuit protection is required for control circuitry, the selection may be made according to columns 4, 5 or 8, 9 or 12, 13 respectively. However, it should be noted that the coordination of the protection for short-circuits is not valid for main and control cables leaving the switchgear assembly.

The coordination of the protection system for the short-circuit protection of cable and wiring in main and control circuitry outside the switchgear assembly, must always be done under due consideration of the loop impedance (refer also to DIN VDE 0100 part 430).

In the case of the co-ordination of a common over-current protection system, the sizing may also be done in terms of overload. Such a coordination is only sensible however, if the protection against indirect contact (e.g. with functional extra-low voltage is achieved by means of safety separated circuits or residual-current protection systems. If this is not the case, the loop impedance must also be taken into consideration for the protection against indirect contact.

In a TN system (network), the loop impedance for the specification of the protection against short-circuits can be found in a similar way to the loop impedance referred to in section 1.4.2.

The "apparent increases" in the cable cross-sections contained in Tables 1.11 and 1.12 are partially determined by the installed switchgear. Switchgear devices are tested at 40 °C with a specified cable cross-section connected, i.e. they are only expected to carry their rated current if conductors exceeding this minimum cross-section are connected. These specified test cross-sections have been incorporated into the T.1 tables of DIN VDE 0660 part 500 A5 (draft), and have been taken into consideration in Tables 1.11 and 1.12.

The main differences between type-tested l.v. switchgear and controlgear assemblies (TTA) and partially type-tested assemblies (PTTA), can be found in Table 1.13 above.

1.3.1.2 Generic specifications

Generic specifications contain basic stipulations which are called for in e.g. construction and installation specifications. In addition, they contain stipulations concerning the protection of personnel, calculating and test procedures, fundamental ratings, etc. Wherever necessary, the Siemens switchboard catalogues also refer to the relevant generic specifications which are met by the equipment.

The most important generic specifications are:

DIN VDE 0100	Specifications for the erection of power installations with rated voltages up to 1000 V In particular, part 410 "Protection measures: Protection against dangerous shock currents" must be noted.
DIN VDE 0102	Calculation of short-circuit current in three-phase networks
DIN VDE 0103	Principles for the dimensioning of power installations with respect to mechanical and thermal short-circuit withstand capabilities
DIN VDE 0106 Part 100	Protection against electrical shock, arrangement of actuation elements in the vicinity of parts dangerous to touch
DIN VDE 0110 Part 1	Insulation co-ordination for electrical equipment in low-voltage installations – basic principles
Part 2	Dimensioning of creepage and clearance distances
DIN VDE 0199	Colours of indicator lights and pushbuttons
DIN VDE 0875	Specifications for the radio interference suppression of equipment and installations
IEC Publication 364	"Electrical installations of buildings"
IEC Publication 73	"Colours of indicator lights and and pushbuttons"
NF/UTE C20-040 (France)	Creepage and clearance distances ("Règles communes aux matériels électriques. Lignes de fuite et distances d'isolement dans l'air")
SEV 3017 (Switzerland)	Creepage and clearance distances"

1.3.1.3 Equipping specifications

Equipping specifications refer to the fitting of equipment in the electrical installation and include, amongst others:

▷ requirements related to the project engineering and design, including the selection of the equipment,

▷ supplementary requirements related to existing construction specifications which take into consideration the use of switchgear and distribution systems in specifically defined applications.

In general, no specific equipping specifications have been adhered to for the standard switchboards and distribution systems listed in the Siemens NV catalogues. The applicable equipping specifications must be determined in the initial project planning phase.

The most important equipping specifications are:

DIN VDE 0100	Specifications for the erection of power installations with rated voltages up to 1000 V, in particular part 726 – cranes and hoists
DIN VDE 0107	Specifications for the erection of power installations in hospitals and rooms used for medical purposes outside of hospitals
DIN VDE 0108 Parts 1 to 8 and suppl. 1 of Part 1	Erection and operation of electrical power installations and emergency power supplies in buildings used for public assembly.
EN 60 0204 Part 1/DIN VDE 0113 Part 1	"Electrical equipment of industrial machines. Part 1: General rules"
DIN VDE 0115	Specifications for electrical traction systems
DIN VDE 0118	Specifications for the erection of electrical installations in underground mining operations
DIN VDE 0160	Specifications for the equipping of electrical power installations with electronic devices
DIN VDE 0165	Erection of electrical installations in hazardous areas (potentially explosive atmospheres)
DIN VDE 0166	Electrical installations and their equipment in hazardous areas (potentially explosive atmospheres)

DIN VDE 0168	Specifications for the erection and operation of electrical installations in surface mining, quarrying and similar operations
IEC Publication 204-1	"The electrical equipping of industrial machines. Part 1: General rules"
BS 2771 Part 1 (Great Britain)	Identical to DIN VDE 0113 Part 1/EN 60 204 Part 1 (The electrical equipping of industrial machines. Part 1: Specification for general requirements)
JIC-EPG-1-30 (USA)	"Electrical standards for general purpose machine tools"
NFPA[1]) 79 (USA)	"Electrical standards for industrial machinery 1987"
NF/UTE C79-130 (France)	Identical to DIN VDE 0113 Part 1/EN 60 204 Part 1 (Appareils électriques industriels – Équipment électrique des machines industrielles – 1re partie: Règles générales)
SEV 1000 (Switzerland)	Regulations for domestic installations with special reference to power tools

For further information, with special reference to equipment specifications, see Section 1.5, page 57.

1.3.1.4 Erection specifications

Erection specifications include:

▷ Requirements related to the erection of electrical installations,
▷ Minimum safety requirements related to the construction of such installations for which no other specifications exist,
▷ Requirements related to the selection of the equipment.

When planning an installation using switchboards and distribution systems from the Siemens NV catalogues, the relevant erection specifications must also be established, since these may contain additional requirements pertaining to erection and commissioning which have not been considered in the standard catalogue designs.

The most important erection specifications are:

DIN VDE 0100	Specifications for the erection of power installations with rated voltages up to 1000 V. Of particular importance is Part 729 – Erection and connection of switch and distribution boards
DIN VDE 0107	Specifications for the erection of power installations in hospitals and rooms used for medical purposes outside of hospitals
DIN VDE 0108 Parts 1 to 8	Erection and operation of electrical power installations and emergency power supplies in buildings used for public assembly
DIN VDE 0118	Specifications for the erection of electrical installations in underground mining operations
DIN VDE 0165	Erection of electrical installations in hazardous areas (potentially explosive atmospheres)
DIN VDE 0166	Electrical installations and their equipment in hazardous areas (potentially explosive atmospheres)
DIN VDE 0168	Specifications for the erection and operation of electrical installations in surface mining, quarrying and similar operations
DIN VDE 0800	Specifications for the erection and operation of telecommunication installations including data processing systems
IEC Publication 364	"Electrical installations of buildings"
NF/UTE C15-100 (France)	Erection and maintenance specifications for low-voltage installations ("Installations électriques à basse tension: Règles (Recueil 1985/1986 incorporé, Édition 2982")
SEV 1000 (Switzerland)	Regulations for domestic installations

---

[1]) NFPA National Fire Protection Association

1.3.1.5 Relevant DIN standards for switchgear assemblies

These standards have a similar character to the generic specifications, and their application in switchgear assemblies has become obligatory insofar as they are not already contained in the construction and equipment specifications.

The most important relevant standards are:

DIN 41 488	Module sizes for cabinets
DIN 40 705	Identification of insulated and bare conductors by colours
DIN 40 719 Part 1 to ..	Circuit diagram documentation
DIN 40 050	Degrees of protection; protection against contact with live parts, protection against ingress of foreign bodies and liquids
DIN 50 010	Definitions of climatic conditions
DIN 50 019 Part 3	Technical climates, statistical climatic models
IEC Publication 529	"Classification of degrees of protection provided by enclosures"
IEC Publication 947-1, App. C	"Degrees of protection of enclosed equipment"
IEC Publication 446	"Indentification of insulated and bare conductors by colours"
BS 3042 (Great Britain)	"Specification for standard test fingers and probes for checking protection against electrical, mechanical and thermal hazard"
NF/UTE C04-200 (France)	Identification of conducteurs ("Repérage des conducteurs" (CEI 152, 391, 446 et HD 324))

In addition to these specifications and standards there are a number of VDI Guidelines which are of particular interest, e.g.

VDI 2853	Technical safety requirements pertaining to the construction, equipping and operation of industrial robots

VDI 2853 is applicable even without additional specific agreement between the manufacturer and the user.

1.4 Protection of persons and material assets

For the protection against dangerous body currents, (protection of personnel) DIN VDE 0660 Part 500 refers in full to DIN VDE 0100 Part 410, which contains the fundamental specifications pertaining to this subject.

1.4.1 Protection against direct contact

In switchgear installations and distribution systems a "protection against direct contact with live parts" (parts which carry a voltage under normal operating conditions), is a fundamental requirement.

1.4.1.1 Protection against electrical shock to DIN VDE 0100

A protection against direct contact is described in DIN VDE 0100 Part 410 and is always to be provided, irrespective of the value of voltage present.

Exception:

This protection is not required in the case of safety extra-low voltage (SELV) up to AC 25 V or DC 60 V (lower values could be specified e.g. in medical applications). However, this exception is e.g. not valid in hazardous locations where a protection against electrical shock is always required.

Protection against direct contact can be classified as total protection or as partial protection. Total protection is achieved by means of insulation, housings, shrouding, cladding and the like. The minimum degree of protection required is determined by the relevant construction, equipment and erection specifications. A partial protection is merely a protection against accidental contact, and not against deliberate or intentional touching. Partial protection is achieved with covers, barriers, guards or similar obstacles.

The nature and extent of the required protective measures are determined by the conditions prevailing at the site of installation. A partial protection is only permitted in electrical operating areas and locked electrical operating areas as defined in DIN VDE 0100, part 731, whereby barriers in front of active parts may only bend by a maximum of 20 mm when subjected to a point force of 500 N at their surface center.

Protection by means of distance in switchgear assemblies is not contained in the specifications DIN VDE 0660 part 500, which however does not exclude the

1 Specifications for low-voltage devices and switchgear assemblies

possibility that switchgear assemblies without any protection against direct contact may be installed in locked electrical operating areas.

1.4.1.2 Protection against electrical shock, positioning of operating elements in the vicinity of parts which are dangerous to touch

The protection against direct contact with live parts in the vicinity of hand actuated devices and operator elements, is an important part of protection against direct contact.

To ensure that the goals of the accident prevention regulations (UVV[1]) as issued by the trade associations for electrical installations and equipment (VBG4[2]) are covered by DIN VDE specifications, the corresponding stipulations are contained in DIN VDE 0106 part 100 (Fig. 1.1). This DIN VDE specification contains the basic principles for the lay-out and design of electrical equipment (e.g. devices, switchgear installations and distribution systems), with respect to protection against direct contact during the "re-establishment of a normal function" (e.g. adjustment, reset operations etc.)

[1] UVV "Unfallverhütungsvorschriften"
[2] VBG4 "Vorschriftenwerk der Berufsgenossenschaften für elektrische Anlagen und Betriebsmittel"

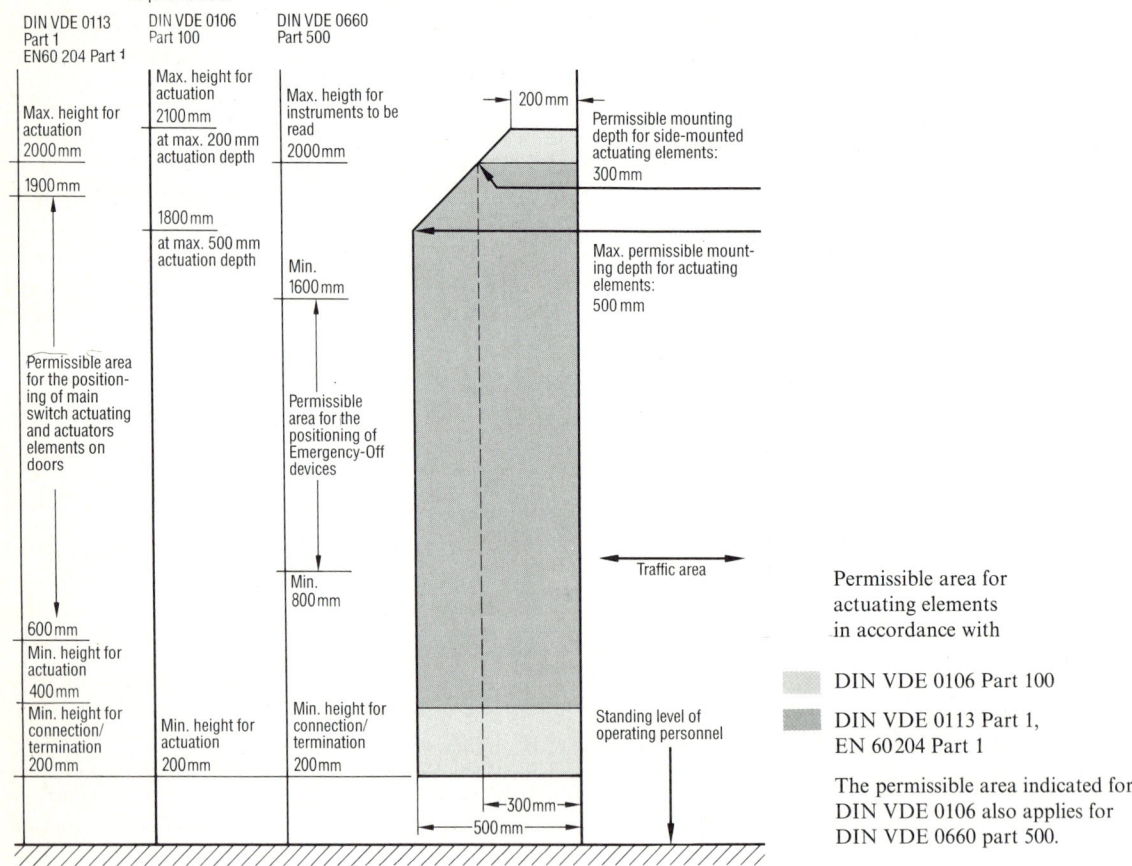

Figure 1.1
Permissible area for the positioning of hand-actuated devices and operator elements in accordance with DIN VDE 0106 part 100 with additional consideration of DIN VDE 0113 part 1 and DIN VDE 0660 part 500

Hand actuated devices and operator elements include:

Actuators
Operation of
    miniature circuit-breakers (m.c.b.),
    circuit-breakers (the previous term motor protection circuit-breakers is no longer defined),
    overload relays, motor and control switches,
    programming switches and keyboards.

Delatching of
    undervoltage or overvoltage relays,
    overload relays, leakage current or voltage operated earth-leakage protection systems,
    fuse monitoring systems,
    latching relays, balanced-beam relays,
    annunciator relays, flag relays.

Resetting of
    air vanes, pressure sensors,
    reclosing lock-out on switchgear devices,
    insulation monitoring systems.

Replacement parts
Replacement of
    screw-in fuses, and miniature scew-in fuses
    indicator lamps and tubes,
    lamps for cubicle lighting.

Undoing and plugging of
    plug-in connectors, plug elements, etc.
    insofar as this represents the re-establishment of a normal function.

Adjustment and setting of
    electronic components e.g. potentiometers,
    time relays, flashing relays, thermostats,
    pressure monitors, measuring relays,
    program cam switches and mechanisms.

Adjustment and setting can only be included under the heading "hand actuated devices and operator elements" under certain circumstances. If a once-only adjustment is required during the commissioning phase, then covers providing protection against finger or back-of-hand contact are not required. If the setting or adjustment serves to re-establish a normal function, however, then DIN VDE 0106 part 100 applies.

For all the cases listed above, it is assumed that the actuation is performed by trained personnel. Actuation by laymen requires a complete protection against direct contact.

HRC fuses do not fall under "hand actuated devices and operator elements". The replacement of HRC fuses is considered as work on the switchgear or switchboard. A protection around fuse switch disconnectors is however recommended, since these fuses are normally drawn and/or replaced under live operating conditions. Suitable covers should be fitted.

Generally speaking, the type tested switchgear assemblies (TTA) listed in the Siemens NV catalogues comply with these stipulations; either by virtue of suitably designed switchgear (e.g. with SIGUT termination technique, see section 5.2.1), by means of suitable covers and barriers, or by the distances between actuating devices/operating elements and live parts (Fig. 1.2 to 1.4).

In addition to the finger and back-of-hand safety area with its corresponding covers, a permissible area for the location of hand actuated devices and operator elements is defined. These requirements are illustrated in Figure 1.1. This figure has been extended to include related requirements from DIN VDE 0113 part 1 and DIN VDE 0660 part 500.

The other aims of the VBG 4, inasmuch as they concern the planner or manufacturer of a switchboard or distribution system, are also met by adherence to the DIN VDE specifications to which reference has been made by the trade associations in their notices (also refer to the appendix of the VBG 4 implementation instructions).

The "Manufacturers' Certificate[1]" to VBG 4, which is requested on an ever increasing scale by the operators of switchgear installations, can be issued for the TTA's in the Siemens NV catalogues. The "VBG 4 Certificate" shown in Figure 1.5 (freely translated from the German version[2]) refers to § 5 para. 4, and certifies that the entire switchboard or equipment including the installation has been checked to DIN VDE 0100 part 600. It is, however, not suitable for a certification of adherence to DIN VDE 0106 part 100.

The following comments on the accident prevention regulations (UVV) serve to clarify a number of points which are frequently misunderstood.

As underwriter for trade and industry, the trade association regards the reduction in the number of accidents as one of its most important goals. For this purpose, special accident prevention regulations pertaining to the various sectors, installations, tools, acti-

---

[1] The "Manufacturers' Certificate" is a certificate of compliance in which the manufacturer and/or the supplier confirms the DIN VDE specifications which have been adhered to.
[2] German version available from Carl Heymann Publishers, Gereonstr. 18–32, 50670 Cologne. Order No. ZH1/293

1 Specifications for low-voltage devices and switchgear assemblier

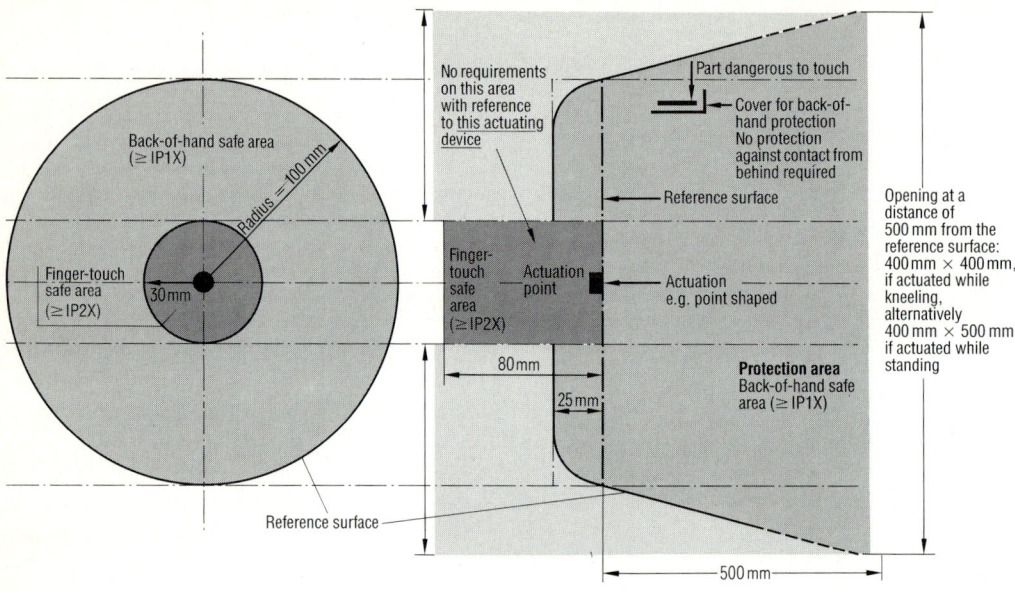

Figure 1.2 Finger and back-of-hand areas with respect to actuating elements

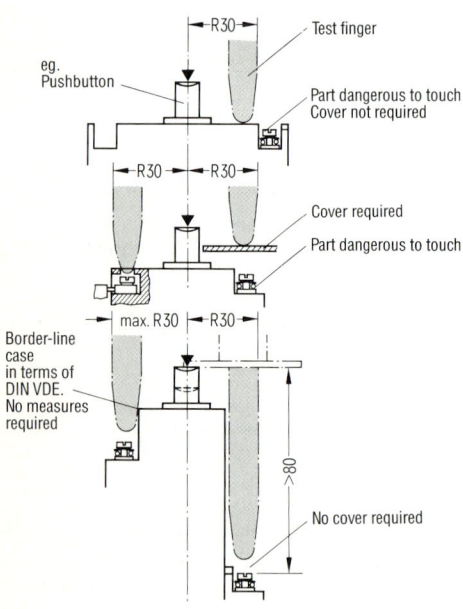

Figure 1.3
Examples of finger-touch protection;
Testing with a straight test finger in accordance
with DIN VDE 0470, part 1

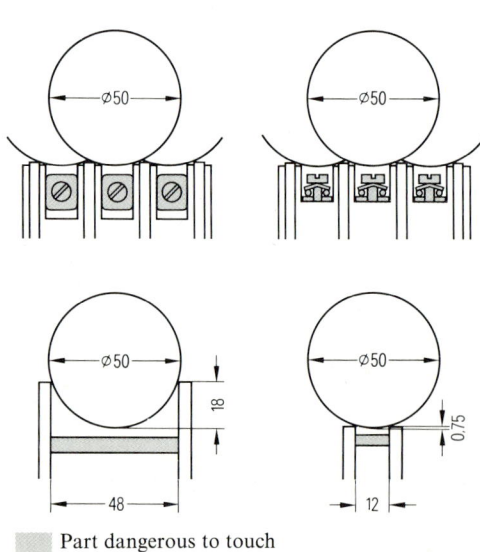

Figure 1.4
Examples of back-of-hand protection

Confirmation

In accordance with § 5 paragraph 4 of the Accident Prevention Regulations for electrical installations and equipment (VBG 4)

To

Address of the purchaser

Hereby it is confirmed that the electrical installation / electrical equipment / electrotechnical equipment of the machine or installation

Exact description of type and location

has been constructed in accordance with the Accident Prevention Regulations for electrical installations and equipment (VBG 4).

This confirmation serves the sole purpose of releasing the contractor from the obligation of inspecting the electrical installation / electrical equipment / electrotechnical equipment of the machine or installation, or having it inspected (§ 5 para. 1, 4 of the VBG 4) before taking it into service for the first time. Warranties and liability claims under civil law are not affected or determined by this confirmation.

Manufacturer or constructor
of the installation / equipment:

Stamp

Place and date   _Signature_

Figure 1.5
Example of a "VBG 4 Certificate" (unofficial translation of original in German – Carl Heymann Publishers KG, Cologne)

vities etc. are issued. These regulations are ratified and imposed by the Federal Department of Labour (Germany). Contravention of the accident prevention regulations can lead to prosecution and punishment by fine.

The "VBG 4 Elektrische Anlagen und Betriebsmittel" (electrical installations and equipment) is primarily aimed at the operator and user, although in certain instances it assumes that the manufacturer has fulfilled the necessary specifications to enable operation in accordance with the accident prevention regulations.

1.4.1.3 Measures to be taken in the modification or extension of switchgear assemblies already in service

In the modification of switchgear assemblies which have been in service for a longer period of time, or in the extension of standard switchgear assemblies from the catalogues, the following points should be noted:

▷ Measures in accordance with DIN VDE 0106 part 100 are only required in the area of hand actuated devices and operator elements which are required for the "re-establishment of a normal function", and only if exposed active parts with a voltage of $>AC\ 50\ V$ or $>DC\ 120\ V$ are present in the finger and back-of-hand safe areas.

▷ It must be verified that the equipment which is to be actuated already meets the specifications of DIN VDE 0106 part 100. If this is not the case, then covers must be provided, or auxiliary devices must be used for the actuation.

▷ It must be verified that the equipment which is to be actuated is so located, that no parts carrying a voltage which is dangerous to touch protrudes into the protection area. If necessary, covers must be provided, auxiliary devices must be used for the actuation or the actuation point must be moved.

▷ If a panel contains equipment which needs to be accessed for the re-establishment of a normal function, then it must be verified that the terminals of equipment built into the panel door are protected from back-of-hand contact. If necessary, suitable covers are to be fitted.

▷ Further, it is recommended that all other devices in the switchgear assembly be checked and if necessary be upgraded with suitable covers or shrouding.

1.4.1.4 "Safety separation" with low-voltage switchgear

The safe separation of circuits with "safety extra-low voltage" (SELV) and/or "functional extra-low voltage" (FELV) from other circuits, to ensure protection from dangerous shock currents, is becoming more and more relevant with the increasing use of electronic systems in power installations.

At present, there are no stipulations in the DIN VDE 0660 specifications for low-voltage switchgear which make specific reference to "safety separation".

In the basic safety specification, DIN VDE 0106 part 101 of 11.86 and the alterations A1, 2.89 (draft), fundamental requirements relating to measures within an item of electrical equipment, are mentioned (see Table 1.14). These are to be taken into consideration in the drafting or revision of safety related standards for electrical equipment.

Since no specific reference is made to "safety separation" in the DIN VDE specifications for low-voltage switchgear, Table 1.14 gives a summary and evaluation of pertinent statements which are made in other related specifications. According to DIN VDE 0106 part 101, a safe separation of circuits is ensured if the transfer of the voltage from one circuit to the other is prevented with adequate certainty. This certainty must be maintained throughout the service life of the item of electrical equipment.

Mention is made of faults, such as a bent or loose conductive part (bent solder pin, broken coil wire, loosened solder joint, fallen out screw) which have to be taken into account. Experience has shown that the possible breakage or cracking of device or housing parts (e.g. a phase barrier between terminals) during transport or installation must also be considered.

The installation conditions under which the Siemens contactors fulfill the requirements for "safety separation" can be found in Table 1.15. For voltages lower than those specified, no restrictions apply in the case of size 00 contactors. For the sizes 0 and 1, too (only contactors 3TH3/3TF3) no restrictions need to be considered up to 415 V. For other items of low-voltage switchgear, investigations are presently being conducted at the respective factories. The results of these tests and the respective recommendations will be published in the l.v. switchgear catalogues. Separate data for contactors and overload relays is available on request. Suitable interface contactor relays are

Table 1.14 Specifications concerning the requirements for "safety separation".

Standard	Topic	Requirements
DIN VDE 0106 part 101, 11.86 paragraph 2, 3, 4.2 and 4.4, part 101 A1, 2.89 (draft)	Safety separation	Safety separation between circuits exists if a single fault does not cause a transfer of the voltage from one circuit to the other. The insulation must have sufficient resistance to ageing for the duration of the expected service life.
	Clearances	The rated impulse withstand voltage of the clearances for increased insulation must be equal to to next higher application class, or to 160% of the impulse withstand voltage of the basic insulation.
	Creepage distances	For the creepage distances, double the distances than those for the basic insulation, apply. Alternatively an additional insulation, having characteristic values the same as those of the basic insulation, may be used.
DIN VDE 0160, 12.90, paragraph 5.7.5.2 (draft)	Clearances and creepage distances	To achieve "safety separation" by means of double or increased insulation, the clearances and creepage distances must be dimensioned as follows: The minimum values must be those pertaining to the pollution degree 2 [1]. For the dimensioning of the creepage distances, the values derived from paragraph 5.7.4.2 of the DIN VDE specifications must be doubled. Clearances and creepage distances which serve to achieve "safety separation" may not be exposed to pollution degrees [1] 3 or 4.
	Alternative: Protective screening	The clearances and creepage distances between the protective screen and the neighbouring circuits must be dimensioned in terms of the basic insulation of these circuits.
DIN VDE 0106 part 1 A1, 4.90 (draft)	Safety separation	The "safety separation" is achieved by: – double or strengthened insulation, – protective screening, – a combination of both. The requirements are similar to those of DIN VDE 0160.

[1] For pollution degrees refer to DIN VDE 0110 parts 1 and 2

Table 1.15
Contactors for "safety separation" to DIN VDE 0106 Part 101, Part 101 A1, 2.89 (draft) and DIN VDE 0160, 12.90 (draft)

Size	Type	Conditions for the "safety separation" [1] of – control circuit (coil) to auxiliary current paths and main current paths – auxiliary current paths to auxiliary current paths and main current paths – main current path to main current path	
		Rated voltage U_n [2]	Use of current paths of the contactors
00	3TF/3TH/3TJ 3TK	127 up to 240 V	The circuits which are to be "safety separated" may not occupy current paths which are immediately adjacent to each other. (eg. one current path must be left unused).
0	3TF/3TH	240 up to 415 V	
	3TB/3TH8	up to 500 V	The circuits which are to be "safety separated" may not occupy current paths which are immediately adjacent to each other or one above the other (including diagonal!).
1	3TB/3TF4	up to 415 V	
2 to 6	3TB/3TC	up to 300 V	The circuits which are to be "safety separated" may not respectively incorporate the NO and NC contacts of the same auxiliary contact block, or the contacts of auxiliary contact blocks mounted adjacent to each other.
2	3TF	up to 415 V	
8 up to 12	3TB/3TC	up to 500 V	
3 up to 14	3TF/3TK	up to 300 V	

[1] An earthed mounting plate or chassis has been assumed. An evaluation of the associated voltages is therefor not necessary.
[2] In the case of a network with earthed star-point, the main current paths may carry $\sqrt{3}$ times the voltage (line-to-line)

1 Specifications for low-voltage devices and switchgear assemblies

recommended in cases where items of switchgear, which do not fulfill the conditions of safety separation, are to be installed.

1.4.2 Protection against indirect contact

The term "protection against indirect contact" refers to measures which must be taken in addition to the normal basic insulation. In the event of an insulation failure, these measures must prevent the possibility of people or domestic animals coming into contact with a dangerous touch voltage on the body of the equipment, alternatively they must prevent the continued existence of touch voltages which are too high.

Touch voltages which are at present considered to be too high in terms of IEC 364-4-41 alt. DIN VDE 0100 part 410 are:

> AC 50 V (r.m.s.), alternatively
> DC 120 V,

whereby both voltages are defined as potentials with respect to earth. In unearthed networks, the voltage which applies is that voltage which can exist between earth and the unaffected conductors, if one of the conductors is shorted to earth.

These voltage limits may be considerably lower for specific applications. For example, in electrical operating areas on farms, or in hospitals etc., a touch vol-

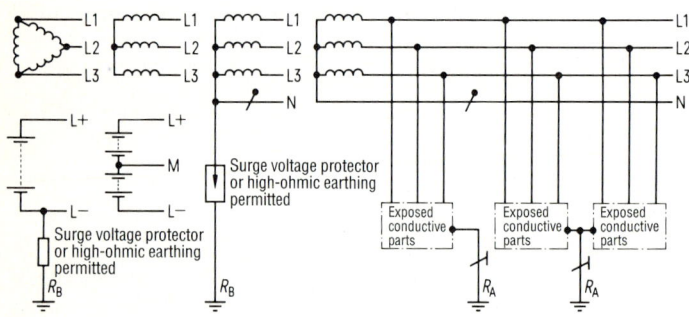

R_A Resistance between earth and the exposed conductive parts
R_B Operational earthing resistance

Figure 1.6   IT System (network)

First letter **I** (Insulation): No direct earthing of a network point

Second letter **T** (Terre): Exposed conductive parts directly earthed (In the case of protection by indication, additional equipotential bonding is required within the area of normal arm's reach.)

Caution when using neutral conductors!

The insulation resistance (dielectric strength) of the equipment in use must be considered!
Overcurrent detection in the neutral conductor is required. All main conductors, as well as the neutral conductor, must be disconnected during switch-off. The neutral conductor may not be disconnected before the main conductors.

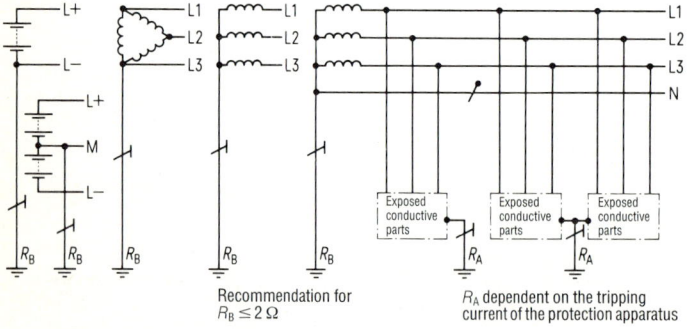

R_A Resistance between earth and the exposed conductive parts
R_B Operational earthing resistance

Figure. 1.7   TT System (network)

First letter **T** (Terre): Direct earthing of a network point required.

Second letter **T** (Terre): Exposed conductive parts directly earthed. (In certain instances, additional equipotential bonding may be required.)

Caution when using neutral conductors!

In the case of protection by means of disconnection through overcurrent protection systems, a detection of overcurrent in the neutral conductor is also required. The neutral conductor may not be disconnected before the main conductors. Without protection in the neutral conductor, switching off of the main conductors in ≤ 200 ms.

Protection against indirect contact

tage above 25 V (AC or DC) is considered to be too high.

Irrespective of the applicable touch voltage limits, protection against indirect contact is always required, i.e. as from 0 V.

DIN VDE 0100 part 410 no longer refers to the "8 protective measures", nor does it distinguish between protective measures with and without protective conductor. The only distinctions which are made, are

▷ measures which serve to ensure protection against direct and indirect contact (Upon closer examination, however, these do not apply, since an additional protection against direct contact is required even at safety extra-low voltages of AC >25 V or DC >60 V), and

▷ measures specific to the network configuration.

In addition to the measures which depend on the network configuration, paragraph 6 also describes the following protective measures:

- total insulation,
- protection by means of local non-earthed equipotential bonding,
- protection by means of non-conducting locations, and
- protection by means of safety separation.

Measures which, in terms of paragraph 4, are designed to fulfill the conditions for protection against direct as well as indirect contact are,

- protection by means of safety extra-low voltage (SELV),
- protection by means of functional extra-low voltage (FELV),
- protection by means of limiting the discharge energy.

For switchboards and distribution systems, the emphasis generally lies on measures which are specific to the network configuration and on total insulation. Thereby, the following points are to be taken into account:

▷ the type of system (network), i.e. whether it is a TT, IT or TN network (Fig. 1.6 to 1.8), as well as

▷ the nature of the protection system, whereby the following protection systems may be used:
  - overcurrent protection systems, such as fuses of the type gL to DIN VDE 0636, miniature circuit-breakers (m.c.b.) with B or C

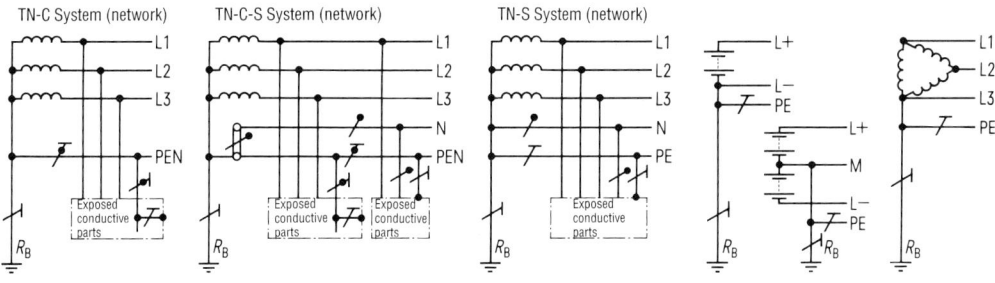

$R_B \leq 2\,\Omega$, alternatively $\dfrac{R_B}{R_E} \leq \dfrac{U_L}{U_0 - U_L} \leq 1:3.4$

R_B Operational earthing resistance
R_E Resistance to earth of foreign conductive parts

Figure 1.8 TN Systems (networks)
First letter **T** (Terre): Direct earthing of a network point required.
Second letter **N** (Neutral): Exposed conductive parts connected to the earthed network point via the protective earth conductor (PE) and/or the protective earth/neutral conductor (PEN). In certain instances, additional equipotential bonding may be required.

  C neutral and protective conductors common all the way as PEN conductor
  C-S neutral and protective conductors partially as common PEN and partially as separate conductors
  S neutral and protective conductors separate all the way

In the case of an earthed phase only a TN-S system (network) is possible. An earthed phase may not be used as a PEN conductor, i.e. separate phase and protective conductors are required for the entire run.

characteristics to DIN VDE 0641 A4,11.88[1]) (other characteristics, e.g. L, were permissible until Dec. '89), and power circuit-breakers to DIN VDE 0660 part 101,

- residual current protection devices to DIN VDE 0664, part 1 and 2[2]),
- insulation monitoring devices to DIN VDE 0413, parts 2 and 8,
- in special cases, also fault voltage protection devices to DIN VDE 0663.

---

[1]) The characteristic B is approximately the same as the type L, the characteristic C approximately the same as type G. Circuit-breakers with characteristics B and C may be used for line protection (Refer also to Section 9.7.3, page 539).

[2]) Network-dependent earth leakage protection devices, e.g. LS/DI are only permissible if the protection is achieved by the LS (circuit-breaker) part.

1.4.2.1 Comments related to the protective measures specific to the network configuration

In IT systems (networks), "protection by disconnection" as well as "protection by indication" may be applied. In both cases, the first fault need not cause immediate disconnection. In the case of "protection by disconnection" a second fault too need not cause immediate disconnection (disconnection under overload condition sufficient). On the other hand, with "protection by disconnection" a second fault must result in immediate disconnection. In this case, the disconnection depends on the type of network which exists after the first fault, i.e whether it is a TT system (bodies individually earthed) or a TN system (All bodies earthed via a common protective conductor).

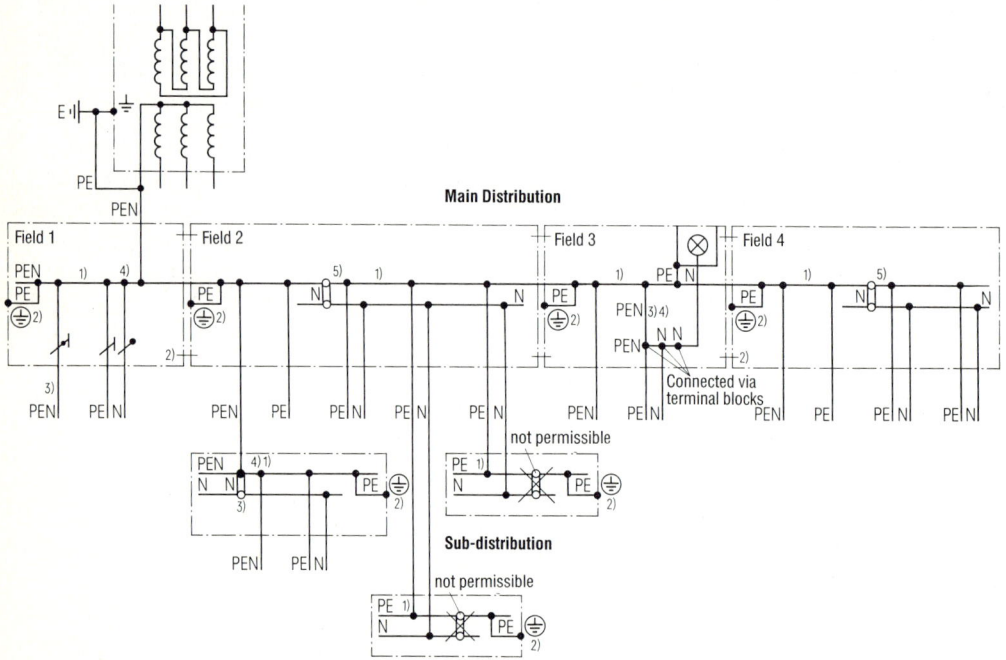

[1]) Connection facility for protective conductors (PE), PEN conductor, neutral conductor and if applicable main equipotential bonding conductor.

[2]) Incorporation of the construction or chassis parts by the fixing of the separately laid PEN or PE busbar by means of suitable connecting elements eg. toothed lock washers, if necessary via seperate protective conductor.

[3]) A combined protective and neutral conductor i.e. a PEN conductor, may only be used at cross-sections ≥ 10 mm² Cu, alternatively ≥ 16 mm² Al.

[4]) In terms of DIN VDE 0100 part 450 paragraph 8.2.5, protective conductors, neutral conductors and PEN conductors may be connected to PEN conductors. A separation into PE and N is only required in the case of cross-sections < 10 mm² Cu, alternatively < 16 mm² Al. After a separation in PE and N, these conductors may not be joined together again, and the neutral conductor may not be earthed.

[5]) The removable link shown here is no longer contained as a requirement in DIN VDE 0100. For practical reasons (measurement of the insulation resistance between the neutral conductor and the protective conductor or earth) it is however recommended that a removable link is used at least on each neutral conductor busbar.

Figure 1.9
The handling of protective, PEN and neutral conductors of differing cross-sections in switchgear assemblies connected to a TN system (network)

Only "protection by disconnection" can be used in TT and TN systems (networks).

In TT systems (networks), dangerous touch voltages >AC 50 V or >DC 120 V must be disconnected within 5 seconds[1] (in special cases within 0.2 s).

In TN systems (networks), dangerous touch voltages in circuits with socket outlets up to 35 A or in circuits with hand-held equipment of the protection class I, must be disconnected within 0.2 s[1]. Faults in circuits with permanently connected equipment must be disconnected in ≤ 5 s[1].

In all network types, an equipotential bonding must be carried out for each building.

The switchgear assemblies of the Siemens NV catalogues can be used in any of the network types. For use in TN systems (networks), however, it must first be established whether the installation will be equipped with 4 or 5 conductors, i.e. with PEN conductor or with PEN and N conductor[2] or with PE and N conductor. Figure 1.9 serves to illustrate these options.

The dimensioning of the protective conductors (PE), PEN conductors and neutral conductors (N) is to be done in accordance with DIN VDE 0660 part 500 Draft A5, as illustrated in Table 1.16.

Since a neutral conductor is not normally used in IT systems (networks), no specific data is provided.

In the Federal Republic of Germany, "protection by disconnection in TN systems (networks)" is the most common form. In the case of this form of protection, the following must be taken into account:

▷ All exposed conductive parts of electrical equipment in a TN-S system (network) must be connected to the star-point of the network transformer by means of a protective conductor. In a TN-C system (network), they must be connected via a protective conductor to the PEN conductor, and finally to the star-point of the network transformer.

[1] New disconnection times are in preparation. Depending on the rated voltage to earth (U_0), they will be between 0.1 and 0.8 s.
[2] In some countries, e.g. in France and to some extent in Italy and Norway, the disconnection of the N conductor is obligatory. This must be taken into account in the planning and ordering stages of a project (4-pole switching devices).

Table 1.16
Dimensioning of protective conductors (PE), PEN conductors and neutral conductors (N) in low-voltage switchgear assemblies (draft DIN VDE 0660 part 500)

Phase conductor	Protective conductor (PE)	PEN Conductor	Neutral conductor (N)
$\geq 0.5 \leq 10$ mm²	as for phase conductor	–	as for phase conductor
$>10 \leq 16$ mm²	as for phase conductor	as for phase conductor	as for phase conductor
$>16 \leq 35$ mm²	min. 16 mm²	min. 16 mm²	min. 16 mm²
$>35 \leq 400$ mm²	50% of the phase conductor cross-section or determined by calculation or testing, however 16 mm² minimum	under consideration of the unbalance current k value of material for higher initial temperature to be taken into account	
$>400 \leq 800$ mm²	min. 200 mm² or determined by calculation or testing, however 16 mm² minimum	up to 30% unbalance minimum 200 mm² above 30% in accordance with Table T1[1] alt. DIN 43670/671 k value of material for higher initial temperature to be taken into account	
>800 mm²	25% of the phase conductor cross-section or determined by calculation or testing, however 16 mm² minimum	up to 30% unbalance minimum 25% of phase conductor above 30% in accordance with Table T1[1] alt. DIN 43670/671 k value of material for higher initial temperature to be taken into account	

[1] Table T1, see draft DIN VDE 0660 part 500 A5, alternatively Tables 1.11 and 1.12 on pages 28 and 29

1 Specifications for low-voltage devices and switchgear assemblies

▷ For cross-sections <10 mm² Cu alternatively <16 mm² Al, the protective and neutral conductors must be laid separately. For cross-sections ≥10 mm² Cu alternatively ≥16 mm² Al, the protective and neutral conductors may be combined[1] in one conductor, i.e. the PEN conductor (refer Fig. 1.9).

▷ As already mentioned, certain disconnection times have to be adhered to. In order to check the disconnection time, it is necessary to measure or to calculate the loop impedance (Z_s).

The condition

$$Z_s \cdot I_a \leq U_0$$

must be fulfilled.

---
[1] To some extent, DIN VDE 0107 specifies that the neutral and protective conductors must be laid separately irrespective of their cross-section. This is also recommended in the case of building installations to DIN VDE 0108. This recommendation is of particular interest for telecommunication installations.

Z_s Loop impedance
I_a Tripping or breaking current of the protection system. For residual-current protection systems $I_a = I_{\Delta N}$
U_0 Voltage to earth

Here follows an example of a calculation of the loop impedance. Alternatively it may be used for the checking of the disconnection conditions (refer to Fig. 1.10):

To simplify the calculation of the total series impedance and the maximum cable length downstream of the last protection system, the calculation form shown in Figure 1.11 may be used. For Section I of the calculation form, the resistance values of the network transformer can be obtained from Table 1.17, whereas the resistance values of connecting cable are given in Table 1.18, a to d (These values refer to the length of cable, including outgoing plus returning i.e. to the distance). For more than three connecting lengths one after the other, Section I of the calculation form must be extended accordingly. The values of the ohmic and the inductive resistances are to be added separately to determine the loop impedances. The branch or feeder circuit to be calculated is described in Section II, whereby the required breaking cur-

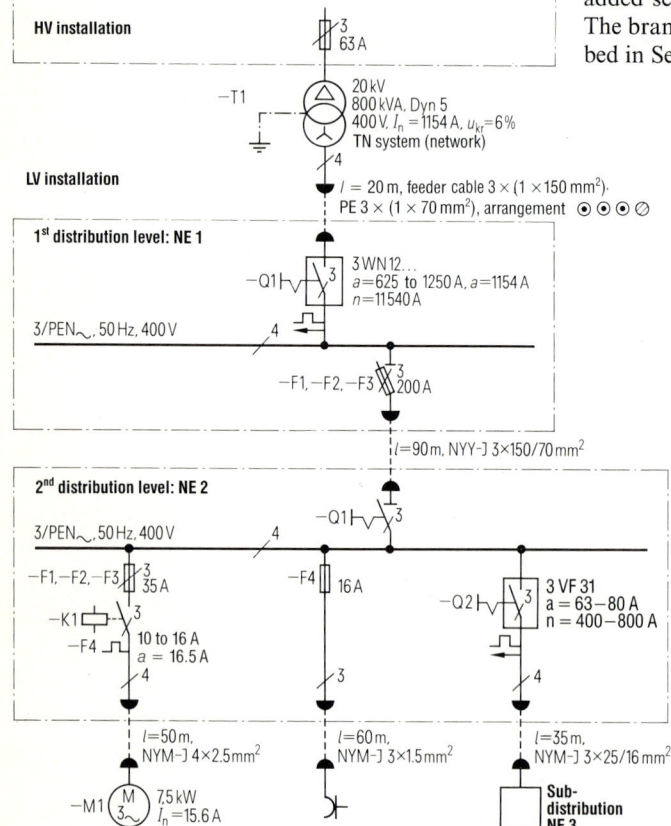

Figure 1.10
Schematic circuit diagram.
Used e.g. for the checking of the protection measure "protection by disconnection" in a TN system (network)

Determination of the loop impedance · Maximum cable lengths

Calculation sheet 1

		R mΩ	X mΩ

Section I — Loop impedance upstream of short-circuit protection device

Cable from: Transformer ___ kVA, U_n = ___ V, I_n ___ A, u_{kr} = ___ %, Vec. grp. ___

Cable from: transformer to 1st distribution
- for I_n = ___ A, cross-section ___ mm², l = ___ m
- $R = \dfrac{R'}{n} \times l$ = ___ mΩ/m × ___ m
- $X = \dfrac{X'}{n} \times l$ = ___ mΩ/m × ___ m

Cable or Busbars from: 1st distribution to 2nd distribution
- for I_n = ___ A, cross-section ___ mm², l = ___ m
- $R = \dfrac{R'}{n} \times l$ = ___ mΩ/m × ___ m
- $X = \dfrac{X'}{n} \times l$ = ___ mΩ/m × ___ m

2nd distribution to 3rd distribution
- for I_n = ___ A, cross-section ___ mm², l = ___ m
- $R = \dfrac{R'}{n} \times l$ = ___ mΩ/m × ___ m
- $X = \dfrac{X'}{n} \times l$ = ___ mΩ/m × ___ m

Section II

- s.c.p.d.[1], type ___ rated current ___ A
- setting of n-release ___ A
- cross-section after s.c.p.d.[1] ___ mm²
- required short-circuit current $I''_{k\,req.}$ ___ A

Sum:
$Z = \sqrt{R^2 + X^2}$
$Z = \sqrt{^2 + ^2}$
$Z = $ mΩ

Section III — Max. permissible cable length after s.c.p.d.[1]

3-phase:
$$L = n \cdot \dfrac{1000 \cdot \dfrac{U_o}{I''_{k\,req.}} - Z}{Z'}$$

AC/DC:
$$L = n \cdot \dfrac{x \cdot s \cdot 10^{-3} \cdot \left(\dfrac{1000 \cdot U_o}{I''_{k\,req.}} - Z\right)}{2} \quad *)$$

L = ___
L = ___ m

Values for 3 phase cable
Z' in mΩ/m for PVC insulated cable NYY with copper conductors

Core cross-section mm²	3½ core cable 4½ core cable	4/5 core cable	single core cable PE = ½ ⊙⊙⊙⊘	single core cable PE = ½ ⊙⊙⊘
0.5	--	107.2	--	--
0.75	--	71.5	--	--
1	--	53.6	--	--
1.5	--	35.7	--	--
2.5	--	21.44	--	--
4	--	13.4	--	--
6	--	8.93	--	--
10	--	5.36	--	--
16	--	3.35	--	--
25	3.11	2.15	3.13	3.12
35	2.23	1.54	2.25	2.24
50	1.56	1.08	1.60	1.59
70	1.12	0.78	1.17	1.16
95	0.84	0.59	0.89	0.88
120	0.67	0.47	0.74	0.72
150	0.55	0.39	0.63	0.61
185	0.45	0.33	0.56	0.53
240	0.37	0.27	0.49	0.46
300	0.31	0.24	0.45	0.42

Z' in mΩ/m for PVC insulated cable NYY with aluminium conductors

Core cross-section mm²	3½ core / 4½ core cable	4/5 core cable	single core PE=½	single core PE=½
16	--	5.70	--	--
25	--	3.64	--	--
35	--	2.60	--	--
50	--	1.83	--	--
70	--	1.31	--	--
95	--	0.97	--	--
120	--	0.77	--	--
150	0.90	0.63	0.96	0.94
185	0.74	0.52	0.80	0.79
240	0.58	--	0.66	0.64
300	--	--	0.57	0.55

Conductivity of copper: $x_{55°}$ = 49.3 m/Ωmm² ; $x_{160°}$ = 37.3 m/Ωmm²

Abbreviations

Abbr.	Unit	Description
l	m	cable length before s.c.p.d.[1]
L	m	Max. permissible cable length after s.c.p.d.[1]
U_o	V	220/230 V in 380/400 V network, 288 V in 500 V network, 380/400 V in 660/690 V network
$I''_{k\,req.}$	A	short-circuit current required by s.c.p.d.[1]
q	mm²	cable cross-section after s.c.p.d.[1]
Z	mΩ	loop impedance from current source (transformer) to the s.c.p.d.[1]
R', X'	mΩ/m	ohmic/inductive resistance of the loop impedance per meter cable of given cross-section
Z'	mΩ/m	loop impedance per meter cable of given cross-section after the s.c.p.d.[1]
n	--	number of parallel conductors of same cross-section

*) 2 if the cross-section of the protective conductor is equal to that of the main conductor
   3 if the cross-section of the protective conductor is equal to half that of the main conductor

[1] s.c.p.d. short circuit protection device

Figure 1.11
Calculation form for the determination of the loop impedance and the maximum cable length.
The data from Section I and II are to be used in the relevant formula for three-phase or single-phase AC or DC of Section III

1 Specifications for low-voltage devices and switchgear assemblies

Example 1: Motor feeder

Determination of the loop impedance upstream of the short-circuit protection device -F1, -F2, -F3 (fuses 35 A) in the second distribution level NE 2

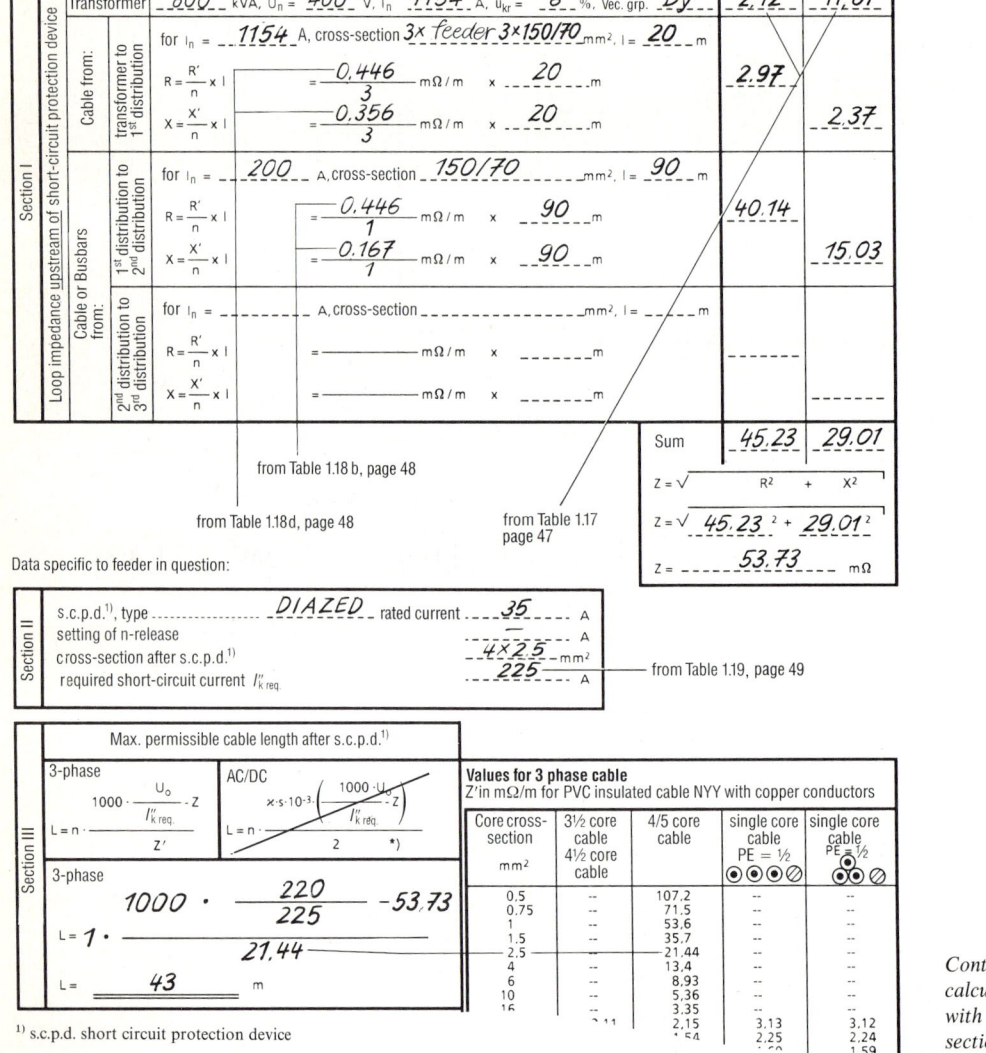

Figure 1.12
Calculation procedure pertaining to the schematic circuit diagram shown in Figure 1.10, page 44

Continuation of the calculation example with 4 mm² cross-section see Figure 1.13, page 52

[1] s.c.p.d. short circuit protection device

rent when using fuses for disconnection times of 5 s or 0.2 s are to be taken from Tables 1.19 and 1.20 respectively (see pages 49 and 50). The values below the bold line, are the values relevant to protection by disconnection. The values above this line refer to breaking currents in the event of short-circuit protection. Tables 1.21a and 1.21b on page 51 apply for the use of miniature circuit-breakers. For moulded case circuit-breakers or power circuit-breakers the values may be found in the corresponding Siemens l.v. switchgear catalogue.

The calculated cable length (Fig. 1.12) is less than the required length of 50 m. To achieve the required

Table 1.17 Three-phase power transformers
Resistance values R, X, Z, of the loop impedance in the event of a single-pole short-circuit, rated primary voltage 12 or 24 kV, rated secondary voltage 400 V

Vector group Yz ... [1]

	Rated power S_{nT} kVA	Rated current I_n A	Short-circuit voltage u_{kr} %	R mΩ	X mΩ	Z mΩ
Oil	50	72	4	58.88	73.30	94.02
	75	108	4	35.50	51.02	62.16
	100	144	4	24.96	39.12	46.40
			6	29.44	62.07	68.70
	125	180	4	18.64	31.91	36.96
			6	22.12	50.16	54.82
	160	231	4	13.60	25.35	28.77
			6	16.00	39.60	42.71

Vector group Dy ... [1]

	Rated power S_{nT} kVA	Rated current I_n A	Short-circuit voltage u_{kr} %	R mΩ	X mΩ	Z mΩ
Cast resin	50	72	4	67.20	107.09	126.43
	75	108	4	38.40	74.91	84.18
	100	144	4	27.20	56.95	63.11
	125	180	4	20.48	46.13	50.47
	160	231	4	14.40	36.69	39.42
Oil/Cast resin	200	289	4	12.24	29.07	31.54
	250	361	4	9.02	23.55	25.22
	315	455	4	6.76	18.83	20.01
	400	577	4	5.08	14.91	15.75
	500	722	4	3.78	12.02	12.60
	630	910	4	2.74	9.61	9.99
Cast resin	100	144	6	28.00	90.27	94.51
	125	180	6	21.25	72.55	75.60
	160	231	6	15.60	56.96	59.06
Oil/Cast resin	200	289	6	13.04	45.41	47.25
	250	361	6	9.79	36.50	37.79
	315	455	6	7.42	29.06	29.99
	400	577	6	5.36	22.99	23.61
	500	722	6	4.03	18.45	18.89
	630	910	6	2.92	14.71	15.00
	800	1154	6	2.12	11.61	11.80
	1000	1443	6	1.62	9.30	9.44
	1250	1804	6	1.22	7.45	7.55
	1600	2309	6	0.96	5.82	5.90
Oil	2000	2886	6	0.92	4.63	4.72
	2500	3608	6	0.68	3.72	3.78
Cast resin	2000	2886	8	0.56	6.27	6.30
	2500	3608	8	0.44	5.01	5.03

[1] For vector groups, refer to Section 9.6.1 on page 535

Table 1.18 Resistance values of PVC insulated connecting cables with copper conductors at **55 °C** (cable temperature)[1]
For PVC insulated cables with *Aluminium* conductors, the resistance value R' must be multiplied by the *factor 1,7*

a) 4 and 5 core cable

Core number × rated cross-section mm²	R' [2] mΩ/m	X' [3] mΩ/m	Z' mΩ/m
4 × 0.5	81.90	0.23	81.90
4 × 0.75	55.74	0.23	55.74
4 × 1	41.18	0.23	41.18
4 × 1.5	27.53	0.23	27.53
4 × 2.5	16.86	0.22	16.86
4 × 4	10.49	0.21	10.49
4 × 6	7.01	0.20	7.01
4 × 10	4.16	0.19	4.16
4 × 16	2.62	0.18	2.63
4 × 25	1.654	0.176	1.663
4 × 35	1.192	0.160	1.203
4 × 50	0.880	0.159	0.894
4 × 70	0.610	0.155	0.629
4 × 95	0.440	0.154	0.466
4 × 120	0.348	0.151	0.379
4 × 150	0.282	0.151	0.320
4 × 185	0.226	0.151	0.272
4 × 240	0.172	0.149	0.228
4 × 300	0.136	0.149	0.202

b) 3½ and 4½ core cable

Core number × rated cross-section mm²	R' [2] mΩ/m	X' [3] mΩ/m	Z' mΩ/m
3 × 25/16	2.135	0.182	2.143
3 × 35/16	1.904	0.183	1.913
3 × 50/25	1.267	0.179	1.279
3 × 70/35	0.901	0.174	0.918
3 × 95/50	0.660	0.168	0.681
3 × 120/70	0.479	0.160	0.505
3 × 150/70	0.446	0.167	0.476
3 × 185/95	0.333	0.163	0.371
3 × 240/120	0.260	0.164	0.307
3 × 300/150	0.209	0.162	0.264

[1] For resistance values at other temperatures, refer to Section 9.4.5, page 529
[2] Base values for outgoing and return conductor (i.e. loop values); in accordance with DIN VDE 0295, Tables 1 and 2, as well as with DIN VDE 0102. Based on average conductor temperature of 55 °C
[3] Base values from "Kabel und Leitungen für Starkstrom", 4th edition, Siemens AG, in accordance with DIN VDE 0102 (loop values).

c) Single core cable

Core number × rated cross-section Main conductor/PEN mm²	R' [2] mΩ/m	X' [3] mΩ/m	Z' mΩ/m
1 × 25/16	2.135	0.390	2.170
1 × 35/16	1.904	0.377	1.941
1 × 50/25	1.267	0.373	1.321
1 × 70/35	0.901	0.366	0.973
1 × 95/50	0.660	0.333	0.739
1 × 120/70	0.479	0.330	0.582
1 × 150/70	0.446	0.327	0.553
1 × 185/95	0.333	0.326	0.466
1 × 240/120	0.260	0.320	0.412
1 × 300/150	0.209	0.328	0.381

d) Single core cable

Core number × rated cross-section mm² Main conductor/PEN mm²	R' [2] mΩ/m	X' [3] mΩ/m	Z' mΩ/m
1 × 25/16	2.135	0.390	2.170
1 × 35/16	1.904	0.377	1.941
1 × 50/25	1.267	0.373	1.321
1 × 70/35	0.901	0.366	0.973
1 × 95/50	0.660	0.362	0.753
1 × 120/70	0.479	0.359	0.599
1 × 150/70	0.446	0.356	0.571
1 × 185/95	0.333	0.354	0.486
1 × 240/120	0.260	0.349	0.435
1 × 300/150	0.209	0.347	0.405

length, either the short-circuit protection may be reduced, or the cross-section may be increased to 4 mm². The increase in cross-section is the preferred method, since thereby the voltage drop is also reduced. The protection measure could also be achieved by the use of an earth-leakage or residual-current protection system.

Continued on page 53

Required short-circuit current

Table 1.19
Required short-circuit current $I''_{k\,req.}$ (breaking current) for line protection fuses to DIN VDE 0636, operation class gL (gG), under consideration of the permissible short-circuit thermal rating (160 °C) of PVC insulated cables with copper conductors and a breaking time of **5 s**

Rated cross-section q mm²	Rated current I_n in A — Required short-circuit current $I''_{k\,req.}$ in A																						
	2	4	6	10	16	20	25	32	35	40	50	63	80	100	125	160	200	224[1]	250	300[1]	315	355[1]	400
0.5	9	19	28	76	180	400																	
0.75			28	55	115	190	370																
1				47	93	137	250	350	890														
1.5					72	100	175	290	550	520													
2.5							88	200	360	350	860	1880											
4								156	225	220	520	890	1750	3200									
6									173	200	330	570	950	1600	3100	6100							
10											260	400	640	1050	1900	3200	6800	4800					
16												351	452	640	1080	1800	3450	3200	5400	6300	10700	8400	
25														573	751	1140	2100	2000	3100	5000	5700	6000	11100
35																995	1380	1350	2000	3700	3500	4300	6100
50																	1310	1300	1580	2300	2450	3100	4100
70																				2050	2070	2300	2950
95																							2650
120																							
150																							
185																							
240																							
300																							
400																							
500																							
2×50																							
2×70																							
2×95																							
to 2×500																							
3×50																							
3×70																							
to 3×500																							
4×50 to 4×500	9	19	28	47	72	88	120	156	173	200	260	351	452	573	751	995	1310	1300	1580	2050	2070	2300	2650

Rated cross-section q mm²	425[1]	500	630	800	1000	1250
16	22000					
25	8800					
35	6800	10100	28000			
50	4900	7150	16000			
70	3200	5000	10050	21000	39000	90000
95	2900	3550	7100	13500	22500	51000
120			5150	9400	15500	35500
150			5100	7400	12000	25000
185				6500	9400	19000
240					8500	15000
300						12300
400						
500				6500	8500	12300
2×50				9000	14500	32000
2×70				6500	10500	19500
2×95					8500	15000
to 2×500					8500	12300
3×50					9400	19000
3×70					8500	13000
to 3×500						12300
4×50 to 4×500	2900	3550	5100	6500	8500	12300

Non-protected range of cross-sections

Protected range of cross-sections, provided that the required short-circuit current $I''_{k\,req.}$ flows.

Where $I''_{k\,req.}$ is not entered for a particular cross-section, the preceeding value for $I''_{k\,req.}$ applies. This value refers to the breaking time of **5 s** in accordance with DIN VDE 0100 part 600.

[1] Fuse not listed by DIN VDE 0636, however available from Siemens

Table 1.20
Required short-circuit current $I''_{k\,req.}$ (breaking current) for line protection fuses to DIN VDE 0636, operation class gL (gG), under consideration of the permissible short-circuit thermal rating (160 °C) and a breaking time of **0.2 s** for protection against indirect contact, referred to PVC insulated cables with copper conductors

Rated cross-section q mm²	\multicolumn{29}{c}{Rated current I_n in A — Required short-circuit current $I''_{k\,req.}$ in A}																												
	2	4	6	10	16	20	25	32	35	40	50	63	80	100	125	160	200	224[1]	250	300[1]	315[1]	355[1]	400	425[1]	500	630	800	1000	1250
0.5	20	40	60	100	180	400																							
0.75				148	191	370																							
1						270	290	550	520																				
1.5								332	367	410	860	1880																	
2.5											578	890	1750	3200															
4												750	1000	1600	3100	6100													
6													1270	1900	3200	6800	4800		8400										
10																1700	2180	3450	3200	5400	6300	10700							
16																	2950	2600	3600	5000	5700	6000	11100	8800	22000				
25																			4400	4800	5000	6100	6800		10100	28000			
35																							6400		8000	16000			
50																										11700	21000	39000	90000
70																										15800	22500	51000	
95																											20500	35500	
120																												28500	
150																													
185																													
240																													
300																													
400																												→	28500
500																												32000	28500
2×50																													
2×70																													
2×95 to 2×500																													
3×50	→	→	→	→	→	→	→	→	→	→	→	→	→	→	→	→	→	→	→	→	→	→	→	→	→	→	→	→	→
3×70 to 3×500	→	→	→	→	→	→	→	→	→	→	→	→	→	→	→	→	→	→	→	→	→	→	→	→	→	→	→	→	→
4×50 to 4×500	20	40	60	100	148	191	270	332	367	410	578	750	1000	1270	1700	2180	2600	2950	3600	4400	4800	5000	6100	6400	8000	11700	15800	20500	28500

Non-protected range of cross-sections

Protected range of cross-sections, provided that the required short-circuit current $I''_{k\,req.}$ flows.

Where $I''_{k\,req.}$ is not entered for a particular cross-section, the preceeding value for $I''_{k\,req.}$ applies. This value refers to the breaking time of **0.2 s** in accordance with DIN VDE 0100 part 600.

Values above the dark line result in shorter breaking times, and are required as protection against prohibitive thermal stressing (≤ 160 °C).

[1] Fuse not listed by DIN VDE 0636, however available from Siemens

Resistance values of PVC cables · Required values of PVC cables

Table 1.21a
Required short-circuit current $I''_{k\,req.}$ (breaking current) of miniature circuit-breakers in A to DIN VDE 0641 A4, characterisic B, under consideration of the permissible short-circuit thermal rating (160 °C) of PVC insulated cables with copper conductors

| Rated cross-section q mm² | \multicolumn{3}{c}{6} | \multicolumn{3}{c}{10} | \multicolumn{3}{c}{16} | \multicolumn{3}{c}{20} | \multicolumn{3}{c}{25} | \multicolumn{3}{c}{32} | \multicolumn{3}{c}{40} | \multicolumn{3}{c}{50¹⁾} | \multicolumn{3}{c}{60¹⁾} |

Rated current I_n in A (column groups above); Rated short-circuit breaking capacity I''_c in kA: 3 | 6 | 10 for each group.

Required short-circuit current $I''_{k\,req.}$ in A:

q mm²	6A/3	6A/6	6A/10	10A/3	10A/6	10A/10	16A/3	16A/6	16A/10	20A/3	20A/6	20A/10	25A/3	25A/6	25A/10	32A/3	32A/6	32A/10	40A/3	40A/6	40A/10	50A/3	50A/6	50A/10	60A/3	60A/6	60A/10
0.5	\multicolumn{15}{l}{Non-protected range of cross-sections for I''_k of 3 kA, 6 kA}												
0.75																\multicolumn{12}{l}{Probably non-protected range of cross-sections. No definitive statement possible, since the required values are not contained in DIN VDE (refer to manufacturer's data)}
1																											
1.5	30			\multicolumn{12}{l}{Non-protected range of cross-sections for 10 kA at the point of installation}												
2.5	+	30	30	50	+	50	80	+	80	100	+	100	125	+	125												
≥4	30	30	30	50	50	50	80	80	80	100	100	100	125	125	125	160	160	160	200	200	200	250	250	250	315	315	
≥6	\multicolumn{27}{l}{Protected range of cross-sections}

¹⁾ Not offered by Siemens

Table 1.21b
Required short-circuit current $I''_{k\,req.}$ (breaking current) of miniature circuit-breakers in A, to DIN VDE 0641 A4, characterisic C, type 5SN... ($I''_{k\,req.} = 10 \cdot I_N$) under consideration of the permissible short-circuit thermal rating (160 °C) of PVC insulated cables with copper conductors

Required short-circuit current $I''_{k\,req.}$ in A (Non-protected range of cross-sections / Protected range of cross-sections):

Rated cross-section q mm²	0.5 A	1 A	1.6 A	2 A	3 A	4 A	6 A	8 A	10 A	16 A	20 A	25 A	32 A	40 A	50 A
0.5	5	10	16	20	30										
0.75															
1						40	60	80	100						
1.5										160	200	250	320	400	
2.5															500
≥4	5	10	16	20	30	40	60	80	100	160	200	250	320	400	500

1 Specifications for low-voltage devices and switchgear assemblies

continuation of the calculation on page 46

Determining the permissible cable length in terms of the protection apparatus

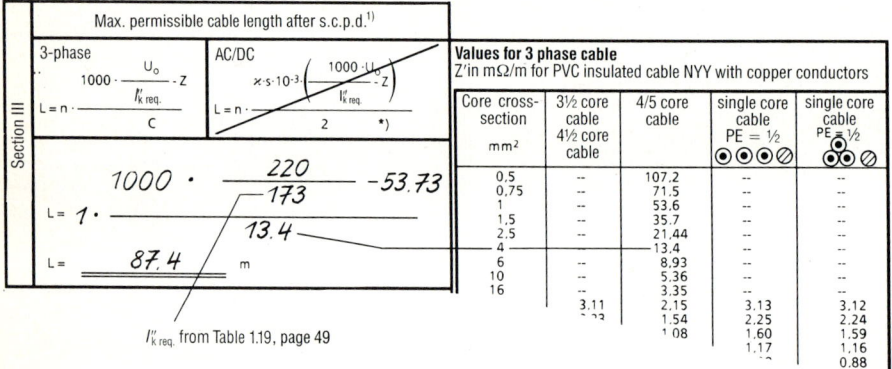

$I''_{k\,req.}$ from Table 1.19, page 49

Thereby, the "protection by disconnection" is ensured for this feeder. Simultaneously, short-circuit protection is achieved.

Figure 1.13 Checking the maximum cable length of cross-section 4 mm²

Example 2: Socket outlet circuit

The loop impedance upstream of the short-circuit protection device −F4 can be taken from Example 1 (Fig. 1.12)

Data specific to feeder in question:

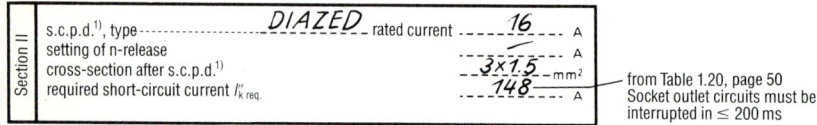

Determination of the permissible cable length downstream of the short-circuit protection device:

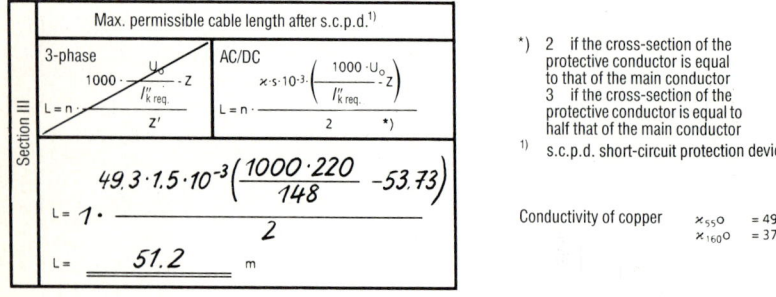

*) 2 if the cross-section of the protective conductor is equal to that of the main conductor
3 if the cross-section of the protective conductor is equal to half that of the main conductor

[1]) s.c.p.d. short-circuit protection device

Conductivity of copper $\varkappa_{55°}$ = 49.3 m/Ωmm²
                          $\varkappa_{160°}$ = 37.3 m/Ωmm²

Thereby, for this feeder too, the "protection by disconnection" and the short-circuit protection is ensured.

Figure 1.14 Calculation procedure pertaining to the circuit diagram shown in Figure 1.10, on page 44

Example 3: Sub-distribution NE 3

The loop impedance upstream of the short-circuit protection device $-Q2$ can be taken from Example 1 (Fig. 1.12)

Data specific to feeder in question:

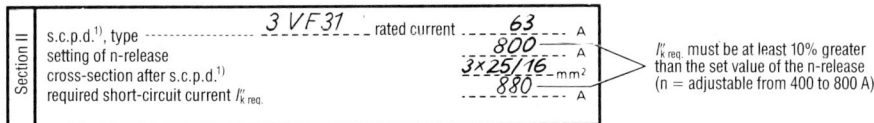

Determination of the permissible cable length downstream of the short-circuit protection device (s.c.p.d.):

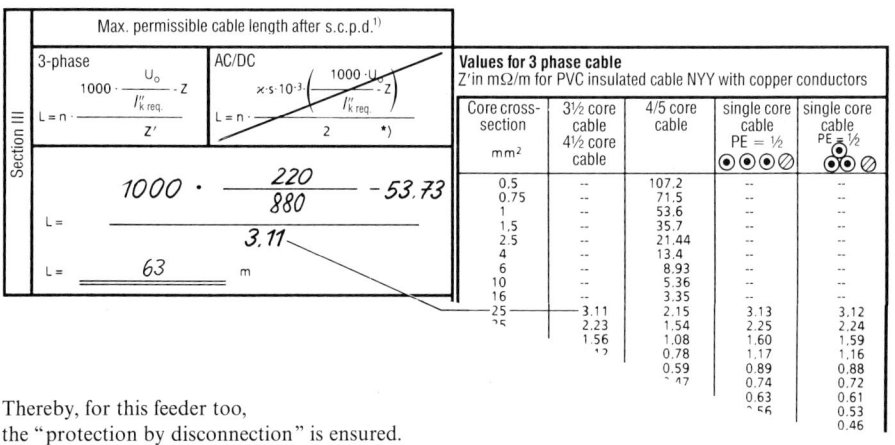

Thereby, for this feeder too, the "protection by disconnection" is ensured.

Figure 1.15 Calculation procedure pertaining to the schematic circuit diagram shown in Figure 1.10, page 44

1.4.3 Protection by total insulation

The use of totally insulated switch and distribution boards is limited through the problems of heat dissipation as well as the size and weight of the equipment to be installed in the enclosures. The totally insulated distribution systems listed in the Siemens NV catalogues are suitable for the widest possible range of applications.

The protection by total insulation (Fig. 1.16) is achieved by means of complete insulation of the equipment. This insulation must take the form of double or reinforced insulation, or must be in addition to the normal operating insulation of the equipment. This is usually achieved by installing the equipment with its standard insulation inside housings or boxes manufactured of insulating material, and closed on all sides.

Criteria for total insulation are:

▷ Only insulated parts may protrude from the insulating housing. If metal parts protrude from the housing, then these must be isolated from the internal active parts and the equipment inside the housing e.g. by means of intermediate insulators.

▷ No protective conductor may be connected to the totally insulated equipment itself, or to equipment or conducting parts installed inside the totally insulating housing. In other words, the connection of bodies/metal parts to a protective conductor is forbidden[1]. The looping through of protective conductors to equipment further downstream is, however, permitted, e.g. via fully insulated PE/PEN busbar links.

▷ Normally, only electrical equipment which is marked with the symbol ▫ is considered to be totally insulated. (exception e.g. cables and conductors).

---
[1] The problem that no protective conductor may be connected to equipment in totally insulated housings, is presently under reconsideration at the DKE

1 Specifications for low-voltage devices and switchgear assemblies

Protection is achieved by:
▷ the use of equipment of protection class II to DIN VDE 0106 part 1; marked with the symbol ▣ to DIN 30 600, e.g. totally insulated housing,

or
▷ the fitting of additional insulation $\geq$ IP 2X [3] around the basic insulation of the equipment,

or
▷ the fitting of strengthened insulation to active parts, e.g. electrical conductors with special insulation such as NSYA

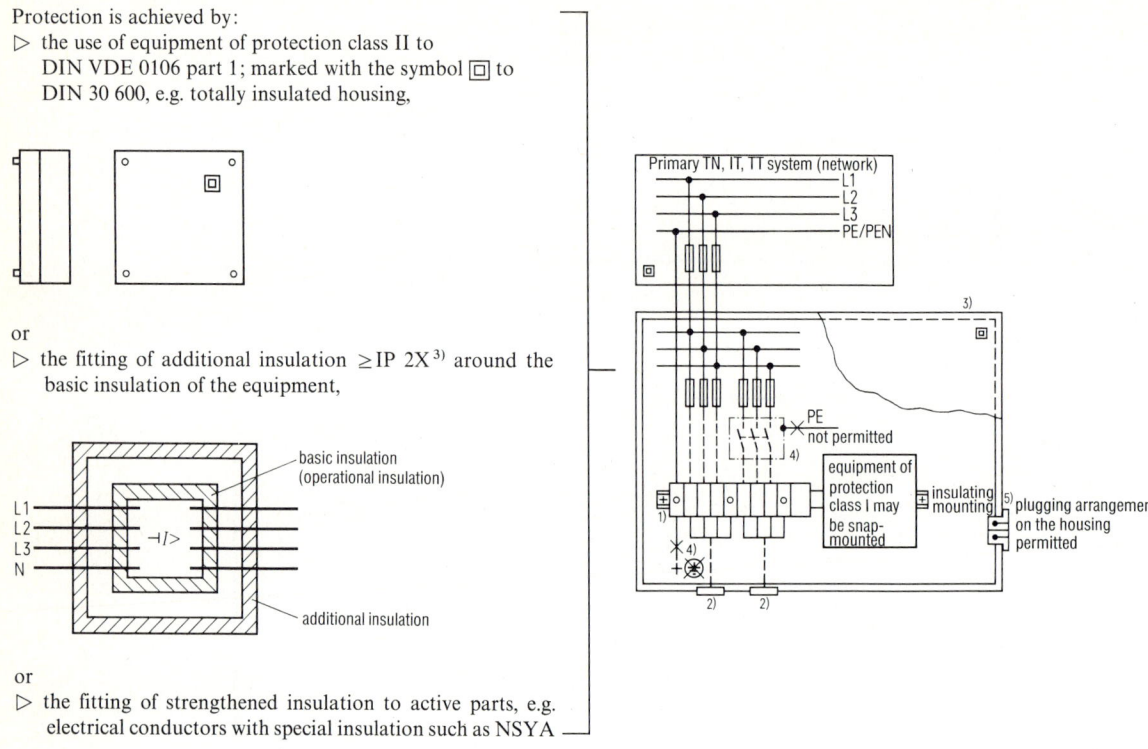

[1] If terminal block mounting rails are used as protective or PEN conductors, they must be marked accordingly (e.g. with green/yellow adhesive tape). These rails must be fitted in such a way as to be insulated from active parts and other conductive parts.
[2] Metal Pg gland adaptors or conduit threads are not permitted. No mounting screws or other conductive parts may protrude through the encapsulating housing, unless they are also totally insulated from the internal live parts and other conductive parts.
[3] In accordance with DIN VDE 0660 part 500, totally insulated equipment must be $\geq$ IP4X.
[4] Termination of protective conductors not permitted.
[5] Plugging arangements on totally insulated housings are presently under contentious discussion!

Figure 1.16   Total insulation of the electrical equipment

1.4.4 Erection and connection of switch and distribution boards

For the compliance with the aforementioned protection measures, the stipulations of DIN VDE 0100 part 729 – Erection and connection of switch and distribution boards – are of particular importance. They ensure the safe movement of operating personnel and require that

▷ a minimum isle width is maintained (refer to Figures 1.17 to 1.20 on pages 55 and 56),

▷ switch and distribution boards with reduced protection against direct contact may only be erected in electrical operating areas and locked electrical operating areas as defined in DIN VDE 0100, part 731,

▷ for isle lengths of $\geq$ 20 m two exits are required, for isle lengths between 6 and 20 m two exits are recommended, and for an isle length of up to 6 m one exit is sufficient.

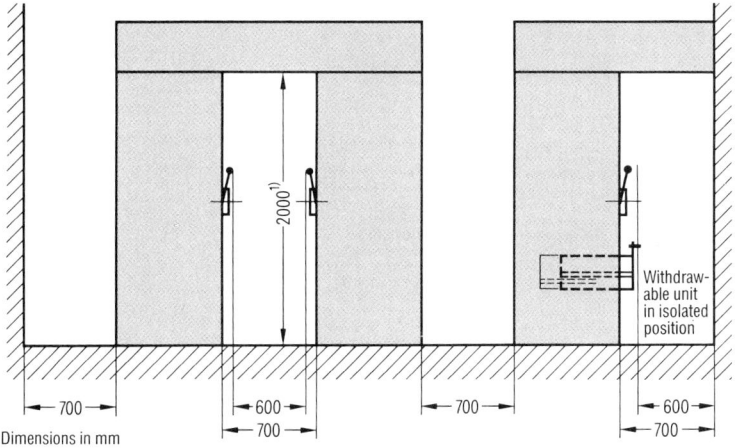

[1] Minimum height for passage under barriers such as e.g. cover plates

Figure 1.17
Minimum isle widths between low-voltage switchboards with a degree of protection $\geq$ IP 2X to DIN 40 050 (for exceptions, see Fig. 1.19 on page 56)

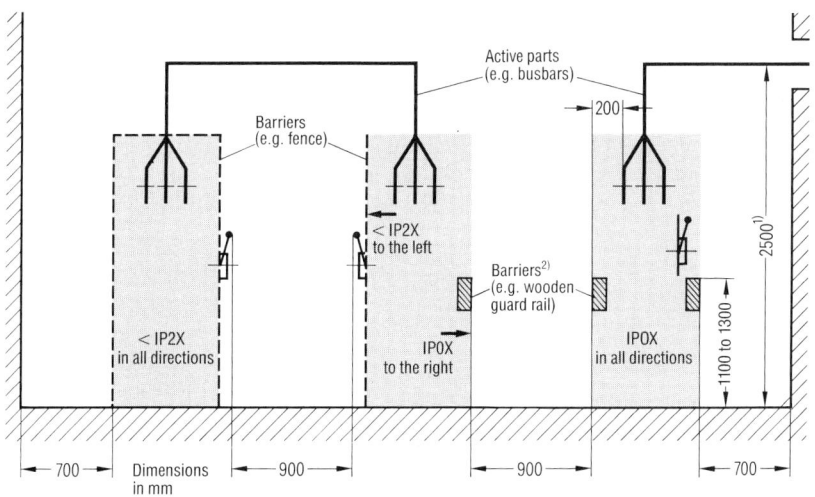

[1] Minimum height for passage under bare active parts
[2] Maximum permissible flexing: 20 mm from rest for a load of 500 N applied midway between supports

Figure 1.18
Isles between low-voltage switchboards with degrees of protection < IP 2X to DIN 40 050 (only permissible in electrical operating areas and locked electrical operating areas, refer DIN VDE 0100, part 731)

1 Specifications for low-voltage devices and switchgear assemblies

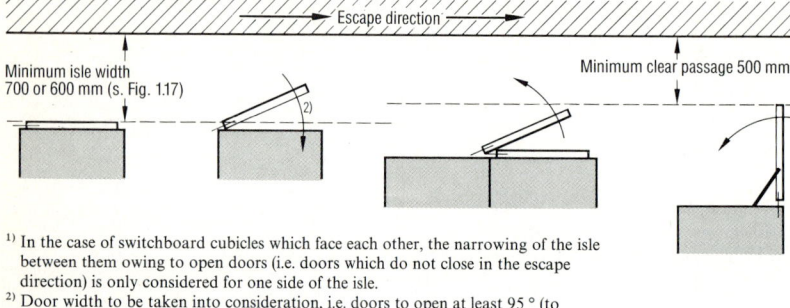

[1] In the case of switchboard cubicles which face each other, the narrowing of the isle between them owing to open doors (i.e. doors which do not close in the escape direction) is only considered for one side of the isle.
[2] Door width to be taken into consideration, i.e. doors to open at least 95 ° (to DIN VDE 0113 part 1/EN 60 204 part 1).

Figure 1.19 Reduced isle widths in the area of open doors (exceptions to Fig. 1.17, page 55)

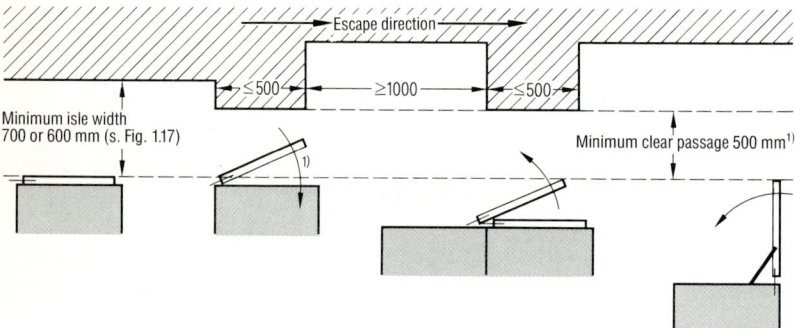

[1] Door width to be taken into consideration, i.e. doors to open at least 95° (to DIN VDE 0113 part 1; EN 60 204 part 1).

Figure 1.20
Reduced isle widths in the area of open doors and/or jutting out building structure.
This supplementary information is not contained in part 729 of DIN VDE 0100.

1.4.5 Clearances and creepage distances

The stipulated clearances and creepage distances to DIN VDE 0110 too, count among the most important generic specifications. The previous specifications DIN VDE 0110 and DIN VDE 0110b as well as the guidelines DIN VDE 0109 were superseded by DIN VDE 0110 parts 1 and 2 in January 1989. The transition time for the previous specifications expired on 31.12.89. As from this date, all planners as well as all the switchgear committees must take the new specifications into account, i.e. as from this date the new stipulations must also be incorporated in the renewal and updating of other related specifications.

In the specifications DIN VDE 0110 and DIN VDE 0110b, basic statements pertaining to the insulation groups (A_0, A, B, C, D) in dependence of the type of room and ambient influences were made.

Minimum values for the clearances and distances paths as well as the mounting distances were assigned to these insulation groups.

In addition, test voltages related to the clearances were specified.

Low-voltage switchgear and their parts built to DIN VDE 0660 conformed to the insulation class C.

Switchgear intended for use on vehicles had to conform to the insulation class D. Material and equipment used primarily in domestic installations, e.g. light switches, socket outlets etc. generally conformed to class B.

In terms of DIN VDE 0660 part 500 all switching devices must possess clearances and creepage distances as specified in the corresponding device specifications. These must be maintained under normal operating conditions. For exposed live conductors (e.g. busbars and wire bridges) the clearances and creepage distances of the associated switching devices must be maintained. This also applies in the case of possible short-circuits. For other clearances and creepage distances, DIN VDE 0110 parts 1 and 2 apply.

The equipment, switchboards and distribution boards listed in the Siemens l.v. switchgear and NV catalogues fulfil the respective conditions.

In terms of DIN VDE 0110 part 1 (Insulation coordination for electrical equipment in low-voltage installations – basic principles) and 2 (Dimensioning of creepage and clearance distances), the following points need to be considered in the determination of the clearances, creepage distances:

– overvoltage category[1],
– pollution degree,
– resistance to tracking[2],
– reference rated voltage or reference rated impulse voltage.

For switchgear and switchgear combinations, the pollution degrees 3 and 4 is usually applied.

---
[1] Only for clearances
[2] Only for creepage distances

1.5 Equipping specifications and relevant standards

Equipping specifications (also known as supplementary specifications) play an important role in the planning of switchboards and distribution boards, since consideration of the construction specifications alone is not sufficient to ensure that additional demands, which may be made by various installation conditions, are met.

The equipping specifications must be taken into consideration when installing switchgear, switchboards and distribution boards under certain defined conditions e.g.

▷ in rooms used for medical purposes (see DIN VDE 0107),

▷ in the equipping of industrial machinery, with which materials are formed, prepared, broken up, mixed, dried, tested or subjected to some other form of process (see DIN VDE 0113 part 1/EN 60204 part 1),

▷ in buildings for public assembly e.g. meeting halls, department stores, lodgings, hospitals etc. (see DIN VDE 0108),

▷ in potentially explosive operating areas (see DIN VDE 0165),

▷ in potentially explosive atmospheres (see DIN VDE 0165 and DIN VDE 0170/0171),

▷ in operating areas threatened by explosive substances (see DIN VDE 0166),

▷ in electrical installations in open-cast mines, stone quarries as well as in underground mining operations (see DIN VDE 0168 and DIN VDE 0118),

▷ for the equipping of nuclear power stations (KTA, i.e. "Kerntechnischer Ausschuß" = commission concerned with nuclear technology).

If the relevant equipment regulations are not taken into account, the resulting alterations and revisions can cause considerable delays as well as additional costs.

The most important topics covered by equiping regulations concern main switches and emergency-off switches to DIN VDE 0113 part 1.

1.5.1 Main switches

Main switches are demanded chiefly by DIN VDE 0113 and IEC 204. In the other specifications, usually only an isolating facility is required.

Electrical equipment of industrial machinery must incorporate a main switch which disconnects the entire equipment from the supply for the duration of cleaning, maintenance and repair operations as well as during long shutdown periods.

The dimensioning of main switches is no longer specified in DIN VDE 0113. Main switches should, however, be dimensioned to carry (not necessarily switch) the sum of the operating currents of all the loads which may be in operation simultaneously. If the main switch is to serve as an emergency-off switch as well, then its switching capacity must at least correspond to that which is required of the emergency-off apparatus.

The characteristics of main switches from Siemens are:

▷ They have disconnector properties, i.e. they reliably disconnect and isolate a circuit in all the current paths when the switch position indicator is in the OFF position, and establish an isolating distance between the contacts which affords sufficient protection for personnel.

▷ As a minimum standard, they are classified as load-break switches, i.e. they are capable of switching equipment ON and OFF during normal operation and they possess a switching capacity which is at least of the same dimension as their rated current.

▷ They can be operated manually, and have only one OFF and one ON position with associated end stops.

▷ They possess a switch position indicator which will only indicate the OFF position once all the contacts have achieved the specified isolating clearance.

▷ The actuating knob or handle for rotary actuation is coloured black. The ON position is marked with I and the OFF position with O. The actuator is lockable in the OFF position (e.g. by means of a padlock). The actuator is to be mounted in such a way as to permit operation from outside the switchboard (not only from behind the door), unless the switch in question is an electrically operated circuit-breaker.

▷ For main switches which are also intended for use as emergency-off devices, red operating handles with a yellow background colour are offered.

▷ The main switch breaks the current in all the non-earthed conductors simultaneously. If a disconnection of the neutral conductor is also required, 4 pole switches must be selected.

▷ All the parts of the main switches which carry a potential when the switch is in the OFF position can be shrouded with terminal covers which are included with the delivery of each main switch.

Remotely controlled circuit-breakers, in a design which does not permit "contact-tapping" may be used as main switches. In such a circuit-breaker the contacts cannot even momentarily "tap" if an attempt is made to close the breaker while its under-voltage release is de-energized.

The Siemens low-voltage switcgear catalogues list a range of main switches for rated operating currents up to 4000 A.

1.5.2 Emergency-off apparatus

The topic "emergency-off" is mainly dealt with in DIN VDE 0113, although it is also covered in other specifications.

For example, statements related to emergency-off apparatus are also found in DIN VDE 0100 parts 460 and 537 as well as in DIN 24981 of 03.89 (draft).

In the event of danger, it must be possible to disconnect every piece of electrical equipment in industrial machinery in such a way as to protect people or machinery from harm or damage. In certain instances, this not merely means the switching off of the machinery. If required, the emergency-off apparatus must initiate the reversing of machine parts to safe park positions; if necessary, by the application of operating voltage.

Two different methods may be applied to meet these requirements:

a) The installation of a single emergency-off switch which interrupts the supply of the relevant circuits. Such a switch may be hand operated or may be remotely operated by the opening of a control circuit.

b) An arrangement in the control circuits which allows the opening of all relevant main circuits by means of a single command. For further information, also refer to Section 4.3.4, page 254 "The use of contactor relays in safety circuits". If more than one contactor is used to switch the same load, e.g.. in the case of reversing and star-delta starters, then the coils of all these contactors must be de-energized.

Emergency-off switches used in accordance with method a) must be capable of breaking the locked rotor current of the largest motor in the machine, plus the sum of the normal operating currents of all the other loads.

The characteristics of emergency-off switches are as follows:

▷ The normally closed contacts are opened by direct force, i.e. they are forced apart for at least some of the contact travel without depending on elastic drive linkages and thereby ensure a reliable and dependable contact separation.

▷ The actuator is of a conspicuous red colour.

▷ The area immediately under the actuator is coloured bright yellow so that an obvious contrast is achieved with the red actuator. This ensures maximum visibility of the switch actuator.

These characteristics ensure that the emergency-off switches fulfil the requirements of EN 60 204 part 1/ DIN VDE 0113 part 1, as well as those of IEC 204-1.

The Siemens l.v. switcgear catalogues offer emergency-off switches for rated operating currents up to 4000 A.

For method b), the individual switching devices (contactors) which are de-energized by a single command need only have a breaking capacity suitable to their respective loads.

Emergency-off command devices are described in DIN VDE 0660 part 200, 07.92.

In addition to the characteristics mentioned for method a) above, such devices must fulfil a number of further criteria e.g.

▷ The actuator must latch in the operated ("emergency-off") position after having been actuated. The delatching may only be possible by hand, or by means of a tool, e.g. a key, directly on the emergency-off command device itself.

The latching must become effective when all the normally-closed switching elements have achieved the minimum opening travel, and all normally-open contacts have closed.

▷ Overlapping (simultaneous contact) of NO and NC contacts is not permissible.

Here follows a number of examples of emergency-off apparatus typically used with method b):

Mushroom-shaped pushbutton, latching, red,
delatching by a turning to the left of the mushroom-shaped actuator,
diameter of actuating surface 30 mm, 40 mm or 70 mm;

Mushroom-shaped push/pull button with
snap-action operation, red,
delatching by pulling,
diameter of actuating surface 50 mm;

Mushroom-shaped pushbutton, red,
latching with CES security lock,
delatching by key,
diameter of actuating surface 40 mm;

Application guidelines

As a rule, larger machines are equipped with separate main switches and emergency-off apparatus. This is sensible, since the actuator of the emergency-off switch must be within easy, quick and safe reach of the operator. The emergency-off switch is therefore located at the point of control, whereas the main switch can be located in the switchboard.

To ensure that the emergency-off apparatus really is easy and rapid in actuation, it is recommended that manually operated emergency-off switches be used only for currents of up to 63 A. Where emergency-off switches of higher current ratings are required, the use of a circuit-breaker equipped with an under-voltage release which is energized via a red mushroom-shaped 3SB1 emergency-off command device (with yellow background), is recommended. The actuation force of the emergency-off apparatus is thus kept to a minimum.

Combined main and emergency-off switches

For machines which do not have a separate control cabinet or switchboard (i.e. smaller machines), and in terms of the above, the two switches would have to be mounted side-by-side. In such cases, one switch may be used as both main and emergency-off switch, provided that all the respective requirements for emergency-off and main switches are met.

Figure 1.21
3SB1 emergency-off command device

1.5.3 Safety disconnecting switches for mechanical maintenance (service/repair switches)

A safety disconnecting switch is a special item of switchgear designed for a specific purpose, e.g. for paint mixing machines into which personnel have to climb to perform cleaning operations between two paint colours. Such switches are used to open the main incoming supply to the machine and must be secured against unauthorized reclosure.

The use of safety disconnecting switches is not stipulated in any of the DIN VDE specifications. The criteria which must be fulfilled by such switches is, however, described in DIN VDE 0100 part 537. Here it is stated that these

▷ should preferably be located in the main feeder of the machine section concerned,

▷ must be capable of switching the full load current of the relevant section,

▷ are to be selected and installed in such a way as to prevent an automatic reclosure,

▷ incorporate a suitable facility to prevent unauthorized reclosure.

In addition, the DIN VDE 0100 part 460 specifications must also be adhered to.

1.5.4 The electrical equiping of industrial machinery scheduled for export to countries outside Germany

If electrical devices are exported as part of the electrical equipment of industrial machinery, not only must the specifications governing the individual devices be considered, but also the relevant equipment and erection specifications applicable at the point of final installation.

The acceptance of the machine and its electrical equipment can be carried out by the local testing authority, a national testing body or its agents, or by a neutral testing company.

Such authorities may test/check:

▷ the types of devices installed,

▷ the design and construction of the control system, in terms of the national and international specifications, and

▷ compliance with local erection specifications, e.g. the connection to the energy supply network.

The various national and international regulations governing the electrical equipment of industrial machinery still differ from one another in certain aspects. In some countries, relevant local specifications do not exist. Whenever electrical equipment is exported as part of industrial machinery, it is imperative to clarify in advance which specifications have to be met, and which deviations from DIN VDE specifications and DIN standards have to be taken into account.

Deviations from these specifications and standards can, for instance, effect individual items of electrical equipment, their arrangement, marking and/or protective measures. A summary is given below of some of the particularities in a number of countries. However, this summary should not and cannot be used as a substitute for the study of the relevant regulations and specifications.

Specifications for individual items of electrical equipment

In most countries, items of electrical equipment which do not carry specific approvals may be used as integral parts of imported machinery. (See section 1.2.1, page 17). For exports to Canada, however, only low-voltage switchgear which carries the CSA approval may be supplied. For exports to U.S.A., it must be clarified with the final buyer whether ⓤ listed[1] components (**U**nderwriter **L**aboratories) must be used in the control systems, or whether ℜ listed[2] ("**R**ecognized **C**omponents") are acceptable. Furthermore, for exports to Canada, and to some extent to U.S.A., only the nationally approved wires and cables may be used. In terms of the respective specifications in U.S.A., Canada, Switzerland and also to some extent in Great Britain, fuses have different dimensions and characteristics from those manufactured according to DIN standards. To ensure the availability of spares and replacement parts, it must be clarified with the customer whether DIN type fuses may be fitted, or whether fuses which are more readily available in the country of destination are to be used.

Note:

The IEC 204-1, which as already mentioned corresponds to EN 60204 part 7 and DIN VDE 0113 part 1, stipulates that only fuses, for which equivalent replacements are available in the respective country, may be used.

---

[1] ⓤ listed devices are permitted for "field and factory wiring"
[2] ℜ listed devices are only permitted for "factory wiring"

Specifications for electrical equipment

The following examples of national, regional and international specifications for the electrical equipment of industrial machinery and machine tools stipulate the manner in which control systems are to be designed and constructed (also refer to Section 1.3.1, page 26):

International specifications

IEC 204-1	"Electrical equipment of industrial machines" General rules,
IEC 204-2	"Electrical equipment of industrial machines" Item designation and examples of drawings, diagrams, tables and instructions.

Regional specifications

EN 60 204-1	"Electrical equipment of industrial machines"; General rules
EN 60 204-1 A 1	"Electrical equipment of industrial machines"; Item designation and examples of drawings, diagrams, tables and instructions

The following national specifications are identical to EN 60 204-1:

– Germany   DIN VDE 0113 part 1
– France   NF/UTE C79-130
– Great Britain   BS 2771 part 1

The rest of the EEC as well as several other European countries are presently bringing their national specifications into alignment with the European standards.

Further national specifications

– Great Britain
   BS/CP 1015   "Electrical equipment of industrial machinery used in trade"
– India
   IS 1356 part 1   "Electrical equipment for machine tools".
– Italy
   CEI 44.1   Electrical equipment for machine tools,
   CEI 44.2   Electrical equipment for machine tools, machines for production lines,
   CEI 44.3   Electrical equipment for machine tools, electronic apparatus.
– Canada
   CSA C22.2-73   "Construction and testing of electrically equipped machine tools",
   CSA C22.2-105   "Electrical equipment for woodworking machines".
– U.S.A.
   JIC EMP-11   "Electrical standards for mass production equipment up to 600 V".

Recommendations for the USA

JIC[1] EGP-1	"Electrical equipment of general purpose machine tools up to 600 V",
NFPA[2] 79	"Electrical regulations for industrial machines",
NEC	"National Electrical Code".

Erection specifications for electrical equipment

The various national specifications for the erection of electrical equipment contain paragraphs of a rather general nature. Where-ever these paragraphs are important for the construction of control assemblies, they are included in the relevant construction specifications for these control assemblies.

In the Federal Republic of Germany, for example, DIN VDE 0100 and DIN VDE 0113 part 1 are applicable for the erection (setting up, connection) of the electrical equipment of industrial machinery. Here, DIN VDE 0100 only covers the erection of the parts up to the supply terminals of the machine, whereas DIN VDE 0113 part 1 covers the erection of the machine including the electrical equipment, e.g. separate control cubicles with connecting cables to the machine parts. If the connecting cables are laid on the building structure, then DIN VDE 0100 becomes applicable again.

---

[1] JIC Joint Industrial Council
[2] NFPA National Fire Protection Association

1.5.5 Colours for pushbuttons, illuminated pushbuttons and indicator lights

The standardized meanings of certain colours have been defined in DIN VDE 0199/3.88 and IEC 73. This has been done to increase the safety to operating personnel as well as to facilitate the correct operation and care of electrical equipment and installations. In addition, DIN VDE 0113 part 1, 02.86 also refers to colours of pushbuttons, illuminated pushbuttons and indicator lights. To a large extent, these specifications are the same.

In Tables 1.22 and 1.23 the colours for pushbuttons and indicator lights, as well as their meanings and some typical applications are given.

For illuminated pushbuttons, the recommended use of colours may be found in Table V of DIN VDE 0113, part 1.

Table 1.22 Colours for pushbuttons and their meanings

Colour	Meaning of colour	Typical application
Red	Operate in case of danger	Emergency stop
	Stop (halt) or Off	Complete shutdown, stop a motor or several motors, stop part(s) of a machine, de-energize a switching device, reset button combined with stop function
Yellow	Intervention	Intervention to suppress abnormal condition or to avoid unwanted change
Green	Start or On	Switch on everything, start a motor or several motors, start part(s) of a machine, energize a switching device
Blue	Any meaning not covered by the colours above	In specific instances, this colour may be assigned a meaning which is not covered by the colours red, yellow or green
Black Grey White	Not assigned any specific meaning	May be used for any meaning, with the exception of "stop" or "off" pushbuttons

Table 1.23
Colours for indicator lights and their meanings

Colour	Meaning	Explanation	Typical application
Red	Danger or Alarm	Warning of possible danger or of conditions which require immediate intervention	Temperature beyond specified (safe) limits, major parts of the equipment stopped owing to the tripping of protection apparatus
Yellow	Caution	Change or impending change of the conditions	Temperature deviating from the normal value, overload condition which may continue for a limited period only
Green	Safety (normal, ready, operating)	Indication of safe and normal operating conditions, or that next stage of process may start	Coolant flowing, machine ready to start
Blue	Special information	Blue can be assigned any meaning except those covered by the above colours red, yellow and green	Selector switch in the setting-up position
White	General information	Any meaning can be assigned; may be used when doubt exists as to application of the colours red, yellow and green, and e.g. as acknowledgement signal	–

The large variety within the Siemens 3SB1, 3SB2 and 3SB3 ranges of command devices offers suitable solutions for all these applications. The selection may be made from the corresponding l.v. switchgear catalogues.

1.5.6 IP degrees of protection against contact with live parts, ingress of foreign bodies and ingress of liquid

In terms of the application of standards governing "switchgear assemblies", a correlation between international and DIN standards for the determination of the necessary degrees of protection, has only been established recently for switch and distribution boards (also refer to Section 7.4.7, page 360).

A number of specifications and standards have not as yet been brought up to date in terms of the currently valid definitions of the IP degrees of protection.

1.5.6.1 IP degrees of protection to DIN and IEC

In all categories of specifications, demands are made on the housings for switchgear and distribution systems with respect to the degree of protection afforded against contact with live parts, against ingress of foreign bodies and against ingress of liquid. These demands are expressed in terms of IP degrees of protection.

The description of the required degree of protection together with the respective IP code numbers are contained in

DIN 40050 July 1980	IP degrees of protection, protection against contact with live parts, ingress of foreign bodies and ingress of liquid for electrical equipment (Table 1.24).

The associated tests (testing equipment, testing procedures and assessments) are laid down in:

DIN 40052	Testing of protection against ingress of foreign bodies. Dust chamber,
DIN 40053	Testing of protection against ingress of water,
DIN VDE 0470 part 1	Testing of protection against contact with live parts. IEC Test Finger.

The contents of these standards correspond with the contents of the IEC Publication 529, including supplements No. 1 and No. 2 "Classification of degrees

Table 1.24
Summary[1] of the IP degrees of protection as outlined in DIN 40050, 7.80, IEC 529, 11.89 and IEC 947-1 app. C

Code digit	First code digit: Degree of protection against contact and ingress of foreign bodies	Second code digit: Degree of protection against ingress of water
0	No particular protection	No particular protection
1	No protection against intentional access, however no access to large body surfaces; protection against ingress of foreign bodies with diameter greater than 50 mm	Protection against vertically falling water drops
2 (2L)	Protection against contact by standard test finger and similar foreign bodies; protection against ingress of foreign bodies with diameter greater than 12.5 mm	Protection against dripping water also when the housing is tilted up to 15° from its normal position
3 (3L)	Protection against contact by test probe and similar foreign bodies with diameter 2.5 mm; protection against ingress of foreign bodies with diameter greater than 12.5 mm	Protection against water spray from an angle of up to 60° from vertical
4 (4L)	Protection against contact by test probe and similar foreign bodies with diameter 1.0 mm; protection against ingress of foreign bodies with diameter greater than 12.5 mm	Protection against water splashed against the enclosure from any direction
5	Limited protection against harmful ingress of dust und dust deposists	Protection against water jets from any direction
6	Complete protection against contact with live parts; protection against ingress of dust	Protection against harmful ingress of water from heavy seas or strong water jets
7	–	Protection against harmful ingress of water during immersion under specified conditions of pressure and time
8	–	Complete protection against ingress of water during submersion

[1] For the official short-form and full definition tests, please refer to the corresponding specifications.

of protection provided by enclosures", which has been declared a Harmonizing Document by CENELEC.

IEC 974-1, appendix C is used for the degrees of protection of low-voltage switchgear and switchboards. This specification is identical to the draft DIN VDE 0660 part 100 of June 1985, appendix C, – degrees of protection of encapsulated devices.

In terms of IEC 947-1, a distinction is made between foreign bodies and protection against the touching of live parts in the first code digit. For the degrees of protection IP **2**., the meaning of the first digit has changed from "protection against ingress of foreign bodies" to "foreign bodies greater than 12.5 mm and is extended by a 3rd code digit, if required. Further details may also be found in the draft of DIN VDE 0660, part 100.

The specifications and standards affected by these revisions are continuously updated in a suitable form.

The following points are still of interest:

▷ The first code digit of an IP degree of protection does not completely describe the protection against contact with live parts. It merely indicates the protection of personnel against contact with parts which are live under normal operating conditions, and against contact with moving parts with the described test instruments.

The complete requirements relating to protection against direct contact with live parts are contained in the applicable construction, equipment and generic specifications. (see Section 1.4.1.1, page 33). For general locations, the minimum IP degree of protection required to achieve protection against direct contact with live parts is IP2X [1].

▷ The IP degree of protection of a piece of electrical equipment alone does not describe sufficient protection under severe operating and ambient conditions (see Section 1.6.2.1, page 72).

1.5.6.2 Degrees of protection according to other national specifications

A summary of national specifications for degrees of protection is given in Table 1.25.

Table 1.25
National specifications for degrees of protection

Country	Specifications	Comment
Finland	E1 (1986), SFS 2972, 1977	–
Norway	NEMKO 22/1952	–
Sweden	SEN 2121-1960	Similar to DIN 40 050, but instead of IP the code letter S is used
Great Britain	BS 5490, 1985	Identical to IEC 529
France	NF/UTE C20-010, 1986	–
Switzerland	SEV 1000, 1985 SEV 3428, 1979	Identical to IEC 529
Spain	UNE 20–324, 1978	–
Canada	CSA, C22.2 No. 94, 1976	Differs significantly from IEC 947-1, App C and IEC 529
USA	UL 508, 1984 NEMA ICS 6, 1983 Definitions of the classifications: NEMA No. 250, 1985	Differs significantly from IEC 947-1, App C and IEC 529
Australia	AS 1939, 1986	–

---

[1] If a degree of protection is indicated by means of an X (e.g. IP2X), then it means that this degree of protection is not relevant to the situation at hand. For example, if the degree of protection against ingress of liquids is not relevant to the protection against direct contact with live parts.

1.6 Operating and ambient conditions

Switchgear listed in the Siemens l.v. switchgear catalogues, and switchgear assemblies (TTA) listed in the NV catalogues, may be installed and operated under the respective operating and ambient conditions as defined in DIN VDE 0660 (also refer to Section 7.4.7., page 360).

1.6.1 Normal conditions

Table 1.26 shows the normal conditions which apply for switchgear and switchgear assemblies in terms of e.g. parts 101, 102 and 103 of DIN VDE 0660.

The terms used are defined in DIN 50010 part 1 follows:

▷ *Indoor climate* is a climate inside rooms which are built in such a fashion that objects are not exposed to the direct influence of an open-air climate.

Note:
Indoor climate can be influenced to a large extent by various heat and moisture sources, dust, chemical influences, e.g. by the presence of people, animals or plants as well as by factory equipment, heating or air conditioning.

▷ *Outdoor climate* is a climate inside structures which are built in such a fashion that objects are not exposed to direct radiation from the sun, precipitation or to wind (if applicable). In all other respects, however, it is the same as an open-air climate.

▷ *Open-air climate* is a climate to which objects are exposed in the open.

In the Siemens l.v. switchgear catalogues, the devices are often designated as being "climate-proof". Years of experience in many different countries where extreme climatic conditions prevail, including e.g. Brazil, India and Indonesia, has confirmed that "climate-proof" equipment can be installed anywhere in the world (see Section 1.6.1.5, page 68).

Abnormal operating and ambient conditions

If abnormal conditions are expected during operation, transportation or storage, e.g.

▷ regular condensation (inside),

▷ higher or lower temperature levels than those specified in DIN VDE 0660,

then these must be made known in the enquiry and ordering stages.

Table 1.26
Normal conditions in terms of DIN VDE 0660 and its parts

	Indoor climate	Open-air climate
Ambient temperature		
Short time peak value	+40 °C	+40 °C
Peak value of 24 hour average	+35 °C	+35 °C
Lowest value	−5 °C	−25 °C [1]
Atmospheric conditions		
Relative humidity	50% [2] at 40 °C	100% at 25 °C
Altitude		
Maximum altitude (m above sea level)	2000 m [3]	2000 m

[1] In temperate climate; in arctic climate −50 °C.
[2] A higher relative humidity may be permissible at lower temperatures, e.g. 90% at 20 °C.
[3] For switchgear >1000 V up to 12000 V, max. 1000 m above sea level.

Where applicable, specific additional information pertaining to abnormal conditions for switchgear and switchgear assemblies is given in the respective catalogues (also refer to Section 1.6.2, page 72).

1.6.1.1 Ambient temperature

The term ambient temperature is understood to mean the temperature of the air, measured under specified conditions, surrounding the switchgear device, switchboard or distribution board. In the case of encapsulated switchgear (e.g. motor starter in enclosure), the ambient temperature is the temperature of the air outside the enclosure or housing. For switchgear devices installed in switch or distribution boards, the ambient temperature is taken to mean the temperature of the air inside the switch or distribution board cabinet (see also Sections 7.4.6 and 7.4.7, pages 351 and 360 resp.).

For switchgear listed in the Siemens l.v. switchgear catalogues, the maximum permissible ambient temperatures are given. In some cases at ambient temperatures above 35 °C a reduction in the current rating, among other possible measures, may be required.

1.6.1.2 Altitude

Two major problems are associated with the installation of low-voltage switchgear at altitudes of over 1800 m above sea level:

▷ A reduced air density results in reduced heat dissi-

1 Specifications for low-voltage devices and switchgear assemblies

Table 1.27
Environmental influences to DIN 50 019 part 1, DIN IEC 721 part 2-1 and the Siemens standard SN 29070 part 1

Designation		DIN 50 019, part 1	DIN IEC 721 part 2-1	Siemens standard SN 29070
Global climate:	Minimum temperature	×	×	×
	Maximum temperature	×	×	×
	Humidity	×	×	×
	Weather factors			×
Indoor installation		–	–	×
Contamination factors		–	–	×

pation by the air surrounding electrical equipment. Since, however, the ambient temperature also drops with increased altitude, it is not usually necessary to de-rate the current carrying ability of the equipment.

▷ A lower air density also results in a reduced breakdown voltage and thus a reduced dielectric strength. This influence can, in most cases, be neglected and is taken into account by the minimum clearances and altitude correction factors in Tables 2a and 2b of DIN VDE 0110 part 1.

Generally speaking, the switching capacity of low-voltage switching devices may be reduced as a result of the lower air density. In critical cases, the supplier of the switchgear should be consulted.

For low-voltage switching devices, the following de-rating factors may be used:

Altitude up to
2500 m $0.93 \cdot I_e$ 3500 m $0.83 \cdot I_e$
3000 m $0.88 \cdot I_e$ 4000 m $0.78 \cdot I_e$

In the case of the 3TF6 vacuum contactor and the 3WS vacuum circuit-breaker, a special version with an adapted magnet or operating mechanism must be used at high altitudes.

1.6.1.3 Environmental influences

In practice, electrical equipment (including machines and devices) is subjected to a great variety of environmental influences. These are described in DIN 50019 part 1, DIN IEC 721 part 2-1 as well as in the Siemens standard SN 29070 (Table 1.27).

DIN 50019 part 1

Climatic models, coding and cartographic representation of open-air climates.

The DIN 50019 assigns a specific code letter to each climatic region on earth which it defines mainly in terms of the temperature and humidity (Table 1.28).

DIN IEC 721

Classification of environmental conditions. Natural influences, temperature and humidity.

The Tables I to IV of DIN IEC part 2-1 define climate groups in terms of extreme maximum and minimum temperatures and humidity (Table 1.29).

The part 3-3, 04/90 defines and classifies the climatic, biological and mechanical environmental conditions, as well as chemically-active materials and particles in 6 tables.

Siemens standard SN 29070

Climatic conditions for materials with and without organic or non-organic coverings, classification of climates, quantification of climatic influences, climatic data (Tables 1.30 to 1.32).

The Siemens standard SN 29070 defines climatic classes which are based on the actual location of the electrical equipment. Here, reference is also made to indoor installation, roofed installations outdoors, as well as the influence of pollution.

Table 1.28
Division of the earth into climatic regions to DIN 50 019 part 1

Code letter	Designation	Code derived from
A	Dry climatic region	aridus
F	Cold climatic region	frigidus
H	Warm, damp climatic region	humidus
M	Coastal climatic region	mare
T	Temperate climatic region	temperatus

Table 1.29 Environmental influences to DIN IEC 721 part 2-1

Climate group	Lower air temperature °C	Upper air temperature °C	Highest air temperature with ≥95% relative humidity °C	Highest absolute humidity g·m^{-3}
Extreme daily averages of air temperature and humidity			
Restricted	−15	+30	+20	17
Extended	−29	+35	+24	22
Common	−45	+35	+31	30
Global	−55	+43	+31	30
Annual extremes of air temperature and humidity			
Restricted	−20	+35	+25	22
Extended	−33	+40	+27	25
Common	−50	+40	+33	36
Global	−65	+55	+33	36
Absolute annual extreme value of air temperature and humidity			
Restricted	−30	+45	+28	25
Extended	−45	+45	+31	30
Common	−60	+45	+37	40
Global	−75	+60	+37	40

Table 1.30 Environmental influences in terms of the Siemens standard SN 29070

Climate class **J1**	Indoors in buildings which have good heat insulation or high thermal capacity; heated or cooled; normally *only* the temperature is monitored, e.g. normal residential rooms, work rooms, offices, shops, telephone exchanges and broadcasting facilities, store rooms for sensitive products.
Climate class **J2**	Indoors in buildings with poor heat insulation or low thermal capacity; heated or cooled; no temperature monitoring, heating or cooling may be non-operational for several days, e.g. unmanned relay, amplification and transformer stations, stables, automobile workshops, manufacturing halls for heavy machinery, hangers.
Climate class **J3**	Indoors in buildings with no particular heat insulation and low thermal capacity; neither heated nor cooled; e.g. telephone booths, building entrance halls, barns, silos, unheated warehouses, sheds, garages.
Climate class **A**	Outdoor room, partially open, but protected from direct radiation from the sun and weather influences; outdoor climate however prevails within the room, possibly with high level of pollution, e.g. covered railway platforms, market halls, bus stop shelters, roofed storage areas.
Climate class **F1**	Outdoor areas with low pollution levels, e.g. rural areas, mountainous regions.
Climate class **F2**	Outdoor areas with high pollution levels, e.g. areas with high traffic concentration, industrial areas, coastal regions and on the open sea.

1.6.1.4 Parameters and pollution levels of the climate classes to IEC 721 and the Siemens standard SN 29070, part 1

The parameters associated with the respective climate classes are shown in Table 1.31. The pollution levels given in Table 1.32. apply for materials with and without organic or non-organic coverings.

1.6.1.5 Protection against corrosion, type of finish for technical products

The standards in Table 1.33 define climatic regions, climate groups and climate classes.

For Siemens products, the terms "normal finish", "normal paint finish" and "increased protection against corrosion" are used.

Table 1.31 Parameters of the climate classes to the Siemens standard SN 29070, part 1

Climatic parameter			Climate class					
			J1	J2[1]	J3[2]	A[3]	F1[3]	F2[3]
Temperature	min.	°C	+5	−25	−40	−50		
	max.	°C	+40	+55	+70	+55		
Relative humidity	min.	%	5	10	10	12		
	max.	%	85	100	100	120		
Absolute humidity	min.	g/m^3	1	0.5	0.1	0.04		
	max.	g/m^3	25	29	35	36		
Air pressure	min.	kPa	70					
	max.	kPa	106[5]			106		
Sun, light and UV radiation		W/m^2	700[6]		1120[6]		1120	
Heat radiation			7)				8)	
Condensation			no	yes[9]		yes		
Precipitation (rain, snow, hail etc.)			no			10)		yes
Freezing, hoar frost			no			yes[9]		yes
Biological influences[14]	Flora		no	9) 11)		11)		13)
	Fauna		no	9) 12)		12)		
Sea air			no			salt mist	no	salt mist
Industrial air			no	see Table 1.32				

[1] The given temperature limits can be encountered at these locations, even if the outdoor temperatures do not reach these levels.

[2] The given temperature limits can be encountered at these locations, even if the outdoor (open-air) temperatures do not reach these levels by a long way. To a large extent, this climate class corresponds to Class 3K6 in DIN IEC 721-3-3 which is generally used as a basis for the design of Siemens low-voltage switchgear.

[3] Excluding the influence in an extremely cold climate.

[4] At locations where direct radiation from the sun or the greenhouse effect (heat accumulation) is not encountered, +40 °C should be applied as the upper limit.

[5] Mining conditions have not been taken into account.

[6] Dependent on the structural or building architecture (e.g. windows).

[7] Heat radiation can occur, e.g. in the vicinity of air heating equipment.

[8] Heat radiation can occur in particular applications, e.g. device dependent.

[9] Not common, but occasional occurence possible.

[10] Depending on the structural or building architecture, wind driven rain, snow, hail etc. is possible.

[11] Occurance of mildew and growth of fungus etc.

[12] Presence of rodents and other harmful animals, with the exception of termites.

[13] Presence of rodents and other harmful animals, including termites.

[14] In accordance with DIN IEC 721-3-3, Siemens low-voltage switchgear complies with the Class 3B2.

Table 1.32 Pollution levels of the climate classes to the Siemens standard SN 29070, part 1

Climatic parameter			Climate class							
			J2	J3		A[1)]		F1	F2	
			Limit value	Mean value	Limit value	Mean value	Limit value	Limit value	Mean value	Limit value
Sulphur dioxide		mg/m^3	0.1	0.3	1.0	5.0	10	0.1	5.0	10
		cm^3/m^3	0.037	0.11	0.37	1.87	3.7	0.037	1.87	3.7
Hydrogen sulphide		mg/m^3	0.1	0.1	0.5	3.0	10	0.01	3.0	10
		cm^3/m^3	0.007	0.07	0.355	2.1	7.0	0.007	2.1	7.0
Chlorine		mg/m^3	0.1	0.1	0.3	0.3	1.0	0.1	0.3	1.0
		cm^3/m^3	0.034	0.034	0.1	0.1	0.34	0.034	0.1	0.34
Ammonia		mg/m^3	0.3	1.0	3.0	10	35	0.3	10	35
		cm^3/m^3	0.425	1.4	4.2	14	50	0.425	14	50
Ozone		mg/m^3	0.01	0.05	0.1	0.1	0.3	0.01	0.1	0.3
		cm^3/m^3	0.005	0.025	0.05	0.05	0.15	0.005	0.05	0.15
Nitric oxide		mg/m^3	0.1	0.5	1.0	3.0	9.0	0.1	3.0	9.0
		cm^3/m^3	0.052	0.26	0.52	1.56	4.68	0.052	1.56	4.68

Note:
The mean values are the expected average levels (long term values). The limit values are peak levels, and do not occur for longer than 30 minutes.

[1)] To a large extent, the values correspond to the Class 3C3 in DIN IEC 721-3-3. Generally speaking, Siemens low-voltage switchgear complies with these conditions.

The term "increased protection against corrosion" replaces the previously common terms "climate proof" and "tropicalized". Table 1.33 indicates the suitable finish for each climate.

In cases of extremely harsh environmental conditions, additional information would be required.

1.6.1.6 Decontaminability

In areas threatened by radiation, (nuclear power stations, atomic laboratories, nuclear medicine facilities), there is a risk that the surfaces of devices and machines may become contaminated by radioactive precipitation. The removal of such precipitation is known as decontamination.

Acid solutions are used in the decontamination process. The surfaces of the devices, switchboards or machines must therefore be of suitable finish to permit the use of such cleaning agents without the risk of damage.

Table 1.33 Protection against corrosion and the decontaminability of technical products

Climatic region, group and class to:			Suitable version	Decontaminability
DIN 50019	DIN IEC-721 part 2-1	SN 29070, part 1		
A, T	Extended	J1, J2, J3 F1	Normal version, Normal finish	Generally no
H, M	Global	A, F2	Increased protection against corrosion, special finish	Generally yes

1.6.1.7 Climatic operating conditions for electronic equipment

Application categories and details concerning the reliability of components in the fields of electronics and communications in accordance with DIN 40040, April 1987

The DIN 40040 standard specifies the application categories for components (capacitors, resistors, diodes) used in the fields of electronics and communications.

The meanings of the first three positions of the category codes are given in Table 1.34.

The 4th position of the category code provides information on the failure quota, the 5th position on the stress duration in hours, the 6th position on vibrational and shock stressing and the 7th position indicates the air pressure at operational altitude.

In terms of DIN VDE 0160, electronic equipment must fulfil the demands of humidity rating G. Generally, programmable logic controllers (PLC's) meet the demands of the humidity rating F.

Application categories for electronic modules and assemblies

Siemens standard SN 26556, October 1978 including Addendum 1 of November 1982.

Application categories and code letters for electronic modules e.g. modules or electronic cards and equipment used in process and control systems, are specified in the SN 26556.

In contrast to the application categories for electronic components, the category codes for modules and assemblies only consist of a single letter which indicates the limit values of temperature and humidity.

One can distinguish between operational and storage/transport temperatures (Table 1.35).

Table 1.34 Application categories for electronic components in accordance with DIN 40040

Code										
1st position:	Lower temperature limit									
	E	−65 °C	G	−40 °C	I	−10 °C	L	+5 °C		
	F	−55 °C	H	−25 °C	K	0 °C				
2nd position:	Upper temperature limit									
	K	125 °C	N	90 °C	R	75 °C	U	60 °C	Y	40 °C
	L	110 °C	P	85 °C	S	70 °C	V	55 °C		
	M	100 °C	Q	80 °C	T	65 °C	W	50 °C		
3rd position:	Influence of humidity, relative humidity:									

	Yearly average	Highest values			Comment
		on 30 days per year continuously[1]	on 60 days per year continuously[1]	on the rest of the days occasionally[2]	
D	≤80%	100%	–	90%	Condensation
E	≤75%	95%	–	85%	Occasional and light condensation permissible[3]
F	≤75%	95%	–	85%	No condensation
G	≤65%	–	85%	75%	
H	≤50%	–	75%	65%	

[1] These days should be distributed throughout the year.
[2] Maintaining the yearly average.
[3] Can e.g. occur as a result of the brief opening of equipment installed in the open.

Table 1.35 Application categories for electrotechnical assemblies in accordance with the Siemens standard SN 26556

Code letter	Temperature range		Condensation temperature t_d [1] and relative humidity U								Condensation	Temperature change	Air pressure [4]
	Lower temperature limit	Upper temperature limit	Yearly average		Highest values								
					on 30 days per year continuously [2]		on 60 days per year continuously [2]		on the rest of the days occasionally [3]				
	°C	°C	U %	t_d °C	U %	t_d °C	U %	t_d °C	U %	t_d °C			mbar (hPa)

Specified values for assemblies in service

A	0	+40	75	17	95	24	–	–	85	20	not permissible	max. 10 °C within 1 hour, max. 1 °C within 3 min	860 to 1080
B	0	+55	75	17	95	24	–	–	85	20	not permissible		
C	0	+70	75	17	95	24	–	–	85	20	not permissible		
D	0	+70	65	17	–	–	85	24	75	20	not permissible		
E	–25	+40	80	20	100	28	–	–	90	23	permissible [5]		
F	–25	+55	80	20	100	28	–	–	90	23	permissible [5]		
G	–25	+70	80	20	100	28	–	–	90	23	permissible [5]		
H	–25	+70	75	17	95	24	–	–	85	20	ocassional, brief, light [6]		

Specified values for assemblies packed for transport and storage

I	–25	+70	75	17 [7]	95	24 [7]	–	–	85	20 [7]	ocassional, brief, light [6]	max. 20 °C within 1 hour	660 to 1080
K	–40	+70	65	17 [7]	–	–	85	24 [7]	75	20 [7]	not permissible		
L	–40	+70	75	17 [7]	95	24 [7]	–	–	85	20 [7]	ocassional, brief, light [6]		

[1] Refer SN 29001.
[2] These days should be distributed throughout the year in a natural way.
[3] Maintaining the yearly average.
[4] The given values correspond to a location up to 1500 m above sea level.
[5] Can e.g. occur as a result of opening and closing of equipment installed in the open air.
[6] Occasional, brief, and light condensation includes cases in which the following conditions are met simultaneously: max. duration of a single condensation: 3 h; frequency of condensation: on average 3 times per year, however no more than 10 per year; shortest period between condensation cycles: 1 day.
[7] The given values correspond to a transport altitude of 3500 m above sea level.

◁ Definition of concepts used in Tables 1.34 and 1.35

Lower temperature limit	The lowest temperature at which the component or assembly may still be operated.
Upper temperature limit	The highest temperature at which the component or assembly may still be operated. The influence of self-generated heat and/or heat from neighbouring devices must be considered.
Influence of humidity	Relative humidity of the air in the vicinity of the component or assembly.
Lower transport and storage temperature limit	The lowest temperature at which transport or storage is still permissible, whereby the component or assembly need not be functional. If no other value is given, –25 °C applies.
Upper transport and storage temperature limit	The highest temperature at which transport or storage is still permissible, whereby the component or assembly need not be functional. If no other value is given, +65 °C applies.

1.6.2 Abnormal conditions for transport, storage and operation

For low-voltage switchgear and switchboards, the term "abnormal conditions" is understood to include:

▷ severe contamination of the surrounding air by dust, smoke, corrosive or radio-active elements, steam or salt,
▷ influence of electric or magnetic fields,
▷ radiation (e.g. radiation from the sun, or radio-active elements),
▷ influence of vegetation and/or vermin and insects,
▷ location in hazardous areas or rooms, and/or
▷ areas subject to strong vibration and/or shocks.

If such conditions or others which may be relevant are to be expected either in operation, transport or storage, it is advisable that they be taken into account at the planning stage.

1.6.2.1 Arduous operating conditions

Switchgear listed in the Siemens l.v. catalogues is intended for use in closed rooms where difficult operating conditions due to dust, corrosive steam or gases are not encountered.

Arduous operating conditions exist, when the surrounding atmosphere supports corrosive action and thereby shortens the service life of the switchgear. This is, for example the case in sulphurous atmospheres or in environments where the air has a high content of carbon dioxide, ammonia or sodium chloride (salt in sea air). Switchgear is subjected to extremely severe conditions in the chemical industry. It is therefore recommended that encapsulation with a suitable degree of protection be used, or that enquiries be made. As a rule, encapsulation meeting the specifications of Class 351 for unprotected equipment, is sufficient.

1.6.2.2 Influence of α, β and γ radiation

The behaviour of materials (changes in their mechanical, electrical or other properties) when exposed to high energy, ionizing radiation is defined as their resistance to radiation.

A distinction is made between α radiation (helium ions) β radiation (electrons) and γ radiation (hard electromagnetic radiation). In practice, mainly γ radiation is encountered.

The resistance to radiation of a material is often expressed in terms of the dose, i.e. the converted energy per kilogram (J/kg) of material during radiation, required to alter the relevant characteristic values of the material by a maximum of 25% or 50%. In practice, however, data concerning the relative changes in the properties of a material (in percent) has little value. It is often of greater importance to know whether a material or a material component meets with a specified minimum requirement under the given conditions.

The effect of ionizing radiation depends on the type of radiation and it's energy (the effective cross-section of α, β and γ rays is in order of 10000:100:1 respectively). They do, however, manifest themselves by the same reaction, viz. the destruction of the chemical bonding through ionization of the atoms and molecules in the material. Radiation affects different materials in different ways.

Metals

No damage is suffered by metal from the effects of radiation (with the exception of neutron radiation) as the ionization of an atom is negated through the free movement of electrons within the material.

Semi-conductors

Semi-conductors are regarded as relatively sensitive to radiation. The additional charge carriers built up during radiation cause a change in the electrical parameters.

Plastics

Under the influence of radiation, plastics are effected in the following ways. The individual properties affected are shown in brackets:

▷ deterioration of the crystal structure in semi-crystaline plastics (reduction in strength),
▷ breaking of the polymer chains, combined with disintegration due to oxidation (reduction in strength, material flow, release of gases),
▷ cross-polymerization (brittleness, hardening),
▷ creation of charge carriers (reduction in insulating properties).

In most cases the dose power has an important influence on the resistance to radiation (in the case of radiation in air), since the diffusion of oxygen from the atmosphere into the material during radiation has an oxidizing effect.

Exact information on the resistance to radiation requires the exact knowledge of the conditions prevailing at the installation site (dose, dose strength and temperature etc.).

Thermoplastics

Thermoplastics exhibit varying degrees of sensitivity to radiation. Thermoplastics which contain fluoride, e.g. TEFLON, exhibit a low resistance to radiation (usable only up to a dose of approximately 10^2 to 10^3 J/kg).

Duroplastics

Duroplastics generally exhibit a higher resistance to the effects of radiation than thermoplastics (some may be used at doses of up to approximately 10^6 to 10^7 J/kg). The filler material plays an important role; inorganic filler generally offers a higher resistance to radiation than organic material.

Elastomers

These are comparable to thermoplastics in respect of resistance to radiation.

Oils and lubricants

Radiation accelerates the ageing and hardening of oils and lubricants.

1.6.2.3 Stressing due to vibrations and impact

Switchgear which is used on piston compressors, vibration screens and similar machine tools or on vehicles, cranes etc., is often subjected to severe vibrations, i.e. oscillations and/or mechanical shocks.

The oscillations are described by their frequency f and their amplitude A. The following relationship applies:

$$f = \frac{1}{2\pi}\sqrt{\frac{a}{A}} = 0.159\sqrt{\frac{a}{A}}.$$

- f oscillation frequency
- a acceleration
- A amplitude of displacement

Information concerning the resistance to vibration of the switchgear is given for the x direction (perpendicular to the mounting plane), y direction (horizontal) and z direction (vertical).

Switchgear should, preferably, be mounted on vibration-free vertical panels. If vibrations cannot be avoided then steps should be taken to reduce the influence of such vibrations, e.g. by supporting the mounting panel on shock absorbers or springs, or through the rotation of the item of switchgear out of the critical plane.

Oscillations produce a continuous mechanical stress, whereas impacts or mechanical shocks produce individual (in some instances repeated) stresses. Mechanical shocks are described according to their form (e.g.. saw tooth, trapezoid or half sinusoidal), the acceleration (usually as a multiple of the acceleration due to gravity g) and the shock duration. Where necessary, please enquire regarding any values not found in the catalogues. By and large, switching devices fulfill Class 3M6 to DIN IEC 721-3-3.

1.6.2.4 Measures against the effects of induced vibration

Induced vibrations are random oscillations caused by a massive load impact such as an earthquake, aircraft crash, explosion or burst pressure wave. Under these circumstances, electrical equipment and installations are subjected to oscillations with various frequency content and acceleration levels, depending on the nature of the load impact. Low-voltage switchboards are generally designed to withstand the effects of earthquake, alternatively the effects of "earthquake superimposed with burst pressure wave". For the case of high frequency load impact, e.g. resulting from aircraft crash, measures in addition to those required for protection against earthquake are generally required. These must serve to lower the level of acceleration, and to reduce the frequency content to values comparable to those typical for earthquake. This can be achieved by a number of constructional measures.

The following could be specified for the switchboard:

▷ stability and soundness,

▷ functional integrity during and after the load impact.

The assessment is either done experimentally on completely equipped switchboards to KWU specifications[1], or in a combined experiment/calculation procedure involving the calculation of the supporting structure in terms of the KWU specifications[2] and the qualification of the installed devices, to the criteria of KWU specifications. The assessment must be documented by means of calculation and/or test reports.

1.6.2.5 Resistance to termites

Very few documents on the subject of resistance to termites or termite proofing have been published, and these contradict each other to some extent. This is understandable, since one can only speak of a relative resistance which, depending on the testing method employed, tends to be judged differently in terms of the practical constraints and local conditions. In general, low-voltage switchgear from Siemens fulfills the conditions set out in the biological Class 3B2 to DIN IEC 721-3-3.

Only steel, concrete, stone or glass can be regarded as being absolutely resistant to termites. Termites generally feed off wood and other celluloid materials. Other materials are eaten without being digested. Sharp edges are attacked first, as smooth surfaces and rounded corners do not offer sufficient purchase for the termites.

Damage can also result from moist termite excrement which can cause creepage paths resulting in leakage currents. Dry excrement can be extremely hard and can lead to mechanical damage.

1.6.2.6 Environmental acceptability

Low-voltage devices from Siemens not only comply with all pertinent standards and specifications, but are also renowned for high quality. To achieve and maintain this reputation, a number of design aspects and selection of materials are specified, including the use of self-extinguishing high quality synthetic compounds. In terms of the present state of the art, a number of these plastics can only be produced with the addition of halogen compounds to act as a flame retarding agent.

The plastics assemblies do not, however, contain the flame retarding agents Bromated Biphenyle or Diphenylether which could produce small quantities of Bromated Furanes or Dioxines.

Furthermore, Siemens low-voltage devices are free from Poly-Chlorinated Biphenyls (PCB) and Asbestos, and do not contain Cadmium as colour pigmentation, stabilizer or surface protection agent.

Consequently, used devices may be disposed of as normal waste without the risk of additional contamination of soil or water.

1.6.2.7 Suppression of radio interference

Radio interference is high frequency disturbance of radio wave reception. It exists if undesirable electromagnetic oscillations are received together with the normal signal by a high frequency channel of an antenna installation or a radio receiver itself via its antenna or antenna input socket, and if it audibly detracts from the sound quality of the main reception.

In the Federal Republic of Germany, the suppression of radio interference produced by electrical devices, machines and installations, which use or produce high frequency, is required by law. Their operation is only allowed if a permit has been issued by the federal post and communications authority (Bundespost). To avoid possible trade barriers within the European community, regulations regarding radio interference suppression are dealt with on an EC level. The DIN VDE specifications for radio interference suppression can only be applied in Germany insofar as their technical content does not deviate from the relevant decrees of the federal minister for post and telecommunications. Still today, the technical requirements of these decrees are not in all cases identical to the stipulations of the DIN VDE specifications. In terms of the present harmonization activities, the current DIN VDE specifications are undergoing continuous changes in order to adapt to the developing EC regulations. To determine which decrees and DIN VDE specifications apply, it is necessary to distinguish between electrical equipment and installations which produce high frequency signals by means of individual electrical operations or repetitive signals at less than 10 kHz, and those which produce or use discreet frequencies or repetitive signals above 10 kHz.

---

[1] KWU specification: Oscillation testing of supporting structures, no. ZXX001/DD/7080.7, and individual devices, no. ZXX001/DD/7080.9 in the field of energy production and transfer for the assessment of the security in event of induced vibrations due to earthquake and similar load impacts

[2] KWU specification: ZXX001/DD/7080.3 Procedure for the calculation of supporting structures in the field of energy production and transfer under consideration of induced vibrations (Siemens AG, department KWU, Erlangen) This specification is based on the draft rules of KTA 2201.4 (KTA = **K**erntechnischer **A**usschuß i.e. commission concerned with nuclear technology) The quoted specifications will be replaced in the foreseeable future by ones with the same titles, but with more stringent requirements, under the codes ZEN000/... instead of ZXX001/... (Publ.: Siemens AG, department KWU, Erlangen). The specifications are based on the rules laid down in: DIN 40046, part 8, 35 and 55 or in future DIN IEC 68 part 2-6 in conjunction with IEEE 344 – 1975.

The currently applicable situation, including exceptions and significant concessions for the use in industry, must be taken from the respective valid legal regulations. At present these are the decrees 523 of 1969, 1044 to 1046 of 1984 and 483 of 1986 in conjunction with the specifications DIN VDE 0871/6.78 "Radio interference suppression of high frequency devices for industrial, scientific, medical (ISM) and similar purposes", DIN VDE 0875 of Nov. 1984 "Radio interference suppression of electrical equipment and installations" with the parts 1, 2 and 3 of December 1988. The parts 1 and 2 of Dec. 1988 are the German versions of the European standards EN 55014 and 55015 of 1987. DIN VDE 0871 is to be replaced by DIN VDE 0875, parts 11, 2A and 3. It is however still valid untill 31.12.1995.

In terms of the new DIN VDE 0875 part 3, the following radio interference levels are defined:

Interference Level G indicates coarse suppression,

Interference Level N is for normal suppression,

Interference Level K is for slight suppression,

Interference Level O indicates devices which cause no radio interference (e.g. immersion heaters).

Radio interference from devices and machines which are to be installed in residential areas must be within the limits specified by part 1, part 2 or alternatively it must be of level N in terms of part 3.

No particular permissible radio interference level is specified for industrial areas. Practical experience has however shown, that if the machines are suppressed to level G, then level N for the neighbouring residential areas is also achieved.

Electrical equipment can be installed in industrial areas without preventative verification provided that the conditions described below are met.

For products which fall within the scope of DIN VDE 0875 part 3 (12.1988), the specification part 3/Dec. 1988 paragraph 10.1.1.2 calls for the following:

"The interference voltage and interference power/field strength of an electrical installation within an associated defined operating area is not restricted, provided provision has been made for the conditions of paragraphs 10.1.1.2.1 and 10.1.1.2.2 of DIN VDE 0875 to be met".

These conditions are:

▷ The interference voltage on the low voltage supply conductors outside the installation, outside the associated factory or industrial area may not exceed the interference level N.

▷ The interference field strength in sound and television ranges in the vicinity of receiver antenna systems which are, or are to be, built in accordance with DIN VDE 0855 part 1, must be so weak that an interference-free reception is ensured when the specified minimum useful field strength is available at the point of installation of the antenna.

If the equipment has the interference level G in terms of DIN VDE 0875 part 3, then the above conditions will be met in practice. No radio interference mark is issued for components such as electric motors, control units, contactors or other low voltage switchgear devices, since these only form part of a machine or installation and cannot perform any function alone. The extent of the necessary interference suppression measures can only be determined once the nature of the load, the switching frequency and the type of encapsulation or housing has been established.

For products which fall within the scope of law 1046 of the federal minister for post and telecommunications (DIN VDE 0871), § 2 paragraph 1.8 states:

"In the case of electronic measurement and control systems incorporated in electrical installations which fall within the scope of DIN VDE 0875 part 3 (12.88), the limit values of the limit value class B only refer to the limits of the associated rooms or operating area."

The operation of these products in the application described above, requires no preventative verification, and is regarded as "generally authorized".

In addition to the above, the "Guidelines of the committee entrusted with the harmonization of statutory specifications concerning electromagnetic compatibility of the member states" are of interest. In terms of these guidelines, issued on the 3rd of May 1989, the so-called "Electromagnetic Compatibility" (EMC) is declared as the minimum standard to which every electrical and electronic device will have to comply if it is brought onto the market or is destined for service within the EC after 1.7.92. Since radio interference suppression forms part of the EMC considerations, it is also covered by these "EMC Guidelines" of the European community. Further changes to the procedures described above are therefore to be expected.

2 Network data and duty types

In the rating and selection of installation components the network data, the network conditions as well as the anticipated operational duty must be taken into account.

2.1 Network data

The nominal voltage, nominal frequency and the behaviour under short circuit conditions are among the most important characteristics of a network. Nominal values are approximate quantity values used to identify the nominal data of supply networks in terms of international agreements.

In contrast, IEC 947-.. (alternatively DIN VDE 0660-..) uses the term "rated" instead of "nominal" to refer to quantity values assigned (generally by the manufacture) for a specific operating condition of a component, device or equipment (also see Section 9.1.2, page 500).

2.1.1 Nominal voltages and frequencies

The DIN IEC 38, 05.87, "IEC Standard Voltages", has been valid in Federal Republic of Germany since May 1987.

This standard contains the international version, IEC 38 – 6th edition of 1983, in unchanged form.

The IEC 38 standard is the result of international agreements aimed at reducing the number of different standard voltages used for electrical supply and traction networks, consumer installations and equipment.

Main circuits

In the field of low-voltage applications, the most significant simplification in the IEC 38 standard is the replacement of the three-phase voltages for energy supply systems, i.e. 220/380 V and 240/415 V, by the single internationally standardized value 230/400 V (see Table 2.1). For the interim period up to the year 2003, tolerence bands for the operational voltages of the networks have been specified. This serves to ensure that electrical equipment selected in accordance

Table 2.1
Nominal voltages in the low-voltage range between 100 and 1000 V to DIN IEC 38, 05.87, for three-phase networks and corresponding loads

Three-phase four-wire or three-wire networks	Single-phase three-wire networks
Nominal voltage V	Nominal voltage V
–	120/240
230/400 [1]	–
277/480 [2]	–
400/690 [1]	–
1000	–

[1] Voltage tolerances in the transition period for countries previously with 220/380 V networks +6%, −10%, alternatively with 240/415 V networks +10%, −6%.
The aim is to achieve 230/400 V ± 10% at the end of the transition period. Bearing the recommended value in mind, 400/690 V +6%, −10% should be used for the existing 380/660 V networks.
The currently valid network voltages, the tolerances and the remaining transition period, must therefore be taken from the relevant national specifications.

[2] Not to be used together with 230/400/690 V.

with the previous voltages may be operated safely for the duration of their normal service life.

The IEC recommendations have already been adopted as a national standard by most of the larger industrial countries, insofar as local conditions permit this. In Germany, it is recommended that the new values are used for all new installations.

A summary of the previously common transmission, distribution and consumer voltages, as well as the distribution systems in networks of the public power reticulation systems in Western and Eastern Europe, Africa, Northern, Middle and South America, Near East, Far East and Australia may be found in the brochure "Supply system voltages and frequencies in countries outside the Federal Republic of Germany" (publ. Siemens AG, order no. A19100-E593-B009-7600), and in Section 9.10 of the appendix on page 549.

The standard DIN 40005, "Nominal frequencies from 16 2/3 up to 10000 Hz" contains the regulations

Nominal voltages in main and auxiliary circuits

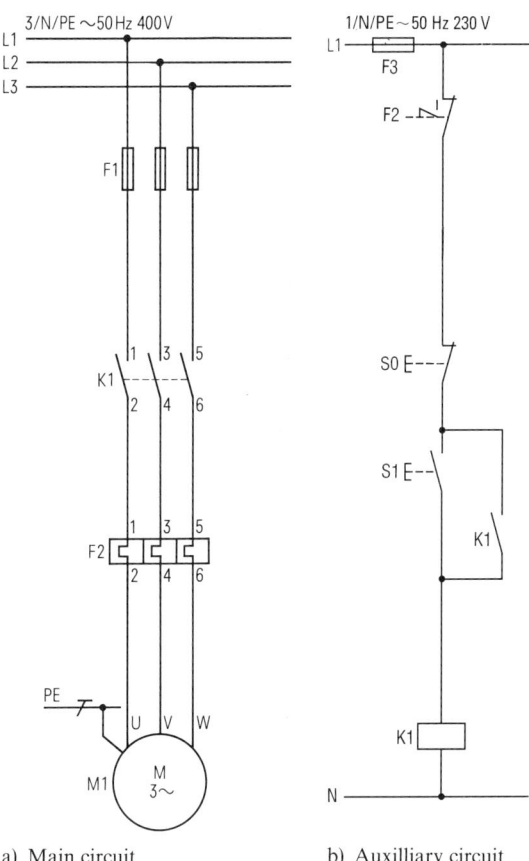

a) Main circuit   b) Auxilliary circuit

Figure 2.1
Main and auxiliary circuits for a motor starter with momentary contact control

Table 2.2
Nominal voltages below AC 120 V and below DC 750 V to DIN IEC 38 for electrical equipment

AC		DC	
Nominal values		Nominal values	
preferred	supplementary	preferred	supplementary
V	V	V	V
–	–	–	2.4
–	–	–	3
–	–	–	4
–	–	–	4.5
–	5	–	5
6	–	6	–
–	–	–	7.5
–	–	–	9
12	–	12	–
–	15	–	15
24	–	24	–
–	–	–	30
–	36	36	–
–	–	–	40
–	42	–	–
48	–	48	–
–	60	60	–
–	–	72	–
–	–	–	80
–	–	96	–
–	100	–	–
110	–	110	–
–	–	–	125
–	–	220	–
–	–	–	250
–	–	440	–
–	–	–	600

for rated frequencies. For power supply systems, this document only provides for the frequency of 50 Hz. Energy supply networks, not only in Europe but also in countries influenced by european technology, are operated at this frequency.

The nominal frequency of 60 Hz is only used in the U.S.A, Canada, in the Middle American countries influenced by American technology, and to some extent in South America e.g. Columbia, Venuzuela, Brazil, as well as in Japan.

In Europe, $16^{2}/_{3}$ Hz is generally the frequency used for a.c. operated railway and traction systems.

Higher frequencies are used in internal networks for motors in certain industrial applications, e.g. 100 to 120 Hz for high speed motors in the textile industry or frequencies of up to 300 Hz in the woodworking industry. The frequency 400 Hz may be found in a number of low-voltage installations. Information regarding the selection of low-voltage switchgear for applications in networks with frequencies higher than 50 Hz, is given in Section 3.7.3, on page 187.

Figure 2.1 shows the circuit diagrams of the main and auxiliary circuits for the direct-on-line starting of a three-phase induction motor (also see Sections 8.1.4 and 8.2).

Table 2.2 provides an overview of nominal voltages below AC 120 V and below DC 750 V to DIN IEC 38 for electrical equipment.

Auxiliary circuits

The standard values of the nominal voltages, to DIN VDE 0660 parts 10., for auxiliary and control circuits are listed in Table 2.3.

2 Network data and duty types

Table 2.3
Standard values of the nominal voltages for auxiliary circuits to DIN VDE 0660, part 10

DC V	AC V
24	24
48	48
110	110
125	–
–	127
220	220
250	–

Operating voltages

In DIN VDE 0100 part 725 the following voltages are recommended:

▷ AC 24, 42, 48, 110, 220[1)] V and
▷ DC 24, 48, 60, 110, 220, 250 V.

Although the preferred and maximum value of alternating voltage is still given in DIN VDE 0113 as AC 220 V, this will be replaced by AC 230 V in terms of IEC 38. This control voltage is not only the most generally preferred value in the Federal Republic of Germany, but also in many countries throughout the world. In addition, a control voltage of 110 V is also common in some countries. In the U.S.A. and in Canada the preferred control voltage is 120 V at 60 Hz. In comparison with lower voltages, the control voltage of AC 230 V offers the following advantages:

▷ higher contact reliability,
▷ lower voltage drop,
▷ conductors with smaller cross-sectional areas.

In terms of DIN VDE 0100 part 725 auxiliary circuits must be so arranged that under any operating condition, the maximum voltage deviation must be within the limits of 0.95 to 1.05 times the nominal voltage. Deviations from these limits are permissible if it is shown that the voltage deviations do not exceed the permissible voltage tolerances of the equipment being operated.

According to DIN VDE 0660 and IEC 947, all control circuits and their functional elements must be so arranged that they will safely perform their function while operating within the voltage limits set out in Table 2.4.

[1)] The preferred value in terms of DIN IEC 38 is 230 V.

Table 2.4
Tolerance limits of operating voltages for general use in terms of DIN VDE 0660. (Other values apply in traction and mining applications)

Equipment	Operation	Operating voltage as a multiple of the rated control supply voltage	
		Lower limit	Upper limit
Switch operating mechanisms	Closing Opening	$0.85 \cdot U_s$ $0.75 \cdot U_s$	$1.1 \cdot U_s$ $1.1 \cdot U_s$
Contactors	Closing Opening	$0.85 \cdot U_s$ [1)] $0.10 \cdot U_s$ (preferably $0.5 \cdot U_s$)	$1.1 \cdot U_s$ [1)] $0.75 \cdot U_s$
Contactor relays	Return to the rest position, also with worn contacts	$0.10 \cdot U_s$ (preferably $0.15 \cdot U_s$)	
Circuit-breakers with shunt trips	Opening	$0.5 \cdot U_s$	$1.1 \cdot U_s$

[1)] Siemens contactors: lower limit $0.8 \cdot U_s$, upper limit $1.1 \cdot U_s$

In terms of DIN VDE 0100 part 725 auxiliary circuits may be connected either directly (galvanically) or via transformers to the main circuit, or they may be totally independent of it.

● Operating voltages dependent on the main circuit:

For auxiliary circuits which have a function related to safety (e.g. with emergency-off switches and limit switches with safety function – see Section 4.3.6.1 on page 260), and which are not galvanically separated from the network, the following applies in terms of DIN VDE 0100:

▷ Auxiliary circuits in earthed networks. The voltage must be taken from between neutral and one phase.
▷ Auxiliary circuits in non-earthed networks. The voltage may be taken from between two phases, if the auxiliary switch serving the safety circuit opens the auxiliary circuit in both poles.

▷ Auxiliary circuits fed via control transformers to DIN VDE 0113 part 1, 02.86.

In terms of this specification, control transformers must be used when

- the machine in question contains more than five electromagnetic operating coils (e.g. in contactors, relays or valves), or
- command, control and indicating devices are located outside the control cabinet, and
- if electronic control and indication circuits are connected.

The transformers must be connected downstream from the main switch, preferably between two phases. This has the advantage, that they may be used for both earthed and non-earthed systems. On the secondary side of the transformer, the auxiliary circuits may be earthed or non-earthed.

If a direct current control voltage is required, this is provided using a power supply with transformer and rectifier.

• Operating voltages independent of the main circuit:

Operating voltages which are not dependent on the main circuit are derived from batteries. In DIN VDE 0113, DC 220 V is recommended. For practical reasons DC 24 V is often selected (12 cells of 2 V).

2.1.2 Short-circuit current

Largest possible (prospective) short-circuit current

The magnitude of the uninfluenced, maximum (prospective) short-circuit current which can occur at the point of installation of switching devices and distribution systems is decisive for their selection in terms of their

▷ Short-circuit strength,
▷ Switching capacity and
▷ Back-up protection (if required).

The current which results from a short-circuit through a negligible impedance at any point between a phase and a protective conductor, is the basis of the protective measures against indirect contact with live parts, as is outlined in DIN VDE 0100 part 410 (also see Section 1.4.2, page 40).

Short-circuit currents are the subject of the "VDE Guidelines for the calculation of short-circuit currents in three phase networks with nominal voltages up to 1000 V" (DIN VDE 0102, 01.90).

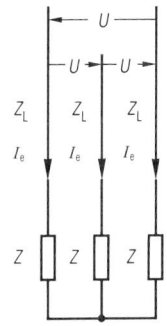

Operating current I_e

$$I_e = \frac{U}{\sqrt{3}(Z_L + Z)}$$

a) Normal operation

Steady-state short-circuit current I_k

$$I_k = \frac{U}{\sqrt{3} Z_L}$$

b) Operation disturbed by short-circuit

Figure 2.2
Rated operating current and steady-state short-circuit alternating current

Under normal conditions, the rated operating current flows in the circuit (Fig. 2.2a). Its magnitude is determined by the rated voltage U, the sum of the impedances of the feeder network Z_L and the load Z. Under short-circuit conditions (Fig. 2.2b) the value of the load impedance Z is removed. The magnitude of the short-circuit current is therefore only dependent on the rated voltage U, the network impedance Z_L and the resistance of the short-circuit point. This resistance (e.g. of the arc in air) can limit the short-circuit current to a large extent. It is, however, not taken into account in short-circuit current calculations.

Transient phenomena

In the event of a short-circuit, the network conditions change from their operational to a short-circuit state. This change, with the exception of the extreme case in which the short-circuit begins exactly at the instant when the current is passing through zero, involves a transition phase similar to that found in every switch-on operation (Fig. 2.3).

The symmetrical alternating current is superimposed on the d.c. component, i_{DC}. At the moment of switch-on, the direct current component i_{DC} is equal to the instantaneous value of the alternating current, but in the opposite direction. The d.c. component i_{DC} decays exponentially with the time constant $\tau = L/R$.

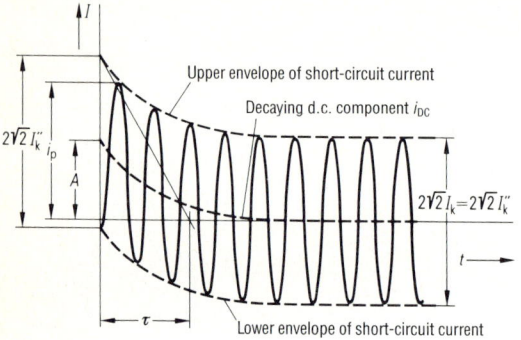

i_{DC} Decaying d.c. component of the short-circuit current
I_k'' Initial symmetrical a.c. short-circuit current
i_p Peak short-circuit current
I_k Steady-state short-circuit current
A Initial value of the d.c. component i_{DC}
τ Time constant of the d.c. component i_{DC}

Figure 2.3
Short-circuit alternating current, for short-circuit remote from generator, according to DIN VDE 0102

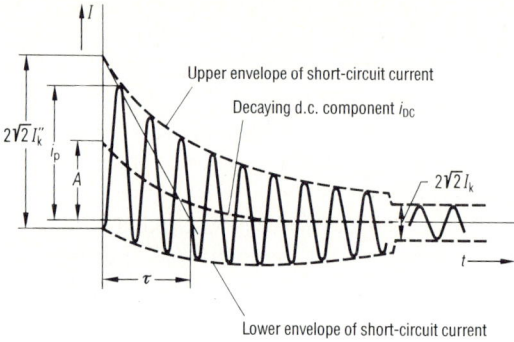

i_{DC} Decaying d.c. component of the short-circuit current
I_k'' Initial symmetrical a.c. short-circuit current
i_p Peak short-circuit current
I_k Steady-state short-circuit current
A Initial value of the d.c. component i_{DC}
τ Time constant of the d.c. component i_{DC}

Figure 2.5
Short-circuit alternating current, for short-circuit near to generator, according to DIN VDE 0102

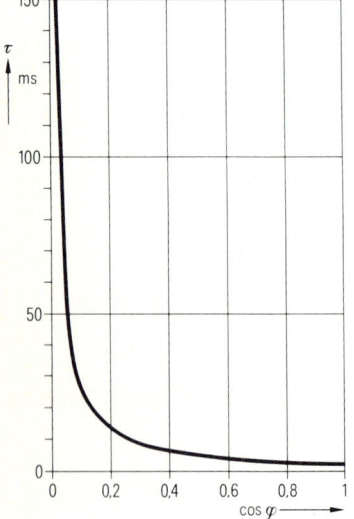

Figure 2.4
Time constant τ of the d.c. component, as a function of the $\cos \varphi$ of the circuit; the values apply for 50 Hz

Figure 2.4 shows the relationship between the time constant τ and the power factor $\cos \varphi$ of the circuit.

To be taken into account in short-circuit calculations:

▷ Subtransient initial a.c. short-circuit current I_k''

This is the effective (r.m.s.) value of the short-circuit current at the instant it occurs. For short-circuits remote from generator, which is usually the case in low voltage networks, the short-circuit current I_k'' remains almost constant for the duration of the short-circuit and can therefore be regarded as being equal to the steady-state short-circuit current I_k. Therefore $I_k'' \approx I_k$.

As a comparison, in the case of short-circuits near to the generator, the decaying initial short-circuit current I_k'' is greater than the steady-state short-circuit current I_k (Fig. 2.5). For the calculation of the subtransient initial short-circuit current I_k'', not only the effective network impedances, but also the initial reactance X_d'' (sub-transient series reactance of the synchronous machines) must be known.

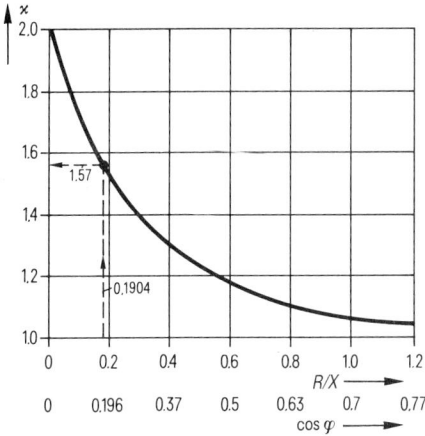

Example

$R = 1.113$ mΩ
$X = 5.8459$ mΩ
$\dfrac{R}{X} = \dfrac{1.113}{5.8459} = 0.1904$

$\varkappa = 1.02 + 0.98 \cdot e^{-3 \cdot R/X}$, gives $\varkappa = 1.57$

Figure 2.6
Factor $\varkappa$ as a function of R/X and $\cos \varphi$

▷ Peak short-circuit current i_p
as the highest instantaneous peak value of the current

This current arises when the short-circuit starts at the instant that the voltage crosses through zero.

$i_p = \varkappa \cdot 2 \cdot \sqrt{2} I_k''$

The factor $\varkappa$ expresses the relationship between the peak short-circuit current i_p and the peak value of the sustained symmetrical short-circuit current $\sqrt{2} \cdot I_k$. It is dependent on the ratio of ohmic resistance to reactive impedance R/X or on the $\cos \varphi$ of the circuit (Figure 2.6). The factor $\varkappa$ may also be calculated numerically with the formula:

$\varkappa = 1.02 + 0.98 \cdot e^{-3 \cdot R/X}$

Power factor $\cos \varphi$ in the short-circuit current path

In low-voltage systems the power factor $\cos \varphi$ of the shorted circuit depends largely on the inductive resistance (reactance) of the feeder transformer. The grea-

Short-circuit current · Types of short-circuits

ter the rated output of the transformer, the greater the inductive component of its impedance, and the lower the power factor $\cos \varphi$ is. This relationship is taken into account in DIN VDE 0660 part 101/IEC 947-2. In these documents a mimimum value for the rated short-circuit making capacity of an a.c. circuit-breaker is given as a multiple of its rated breaking capacity ($I_{cm} = n \cdot I_{cn}$). See Section 3.1.2, page 104 for more information.

2.1.3 Types of short-circuits

In a three-phase network, short-circuits may occur between three phases, two phases or between one phase and the neutral conductor or earth (Table 2.5).

The highest currents are caused by three pole and single pole short-circuits directly on the terminals of the low-voltage side of the transformer (if the transformer is the only source of current for the short-circuit).

Single source three pole short-circuit remote from the generator

● Initial a.c. short-circuit current I_k''

From Figure 2.7,

$$I_k'' = \dfrac{c \cdot U_{nT}}{\sqrt{3} Z_k} = \dfrac{c \cdot U_{nT}}{\sqrt{3} \sqrt{R_k^2 + X_k^2}}. \quad (2.1)$$

c factor for determining the equivalent short-circuit driving voltage

U_{nT} rated voltage of the transformer l.v. winding

For the calculation of the highest short-circuit current, use

$c_{max} = 1.0$ for 230/400 V, alternatively 1.05 for other voltages

$R_k = R_{Qt} + R_T + R_L$; the sum of the series resistances per conductor from Figure 2.7 (at conductor temperature 20 °C)

$X_k = X_{Qt} + X_T + X_L$; the sum of the series reactances per conductor from Figure 2.7

● Peak short-circuit current i_p

$$i_p = \varkappa \cdot \sqrt{2} I_k'' \quad (2.2)$$

The factor $\varkappa$ is derived from Figure 2.6 for the value on the abscissa $R/X = R_k/X_k$, or it is numerically calculated

2 Network data and duty types

Table 2.5 Types of short-circuits and the resultant a.c. short-circuit currents in three-phase networks

Type of short-circuit	Short-circuit at the transformer terminals			Short-circuit along the length of the conductors		
		Short-circuit current	Relationship		Short-circuit current	Relationship
Three-phase		$I_{k3} = \dfrac{U}{\sqrt{3} \cdot Z}$	$\dfrac{I_{k3}}{I_{k3}} = 1$		$I_{k3} = \dfrac{U}{\sqrt{3} \cdot Z}$	$\dfrac{I_{k3}}{I_{k3}} = 1$
Two-phase		$I_{k2} = \dfrac{U}{2 \cdot Z}$	$\dfrac{I_{k2}}{I_{k3}} = 0.87$		$I_{k2} = \dfrac{U}{2 \cdot Z}$	$\dfrac{I_{k2}}{I_{k3}} = 0.87$
Single-phase (earth fault)		$I_{k1} = \dfrac{U}{\sqrt{3} \cdot Z}$	$\dfrac{I_{k1}}{I_{k3}} = 1$		$I_{k1} = \dfrac{U}{\sqrt{3} \cdot (Z + Z_N)}$	$\dfrac{I_{k1}}{I_{k3}} \leq 0.5$

U Line-to-line voltage
Z Impedance (resistance and reactance) of a phase
Z_N Impedance (resistance and reactance) of the neatral conductor

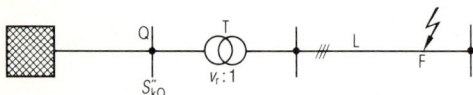

a) Network circuit diagram

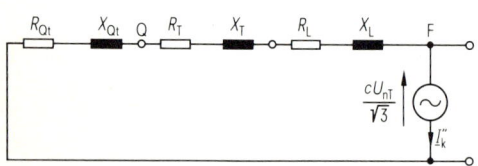

b) Equivalent circuit including equivalent voltage source

Q Point of connection to the network
S''_{kQ} Initial symmetrical short-circuit power at the point of connection to the network
v_r Rated voltage ratio at the principle tapping
F Point of fault
R_{Qt} Equivalent resistance of the network feeder
R_T Equivalent resistance of the transformer
R_L Equivalent resistance of the conductors
X_{Qt} Equivalent reactance of the network feeder
X_T Equivalent reactance of the transformer
X_L Equivalent reactance of the conductors

Figure 2.7
Example for the calculation of the subtransient initial short-circuit current I''_k for a single source three-phase short-circuit remote from the generator

Single source two-phase short-circuit remote from the generator

● Initial a.c. short-circuit current:

$$I''_{k2} = \frac{\sqrt{3}}{2} I''_k. \tag{2.3}$$

I''_k is determined by the equation (2.1), using the values:

For the calculation of the highest short-circuit current, use

$c_{max} = 1.0$ for 230/400 V, alternatively 1.05 for other voltages, and

R_L at conductor temperature 20 °C

For the calculation of the lowest short-circuit current, use

$c_{min} = 0.95$ for 230/400 V, alternatively 1.0 for other voltages, insofar as other DIN VDE specifications do not stipulate lower values, and

R_L at conductor temperature 80 °C.

● Peak short-circuit current:

$$i_{p2} = \varkappa \cdot \sqrt{2} I''_{k2}. \tag{2.4}$$

The factor $\varkappa$ is derived from Figure 2.6, or numerically calculated. Thereby, the relationship of R/X applicable for a three-phase short-circuit may be assumed.

$$I_{k1}'' = \frac{\sqrt{3} \cdot c \cdot U_{nT}}{\sqrt{(2R_{Qt} + 2R_T + 2R_L + R_{0T} + R_{0L})^2 + (2X_{Qt} + 2X_T + 2X_L + X_{0T} + X_{0L})^2}} \qquad (2.5)$$

Single source single-phase short-circuit to earth remote from the generator

For transformers with the vector groups Dy, Dz, Yy and Yz in which the star point is only earthed on the low-voltage side, the above equation (2.5) applies. Herein:

I_{k1}'' Initial a.c. short-circuit current of a single-phase fault
c Factor for determining the equivalent short-circuit driving voltage
U_{nT} Rated voltage of the transformer l.v. winding (e.g. 400 V)
R_{0T} Resistance of transformer in the zero phase-sequence network
R_{0L} Resistance of the conductors in the zero phase-sequence network
X_{0T} Reactance of transformer in the zero phase-sequence network

For further explanations, see Figure 2.7.

For the calculation of the highest short-circuit current, use

$c = 1.0$ for 230/400 V, alternatively 1.05 for other voltages

R_L, R_{0L} at conductor temperature 20 °C

For the calculation of the lowest short-circuit current, use

$c_{min} = 0.95$ for 230/400 V, alternatively 1.05 for other voltages insofar as other DIN VDE specifications do not stipulate lower values, and

R_L, R_{0L} at conductor temperature 80 °C.

The equation (2.5) does not apply for transformers with vector group Yy in which the star point is earthed on both the high and low-voltage sides, or when neutral earthing transformers are used. In these cases, calculation is done in accordance with DIN VDE 0102, 01.90 (method of symmetrical components).

2.1.3.1 Contribution to the short-circuit current by connected motors

In terms of DIN VDE 0102, a short-circuit calculation must include the portion contributed by connected motors. To this end, synchronous motors and condensors are treated as generators.

The contribution from asynchronous induction motors with rated voltage up to 1000 V may be ignored, if the sum of the rated currents of the motor group ΣI_{nM} is less than 1% of the three-pole short-circuit current I_{k3}'' without motors, i.e.

$\Sigma I_{nM} < 0.01 \, I_{k3}''$ (without motor contribution)

Motors which are not connected in terms of the circuit diagram (interlocking), or owing to their function in the process (reverse drives), are not relevant to the short-circuit calculation and are not taken into account.

The calculation of the highest short-circuit currents produced by asynchronous motors in the event of a short-circuit across their terminals is achieved by means of the following equations:

- Initial a.c. short circuit current:

$$I_{k3M}'' = \frac{U_{nT}}{\sqrt{3} X_M}. \qquad (2.6)$$

- Peak short-circuit current:

$$i_{pM} = \varkappa_M \cdot \sqrt{2} I_{k3M}'' \qquad (2.7)$$

U_{nT} Rated voltage of the transformer l.v. winding (e.g. 400 V)

$$X_M = \frac{1}{I_{an}/I_{nm}} \cdot \frac{U_{nT}}{\sqrt{3} X_M}$$

$\varkappa_M =$ 1.4 corresponding to $R_M/X_M = 0.3$ (fictitious value)
U_{nM} Rated voltage of the motor (e.g. 400 V)
I_{an} Locked rotor current of the motor
I_{nM} Rated current of the motor
X_M Short-circuit reactance of the motor

If several motors of various ratings are connected to a common low-voltage busbar system, the motors and their feeder cables may be combined as a mathematical equivalent circuit of a single motor to simplify the calculation.

For the equivalent circuit, use

X_M as per given formula above,
I_{nM} the sum of the rated currents of all the single motors in the motor group,
I_{an}/I_{nM} $= 5$, and
$\varkappa_M$ $= 1.4$ corresponding to $R_M/X_M = 0.3$ (see Fig. 2.6).

In these recommended guide values, the influence of the feeder cables to the individual motors has been taken into account.

Example

Given: I''_{k3} (without motors) = 20 kA;
I_{k3M} (motor group) = 400 A; U_{nT} = 380 V

Required: I''_{k3} (with motors)

Calculation:

$$X_M = \frac{1}{5} \cdot \frac{380}{\sqrt{3} \cdot 400} = 0.109695 \; \Omega$$

$$I''_{k3M} = \frac{380}{\sqrt{3} \cdot 0.109695} = 2000 \; A$$

Result: I''_{k3} (with motors) = 22 kA

Owing to the contribution of the motor group, the a.c. short-circuit current is increased from 20 to 22 kA.

2.1.4 Effects of short-circuit currents

Short-circuit current subjects the system components to dynamic and thermal stresses. The design and dimensioning of power installations for mechanical and thermal short-circuit strength are contained in the DIN VDE 0103 specification of April 1988.

Dynamic effect

Two parallel conductors either attract or repel each other, depending on whether the currents in them flow in the same or in opposite directions (Fig. 2.8).

As a first approximation, the force F between two long rigid parallel conductors can be expressed as:

$$F = 0.2 \, i_1 \cdot i_2 \frac{l}{a} \qquad (2.8)$$

where,

F Force in N
i_1, i_2 Instantaneous value of the currents in the conductors in kA
l Length of the conductors (distance between busbar supports) in cm
a Distance between the centres of the conductors (busbars spacing) in cm

For the force F produced by the peak short-circuit current i_{p2} as a function of the distance between conductors a_s, and for a two-pole short-circuit, the following applies (also refer to Section 9.8.2, page 547):

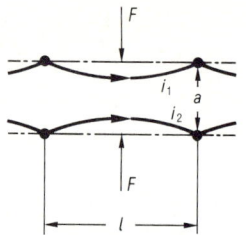 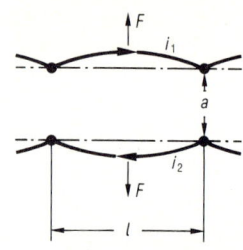

a) Current flowing in the same direction; the conductors attract each other

b) Current flowing in opposite directions; the conductors repel each other

Figure 2.8
Effect of the force F, produced by the short-circuit current, on parallel conductors (exaggerated representation)

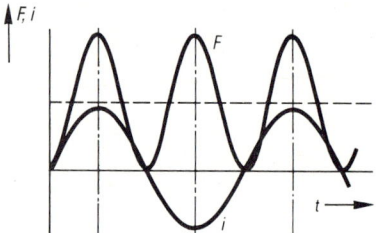

Figure 2.9
Alternating current i causes a pulsating force F, at double the frequency of the current itself

for parallel rigid conductor section (several busbars per phase),

$$F = 0.2 \cdot \left(\frac{i_{p2}}{n}\right)^2 \cdot \frac{l}{a_s}. \qquad (2.9)$$

F Force on the busbar section in N
i_{p2} Short-circuit current (two-pole short-circuit) in kA
l Distance between supports
n No. of busbar sections
a_s Effective distance between busbar sections (phases)

In the case of one busbar per phase, the effective distance between busbar sections a_s is the same as the distance between conductors a.

If there is more than one busbar per phase, a_s is not equal to a, but is dependent on

▷ busbar cross-section (width/thickness)
▷ number of busbars
▷ busbar shape

Corresponding values for a_s, for square and U-shaped busbars may be found in Table 1 or Table 5 of DIN VDE 0103. Alternatively, a_s must be calculated.

In three-phase networks, the middle busbar is usually subjected to the greatest stress in the event of a three pole short-circuit.

In a. c. circuits, the current flows in a sinusoidal wave form. Therefor, the magnitude of the force changes with the square of a sinusoidal wave. If the short-circuit condition lasts for longer than half a wave cycle (10 ms at 50 Hz), then the conductors are subjected to mechanical stresses which oscillate at a frequency twice that of the supply (Fig. 2.9).

Thermal effect

Short-circuit currents cause extreme heating of the conductors. The extent of this heating depends on the square of the effective short-circuit current I_k and its duration ($I^2 \cdot t$ value).

2.1.5 Diagrams for determining resistance values and short-circuit currents, computer program "KUBS" for short-circuit calculations and product selection

The three-pole short-circuit current at the point of installation of a piece of switchgear, can be estimated by means of the "nomograms for determining resistance values and short-circuit currents" (Fig. 2.10 to 2.12), or with the aid of a calculation sheet (Fig. 2.13).

The computer program "KUBS" provides an exact calculation according to DIN VDE 0102. In addition to the maximum three-pole short-circuit currents, it also provides the maximum two-pole and the smallest single-pole short-circuit values.

By using the diagrams (Fig. 2.10 to 2.12), it is possible to estimate the value of the initial a.c. short-circuit current I_k'' for the rated voltages 220/230 V, 380/400 V and 500 V. Thereby, the following assumptions are made:

▷ network is supplied via transformers only
▷ the primary voltage on the transformer remains constant.

The reactances X and resistances R for various cross-sections of open lines and cables can be derived from the diagrams. The same applies for the resistance R_T and reactance X_T corresponding to the various reactance voltages u_x of the transformers.

Table 2.6
Reactance voltage u_x of transformers with $u_{kr}=4\%$, $u_{kr}=6\%$ and various values of u_{Rr}

u_{Rr} in %	u_x in % with $u_{kr}=4\%$	u_x in % with $u_{kr}=6\%$
1	3.87	5.92
1.1	3.85	5.89
1.2	3.82	5.88
1.3	3.78	5.86
1.4	3.75	5.83
1.5	3.71	5.81
1.6	3.67	5.78
1.7	3.62	5.75
1.8	3.57	5.72
1.9	3.52	5.69
2	3.46	5.66
2.1	3.40	5.62
2.2	3.34	5.58
2.3	3.27	5.54
2.4	3.20	5.50
2.5	3.12	5.45

The reactance voltage (Table 2.6) is determined by:

$$u_x = \sqrt{u_{kr}^2 - u_{Rr}^2}. \tag{2.10}$$

u_x Reactance voltage of the transformer in %
u_{kr} Impedance voltage of the transformer in %
u_{Rr} Resistive voltage drop of the transformer in %

The values for u_{kr} and u_{Rr} are given in the Siemens transformer catalogues (a number of values are also listed in Section 9.6, page 532).

From the Figures 2.10 to 2.12, therefor, the effective value of the initial symmetrical short-circuit current I_k'' as a function of the geometrical sum of the reactances and resistances, may be found. The values thus determined all have a safety factor to a varying degree, since the short-circuit current is limited to a significant extent by contact resistances, current loops and current displacement effects. These factors are not been taken into account by the use of the diagrams.

2 Network data and duty types

Example for 230 V networks

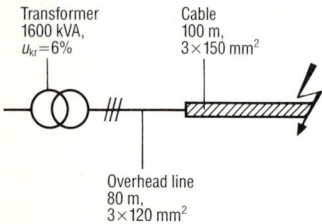

Rated output of the transformer: $S_{nT} = 1600$ kVA

Overhead line: Copper, length 80 m cross-section 3×120 mm²

Cable: Copper, length 100 m cross-section 3×150 mm²

Impedance voltage: $u_{kr} = 6\%$

Resistive voltage drop: $u_{Rr} = 1\%$

Rated operating voltage on the low-voltage side: 230 V

Reactance voltage: (also see Table 2.6)
$u_x = \sqrt{u_{kr}^2 - u_{Rr}^2} = \sqrt{6^2 - 1^2} = 5.92\%$

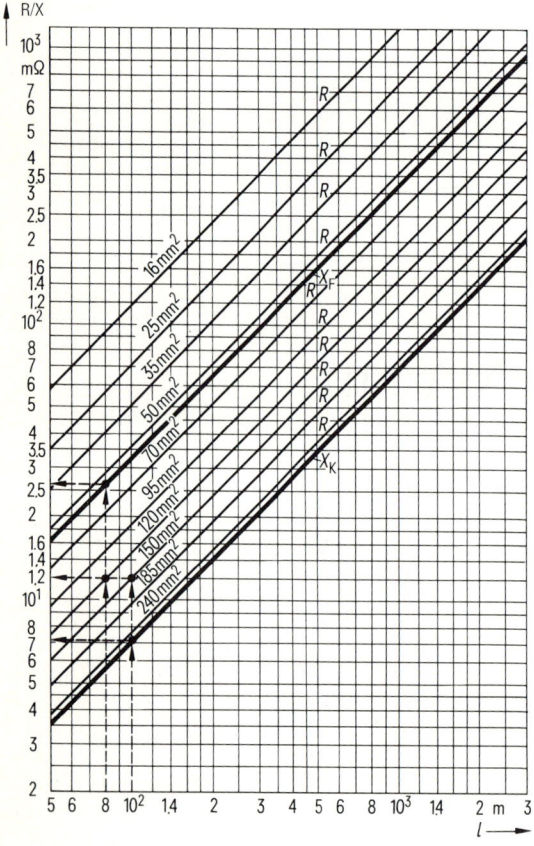

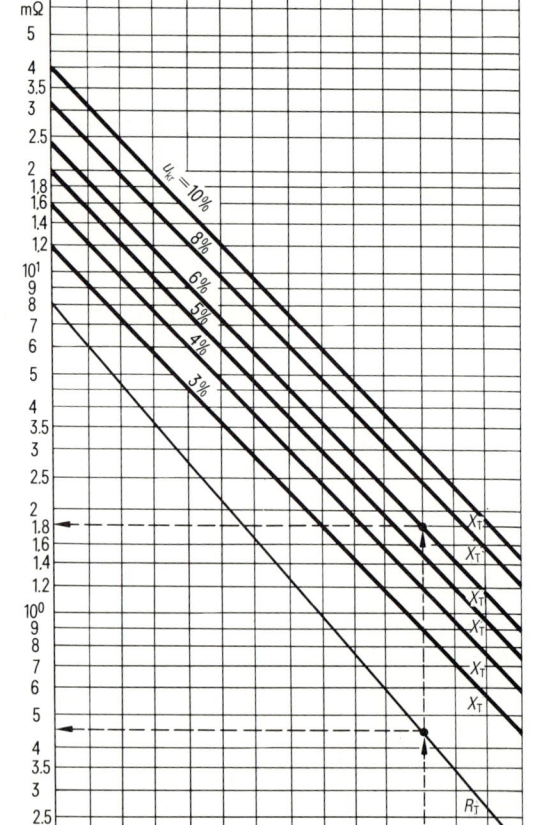

R Resistance at 20 °C
X_F Reactance of overhead lines
X_K Reactance of cables

(In the case of *aluminium* conductors, the resistance values taken from the diagram must be multiplied by *1.7*)

R_T Resistance X_T Reactance

a) Resistances and reactances of overhead lines and cables

b) Resistances and reactances of transformers

Figure 2.10 Nomograms to determine the short-circuit current in 220/230[1]) V networks

Table 2.7
Table of the resistances and reactances from Figures 2.10a and b

	Resistance R mΩ	Reactance X mΩ
Overhead line and Cable from Fig. 2.10a	12.0 12.0	26.0 7.2
Transformer from Fig. 2.10b	$0.45 \cdot 1.09$ [1)]	$1.8 \cdot 1.09$ [1)]
Total impedance of the short-circuit path	24.5	35.2

The total impedance of the short-circuit path is the arithmetic sum of the individual resistive and reactive impedances respectively (Table 2.7).

Reading off the corresponding initial symmetrical short-circuit current

By using the total resistance and reactance values determined above, the short-circuit current can be read from Figure 2.10c. The intersection between the vertical resistance and horizontal reactance lines gives the magnitude of the initial symmetrical short-circuit current I''_k. If this point lies between the curves for I''_k, then the actual value of the initial symmetrical short-circuit current I''_k is obtained by interpolation.

Thus, the initial symmetrical short-circuit current for the resistances and reactances determined is approximately $I''_k = 3000$ A.

[1)] For 230 V networks, the resistive and reactive impedances shown in Figure 2.10b are shifted upward by approximately 9%.

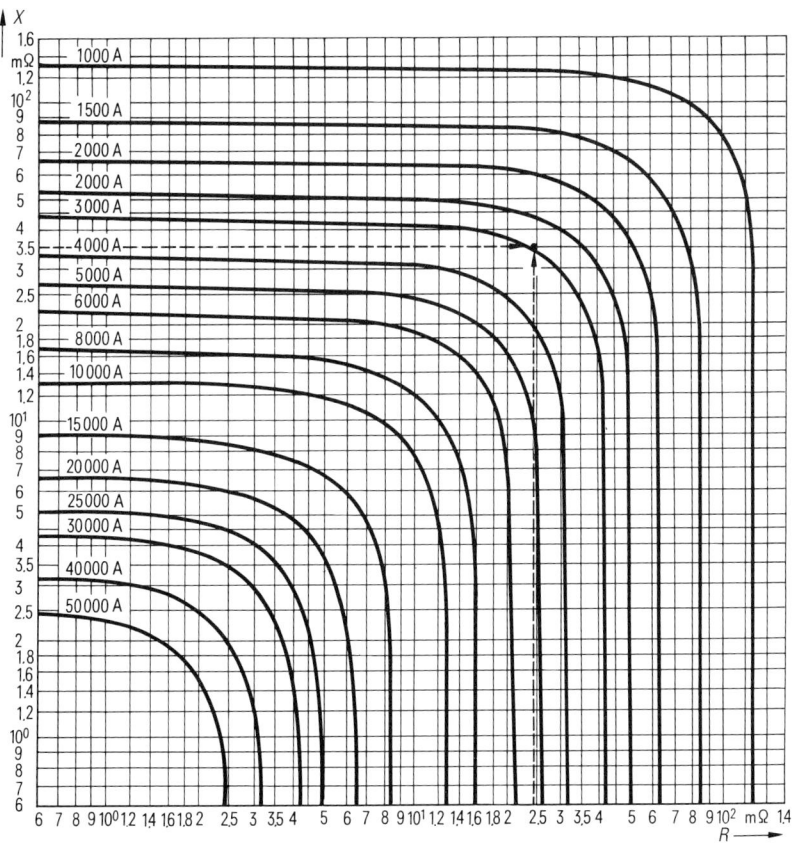

R Resistance
X Reactance

c) Initial symmetrical a.c short-circuit current I''_k (r.m.s. value) as a function of the total impedance of the short-circuit path

2 Network data and duty types

Example for 400 V networks

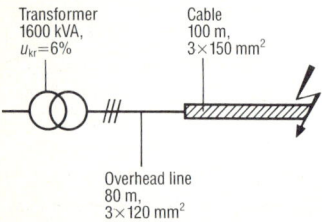

Rated output of the transformer: $S_{nT} = 1600$ kVA

Overhead line: Copper, length 80 m, cross-section 3×120 mm^2

Cable: Copper, length 100 m, cross-section 3×150 mm^2

Impedance voltage: $u_{kr} = 6\%$

Resistive voltage drop: $u_R = 1\%$

Rated operating voltage on the low-voltage side: 400 V

Reactance voltage (also see Table 2.6):
$$u_x = \sqrt{u_{kr}^2 - u_{Rr}^2} = \sqrt{6^2 - 1^2} = 5.92\%$$

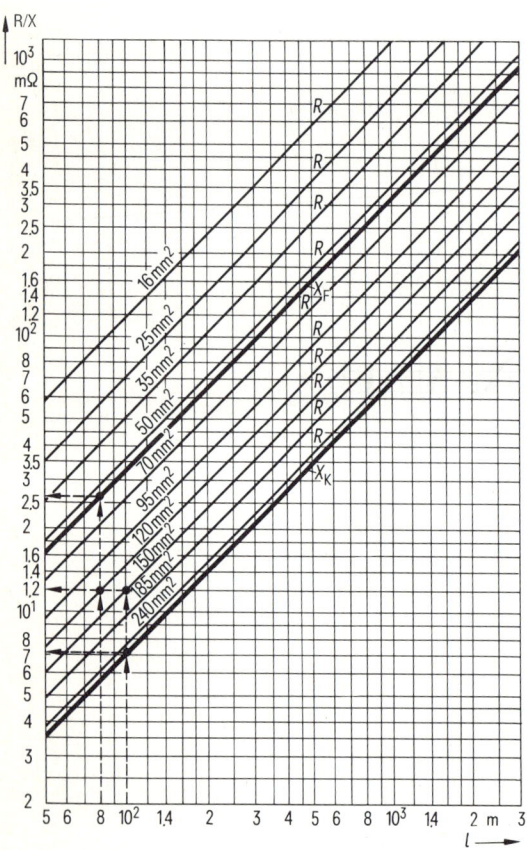

R Resistance at 20 °C
X_F Reactance of overhead lines
X_K Reactance of cables

(In the case of *aluminium* conductors, the resistance values taken from the diagram must be multiplied by 1.7)

a) Resistances and reactances of overhead lines and cables

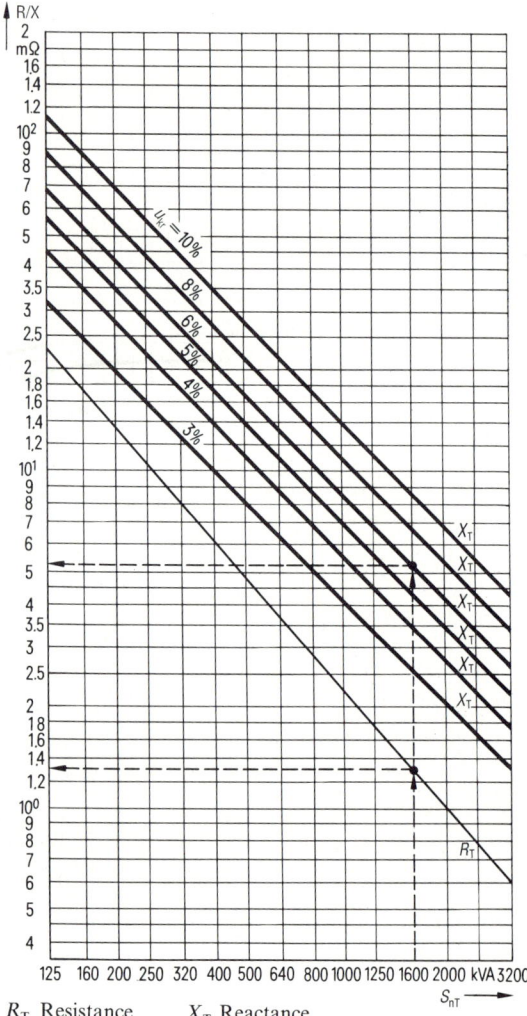

R_T Resistance X_T Reactance

b) Resistances and reactances of transformers

Figure 2.11 Nomograms to determine the short-circuit current in 380/400[1] V networks

Table 2.8
Table of the resistances and reactances from Figures 2.11a and b

	Resistance R mΩ	Reactance X mΩ
Overhead line and Cable from Fig. 2.11a Transformer from Fig. 2.11b	12 12 $1.3 \cdot 1.1$ [1]	26 7.2 $5.2 \cdot 1.1$ [1]
Total impedance of the short-circuit path	25.4	38.9

The total impedance of the short-circuit path is the arithmetic sum of the individual resistive and reactive impedances respectively (Table 2.8).

Reading off the corresponding initial symmetrical short-circuit current

By using the total resistance and reactance values determined above, the short-circuit current can be read from Figure 2.11c. The intersection between the vertical resistance and horizontal reactance lines gives the magnitude of the initial symmetrical short-circuit current I_k''. If this point lies between the curves for I_k'', as in this example, then the actual value of the initial symmetrical short-circuit current I_k'' is obtained by interpolation.

Thus, the initial symmetrical short-circuit current for the resistances and reactances determined is approximately $I_k'' = 4700$ A.

---
[1] For 400 V networks, the resistive and reactive impedances shown in Figure 2.11b are shifted upward by approximately 10%.

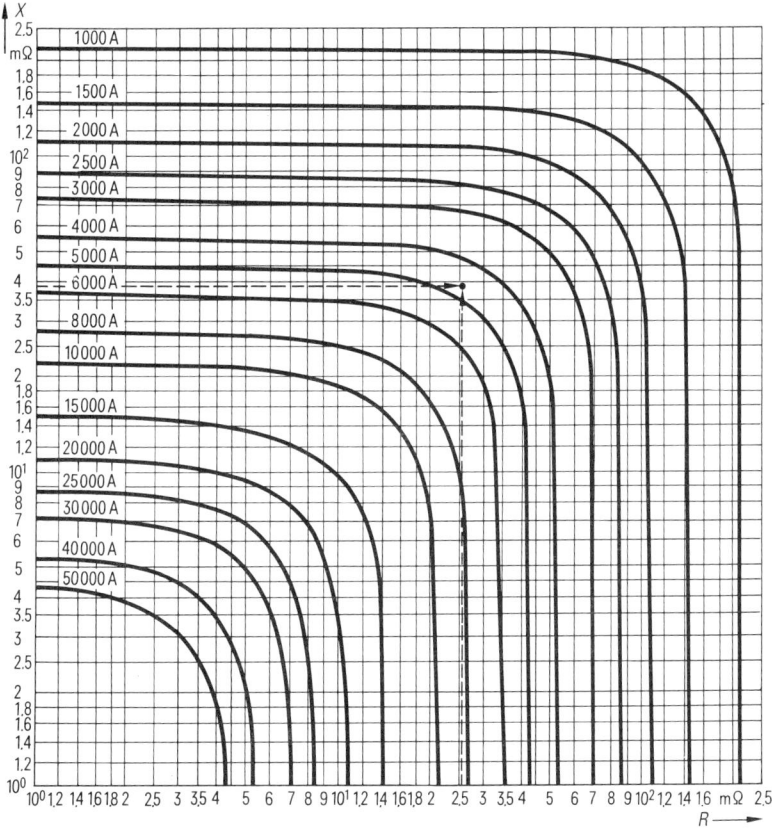

R Resistance
X Reactance

c) Initial symmetrical a.c short-circuit current I_k'' (r.m.s. value) as a function of the total impedance of the short-circuit path.

2 Network data and duty types

Example for 500 V networks

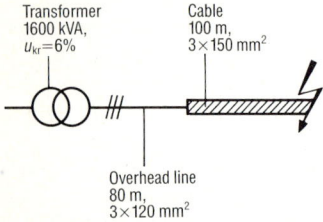

Rated output of the transformer: $S_{nT} = 1600$ kVA

Overhead line: Copper, length 80 m, cross-section 3×120 mm^2

Cable: Copper, length 100 m, cross-section 3×150 mm^2

Impedance voltage: $u_{kr} = 6\%$

Resistive voltage drop: $u_{Rr} = 1\%$

Rated operating voltage on the low-voltage side: 500 V

Reactance voltage: (also see Table 2.6)
$$u_x = \sqrt{u_{kr}^2 - u_{Rr}^2} = \sqrt{6^2 - 1^2} = 5.92\%$$

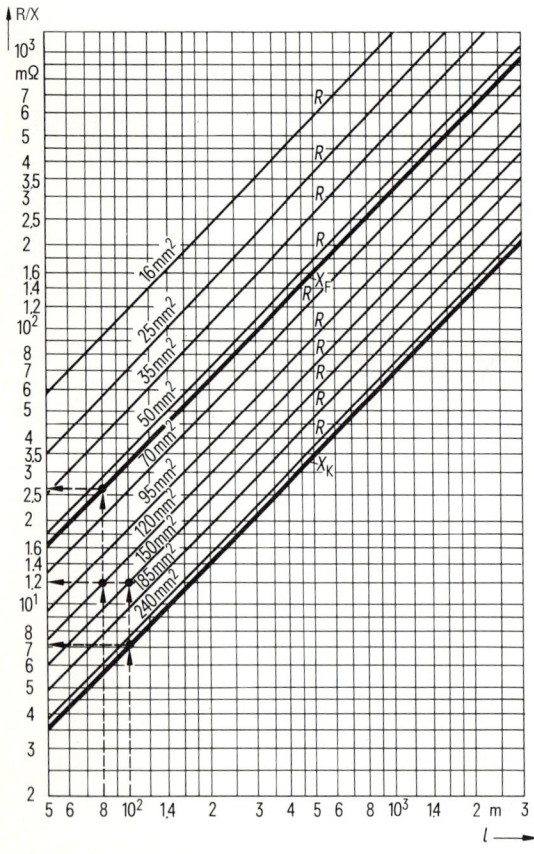

R Resistance at 20 °C
X_F Reactance of overhead lines
X_K Reactance of cables

(In the case of *aluminium* conductors, the resistance values taken from the diagram must be multiplied by 1.7)

a) Resistances and reactances of overhead lines and cables

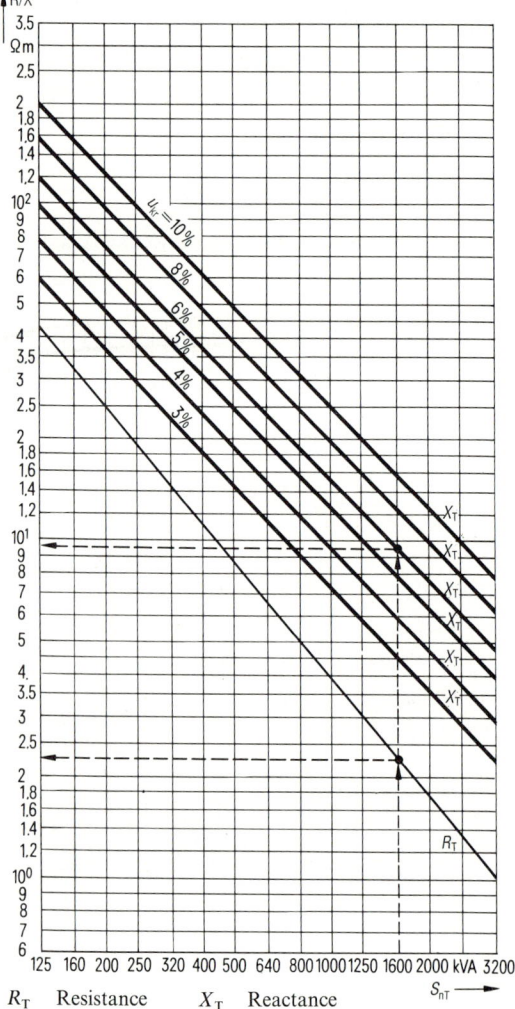

R_T Resistance X_T Reactance

b) Resistances and reactances of transformers

Figure 2.12 Nomograms to determine the short-circuit current in 500 V networks

Table 2.9
Table of the resistances and reactances from Figures 2.12a and b

	Resistance R mΩ	Reactance X mΩ
Overhead line and Cable from Fig. 2.12a	12 12	26 7.2
Transformer from Fig. 2.12b	2.2	9.5
Total impedance of the short-circuit path	26.2	42.7

The total impedance of the short-circuit path is the arithmetic sum of the individual resistive and reactive impedances respectively (Table 2.8).

Reading off the corresponding initial symmetrical short-circuit current

By using the total resistance and reactance values determined above, the short-circuit current can be read from Figure 2.12c. The intersection between the vertical resistance and horizontal reactance lines gives the magnitude of the initial symmetrical short-circuit current I_k''. If this point lies between the curves for I_k'', as in this example, then the actual value of the initial symmetrical short-circuit current I_k'' is obtained by interpolation.

Thus, the initial symmetrical short-circuit current for the resistances and reactances determined is approximately $I_k'' = 5800$ A.

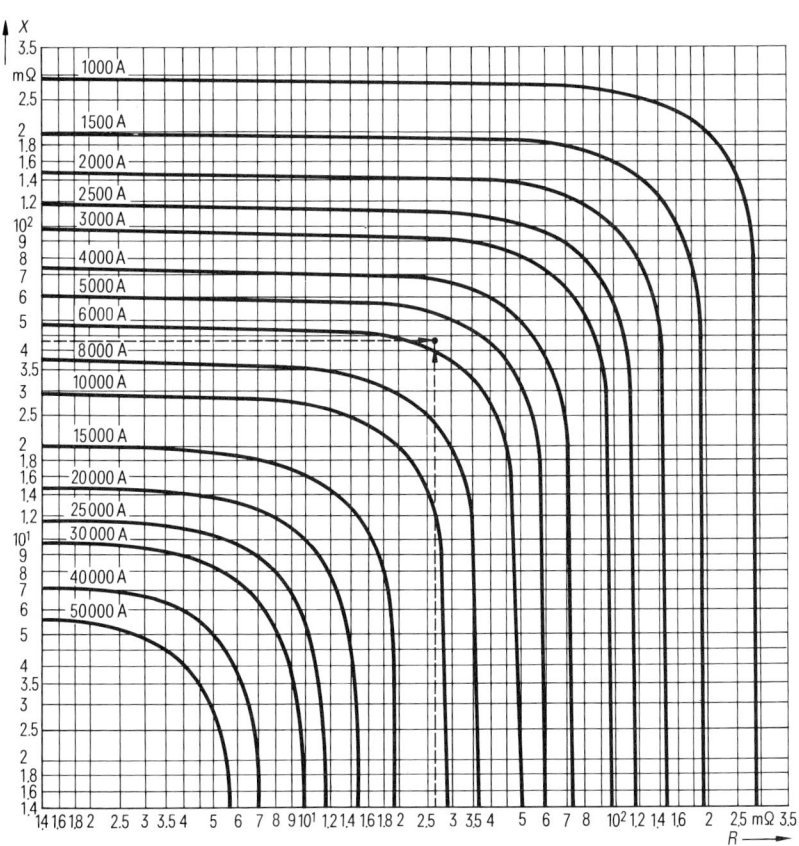

R Resistance
X Reactance

c) Initial symmetrical a.c short-circuit current I_k'' (r.m.s. value) as a function of the total impedance of the short-circuit path

SIEMENS

Calculation of short-circuit currents in low-voltage networks to DIN VDE 0102

Three-phase short-circuit fed from a network via a transformer

Customer/location: *Günther / Nürnberg*
Network point: *A*

Rated voltage of the transformer low-voltage side
U_{rTUS} = 231, **400**, 525 or .... V, **50** or .... Hz

Network data				Calculation process	$U_{rTUS}^2 = 231^2 = 53.4 \cdot 10^3$ $400^2 = 160 \cdot 10^3$ $525^2 = 275.6 \cdot 10^3$	Resistance R [mΩ]	Reactance X [mΩ]
a) Network	S_{kQ}'' ...... MVA (req. from power utility) **250** MVA (actual value) U_{nQ} ...... kV			Referred to low-voltage side: $Z_{Qt} = \frac{1.1 \cdot U_{nQ}^2}{S_{kQ}'' \cdot 10^3} \cdot \left(\frac{U_{rTUS}}{U_{rTOS}}\right)^2 = \frac{1.1 \cdot 160 \cdot 10^3}{250 \cdot 10^3} = 0.704$ mΩ $X_{Qt} = 0.995 \cdot Z_{Qt} = 0.995 \cdot 0.704$ $R_{Qt} = 0.1 \cdot X_{Qt} = 0.1 \cdot 0.7004$		0.07	0.7004
b) Transformer	S_{rT} kVA	u_{kr} %	u_{Rr} %	$Z_T = u_{kr} \cdot \frac{U_{rT}^2}{S_{rT} \cdot 100\%} = 6 \cdot \frac{160 \cdot 10^3}{2000 \cdot 100} = 4.8$ mΩ $X_T = \sqrt{Z_T^2 - R_T^2} = \sqrt{4.8^2 - 0.84^2} = \sqrt{22.33} = 4.7255$ $R_T = u_{Rr} \cdot \frac{U_{rT}^2}{S_{rT} \cdot 100\%} = 1.05 \cdot \frac{160 \cdot 10^3}{2000 \cdot 100} = 0.84$		0.84	4.7255
	1. **1000**	**6**	**1.05**	If necessary, obtain from catalogue			
	2. **1000**	**6**	**1.05**				
	3.						
	Total **2000** S_{rT}	**6**¹⁾	**1.05**¹⁾ Mean values ¹⁾ refer overleaf				
c) Cable or overhead line	n^* **4** × **240** mm² q $l = \underline{6}$ m Cu $x'^{**} \approx 0.08$ mΩ/m (cable) $x'^{**} \approx 0.33$ mΩ/m (overhead line) ϱ refer overleaf			$R_{L_1} = \frac{l \cdot \varrho}{q \cdot n} \cdot 10^3 = \frac{6 \cdot 1}{240 \cdot 4 \cdot 56} \cdot 10^3 = 0.111$ mΩ $X_{L_1} = x' \cdot \frac{l}{n} = 0.08 \cdot \frac{6}{4} = 0.12$ mΩ		0.111	0.12
d) Busbars in the switchboard	Busbars per phase **1** × **80** × **10** mm thus $q =$ **800** mm²/Ph $l =$ **1.5** m Cu $x'^{**} \approx 0.12$ mΩ/m ϱ refer overleaf			$R_{L_2} = \frac{l \cdot \varrho}{q} \cdot 10^3 = \frac{1.5 \cdot 1}{800 \cdot 56} \cdot 10^3 = 0.033$ mΩ $X_{L_2} = x' \cdot l = 0.12 \cdot 1.5 = 0.18$ mΩ		0.033	0.18
e) Distribution cable (wire) *not applicable*	n^* × mm² q $l =$ m Cu/Al/Fe $x'^{**} \approx 0.08$ mΩ/m ϱ refer overleaf			$R_{L_3} = \frac{l \cdot \varrho}{q \cdot n} \cdot 10^3 = \underline{} \cdot 10^3 = \underline{}$ mΩ $X_{L_3} = x' \cdot \frac{l}{n} = 0.08 \cdot \underline{} = \underline{}$ mΩ		/	/
f) *Distribution busbars (copper)*	*Busbars per phase* **1·30·10** mm *thus* $q =$ **300** mm² $l =$ **1** m $\varkappa =$ **0.12** mΩ/m			$R = \frac{l \cdot q}{q} \cdot 10^3 = \frac{1 \cdot 1}{300 \cdot 56} \cdot 10^3 = 0.059$ mΩ $X = \varkappa' \cdot l = 0.12 \cdot 1 = 0.12$ mΩ		0.059	0.12

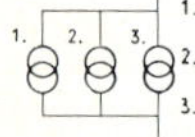

* n = number of parallel conductors per phase
** x' = values valid for 50 Hz – for other frequencies, convert proportionately

Total: $R_k = 1.113$ $X_k = 5.8459$

$Z_k = \sqrt{R_k^2 + X_k^2} = \sqrt{1.113^2 + 5.8459^2} = \sqrt{35.4133} = 5.9509$ mΩ

The maximum initial symmetrical three-phase short-circuit current at the point in question (prospective current):

$I_k'' = \frac{U_{rT}}{\sqrt{3} \cdot Z_k} = \frac{400}{\sqrt{3} \cdot 5.9509} = \underline{38.81}$ kA_eff

The max. prospective peak short-circuit current at the point in question:

i_p [peak value] = $\sqrt{2} \cdot \varkappa \cdot I_k'' = \sqrt{2} \cdot 1.57 \cdot 38.81 = \underline{86.17}$ kA

$\varkappa$ refer overleaf

U_{rTUS} = Rated voltage of the transformer low-voltage side (e.g. 400 V, see above)
U_{rTOS} = Rated voltage of the transformer high-voltage side

Network data determined by/on: *Guillen / 20.6.91*
Calculated by/on: *Dommisse / 28.6.91*

① This assumption is approximately correct (~), if each transformer has a connecting line or cable. If not, the assumption is correct (=).

Figure 2.13
Calculation sheet. Example for the calculation of the three-pole short-circuit currents in a low-voltage feeder circuit

Calculation sheet

The calculation sheet "Calculation of the short-circuit currents in low-voltage networks according to DIN VDE 0102, January 1990" (Fig. 2.13) may be ordered from Siemens AG, Erlangen (order no. H30-E 1056-N41 – German).

The approximate mean value of the impedance voltage u_{kr} for several transformers operating in parallel $u_{kr\,mean}$ can be calculated as follows ($u_{Rr\,mean}$ is calculated in a similar way):

$$U_{kr\,mean} = \frac{S_{n_1} + S_{n_2} + S_{n_3} + \cdots}{\frac{S_{n_1}}{u_{z_1}} + \frac{S_{n_2}}{u_{z_2}} + \frac{S_{n_3}}{u_{z_3}} + \cdots}. \quad (2.11)$$

The specific resistances ϱ (at 20 °C) for

Copper (Cu): $\quad \varrho_{Cu} = \frac{1}{57} \frac{\Omega \, mm^2}{m},$

Aluminium (Al): $\varrho_{Al} = \frac{1}{34} \frac{\Omega \, mm^2}{m},$

Iron (Fe): $\quad \varrho_{Fe} = \frac{1}{8} \frac{\Omega \, mm^2}{m}.$

Computer program "KUBS" for short-circuit calculations and product selection

The name of the computer program KUBS is derived from the German "**K**urzschlußstromberechnung, **B**ack-up-Schutz und **S**elektivität", which means "short-circuit calculation, back-up protection and selectivity" (updated English version in preparation). It calculates the maximum single and three-pole, as well as the smallest single-pole short-circuit currents in low-voltage radial networks. The program is based on the instructions in DIN VDE 0102 part 2, for the calculation of a single-source short-circuit remote from the generator.

The following hardware (example) is required to run KUBS:

▷ Computer: Siemens PCD-3T, PCD-4T or other AT-compatible computer with at least one $3\frac{1}{2}''$ or $5\frac{1}{4}''$ floppy disk drive.

▷ Operating system: MS-DOS 2.11 or higher
▷ Printer: Siemens PT-88, or compatible
▷ Plotter: Sharp CE 515P, or compatible (optional)
▷ Laser printer: Kyocera F 1200, or compatible (optional)

In practice, all the data required for the calculation of short-circuit currents is not always available, e.g. the initial symmetrical short-circuit apparent power, S_k'', of the medium-voltage network in the case of an infeed via transformer. In this case, for a 20 kV network, for example, KUBS suggests a value of 500 MVA. For the characteristic values of the transformer, the program provides the relevant reactance voltage u_{kr} and resistive voltage drop u_{Rr} by means of empirical formulae, or adopts the values for the calculation from the corresponding DIN standard power transformer. The programm also allows for the use of an equivalent single transformer model to represent a parallel connection of several feeder transformers.

Often, it is required to calculate the short-circuit current for a point in the network, using a known fault level at some point upstream as a starting base value. This is the case, for example, in larger networks in which the calcultion needs to be carried out in several stages. For this, the program offers the possibility to select a starting point. Apart from the three pole symmetrical short-circuit current, the program needs a number of further parameters to continue the calculation: either the asymetrical peak short-circuit current i_p, or the power factor cos φ, or the ratio R/X. If none of these values is known, the power factor cos φ used for the switching capacity testing of power circuit-breakers (to DIN VDE 0660 part 101), in relation to the short-circuit current, may be called up by the program (see Section 3.1.2, page 104). Also, the rated operating voltage U_e and rated system frequency f, must be known. The results obtained are sufficiently accurate for most applications encountered in practice.

The network data is entered in a dialog with the computer. The program numbers the network points sequentially. It is recommended that the network diagram be sketched by hand first, and that the points defined and numbered by the program are entered on this sketch as one proceeds through the calculation.

Point 1 refers to the secondary terminals of the transformer, or the generator terminals or the source point of the short-circuit current. By using the function "Continue", a further feeder may be attached in series to the previously calculated network point. By selecting the function "New Branch", any already calculated point may be treated as a new starting point for

2 Network data and duty types

a further feeder. The user is requested to "Enter the network point" at which the further distribution is to be attached.

Even if the initial information is not very extensive, by means of these functions it is possible to process the network data in virtually any sequence.

The conductor types (cable, busbars or overhead lines) and their material (copper or aluminium) can be entered for each feeder. If the geometrical arrangement of the busbars or overhead lines is known, the program can use this information to determine the inductive reactance of the conductors. Impedance values from tables may also be entered individually.

A number of filters and plausibility checks prevent the input of forbidden values. On the other hand, the program does not reject all values which are theoretically possible, even though they may appear to be technically unrealistic or non-sensical.

Corrections are possible even after the network calculation has been completed, e.g. if the network has changed, or if an input error is discovered. Thus the data of every part of the network, including the infeed, may be changed. The insertion or deletion of feeders at any point in the network diagram can be implemented. All these changes cause the calculations of the downstream network points to be updated. These features are particularly useful for carrying out "what if" calculations during network design.

The program function "Breaker" enables the optimal selection of a circuit-breaker for each specified point and application in the network. This selection takes into account the continuous current rating of the circuit-breaker at the specified ambient temperature, its rated short-circuit breaking capacity at the given rated voltage U_e of the network, and the setting range of the overload release.

To enable the program to recommend a circuit breaker, it is only necessary to enter the required continuous current and the number of the network point in question. The rated voltage and the prospective short-circuit current (fault level) are automatically taken into account.

Even the back-up protection by means of an upstream circuit-breaker is taken into consideration by the selection feature. The tables from the Siemens brochure "Discrimination and back-up protection in fuseless low-voltage feeders" (order no. E-20001-P285-A372-X7600), is used a basis for this.

The program indicates the degree of discrimination between the upstream und downstream circuit-breakers. If the given discrimination limits ("Sel" for selectivity limit) are not suitable for the application in question, the program may be called upon to suggest further alternatives.

The function "Recall Data" enables the display of the data entered and the short-circuit current calculated for each and every network point (see Fig. 2.14).

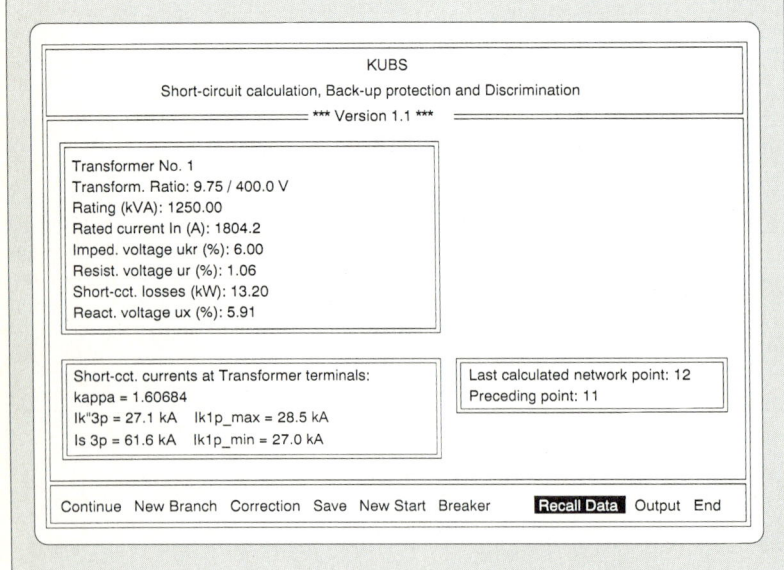

Figure 2.14
The function "Recall Data" may be used to check the input data and the calculated short-circuit currents for each network point

Short-circuit calculation using "KUBS"

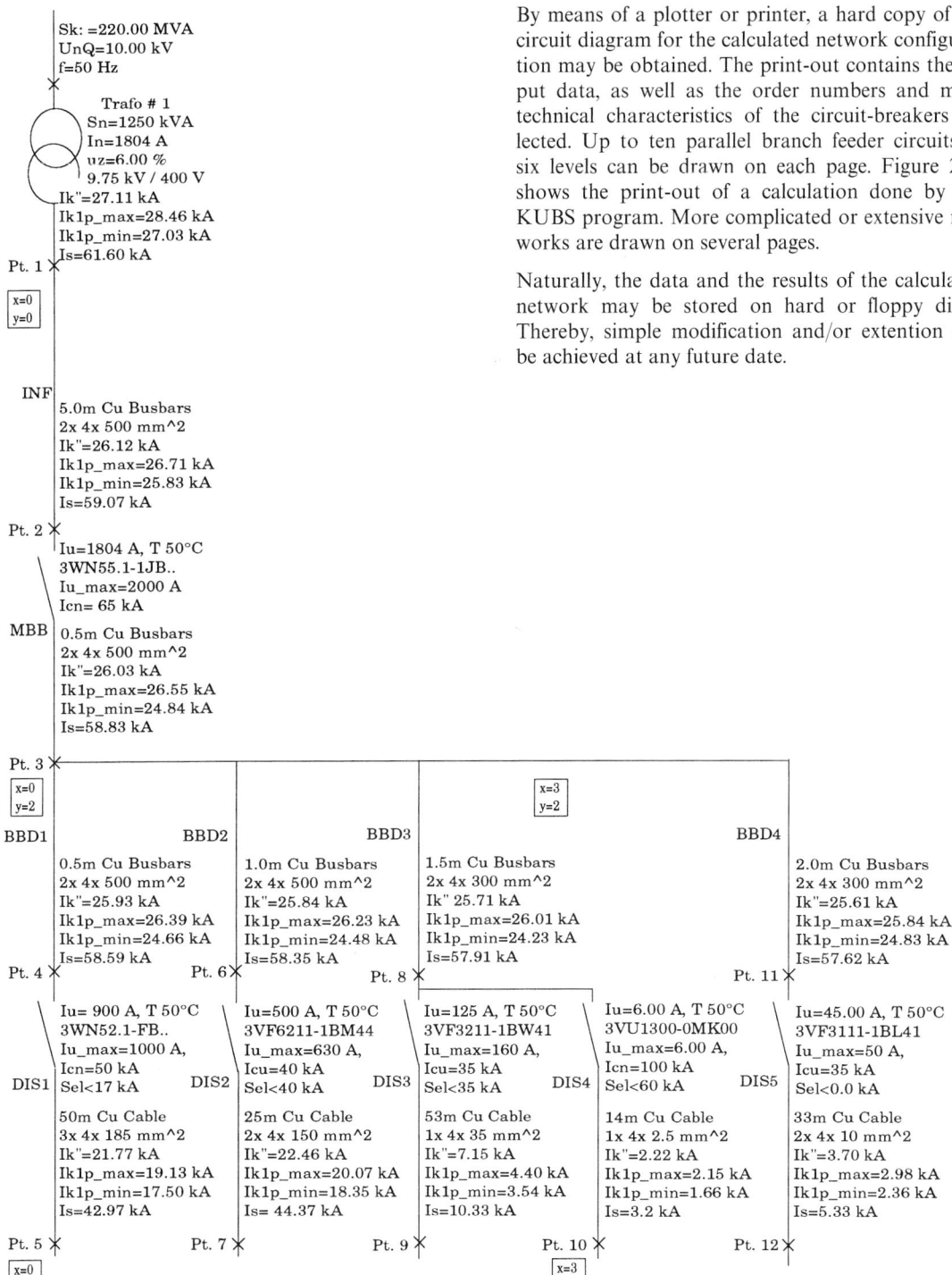

By means of a plotter or printer, a hard copy of the circuit diagram for the calculated network configuration may be obtained. The print-out contains the input data, as well as the order numbers and main technical characteristics of the circuit-breakers selected. Up to ten parallel branch feeder circuits in six levels can be drawn on each page. Figure 2.15 shows the print-out of a calculation done by the KUBS program. More complicated or extensive networks are drawn on several pages.

Naturally, the data and the results of the calculated network may be stored on hard or floppy disks. Thereby, simple modification and/or extention can be achieved at any future date.

Figure 2.15 Typical print-out from the KUBS program

2.1.6 Influence of the transformers and conductors on the short-circuit current

For a given operating voltage, the magnitude of the short-circuit current depends on the sum of the impedances between the generator and the point of short-circuit.

In low-voltage networks, these impedances are dominated by those of the transformers, transmission lines and cables. The additional resistive and inductive impedances in the current path, e.g. contact resistances and inductive influences of neighbouring metal structures, are either difficult to ascertain or, in fact, remain unknown for the calculation.

Figure 2.16 indicates the expected value of short-circuit current I_k'' as a function of the rated apparent power S_{nT} and the impedance voltage of the transformers u_{kr}, for a given rated voltage. The short-circuit currents indicated, refer to a full three pole short-circuit directly across the low-voltage terminals of the transformer.

The extreme limiting effect of the transmission lines and cables on the short-circuit current is illustrated by means of an example of a distribution system in Figure 2.18 (also refer to Section 9.8.1, page 546). More extensive information may be found in the brochure "Overload and short-circuit protection in low-voltage installations" (order no. A19100-E732-A326-X-6700).

Smallest short-circuit current

For protection by disconnection, e.g. in TN systems, it is necessary to determine the smallest short-circuit current which will flow in the event of a short-circuit between a phase and the PEN conductor at its furthest point from the supply. This is to establish whether the short-circuit protection apparatus will cause the current to be interrupted within the prescribed time.

In DIN VDE 0100 part 410, paragraph 6.1.3, the following values have been specified for consumer installations:

▷ maximum 0.2 seconds for circuits with socket outlets and rated currents up to 35 A,
▷ maximum 0.2 seconds for all other circuits associated with equipment of the protection class I, if they are continuously hand-held or are enclosed by the hand, e.g. operating switches on machine tools
▷ maximum 5 seconds for all other circuits (also see Section 1.4.2.1, page 42)

The following must be true:

$Z_s \cdot I_a \leq U_0$

Z_s Impedance of the fault current loop
I_a A.c. operating current which trips the automatic opening of the circuit
U_0 Voltage to earthed conductor

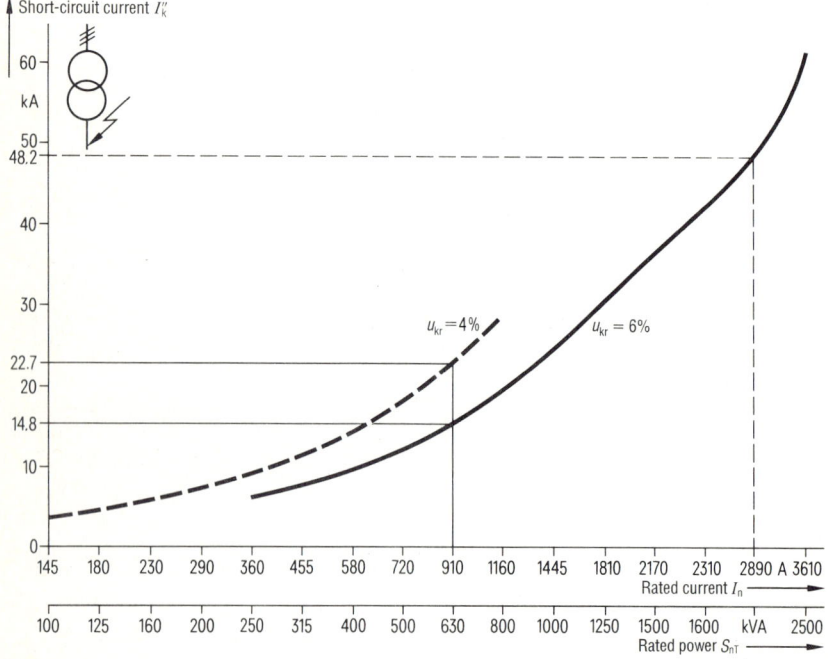

Figure 2.16
Initial a.c. short-circuit current I_k'' of transformers (400 V, 50 Hz) as a function of the apparent power S_{nT} and the impedance voltage u_{kr}

Influences on the short-circuit current

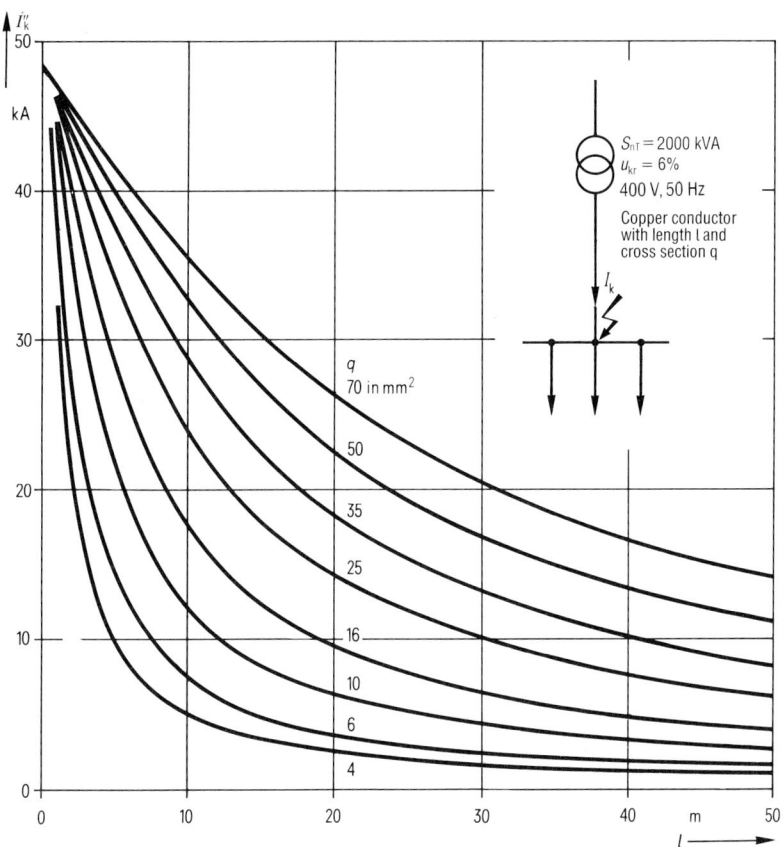

Figure 2.17
Initial a.c. short-circuit current I_k'' as a function of cable or conductor length l and cross-section q (example)

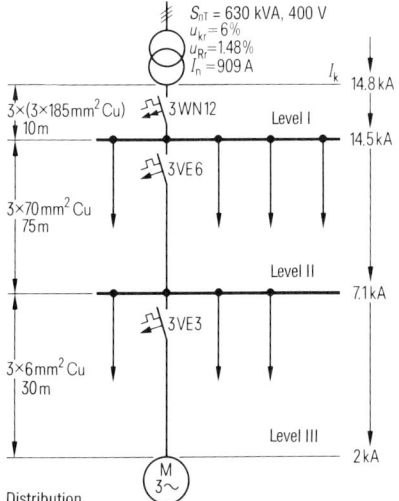

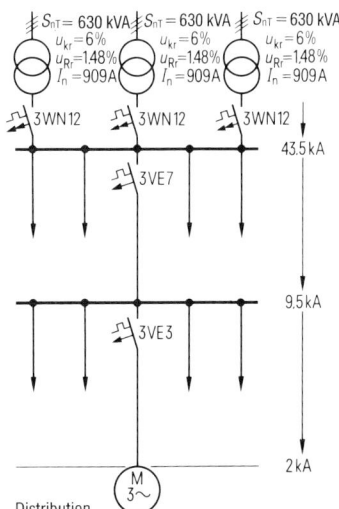

a) Infeed via a single transformer

b) Infeed via three transformers

Figure 2.18
Limiting of the short-circuit current by the transmission lines and cables in a distribution system

2 Network data and duty types

The smallest single-phase short-circuit current in TT and TN-C networks can be determined by means of the short-circuit calcutation program "KUBS" (see page 93).

The program also recommends circuit-breakers incorporating electromagnetic or electronic instantaneous releases with operating currents which are below the smallest single-pole short-circuit current at the point of circuit-breaker installation, i.e. it recommends circuit-breakers which fullfil the above requirement.

2.2 Duty types

Low voltage devices and asynchronous three-phase induction motors, with the exception of motors for special applications, e.g. servo-drives, are dimensioned for continuous operation at their rated output. To a large extent, however, motors are used in applications where the operating duty deviates considerably from that of continuous operation. Since in these cases the power rating of the drives and the associated switchgear for the optimal use of power reserves differs from that associated with normal continuous duty, it is necessary to specify exact discriptions of the various duty types[1]. To simplify the communication between user and manufacturer, the various possible duties have been divided into 9 main categories in DIN VDE 0530 part 1.

Several of the applications found in practice are covered by the duties S1 to S9. If the actual loading differs from these, an exact discription of the duty cycle must be given. Alternatively a duty which represents an equally severe loading must be specified.

2.2.1 Continuous operation duty

Continuous operation duty (S1)

This refers to operation under constant load condition (Fig. 2.19), e.g. at rated power, of sufficient duration for thermal equilibrium to be attained.

Plant components must be so dimensioned that they can conduct the constant load current for an unlimited period of time without the need for intervention

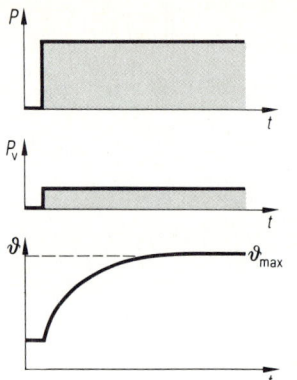

Figure 2.19
Duty type S1:
Continuous operation

and without the permissible temperature limits of the components being exceeded (also refer to Section 4.3.5.2, page 259).

2.2.2 Short-time operation duty

Short-time operation duty (S2)

This also refers to operation under constant load condition (Fig. 2.20), however for a period of time which is not long enough for thermal equilibrium to be attained, and followed by an interval of sufficient duration to allow the machine temperature to be within 2 K of the coolant temperature.

Designation: S2 and by the duration of operation and power rating eg. S2: 20 min, 15 kW.

During short-time operation duty (in German, also referred to as "KB") at a contant rated operating current I_e, the maximum permissible temperature of the plant components e.g. the switchgear, is not reached during the load period.

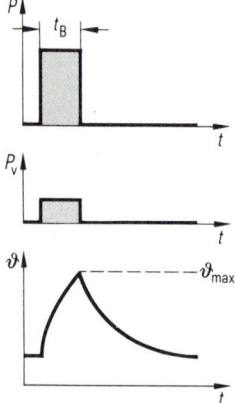

Figure 2.20
Duty type S2: Short-time operation

---

[1] Further information (also for design and dimensioning purposes) may be found in the "Project Manual" section of the Siemens M10 catalogue as well as in the special issue "Planning of Standard Motors". For the selection of 3TF, 3TK and 3TH contactors in short-time and intermittent periodic duty, see Sections 3.7.7 to 3.7.9, page 197 onward.

Continuous, short-time and intermittent duty

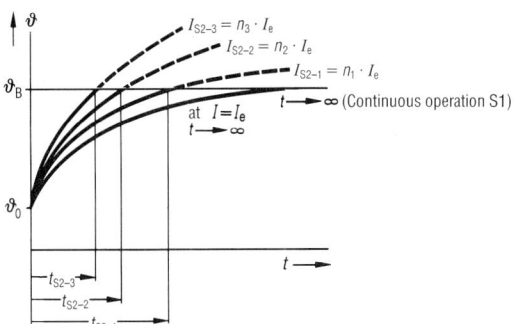

I_e Rated operating current
I_{S2} Current in short-time operation duty
n Multiples of I_e
t_{S2} Permissible operating times at I_{S2}
ϑ_B Permissible temperature limit
ϑ_0 Ambient temperature

Figure 2.21 Temperature-rise characteristics of switchgear

The intervals between the loading periods are long enough to allow the item of switchgear to cool down to the ambient temperature level. Thus, in the case of S2 duty, an item of switchgear may be loaded with a current $I_{S2} > I_{eAC1}$ without its temperature limit being exceeded. The load period t_{S2} during S2 duty, is simultaneously the heating period in which the item of switchgear may reach its maximum permissible temperature. The greater the load current I_{S2}, the shorter the permissible load period t_{S2} (heating up period) must be. The maximum load current may not exceed the dynamic current carrying capacity of the switchgear (e.g. cause contact lifting). The permissible load period t_{S2} is also reduced if the ambient temperature increases. Generalized temperature rise characteristics for switchgear are given in Figure 2.21.

2.2.3 Intermittent periodic duty

Intermittent periodic duty without the influence of the starting sequence (S3)

This refers to operation which consists of a number of identical load cycles (Fig. 2.22), each comprising a period under constant load and an interval. The starting current does not have a significant effect on the heat rise.

Designation: By S3, the load duration (t_B), the cycle time (t_S) and the power rating, e.g. S3: 15 min/60min, 20 kW, or by S3, the cyclic duration factor t_r in percent (c.d.f.) and the cycle time, e.g. S3: 25%, 60 min, 20 kW.

In terms of DIN VDE 0530, the duty cycle for motors during intermittent periodic duty is given as 10 min

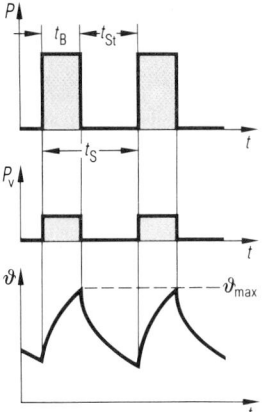

Figure 2.22
Duty type S3:
Intermittent periodic duty without the influence of the starting sequence

$$t_r = \frac{t_B}{t_B + t_{St}} \cdot 100 \text{ (in \%)}$$

if no other stipulation is made. This duty cycle is to be regarded as the maximum value in practice.

During intermittent periodic duty (in German, also referred to as "AB" or **A**ussetz**b**etrieb) the items of switchgear are switched on and off in a cyclic manner. The cycle times are too short for thermal equilibrium to be achieved within either the load period or the interval.

The intermittent periodic duty is designated by the relative load duration and the cycle time, or alternatively by the cyclic duration factor in percent (c.d.f.)

c.d.f. $= t_B/t_S \cdot 100$ (in %)

where, the cycle time t_S is the sum of the load duration t_B and the current-free interval t_{St}; $t_S = t_B + t_{St}$.

In terms of DIN VDE 0660 part 100, the preferred values of cyclic duration factor are 15%, 25%, 40% and 60%.

In practice, intermittent periodic duty is a series of on-load and off-load periods, as shown for example in Figure 2.23. A number of on-load periods of varying durations with intervals of varying lengths between them are repeated cyclically. The cycle duration factor for the whole duty cycle is derived from the equation:

$$\text{c.d.f.} = \frac{\Sigma t_B}{\Sigma t_B + \Sigma t_{St}} \cdot 100 \text{ (in \%)}. \tag{2.12}$$

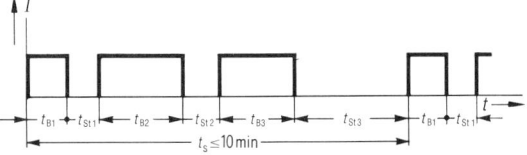

Figure 2.23
Intermittent periodic duty with a duty cycle comprising load durations and intervals of varying lengths

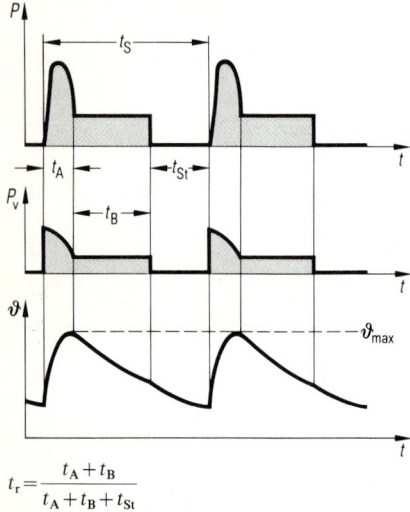

$$t_r = \frac{t_A + t_B}{t_A + t_B + t_{St}}$$

Figure 2.24
Duty type S4: Intermittent periodic duty with influence of the starting sequence

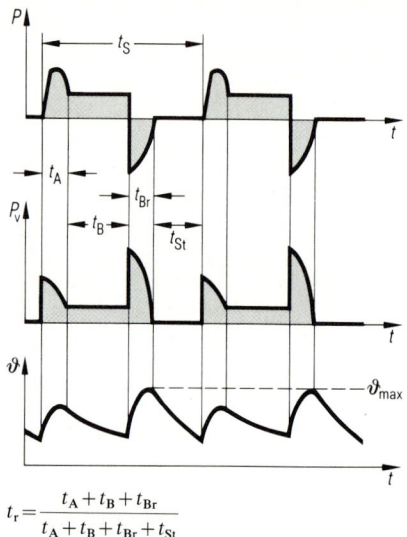

$$t_r = \frac{t_A + t_B + t_{Br}}{t_A + t_B + t_{Br} + t_{St}}$$

Figure 2.25
Duty type S5: Intermittent periodic duty with influence of the starting sequence and dynamic braking

Intermittent periodic duty with influence of the starting sequence (S4)

This refers to operation comprising a sequence of identical duty cycles (Fig. 2.24) each consisting of a significant run-up time (t_A), a time at constant load (t_B) and an interval (t_{St}).

Additional information regarding the moment of inertia and the load torque during starting are required.

Designation: By S4, the cyclic duration factor in percent, the number of starts per hour and the power rating, e.g S4: 40%, 520 starts per hour, 30 kW.

Intermittent periodic duty with influence of the starting sequence and electric braking (S5)

This refers to operation comprising a sequence of identical duty cycles (Fig. 2.25) each consisting of a significant run-up time (t_A), a time at constant load (t_B), a period of rapid electric braking (t_{Br}) and an interval (t_{St}).

Additional information regarding the moment of inertia and the load torque during starting and stopping are required.

Designation: The same way as for S4, however the type of braking must be specified, e.g. S4: 30%, 250 starts per hour, dynamic plug braking, 50 kW.

Continuous duty with intermittent loading (S6)

This refers to operation comprising a sequence of identical duty cycles (Fig. 2.26), each consisting of a time at constant load (t_B) and a time at no-load (t_L). There are *no* intervals.

Designation: The same as for S3, e.g. S6: 30%, 40 min, 85 kW.

Note

The cycle time is generally so short, that thermal equilibrium is *not* reached.

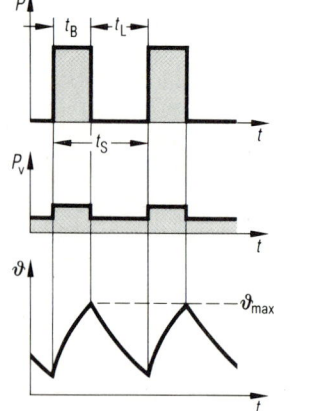

$$t_r = \frac{t_B}{t_B + t_L}$$

Figure 2.26
Duty type S6:
Continuous duty with intermittent loading

Continuous-operation periodic duty with starting and electric braking (S7)

This refers to operation comprising a sequence of identical duty cycles (Fig. 2.27) each consisting of a significant run-up time (t_A), a time at constant load (t_B), and a period of rapid electric braking (t_{Br}). There are *no* intervals.

Additional information regarding the moment of inertia and the load torque during starting and stopping are required.

Designation: The same as for S5, however without t_r, e.g. S7: 12 kW, 500 reversing operations per hour.

Continuous-operation periodic duty with regular load/speed changes (S8)

This refers to operation comprising a sequence of identical duty cycles (Fig. 2.28). Each of these cycles consists of a period at constant load and fixed speed (t_B) followed by one or more periods at different load and associated speed.

Designation (and additional information): As for S5, however for each speed.

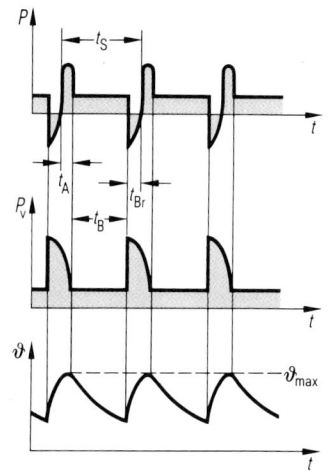

Figure 2.27
Duty type S7: Continuous-operation periodic duty with starting and electric braking ($t_r = 1$)

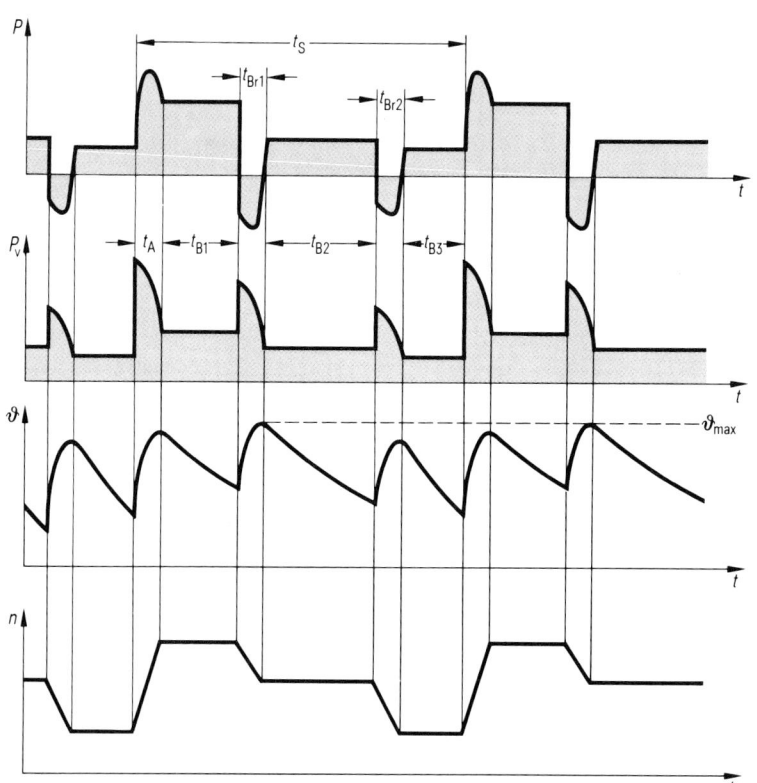

$$t_{r1} = \frac{t_A + t_{B1}}{t_A + t_{B1} + t_{Br1} + t_{B2} + t_{Br2} + t_{B3}}$$

$$t_{r2} = \frac{t_{Br1} + t_{B2}}{t_A + t_{B1} + t_{Br1} + t_{B2} + t_{Br2} + t_{B3}}$$

$$t_{r3} = \frac{t_{Br2} + t_{B3}}{t_A + t_{B1} + t_{Br1} + t_{B2} + t_{Br2} + t_{B3}}$$

Figure 2.28
Duty type S8: Continuous-operation periodic duty with regular load/speed changes

2 Network data and duty types

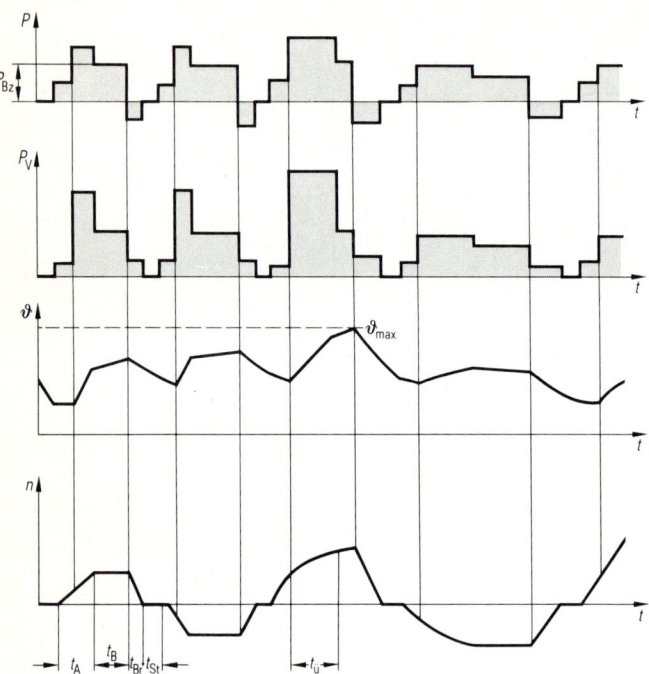

Figure 2.29
Duty type S9: Continuous-operation periodic duty with non-regular load/speed changes

Continuous-operation periodic duty with non-regular load/speed changes (S9)

This refers to operation during which the load and speed generally vary in a non-periodic way within the permissible operational limits (Fig. 2.29). Load peaks, which may be well above the rated value, often occur in this type of duty.

Note

In this duty type, a suitably selected continuous load rating must be used as the basis value for the duty cycle.

Summary of short-time and intermittent operation

Upon closer inspection of the duty types S2 to S9, one can recognize that they may be divided into two groups. These are,

▷ duty types which, in comparison with continuous operation S1, allow an increase in the power rating (S2, S3, S6), and

▷ duty types which, in comparison with continuous operation S1, demand a derating of the switchgear (S4, S5, S7, S8, S9).

2.2.4 Operation at inconsistent loading

One of the most frequent deviations from the duty types defined by DIN VDE 0530 is that the required output during the load cycle is not constant. In such cases, the power P (current I, torque M) is replaced by a average value P_{mi} (current I_{mi}, torque M_{mi}). It is the root-mean-square value (r.m.s.) of the individual loads. In this method of calculation, the maximum torque may not exceed 80% of the breakdown torque.

If the largest required output power differs from the smallest by more than the factor 2, then the average power becomes too inaccurate. In such cases, one must calculate with the average current.

This method of determining the average is not suitable for the duty type S2 – further details have to be obtained from the respective manufacturers.

If the load current during short-time duty does not have a constant value during the entire load duration t_{S2}, then the average value I_{mi} which determines the temperature rise must be calculated.

Figure 2.30 shows an example of a non-constant load current during short-time operation. The load current $I_{S2} = f(t)$ is replaced by steps of constant current I_n for the durations t_n. If the load current curve is subdivided into n segments in this way, then the root-mean-square value I_{qmi} can be determined as follows:

$$I_{qmi} = \sqrt{\frac{I_1^2 t_1 + I_2^2 t_2 + \cdots + I_n^2 t_n}{t_1 + t_2 + \cdots + t_n}}.$$

The resulting r.m.s. value I_{qmi} determines the thermal loading of the switching device during the short-time duty, i.e. $I_{qmi} = I_{S2} = n \cdot I_{eAC1}$.

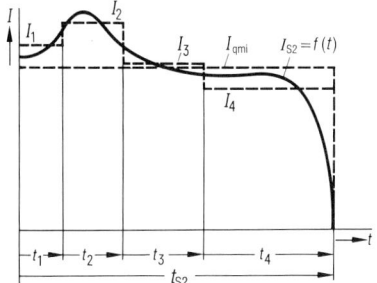

Figure 2.30
Non-constant load current in short-time duty during the load duration t_{S2}

3 Selection criteria for low-voltage switchgear in main circuits

In general, low-voltage switching devices may be categorized according to their function, i.e.

▷ switchgear for main circuits, and
▷ switchgear for auxiliary circuits.

This chapter deals with selection criteria for switchgear in main circuits. The following chapter deals with selection criteria in auxiliary circuits.

3.1 Network and operating conditions

3.1.1 Rated voltage and rated frequency

The network voltage and network frequency determine the selection of switchgear in terms of their

▷ Rated insulation voltage U_i and
▷ Rated operational voltage U_e

The rated insulation voltage U_i is the standardized value of voltage, for which the insulation of an item of switchgear, or of its accessories, is dimensioned (also refer to Section 1.4.5, page 56).

The rated operational voltage U_e of an item of switchgear is the voltage upon which, for example, the characteristic values of switching capacity are based in the case of circuit-breakers, or upon which the duty type and utilization categories are based in the case of motor starters and contactors.

The most common values of rated voltages for switchgear are shown in Table 3.1. In the case of three-phase systems, these values apply to the line-to-line or delta voltage of the network.

An item of switchgear may have a number of values of rated voltage each relating to specific characteristics such as e.g. switching capacity or service life.

In the catalogues the values of U_i and U_e are given for every item of switchgear.

Siemens low voltage switchgear for a.c. application is usually dimensioned for a frequency of 40 to 60 Hz. The characteristic data may be different at other system frequencies.

Table 3.1 Rated voltages for switchgear.
Values printed in bold type are preferred values. (Alignment with DIN IEC 38 has not yet been achieved, also see section 2.1.1, page 76)

Direct current V	Alternating current V	Direct current V	Alternating current V
24	24	1200	1000
60	60	**1500**	
110	125	**2400**	
220	**220** (230)	**3000**	3000
440	**380** (400)		6000
660	500		10000
750	**660** (690)		

For the influence of network frequency on the various switchgear functions, see Section 3.7.3, on page 187.

For the influence of harmonic currents on the tripping performance of current-dependent overcurrent releases and relays see Section 3.7.3.4, on page 189.

For the use of a.c. contactors in square-wave a.c. systems see Section 3.7.5, on page 194.

3.1.2 Rated short-circuit strength and rated switching capacity

The prospective short-circuit current at the point of installation is a decisive factor in the selection of switching devices in terms of their

▷ short-circuit strength,
▷ rated switching capacity, or if applicable in consideration of their
▷ back-up protection.

The dynamic short-circuit strength and the rated peak short-circuit current I_s as such will in future no longer be contained in the product standards. In terms of DIN VDE 0660 part 100 I_s corresponds to the selectivity limit current, and in terms of DIN VDE 0102, 01.90 to the peak short-circuit current i_p. The behaviour of switchgear under short-circuit conditions is expressed in the currently valid specifications and in the IEC publications (e.g. IEC 947-2) by means of the following:

▷ Rated short-time withstand current I_{cw},
▷ Rated short-circuit making capacity I_{cm},
▷ Rated conditional short-circuit current, dependent on protection by fuses,
▷ Rated conditional short-circuit current, dependent on protection by circuit-breaker,
▷ Various short-circuit breaking categories, e.g. I_{cn}, I_{cs}, I_c, I_{cu}. For characteristic quantities and units of electricity, refer to Section 9.1.2, page 500.

The permissible thermal short-circuit load capability is given as the rated short-time current. This is the permissible current, which can flow through an item of switchgear for a given time without causing damage.

Where applicable, the values for the rated short-circuit making capacity, rated short-circuit breaking capacity and rated short-time current are given in the Siemens l.v. switchgear catalogues.

In the case of circuit-breakers, for example, the rated short-circuit making capacity is that short-circuit current onto which the circuit-breaker can be closed, at rated operating voltage +10%, rated frequency and a specified power factor. It is stated as the maximum peak value of the prospective short-circuit current, and is at least equal to the rated short-circuit breaking capacity multiplied by the factor n, from Table 3.2 (see also Section 2.1.2, page 79).

The rated short-circuit breaking capacity I_{cn} is the short-circuit current which a circuit-breaker can break given rated operating voltage +10%, rated frequency and a specified power factor. It is expressed as the r.m.s. value of the a.c. current component.

The stated switching capacity is based on the test sequence O-t-CO-t-CO. In addition, the switching capacity may be stated in terms of the shortened test sequence O-t-CO (for an explanation of O, t and C, refer to Table 3.3).

Previously, IEC Recommendations 157-1, 1973 specified short-circuit categories for circuit-breakers which indicated, how many times the device must be capable of switching its rated making and breaking capacity, and in what condition it must be after the stated test sequence (see Table 3.3). Today, the requirements, terms and specifications for low-voltage switchgear are defined in IEC 947-1, 1988 ("Low-voltage switchgear and controlgear, Part 1: General rules"). The standards specifically related to circuit-breakers are contained in IEC 947-2, 1989 and DIN VDE 0660 part 101, 07.92. Table 9.4 on page 502 shows a comparison between the previously valid short-circuit categories to IEC 157-1 and the currently applicable values according to IEC 947-2.

Table 3.2
Rated short-circuit making capacity I_{cm} as a function of rated short-circuit breaking capacity I_{cn}

Rated short-circuit breaking capacity I_{cn} A	Power factor $\cos \varphi$	Time constant ms	Rated short-circuit making capacity (minimum value) $I_{cm} = n \cdot I_{cn}$
$I_{cn} \geq 1500$	0.95	5	$1.41 \cdot I_{cn}$
$1500 < I_{cn} \geq 3000$	0.9	5	$1.42 \cdot I_{cn}$
$3000 < I_{cn} \geq 4500$	0.8	5	$1.47 \cdot I_{cn}$
$4500 < I_{cn} \geq 6000$	0.7	5	$1.53 \cdot I_{cn}$
$6000 < I_{cn} \geq 10000$	0.5	5	$1.7 \cdot I_{cn}$
$10000 < I_{cn} \geq 20000$	0.3	10	$2.0 \cdot I_{cn}$
$20000 < I_{cn} \geq 50000$	0.25	15	$2.1 \cdot I_{cn}$
$50000 < I_{cn}$	0.2	15	$2.2 \cdot I_{cn}$

Table 3.3
Short-circuit breaking capacity categories to DIN VDE 0660, part 101

Short-circuit category	Rated ultimate short-circuit breaking capacity I_{cu}	Rated service short-circuit breaking capacity I_{cs}
Switching cycle [1]	O–t–CO [2]	O–t–CO–t–CO [2]
Condition of the circuit-breaker after the short-circuit test	A temperature rise test is to be performed with thermal rated current. Adjacent isolating material may not be damaged. Limited shift of the overload release characteristic curve is permissible.	The breaker shall be able to conduct its rated thermal current without prior maintenance (temperature rise test only necessary in case of doubt). Shift of the overload characteristic curve is not permissible

[1] Also refer to Section 9.1.3, page 502

[2] O   Open
CO   Close Open
t   Pause (t = time)

3.1.3 Rated currents

The rated duty, such as continuous operation, intermittent periodic duty or short-time operation, is decisive for the selection of an item of switchgear in terms of its rated currents.

The following rated currents are defined in terms of thermal effects:

▷ conventional free-air thermal current I_{th},
▷ rated uninterrupted current I_u,
▷ rated operational current I_e.

Conventional free-air thermal current I_{th},
Rated uninterrupted current I_u

The conventional free-air thermal current I_{th}, or I_{the} for motor starters in enclosures, is defined as an 8 hour current in terms of DIN VDE 0660, part 100, part 102, part 104 and part 107. It is the maximum current which can be conducted for this period of time during which no switching operations are carried out, without the need for intervention and without the temperature limit being exceeded. Accordingly, the rated uninterrupted current I_u can be conducted for an unlimited period of time.

For adjustable thermal overload relays and releases, the rated current is the highest value of current to which the device can be adjusted.

Rated operational current I_e

The rated operational current I_e is the current which is determined by the conditions under which the item of switchgear is used, the rated operational voltage, the network frequency, the switching capacity, the rated duty type, the utilization category, the required contact life and the necessary degree of protection. An item of switchgear may have several rated operational currents.

3.2 Switching tasks and conditions

3.2.1 Switching tasks

In general, the following switching tasks need to be performed in any switchboard or electrical installation:

▷ Disconnection,
▷ No-load switching,
▷ Load switching,
▷ Motor switching and
▷ Power switching.

Special switchgear devices exist for each of these tasks. However, items of switchgear may be built to perform more than one of the tasks.

3.2.1.1 Disconnection

Disconnectors are switches, which isolate or open a circuit in all of the current paths (poles) and which provide an absolutely reliable switching state indication.

During repair and maintenance activities on potentially live parts of electrical plant and equipment, a potential free condition must be established and maintained for the duration of the work (Accident Prevention Regulations VBG No. 4, see Section 1.4.1.2 on page 34).

In order to establish the potential free condition, the plant and equipment must be isolated from the network. Switches which meet the requirements for disconnectors are offered in the relevant Siemens l.v. switchgear catalogue. For disconnectors, the specified clearances to DIN VDE 0110 (see Section 1.4.5, p. 56) must not only exist between the current paths and the adjacent earthed parts, but also between the opened contacts, i.e. the opening distances (Fig. 3.1)

If the potential free condition is to be guaranteed for the duration of the repair and maintenance activity, then a disconnector with a drive mechanism which can be locked in the "Off" position, e.g. with one or more padlocks, should be selected.

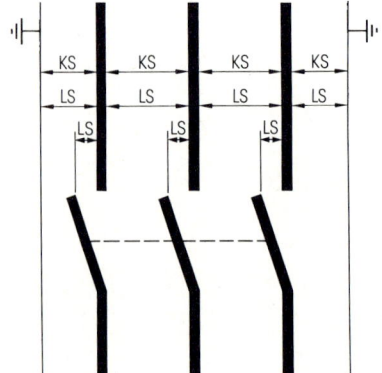

KS Creepage distance or tracking path
LS Clearance

Figure 3.1
Clearances and creepage distances for disconnectors (schematic representation)

3.2.1.2 Off-load switching

The term off-load switching is understood to mean the opening or closing of a circuit, in which either no current is flowing before and after the switching operation, or only a very small current (e.g. owing to line capacitance) flows through the contacts at the moment of opening.

The use of off-load switches pre-supposes that accidental switching under load, e.g. under fault conditions, is always prevented by other measures. If this were not the case, damage to the item of switchgear and therefore to the plant could result, but above all, operating personnel could be exposed to the dangers of switching arcs. For this reason, off-load switches are only rarely used nowadays, and only in special applications e.g. in heavy current installations. In all other cases, on-load disconnectors are selected as a minimum.

3.2.1.3 On-load switching

On-load switch-disconnectors are capable of switching the rated current of equipment or sections of an electrical installation under fault-free conditions. By definition, on-load switch-disconnectors can also make and break overload currents.

In terms of DIN VDE 0660, their switching capacity lies between 1.5 to $10 \cdot I_e$. More specifically,

AC-21: 1.5·rated operating current I_e,
AC-22: 3·rated operating current I_e,
AC-23: 6 to 10·rated operating current I_e.

The switching capacity of on-load switch-disconnectors and fuse switch disconnectors is given in the relevant Siemens l.v. switchgear catalogue. It is stated as a multiple of the rated current. In addition, since the Siemens switch-disconnectors feature a very high making capacity, no danger to personnel will result from the unintentional closing onto an existing short-circuit (also refer to Section 3.3.4.2, page 119).

3.2.1.4 Motor switching

Motor switches are items of switchgear specifically designed for the switching of motors. Their switching capacity meets the demands made by the various types of motors and operating duty types e.g. inching and plugging (see Section 3.2.2.1, page 111).

Although on-load switch-disconnectors and fuse switch disconnectors, as well as switch-disconnectors with fuses (fuse-combination units, combined fuse switches or c.f.s. units) do have motor-switching capacity, they are not intended for the operational switching of motors. Contactors and motor control switches with AC-3 switching capacity are more suitable for this application.

Like circuit-breakers, fuse switch disconnectors and switch-disconnectors with fuses also possess a short-circuit making capacity.

3.2.1.5 Power switching, switching of short-circuits

Circuit-breakers are items of switchgear, which can be used to switch not only load, motor and overload currents, but also for the interruption of short-circuit currents.

In terms of the arc quenching technique employed, one can distinguish between two basic types or designs of circuit-breakers:

▷ zero-point quenching circuit-breakers,
▷ current-limiting circuit-breakers.

Zero-point quenching circuit-breakers

In circuit-breakers of this type, the arc drawn between the circuit-breaker contacts during opening is extinguished when the alternating current passes through the zero point (Fig. 3.2).

The high current peaks of short-circuit currents produce extremely strong forces of repulsion between

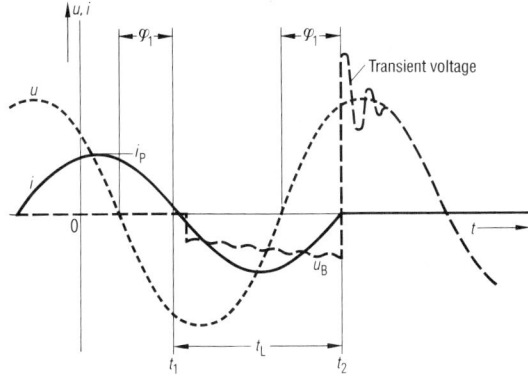

t_1 Parting of the contacts
t_2 End of the breaking process
t_L Arc duration
u_B Arc voltage

Figure 3.2
Current and voltage during a short-circuit interruption by a zero-point quenching circuit-breaker

3 Selection criteria for low-voltage switchgear in main circuits

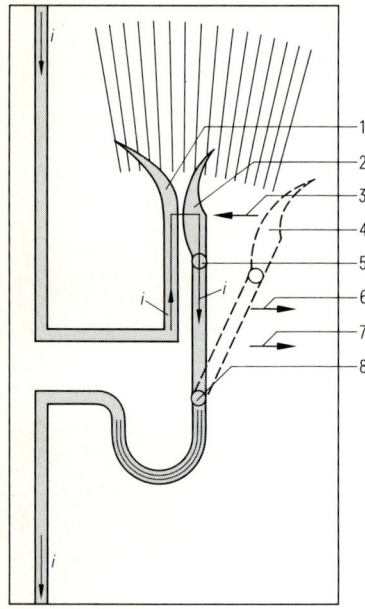

1 Fixed contact
2 Moving contact
3 Direction of the contact pressure reinforcement
4 Moving contact, opened
5 Fulcrum of the moving contact assembly in the closed position
6 Direction of repulsive force of owing to current in parallel conductors
7 Direction of force from the contact pressure spring
8 Hinge point of the moving contact assembly during opening or closing

Figure 3.3
Switching elements of a discriminative zero-point quenching circuit-breaker with current rating above approx. 160 A (schematic illustration)

the contacts. The circuit-breakers therefore either have a correspondingly high mechanical contact pressure, or they employ an electrodynamic technique which reinforces the contact pressure during short-circuit conditions, to prevent parting of the contacts before the electromagnetic release has operated. The electrodynamic method for increasing the contact pressure is used particularly in discriminative type circuit-breakers with current ratings higher than 160 A, where an opening delay under short-circuit conditions is required.

Refer to Figure 3.3. As a result of the force of repulsion which is induced between two parallel conductors carrying current in opposite directions, the lon-ger moving contact lever arm is repelled from the fixed contact assembly in the event of a short-circuit. This force of repulsion is converted into a force in the opposite direction on the shorter moving contact lever arm by the presence of a fulcrum (point of support), and causes an increase in the contact pressure. This fulcrum is released by the opening command, and the repulsive force then assists the opening movement while the current flows via the switching arc.

Current-limiting circuit-breakers

The term "current-limiting" refers to the technique by which the peak value i_p of a prospective short-circuit current, is limited to a smaller let-through or cut-off current i_D. This current-limiting can be achieved by circuit-breakers in a number of ways.

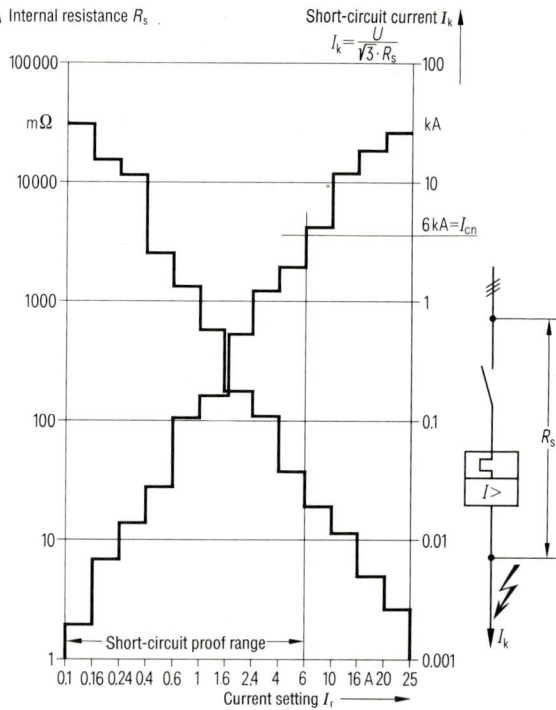

I_{cn} Rated short-circuit breaking capacity
I_r Current setting on the current-dependent delayed overload release

Figure 3.4
Short-circuit withstand capacity by virtue of circuit-breaker internal resistance R_S;
Example: 3VU13 circuit-breaker at 400 V

Circuit-breakers with a high internal resistance

In circuit-breakers with low overload-current settings, the combined resistance in each pole, of the bimetal heating coil and the winding of the instantaneous electromagnetic release, is relatively high. It may be so high, that every short-circuit current I_k will be limited to a value which the circuit-breaker will not only withstand, in terms of the thermal and dynamic stresses, but will also be able to interrupt. The circuit-breaker is then said to be "short-circuit proof". It may be installed at points in the network where the prospective short-circuit current level may be higher than 80 kA. The setting ranges of the thermal overload release for which the above holds true, is dependent upon the inherent breaking capacity of the circuit-breaker.

This in turn is dependent on the system voltage and the power factor, and therefore the degree to which the circuit-breaker is "short-circuit proof" will be different at various system voltages.

Figure 3.4 illustrates the short-circuit proof range of the 3VU13 circuit-breaker for a system voltage of 400 V. It can be seen that even in the event of a full short-circuit across the terminals of the circuit-breaker, the internal resistance R_s of the overload setting ranges up to 6 A, will limit the short-circuit current I_k to a value smaller than the inherent breaking capacity of 6 kA.

Circuit-breakers with extremely short opening time and high arc voltage

This type of current-limiting circuit-breaker incorporates a release and opening mechanism which differs fundamentally from that of the zero-point-quenching type. The opening of the contacts is not initiated by the relatively slow process of unlatching the switch mechanism.

As in the case of fuses, two conditions are met:

▷ the circuit is opened, before the peak of the prospective short-circuit current is reached, and
▷ a large resistance, in the form of an arc voltage, is introduced into the circuit or current path.

The rapid opening of the contacts is achieved by an arrangement, as can be seen in Figure 3.6 for smaller circuit-breakers, which causes the moving contact arm to be struck away from the fixed contact assembly by a moving armature of the electromagnetic release.

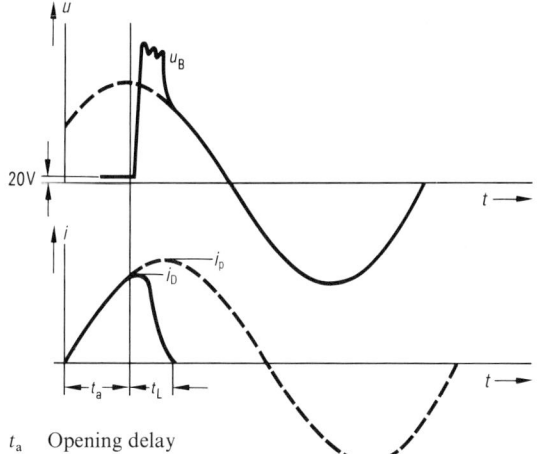

t_a Opening delay
t_L Arc duration
u_B Arc voltage
i_D Let-through current
i_p Peak of prospective short-circuit current

Figure 3.5
Current and voltage during a short-circuit interruption by a current-limiting circuit-breaker

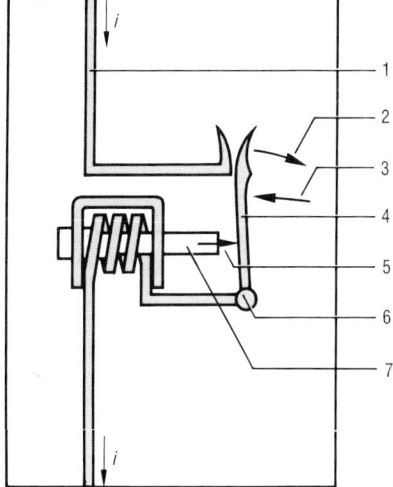

1 Fixed contact assembly
2 Direction of repulsive force between contacts
3 Direction of force from contact pressure spring
4 Moving contact assembly
5 Direction of impact from moving armature
6 Hinge point of moving contact assembly
7 Moving armature

Figure 3.6
Switching elements of a current-limiting circuit-breaker for rated continuous currents up to approx. 63 A (schematic illustration)

3 Selection criteria for low-voltage switchgear in main circuits

In larger circuit-breakers the switching elements form a current loop (Fig. 3.7). The large force of repulsion which is induced between the fixed and moving contact assemblies at the onset of a short-circuit current, drives the moving contact from the fixed one. This force of repulsion increases with an increase in the current i.

A large arc voltage is produced in a very short time by means of the specially shaped contact pieces, and a suitably designed arc chamber, which cause the arc to move away from the contact surfaces and into the arc chamber at very high speed. Here the arc is divided into several smaller arcs and intensively cooled by the arc-splitting plates.

In general, for voltages up to 400 V, current-limiting circuit-breakers normally have a higher short-circuit breaking capacity than zero point quenching circuit-breakers of the same current rating.

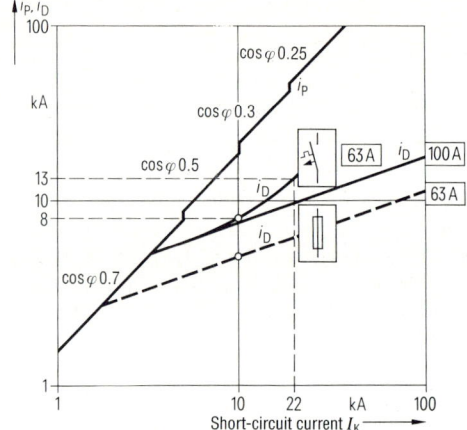

i_D Let-through currents
i_p Prospective short-circuit current (peak value)

Figure 3.8
Comparison of current-limiting at 400 V, 50 Hz of a 63 A circuit-breaker with that of 63 A and 100 A HRC fuses. E.g. at 10 kA, i_D fuse = 7.5 kA, and i_D circuit-breaker = 8 kA

It should be pointed out that they do not possess an "unlimited" breaking capacity, as breaking capacity and let-through current are dependent on voltage.

Figure 3.8 illustrates the current-limiting performance of a 63 A circuit-breaker at 400 V, 50 Hz in comparison with 63 A and 100 A HRC fuses, type 3NA, of utilization category gL. Owing to the high motor starting currents, the current rating of the fuses must be higher than that of the motor, i.e. for a 30 kW motor, a 63 A circuit-breaker or 100 A fuses may be selected (also refer to Section 3.4.3, page 125).

3.2.2 Switching of components in an electrical installation

This section deals with the conditions which exist for an item of switchgear during the switching of the most common components found in an electrical installation. These components are:

▷ Low-voltage motors,
▷ High-voltage motors,
▷ Capacitors,
▷ Electric heating equipment,
▷ Lighting systems,
▷ Low-voltage transformers.

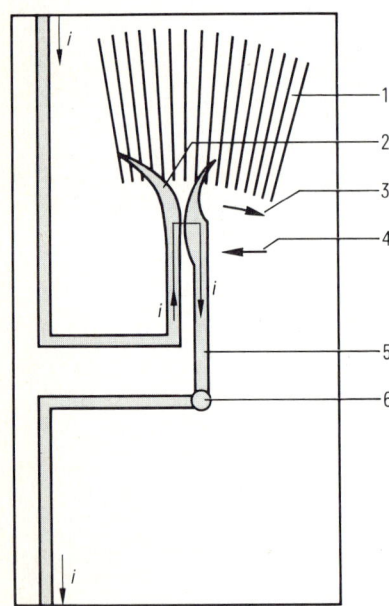

1 Arc-splitting plates
2 Fixed contact
3 Direction of repulsive force and contact lifting force
4 Direction of force from contact pressure spring
5 Moving contact arm
6 Hinge point of moving contact arm

Figure 3.7
Switching elements of a current-limiting circuit-breaker for rated operational currents above approx. 200 A (schematic illustration)

3.2.2.1 Starting of low-voltage motors

Starting of slip-ring motors

In terms of DIN VDE 0660 Part 301 a distinction is made between:

▷ Half-load starting, average torque 0.65 to 0.7 times rated torque, e.g. elevators,

▷ Full-load starting, average torque 1.3 to 1.5 times rated torque, e.g. presses, lathes,

▷ Heavy starting, average torque 1.7 to 2 times rated torque, e.g. centrifugal pumps, crushers.

By switching resistors into the rotor circuit, the starting current can be reduced and the starting torque can be controlled between zero and breakdown torque. Starter resistances and starter combinations are available for this purpose, and are described in Section 3.7.12, page 210.

Three-phase motors with slip-ring rotors are often used in crane applications. For this application, the utilization category AC-2 to DIN VDE 0660 part 102 applies.

The utilization category AC-2 reflects the typical operating conditions which prevail in fully loaded slip-ring motors during starting as well as during switching off while still in the starting sequence, during plug braking and in reversing and jogging duty.

Typical crane duty (mixed), however, represents a lower loading since switching off also occurs during the steady state operation and at partial loads.

Details for the various load conditions can be found in the Siemens H1 and H3 catalogues. Here the increased switching frequency associated with crane duty is also given special consideration.

Starting of squirrel-cage motors

Direct-on-line starting (d.o.l.)

When a squirrel-cage motor is switched on directly, the starting current I_{an} flows after the decay of the initial transient inrush current. The value of the starting current I_{an} is *independent* of the load torque, whereas the starting time t_{an} is *a function* of the load torque, the moment of inertia of the whole drive and the accelerating torque. In the case of a longer starting time one speaks of heavy-duty starting (Fig. 3.9).

Typical values:

▷ Starting inrush current peak (maximum value) $I_s = 2\sqrt{2} \cdot I_{an}$,

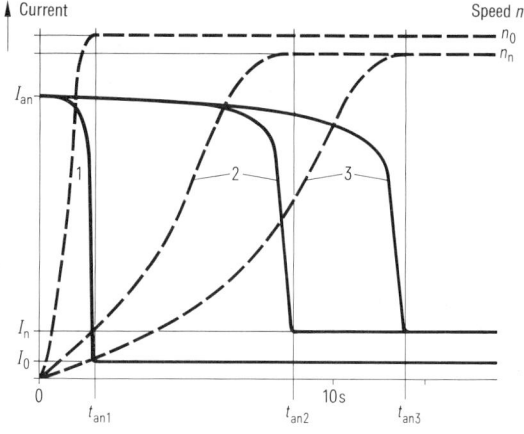

n_0	Speed at no load	1	Starting without load
n_n	Rated speed	2	Normal starting
I_{an}	Starting current	3	Heavy-duty starting
I_n	Rated current		
I_0	No-load current		
t_{an}	Starting times		

Figure 3.9
Current and speed curves during the starting of squirrel-cage motors

▷ Starting current (r.m.s. value) $I_{an} = 4$ to $8.4 \cdot I_n$, in exceptional cases also up to $13 \cdot I_n$,

▷ No-load current $I_0 = 0.95$ to $0.2 \cdot I_n$ (lower for increase in motor size and higher for increase in the number of poles),

▷ Run-up time t_{an},
under normal operating conditions $t_{an} < 10$ s,
for heavy-duty starting $t_{an} > 10$ s, (monitoring of the motor temperature rise necessary).

Even higher in-rush current peaks are to be expected in motors used in reversing applications.

If a motor is switched on under locked-rotor conditions, then the maximum starting current (locked-rotor current) will flow.

A number of methods exist for reducing the starting current of a motor.

Star-delta starting

In all forms of starting in which the voltage on the motor windings is lowered, the torque is decreased with the square and the current is decreased as a linear function of the voltage reduction.

In star-delta starting of three-phase asynchronous motors, the starting current is reduced to 1/3 of the direct-on-line starting current. This value results from the linear reduction of the current corresponding to the $1/\sqrt{3}$ reduction in the voltage across each winding, as well as the $1/\sqrt{3}$ relationship between the line current and the winding current in the star connection.

This form of starting is however only suitable for three-phase asynchronous motors of which the rated voltage for delta starting is the same as the network voltage, and the winding ends are brought out separately to the terminal box.

Owing to the $1/\sqrt{3}$ reduction in winding voltage, the motor torque is also reduced to approximately 1/3 of the direct-on-line value. ($\approx 0.33 \cdot M_n$ for motors up to approx. 15 kW at 400 V, 50 Hz and $\approx 0.29 \cdot M_n$ for motors above approx. 18.5 kW). This achieves a gentle acceleration of the drive.

Although it is commonly known that star-delta starters are only suitable for the starting of motors in applications where the load torque remains low during the starting sequence, it does happen occasionally that star delta starters are used incorrectly. This is the case, for example, if the load torque of the driven machine during the run-up period of the motor, is or will become so large, that the motor cannot attain at least 90% of its rated speed whilst still in the star mode.

The accelerating torque of the drive may be improved through the selection of a motor with a higher torque or rotor class, e.g. KL 16. Torque class KL 16 means that the motor can successfully be started in the direct-on-line connection, even at 5% undervoltage, with a load torque of up to 160% rated torque.

Driven machines which could cause problems during the starting sequence as a result of their particular torque characteristics include fans, centrifugal pumps ($M_n \sim n^2$), compressors and other machines with similar torque/speed curves, since in these cases the speed drops drastically during the change-over period from star to delta (≥ 50 ms).

At the instant when the delta contactor is closed, current peaks result from the transients in the motor. These may be exaggerated by an unfavourable position of the (rotating) rotor field in relation to the network frequency. These current peaks can be higher than those which occur during direct starting of the stationary motor in the delta configuration. In extreme cases, the making capacity of the delta contactor may be exceeded. This, in turn, could cause contact welding.

A reduction in the transient currents which arise during the switch-over from star to delta, may be achieved by the use of the correct wiring diagram for the main circuit (refer to the basic circuit diagrams in Section 8.3, page 444). This not only reduces the loading of the current paths in the contactors, but also lowers the dynamic stressing of the winding head in the motor.

For circuit configuration and contactor selection, see Section 8.3.1, page 444.

Closed-transition star delta starting

To ensure that the transient currents which arise during switch-over from star to delta are kept as small as possible, a fourth contactor (transition contactor) may be used to maintain the connection between the motor windings and the network via resistors. This is done in such a way that the change-over occurs without the disconnection of the motor from the supply. The current peaks are suppressed to a great extent, typically in the case of a motor which does not achieve sufficient speed before change-over to delta.

For circuit configuration and switchgear selection, see Section 8.3.2, page 454.

Motor starting via single-pole resistor (KUSA starter)

This circuit configuration is also known as a "KUSA starter" (KUSA from the German "Kurzschlußläufer-Sanftanlauf", meaning soft start of a short-circuited or cage rotor). Owing to the voltage drop across a resistor in one line, terminal voltage on the motor is initially asymmetrical. This effectively produces a second torque in opposition to the main torque which results in a reduction of the total torque. The starting current in the two remaining lines without resistors is, however, not reduced. It is, in fact somewhat higher than the value which would be expected with direct-on-line starting. After the run-up period, the resistance is short-circuited.

For circuit configuration and switchgear selection, see Section 8.3.6, page 454.

Motor starting via three-pole resistances

The starting current of the motor produces a voltage drop across the resistors, so that the motor starts with a correspondingly reduced terminal voltage. The starting current is decreased in direct proportion to the voltage reduction. In single step starters, the resistances are dimensioned to limit the starting current to approximately 3 times the motor rated current. In two step starters the starting current can be limited to approximately 1.5 to 2 times the rated motor current, whereby the starting torque is then also very low.

For circuit configuration and switchgear selection, see Section 8.3.7, page 456.

Autotransformer starting of squirrel-cage motors

During the run-up period, the motor is connected to the tappings of a transformer in the economy star configuration. Usually the motor is connected to about 70% of its rated voltage. In this case both the starting torque and the starting current are reduced to about 49% ($M \sim u^2$) of the values for direct on-line starting. This has the advantage, that the network is relieved during the run-up period. In addition, the motor requires only 3 connecting cables as opposed to 6 when using star delta starting.

For circuit configuration and switchgear selection, see Section 8.3.9, page 460.

Three-phase reverser circuits

Three-phase reverser circuits are used for reversing operations, when the direction of rotation of the motor is changed rapidly by the reversal of the phases. During the phase reversal, a residual magnetic field and an associated high residual voltage exist in the motor windings. This leads to additional change-over transient current peaks of up to $20 \cdot I_n$.

It should be noted that a sufficiently long change-over period is required to ensure that the arcs drawn between the contacts of the "opening switch" are quenched before the contacts of the "closing switch" touch. Contactor-type reversers must be electrically interlocked via their respective NC auxiliary contacts to achieve this. In addition, the contactors may be mechanically interlocked. The mechanical interlock insures that both contactors cannot be closed at the same time, even in the event of wiring faults or incorrect control signals (e.g. during commissioning), or under conditions of severe vibration. Thus, it protects against line-to-line short-circuits.

For circuit diagram, see Section 8.2.4, page 420. Selection of the switchgear: as for direct-on-line starting.

Control of pole-changing three-phase motors

Under constant frequency conditions, the speed of an induction motor is determined by its number of poles. If the same motor is to be operated at more than one speed, then its stator winding is so configured that it is possible to connect the motor to the network in different ways, to change its pole number.

▷ Version 1: Dahlander connection; two speeds (speed ratio 1:2)

   Different output ratings result for each of the 2 speeds. For a reduction of the starting current, the motor may be started at the lower speed by means of a star-delta connection. For this, nine winding tails must be brought out to the terminal box of the motor, and the star point must be accessible.

▷ Version 2: Two separate windings; two speeds

   Different output ratings result for each of the 2 speeds.

▷ Version 3: Dahlander connection and one separate winding; three speeds

   The Dahlander winding, with its speed relationship of 1:2, is supplemented by the extra winding. Its speed may lie below, between or above the two Dahlander speeds.

▷ Version 4: Two Dahlander connections; four speeds

   The speeds may follow each other sequentially in the order of each Dahlander winding, or may be used in any other order.

Pole-changing three-phase motors may be controlled by means of motor-control switches or contactors.

For circuit configuration and switchgear selection, see Section 8.2.5, page 422 et seq.

Inching or jogging

The term "inching" or "jogging" refers to repeated short duration starts of a motor, in order to achieve small movements. This means that the motor current is switched off during run-up. Inching operations are common in applications such as the setting up of machine tools and load elevators.

Inching duty causes severe erosion of contact material due to the arcing between the contacts of the switching device. The selection of motor switches for inching duty must therefore be made in terms of their AC-4 rating. For mixed operation AC-3/AC-4, the expected contact life of the contactor may be estimated (see Section 3.7.7, page 199).

3.2.2.2 Switching of high-voltage motors

High-voltage motors up to a rated output of 6500 kW with rated voltage in the range of 3.6 to 12 kV are mostly switched in AC-2 and AC-3 duty by means of high-voltage vacuum contactors.

The most important characteristics of the Siemens 3TL6 high-voltage vacuum contactors (Fig. 3.10) commonly used for these applications are:

▷ one frame size for the rated voltages 3.6, 7.2 and 12 kV; thereby, the same space requirements, uniform connection and standard installation parts in all the switch cubicles;

▷ centrally located terminal rail for all auxiliary and control connections;

▷ positively driven operation of the auxiliary contacts for reliable monitoring of the main contacts;

▷ high contact reliability of the auxiliary contacts through the use of double moving contact bridges; this also ensures the reliable operation with electronic control systems;

▷ d.c. magnet system in economy configuration for a.c. and d.c. excitation, with high mechanical service life and low pull-in and hold-in power consumption;

▷ possibility of continuous closed state by means of a mechanical latch on the magnet system;

▷ lock-out mechanism to prevent unintentional closing of the contactor in the event of severe vibration or shock;

▷ mechanical interlocking of two contactors for reversing operation, to prevent the simultaneous closing of both contactors;

▷ high degree of operational reliability even under extreme ambient conditions (dust, moisture, aggressive vapours) as a result of the increased dielectric strength;

▷ free of maintenance for the entire service life.

Technical data, as well as ordering details may be found in the Siemens HG 11 catalogue.

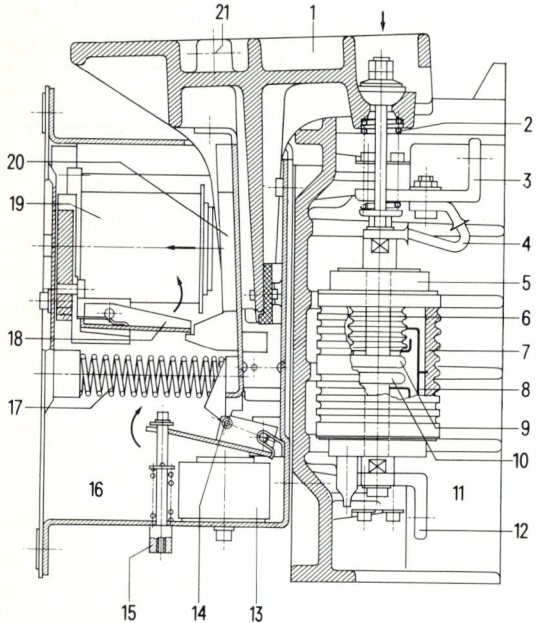

1 Rocker assembly
2 Supplementary contact pressure spring
3 Upper main terminal
4 Flexible connector
5 Vacuum tube
6 Metallic bellows
7 Ceramic insulator
8 Metallic shield
9 Moving contact
10 Fixed contact
11 High-voltage section
12 Lower main terminal
13 Tripping magnet
14 Mechanical latch
15 Tripping pin
16 Low-voltage section
17 Opening spring
18 Mechanical lock-out
19 Solenoid operating mechanism
20 Armature plate
21 Fulcrum of the rocker assembly

Figure 3.10
Sectional view of the 3TL6 high-voltage vacuum contactor

3.2.2.3 Switching of capacitors

When capacitors are switched into a circuit, they are charged during a transient period to their final value. During this period, high current peaks, at frequencies in the order of a few hundred to a few thousand Hertz may be produced. These, naturally, impose particular demands on the switchgear. The amplitude and frequency of these switch-on currents are determined by the capacitance of the capacitor, the reactances in the associated circuit and supply lines, and the point of switch-on with respect to the wave-form of the network voltage.

Switching in of single capacitors

When a single capacitor of a particular reactive power rating is switched into the network, the switch-on current is mainly determined by the size of the feeder transformer and the network impedance between the transformer and the capacitor terminals.

The stresses on the switchgear increase with:

▷ an increase in capacitor reactive power rating,

▷ an increase in the power rating of the feeder transformer, and thus an increase in the short-circuit reactance value,

▷ a decrease in the impedance of the incoming cables.

Switching in of capacitors in a power factorcorrection system

When groups of single capacitors, or so-called capacitor banks, are switched, e.g. in power factor correction systems, the switching conditions are particularly severe for the switchgear. This is because the capacitors already connected to the network, represent an additional energy source.

The switch-on current peaks are limited by the impedance of the incoming cables and, in favourable cases, also by the inductance of the capacitors themselves and the interconnecting leads between the respective capacitor banks.

The stresses imposed on the switchgear are therefore determined by:

▷ the ratio between the reactive power of the capacitor bank being switched in, and those already connected to the network,

▷ the length and path, i.e. the resistance and reactance, of the connecting cables between the individual capacitor banks, and

▷ the inherent inductivity of the capacitors themselves, which e.g. in the case of low-loss canister-type capacitors is negligibly small.

To achieve the switching of capacitors without major expense, it is often necessary to take additional measures. For example, there is a risk of contact welding if motor contactors are used without appropriate derating or at reduced switching frequency. Here, the permissible switching frequency of the capacitor bank may be increased, e.g. by the use of additional inductances, or coils, in the individual bank feeder cables.

By the use of special capacitor contactors or capacitor contactor combinations a higher switching frequency and minimal disturbance of the three-phase supply network can be achieved. These contactors switch the capacitors into the network via pre-charge resistors, so that the switch-on current peaks are limited to a low value.

For contactors suitable for switching capacitors see Section 3.7.6.2, page 195.

3.2.2.4 Switching of electric heating equipment

Electric heating equipment is used, for example, in room heating systems, resistance ovens and kilns, and in air conditioning plants.

For wire wound resistance elements the switch-on current may be 1.4 times the rated current. A 10% increase in the network voltage results in a corresponding increase in the current drawn by resistance elements. This should be taken into account when selecting switchgear in terms of its rated operating current. The utilization categories which apply in this case are AC-1 for alternating current, and DC-1 for direct-current systems. For the switching of resistance elements any switch with the corresponding load breaking capacity may be selected.

Heating circuits are often operated as single-phase installations. Generally, multi-pole switching devices with parallel connected current paths to increase the permissible load current, are used.

For the loading of switchgear with parallel connected current paths see Section 3.7.1, on page 185.

3.2.2.5 Switching of lamps in lighting installations

In lighting installations, switchgear can be particularly stressed by high inrush currents.

Incandescent lamps

The inrush current of incandescent lamps can be as high as 15 times the rated current. The making capacity of the contactor must therefore be at least this high.

Fluorescent lamps

During the normal starting phase, fluorescent lamps without power factor correction or those in a twin-lamp lead-lag connection, draw pre-heating currents which may attain values of about twice the rated current for a short period of time. The lamp current may not exceed the AC-1 rated operational current I_e of the associated contactor.

In the case of parallel compensated fluorescent lamps, on the other hand, inrush currents of 20 times the capacitor rated current can occur. In the selection of the contactor, the permissible capacitor rating for single compensation and a typical cable impedance of 0.8 Ω, should be taken into account.

In the case of fluorescent lamps with electronic ballasts, the starting current reaches an average of 10

Table 3.4 Inrush and starting currents of lamps. Selection criteria for contactors
(for the selection of contactors for the switching of lamps, see Tables 3.30 to 3.32, page 206 et seq.)

Lamp type	Inrush current I [1] A	Starting time min	Starting current I_{an} A	$\cos \varphi$	Calculation parameters for contactors in terms of utilization categories AC-1/DC-1 and AC-3	the switching capacity for capacitive load (AC-6b)
Incandescent lamps	$15 \cdot I_{eL}$	–	–	1	$I_{eL} \leq 0.83 \cdot I_{eAC-3}$ and $\leq I_{eAC-1}$	–
Fluorescent lamps and compact fluorescent lamps [2] with choke ballast						
– non-p.f. corrected	$\approx 2.0 \cdot I_{eL}$	–	–	0.4–0.6	$I_{eL} \leq I_{eAC-1}$	–
– p.f. corrected	$\approx 20 \cdot I_{eL}$	–	–	0.9	$I_{eL} \leq I_{eAC-1}$ and $R_{Ltg} \geq 0.8\,\Omega$	$C_{perm} \leq 11.1 \cdot I_{eAC-3}\,\mu F$
– lead/lag connection	$1.0 \cdot I_{eL}$	–	–	0.9	$I_{eL} \leq I_{eAC-1}$	–
Luminaires with fluorescent lamps and electronic ballast units						
– a.c. operation	$10 \cdot I_{eL}$	–	–	0.9	$I_{eL} \leq 0.7 \cdot I_{eAC-3}$	–
– d.c. operation	$10 \cdot I_{eL}$	–	–	0.9	$I_{eL} \leq 0.7 \cdot I_{eAC-3}$ and $\leq I_{eDC-1}$	–
High-pressure mercury vapour lamps (HQL)						
– non-p.f. corrected	$20 \cdot I_{eL}$	3–5	$2 \cdot I_e$	0.4–0.6	$I_{eL} \leq 0.5 \cdot I_{eAC-1}$ and, with only one lamp $\leq 0.88 \cdot I_{eAC-3}$	–
– p.f. corrected	$20 \cdot I_{eL}$	3–5	$2 \cdot I_e$	0.9	$I_{eL} \leq 0.5 \cdot I_{eAC-1}$ and, with only one lamp $\leq 0.88 \cdot I_{eAC-3}$ and $C_{perm} \leq 11.1 \cdot I_{eAC-3}\,\mu F$	–
Metal-halide lamps (HQI)						
– non-p.f. corrected	$20 \cdot I_{eL}$	5–10	1.7 to $2.2 \cdot I_e$	0.4–0.5	$I_{eL} \leq 0.5 \cdot I_{eAC-1}$ and, with only one lamp $\leq 0.88 \cdot I_{eAC-3}$	–
– p.f. corrected	$20 \cdot I_{eL}$	5–10	1.7 to $2.2 \cdot I_e$	0.9	$I_{eL} \leq 0.5 \cdot I_{eAC-1}$ and, with only one lamp $\leq 0.88 \cdot I_{eAC-3}$ and $C_{perm} \leq 11.1 \cdot I_{eAC-3}\,\mu F$	–
High-pressure sodium-vapour lamps (NAV)						
– non-p.f. corrected	$20 \cdot I_{eL}$	5–10	1.7 to $2.2 \cdot I_e$	0.4–0.5	$I_{eL} \leq 0.5 \cdot I_{eAC-1}$ and, with only one lamp $\leq 0.88 \cdot I_{eAC-3}$	–
– p.f. corrected	$20 \cdot I_{eL}$	5–10	1.7 to $2.2 \cdot I_e$	0.9	$I_{eL} \leq 0.5 \cdot I_{eAC-1}$ and, with only one lamp $\leq 0.88 \cdot I_{eAC-3}$ and $C_{perm} \leq 11.1 \cdot I_{eAC-3}\,\mu F$	–

[1] Expressed as a multiple of the lamp rated current   [2] e.g. OSRAM DULUX

times the rated current of the lamp. The contactor should be selected in accordance with Table 3.4, and Tables 3.30 to 3.32 on pages 206 to 208. Alternatively, the tables in the Siemens l.v. switchgear catalogue may be used.

*High pressure mercury-vapour
and metal halide lamps*

These lamps are switched on via ballast units in the form of series inductors and, in the case of the metal halide lamps, with the aid of ignition devices. During the initial 3 to 5 minutes after switch-on, and before the lamps reach their normal operating condition, they draw an almost purely inductive current, which can be as high as approximately double their rated current. For metal halide lamps, the AC-1 rated operating current I_e of the contactor must be at least equal to this starting current.

High pressure sodium vapour lamps

During the starting time of approximately 5 to 10 minutes, an inductive current flows which is approximately 1.7 to 2.2 times the rated current. This starting current value should be used as a basis for selecting the contactor according to its rated operating current I_e.

*High pressure mercury vapour lamps,
metal halide lamps
and sodium vapour lamps with power factor correction*

If these lamps are operated with power factor correction, it is necessary to take the high inrush currents associated with the switching of capacitors into account when selecting the contactor. However, since the inductive starting current is partly compensated by the capacitor, it may be possible to use a smaller contactor than would be selected for uncompensated lamps, provided that the actual inrush current of the capacitor in relation to the making capacity of the contactor allows this.

Table 3.4 provides a summary of the inrush and starting currents of the various lamps, and provides a guide to the criteria for selection of the contactors.

Selection of Siemens 3TF, 3TK and 3TB contactors for the switching of lamps, see Section 3.7.10, pages 206 to 208.

3.2.2.6 Switching of low-voltage transformers

A very high current peak of extremely short duration is produced when a low-voltage transformer is switched on. These inrush currents, which are caused by the initial build-up of the magnetic field, may be as high as 30 times the rated current of the transformer. The inrush currents are, however, different for each transformer type. They are dependent on the position of the transformer winding, the characteristic values of the magnetic circuit and particularly, on the phase angle of the voltage at the instant of switch-on. Contactors selected for the switching of low-voltage transformers must have a sufficiently high making capacity, to avoid contact welding during switch-on.

For selection of Siemens 3TF, 3TK and 3TB contactors for the switching of transformers, see Section 3.7.11, page 210 and the relevant Siemens l.v. switchgear catalogue. In addition, the computer program ASIST is available to facilitate selection (also refer to Section 8.13, page 496).

3.3 Switching frequency and service life

The switching frequency required of a switching device, is associated with its function in the circuit. Depending on the application, the switch is operated more, or less frequently. For example, plant sections are only seldom disconnected from the network for overhaul and maintenance purposes, while groups of machine tools may be disconnected either daily or weekly according to planned stoppages and shift breaks. The machine tools themselves may be switched on and off on an hourly or more frequent basis, and automated machine tools may require as many as 3000 switching operations per hour (Also refer to Section 9.9, page 548 for the max. number of switching cycles).

It is therefore extremely important, that the switchgear selected and installed is suited to the switching frequency and service life required.

3.3.1 Permissible switching frequency

The permissible switching frequency is expressed in terms of the permitted number of switching cycles or make/break operations per hour. It is dependent on the rated operating current of the particular utilization category in question (see Section 3.3.4, page 118).

Specific data on switching frequency in the various utilization categories is given in the Siemens low-voltage switchgear catalogues.

Theoretically, the maximum permissible switching frequency is the no-load switching frequency of the device. In comparison to continuous operation, however, a high switching frequency causes increased heating. The reasons for this include the repeated high starting currents after closing, as well as the heat caused by the switching arcs during opening.

In practice, the maximum switching frequency is determined by the nature of the load, the utilization category and the ambient temperature. The permissible switching frequency can be calculated for each utilization category as a function of breaking current, rated operating voltage and ambient temperature. For contactors, the following applies:

$$Z' = Z \cdot \frac{I_e}{I'} \cdot \left(\frac{400\text{ V}}{U'}\right)^{1.5} \cdot \frac{55\text{ °C}}{\vartheta_u}$$

(for $55\text{°C} \geq \vartheta_u \geq 70\text{ °C}$)

Z' Permissible switching frequency
Z Switching frequency in terms of the utilization category
I' Actual operating current
I_e Rated operating current in terms of the utilization category
U' Actual operating voltage
ϑ_u Ambient temperature

Hereby, one must take care that the no-load switching frequency is not exceeded. Also, the permissible switching frequency of contactors and circuit-breakers may further be restricted by series-connected overload protection devices (e.g. thermally delayed overload relays and tripping units). Further details may be found in Section 3.4.3.4 on page 131.

3.3.2 Mechanical service life

The mechanical service life of an item of switchgear is expressed in terms of the number of make-break operations, which can be performed under no-load conditions. It is determined by the friction between the mechanical moving parts in the device. The greater the forces which have to be overcome, the larger the drive mechanism and the stressing on the materials will be. Disconnectors and circuit-breakers for high currents operate with high contact pressure and large masses. Their mechanical service life is therefore limited to a relatively low value. For longer mechanical service life, specific items of switchgear (e.g. contactors) are available which operate with relatively lower contact pressures.

3.3.3 Electrical service life

The electrical service life of an item of switchgear is determined by the number of switching cycles it can perform before its contacts are worn out (contact material eroded away). The contacts are stressed during the opening and closing operations under load. Contact erosion is mainly caused by the breaking of arc currents which burn between the contact pieces during the opening operation. During the closing operation, the moving contact pieces may be subjected to bounce. If this is the case, the contacts are subjected to arc erosion during the closing operation as well. The degree of contact erosion is dependent on the switching duty and therefore on the voltage, current and the arc duration.

3.3.4 Selection in terms of utilization categories

In terms of DIN VDE 0660 part 102, 107 and 200 (Table 3.5), the application and associated stressing of contactors, disconnectors, load-break switches and fuse combination units can be defined by a utilization category in connection with the given rated operating current, or motor power rating, and the rated voltage. Often, several different utilization categories, defined by the respective operating current, voltage and switching function may apply for a particular item of switchgear. The rated making and breaking capacities, as well as the electrical service life for the various utilization categories at the specified values of current, voltage and power factor, are given in Tables 3.6 and 3.7.

3.3.4.1 Selection of contactors

In terms of DIN VDE 0660 part 102, only the utilization categories AC-1 to AC-4 and DC-1 to DC-4 applied for the main contacts of contactors up to the end of 1992 (see Table 3.5). They have subsequently been extended by the categories AC-5a/b to AC-8a/b and DC-6.

The rated *making* capacity of a contactor is that value of current which can be switched on under given conditions without the contacts welding, without excessive contact erosion and without the occurrence of excessive arcing.

The rated *breaking* capacity of a contactor is that value of current which can be switched off under given conditions without excessive contact erosion and without the occurrence of excessive arcing.

Service life · Utilization categories

Table 3.5 Utilization categories to DIN VDE 0660, parts 102, 107 and 200 and to IEC 947 parts 3, 4 and 5

Alternating current		Direct current		DIN VDE 0660 part	IEC 947 part
Utilization category	Typical application	Utilization category	Typical application		
AC-1	Non-inductive or slightly inductive loads, resistive furnaces	DC-1	Non-inductive or slightly inductive loads, resistiv furnaces	102	4
AC-15 [1]	Control of electromagnetic loads (>72 VA)	DC-13 [1]	Control of electromagnets	200	5
AC-2	Slip-ring motors: starting, switching off	–	–	102	4
AC-20 [2)3)]	Connecting and disconnecting under no-load conditions	DC-20 [2)3)]	Connecting and disconnecting under no-load conditions	107	3
AC-21 [2)]	Switching of resistive loads including moderate overloads	DC-21 [2)]	Switching of resistive loads including moderate overloads	107	3
AC-22 [2)]	Switching of mixed resistive and inductive loads, including moderate overloads	DC-22 [2)]	Switching of mixed resistive and inductive loads, including moderate overloads (e.g. shunt-wound motors)	107	3
AC-23 [2)]	Switching of motors or other highly inductive loads	DC-23 [2)]	Switching of highly inductive loads (e.g. series-wound motors)	107	3
AC-3	Squirrel-cage motors: starting, switching off motors during running [6)]	DC-3	Shunt-wound motors: starting, plugging [4)], reverseing [4)], inching [5)], dynamic braking [2)]	102	4
AC-4	Squirrel-cage motors: starting, plugging [4)], reverseing [4)], inching [5)]	–	–	102	4
AC-5a	Switching of electric discharge lamp controls	DC-5	Series-wound motors: starting, plugging [4)], reverseing [4)], inching [5)], dynamic braking	102	4
AC-5b	Switching of incandescent lamps				
AC-6a	Switching of transformers	DC-6	Switching of incandescent lamps	102	4
AC-6b	Switching of capacitor banks				
AC-7a	Slightly inductive loads in household appliances and similar applications				
AC-7b	Motor loads for household applications				
AC-8a	Hermetic refrigerant compressor motor control with manual resetting of overload releases [7)]				
AC-8b	Hermetic refrigerant compressor motor control with automatic resetting of overload releases [7)]				

[1)] In DIN VDE 0660 (07.92), AC-11 is changed to AC-15 and DC-11 to DC-13 respectively.
[2)] The suffix "A" or "B" is added for "frequent operation" or "infrequent operation" respectively.
[3)] The use of these utilization categories is not permitted in the United States of America.
[4)] Plugging or reversing of the motor is the rapid braking or changing of the direction of rotation by means of swapping two of the three supply line phases while the motor is running.
[5)] Inching (or jogging) is understood to mean the single or repeated short-time switching-on of a motor to achieve small movements in the driven machine.
[6)] The devices may be used for occasional inching (jogging) or plugging for limited periods of time; thereby the number of such operations should not exceed five per minute or more than ten in a ten minute period.
[7)] A hermetic refrigerant compressor motor is understood to be a combination consisting of a compressor and a motor, both of which are enclosed in the same housing, with no external shaft or shaft seals, the motor operating in the refrigerant.

The electrical service life of a contactor characterizes its resistance to electrical wear. It is determined by the number of switching cycles (make/break operations) the contactor can perform under the operational load conditions without the need for maintenance or the replacement of parts.

In Section 3.7.7 (page 197), the selection of 3TF contactors in terms of the utilization categories AC-1, AC-3 and AC-4, as well as the selection in terms of contact service life under mixed operating conditions, is illustrated.

In addition to this, the computer program "ELLE" is available for the calculation of the electrical service life of contactors under mixed operating conditions and for the selection of the most suitable Siemens contactor for each particular application (Order No. E86010-D1802-A117-A2).

3 Selection criteria for low-voltage switchgear in main circuits

Table 3.6
Verification of the making and breaking capacities (a) and the electrical service life (b) of contactors in main circuits in terms of the utilization categories to DIN VDE 0660 part 102 (alt. IEC 947-4-1). For the verification of the switching capacities in auxiliary circuits, see Section 4.1.2, Table 4.2, page 231

I	Making current
I_c	Breaking current
I_e	Rated operational current
U	Applied voltage
U_e	Rated operational voltage
U_r	Power-frequency or d.c. recovery voltage
L/R	Time constant of the test circuit
$\cos\varphi$	Power factor of the test circuit

a) *Verification of the rated making and rated breaking capacities*

Alternating current				Direct current			
Utilization categoriy	Make and break conditions			Utilization categoriy	Make and break conditions		
	I_c/I_e [7]	U_r/U_e	$\cos\varphi$ [8]		I_c/I_e [7]	U_r/U_e	L/R [9] ms
AC-1	1.5	1.05	0.80	DC-1	1.5	1.05	1.0
AC-2	4.0 [1]	1.05	0.65 [1]				
AC-3 [2]	8.0	1.05	[3]	DC-3	4.0	1.05	2.5
AC-4 [2]	10.0	1.05	[3]				
AC-5a	3.0	1.05	0.45	DC-5	4.0	1.05	15.0
AC-5b	3.05 [4]	1.05	[4]				
AC-6a	[5]	[5]	[5]	DC-6	1.5 [4]	1.05	[4]
AC-6b	[6]						
AC-7a	1.5	1.05	0.80				
AC-7b	8.0	1.05	[3]				
AC-8a	6.0	1.05	[3]				
AC-8b	6.0	1.05	[3]				

b) *Verification of the electrical service life*

Alternating current

Utilization categoriy	I_e	Make conditions			Break conditions		
	A	I/I_e [7]	U/U_e	$\cos\varphi$ [8]	I_c/I_e	U_r/U_e	$\cos\varphi$ [8]
AC-1	all values	1	1	0.95	1	1	0.95
AC-2	all values	2.5	1	0.65	2.5	1	0.65
AC-3	$I_e \leq 17$	6	1	0.65	1	0.17	0.65
	$I_e > 17$	6	1	0.35	1	0.17	0.35
AC-4	$I_e \leq 17$	6	1	0.65	1	0.17	0.65
	$I_e > 17$	6	1	0.35	1	0.17	0.35

Direct current

				L/R [9] ms			L/R [9] ms
DC-1	all values	1	1	1	1	1	1
DC-3	all values	2.5	1	2	2.5	1	2
DC-5	all values	2.5	1	7.5	2.5	1	7.5

[1] The values shown apply for contactors in the stator circuit. For contactors in the rotor circuit, the test is made with a current of four times the rated rotor operational current and a power factor of 0.65.
[2] The make conditions for utilization category AC-3 ($I/I_e = 10$) and AC-4 ($I/I_e = 12$) must also be verified. The verification may be carried out during the make and break test, but only with the manufacturer's agreement. 25 operating cycles shall be made at a control supply voltage equal to 110% of the rated control supply voltage U_s and 25 operating cycles at 85% of U_s.
[3] 0.45 for $I_e \leq 100$ A; 0.35 for $I_e > 100$ A
[4] The tests are to be carried out with an incandescent light load.
[5] To be derived from the test values for AC-3 or AC-4 in accordance with Table VIIb of the specification (IEC 947-4-1).
[6] Capacitive ratings may be derived from capacitor switching tests or may be assigned on the basis of established practice and experience. As a guide, reference may be made to the equations given in Table VIIb of the specification.
[7] In a.c. the conditions for making are expressed in r.m.s. values but it is understood that the peak value of asymmetrical current, corresponding to the power factor of the circuit, may assume a higher value.
[8] Tolerance for $\cos\varphi$: ± 0.05.
[9] Tolerance for L/R: $\pm 15\%$.

Table 3.7
Verification of the making and breaking capacities (a) and the electrical service life (b) of disconnectors, load switches, switch-disconnectors and fuse-combination units to DIN VDE 0660 part 107 and IEC 947-3

I Making current
I_c Breaking current
I_e Rated operational current
U Applied voltage
U_e Rated operational voltage
U_r Power-frequency or d.c. recovery voltage (between the supply side terminals of the switching device)

a) *Verification of the rated making and rated breaking capacities* [1]

Type of current	Utilization category		I_e A	Making [2]			Make-break		
				I/I_e	U/U_e	$\cos\varphi$	I_c/I_e	U_r/U_e	$\cos\varphi$
Alternating current	AC-20 A [3]	AC-20 B [3]	all values	–	–	–	–	–	–
	AC-21 A	AC-21 B	all values	1.5	1.05	0.95	1.5	1.05	0.95
	AC-22 A	AC-22 B	all values	3	1.05	0.65	3	1.05	0.65
	AC-23 A	AC-23 B	$0 < I_e \leq 100$ A	10	1.05	0.45	8	1.05	0.45
			$100\,\text{A} < I_e$ [1]	10	1.05	0.35	8	1.05	0.35
						L/R ms			L/R ms
Direct current	DC-20 A [3]	DC-20 B [3]	all values	–	–	–	–	–	–
	DC-21 A	DC-21 B	all values	1.5	1.05	1	1.5	1.05	1
	DC-22 A	DC-22 B	all values	4	1.05	2.5	4	1.05	2.5
	DC-23 A	DC-23 B	all values	4	1.05	15	4	1.05	15

b) *Verification of the electrical service life* [4]

Type of current	Utilization category		I_e A	Making [2]			Breaking		
				I/I_e	U/U_e	$\cos\varphi$	I_c/I_e	U_r/U_e	$\cos\varphi$
Alternating current	AC-20 A	AC-20 B	–	–	–	–	–	–	–
	AC-21 A	AC-21 B	all values	1	1	0.95	1	1	0.95
	AC-22 A	AC-22 B	all values	1	1	0.80	1	1	0.80
	AC-23 A	AC-23 B	all values	1	1	0.65	1	1	0.65
						L/R ms			L/R ms
Direct current	DC-20 A	DC-20 B	–	–	–	–	–	–	–
	DC-21 A	DC-21 B	all values	1	1	1	1	1	1
	DC-22 A	DC-22 B	all values	1	1	2	1	1	2
	DC-23 A	DC-23 B	all values	1	1	7.5	1	1	7.5

[1] Table III of IEC 047-3 specifies 5 operating cycles (exception: for AC-23, 100 A $< I_e$ only 3 cycles are prescribed).

[2] For a.c. the making current is expressed by the r.m.s. value of the current, but it is understood that the peak value of asymmetric current, corresponding to the power factor of the test circuit, may assume a higher value.

[3] If the switching device has a making and/or breaking capacity, then the verification must be made at the values of voltage, current and power factor specified by the manufacturer. The number of operating cycles for the verification of the electrical service life must correspond to those given for the particular rated operational current in the specification. The use of these utilization categories is not permitted in the United States of America.

[4] In addition to these parameters, the specification stipulates the number of operating cycles for the operational performance test as a function of the rated operational current I_e. A distinction is made between the categories A and B in terms of the total number of operating cycles with and without current respectively (Table IV of IEC 947-3).

3.3.4.2 Selection of load switches, disconnectors, on-load disconnectors and combination fuse switch units

For these devices, the utilization categories AC-20 to AC-23 and DC-20 to DC-23 in terms of DIN VDE 0660, part 107, still apply in principle. The latest specification (e.g. IEC 947-3) distinguish between frequent operation (AC-20 A to AC-23 A and DC-20 A to DC-23 A) and infrequent operation (AC-20 B to AC-23 B and DC-20 B to DC-23 B).

The rated *making* capacity of an on-load disconnector is that value of current which the device can switch on under given conditions.

The rated *breaking* capacity of an on-load disconnector is that value of current which the device can switch off under given conditions.

The electrical service life characterizes the resistance to electrical wear. It is determined by the number of switching cycles (make/break operations) which can be performed under the operating load conditions without the need for maintenance or the replacement of parts.

3.4 Protection against overcurrent and excessive temperature rise

One can distinguish between the protection against

▷ overload currents
▷ the effects of short-circuit currents, and
▷ excessive temperature rise.

The protection against overload currents is generally performed by current-dependent, thermally or electronically delayed overload relays and tripping devices.

The protection against the effects of short-circuit currents is usually performed by instantaneous electromagnetic or electronic relays and tripping devices, or fuses. In the case of discriminative short-circuit protection, short-time delayed electromagnetic or electronic relays and tripping devices, or suitably selected fuses, are used.

The direct protection against excessive temperature rise is commonly achieved by thermistor protection devices.

Table 3.8 gives an overview of the protection devices used in low-voltage installations.

3.4.1 General functions

3.4.1.1 Protection against overload

Plant components can become overloaded, when operational overcurrents are present for excessive periods of time, or if the equipment, e.g. motors and/or cables are incorrectly dimensioned. These overload currents raise the temperature of motor windings and of cables above the permissible levels, and shorten the service life of their insulation.

Table 3.8
Protection devices used in low-voltage installations

Protection device	Used for protection against		
	overload	short-circuit	overheating
Fuse			
operation class gL [1]	× [3]	×	—
operation class aM [2]	—	×	—
Circuit breaker with			
a-release	×	—	—
z-release	—	×	—
n-release	—	×	—
Miniature circuit-breaker (m.c.b.) [4]	×	×	—
Current-dependent thermally delayed overload relay	×	—	—
Electronically delayed overload relay	×	—	—
Instantaneous electromagnetic overcurrent relay	×	×	—
Thermistor protection device	× [5]	—	×

a Current-dependent delayed release
z Short-time delayed release
n Instantaneous electromagnetic release

[1] General purpose fuses for cable and conductor protection
[2] Accompanied fuses for the protection of switching devices
[3] Not for motors
[4] Miniature (line protection) circuit-breakers to DIN VDE 0641 are circuit-breakers with current dependent delayed overload and instantaneous short-circuit releases
[5] Not for rotor critical motors

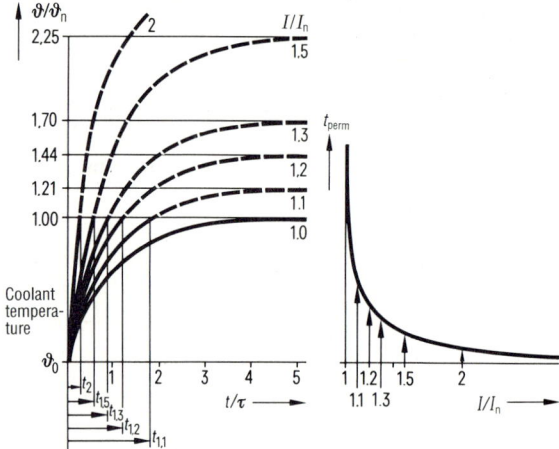

a) Temperature rise for loading with rated current ($1.0 \cdot I_n$) and with various overcurrents ($1.1 \cdot, 1.2 \cdot, 1.3 \cdot, 1.5 \cdot, 2 \cdot I_n$)

b) Permissible loading

ϑ_n Permissible rated temperature limit
ϑ_0 Coolant temperature
t_{perm} Permissible loading time
τ Thermal time constant

Figure 3.11
Permissible thermal loading of a body, if its maximum permissible steady state temperature is not to be exceeded

The higher the overload current, the sooner the permissible temperature is reached and thus the shorter the permissible loading time (Fig. 3.11 a). The permissible loading curve may be obtained by entering the permissible loading times in the time/current diagram (Fig. 3.11 b).

The task of the overload protection is to allow normal operational overload currents to flow, but to interrupt these currents before the permissible loading period is exceeded.

3.4.1.2 Protection against the effects of short-circuits

Short-circuits may be caused by insulation damage or failure, or by incorrect switching operations. Short-circuits are mostly associated with electric arcs.

Short-circuit arcs can destroy equipment and endanger operating personnel. Short-circuit currents severely stress the cables and plant components through which they flow. These stresses are of a thermal and mechanical (dynamic) nature.

The task of the short-circuit protection is to restrict the dangerous effects of short-circuits to a minimum, in that short-circuit currents are detected and interrupted within milliseconds. An additional current limiting effect can be of great advantage.

3.4.1.3 Protection against excessive temperature rise

In contrast to overload protection, temperature sensing protection devices (e.g. thermistor protection) respond directly to the actual relevant temperature inside the equipment (e.g. on motor or transformer windings).

The task of thermistor protection is to allow normal operational overtemperature to occur, but to cause the equipment to be switched off before the permissible temperature limit is exceeded.

Prior to switch-off, an alarm may be given at a specified value of overtemperature, so that preventative measures may be taken to avoid excessive thermal stressing.

3.4.2 Specifications

3.4.2.1 Specifications for overload protection devices

Stipulations for overload protection devices are contained in DIN VDE 0660 part 102 for motor starters, and in DIN VDE 0660 part 101 for circuit-breakers (see Table 3.9 below). If circuit-breakers are used as motor starters, then the characteristic data according to DIN VDE 0660 part 102, applies for the overload protection (see Table 3.10). For overload protection devices, IEC 947-4-1, 05.90 must be taken into account. For circuit-breakers, IEC 947-2, 01.89 (alt. DIN VDE 0660 part 202, 07.92) applies.

In the case of motor starters, a distinction is made between overload relays of Type 1 and Type 2. An overload relay is of

Type 1, if its current setting refers to the rated current of the motor, and of

Type 2, if its current setting is the operating current limit.

Siemens overload relays and circuit-breakers are of the Type 1.

For Factor A times the current setting I_r, and starting from the cold state, the release may not trip within

Table 3.9 Test currents and tripping times for circuit-breakers to DIN VDE 0660 Part 101

Overload release	Current setting I_r	Test current as a multiple of the current setting I_r		Tripping time t	Ambient temperature ϑ_u (reference value)
		Factor A	Factor B		
without temperature compensation	$I_r \leq 63$ A	1.05	1.3	1 h	+20 or +40 °C unless otherwise specified by the manufacturer
	$I_r > 63$ A	1.05	1.3	2 h	
with temperature compensation	$I_r \leq 63$ A	1.05 1.05 1.00	1.3 1.4 1.3	1 h 1 h 1 h	+20 °C −5 °C +40 °C
	$I_r > 63$ A	1.05 1.05 1.00	1.25 1.35 1.25	2 h 2 h 2 h	+20 °C −5 °C +40 °C

Table 3.10
Test currents and tripping times (limits of operation) for three-pole current-dependent delayed overload relays to DIN VDE 0660 part 102 (alt. IEC 947-4-1)

Overload relay	Test current as a multiple of the current setting I_r [1]				Tripping time t	Ambient temperature ϑ_u (reference value)
	Factor A	Factor B	Factor C	Factor D		
a) All poles equally loaded						
without temperature compensation	1.00	1.2	1.5	7.2	2 h	+40 °C [2]
with temperature compensation	1.05 1.05	1.2 1.3	1.5 1.6	7.2 7.3	2 h	+20 °C or +40 °C −5 °C
b) Only two poles loaded, without phase-loss sensitivity						
	Factor A	Factor B				
without temperature compensation	1.00 (3 poles loaded)	1.25 (2 poles loaded) 0 (1 pole loaded)			2	+40 °C
with temperature compensation	1.00 (3 poles loaded)	1.32 (2 poles loaded) 0 (1 pole loaded)			2 h	+20 °C
c) Poles not equally loaded, with phase-loss sensitivity						
with temperature compensation	1.00 (2 poles loaded) 0.90 (1 pole loaded)	1.15 (2 poles loaded) 0 (1 pole loaded)			2 h	+20 °C

[1] The values apply for all possible combinations of the poles.
[2] The reference value for the ambient temperature may be any value between −5 °C and +40 °C; preferred values +20 °C and +40 °C.

the tripping time t. Following this test, the current is increased to Factor B times the current setting. At this higher current, the release must trip within the tripping time t.

If only two poles are loaded, the Factor B may be 10% higher than in the case of three pole loading.

The characteristic data for the tripping of three-pole current-dependent thermally delayed overload relays to DIN VDE 0660 part 104 and IEC 947-4-1 is shown in Table 3.10.

For the overload relay or starter mounted in its enclosure, and loaded with a current of Factor A times the current setting (starting from the cold state), tripping may not occur in less than 2 hours. When the current is subsequently increased to Factor B times the current setting, the relay must trip within 2 hours. For Class 10 A, 10, 20, and 30 overload relays, loaded at Factor C times the current setting, tripping must occur within 2, 4, 8 or 12 minutes respectively (starting from thermal equilibrium at a load current equal to the current setting). At a load current of Factor D times the current setting, tripping must occur within the limits of the tripping time T_p specified for the appropriate trip class (starting from the cold state):

Class 10 A $2\,\text{s} < T_p \leq 10\,\text{s}$
Class 10 $4\,\text{s} < T_p \leq 10\,\text{s}$
Class 20 $6\,\text{s} < T_p \leq 20\,\text{s}$
Class 30 $9\,\text{s} < T_p \leq 30\,\text{s}$

3.4.2.2 Service factors in terms of UL and CSA specifications

The tripping current limits of an overload protection device are fixed in terms of the service factor (SF) of the motor. The SF gives the factor (albeit at a reduced service life in comparison to operation at SF = 1) by which the motor current and power may be increased. One can distinguish between SF ≥ 1.15 and SF < 1.15. In the case of motors with SF ≥ 1.15, the tripping current limit at symmetrical three-pole loading is 1.25 times the current setting I_r. For motors with SF < 1.15, the tripping current limit is $1.15 \cdot I_r$.

For a two pole load of 200% the maximum tripping current limit, the relay must trip within 8 minutes; at 600% within 30 seconds. Both tests are done starting from the cold state.

The Siemens current-dependent thermally delayed overload relays for Class 10 applications are suitable for use with motors having SF ≥ 1.15 without the need for a trip current setting correction. In the case of motors with SF < 1.15, the trip current setting must be adjusted to 0.92 times the motor rated current.

Thermally and electronically delayed relays operating in conjunction with current transformers are suitable for use with motors having SF < 1.15 without the need for a trip current setting correction. For SF ≥ 1.15, the trip current setting must be adjusted to 1.09 times the motor rated current.

3.4.2.3 Specifications for overcurrent protection devices

Circuit-breakers are equipped with instantaneous electromagnetic tripping units or releases to enable them to open immediately in the event of a short-circuit. Within the framework of the DIN VDE 0660 part 101 07.92 and IEC 947-2, merely the permissible tolerances for the operating currents of such tripping units are specified, and not the absolute values. The operating currents may be fixed or adjustable, and normally have a value of 5 to 10 times the circuit-breaker rated current for line protection. For motor protection, the operating current of an instantaneous tripping unit in a circuit-breaker is chosen to be much higher than this.

3.4.2.4 Specifications for temperature dependent protection devices

The specifications for temperature dependent protection devices using PTC thermistors are contained in the IEC publication IEC 34-11-2 section 1 and DIN VDE 0660 part 302, 02.87 " Thermal motor protection. Thermal detectors and tripping units" as well as in IEC 34-11-2 section 2 and DIN VDE 0660 part 303, 02.87 "Thermal motor protection. PTC thermal detectors and tripping units" (also see Section 1.2.1 page 17). DIN VDE 0660 part 303, contains a summary of the characteristic data for matching "Mark A" detectors with "Mark A" tripping units.

Resistance/Temperature characteristic of the "Mark A" detector

The change in resistance with temperature of each individual PTC thermistor must fulfill the following conditions (referred to the rated operating temperature TNF):

a) Resistance ≤ 550 Ω at a temperature of (TNF − 5) °C and a reference d.c. voltage of ≤ 2.5 V

b) Resistance ≥ 1330 Ω at a temperature of (TNF + 5) °C and a reference d.c. voltage of ≤ 2.5 V

c) Resistance ≥ 4000 Ω at a temperature of (TNF + 15) °C and a reference d.c. voltage of ≤ 7.5 V

d) Resistance ≤ 250 Ω at any temperature between − 20 °C and (TNF − 20) °C and a reference d.c. voltage of ≤ 2.5 V.

The preferred configuration consists of three detectors connected in series. If more than three detectors are to be connected in series, then the resistance of the individual detectors must be such that the total resistance value at a temperature of − 20 °C to (TNF − 20) °C does not exceed 750 Ω (also see Section 3.4.3.5, page 138.

Characteristics of the "Mark A" tripping units

If a thermistor tripping unit is to operate under normal conditions in accordance with DIN VDE 0660 part 302/02.87, then the following stipulations must be fulfilled

a) At detector circuit resistance of ≤ 750 Ω, it must be possible to switch on or reset the tripping circuit of the unit.

b) The unit must trip if the detector circuit resistance is increased from 1650 Ω to 4000 Ω.

c) If the detector circuit resistance is decreased from 1650 Ω to 750 Ω, the tripping unit must reset or it must be possible to switch on or reset the tripping circuit of the unit.

d) If a resistance of 4000 Ω is connected across each pair of detector circuit terminals, the voltage across each pair of these terminals may not exceed 7.5 V (d.c. or a.c. peak value) if the unit is operated at its rated voltage.

e) There shall be no significant change in the operating values of the tripping unit for a detector circuit capacitance of up to 0.2 µF.

3.4.3 Protection devices

3.4.3.1 Fuses

Time-current characteristics

The melting time (pre-arcing time) of a fuse is normally illustrated as a function of current in a time-current diagram having a logarithmic scale. The melting time-current characteristic lies between two asymptotes. The first (vertical), is the minimum mel-

3 Selection criteria for low-voltage switchgear in main circuits

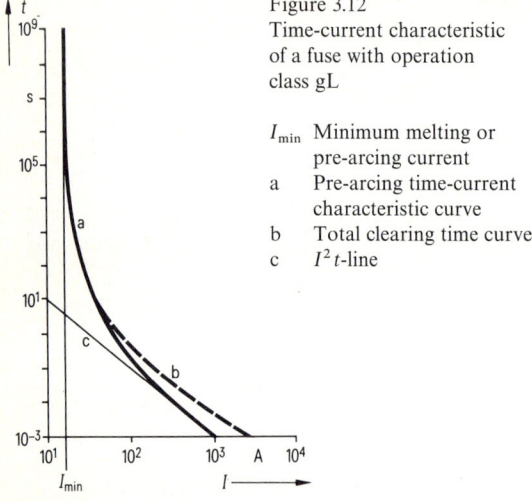

Figure 3.12
Time-current characteristic of a fuse with operation class gL

I_{min} Minimum melting or pre-arcing current
a Pre-arcing time-current characteristic curve
b Total clearing time curve
c $I^2 t$-line

ting current which only just causes the fuse element to melt through, and the second is a diagonal line of equal current heat values (I^2t line), based on higher short-circuit currents, and representing the constant melting heat I^2t of the element (see Fig. 3.12).

The position of the characteristic is determined mainly by the heat transfer from the fuse element to outside. DIN VDE 0636 specifies the ranges of time and current within which a fuse characteristic must lie. These relatively wide time and current ranges are necessary because of the nature of the manufacturing tolerances.

A ±10% deviation of the time-current characteristic on the current axis is permissible in terms of DIN VDE 0636. The Siemens HRC fuses of operation classes gL and aM deviate from the specifications by less than 5%, which improves discrimination performance.

Up to approximately $20 \cdot I_n$, the melting time characteristic is the same as the total clearing time-current characteristic. At higher short-circuit currents the two characteristics separate. The difference is due to the corresponding arc-quenching times which, apart from power factor, depend largely on the operating voltage and the prospective short-circuit current value.

The mean virtual pre-arcing time vs. prospective short-circuit current characteristics at an ambient temperature of (20±5) °C are given in the data sheets of the relevant Siemens catalogues and technical documentation. They apply to fuses without pre-loading (also refer to Section 9.7, page 536).

Current limiting

At very large prospective short-circuit currents, a fuse will rupture before the current can reach its peak value i_p (Fig. 3.13). The highest instantaneous value of the current, which is reached during the interruption procedure, is known as the cut-off current or let-through current i_D. The current limiting effect is illustrated in the relevant catalogues in the form of let-through (or cut-off) current diagrams (Fig. 3.14). Current limiting becomes effective as from the short-circuit current I_k at which the let-through current line i_D intersects the line of the prospective unlimited peak short circuit current i_p.

Example

System voltage	500 V, 50 Hz
Effective (r.m.s.) value of the prospective short-circuit current I_k (derived from calculated initial a.c. short-circuit current I_k'')	50 kA
peak short-circuit current i_p with largest d.c. component	105 kA
proposed fuse: HRC fuse link, size 00, operation class gL	
rated current	100 A
peak value of let-through current i_D (from let-through current diagram)	12 kA

The 100 A fuse limits the prospective short-circuit current to 11.5% of its expected value ($i_D = 0.115 \cdot i_p$).

Breaking (or rupturing) capacity

Owing to their current limiting capabilities (see Fig. 3.14), fuses are able to interrupt very large short-circuit currents within a very confined quenching volume, i.e. within the body of the fuse. The greater the current limiting effect is, the greater is the breaking capacity for the same arc-quenching volume. For example, Siemens LV-HRC fuses are capable of interrupting short-circuit currents of at least up to 120 kA at cos $\varphi = 0.1$. This exceeds the value of 50 kA required by DIN VDE 0636 for a.c. networks by far; even in 690 V networks.

The effective current limiting and corresponding high rupturing capacity are special properties of fuses which render them indispensable in the field of short-circuit protection.

Fuses · Current limiting

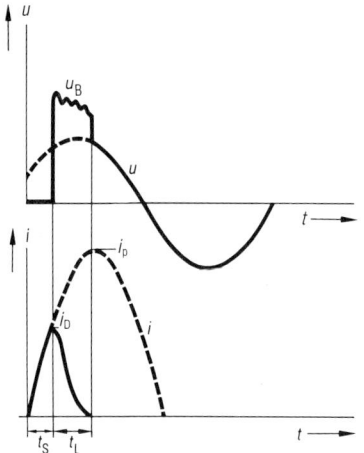

i_p Peak of prospective short-circuit current
i_D Let-through current peak
t_s Melting time
t_L Quenching time
u_B Arc voltage

Figure 3.13
Oscillograph of short-circuit interruption by a fuse

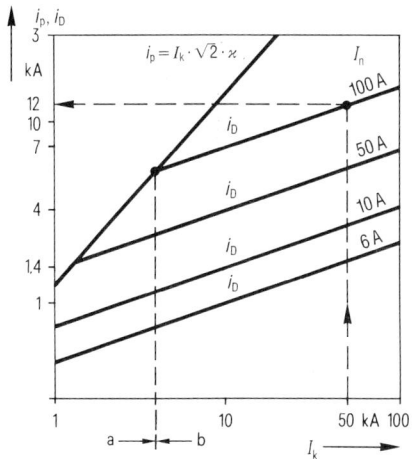

I_k Sustained short-circuit current	a Non current-limiting range
I_n Rated current of the fuse	b Current-limiting range
i_D Peak of let-through current	
i_p Peak of prospective short-circuit current	

Figure 3.14
Let-through current diagram for size 00 HRC fuses, operation class gL, rated current 100 A

Thermal values (I^2t values; Joule values)

The amount of Joulean heat, produced by the current flowing through the fuse, which is required to melt the fuse element is known as the melting or pre-arcing I^2t_s value. The total heat which is absorbed up until the arc is quenched is called the rupturing or total I^2t value. Here too, DIN VDE 0636 specifies the limits for these values. Both the pre-arcing I^2t_s values and the total I^2t values of Siemens LV-HRC fuses lie within these limits.

The I^2t values of fuses determine the degree of selectivity (or discrimination) between fuses at higher currents and the extent to which the protected equipment, e.g. a cable, is stressed thermally (also refer to Section 3.4.4.4, page 144).

Influence of ambient temperature

In practice, fuses are subjected not only to ambient temperatures of various values, but also to changing temperatures. In the case of Siemens LV-HRC fuses, ambient temperatures in the range of $-5\,°C$ to $+45\,°C$ have a negligible effect on the time-current characteristic. In addition, they are capable of carrying their full rated current continuously at an ambient temperature of $+55\,°C$, and for at least 24 hours at $+65\,°C$.

Rated power loss

Low power losses not only contribute to more economical operation but also produce less heat, which is especially advantageous considering the trend to ever smaller installation volumes. A further benefit of a lower temperature rise is the reduction in thermal loading of the external fuse connections (e.g. blades), the fuse base and the connected cables or conductors.

The rated power losses of Siemens LV-HRC fuses are well below the maximum values specified by DIN VDE 0636 for the largest current rating in each fuse size.

Resistance to ageing

Fuses are often used for the short-circuit protection of motor feeder cables. In such cases, regular starting currents of 8 to 12 times the rated operating current of the motor must be carried by these conductors for short periods. If correctly selected, the characteristics of Siemens LV-HRC fuses are not affected by these short time overloads. Even operational overloads of longer duration do not adversely effect their long-term protection properties.

3 Selection criteria for low-voltage switchgear in main circuits

Table 3.11
Classification of low-voltage fuses in terms of their use and operating characteristics to DIN VDE 0636 part 1/12.83

Function class, designation	Continuous current up to	Rupturing current	Operation class, designation	Protection of
General purpose fuses (full range)				
g	I_n	$\geq I_{min}$	gL/gG [1] gR gB	Cables and conductors, semi-conductors, mining installations
Accompanied fuses (partial range)				
a	I_n	$> 4 \cdot I_n$ $\geq 2.7 \cdot I_n$	aM aR	Switching devices, semiconductors

I_{min} Minimum rupturing current (melting current)

Classification

Low-voltage fuses are classified according to their use and operating characteristics and according to their construction. An overview of the classification of fuses according to their characteristics is given in Table 3.11.

The function class [2] **g** (general purpose fuses), contains wide range fuses, which can carry currents of up to at least their rated current continuously, and which are able to interrupt currents from the lowest pre-arcing current up to their rated rupturing capacity. The gL [1] fuses, for cable and busbar protection, fall into this group.

The function class [2] **a** (accompanied fuse), contains partial range fuses, which can carry currents of up to at least their rated current continuously and which are able to interrupt currents from a specific multiple of their rated current up to their rated rupturing capacity.

The switchgear protection fuses aM fall into this group. Their rupturing current begins at approximately 4 times the rated current and they therefore serve only for short-circuit protection.

A comparison of the pre-arcing time-current characteristics of two 200 A LV-HRC fuses with operation

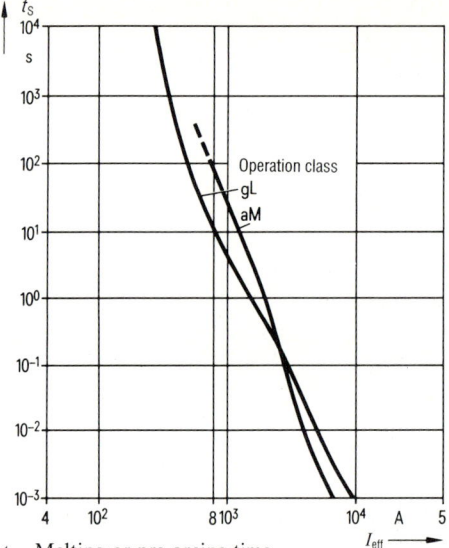

t_s Melting or pre-arcing time

Figure 3.15
Pre-arcing time-current characteristics of LV-HRC fuses, rated current 200 A, operation classes gL and aM

class gL and aM respectively, are shown in Figure 3.15.

3.4.3.2 Circuit-breakers

The main function of circuit-breakers is the protection against the effects of short-circuits, although they are also used for protection against the dangers of overload and earth fault currents. Depending on the application, they incorporate tripping devices or releases. These may be an integral part of the breaker and be installed directly in the current paths (e.g. overload and short-circuit releases). Alternatively, they may be of modular design for retrofitting to the basic unit, and may require an external electrical signal from a relay or other switch for their operation (e.g. undervoltage release or shunt release controlled by the signal from a thermistor protection tripping unit). An overview of the internal releases and external relays commonly used in conjunction with circuit-breakers, is given in Table 3.12.

For further information on earth-leakage and earth-fault protection, refer to Section 3.6, page 183.

Rated short-circuit switching capacity

The rated short-circuit breaking capacity I_{cn} and the rated short-circuit making capacity I_{cm} of a circuit-breaker must be at least equal to the initial symmetrical short-circuit current I''_k which can be produced

---
[1] Operation class gG (general purpose) replaces the characteristic gL in the new edition of DIN VDE 0636
[2] Also known as utilization class

Table 3.12
Protective functions of internal releases and external relays used in conjunction with circuit-breakers

Function	Releases	Relays
Overload protection	Overload releases current/time-dependent thermally or electronically delayed	Overload relays current/time-dependent thermally or electronically delayed
		Thermistor protection tripping devices
Short-circuit protection	Overcurrent releases instantaneous electromagnetic or electronic	Overcurrent relays instantaneous electromagnetic
Discriminative short-circuit protection	Overcurrent releases short-time delayed electromagnetic or electronic	–

at its point of installation. If this is not the case then back-up fuses must be installed on the line side of the circuit-breaker. The maximum permissible back-up fuse ratings for each circuit-breaker are given in the relevant Siemens l.v. switchgear catalogue.

Rated peak withstand current and rated short-time withstand current

Circuit-breakers intended for use in a discriminative (selective) short-circuit protection chain, must be capable of withstanding the initial uninfluenced short-circuit current peak i_p, and must be able to conduct the short-circuit current during the specified short-time delay period until the current is interrupted by the breaker itself or by a downstream counterpart. Data on the rated peak withstand current and the rated short-time withstand current (1 s current) for circuit-breakers is given in the relevant Siemens l.v. switchgear catalogue. In the case of alternating current, the r.m.s. value of the uninfluenced a.c. component of the short-circuit current is used as a basis for the determination of the circuit-breaker suitability.

Overload releases

Thermally or eletronically delayed overload releases may be adjustable within a given current range, or may be fixed at a specific value. Adjustment is usually achieved with a twist knob, lever or rating plug.

The tripping characteristic itself is usually fixed, and is merely shifted on the time axis with the adjustment of the tripping current I_r. Thus, the tripping characteristics of circuit-breakers and overload relays are expressed in terms of multiples of the trip current setting I_r. These characteristics may be found in the relevant Siemens catalogues (see Fig. 3.16b on page 130).

Electronically delayed overload releases and relays, usually allow the setting of the tripping time for a current of 6 times I_r.

The tripping currents and delay times of current-dependent delayed thermal overload releases and relays are stipulated in the DIN VDE specifications and IEC publications. In the case of overload relays, one distinguishes between two types, i.e.

Type 1 characterized by a current setting which corresponds to the rated current of the associated motor, and

Type 2 characterized by a current setting which corresponds to the operating current limit.

Type 1 is most commonly used.

Short-circuit releases

Electromagnetic short-circuit releases may be either fixed or adjustable, where-as the electronic short-circuit releases of Siemens circuit-breakers are always adjustable.

Table 3.13 gives an overview of the tripping current ranges of short-circuit releases in circuit-breakers. According to DIN VDE 0660, the actual current at which the circuit-breakers trips, may deviate by $\pm 20\%$ from the set value.

Mechanical reclosing lock-out

A mechanical reclosing lock-out is used to prevent a circuit-breaker, which has tripped owing to a short-circuit, from accidentally being re-closed onto the faulty circuit. When the short-circuit release operates and trips the breaker, the reclosing lock-out mechanically latches the tripping mechanism in the tripped position. Should an attempt be made to reclose the breaker, before the reclosing lock-out has been reset, the switch mechanism will not engage.

In moulded case circuit-breakers (m.c.c.b.'s), the operating mechanism moves to a tripped position after a release has operated. The circuit-breakers can only be reclosed after the operating mechanism or toggle has been moved from the tripped to the off position.

3 Selection criteria for low-voltage switchgear in main circuits

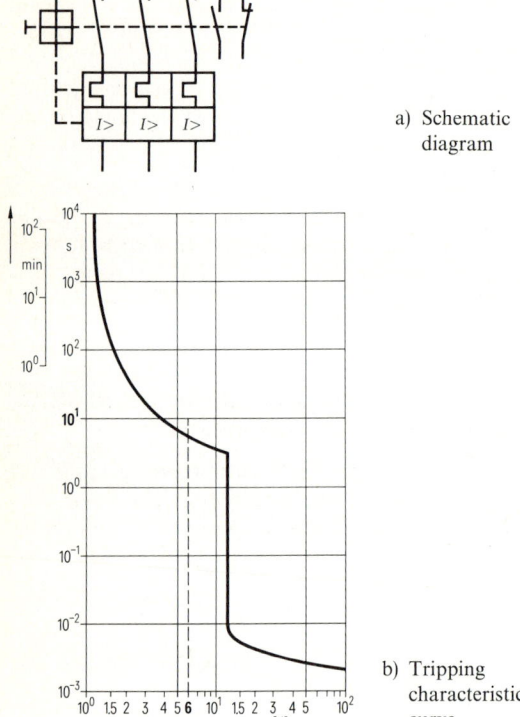

a) Schematic diagram

b) Tripping characteristic curve

Figure 3.16
Circuit-breaker with a current-dependent delayed overload release and an instantaneous electromagnetic short-circuit release (a and n releases)

Table 3.13
Tripping current ranges of short-circuit releases
(to DIN VDE 0660, part 101)

Application, preferred device for interrupting short-circuit current	Type of delay	Tripping ranges of the current dependent delayed overload release as a multiple of the current setting I_r
Circuit-breaker for generator protection	Instantaneous or short-time delayed	Approx. 3 to $6 \cdot I_r$
Circuit-breaker for cable protection	Instantaneous	Approx. 6 to $12 \cdot I_r$
Circuit-breaker for motor protection	Instantaneous or short-time delayed [1]	Approx. 8 to $15 \cdot I_r$

[1] Short-time delayed e.g. to avoid tripping resulting from current surges (rush currents)

Undervoltage releases

An undervoltage release (u.v.r.) is used to sense the presence of a voltage (usually the supply voltage), for interlocking purposes and/or for remote tripping.

The release must trip the breaker if the rated control supply voltage U_c drops to a value of between 0.35 and 0.7 times its rated operating voltage U_s.

If the operating voltage of the u.v.r. is derived from the network, then the breaker trips instantaneously in the event of network supply voltage loss or a severe sag.

Delayed undervoltage releases are used in weak networks where short duration voltage dips tend to cause nuisance tripping of the breaker. The delay, which is typically selectable as 1.2 or 3 seconds, is usually achieved by an auxiliary delay unit.

By the opening of the connection between the u.v.r. and the drop-out delay unit, instantaneous tripping of the circuit-breaker may be achieved (e.g. operational switch-off, emergency switch-off).

Shunt releases (also known as shunt trips)

A shunt release is used for the remote tripping of a circuit-breaker. Its operating voltage lies between 0.5 and 1.1 times its rated operating voltage U_s.

Meshed network releases (shunt release with energy storage)

The meshed network release is a shunt release with an operating range of 0.1 to 1.1 times U_s. It usually consists of a normal shunt release in combination with a separate energy storage device. The capacitance of the energy storage unit is so dimensioned, that if fully charged, the release will remain operative for 4 to 5 minutes after a total loss of the network supply voltage (also refer to Section 3.4.6.3, page 174).

3.4.3.3 Line-protection (miniature) circuit-breakers

Fields of application

Miniature circuit-breakers (m.c.b.'s) are used for the overload and short-circuit protection of cables, wires and conductors in electrical installations and distribution systems.

In TN and TT systems (networks) with switch-off by overload protection equipment, the m.c.b.'s also prevent the continued existence of excessively high touch voltages in the event of a fault.

Miniature circuit-breakers used in an industrial environment, in high-rise building distribution systems or in trade and contracting applications are also often called upon to protect pieces of electrical equipment or small machines.

Especially in these applications, the demands made on miniature circuit-breakers far exceed those of household and domestic installations. Line protection circuit-breakers with auxiliary contacts and m.c.b.'s for a.c./d.c. usage (with or without auxiliary contacts) meet several of these demands.

Line protection circuit-breakers have a fixed overload release for a each rated current. This means that they are not ideal for the protection of motors. Unlike circuit-breakers for motor protection and overload relays, they cannot provide an accurate protection characteristic which has been optimized for motor behaviour. Line protection circuit-breakers should therefore only be used to provide a rough form of protection for motors in cases where a more accurate protection is not required.

The Siemens a.c. miniature circuit-breakers up to 63 A are suitable for use in all three-phase networks with a system voltage up to 240/415 V. The a.c./d.c. m.c.b.'s are, in addition, also suitable for all d.c. systems up to a rated voltage of 440 V, depending on the number of poles involved. Furthermore, they are suitable for use in single or multi-phase networks, as single or multi-pole devices, and for frequencies other than 50 Hz.

Switching capacity

Particular demands are made on miniature circuit-breakers with respect to switching capacity.

For this reason, rated switching capacity classes are defined in IEC 898, DIN VDE 0641 part 11, 08.92 and CEE publication 19, 2nd Edition (Table 3.14). Not all the rated switching capacity classes were adopted by the new DIN VDE 0641 A4/11.88.

The Siemens N-System a.c. miniature circuit-breakers of the type 5SQ have a rated switching capacity of 3000 A, the types 5SN and 5SX of 6000 A and the series 5SL of 10 000 A.

In order to define the extent of line protection under short-circuit conditions and the selectivity with respect to a back-up fuse, m.c.b.'s up to 16 and 32 A are divided into three groups, or current limiting classes, according to the degree to which they limit a short-circuit current upon opening. The permissible let-through (or cut-off) I^2t values for line protection miniature circuit-breakers up to 16 A, and from 16 to 32 A, may be found in the DIN VDE specifications.

The testing requirements for the rated switching capacity according to BS 3871 are similar to those of DIN VDE 0641 and CEE Publ. 19. The switching capacity classes differ to some extent.

In terms of BS 3871, Siemens N-System a.c. miniature circuit-breakers of the type 5SN and 5SX2 can be grouped with the switching capacity class M6, whereas the types 5SL belong to the class M9.

Depending on the testing parameters used, other standards, specifications and recommendations may classify the breakers as having higher switching capacities (e.g. UL). This should be borne in mind whenever line protection miniature circuit-breakers are used.

3.4.3.4 Overload relays

Overload relays are used to protect electrical equipment, such as three-phase motors and transformers, from overheating.

Overheating of a motor can, for example, be caused by a mechanical overloading at the output shaft, by the drawing of an asymmetrical operating current, as a result of an asymmetrical network supply voltage or the loss of one phase, or owing to a blocked rotor. In all these cases, the increased current drawn by the motor is monitored in all the phases by the overload relay.

Overload relays are current-dependent protection devices. The time-current tripping characteristic is usually achieved using the bimetal principle, the melting of a eutectic material or electronically. Only relays of the bimetallic and electronic kind are dealt with in this book (refer to Section 9).

Overload relays with bimetal strips

Current-dependent thermally delayed overload relays generally have three bimetal strips. These are heated

Table 3.14 Rated switching capacity classes

To IEC 898 and CEE Publ. 19, 2nd edition	To DIN VDE 0641 part 11, 08.92
1 500 A	—
3 000 A	3 000 A
4 000 A	—
6 000 A	6 000 A
10 000 A	10 000 A

indirectly by the motor current which flows through heater elements or windings.

For motor currents above 180 A, the secondary current of a current transformer is used to heat the bimetal strips. On the one hand, this reduces the losses, while on the other hand increasing the short-circuit withstand capability. Figure 3.17 illustrates the construction and operating principles of a bimetallic overload relay.

As the bimetal strips are heated by the motor current flowing through the heater windings L1/T1, L2/T2 and L3/T3, the strips bend and the tripping bar is moved. The tripping bar operates the tripping lever via the ambient temperature compensating strip. The tripping lever, in turn, transfers the force to the spring-loaded snap-action moving contact. Prior to tripping, the NC contact (95/96) is closed, and the NO contact (97/98) is open.

In the event of an overload, the tripping lever pushes the snap-action moving contact so that the NC contact (95/96) is opened and the NO contact (97/98) closed. A mechanical switch position indicator signals the tripped state.

If the reset selector is in the "Manual Reset" position, the snap-action moving contact is pushed beyond its dead-centre position (reclosing lock-out). After sufficient cooling of the bimetal strips, the reset button must be used to push the snap-action moving contact back over its dead-centre position to reset the relay.

If the reset selector is in the "Automatic Reset" position, the NC contact is opened and the NO contact closed, even though the snap-action moving contact is not moved beyond its dead-centre position. In this case, the moving contact returns to its original position automatically once the bimetal strips have cooled down.

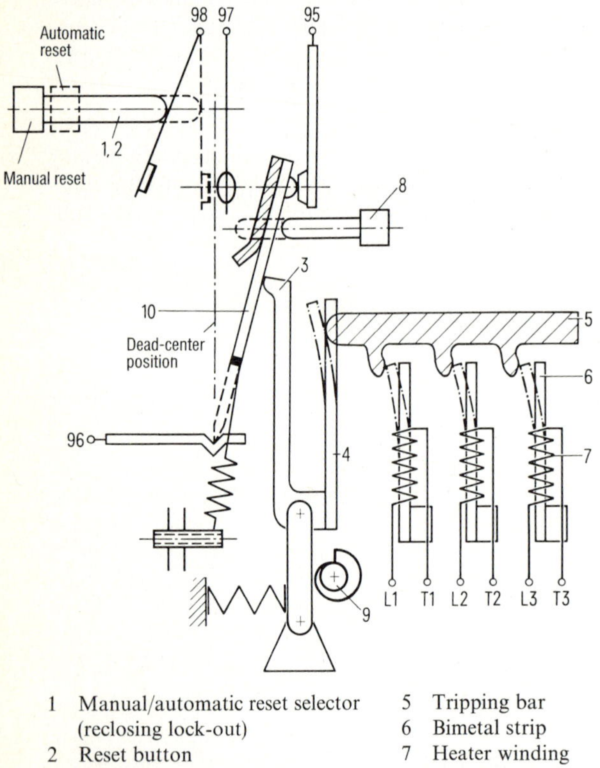

Setting of the overload relay

The overload relays have either a knob or lever which can be adjusted steplessly within the setting range to the desired current setting I_r. Either the knob must be turned so that the pointer on the body of the relay lines up with a value or division line on the scale of the knob, or the lever must be moved to the desired current value on the setting scale. In the case of an overload relay with current transformer, when the units are mounted separately from one another, the special dot mark must be used for the setting. This provides a compensation for the different thermal conditions at the relay and current transformer.

Recovery times

After tripping, thermally delayed overload relays require a specific period in which to cool down. Only after this period, can the relay be reset again. This time period is known as the recovery time.

The actual recovery time depends on the value of current which caused the overload relay to trip, and on the time-current tripping characteristic or, in the case of electronic relays, on the recovery time value stored in memory of the relay itself. The recovery times of thermal overload relays may be read off from

1 Manual/automatic reset selector (reclosing lock-out)
2 Reset button
3 Tripping lever
4 Ambient temperature compensation strip
5 Tripping bar
6 Bimetal strip
7 Heater winding
8 Test button
9 Setting screw
10 Switch rocker

Figure 3.17
Operating principle of a three-pole overload relay with temperature compensation, selectable reclosing lock-out and test button

Construction and operating principles of overload relays

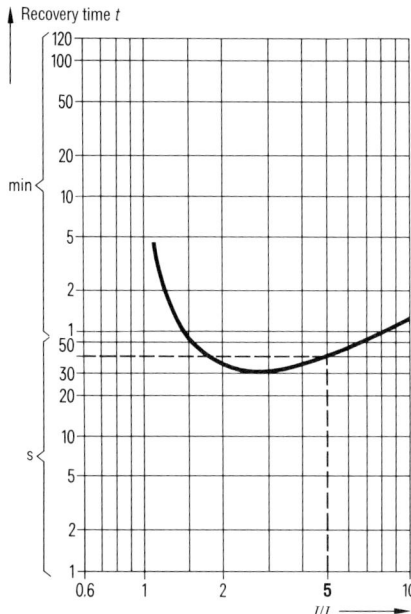

Figure 3.18
Recovery times of thermally delayed overload relays as a function of the tripping current and the trip-current setting

the diagram in Figure 3.18. For example, after a tripping operation which occurred at five times the value of the trip-current setting I_r, the recovery time will be approximately 40 seconds.

The recovery time forces a stoppage of operation, which contributes to the time required for the motor to cool down after the overload trip. The recovery time of the relay is however not always sufficient to allow adequate cooling prior to renewed starting of the motor. This will depend on the particular motor in question, the ambient conditions, and the operating duty.

Protection against inadvertent reclosure

Siemens overload relays are equipped with a reclosing lock-out facility ("Manual Reset") to prevent the inadvertent automatic reclosure of the relay after the recovery time has expired. The auxiliary contacts of the overload relay only return to their non-tripped state when the reset button is depressed, and only then can the associated contactor be re-energized.

Should an overload condition occur while the reset button is held in the depressed position, the relay will still trip. All Siemens overload relays incorporate this trip-free feature.

The reclosing lock-out may also be switched from "Manual Reset" to "Automatic Reset". The relays are, however, always delivered with the lock-out in the manual position.

For safety reasons, overload relays in the automatic mode should only be used in circuits employing momentary contact control (also known as 3-wire control) for the contactor actuation (Fig. 3.19a). Although the NC contact returns to its original state after the recovery time, the control circuit remains open owing to contact S1.

Overload relays in the manual reset mode may also be used in continuous contact control circuits (also known as 2-wire control; see Fig. 3.19b). The NC auxiliary contact of the relay, and therefore the coil circuit of the contactor, remains open after a tripping

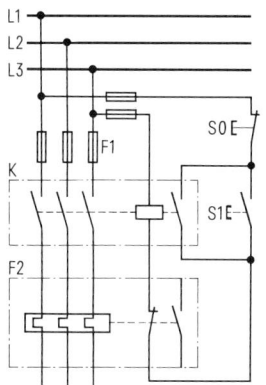
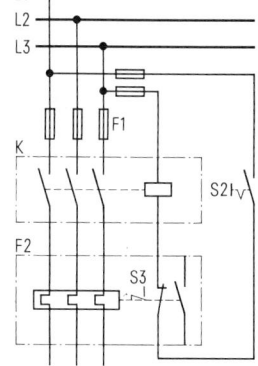

a) Momentary-contact operation, relay without reclosing lock-out (automatic reset)

b) Maintained-contact operation, relay with reclosing lock-out (manual reset)

F1 Fuse
F2 Thermally delayed overload release
K Contactor
S0 Off button
S1 On button
S2 Switch with maintained contact
S3 Button to release the reclosing lock-out

Figure 3.19
Application examples showing overload relays with automatic and manual reset

operation, even though S2 (e.g. a contact of a pressure monitor, position switch, thermostat etc.) may close again.

Siemens thermal overload relays are fitted with a mechanical indicating pin (green) to show that the relay and its auxiliary contacts are in the tripped state. For a remote indication of this condition, the NO contact may be used. For remote resetting of an overload relay, an electrical reset module may be snapped over the reset button.

Test feature

The correct operation of the auxiliary contacts of an overload relay in the non-tripped state may be checked by depressing the test button. This causes the opening of the NC contact (95-96) and the closing of the NO contact (97-98). In this way, the correct wiring of the control circuit may also be checked.

The test button may also be used as an off button to drop out the contactor and thereby switch off the motor.

If the overload relay is in the manual reset mode, then it must be reset again after the test button has been depressed and released. In the automatic reset mode, the relay is reset the moment the test button is released.

Overload relays of the type 3UA7 have a combined off/reset button. The NC contact is held open for as long as this button is depressed.

Time-current tripping characteristics

The tripping characteristics of an overload relay give the tripping time as a function of the tripping or operating current I_r, which is expressed as a multiple of the current setting (Fig. 3.20). They are usually provided for symmetrical three-pole and for 2-pole loading from the cold state.

The smallest current which can cause a tripping operation, is known as the minimum tripping current. According to DIN VDE 0660 part 104, this current must lie within specified limits (also refer to the specifications for overload relays in Section 3.4.2.1, page 123).

The tripping class specifies the tripping time limit for a symmetrical 3-pole loading at 7.2 times the current setting from the cold state. At this load, a relay of Class 10 must trip within 10 seconds and a relay of Class 30 within 30 seconds.

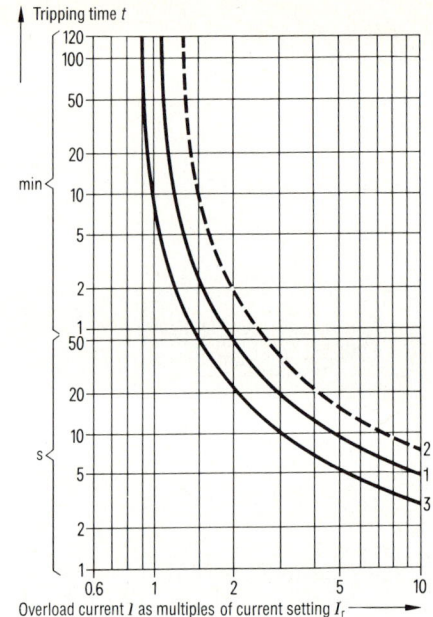

1 Symmetrical three-pole loading from the cold state
2 Two-pole loading from the cold state, relay without phase failure sensitivity
3 Two-pole loading from the cold state, relay with phase failure sensitivity
I_r Current setting of the overload relay

Figure 3.20
Tripping characteristic of a three-pole overload relay for normal starting conditions (mean values)

Class 10 relays are intended for normal starting conditions, where-as Class 30 relays should be selected for heavy duty starting of three-phase motors.

Single-pole loading

If single-phase or d.c. loads are to be protected by a 3-pole thermally delayed overload relay, then it must be ensured that all three bimetal strips of the relay are heated by the load current. For this reason, all three current paths, or poles, of the relay must be connected in series. If this is the case, curve 1 of Figure 3.20 applies.

Behaviour at operating temperature

A motor running at its operating temperature naturally has far lower thermal reserves than a cold motor. This is taken into account by the overload relay. If the set current I_r flows through the overload relay for a longer period of time, then the tripping times are reduced to about 25% of the cold state values.

Accuracy of the tripping times

Manufacture, material and calibration tolerances cause the actual tripping times to lie within a scatter band from which the published tripping characteristics are statistically derived.

Tables of the actual tripping times, with an accuracy tolerance of ±20%, are given in the relevant Siemens l.v. switchgear catalogues for each setting range at load currents of 3 to 8 times the current setting I_r.

Temperature compensation

The temperature compensation feature reduces the effect of the ambient temperature on the tripping behaviour of the overload relay. All Siemens thermal overload relays are fitted with temperature compensation, as standard (see Fig. 3.17, page 132). The temperature compensating strip ensures that the minimum tripping current of the relay lies within the limits specified by DIN VDE 0660 part 104 for a range of ambient temperature from −25 °C to +55 °C.

For lower temperatures between −25 °C and −40 °C, specially calibrated overload relays can be supplied.

For higher temperatures in the region of +55 °C to +80 °C, the maximum value of the setting range is to be reduced by the respective factor from Table 3.15.

Phase-loss sensitivity (single-phasing protection)

The tripping characteristic of a overload relay is based on the assumption that all three bimetal strips are simultaneously heated by the same current.

If one phase is interrupted (single-phasing) and therefore only two of the three bimetal strips are heated, then these two remaining strips alone must produce the necessary force on the tripping mechanism for the tripping operation. This requires a higher current, or alternatively a longer tripping time (curve 2 in Fig. 3.20).

The motor could be damaged if it draws this higher current for a long period of time. To ensure the adequate protection of the motor even in the event of an asymmetrical supply or the loss of one phase, overload relays are designed to have phase-loss sensitivity.

Overload relays with phase-loss sensitivity are fitted with an additional differential tripping bar (designated tripping bar II in Fig. 3.21 below). The two tripping bars are connected via a hinged lever.

For an operational 3-pole symmetrical load of 1.0 times the current setting I_r, all three bimetal strips bend by the same amount and their ends are displaced by the distance a. Both tripping bars are moved the same distance a, so that the vertical position (in the illustration) of the hinged lever is maintained. The relay does not trip.

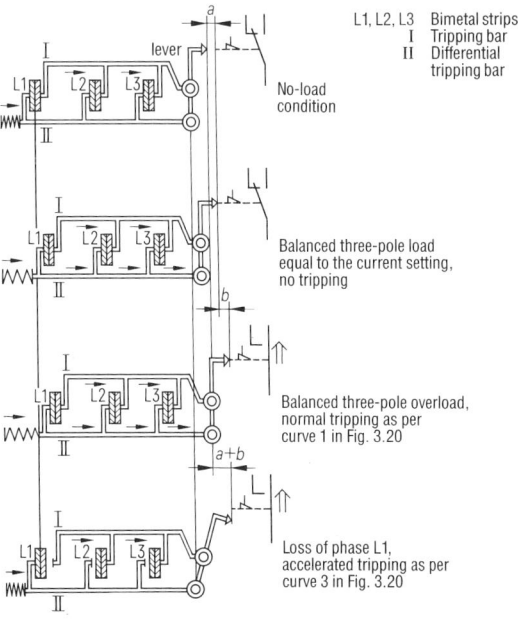

Figure 3.21
Operating principle of a thermally delayed overload relay with phase-loss sensitivity

Table 3.15
Derating factors for thermally delayed overload relays at ambient temperatures above 55 °C.

Ambient temperature °C	Derating factor for the upper setting value
+55	1.0
+60	0.94
+65	0.88
+70	0.82
+75	0.74
+80	0.67

In the event of a 3-pole symmetrical overload, the bimetal strips bend to a greater degree, and their ends are displaced by the distance b. Again, the vertical position of the hinged lever is maintained owing to the similar displacement of all three bimetal strips. The relay trips in accordance with the tripping characteristic for 3-pole symmetrical loading (curve 1 of Fig. 3.20).

If the phase L1, for example, is disconnected during operation then the respective bimetal strip will cool down and its end will return to the cold position. This will cause the tripping bar II and the lower hinged point of the hinged lever to be moved to their original positions as well (lowest illustration in Fig. 3.21). The heated bimetal strips L2 and L3, on the other hand, continue to force the tripping bar I in the direction of the tripping point. Owing to this differential movement of the tripping bars, the hinged lever turns about its lower hinge point, and its end moves towards the tripping point at a greater rate determined by the leverage ratio. Accelerated tripping occurs when the point of the hinged lever has covered the distance $a+b$ (curve 3 of Fig. 3.20).

Electronic overload relays

In the case of electronic overload relays, the actual load or motor current in each phase is measured via current transformers. The secondary current is converted to a proportional voltage, rectified and fed to a microprocessor by means of an analog-to-digital converter. Here the signal is processed in a corresponding programme, and an impulse is given to the output relay in the event of an overload. The operational features and properties of the Siemens 3UB1 electronic overload relay are described below.

Setting of the overload relay

Once the fixing screw has been loosened, the dial on the overload relay may be adjusted to the required value in the setting range. The dial either has an absolute scale in amps, or it is calibrated in percent.

The yellow LED may be used as an adjustment aid. This starts flashing when the instantaneous value of the load current is equal to 100% of the current setting I_r. As soon as the load current increases to 110% of the set value, the yellow LED stops flashing and glows continuously, indicating an overload condition and pending trip. The tripped condition is indicated by a red LED.

Tripping class

By means of a slide switch on the front of the relay, any one of six tripping classes may be selected (viz Class 5, 10, 15, 20, 25, and Class 30). The relays leave the Siemens factory with Class 10 selected.

Manual (Hand) or Automatic reset

The 3UB1 overload relays are delivered in the "Hand Reset" mode. By means of a recessed selector button, the mode may be changed to "Automatic Reset".

For application examples, refer to Figure 3.19, on page 133.

Test and Reset facilities

The correct operation of the overload relay can be checked by pressing the combined test/reset button. If the button is held depressed, an overload condition is simulated, and the relay must trip (red LED on) after a specific time determined by the selected tripping class (e.g. after 6 seconds in Class 10 or after 17 seconds in Class 30). While the button is held depressed, the yellow LED glows and indicates the test operation.

A tripped overload relay can be reset by pressing and immediately releasing the test/reset button. This causes the contacts of the output relay to return to their normal states and the red LED to extinguish. This operation is, however, only possible once the recovery time has expired.

Recovery time

The recovery time is intended to provide a minimum cooling period for the motor after an overload trip. The recovery times are fixed values programmed into the microprocessor, and have been selected according to the setting range of the overload relay to correspond roughly to the size and hence the cooling time of the associated motor. The actual recovery times are given in the relevant Siemens l.v. switchgear catalogue.

If the control supply voltage to the relay is interrupted during the recovery period, then the full period will have to expire again once the relay is reconnected, before it will reset.

Tripping characteristic

In the Class 10 setting, and starting from the cold state, tripping takes place within 8 seconds for a symmetrical three-phase load current of 7.2 times the current setting. Under the same conditions, but in the

Class 30 setting, tripping occurs within 24 seconds. Refer to Section 3.4.2.1 on page 124 for definitions of the tripping class and tripping-time limits to IEC 947-4-1.

In the minimum tripping current region, tripping takes place at between 105 and 115% of the current setting (Fig. 3.22). Under overload conditions, the deviation from the predicted tripping times is max. ±5%. Thus, the 3UB1 electronic relays also meet the demands of UL and CSA.

Behaviour at operating temperature

The overload relay registers the value and duration of the load current. Depending on the load prior to the overload condition, the overload tripping times are reduced below those which apply for loading from the cold state.

For example, if the load current prior to the onset of the overload was 1.0 times the set current I_r, for a given period of time, then the tripping times are reduced to 30% of the cold state values.

Asymmetrical loading

In the event of even a relatively small degree of asymmetry in the supply voltage, three-phase induction motors react by drawing large asymmetrical or unbalanced currents.

The losses which are caused by the resulting opposing magnetic field in the motor, lead to a increased temperature rise in stator and rotor windings.

Electronic relays can sense this unbalanced load current, and will switch to the accelerated tripping characteristic for two pole loading (Fig. 3.22) if the degree of current unbalance should exceed 40%. In this way, an overloading of the motor in the event of a single phasing condition (loss of one phase), is also prevented.

Single-pole loading

The electronic overload relays are designed for the protection of three-phase induction motors. If these relays are to be used for the protection of single-phase motors, it must be insured that the microprocessor receives the same signals as for three pole loading. For this reason, the current transformer of the overload relay must be connected in a special way in accordance with the operating instructions.

Behaviour of the auxiliary contacts in the event of control supply voltage loss

A recessed selector button on the front of the 3UB1 electronic overload relay allows the choice of two different reactions to control supply voltage loss. These are:

▷ Position "M" The overload relay will adopt the tripped state when the supply voltage is removed, and will return to the state it had before supply loss when the voltage is re-applied (M = monostable). A short interruption of the supply voltage, up to 200 ms, will not cause the auxiliary contacts to change state. This operating mode is selected for operation in installations in which the control voltage is not separately monitored.

▷ Position "B" The overload relay will maintain its state of "tripped" or "not tripped" when the supply voltage is removed (B = bistable). The relay will only trip in the event of an overload and if the control voltage is present. This operating mode is selected for operation in installations in which the control voltage is separately monitored.

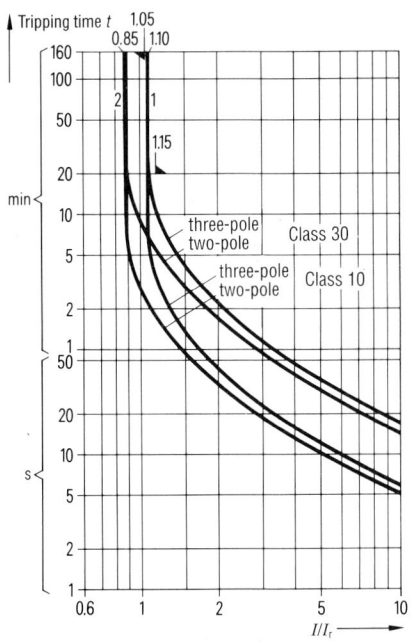

1 Three-pole balanced loading from the cold state
2 Two-pole unbalanced loading from the cold state
I_r Current setting

Figure 3.22
Tripping characteristic of an electronic overload release in Class 10 and Class 30 mode, respectively

3 Selection criteria for low-voltage switchgear in main circuits

Temperature range

The 3UB1 overload relays can be used at ambient temperatures from −25 °C to +55 °C. If ambient temperatures outside this range are expected, then appropriate separate heating or forced cooling must be employed.

3.4.3.5 Thermistor motor protection devices

In contrast to overload relays which respond to the load current and cause a switch-off if an excessively high current flows for a prolonged period of time, thermistor motor protection devices sense the actual temperature in the protected machine (e.g. in the motor windings) by means of thermal detectors (thermistors).

In general, PTC thermistors are used as thermal detectors, although NTC thermistors and resistor elements are also used in special cases.

Thermistor motor protection devices with PTC thermistors

PTC thermistors are semi-conductor elements with a very high positive resistance/temperature coefficient. For a temperature change of only 10 K around their rated operating temperature, their resistance increases more than ten-fold (Fig. 3.23). Owing to this response at a particular temperature (TNF), these detectors are generally used in the protection of series manufactured standard motors, in which the permissible temperature limits and thermal characteristics are known in advance. Also refer to Section 3.4.2.4 on page 125.

Usually, three PTC thermistors in series (i.e. one for each phase) are built into the overhang of the stator winding at the coolant air exhaust end of the motor. The thermistors are selected to have a rated operating temperature (TNF) to correspond with the insulation class, type and construction of the motor. If the temperature at one or more of the thermistors exceeds this value, then a switching signal will be generated in the separate tripping unit as a result of the sharp increase in the detector circuit resistance. This switching signal may be used to switch off the motor or to initiate an alarm. Should both tripping and prior alarm signal be required, then an additional detector circuit with the corresponding rated operating temperature must be installed in the motor winding. Either an additional tripping unit or a unit which can evaluate two detector circuits must then be used.

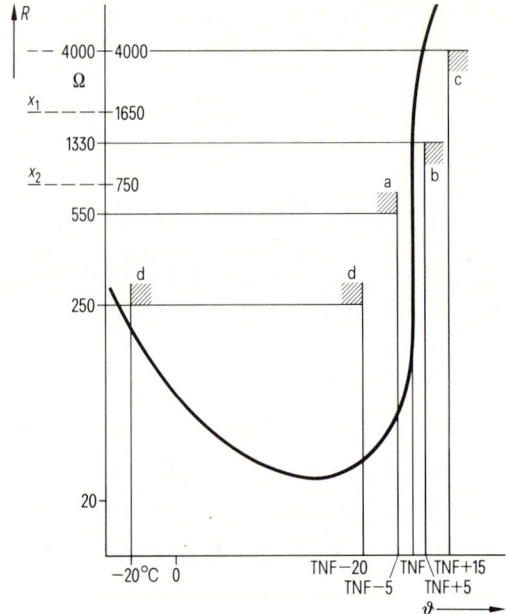

R	Resistance of detector circuit
ϑ	Temperature
x_1	Tripping range of the tripping units
x_2	Resetting range of the tripping units
TNF	Rated operating temperature (tolerance limits in °C)

Figure 3.23
Resistance/temperature characteristic of a Type A PTC thermistor to DIN VDE 0660 part 303

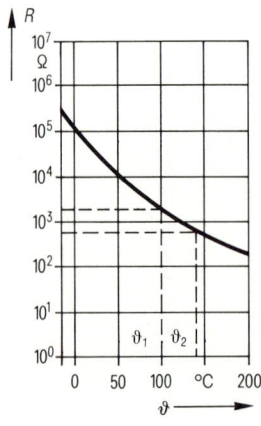

ϑ_1, ϑ_2 Adjustable tripping values of the tripping unit
R Resistance of the detector circuit

Figure 3.24
Resistance/temperature characteristic of an NTC thermistor

Motors incorporating PTC thermistors may be selected from the Siemens M 11 catalogue.

Several different versions of tripping units are available for motor protection with PTC thermistors:

– for one detector circuit, with automatic reset
– for one detector circuit, with manual reset and test function
– for one detector circuit, with manual/automatic/remote reset, test function and detection of short-circuit in the detector circuit
– for two detector circuits (warning/alarm and tripping) with manual/automatic/remote reset and test function
– for six detector circuits, with manual reset, test function and optical indication of detector circuit causing trip (for multiple motor protection).

Furthermore, one can choose between versions which either maintain their switching state after loss of supply voltage or which trip upon supply voltage loss (and reset automatically once supply voltage is re-applied).

For technical and selection data on thermistor protection units, refer to the relevant Siemens l.v. switchgear catalogue.

For circuits using thermistor protection units and PTC thermistors, see Section 8.4, page 462.

Thermistor motor protection devices with NTC thermistors

NTC thermistors are semi-conductor elements with a negative resistance/temperature coefficient. They are generally used for the protection of motors which are manufactured on a one-off basis.

In contrast to that of metal the resistance of an NTC thermistor decreases with increase in temperature (Fig. 3.42). Thermistor motor protection with NTC detectors is mainly used when the temperature characteristics of the motor are not known prior to manufacture and testing, and the optimal response temperature of the protection system can only be determined once the machine is complete.

A tripping unit is offered on which two different operating temperatures, i.e. one for warning or alarm and one for tripping, may be set.

Technical data on the NTC detectors and their associated tripping units as well as information on the setting of the operating temperatures may be found in the Siemens l.v. switchgear catalogue. Basic circuit diagrams for thermistor motor protection using NTC detectors are given in Section 8.4.2, see page 468.

3.4.3.6 Instantaneous electromagnetic overcurrent relays

The 3UG1 instantaneous electromagnetic overcurrent relays protect cables, slip-ring motors and electrical devices from overload and short-circuit current. They are generally used in conjunction with circuit-breakers. Their rated operating current ranges from 0.4 A to 1300 A. If the set operating value is higher than the continuous operating current, then an additional overload protection is required.

3.4.4 Switchgear combinations

In principle, every switching device must be provided with short-circuit protection if its switching capacity is lower than the prospective short-circuit current at its point of installation.

One can distinguish between two possibilities:

1. The short-circuit protection device itself has a breaking capacity which is equal to or higher than the prospective short-circuit current at the point of installation.

2. The combination of the switching device and the short-circuit protection device fulfill the necessary requirements relating to the breaking capacity. This is the most common solution since the overall switchgear costs are generally lower.

3.4.4.1 Switchgear combinations with fuses

Switchgear combination "Circuit-breaker with fuses"

If the prospective short-circuit current at a point in the network exceeds the rated breaking capacity of a circuit-breaker to be installed at this point, then back-up fuses may be connected on the line side of the breaker. To ensure that the contact system of the circuit-breaker is not damaged during a short-circuit, the properties of both short-circuit protection devices i.e. fuses and circuit-breaker, should be compared carefully. The maximum permissible ratings of back-up fuses are given in the relevant Siemens catalogues.

The fuses must operate and interrupt the short-circuit current at a level which does not exceed the rated breaking capacity of the circuit-breaker (Fig. 3.25).

A specific protection range is assigned to each of the elements in the switchgear combination. Protection

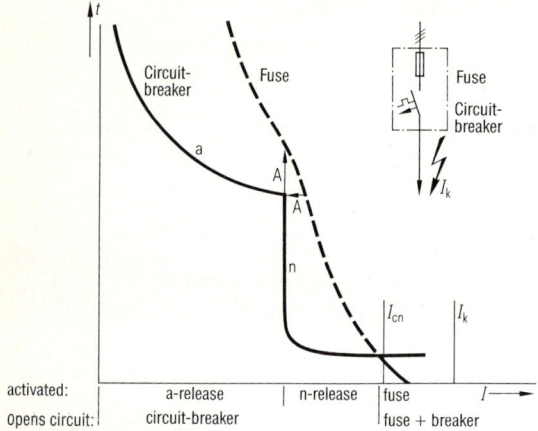

a   Current-dependent delayed overload release
n   Instantaneous electromagnetic short-circuit release
I_{cn} Rated short-circuit breaking capacity
I_k Prospective sustained short-circuit current at the point of installation
A   Distance between characteristic curves

Figure 3.25
Switchgear combination "circuit-breaker with back-up fuses"

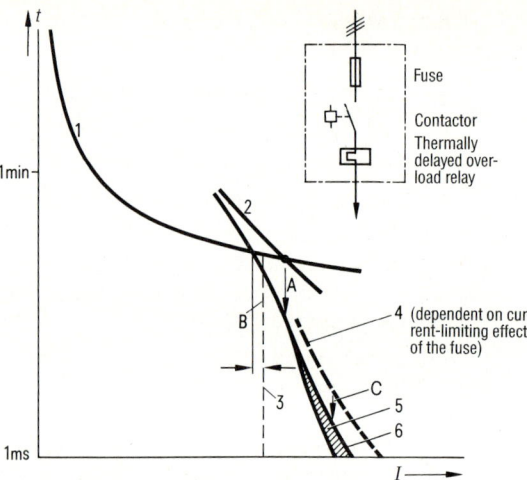

1   Tripping characteristic of the current-dependent thermally delayed overload relay
2   Destruction curve of the thermal overload relay
3   Breaking capacity of the contactor
4   Characteristic curve of the contactor for easily separable welding of the contacts
5   Pre-arcing time characteristic of the fuse, operation class aM
6   Total clearing-time characteristic of the aM fuse
A, B, C   Safety margin for reliable short-circuit protection

Figure 3.26
The tripping characteristic curves of the switchgear combination "contactor, overload relay and fuses"

against overload currents is provided by the current-dependent delayed overload release (a-release), where-as short-circuit currents up to approximately the rated breaking capacity of the circuit-breaker are sensed by its instantaneous electromagnetic short-circuit release (n-release).

All overload and short-circuit currents up to approximately the rated breaking capacity of the circuit-breaker are interrupted by the breaker alone, thereby offering the specific advantages of interrupting the circuit in all poles as well as the possibility to re-close the circuit immediately after the opening operation.

Only short-circuits higher than the breaking capacity of the circuit-breaker are the sole responsibility of the fuses. Even in these cases, however, the circuit-breaker guarantees an opening of all the poles, since its n-release will be operated by the let-through current. This also applies when only one of the three fuses rupture owing to e.g. a single pole earth fault (also refer to page 145).

**Switchgear combination
"Contactor, overload relay and fuses"**

Contactors are mainly used for the switching of motors. The overload protection for the motor, the motor feeder cables and the contactor is provided by the overload relay. The fuses provide the short-circuit protection for the switching devices as well as the feeder cables. Care must be taken to ensure the correct co-ordination between the assigned operating ranges and the respective protection properties of all the components in the switchgear combination (Fig. 3.26).

▷ The time-current characteristics of the overload relay and the fuses must be such as to allow sufficient time for the motor to run up to speed.

▷ The fuses must protect the overload relay from destruction by overload currents which exceed approximately 10 times its rated current.

▷ The fuses must interrupt currents which are too large to be switched off by the contactor (currents greater than approx. 10 times the rated operating current of the contactor).

▷ The fuses must protect the contactor in the event of a short-circuit. Generally, this protection can be regarded in two ways: either it is specified that no contact welding may occur, or light easily separable welding between contact surfaces may be permitted.

With regard to the latter approach, IEC 947-4-1, 05.90, paragraph 8.3 applies. In this new specification the permissible damage of the devices, resulting from a short-circuit, is classified in the following modified form:

Coordination Type "1"

No danger may be caused to persons on neighbouring parts of the installation.

The destruction of the contactor and the overload relay is permissible. If necessary, the contactor and/or the overload relay must be repaired or replaced. The equipment need not be suitable for further service after the short-circuit condition.

Coordination Type "2"

No danger may be caused to persons on neighbouring parts of the installation.

The overload relay may not suffer any damage. Welding of the contacts in the contactor is only permissible if such welding can be broken easily and the contacts separated. A replacement of the contactor must not be necessary. The combination must be suitable for further service after the short-circuit condition.

These stipulations are also contained in DIN VDE 0660 part 102 (07.92).

The new IEC stipulations replace the previously valid degrees of damage, types "a" to "c".[1]

3.4.4.2 Switchgear combinations without fuses

**Back-up protection
(Circuit-breakers in cascade)**

If two circuit-breakers with identical opening times and arc quenching techniques are connected in series in the current path, then they will open simultaneously in the event of a short-circuit at point B on the

---
[1] Type a Destruction and replacement of parts or a complete switching device.
  Type b Contact welding and a permanent change in the tripping characteristic of the overload relay.
  Type c Contact welding and no permanent change in the tripping characteristic of the overload relay.

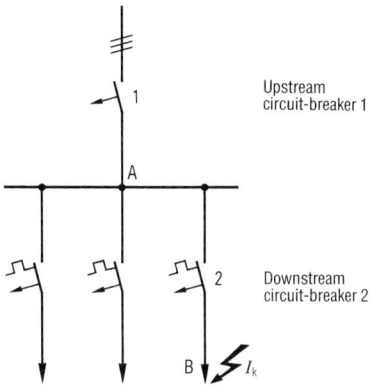

a) Illustration of the back-up protection circuit (cascade connection)

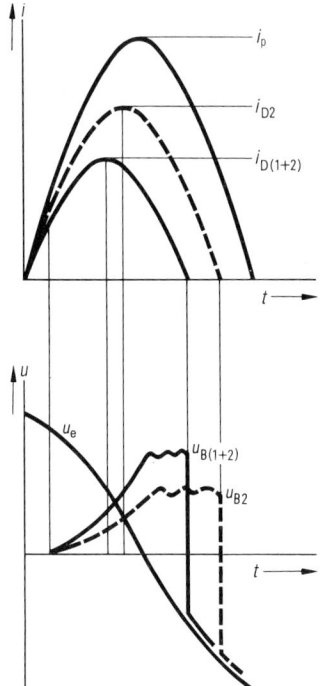

b) Arc voltages in the cascade during short-circuit interruption

U_e Driving voltage (rated operational voltage)
u_{B2} Arc voltage of downstream breaker 2
i_{D2} Let-through current of downstream breaker 2
$u_{B(1+2)}$ Sum of the arc voltages of the upstream breaker 1 and the downstream breaker 2
$i_{D(1+2)}$ Actual resultant let-through current

Figure 3.27
Illustration of a cascade connection and its mode of operation

load side of the downstream breaker (Fig. 3.27a), provided that the short-circuit current is greater than the discrimination (or selectivity) limit of the combination. The short-circuit current is detected by two series connected arc quenching devices; each contributing to the final quenching of the arc and interruption of the fault current. As a result, it is permissible for the downstream circuit-breaker to have a breaking capacity which is lower than the prospective short-circuit current at its point of installation.

Figure 3.27 illustrates the principle of a cascade connection and its method of operation. The rated current of the upstream circuit-breaker (1) is determined by the required operating current.

This distribution circuit-breaker (1) may also act as the main switch for several feeders. The operating current of its instantaneous overcurrent release (n) is set to a high value; if possible, to the breaking capacity of the downstream circuit-breaker. The feeder circuit-breaker (2) is responsible for the overload protection and also interrupts smaller short-circuit currents, resulting from e.g. phase-to-frame faults, insulation breakdown or remote faults at the end of long conductor or cable runs, by itself.

Only in the event of a greater short-circuit current, as would result from a full short-circuit close to the feeder circuit-breaker (2), would the upstream breaker (1) be called upon to open as well (also refer to Section 3.4.6.2, page 172).

Switchgear combination "contactor with circuit-breaker"

The circuit-breaker (with a- and n-releases) is responsible for the overload and short-circuit protection, whereas the contactor performs the switching function. In this case the requirements which must be met by the circuit-breaker (Fig. 3.28) are the same as those which apply to the fuses in the combination "contactor, overload relay and fuses (see page 140).

Switchgear combination "contactor, overload relay and circuit-breaker" (breaker with adjustable short-circuit n-release)

The overload protection is provided by the overload relay in conjunction with the contactor. Short-circuit protection is provided by the circuit-breaker.

The operating current of the instantaneous overcurrent release "n" is set to the lowest possible value which the starting conditions of the motor will allow,

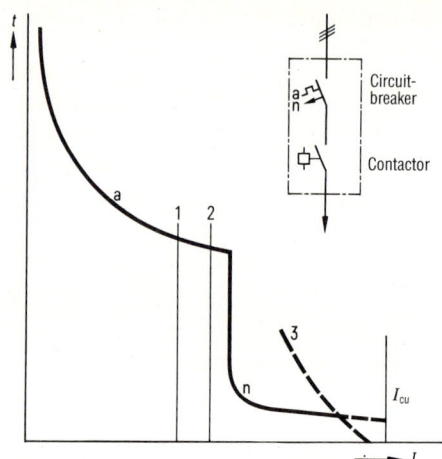

1 Breaking capacity of the contactor
2 Making capacity of the contactor
3 Characteristic curve of the contactor for easily separable welding of the contacts
a Characteristic curve of the current-dependent delayed overload release
n Characteristic curve of the instantaneous electromagnetic short-circuit release
I_{cu} Rated ultimate short-circuit breaking capacity

Figure 3.28
Switchgear combination "contactor with circuit-breaker"

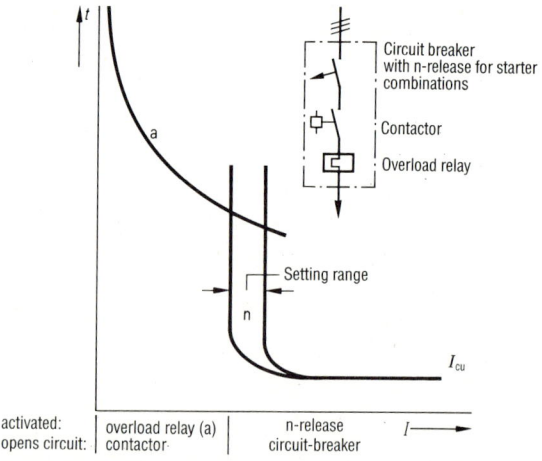

a Characteristic curve of the current-dependent thermally delayed overload relay
n Characteristic curve of the adjustable instantaneous short-circuit release

Figure 3.29
Switchgear combination "contactor, overload relay and circuit-breaker" (breaker with adjustable short-circuit n-release)

whilst also ensuring that even small short-circuit currents are detected and interrupted instantaneously (Fig. 3.29). This switchgear combination has the advantage that one can ascertain whether a short-circuit or an overload condition caused the motor to be switched off.

In the event of an overload, the contactor is opened whereas a short-circuit causes the circuit-breaker to trip. In addition, the starter circuit-breaker offers the advantages of 3 pole disconnection and the ability to reclose the circuit immediately after the short-circuit interruption. The switchgear combination incorporating a starter circuit-breaker (MCP = Motor Circuit Protector) is becoming increasingly popular as the usage of fuses in industrial motor circuits decreases.

For the short-circuit protection of the contactor, the same conditions apply as for the switchgear combination "contactor with circuit-breaker".

3.4.4.3 Switchgear combinations with thermistor motor protection devices

In cases where the motor current is not directly related to the winding temperature, overload protection by means of overload relays or releases may not be sufficient. This condition is caused by

– high switching frequency,
– irregular short time duty,
– restricted cooling,
– increased ambient temperature.

In these cases switchgear combinations comprising thermistor motor protection devices are used. De-

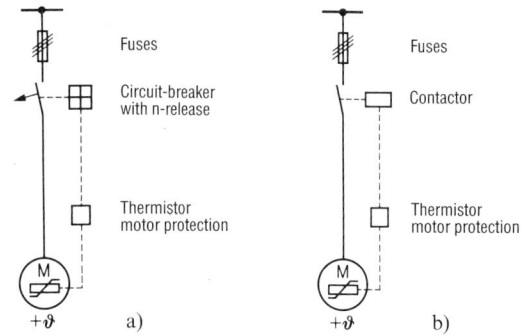

Figure 3.30
Switchgear combinations with thermistor motor protection devices and *without* additional overload protection (single-line diagrams)

pending on the overall design concept of the installation, the combinations may be with or without fuses in terms of Sections 3.4.4.1 and 3.4 4.2 above.

The quality of the protection which can be achieved depends on whether the motor to be protected is "stator critical" or "rotor critical" (see Section 3.4.5.1, page 150). The operating temperature of the thermistors, the heat transfer time constant and the position of the temperature detectors in the motor winding also contribute significantly to the quality of protection. As a rule, these factors are determined by the motor manufacturer.

"Stator critical" motors can be protected sufficiently from overload and overtemperature by means of thermistor motor protection device alone, without

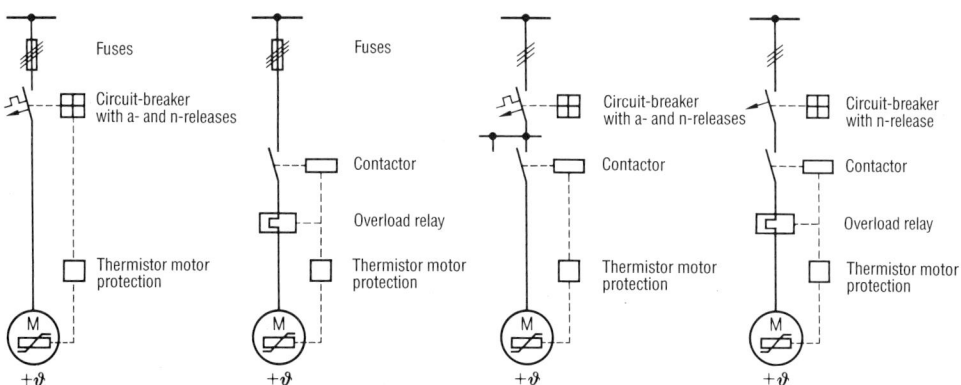

Figure 3.31
Switchgear combinations with a thermistor motor protection device, and *with* additional overload protection (single-line diagrams)

the need for an overload relay. The short-circuit and overload protection of the feeder cables must be ensured by fuses and a circuit-breaker (Fig. 3.30a) or by fuses alone (Fig. 3.30b).

"Rotor critical" motors can only be protected adequate, particularly under locked rotor conditions, if an additional overload relay or release is provided. The overload relay or release thereby also provides the overload protection for the feeder cables (Fig. 3.31).

3.4.4.4 Protection properties of switchgear combinations

Comparison between protection properties of fuses and circuit-breakers

In the time-current diagram, Figure 3.32, the pre-arcing time characteristic (a) of a fuse of operation class gL and the tripping characteristic of a circuit-breaker (b) are shown. The current rating of the fuse and the set operating current of the current-dependent delayed overload release of the circuit-breaker are the same.

The minimum rupturing current of the fuse lies e.g. between 1.3 and 1.6 times the rated current, whereas the minimum operating current of the overload release lies between 1.05 and 1.2 times the set current. In the case of an adjustable overload release, the current setting and therefore the minimum operating current can be matched more closely to the continuous load capabilities of the protected object than is the case with fuses, where the discrete current ratings allow only approximate matching with the load. The minimum rupturing current of the fuses provide sufficient overload protection for the cables and contactors. To permit the normal starting of the motor however, fuses with a pre-arcing characteristic shown as a' must be selected.

In the overload range the pre-arcing current characteristic of the fuse is steeper than the tripping characteristic of the overload release. This is beneficial for the overload protection of cables and conductors. For the overload protection of motors however, a "slower" tripping characteristic is required.

In the short-circuit current range the circuit-breaker will respond faster than the fuse at currents which are marginally greater than the set operating current of the instantaneous short-circuit release. Larger currents are interrupted faster by the fuse. In the case

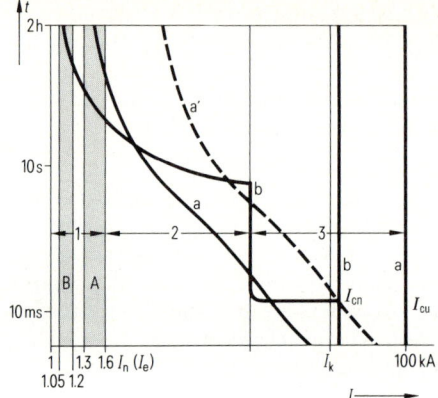

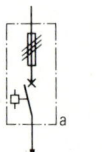 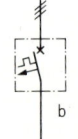

1 Minimum rupturing-/tripping-current range
2 Overload current range
3 Short-circuit current range
A Test range for fuse current rating
B Test range for minimum tripping current of the circuit-breaker
I_{cn} Rated short-circuit breaking capacity
I_{cu} Rated ultimate short-circuit breaking capacity

Figure 3.32
Characteristic curves and breaking capacities of a fuse (a) and a circuit-breaker (b) with a- and n-releases (an-release)

of very high short-circuit current values, the fuse limits the peak let-through current.

This results in the extremely high breaking capacity of more than 100 kA at an operational a.c. voltage of 690 V. In comparison, the rated breaking capacity I_{cn} of circuit-breakers depends on a number of factors such as the rated operating voltage U_e and the construction of the breaker itself.

Table 3.16 provides a comparison between the protective properties of fuses and circuit-breakers. Section 3.4.5, as well as Tables 3.17 and 3.18 (pages 147, 148) provide a comparison between the protective properties of switchgear combinations comprising several low-voltage devices. Information related to discrimination (selectivity) between low-voltage protection devices is given in Section 3.4.6 (page 163).

Table 3.16 Comparison of the protective properties of fuses and circuit-breakers

Property	Fuse	Circuit-breaker
Breaking capacity with alternating current	>100 kA, 690 V	$f(I_n, U, \text{construction}^{1)})$
Current-limiting	$f(I_n, I_k)$	$f(I_n, I_k, U, \text{construction}^{1)})$
Additional arcing space	none	$f(I_n, U, I_k, \text{construction}^{1)})$
Outwardly recognizable indication of serviceability	yes	no
Operational safety	at extra cost [2]	yes
Remote switching	no	yes
Automatic disconnection of all poles	at extra cost [3]	yes
Possibility of remote indication	at extra cost [4]	yes
Possibility of interlocking	no	yes
Reclosing ability after		
Overload interruption	no	yes
Short-circuit interruption	no	f (condition)
Downtime after interruption	yes	f (condition)
Maintenance costs	no	f (no. of operations and condition)
Discrimination	without extra cost	at extra cost
Replacement	yes [5]	same make/model
Short-circuit protection		
Cable and conductors	very good	good
Motor	very good	good
Overload protection		
Cables and conductors	adequate	good
Motor	not possible	good

[1] Construction can refer to: quenching technique, short-circuit proof owing to internal resistance, mechanical design
[2] For example, with the aid of enclosed fuse switch disconnectors with rapid closing feature
[3] By means of fuse monitoring and an associated circuit-breaker
[4] By means of fuse monitoring
[5] Since the sizes are standardized

Comparison between the required protective properties of back-up fuses and miniature circuit-breakers for line protection

In Germany, the pre-conditions for connection to the electricity supply, TAB[1], require that the switching capacity of the miniature circuit-breakers be 6 kA. This switching capacity ensures that virtually all faults are cleared safely by the miniature circuit-breaker alone.

In the event of a higher short-circuit current, the back-up fuse and miniature circuit-breaker operate together. As an example, Figure 3.33a shows the minimum pre-arcing characteristic of the max. permissible 100 amp fuse (acc. DIN VDE 0636), and the max. permissible let-through $\int i^2 dt$-characteristic [2] of a miniature circuit-breaker with current-limiting class 3 (to DIN VDE 0641). For short-circuit currents which lie below the intersection between the two characteristic curves, the conditions for discrimination are fulfilled, i.e. the miniature circuit-breaker trips and interrupts the short-circuit current, while the fuse remains intact.

At higher values of short-circuit current (higher than the value at the point of intersection) the back-up fuse and the miniature circuit-breaker interrupt the short-circuit current together, acting as a single overcurrent protection unit. The let-through $\int i^2 dt$ characteristic curve of the combination (Fig. 3.33b) approaches the total let-through $\int i^2 dt$ characteristic curve of the back-up fuse asymptotically with increase in short-circuit current.

[1] "Technische Anschlußbedingungen der EVU"
[2] The let-through $\int i^2 dt$ is the Joule energy value with which the current path is thermally loaded during the short-circuit.

3 Selection criteria for low-voltage switchgear in main circuits

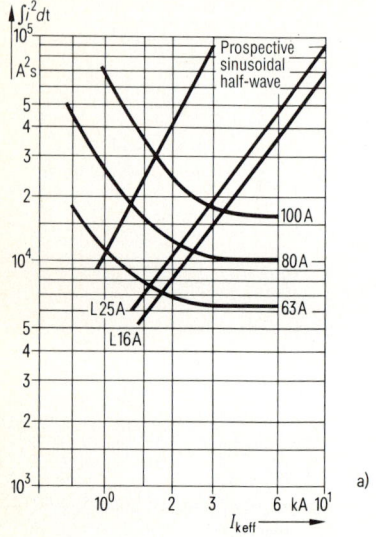
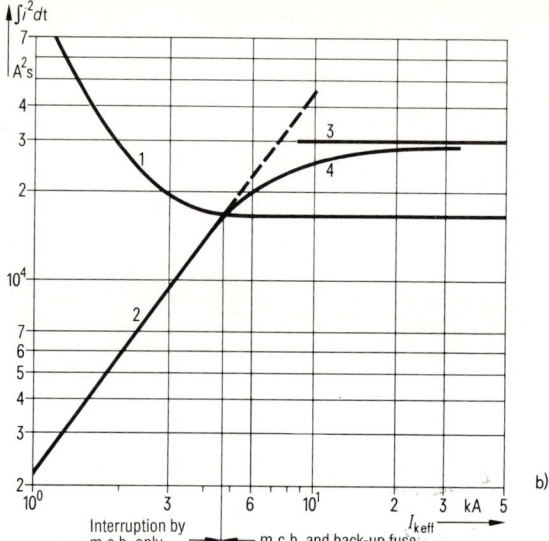

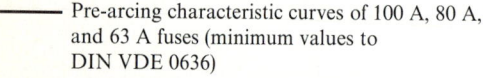 Pre-arcing characteristic curves of 100 A, 80 A, and 63 A fuses (minimum values to DIN VDE 0636)

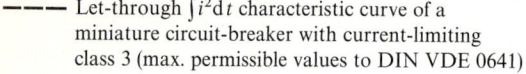 Let-through $\int i^2 dt$ characteristic curve of a miniature circuit-breaker with current-limiting class 3 (max. permissible values to DIN VDE 0641)

1. Pre-arcing characteristic of the 100 A fuse
2. Let-through $\int i^2 dt$ characteristic of a miniature circuit-breaker
3. Total let-through $\int i^2 dt$
4. Let-through $\int i^2 dt$ characteristic of the overcurrent protection combination miniature circuit-breaker with back-up fuse

Figure 3.33
$\int i^2 dt$ Characteristic curves (a) and let-through-$\int i^2 dt$ characteristic curves (b) of the overcurrent protection combination "miniature circuit-breaker 6000/3 with back-up fuse"

Advantage of a higher switching capacity

If one considers the combined characteristic curve "100 A back-up fuse and miniature circuit-breaker" in Figure 3.33b, it is clear, that as from approx. 5 kA, the $\int i^2 dt$ value only increases by a small amount. The additional loading of the miniature circuit-breaker at higher short-circuit current is minimal. It is protected by the fuse, if its own switching capacity is 6 kA. This means, that a higher switching capacity is not required of the m.c.b., since it would not bring additional benefits.

Electrical contractors and operators of consumer installations connected to the public network are not always fully aware of the relationship between the quality of the short-circuit protection and breaking capacity. A higher breaking capacity does not necessarily mean an improved quality of the short-circuit protection, as the above example illustrates.

3.4.4.5 Comparison between the protection properties of switchgear combinations

Table 3.17 shows a comparison between the protection properties of fuses, circuit-breakers, overload relays and thermistor motor protection devices, as well as that of their combinations.

Table 3.17 Comparison between the protection properties of switchgear combinations (single-line diagrams)

Protected object and switching frequency	Protection systems						
	Fuses / Circuit-breaker / Contactor / Overload relay / Thermistor motor protection						
Overload protection – Cable – Motor (stator critical) – Motor (rotor critical)		+ + + +[1] + +[1]	+ + + + + +	+ + + +	+ + + +	+ + + + + +	+ + + + + +
Short-circuit protection – Cable – Motor		+ + + +	+ + + +	+ + + +	+ + + +	+ + + +	+ + + +
Switching frequency		–	+ +	–	+ +	–	+ +

Protected object and switching frequency	Protection systems						
	Circuit breaker / Contactor / Overload relay / Thermistor motor protection						
Overload protection – Cable – Motor (stator critical) – Motor (rotor critical)		+ + + +[1] + +[1]	+ + + + + +	+ + + + + +	+ + + + + +	+ + + +[1] + +[1]	+ + + + +
Short-circuit protection – Cable – Motor		+ +	+ +	+ +	+ +	+ +	+ +
Switching frequency		+	+	+	+	–	–

+ + very good + good – low

[1] protection with minor limitations in the case of phase failure

3 Selection criteria for low-voltage switchgear in main circuits

Table 3.18 Power distribution circuit with fuses *and* circuit-breakers

No.	Type of circuit-breaker	Order No.	Rated short-circuit breaking capacity (I_{cn})	Type of overload release: a Adjustable	Fixed	z Adjustable	Fixed	n Adjustable	Back-up fuse $I_{cn} >$ 100 kA	Tripping characteristic
Incomer circuit-breaker										
1	Circuit-breaker for discriminative protection	3WN	$\geq I_{k1}$	×	–	×	–	×	×	
Distribution circuit-breaker										
2	Fuses and circuit-breaker for distribution protection	3NA 3VF	$\geq I_{k2}$ $\leq I_{k2}$	– –	– ×	– –	– –	– ×	× –	
Load (consumer) circuit-breaker										
3	Fuses and circuit-breaker for motor protection	3NA 3VU	$\geq I_{k3}$ $\leq I_{k3}$	– ×	– –	– –	– ×	– –	× –	
4	Fuses and d.o.l. starter	3ND 3NA 3TW	$\geq I_{k3}$ $\geq I_{k3}$ $\leq I_{k3}$	– – ×	– – –	– – –	– – –	– – –	× × –	

[1]) I_{cn} Rated short-circuit breaking capacity

3.4.4.6 Selection of circuit-breakers for power distribution systems with or without fuses

Power distribution and control circuits can be built with or without the use of fuses.

Distribution systems with fuses

The most common classical designs of distribution systems with fuses (Table 3.18), actually use a combination of circuit-breakers and fuses. Here, each protection device has its own specific task to perform.

The incoming circuit-breaker is responsible for the overload and discriminative (or selective) short-circuit protection of the transformer and the distribution system. The Siemens 3WN circuit-breakers are ideal for this task.

The switchgear combination comprising fuses and distribution protection circuit-breaker is responsible for the overload and short-circuit protection of the cables and conductors leading to the sub-distribution levels. The switchgear combinations comprising fuses and motor protection circuit-breakers as well as those comprising fuses, contactor and overload relay provide the overload and short-circuit protection for the motor feeder cables and the motors themselves.

The respective protection ranges of the fuses and the switchgear devices are described in Section 3.4.3 on page 125 et seq.

Power distribution circuits with or without fuses

Table 3.19 Power distribution system with circuit-breakers but *without* fuses

No.	Type of circuit-breaker	Order No.	Rated short-circuit breaking capacity (I_{cn})	Type of overload release: a Adjustable	Fixed	z Adjustable	Fixed	n Adjustable	Tripping characteristic
Incomer circuit-breaker									
1	Circuit-breaker for discriminative protection	3WN	$> I_{k1}$	×	–	×	–	×	
Distribution circuit-breaker									
2	Circuit-breaker for distribution protection	3VF	$> I_{k2}$	–	×	–	–	×	
3	Circuit-breaker for discriminative protection	3WN	$> I_{k2}$	×	–	×	–	–	
Load (consumer) circuit-breaker									
4	Circuit-breaker for motor protection	3VU	$> I_{k3}$	×	–	–	×	–	
5	Circuit-breaker and d.o.l. starter	3VU 3TW	$> I_{k3}$ –	× ×	– –	– –	× –	– –	

[1] I_{cn} Rated ultimate short-circuit breaking capacity

Power distribution systems without fuses

In distribution systems without fuses (Table 3.19), distribution protection circuit-breakers are used for the short-circuit and overload protection of the feeder circuits. Motor protection circuit-breakers alone, or in combination with contactors as starter combinations, are used for the load switching.

The protection ranges of the switchgear combinations comprising circuit-breaker, contactor and overload relay are described in Section 3.4.4.2, on page 141 et seq.

For technical data on circuit-breakers, please also refer to the relevant Siemens catalogue for low-voltage switchgear.[2]

[2] Selection Tables for the devices in switchgear combinations "circuit-breaker, contactor and overload relay" for use in motor feeder circuits without fuses under normal, heavy and star-delta starting conditions, are given in the Siemens booklets "Motor feeder circuits without fuses" (Order No. A19100-E732-260-V1, German ed.) and "Overload and short-circuit protection in low voltage installations" (Order No. A19100-E732-A326, German ed.).

3.4.5 Protection of plant components

3.4.5.1 Protection of three-phase induction motors

The durability of the materials used for insulation depends on the thermal, electrical and mechanical loading to which they are exposed, on the presence of harmful environmental influences (e.g. humidity), and the detrimental effects of chemical substances/contamination. Under normal operating conditions, the insulation materials in motors achieve a satisfactory service life, provided that the temperature limit associated with the particular insulation class to DIN VDE 0530 is not exceeded.

Causes of thermal overload

One or more of the following conditions can cause a motor to become thermally overloaded:

▷ Increased losses owing to the type of operation, i.e. a load torque which is too high in continuous operation, an on-load factor in intermittent duty which is too high, run-up and/or braking periods which are too long owing to an excessive moment of inertia, or a switching frequency which is too high;

▷ Blocking of the motor during operation;

▷ Connection and switching faults;

▷ Increased losses owing to the nature of the supply network, i.e.

excessive deviations of the network frequency or voltage from the rated values, or

loss of a supply phase (known as phase failure or single phasing);

▷ Insufficient cooling, i.e.

coolant temperature too high,

altitude of installation site too high (lower air density at altitudes over 1000 m above sea level) and

restriction of the coolant flow.

Scope of protection of offered by motor protection equipment

Table 3.20 shows the scope of protection offered by each of the motor protection techniques dealt with in this handbook. As can be seen, a circuit-breaker or the switchgear combination "contactor, overload relay and fuses" offers sufficient overload protection under normal operating conditions. Under abnormal operating conditions, such as excessively long run-up and braking periods, excessive switching frequency or single-phase operation, stator critical motors can be adequately protected using thermistor motor protection devices. In addition, the thermistor motor protection saves the motor windings from overheating even if this is not caused by overcurrent, e.g. if it is due to a restricted coolant flow.

In the case of rotor critical motors, an overload relay or release is required in addition to the thermistor motor protection (also refer to Section 3.4.4.3, on page 143).

Protection of three-phase induction motors with power factor correction capacitors

For power factor correction of the motor, capacitors are often connected directly to the motor terminals and are switched in and out of circuit by the motor contactor (Fig. 3.34). In this way, predominantly active current I_w is drawn from the network. This current flows through the switching device and overload relay or release. These must therefore be set to the active current value.

The current setting I_r can be determined with sufficient accuracy by the equation:

$$I_r = \frac{I_n \cdot \cos\varphi}{0.9}$$

where the values for rated current I_n and $\cos\varphi$ are taken from the rating plate of the motor, and correction to $\cos\varphi = 0.9$ is assumed (typical).

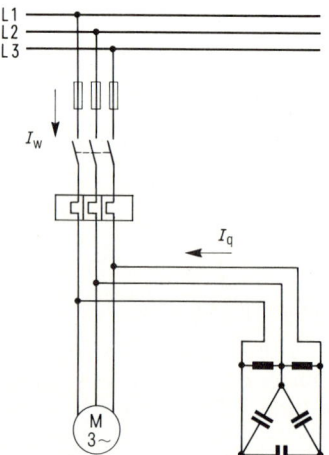

I_w Active current
I_q Reactive current

Figure 3.34
Overload protection of a three-phase induction motor with power factor correction

Protection of three-phase induction motors

Table 3.20 Scope of protection offered by motor protection equipment

Motor protection equipment → Abnormal operating conditions ↓	Contactor and overload relay with phase-loss sensitivity, fuses	Circuit-breaker with delayed current dependent and instantaneous electromagnetic overcurrent releases (an)	Circuit-breaker with instantaneous electromagnetic overcurrent release (n), thermistor motor protection	Contactor, thermistor motor protection, fuses	Circuit-breaker with delayed current-dependent and instantaneous electro-magnetic overcurrent releases (an), thermistor motor protection	Contactor, overload relay, thermistor motor protection, fuses
Increased motor losses during operation						
Continuous overloading	●	●			●	●
Run-up and braking times too long	◐	◐	●	●	●	●
Irregular intermittent duty	○³⁾	○³⁾	●	●	●	●
Switching frequency too high	○³⁾	○³⁾	●	●	●	●
Increased motor losses under fault conditions						
Single phasing and current unbalance	●	●²⁾	●	●	●	●
Voltage and frequency fluctuations	●	●	●	●	●	●
Locked rotor	●	●	●	●	●	●
Starting under locked rotor conditions: stator-critical motors	●¹⁾	●¹⁾	●¹⁾	●¹⁾	●¹⁾	●¹⁾
Starting under locked rotor conditions: rotor-critical motors			○¹⁾	○¹⁾	●¹⁾	●¹⁾
Impaired cooling						
Increased ambient temperature	○	○		●	●	●
Obstructed coolant flow	○	○		●	●	●

○ No protection ◐ Limited protection only ● Full protection

¹⁾ For high-voltage motors only in conjunction with rotor temperature monitoring
²⁾ For overload releases with phase-loss sensitivity
³⁾ Not applicable owing to nuisance early tripping during run-up

Example

Three-phase asynchronous induction motor: 4-pole (1500 r.p.m.); rated output 22 kW at 400 V, 50 Hz; rated current $I_n = 45$ A; power factor $\cos \varphi = 0.83$. The power factor is corrected to $\cos \varphi = 0.9$.

The operating current on the overload protection device must be set to

$$I_r = \frac{45 \text{ A} \cdot 0.83}{0.9} = 41.5 \text{ A}.$$

Protection of pole-changing three-phase induction motors

A pole-changing three-phase motor with a single winding (Dahlander connection), two speeds and one direction of rotation normally has a different power rating for each speed. The operating current drawn by the motor depends on the speed which has been selected. It is therefore necessary to provide a separate overload relay for each of the speed settings.

This also applies for the following pole-changing motors:

▷ Pole-changing three-phase motors with two separate windings, three speeds, and one direction of rotation. In this case a total of three overload relays are necessary; each set to the respective motor currents of the three speeds.

▷ Pole-changing three-phase motors with two separate windings, four speeds, one direction of rotation (both windings in Dahlander connection). An overload relay is required for each speed of rotation; i.e. four are required.

For basic circuit diagrams of pole-changing motors, refer to Section 8.2.5 from page 422 et seq.

Protection of three-phase asynchronous motors under heavy starting conditions

If an overload protection device which has been selected for normal operating conditions trips during run-up of the motor, then heavy starting conditions may be present. This would typically apply if the motor requires a run-up time longer than $t_{an} = 10$ s. Typical examples of such drives are centrifuge pumps or high speed fans with large diameter and inertia, as well as machines which have to be started under load, such as conveyer belts or coal crushers. Either thermally or electronically delayed overload relays, which have been specifically designed for heavy duty starting, are used for the protection of such drive motors.

Naturally, all the other items of switchgear and cables in the circuit must also be selected in terms of the increased demands of heavy duty starting. Please refer to the Siemens l.v. switchgear catalogues for corresponding derating factors and selection data.

Heavy duty starting with thermally delayed overload relays

The bimetal strips of these special relays are heated via saturating current transformers. The current transformers have a low saturation factor and transform the motor current linearly up to approx. 2 times the rated current of the relay (upper setting value). For higher overload currents the core of the current transformer becomes saturated and the secondary current no longer increases with an increase in primary c.t. current. The heating of the bimetal strips is correspondingly lower, resulting in the tripping characteristic shown in Figure 3.35. The position and gradient of this characteristic depends on whether the

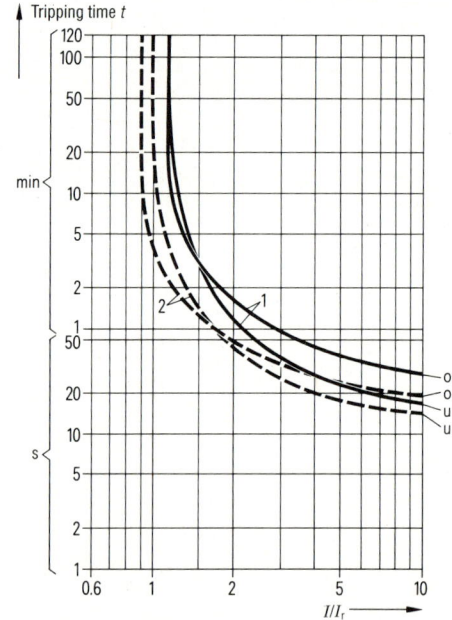

o Upper setting value $= I_r$
u Lower setting value $= 0.65 \cdot I_r$
I_r Current setting
1 three-pole balanced loading from the cold state
2 two-pole balanced loading from the cold state

Figure 3.35
Tripping characteristic of a current-dependent thermally delayed overload relay for heavy duty starting (mean values)

set current I_r corresponds to the lower or upper value of the setting range of the relay.

This difference is due to the fact that the core of the current transformer is saturated to a greater extent by higher currents, i.e. I_r at the upper value of the setting range. As can be seen from Figure 3.35 on page 152, the tripping time for a current $6 \cdot I_r$ is approximately 20 to 30 seconds.

This corresponds to a tripping characteristic known as Class 30. The current transformer of the thermally delayed overload relay for heavy duty starting should be mounted separately from the associated contactor.

Heavy duty starting with electronic overload relays

In the case of these overload relays, the motor current in each phase is monitored via current transformers. For lower current values these current transformers are small enough to be incorporated in the housing of the tripping unit. For larger motor currents an additional three-phase current transformer is usually mounted directly onto the motor contactor.

Current transformers having a high overcurrent factor are used. These ensure that the motor current is transformed linearly up to approximately 10 times the rated current of the overload relay (upper value of setting range). Depending on the run-up (or starting) time of the drive, the required tripping class (Class 15 to Class 30, see Fig. 3.22, page 137) can be selected. The tripping time remains constant throughout the whole setting range.

As can be seen in Figure 3.22 on page 137, the tripping time for 3-pole balanced loading from the cold state at $6 \cdot I_r$, is exactly 29 seconds when Class 30 has been selected on the overload relay.

Extreme heavy duty starting with an electronic overload relay

In the event of a motor run-up time being longer than the tripping time associated with the tripping Class 30, one speaks of extreme heavy duty starting. In such cases an overload relay for heavy duty starting would trip too quickly to allow successful run-up of the motor. Two possible solutions for extreme heavy duty starting using electronic overload relays, are described below. In both cases the relays are switched out of circuit for a specific period of time. This is done by means of a bridging circuit.

Bridging time

The bridging time t_B required for successful run-up of the motor can be calculated from the run-up time t_{an} minus the relay tripping time t_K, based on starting from the cold state with a load corresponding to the starting current of the motor:

$$t_B = t_{an} - t_K$$

The tripping time t_K is determined assuming that the motor run-up current is the same as the starting current and that it remains constant for a time t_K.

During the bridging time t_B, the motor current is not monitored by the overload relay. Should the motor be blocked during this time, the relay will sense the locked rotor current after the time t_B has expired. After this point in time, the tripping of the motor is determined by the overload relay tripping characteristic. The motor remains loaded with the locked rotor current for a period of time corresponding to the run-up time t_{an}.

▷ Bridging circuit for cases in which the run-up time/current is within the permissible load limits of the current transformers:

In this circuit, and under locked rotor conditions, the full motor starting current will flow through the current transformers of the electronic overload relay for the entire bridging period t_B. This circuit can therefore only be used, if neither relay nor associated current transformer become thermally overloaded under these conditions, i.e. only if the intersection between the starting current and starting time lie below the permissible loading curve of the current transformer/relay (see Fig. 3.37 overleaf).

Electronic overload relays in which the auxiliary contacts maintain their switched state in the event of supply voltage loss (bistable setting), are used (see Section 3.4.3.5, page 137).

The motor current can also flow during the time in which the power supply to the overload tripping device is switched off (e.g. in the event of control voltage failure). For this reason a bridging circuit as is shown in Figure 3.36 is required. In this configuration, the power supply to the tripping unit is only connected once the bridging time t_B has expired.

Mode of operation:

Actuation of the button S2 causes the motor contactor K1 and the time relay K2 to be energized. The sealing-in circuit of K1 is closed. Both the time relay

3 Selection criteria for low-voltage switchgear in main circuits

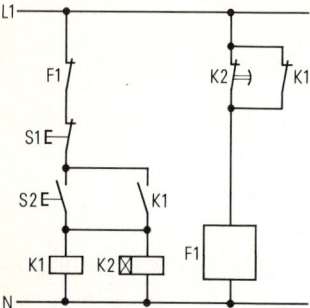

S1 Off pushbutton
S2 On pushbutton
K1 Motor contactor with 1 NO + NC
K2 Time relay with 1 NO delayed contact
F1 Tripping unit of the electronic overload relay in the operating mode "no tripping in the event of power loss"

Figure 3.36
Bridging circuit for use when the run-up time/current are within the permissible loading limits of the overload relay and its current transformer. See Figure 3.37 below

contact K2 and the NC auxiliary contact K1 of the motor contactor, are opened. Thereby, the power supply to the overload relay F1 is switched off. The output contact of the electronic overload relay F1 does not change state (see Section 3.4.3.5, page 137).

After the time t_B set on time relay K2 has expired, its delayed contact is closed. This restores the power supply to the relay F1 and monitoring of the load current can begin.

In the event of a tripping operation, the output relay contact of F1 is opened. The motor contactor K1 and the time relay K2 are switched off. K2 opens and the NC contact K1 closes, thus maintaining the power supply to the overload relay F1.

After the reset time has elapsed, the overload relay may be reset and the motor restarted.

▷ Bridging circuit for cases in which the run-up time/current exceeds the permissible loading of the current transformer:

In this circuit (see Fig. 3.38), no current flows through the current transformer or the overload relay during the bridging time t_B. Thus, they cannot become thermally overloaded during this time. The primary circuit of the overload relay/current transformer combination is bridged out by the bridging contactor K2 during the bridging time t_B.

The motor is started via the contactor K2. Contactor K1 is open. Thus, no current flows via the electronic overload relay.

After the bridging time t_B set on the time relay has expired, the motor load is changed over from contactor K1 to K2 (normal operation). The contactor K2 may, however only open once the contactor K1 has closed.

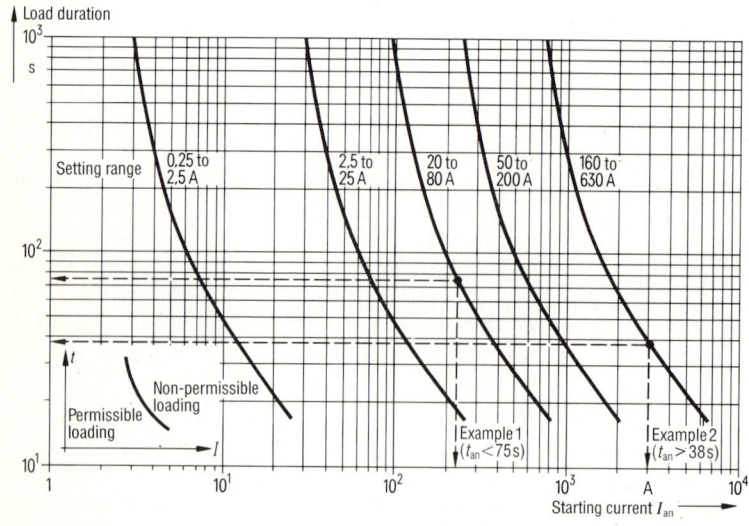

Figure 3.37
Permissible load characteristic curve of the electronic overload relay 3UB1 and associated current transformer

Protection of three-phase induction motors

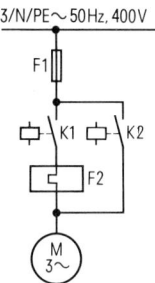

3/N/PE~50Hz, 400V

F1 Short-circuit protection (e.g. fuses)
F2 Overload relay
K1 Motor contactor
K2 Bridging contactor

Figure 3.38
Bridging circuit for cases in which the run-up time/current exceeds the permissible loading of the overload relay and current transformer. See characteristic curve, Fig. 3.37

Examples for determining the permissible loading in cases of extreme heavy duty starting are given in Table 3.21

By using the loading characteristic curves, and depending on the run-up time and the starting current of the motor, one can decide which of the two bridging circuits would be most suitable.

In addition to the overload relay, all the other elements in the circuit including motor, switchgear and feeder cables must be dimensioned for extreme heavy duty starting.

Protection of three-phase induction motors under conditions of high switching frequency

Permissible switching frequency

If overload relays are to be used in combination with contactors at higher switching frequency, then particular care should be taken to ensure the suitability and/or the correct adjustment of the overload relay. The permissible switching frequency depends on the following operational data:

Rated operational current of the motor	I_e,
Starting current of the motor	I_{an},
Run-up time of the motor	t_{an}
ON-time	t_{ED};
Duty cycle	t_s

Example

For $t_{ED} = 20$ s and $t_{an} = 2$ s, approx. 40 switching operations may be carried out per hour. In this case the duty cycle time t_s would be

$$t_s = \frac{3600 \text{ s}}{40} = 90 \text{ s.}$$

Table 3.21
Examples for determining the permissible loading of the current transformer during extreme heavy duty starting.

			Example 1	2
1.	Motor data: Rated motor current I_n A Factor I_{an}/I_n Run-up time t_{an} s		33 6.8 45	530 5.8 55
2.	Selection of the setting range corresponding to the rated motor current I_n A		20–80	160–630
3.	Calculation of the absolute value of starting current I_{an} $I_{an} = I_n \cdot \dfrac{I_{an}}{I_n}$ A		33×6.8 224	530×5.8 3074
4.	Find the starting current on the x-axis of Figure 3.37 and read off the permissible load duration on the y-axis s		75	38
5.	Is the permissible load duration longer than the run-up time? If yes: Overload relay may be used with bridging circuit as per Figure 3.36 If no: Overload relay must be bridged out as per circuit in Figure 3.38		× ($t_{an} < 75$ s) × ($t_{an} > 38$ s)	

If a three-phase induction motor cannot be protected effectively by means of an overload relay owing to high switching frequency, thermistor motor protection devices should be used (also refer to Section 3.4.3.6, page 138).

Protection of rotor critical three-phase induction motors

In the case of three-phase induction motors up to a rating of approx. 15 kW, the stator winding reaches the permissible temperature limit under locked rotor conditions sooner than the rotor does (stator critical). This temperature can be sensed effectively by means

155

3 Selection criteria for low-voltage switchgear in main circuits

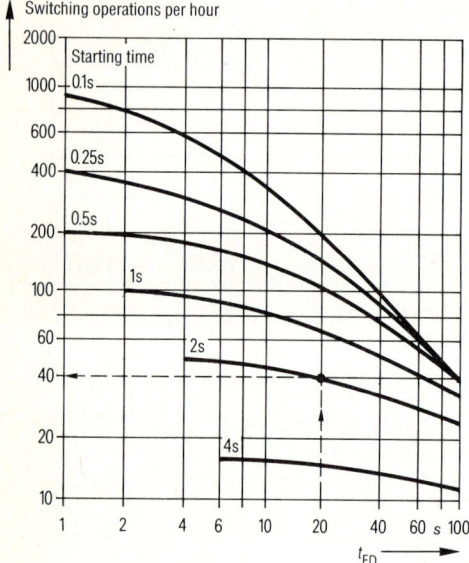

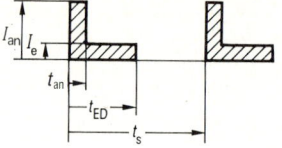

I_{an} Starting current of the motor
I_e Rated operating current of the motor
t_{an} Starting (or run-up) time of the motor
t_{ED} On-load period
t_s Operating cycle time

Figure 3.39
Permissible switching frequency of a thermal overload relay at $I_{an} = 6 \cdot I_e$

of PTC thermistors in the stator winding. Thus, one can ensure that the rotor will not become overheated.

Motors with a higher rating are generally "rotor critical". In this case, the locked rotor reaches the permissible temperature limit earlier than the stator does, i.e. sooner than the temperature detectors embedded in the stator winding can reach their rated operating temperature (TNF). This is particularly true if the motor is started from cold with a locked rotor. Here, the overtemperature of the rotor can be substantially higher than that of the stator. This causes a significant reduction in the expected service life of the motor.

To ensure adequate protection of rotor critical motors under such extreme conditions, the motor current is monitored by a three-pole thermal overload relay in addition to the PTC thermistors embedded in the stator winding. The overload relay will ensure tripping of the motor under locked rotor conditions before the rotor becomes overheated.

The scope of protection achieved for various motors by the use of thermistor motor protection devices with PTC thermistors, is given in the Siemens M11 catalogue.

Protection of three-phase induction motors in reversing operation

In each direction of rotation, the motor draws the same current. The overload relay is therefor connected in the motor feeder cable and is adjusted to the rated operating current of the motor (For basic circuit diagrams see Section 8.2.4, page 420). This also applies for braking by means of plugging (braking by phase reversal) in conjunction with a brake monitor or zero-speed switch.

Protection of three-phase induction motors with star-delta starting

In the star connection the motor draws only $1/\sqrt{3}$ times its rated current from the network. In the delta connected mode, and under rated load conditions, the full rated current flows. The current flowing through the actual motor winding, however is the same in both cases. For this reason, the overload relay is connected directly in the winding circuit and is set to the winding rated current, i.e. at $1/\sqrt{3}$ times (0.58 times) the motor rated current. In this way a single overload relay provides the motor protection in both the star and delta connections (for basic circuits see Section 8.3, page 444 et seq.).

Protection of high-voltage motors

Often, high-voltage motors are protected against overload by means of low-voltage overload relays (e.g. 3UA or 3UC as well as 3UB1 electronic overload relays), connected in a secondary circuit. The secondary circuit relay must be connected via the main current transformer. This current transformer must have sufficient short-circuit withstand capacity, alternatively it must be connected at the star point of the motor. The main current transformer may have an accuracy class of 3% and should have a rated overcurrent factor ≥ 10.

Its power rating must be matched to the power consumption of the overload relay together with the associated cable losses, to ensure that the tripping characteristic is maintained.

The overload relay and main current transformer must be selected to ensure that the secondary current flowing for motor rated current in the primary circuit, falls within the setting range of the overload relay.

Example

Rated motor current 210 A,
main current transformer 400/5 A, ratio = 80, therefore, secondary current at motor rated current in the primary will be:

$$\frac{210}{80} = 2.63 \text{ A}.$$

An overcurrent relay with a setting range from 2.6 to 4 A (setting = 2.63 A) may be selected.

In cases of high-voltage motors under difficult starting conditions (high moment of inertia, several successive starts), monitoring of the rotor winding temperature by means of a radiation thermometer is recommended.

High-voltage HRC fuses or a circuit-breaker may be used for short-circuit protection.

In the case of high-voltage fuses, the let-through current i_D may not exceed 63 kA. This also applies for the total let-through current i_{D2} of two parallel connected high voltage fuses ($i_{D2} = 1.6 \cdot i_{D1}$).

High-voltage fuses suitable for motor circuits are listed in the Siemens HG 12 catalogue.

Circuit-breakers may be used for the short-circuit protection of larger motors for which no suitable high voltage fuses are available. Owing the longer total opening time of the circuit-breakers (maximum permissible 120 Milliseconds), the sustained short-circuit current I_k may not exceed 20 kA (r.m.s. value).

Protection of explosion proof three-phase induction motors with type of protection "increased safety" (EEx e)

For the overcurrent protection of motors with type of protection, "increased safety" (EEx e) the following specifications and regulations apply:

▷ European Norm EN 50019 of May 1978, Electrical apparatus for potentially explosive atmospheres, increased safety "e",

▷ DIN VDE Regulations for electrical apparatus in firedamp and potentially explosive areas, DIN VDE 0170/0171 part 6, 5.78 as well as draft 07.89;

▷ DIN VDE regulations, Installation of electrical apparatus in potentially explosive atmospheres, DIN VDE 0165, 02.91.

In addition, the PTB test regulations (PTB, *Physikalisch-technische Bundesanstalt*, Braunschweig), volume 3 for explosion protected machines with type of protection "increased safety" (Ex)e of 1978, must also be taken into account.

It is imperative that the motor be protected against excessive temperature rise as a result of overloading. The following protection apparatus, among others, may be considered:

▷ Overload protection systems with current-dependent delayed tripping,

▷ Systems for direct temperature monitoring by means of temperature detectors.

Protection by current-dependent delayed overload releases of circuit-breakers and overload relays

The rated breaking capacity of the circuit-breaker, in terms of AC-3 and AC-4 duty to DIN VDE 0660, must be at least equal to the starting current I_{an} of the machine.

The protection apparatus must operate in all poles. The setting current I_r of the overload release or relay must not be greater than the rated current of the machine. In networks which are not solidly earthed and with rated voltages above 1 kV, protection in two phases is sufficient.

The tripping characteristics for releases and relays must be available at the point of installation. These tripping characteristics should indicate the tripping times based on three-pole loading from the cold state at a room temperature of 20 °C, as a function of 3 to 8 times the current setting I_r. The actual tripping times may not deviate from those given in the tripping characteristics by more than $\pm 20\%$.

Releases and relays for the protection of squirrel-cage machines must not only be selected in terms of the rated current I_n. Their selection must also take into account the tripping time according to the cold-state tripping characteristic, for the ratio I_{an}/I_n of the mo-

Table 3.22
Selection of overload relays for motors with type of protection increased safety "e" (examples).

Motor, 1500 r.p.m., 4-pole, 50 Hz from Siemens M 11 catalogue				Thermal overload relay from Siemens NS 2 catalogue		
P_n at T3(G3) kW	I_n at 380 V A	I_{an}	t_E at T3(G3) s	t_A at I_{an} s	Setting range A	I_r A
15	30	$5.9 \cdot I_n$	11	8.2	20 to 32	30
84	154	$6.4 \cdot I_n$	10	9.2	135 to 160	154
115	210	$6.7 \cdot I_n$	13	8.5	160 to 250	210

T3(G3)	Temperature class	t_A	Tripping time
I_{an}	Starting current	t_E	Safe locked-rotor time
I_r	Current setting		

tor to be protected. This tripping time t_A may not be greater than the safe locked-rotor time t_E as given on the motor rating plate. Table 3.22 contains a number of relevant examples.

The t_E-time is the time taken for the motor to reach maximum permissible temperature under locked rotor conditions, starting from the operationally warm condition.

Figure 3.40 shows the minimum values for the t_E-time of motors to DIN VDE 0170/0171 and the V.I.K. recommendations, as a function of starting current ratio I_{an}/I_n. (V.I.K. = Vereinigung Industrielle Kraftwirtschaft; transl. Power Industry Association)

For motors, the protection systems must provide motor protection in the event of phase failure. For example, current-dependent delayed thermal overload relays or releases are suitable if they incorporate phase-loss sensitivity to DIN VDE 0660 part 104.

In general, motors with current-dependent protection may only be used for continuous operation with light and less frequent starts during which temperature rise owing to motor starting is minimal. Motors for heavy duty starting or for high switching frequency may only be used with specially adapted protection apparatus which ensures that the maximum permissible temperatures are not exceeded.

Even during starting, the temperature limits may not be exceeded.

Note

Heavy duty starting is defined as the conditions under which a protection system suitable for normal starting conditions would trip during the run-up time of the motor. Generally, this is the case when the run-up time is greater than approx. 1.7 times the t_E-time.

Slip-ring motors with windings for type of protection increased safety "e", must be provided with additional instantaneous overcurrent protection releases or relays. These must be set at a current marginally above the expected starting peaks, but no higher than 4 times the rated current.

Thermistor motor protection

If overload protection is to be provided exclusively by direct temperature monitoring apparatus using temperature sensors, the version of the machine must be specially tested and certified as being suitable for this form of protection.

The Siemens 3UN2 thermistor protection tripping devices comply with DIN VDE 0660 part 303, 02.87 and are used in conjunction with PTC temperature detectors to DIN 44081/44082 for the direct tempera-

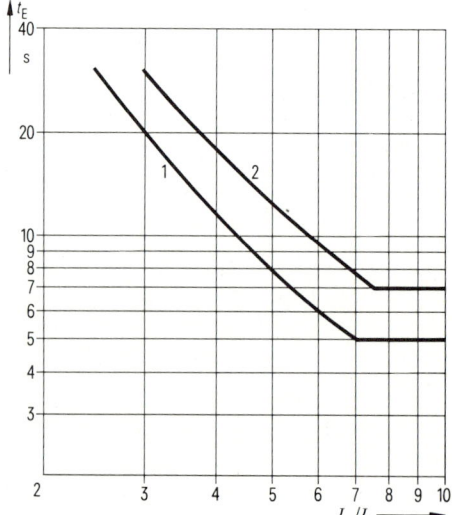

1 To DIN VDE 0170/0171
2 Recommendation of the Vereinigung Industrielle Kraftwirtschaft (V.I.K.)

Figure 3.40
Minimum value) of the safe locked-rotor time (t_E-time) as a function of the starting current ratio I_{an}/I_n

ture monitoring of explosion protected motors with type of protection increased safety EEx e. Furthermore they carry the relevant test mark.

In accordance with the relevant specifications, Siemens offers the following devices and documentation for the protection of motors with type of protection increased safety EEx e.

▷ Three-pole circuit-breakers for motor protection,
▷ Three-pole overload relays with phase-loss sensitivity,
▷ Electronic overload relays,
▷ Overload relays for heavy duty starting,
▷ Instantaneous electromagnetic overcurrent relays,
▷ Thermistor motor protection devices and
▷ Tripping characteristics,
▷ PTB test reports.

Further monitoring and protection methods

Further monitoring and protection possibilities for motors include

▷ Standstill monitoring:
   For drives equipped with tachogenerators, the evaluating unit 6KB4301 is recommended for monitoring the standstill condition and speeds near zero.
▷ Speed monitoring:
   For speed monitoring of machines from approx. 0 up to 6000 r.p.m., the evaluating unit 6KB4202 may be used in conjunction with an impulse generator.
▷ Monitoring of the braking procedure:
   The 6KB4110 speed monitor (brake monitor) is used in conjunction with contactors for the rapid stopping of electric drives by the plugging or reverse-current braking technique.
▷ Conveyer belt monitoring:
   The 6KB4111 conveyer belt monitors shut down a drive if the conveyer belt is slipping badly, is jammed, or when the speed of the drive deviates excessively from its rated value.
▷ Pressure monitoring:
   The 6KC3 electronic pressure-operated switches are used for monitoring the pressure in pneumatic control systems normally used e.g. in automated production and packaging machines.

The product spectrum includes switches for one underpressure range and four overpressure ranges. Each range is characterized by a maximum adjustable switching threshold referred to the ambient barometric pressure. The switches are suitable for use with all dry, nonaggressive gaseous media. These switches are further characterized by low switching hysteresis and a high frequency of operation. They are insensitive to pressure shock waves and overloading within the stated limits.

Further monitoring methods may be found in the Siemens l.v. switchgear catalogues. Basic circuit diagrams are shown in Section 8.5, see page 470.

3.4.5.2 Protection of conductors and cables outside the switchgear combinations

The protection of conductors and cables against excessive heating is generally based on the specification DIN VDE 0100 part 430/6.81 "Installation of power plant with rated voltages up to 1000 V".

In terms of these specifications, the overcurrent protection systems used in conjunction with conductors and cables must protect against excessive heating which may be caused by operational overloading as well as by short-circuit currents of varying severity.

Table 3.23 gives an overview of the overcurrent protection apparatus which may be used for this purpose.

Current-carrying capacity

The current-carrying capacity of insulated conductors and cables in air (not buried in the ground) is given in the DIN VDE specifications. Also refer to Section 9.4.1, page 517.

Protection against overload

The co-ordination of fuses or circuit-breakers for cable protection to the rated conduction cross-sectional area, applicable for insulated conductor and cables in air and ambient temperatures up to 30°C, is given in Section 9.4.2 on page 521.

Protection against short-circuit

The protection against short-circuit must be made in accordance with constraints discussed in Section 1.4.2.1 (refer to page 42). The specification DIN VDE 0100 part 430, also contains information for the determination of the required short-circuit protection.

Table 3.23
Overcurrent protection equipment for conductors and cable; relevant specifications and protection range

Overcurrent protection equipment	Specifications Standards	Protection: Overload	Short-circuit
Line protection fuses, operation class gL	DIN VDE 0636, IEC 269	×	×
Switchgear protection fuses, operation class aM	DIN VDE 0636	–	×
Miniature circuit-breaker (line-protection)	DIN VDE 0641 IEC 898	×	×
Circuit-breaker with overload and instantaneous over-current releases (an)	DIN VDE 0660 part 101 IEC 947-2	–	×
Circuit-breaker for starter combinations with instantaneous over-current releases (n)	DIN VDE 0660 part 101 IEC 947-2	–	×
Switchgear combinations comprising: Contactor and overload relay with gL or aM back-up fuses	DIN VDE 0636 IEC 269 DIN VDE 0660 part 104	×	×
Contactor and overload relay with circuit-breaker for starter combinations	DIN VDE 0660 part 101 IEC 947-2 DIN VDE 0660 part 104 IEC 947-4	×	×

Protection by means of a common overcurrent protection device

If the breaking capacity of the selected protection apparatus for overload protection is equal to or greater than the prospective short-circuit current at the point of installation, it may also provide the short-circuit protection for the downstream conductors and/or cables. For particular cable protection and power circuit-breakers, especially those without short-circuit current limiting, this does not apply for

the complete range of short-circuit currents which may occur at the said point of installation. A careful check prior to selection is therefor required.

Protection of the phases (L1, L2, L3, L+, L−) and the neutral or M conductor

Overcurrent protection must be provided for all the main of phase conductors (L1, L2, L3, L+, L−). It must provide for the switching off of the conductor in which the overcurrent occurs, but must not necessarily switch off the remaining active conductors. If however, the switching off of a single main conductor would cause a danger (e.g. with three-phase induction motors) suitable provision must be made to switch off all active conductors (refer to Section 3.6, page 183).

In earthed networks, protection of the neutral conductor may be omitted when its cross-sectional area is at least equal to that of the phase or main conductors. For smaller cross sections, an overcurrent in the neutral must cause a switching off of the main conductors but not necessarily of the neutral conductor itself. A sensing of overcurrent in the neutral conductor may be omitted if the neutral conductor is protected against a short-circuit by the protection apparatus in the main or phase conductors, i.e. if under normal operating conditions the current in the neutral conductor is smaller than its current carrying capacity.

3.4.5.3 Protection of transformers

Power transformers in radial networks

Transformers may be overloaded, whereby the degree and permissible duration of the overload depends on the preceding continuous loading and on the coolant temperature (IEC publication 354).

Circuit-breakers are most commonly used for the protection of the low-voltage circuit. Occasionally fuses are selected for transformers of smaller output rating.

For transformer protection, the circuit-breaker must have a switching capacity which is equal to or greater than the prospective short-circuit current at the terminals of the low-voltage winding. Its overload release is set to the rated current of the transformer.

For short-circuit protection, an instantaneous electromagnetic overcurrent release is used. In applications calling for discriminative operation with respect to downstream circuit-breakers or fuses, the circuit-

Protection of transformers

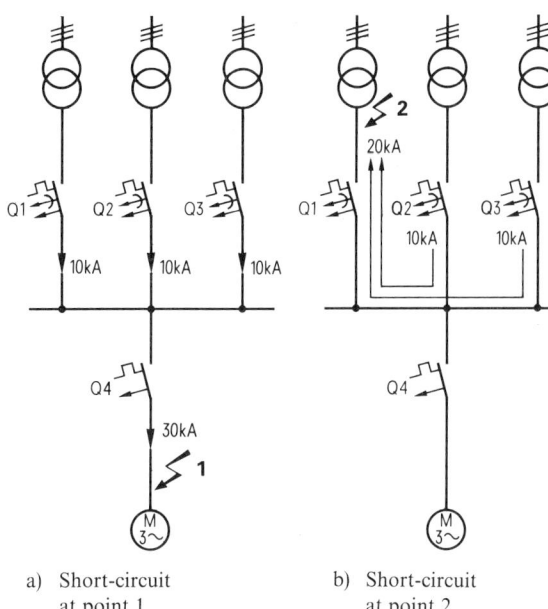

a) Short-circuit at point 1

b) Short-circuit at point 2

Figure 3.41
Short-circuit current for three parallel connected transformers of the same output rating

breaker typically incorporates an additional short-time delayed overcurrent release.

Protection of two parallel-connected transformers of equal rating

For the protection of two transformers having the same output rating and connected in parallel, two separate circuit-breakers are installed (Q1, Q2). Their switching capacity must correspond to the prospective short-circuit current of one transformer.

Protection of three parallel-connected transformers of the same output rating

In the event of a short-circuit at point 1 (Fig. 3.41), the short-circuit current flowing through each circuit-breaker corresponds to that of its respective transformer. For a short-circuit at the point 2, however, the short-circuit current flowing through circuit-breaker Q1 corresponds to the sum of the short-circuit currents of the other two transformers. Accordingly, the circuit-breakers must have double the breaking capacity. In all cases, the circuit-breaker Q4 must have a breaking capacity corresponding to the sum of all three the transformer short-circuit currents.

In such installations, more effective control over short-circuit currents may be achieved by the use of bus-coupler circuit-breakers between the transformers.

Power transformers in meshed networks

In a meshed network with multiple section operation, i.e. fed by multiple high voltage lines and transformers, faults occurring in a transformer station or on a high-voltage line must be isolated to prevent the low-voltage network feeding back to the fault point. For this application a network master relay and special circuit-breakers for mesh-connected systems are used. These are three-pole power circuit-breakers with or without overcurrent release, but incorporating a "meshed-network shunt trip" This is a shunt trip with energy storage to ensure tripping even after the control voltage supply has collapsed (also known as an fc-release). In the event of a fault as described, the circuit-breaker receives its tripping command from the network master relay via the meshed-network shunt trip. The tripping command is given by a normally open contact in the network master relay which then closes the operating circuit of the meshed-network shunt trip.

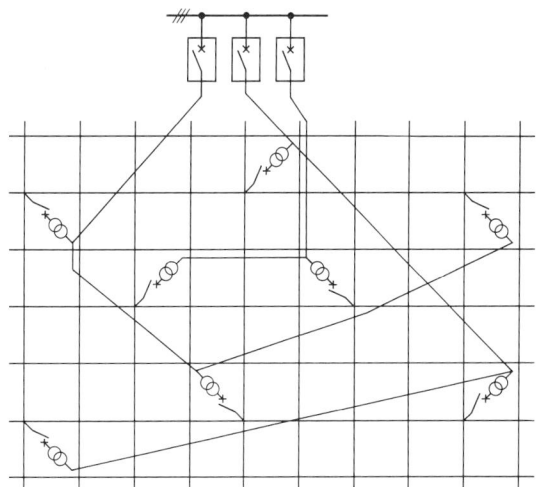

Figure 3.42
Example of a meshed network with multiple feeds

161

3 Selection criteria for low-voltage switchgear in main circuits

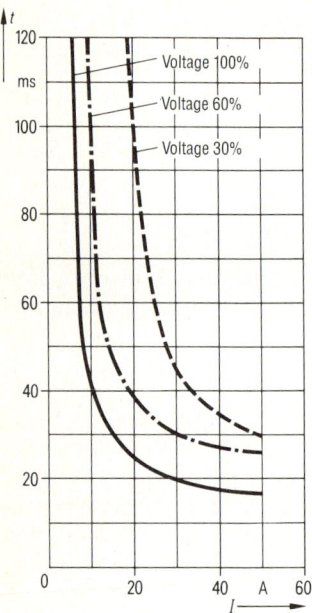

Figure 3.43
Tripping characteristic of a network master relay type 7RM at normal set value (6 A)

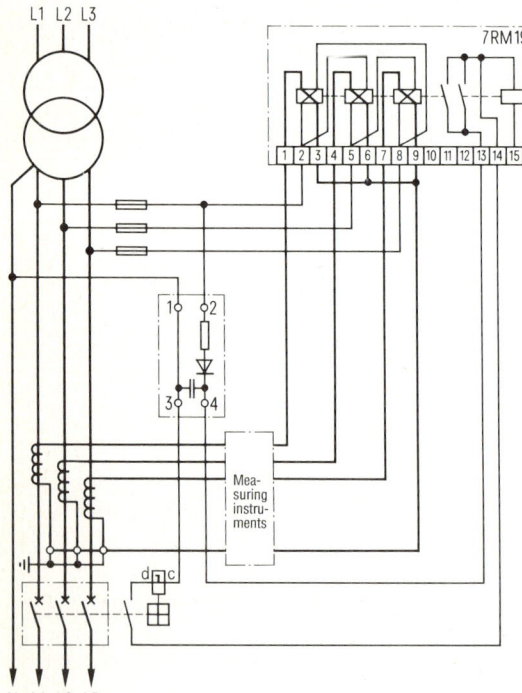

Figure 3.44
Example of a circuit diagram incorporating a circuit-breaker for mesh-connected systems in conjunction with a network master relay for tripping via a stored energy device

Circuit-breakers for mesh-connected systems

When specifying the switching capacity for a circuit-breaker to be used in a meshed network, it must be taken into account that the highest prospective short-circuit current can be expected between the terminals of the transformer and the breaker itself. In this case the sum of the short-circuit currents of all the feeder points will flow through the meshed network and the circuit-breaker to the point of short-circuit. This current may be greater than the prospective short-circuit current of the associated transformer.

Network master relay

Network master relays are used in conjunction with circuit-breakers for mesh-connected systems. They provide rapid and selective isolation of damaged high voltage feeder cables between mesh-connected low-voltage networks. The relay senses the reversal in the direction of the energy flow if a short-circuit in the high voltage feeder cable causes high currents to flow from the meshed network through the associated transformers to the point of fault.

To avoid nuisance tripping however, at rated voltage the network master relay will permit transient currents of up to its rated current to flow before a tripping operation is initiated (adjustable by means of spring tensioning from 2 A to 6 A). Figure 3.43 shows the tripping characteristics for a normal setting at 6 A.

Figure 3.44 shows the basic circuit diagram of a circuit-breaker for mesh-connected systems used in conjunction with a network master relay. If the auxiliary contact (terminals 13, 14) of the 7RM network master relay operates in series with a shunt trip for 230 V 50 Hz and the energy storage device, the contact closes onto the pull-in current of approx. 2 A at DC 300 V. The auxiliary contact opens under no-load, and therefore does not require a particularly high breaking capacity.

If however, tripping is to carried out with the full AC 230 V system voltage, then the built-in auxiliary relay with a current coil for tripping currents of 0.75 to 12 A, must be connected into the circuit (terminals 13, 15). This quick-acting auxiliary relay provides the necessary switching capacity within milliseconds of the signal by the network master relay.

These contacts do not require particularly high breaking capacity, since an auxiliary contact on the circuit-breaker performs the interruption of the shunt trip current.

For further technical details of circuit-breakers used in mesh-connected networks, please refer to the Siemens l.v. switchgear catalogue. For network master relays refer to the Siemens R catalogue.

3.4.5.4 Protection of capacitors

In terms of DIN VDE 0560 part 4, capacitor units must be capable of continuous operation at a current smaller than or equal to 1.3 times the current which flows upon application of the sinusoidal rated voltage at rated frequency.

For this reason, capacitor units are generally not provided with overload protection.

Capacitors can only be overloaded in networks which contain harmonic-producing elements (e.g. generators and drives with static convertors). In combination with the series connected transformer and the short-circuit reactance of the feeding network, the capacitors form a parallel oscillating circuit. Resonance phenomena will be experienced when the natural frequency of this oscillating circuit corresponds to the frequency of a harmonic current created by the converter. To avoid resonance, the capacitors must be equipped with suitable chokes. The combination of choke and capacitor forms an LC tuned circuit with a resonance frequency which is lower than the frequency of the lowest order harmonic in the load current (250 Hz).

Thus, the capacitor/choke combination represents an inductive load for all harmonic currents contained in the load current, and resonance with the network reactance is not possible.

Another possibility is to minimize the harmonic content of the network currents by means of filter circuits. Filter circuits are also resonance elements, but unlike choked capacitors they are finely tuned to the frequencies of the resonance currents to be removed. At their tuned frequency they have an impedance which is virtually 0.

Further details regarding the "choking" of capacitors and the use of filter circuits may be found in Section 7.4.8.6 (see page 373).

If thermally delayed relays are used for protection against overcurrents, then the tripping value may be set at 1.3 to 1.43 times the capacitor rated current. This factor takes into account the permissible tolerance of capacitor current which may be $1.1 \cdot 1.3 = 1.43$ times the capacitor rated current.

In the case of overcurrent relays or releases which are operated via current transformers, premature tripping may occur. This is caused by the different transformer ratios which apply for currents at higher than rated frequency, e.g. at harmonic frequencies.

Low-voltage HRC fuses, of operation class gL, are most commonly used for the short-circuit protection of capacitor units.

To avoid a rupturing of the fuses in the overload range and during switching of the capacitors, the fuse rating is normally selected to be 1.6 to 1.7 times the value of the capacitor rated current.

3.4.6 Discriminative protection (selectivity)

In the past one has tended to focus on the switching capacity and current limiting capabilities of the individual switching devices themselves. More recently the emphasis has shifted to the installation as a whole. The operational reliability and safety of the low voltage installation is not only determined by the performance of the individual devices, but by the coordinated operation and careful matching of all the system components.

An optimal coordination between switching and protection devices depends on the discriminative operation of conductor and device protection systems. Each feeder circuit and each switching device should have the best possible protection against overload and short-circuit, but should not loose their power supply as a result of switching operations carried out to protect other parts of the installation.

By means of a suitable network and protection configuration, the effects of short-circuits must be kept to a minimum. In addition, the degree to which a fault in one part of the installation effects the other parts must be restricted.

Thus, the overcurrent protection system must operate in a "selective" or discriminative manner in the event of a fault, i.e. the point of fault must be localized. Discriminative protection means that only those protection devices which are located closest to the fault will operate, and this in the quickest manner possible, so that no widespread shutdown results and no loads are unnecessarily disconnected from the supply.

The prerequisite for the discriminative operation of

different protection devices connected in series, and therefore for the reliability of supply demanded by the consumers, is the coordination of the tripping and release criteria.

Depending on the voltage level, the network type and the characteristics of the load itself, a number of different protection devices may be considered for short-circuit protection. In low and medium voltage networks, current-dependent and current-independent releases or relays, as well as fuses are used. The basic shapes of the characteristic tripping curves for these protection devices are shown in Figure 3.45. The various possibilities for shifting these curves are indicated by arrows.

The discriminative behaviour of protection devices with respect to one another must be checked by a comparison of their respective time-current characteristics. Depending on the voltage level and the switchgear types, the release times and the tripping currents of series connected protection devices must differ by a corresponding safety margin. (In special cases, e.g. in the primary and secondary protection of a transformer, an overlapping of these values may be permitted). If these conditions cannot be met completely, the non-discriminative behaviour of the protection devices must be recognizable and it must be possible to determine the extent to which the network is effected in the event of a fault.

Coordinated protection is of particular importance in industrial and building networks. In spite of a relatively simple network configuration (radial network), these systems often incorporate more than one voltage level and/or several levels of distribution, and therefore require a great number of protection devices. The series connected protection devices between the supply and the load, i.e. the devices through which a short-circuit current will flow for a fault at the load end, and therefore the devices which must operate in a discriminative manner (e.g. graded w.r.t. tripping time), are determined by the selected distribution circuit. The main protection devices must thereby be carefully matched to the protection switchgear in the sub-distribution branches of the network.

3.4.6.1 Discrimination in radial networks

The following devices may be series connected (listed in direction of power flow):

▷ Fuse with downstream fuse,
▷ Circuit-breaker with downstream circuit-breaker,
▷ Circuit-breaker with downstream fuse
▷ Fuse with downstream circuit-breaker,
▷ Several parallel incoming feeders (with or without bus-coupler(s)) with downstream circuit-breaker or fuse,
▷ HV HRC fuse with downstream LV fuse,
▷ HV HRC fuse with downstream circuit-breaker.

Discrimination between series connected fuses

The incoming cables and the feeders branching from the busbars carry different operating currents and therefore have differing cross-sections. For this reason, they are be protected by fuses of unequal ratings. In the event of a fault, however, the same short-circuit current could flow through both fuses.

As a first principle, the following applies:
Series connected fuses will behave in a discriminative manner provided that their respective melting-time characteristic curves are sufficiently far apart, or more accurately, the tolerance bands of these curves do not touch (see Fig. 3.46).

For large short-circuit currents however, this condition alone is no longer adequate. In such cases discrimination can only be guaranteed if the joule heat value of the current (I^2t value) during the melting and the quenching time of the smaller downstream fuse is smaller than the melting joule heat (I^2t_s value) required to rupture the larger upstream fuse. In general, this is the case when the ratio between fuse ratings is greater than, or equal to 1:1.6. For example, conditions for discrimination are met when a 100 A and a 160 A fuse are connected in series.

Details on the discrimination between LV fuses in

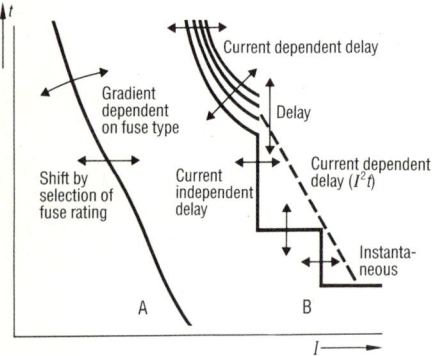

Figure 3.45
Characteristic curves of a fuse (A) and a circuit-breaker with overload releases (B) in a low-voltage network

Discrimination in radial networks

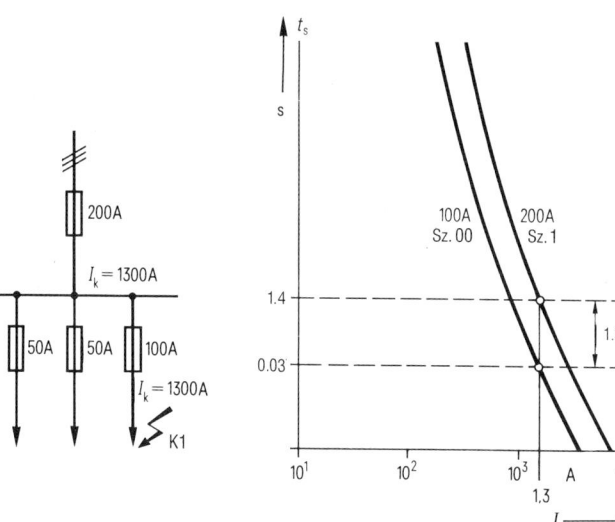
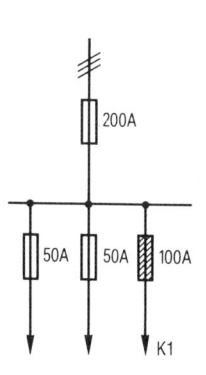

Figure 3.46
Discriminative behaviour of two series connected fuses (example)

t_s Virtual pre-arcing time (melting time)

a) Single-line diagram

b) Pre-arcing (melting) times t_s for $I_k = 1300$ A

c) Discriminative (selective) isolation of the point of short-circuit K1

radial networks at operating voltages up to 400 V and up to 690 V, as well as discrimination in meshed networks, may be found in the technical documentation for LV fuses (e.g. Siemens I2.2 catalogue). Also refer to Section 9.7.5 on page 543.

Discrimination between series connected circuit-breakers

Discrimination by grading of the operating currents of instantaneous overcurrent releases (current-based discrimination)

In the series connection of circuit-breakers, one refers to the overload and short-circuit protection as being "discriminative" when only the circuit-breaker located immediately upstream from the point of fault opens in the event of an overload or short-circuit.

Without resorting to specific testing, one can only be sure of successful discrimination through the grading of instantaneous overcurrent release tripping currents, if the prospective short-circuit currents at the points of installation of the circuit-breakers are of sufficiently different values. The tripping current of the instantaneous overcurrent release (n) in the upstream circuit-breaker must be so adjusted, that its tripping value is higher than the peak of the greatest possible short-circuit current which can arise at the point of installation of the downstream breaker (Fig. 3.47).

In this case, the protection in the event of a full short-circuit on the incoming terminals of the downstream circuit-breaker is performed by the current-dependent delayed overload release (a) of the upstream breaker.

If the downstream circuit-breaker is of the current limiting type, the tripping current of the upstream breaker must be set at a value exceeding the peak let-through current of the downstream unit.

If the prospective short-circuit currents at the points of installation of upstream and downstream circuit-breakers are virtually the same, then discrimination can only be achieved up to a specific value of short-circuit current (discrimination limit). Refer to Section 3.4.6.2, on page 172).

The design engineer must check whether the resulting long tripping times are permissible in terms of the applicable specifications (also refer to Sections 2.1.2 and 3.1.2, pages 79 and 104 resp.)

The fulfilling of specifications can be facilitated by the use of discrimination tables. These tables frequently provide discrimination limits which are much higher than the tripping current of the upstream breaker.

Another possibility for ensuring discrimination in the event of similar prospective short-circuit currents at the two points of circuit-breaker installation, is the use of "time-based discrimination".

3 Selection criteria for low-voltage switchgear in main circuits

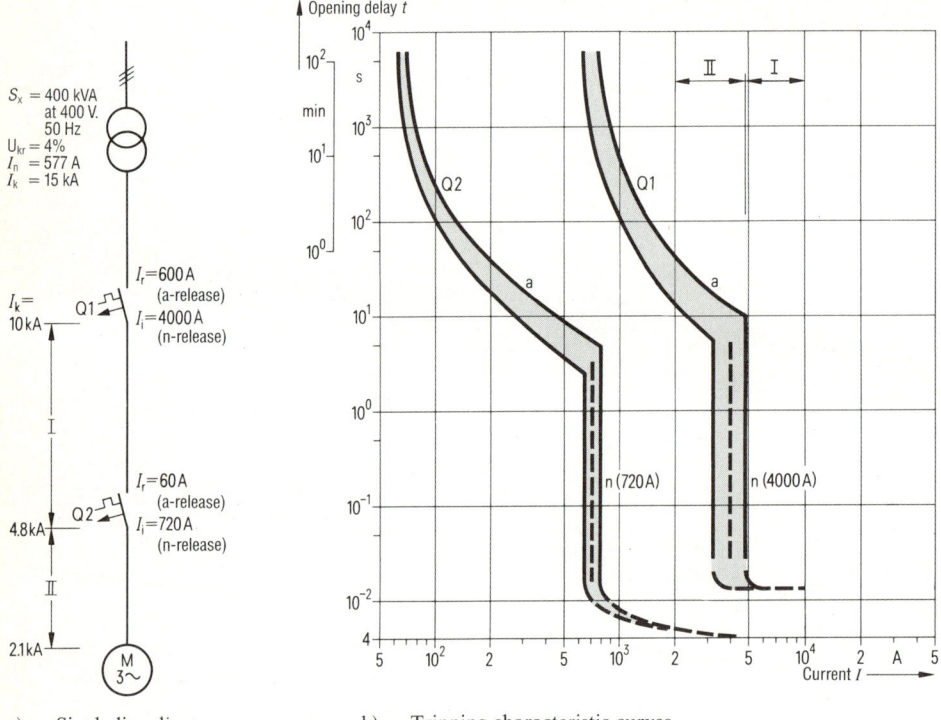

a) Single-line diagram b) Tripping characteristic curves

Q2 Circuit-breaker for motor protection (current-limiting)
Q1 Circuit-breaker (zero-point quenching)

a Current dependent delayed overload release
n Instantaneous electromagnetic overcurrent release

Figure 3.47
Current-based discrimination between two series connected circuit-breakers installed at points in a circuit having differing prospective short-circuit currents (example)

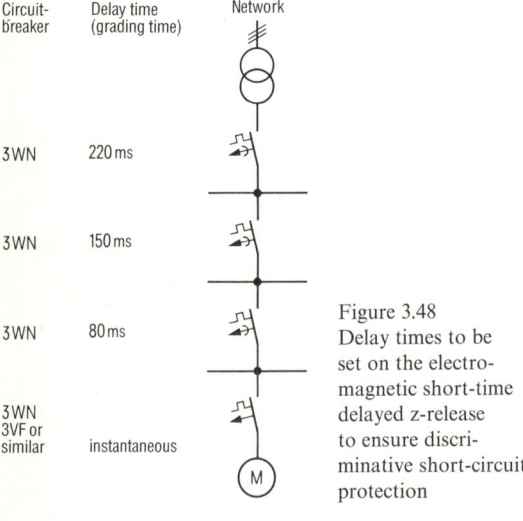

Figure 3.48
Delay times to be set on the electromagnetic short-time delayed z-release to ensure discriminative short-circuit protection

Discrimination by the use of short-time delayed overcurrent releases (time-based discrimination)

To ensure discriminative behaviour in the case of similar prospective short-circuit currents at the points of circuit-breaker installation, the technique of time-based discrimination may be employed (Fig. 3.48 shows an example of the grading times for discriminative short-circuit protection). For this type of protection, the upstream circuit-breaker requires a short-time delayed overcurrent release (z), so that in the event of a fault only the downstream circuit-breaker will operate to disconnect the affected part of the installation from the network. Both the tripping delays and the tripping currents of the respective overcurrent releases are graded in relation to each other.

The example in Figure 3.49 shows the single-line diagram of three series connected circuit-breakers as well as the corresponding grading diagram.

Discrimination in radial networks

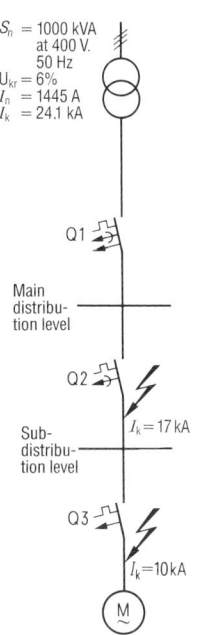 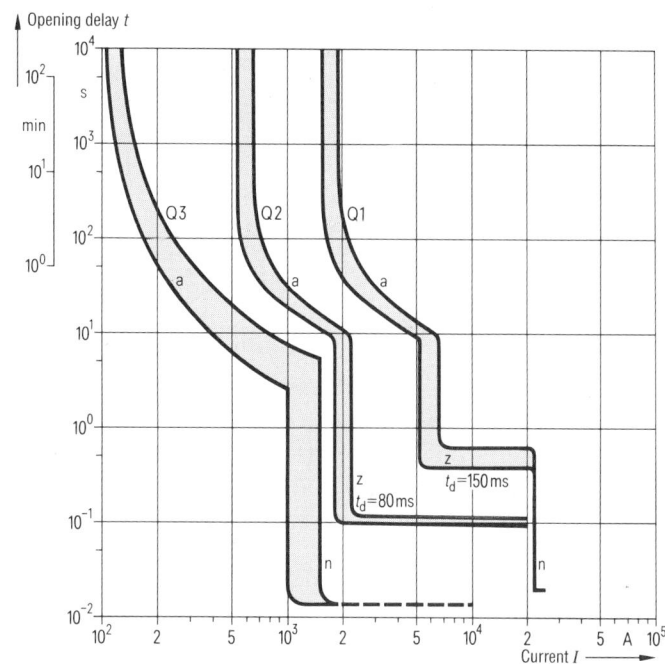

a) Single-line diagram
b) Grading diagram (tripping characteristic curves)

Q1 Circuit-breaker
Q2 Circuit-breaker for distribution protection
Q3 Circuit-breaker for motor protection

a Current dependent delayed overload release
n Instantaneous electromagnetic overcurrent release
z Short-time delayed overcurrent release

Figure 3.49 Discriminative behaviour of three series connected circuit-breakers

The required grading time which must allow for tolerances in all the tripping characteristics, depends on the operating principle of the release and on the construction of the breaker itself.

In the case of electronic short-time delayed overcurrent releases (z releases), a grading time of approx. 70 to 100 ms between each circuit-breaker is normally sufficient to allow for all tolerance bands.

The tripping current of the short-time delayed overcurrent release must be set to at least 1.25 times the value set on the downstream circuit-breaker.

A time setting range up to 500 ms permits grading of up to 7 series connected circuit-breakers.

To reduce the stresses in the event of a complete short-circuit directly at the upstream breaker, it may be equipped with an instantaneous short-circuit release in addition to the short-time delayed release.

The tripping current of this instantaneous release (n) must be such that the release only operates in the event of a complete short-circuit close to the circuit-breaker load terminals, and does not disturb the time grading.

Short-time grading control (ZSS)

To avoid the unwanted long tripping times which can arise when several circuit-breakers are connected in series, Siemens developed the micro-processor based "short-time grading control" (ZSS = Zeitverkürzte Selektivitäts-Steuerung) for the 3WN air circuit-breakers.

This feature allows a reduction in the tripping delay to max. 50 ms for the circuit-breaker installed immediately upstream from the point of fault.

Figure 3.50 shows the operating principle of the ZSS short-time grading control.

3 Selection criteria for low-voltage switchgear in main circuits

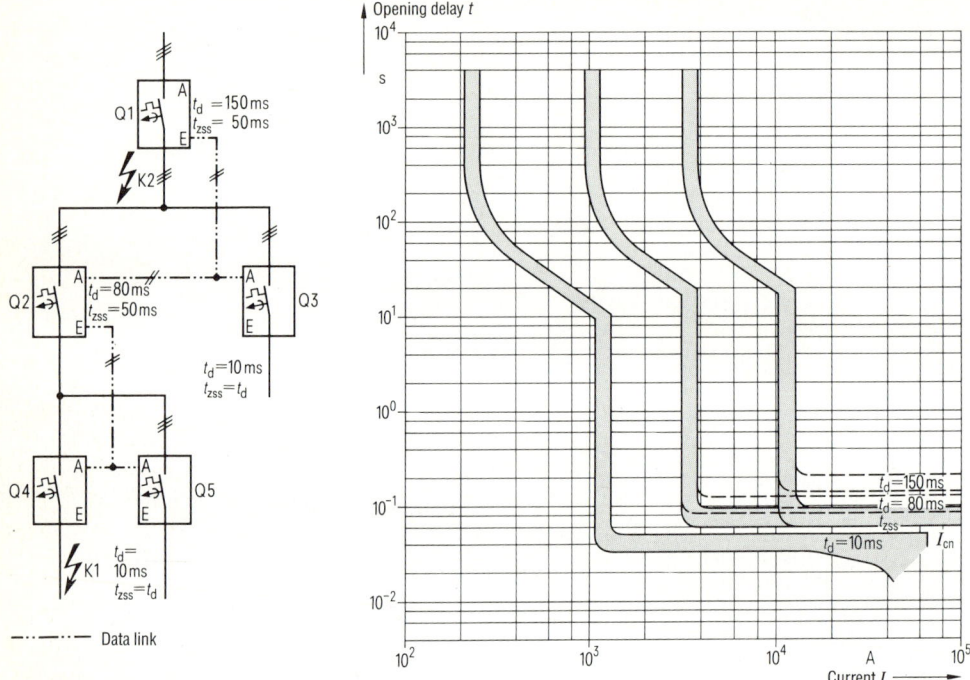

Operation:
A short-circuit at point K1 is sensed by Q4, Q2, and Q1. If ZSS is activated Q2 is temporarily blocked by Q4, and Q1 is temporarily blocked by Q2. Since Q4 does not receive a blocking signal, it trips within 10 ms. A short-circuit at point K2 is only sensed by Q1, and since it does not receive a blocking signal, it trips within 50 ms. Without ZSS, tripping would only take place after 150 ms.

t_d Maximum duration of blocking signal, alternative delay time with ZSS deactivated
t_{zss} Delay time of the circuit-breaker if it senses a short-circuit current and does not receive a blocking signal
A   Output; sends blocking signal
E   Input; receives blocking signal
I_{cn} Rated short-circuit breaking capacity

Figure 3.50   Operating principle of short-time grading control (ZSS) with series or parallel connected circuit-breakers

Discrimination between a circuit-breaker and a downstream fuse

In the overload range and up to the tripping current I_{An} of the instantaneous overcurrent release in the circuit-breaker, conditions for discrimination are met if the time-current characteristic curve of the fuse (including its tolerance band) does not touch the tripping curve of the current-dependent delayed overload release in the breaker. This must also hold true if the circuit has been under load and the fuses/overcurrent release are at operating temperature (Fig. 3.51).

For prospective short-circuit currents which would equal or exceed the tripping value of the instantaneous overcurrent release, discrimination is only possible if the fuse limits the actual short-circuit current during rupturing to a let-through value which is lower than the tripping current of the release (Fig. 3.52). This can only be expected if the fuse rating is much lower than the rated current of the circuit-breaker.

Practical experience has shown, that short-circuit discrimination is ensured if the delay time t_v of the short-time delayed overcurrent release shifts the tripping curve to at least 100 ms above the time-current characteristic curve of the fuse (Fig. 3.52).

Discrimination between a fuse and a downstream circuit-breaker

Here too, discrimination is ensured in the overload range if the tripping curve of the current-dependent delayed overload release does not touch the time-current characteristic curve of the fuse (Fig. 3.53).

Discrimination in radial networks

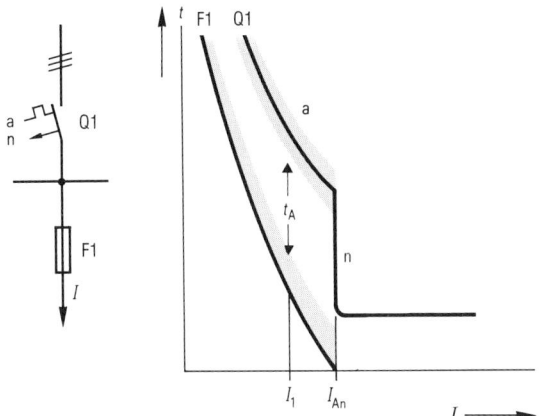

- F1 Fuse
- Q1 Circuit-breaker
- a   Current dependent delayed overload release
- n   Instantaneous electromagnetic overcurrent release
- t_A Safety margin
- I_{An} Tripping current of the n-release

Figure 3.51
Discrimination in the **overload range** between a circuit-breaker and a downstream fuse. The time-current characteristic curves (including tolerance bands) do not touch each other

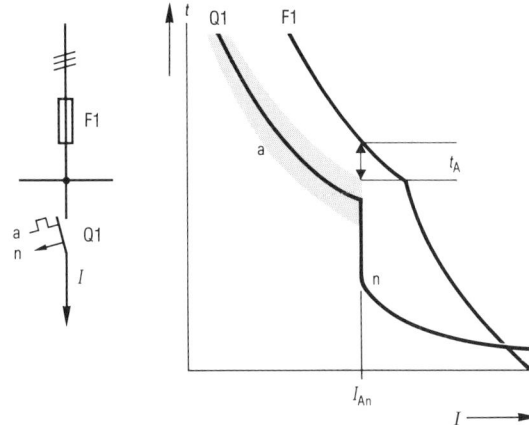

- Q1 Circuit-breaker
- F1 Fuse
- a   Current dependent delayed overload release
- n   Instantaneous electromagnetic overcurrent release
- t_A Safety margin
- I_{An} Tripping current of the n-release

Figure 3.53
Discrimination in the **overload range** between a fuse and a downstream circuit-breaker. The time-current characteristic curves (including tolerance bands) do not touch each other

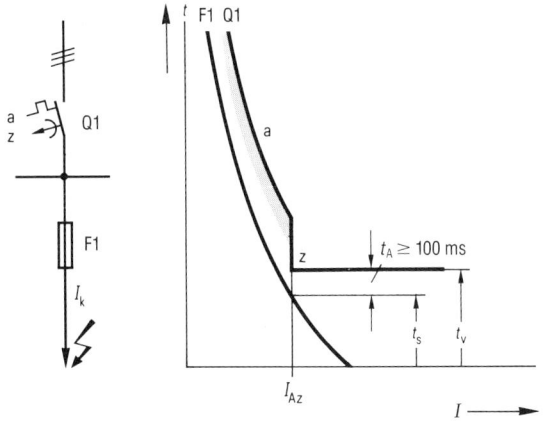

- a   Current dependent delayed overload release
- z   Short-time delayed overcurrent release
- t_A Safety margin
- I_{Az} Tripping current of the z-release
- t_s Pre-arcing time of the fuse (melting time)
- t_v Delay time of the z-release

Figure 3.52
Discrimination in the **short-circuit range** between a circuit-breaker and a downstream fuse

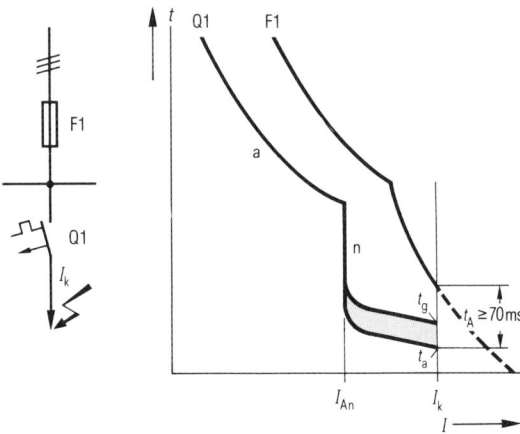

- t_a Opening delay of the circuit-breaker
- t_g Total clearing time of the circuit-breaker

For further explanations, refer to Figure 3.53

Figure 3.54
Discrimination in the **short-circuit range** between a fuse and a downstream circuit-breaker

In the event of a short-circuit, it must be taken into account that the current continues to heat the fuse element during the arcing time of the circuit-breaker.

In practice, it is sufficient of the melting time-current characteristic curve of the fuse lies at least 70 ms above the tripping characteristic curve of the instantaneous overcurrent release of the circuit-breaker (Fig. 3.54).

Discrimination in the case of two or more parallel incoming feeders

In the case of two or more parallel feeders, the prospective short-circuit current (fault level) does not continue to increase as one moves up through the circuit towards the point of supply, but is shared between the incomers. The conditions for discrimination are thereby favoured.

Two identical incoming feeders

In the case of a short-circuit at point K1 below circuit-breaker Q3 (Fig. 3.55a), the total short-circuit current of 20 kA will flow through Q3, whereas the feeder circuit-breakers Q2 and Q3 will conduct only half each, viz. 10 kA.

In the grading diagram (Fig. 3.55b) the tripping characteristic curves of the feeder circuit-breakers Q1 and Q2, must therefore be referred to the current flowing through circuit-breaker Q3.

Since the total short-circuit current is shared equally between the two incoming feeders, the characteristic tripping curve for the circuit-breakers Q1 & Q2 must be shifted to the right on the current axis by the factor 2.

Three identical incoming feeders

If more than two feeder transformers are connected in parallel (Fig. 3.56), then the circuit-breakers Q1, Q2 and Q3 are equipped with instantaneous overcurrent releases (n) in addition to their short-time delayed overcurrent releases (z). The tripping current setting of these n releases is higher than the prospective short-circuit current of the corresponding transformer.

Thus, for a short-circuit at point K1 between the transformer and the circuit-breaker (Fig. 3.56b), the sum of the parallel currents flowing through Q2 and Q3 will cause Q1 to trip instantaneously. The circuit-

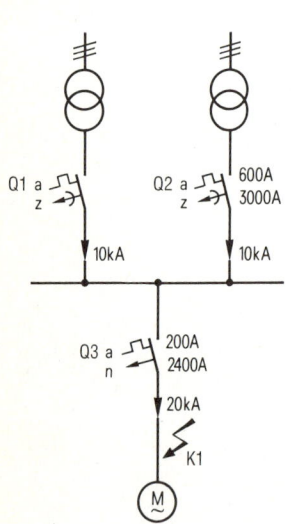 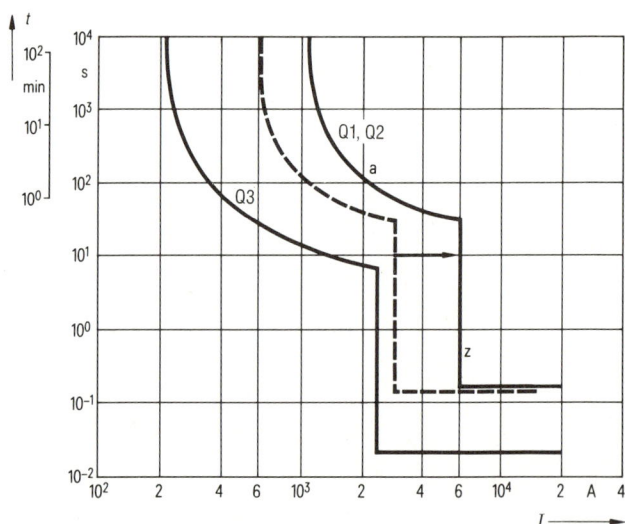

a) Single-line diagram
b) Grading diagram

Q1, Q2 Feeder circuit-breakers
Q3 Circuit-breaker for motor protection

a Current dependent delayed overload release
z Short-time delayed overcurrent release

Figure 3.55 Distribution system with 2 simultaneously connected feeder transformers of the same power rating

breakers Q2 and Q3 remain closed and maintain power supply to the installation.

Parallel feeders connected via a bus-coupler circuit-breaker

The selection of the switchgear for the outgoing branch circuits, and the discriminative behaviour of the distribution circuit as a whole, depend on whether the bus-coupler circuit-breaker is of the zero-point quenching type, i.e. without current limiting, or a current limiter.

High speed current-limiting bus-coupler circuit-breakers relieve the sub-distribution circuits from the effects of summated peak short-circuit currents i_p and therefore permit the use of less robust and more economical circuit-breakers.

Discrimination and undervoltage protection

When a short-circuit occurs, the network voltage at the point of short-circuit collapses.

The remaining voltage depends on the value of fault resistance. In the case of a full short-circuit, the fault resistance, and therefor the voltage at the point of short-circuit, is practically equal to zero. In general, however, severe arcing occurs during short-circuits. The arc voltage will typically be 30 to 70 V. The network voltage along the current path back to the energy source drops to a low value while the short-circuit current is flowing. The value of this remaining voltage is determined by the conductor resistance and therefor by the distance to point of fault. Figure 3.57 shows the behaviour of the voltage in a low-voltage distribution system during the existence of a full short-circuit.

For a short-circuit at point K1 (Fig. 3.57a), the operating voltage U on the busbars of the sub-distribution Section drops to 0.13 times the rated operational voltage U_e, while the voltage on the main distribution busbars drops to $0.5 \cdot U_e$. In this case, the circuit-breaker Q3 installed directly above the point of fault, trips. Depending on the breaker size and type, the total breaking time will be up to 30 ms for a zero point quenching circuit-breaker, and a maximum of 10 ms in the case of a current limiter.

For a short-circuit at point K2 (Fig. 3.57b), circuit-breaker Q2 trips. It is equipped with a short-time delayed overcurrent release (z-release). The delay time will be at least 70 ms. During this time the network voltage on the main distribution busbars drops to $0.13 \cdot U_e$.

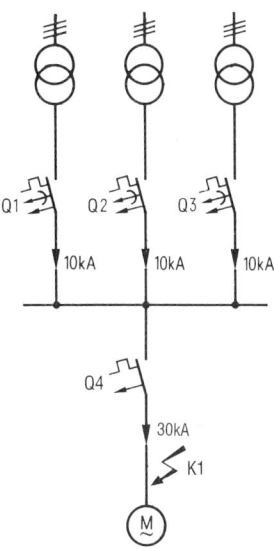

a) Short-circuit in the motor feeder cable: circuit-breaker Q4 trips instantaneously, incomer circuit-breakers Q1, Q2, Q3 remain closed

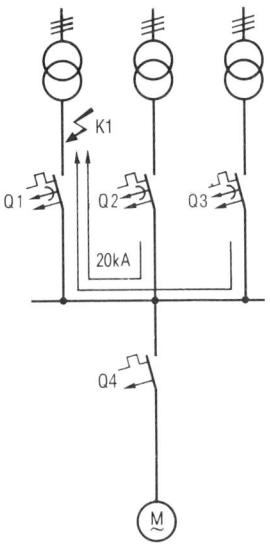

b) Short-circuit between a transformer and the associated circuit-breaker: circuit-breaker Q1 trips instantaneously, circuit-breakers Q2, Q3 remain closed

Q1, Q2, Q3   Feeder circuit-breakers
Q4   Circuit-breaker for motor protection

Figure 3.56
Distribution system with three parallel connected feeder transformers of the same output power rating

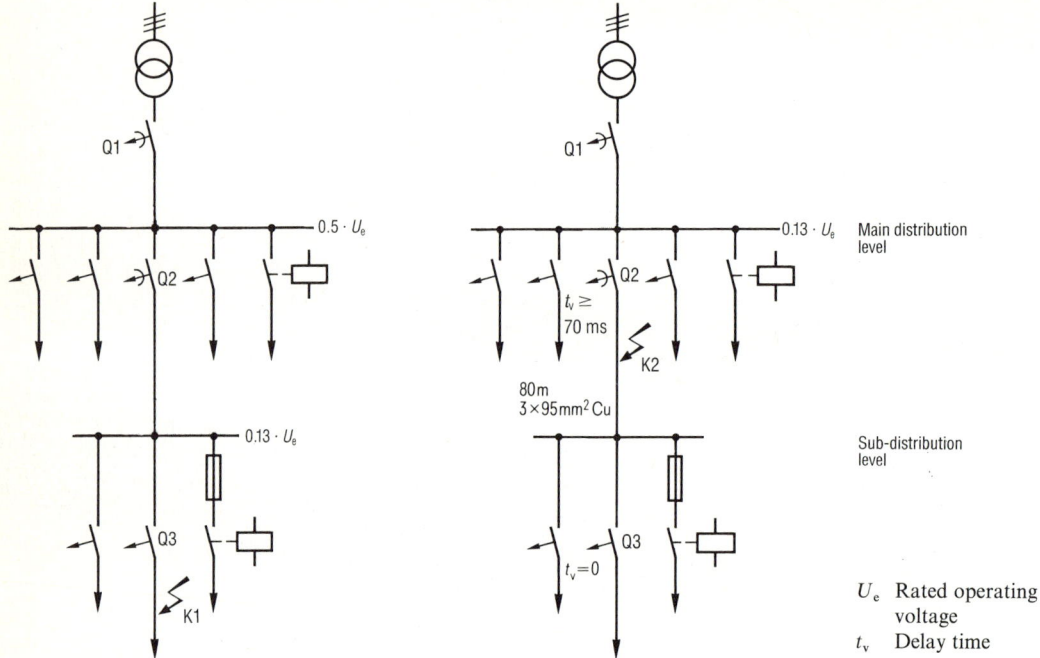

a) Short-circuit in the sub-distribution system   b) Short-circuit in the main distribution system

Figure 3.57
Resultant voltages during the existence of a short-circuit in a low-voltage system with main and sub-distribution levels

If the network voltage drops to between 0.7 and 0.35 times the rated value, and if this condition exists for longer than approx. 20 ms, then all circuit-breakers fitted with undervoltage releases will be tripped. In the same way, all contactors will drop out if the rated control supply voltage drops to below 75% of its rated value for longer than 5 to 30 ms.

Discriminative overcurrent protection therefore requires that undervoltage releases and contactors be equipped with drop-out delay devices. This may not be necessary in the case of current limiting circuit-breakers, in which the total breaking time is 10 ms or less.

3.4.6.2 Use of discrimination tables

If the prospective short-circuit currents at the various points of circuit-breaker installation are similar, a grading of the tripping currents of the instantaneous overcurrent releases permit discrimination only up to a specific short-circuit current value. This current is known as the "discrimination limit" (Fig. 3.58).

If a short-circuit calculation (e.g. to DIN VDE 0102) is done for the point of installation of the downstream breaker, and the value of prospective short-circuit current is found to be less than the discrimination limit given in the tables for the selected combination, then discrimination is guaranteed for all possible short-circuits at or downstream from this point in the distribution circuit.

If the prospective short-circuit current at this point of installation is found to be higher than the discrimination limit, then discriminative operation by the downstream breaker will only be guaranteed for short-circuit currents up to the value given in the tables. Since the probability that the maximum short-circuit current will flow is generally small, the specifier may decide that this value is sufficiently high. If not, a switchgear combination with a higher discrimination limit must be selected.

Discrimination tables · Back-up protection

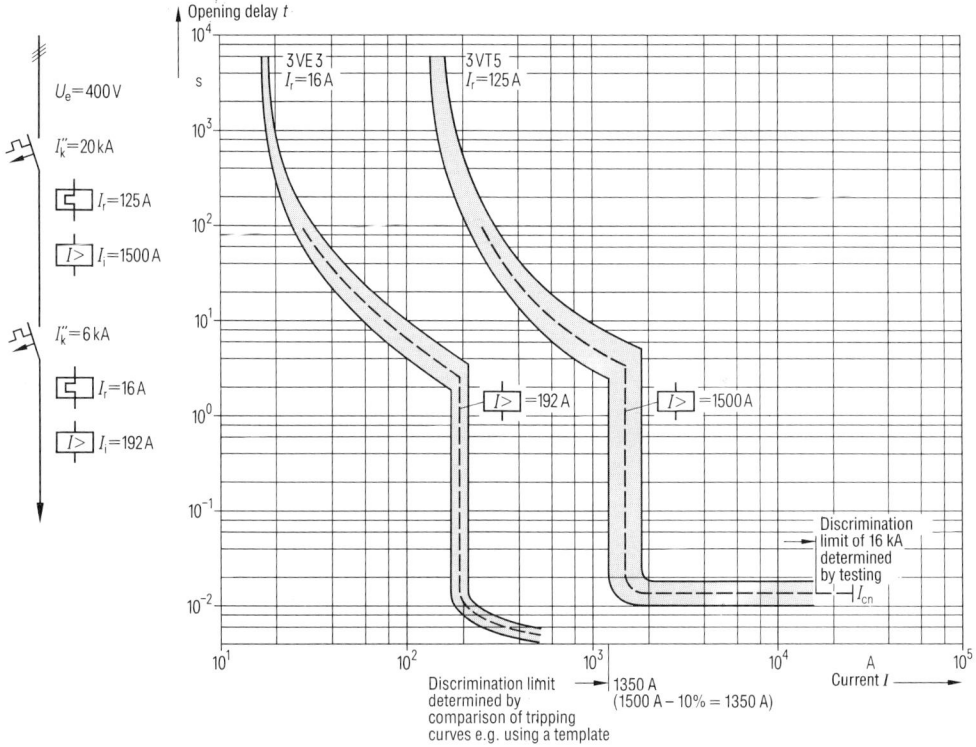

Figure 3.58
Discrimination limit illustrated by the comparison of tripping characteristic curves

Back-up protection

The DIN VDE specifications and the IEC publications permit back-up protection of an item of switchgear by one of the upstream protection devices with corresponding short-circuit switching capacity, provided that both the feeder circuit and the downstream protection device are sufficiently protected. This means, that through the use of back-up protection, the downstream circuit-breaker is permitted to have a breaking capacity which is smaller than the prospective short-circuit current at its point of installation. This method allows a certain degree of freedom in the system design, and can be particularly advantageous in the optimizing of component protection and for economic considerations.

In practice it means, that at a point in the network with a prospective short-circuit current of e.g. 50 kA, a less expensive protection device with only 20 kA short-circuit breaking capacity may also be used if back-up protection is provided by an upstream protection device in such a way that the combination has a breaking capacity equal to, or in excess of, 50 kA.

To some extent, the assessment of back-up protection for power circuit-breakers and for line protection circuit-breakers (m.c.b.) to DIN VDE 0641 is still based on the switching capacity tests for the determination of I_{cn} to DIN VDE 0660 part 1 (alt. IEC 947-2).

The limit values for discrimination and back-up protection for Siemens switchgear are given in the tables contained in the booklet "Discrimination and back-up protection in fuseless low-voltage feeders" (order number E20001-P285-A372-X-7600), as well as in the Siemens booklet "Overload and short-circuit protection in low-voltage installations" (order number A19100-E732-A326, German ed.).

3.4.6.3 Discrimination in meshed networks

In a meshed network, the discrimination must fulfill two conditions:

1. Only the cable on which the short-circuit has occurred may be isolated from the network.
2. In the event of a short-circuit on the terminals of an incoming feeder transformer, only the point of fault may be isolated from the network.

Node point fuses

In a low-voltage meshed network, normally cables of the same cross-section, and therefore LV fuses with operation class gL of the same type and current rating, are used. The fuses are located at the node points in the network (Fig. 3.59).

Should a short-circuit occur in one of the meshed network cables at point K1, the short-circuit currents I_{k1}, I_{k2} and I_{k3} flow from the feeder points a and b to the point of short-circuit. The values of these short-circuit currents depend on cable resistances and on the nature of the feeder points. They may therefor vary greatly. Figure 3.59 shows a simplified illustration of node points created by three cables.

Discrimination is provided if the fuse carrying the sum of currents, I_{k1} plus I_{k2}, ruptures while the fuses carrying only a portion of the short-circuit current, i.e. I_{k1} or I_{k2}, remain intact.

In the case of Siemens LV HRC fuses, the current value ratio $I_{k1}/(I_{k1}+I_{k2})$ permissible for higher short-circuit currents, is 0.8.

Meshed network circuit-breakers and node point fuses

In the incoming feeder points of a meshed network fed by several high-voltage lines, meshed network circuit-breakers are generally installed between the transformer and the node point (Fig. 3.60).

If a short-circuit occurs on the high-voltage side of the transformer (point K1) or between the transformer and the meshed network circuit-breaker (point K2), then the HV fuse on the high-voltage side ruptures. On the low-voltage side, power is fed back through the meshed network circuit-breaker to the point of fault. The shunt trip of the circuit-breaker receives the trip command from the network master relay. In this way the point of fault is selectively isolated from the network (see Section 3.4.5.3, page 160).

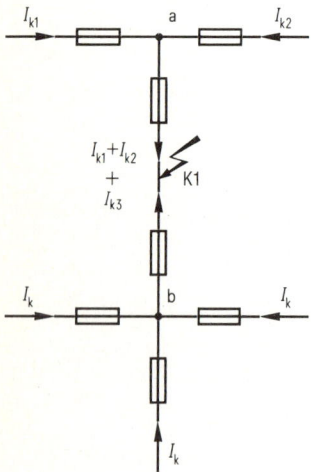

Figure 3.59
A short-circuited cable fed from two node points a and b

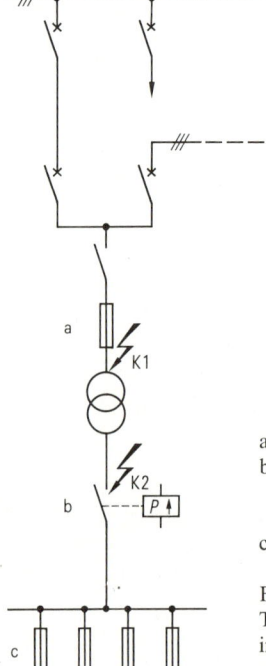

a   HV HRC fuses
b   Meshed network circuit-breaker with network master relay
c   Node point fuses

Figure 3.60
Typical configuration of an incoming feeder point in a low-voltage meshed network (single-line diagram)

3.5 Protection against overvoltage

3.5.1 Overvoltage transients associated with vacuum switchgear

Causes of overvoltages

Irrespective of the arc quenching method employed (i.e. also with SF6 switching technology), switching overvoltages result whenever an inductive alternating current is switched off abruptly before it passes through a natural zero point. The rate-of-rise and the amplitude of these voltages depend on the network and load side conductor configuration (busbar inductance and cable capacitance), the value of the current to be interrupted and on the arc-quenching method. (Fig. 3.61). The energy E stored on the load side results mainly from the chopping current I_{ab} and the inductivity of the load L_a:

$$E = \tfrac{1}{2} L_a \cdot I_{ab}^2.$$

After the separation from the network side, the periodic charging/discharging of the energy stored in the load through the associated stray capacitance C_a, causes a voltage of u_a with maximum value $U_ü$, at the contact gap (Figure 3.62). The energy can be expressed as

$$E = \tfrac{1}{2} C_a \cdot U_ü^2.$$

Ignoring the cable losses, this results in an overvoltage of roughly

$$U_ü = \sqrt{\frac{L_a}{C_a}} \cdot I_{ab},$$

Derived from

$$\tfrac{1}{2} C_a \cdot U_ü^2 = \tfrac{1}{2} L_a \cdot I_{ab}^2.$$

The associated transient frequency equals

$$f = \frac{1}{2\pi \cdot \sqrt{L_a \cdot C_a}}.$$

There are three basic causes of overvoltages:

1. Chopping effect (premature break-off of current)

The negative resistance of the switching arc can lead to high frequency oscillating currents within the switching gap of the vacuum tube. If this high frequency current passes through zero, then the current at operating frequency will be quenched before it passes through its own natural zero point (Fig. 3.62).

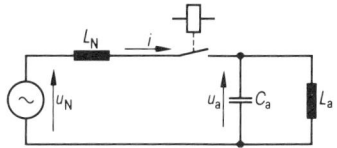

u_N Line-earth voltage, at operating frequency
u_a Line-earth voltage on the load side of the switch
i Load current
L_N Network inductivity
L_a Equivalent inductance of the load side
C_a Equivalent capacitance of the load side

Figure 3.61
Single-line equivalent circuit for the interruption of inductive currents

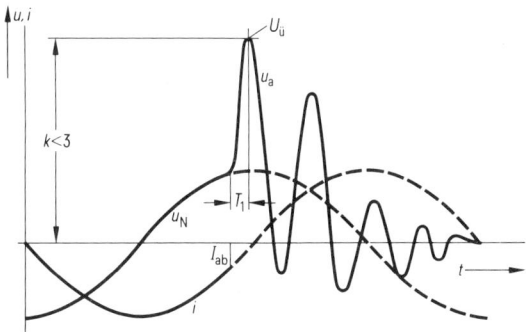

u_N Line-earth voltage, at operating frequency
u_a Line-earth voltage on the load side of the switch
i Current at operating frequency
I_{ab} Chopping current
$U_ü$ Maximum overvoltage on the load side
T_1 Unipolar wavefront time
k Overvoltage factor

Figure 3.62
Voltage/current waveform during the interruption of inductive loads

Even in the event of the most unfavourable network and load circuit parameters, the amplitude and rate-of-rise (unipolar front time T_1) of the possible resulting switching overvoltages will not endanger the insulation of the load. This is because the maximum overvoltage (k<3) lies within the specified dielectric strength of the associated load circuit (k<5 for SF6 contactors).

2. Multiple restriking

If the switching arc re-ignites in the contact gap during the opening of the contacts and after the arc has been quenched once, because the dielectric strength of the switching gap does not build up fast enough (e.g. owing to insufficient contact distance) in relation to the rise in voltage, then multiple restriking results. This phenomena is initiated by the high frequency transient current in the oscillating circuit comprising L_s and C_a. If this transient current (at t_2 to t_3, t_4 to t_5 etc.) is repeatedly quenched by the contact gap as the contacts move further apart (at t_3, t_5 etc.), the cycle of quenching and restriking will continue until either the dielectric strength of the contact gap is sufficiently high, or until the load current at operating frequency starts flowing again (standing arc). See Figure 3.63.

During this repeated quenching and restriking cycle, overvoltages with short bipolar wavefront time (T_2 of a few μs) and overvoltage factor $k \geq 7$, can result (Fig. 3.64).

Multiple re-ignition also sometimes occurs during *closing*, as a result of pre-striking.

Switchgear with a high closing speed (≈ 0.3 mm/ms), e.g. vacuum contactors, limit the overvoltage created during closing to uncritical values (k < 3).

Switching overvoltages caused by multiple restriking do not only occur when switching is done under vacuum, but are also found, albeit to a lesser degree ($k < 3$), e.g. in SF6 contactors.

3. Virtual current chopping

In a three-phase system, multiple restriking occurs primarily in the pole that clears first. The current waveform in the other two poles can partially be influenced by the multiple restriking occurring in the first pole to clear (Fig. 3.65).

Certain network and load-side configurations, as well as particular load operating conditions, permit cross-coupling effects. These cause superimposed high frequency currents which can force unwanted zero

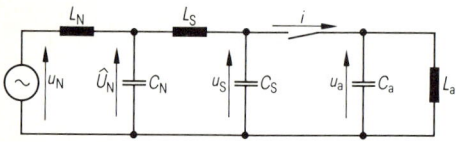

$\hat{U}_N$ Peak value of the line-earth voltage
u_s Line-earth voltage on the network side of the switch (at frequency f_a)
u_a Line-earth voltage on the load side of the switch (at frequency f_a)
i Load current
L_N Network inductivity
L_S Busbar inductivity (including coupled inductance)
C_N Network capacitance
C_S Busbar capacitance

$L_N < L_a$ (the network short-circuit power, determined by L_N, is much greater than that of the load which is a function of L_a);
$C_N < C_a$ (C_N is the total capacitance of feeder network and therefore much greater than the capacitance C_a of the individual cables between the switch and the load);
$L_S < L_N$,
$C_S < C_N$

Figure 3.63
Single line equivalent circuit for the restriking during opening

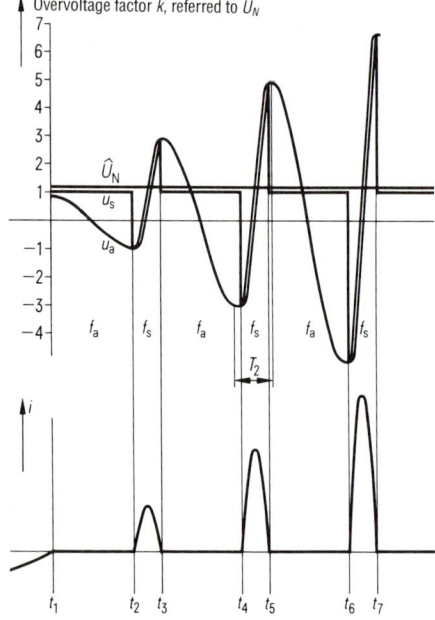

f_a Frequency, dependent on L_a and C_a
f_s Frequency, dependent on L_s and C_s
T_2 Wavefront time (bipolar)

Figure 3.64
Theoretical voltage and current waveforms during multiple restriking

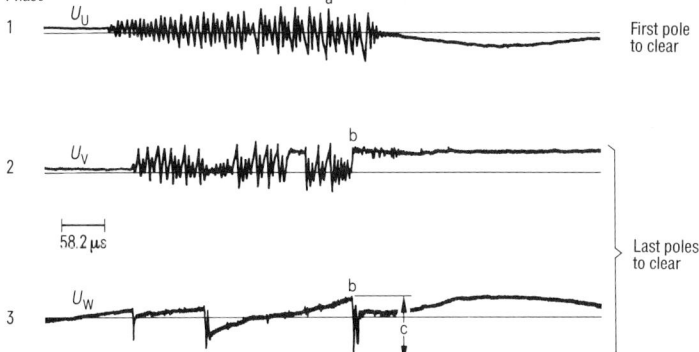

Figure 3.65
Voltage wave measured at the motor terminals with multiple restriking and virtual chopping. Phase 2 is fitted with an overvoltage limiter

crossings in the poles which are last-to-clear. This phenomenon is known as "virtual chopping".

These current zero crossings may also cause multiple restriking. The resulting overvoltages may achieve values of $k \geq 10$ with short bipolar wavefront times (T_2 of a few microseconds).

3.5.2 High voltage vacuum contactors, type 3TL6, for the switching of three-phase inductive motors with slip-rings or short-circuited rotors over 1 kV up to 12 kV

The transients in the circuit depend not only on the network and conductor configuration, but also on the operating condition of the motor at the instant of switching.

Owing to the damping influence of the back-e.m.f. created in a motor, no overvoltages are caused during the operational switch-off with a 3TL voltage vacuum contactor after the motor has run up to speed, or while it is running at no load. Already after the first arc-quenching, the dielectric strength of the contact gap rises at a greater rate than the amplitude of the recovery voltage. This is typical for vacuum contactors and no special overvoltage protective circuitry is required.

Overvoltages and their effects

Three-phase motors with starting current up to 600 A

In the special case of "switching off a motor during start-up or under locked rotor conditions" the back-e.m.f. of the motor is still small or equal to zero. Therefore, multiple restriking (also caused by virtual chopping) may occur.

During restriking, bipolar surge voltages with high peak values are created (Fig. 3.64) and these spread out from the contactor in the form of travelling waves. They are split between the network and load sides in terms of the respective wave resistance. Since there are generally several cable branches on the network side and only one cable between the contactor and the motor (and assuming the same cable type), a portion of $n/(n+1)$ of the travelling wave will run towards the motor.

The wave resistance of the cable lies in the range of 10 Ω, whereas the wave resistance of the motor is of the order of 1000 Ω. Therefore, the wave is reflected at the join between motor and cable. The reflection increases the overvoltage at the motor terminals by a factor ≤ 1.7.

The stressing of the motor insulation depends on both the amplitude of the voltage and the wavefront time of the travelling wave.

On the one hand, a travelling wave (surge wave) entering the motor stresses the main insulation between the winding and the iron slots of the stator. On the other hand, it is split up between the individual winding coils in the motor, whereby the first coil (line-end coil) is stressed to the greatest degree.

In high-voltage motors, the dielectric strength of the

motor is determined to a large extent by the permissible stressing of the line-end coil insulation.

Even line-end coils with reinforced insulation cannot withstand the severe voltage stressing caused by transient waves having very short wavefront times, especially if the overvoltage is increased by reflection effects.

Protective measures are therefore generally required.

Three-phase motors with starting currents higher than 600 A

Motors of higher power rating are less sensitive to switching overvoltages caused by multiple restriking. The reasons for this lie firstly in the construction of the line-end coils in the stator winding, and secondly in the current/voltage waveform during switch-off of the motor under locked rotor or partially run-up conditions.

Motors of higher power rating are wound in such a way that each slot in the stator contains only two conductors, one above the other. The insulation between conductors is therefore only stressed with the voltage difference between these windings (volt per winding). During switch-off of starting currents higher than 600 A, the arc-quenching procedure spans several half cycles without premature current chopping. The arc is finally quenched during a current zero at operating frequency. Overvoltages seldom occur and if so in a weaker form ($k<3$).

The reinforced insulation of the line-end winding provides sufficient safety. Protection circuits are therefore not necessary for motors of larger power rating.

Protective measures for three-phase inductive motors between 1 and 12 kV

Multiple restriking is expected in the range indicated by the shaded line in Figure 3.66. Special overvoltage limiters, type 3EF, for use in industrial cable networks have been developed for the protection against overvoltage in high-voltage motors of lower power rating. These units comprise a zinc oxide varistor in series with an arc gap. By means of the arc gap, the varistor is only used if the voltage peaks exceed a specified value. Under normal operating conditions the varistor is isolated. In this way it is possible to set the limiting voltage so low that the overvoltage factor of $k=3$, is not exceeded (Fig. 3.67).

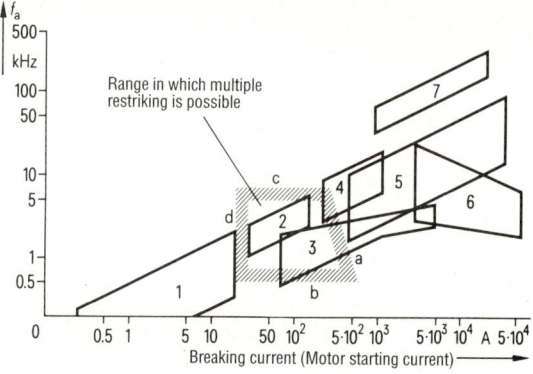

1. Switching of unloaded transformers
2. Switching of neutral earthing transformers with arc suppression coils
3. Switching of motors during starting
4. Switching of compensating coils
5. Switching of transformers with a short-circuit on the secondary side
6. Switching of network short-circuits
7. Switching of short-circuited chokes

Figure 3.66
Examples of specific switching parameters in inductive circuits to illustrate the range in which multiple restriking is possible

The 3EF overvoltage limiters restrict the amplitude of the voltage spikes occurring during multiple restriking, but do not change the rate-of-rise of these spikes. For motors with insulation level $k \geq 6$, the 3EF overvoltage limiters offer optimal protection and the most economical solution. The service life of the limiters is practically unlimited.

The overvoltage limiters also protect against possible voltage spikes caused by virtual current chopping in the last pole to clear which result from the cross-coupling of multiple restriking in the pole which clears first. In particular cases, if the network capacitance is very small in comparison to that of the load side, additional damping measures with RC elements may become necessary.

This is especially true for older motors in which the quality of the winding insulation is unknown. By means of careful and coordinated selection of the components, RC elements used in conjunction with overvoltage limiters may achieve complete suppression of multiple restriking. For the selection of the

Protection against overvoltage

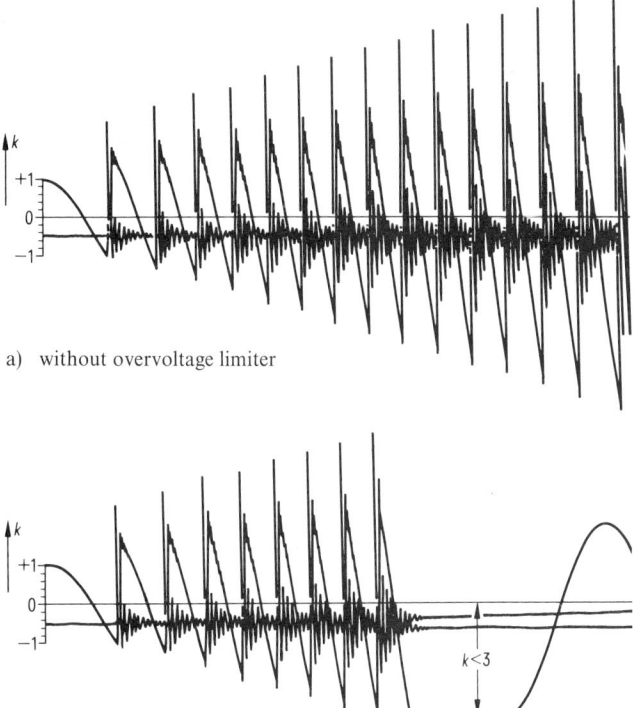

Figure 3.67
Voltage waveform at the motor terminals during multiple restriking

a) without overvoltage limiter

b) with overvoltage limiter

A Response voltage of the limiter

components in such a tuned the RC circuit, the complete installation network data such busbar inductance (including that of neighbouring branches), load side cable type, cable lengths and motor sizes must be known.

The overvoltage limiters and RC elements are connected between the individual phase conductors of the load circuit and an earthed star point.

Examples of circuits for overvoltage protection at rated voltages from 3.6 to 12 kV

A cable length of $l = 100$ m has been assumed for the circuit examples shown in Figure 3.68. Space considerations normally dictate that the protection elements are installed within the switching cubicle near to the high-voltage vacuum contactor. Generally speaking, this location is acceptable for the functioning of the protection circuit. In AC-4 operation of the motor and unfavourable cable configuration (for cable lengths over 100 m the difference between the wave resistance of the feeder cable and that of the

3 Selection criteria for low-voltage switchgear in main circuits

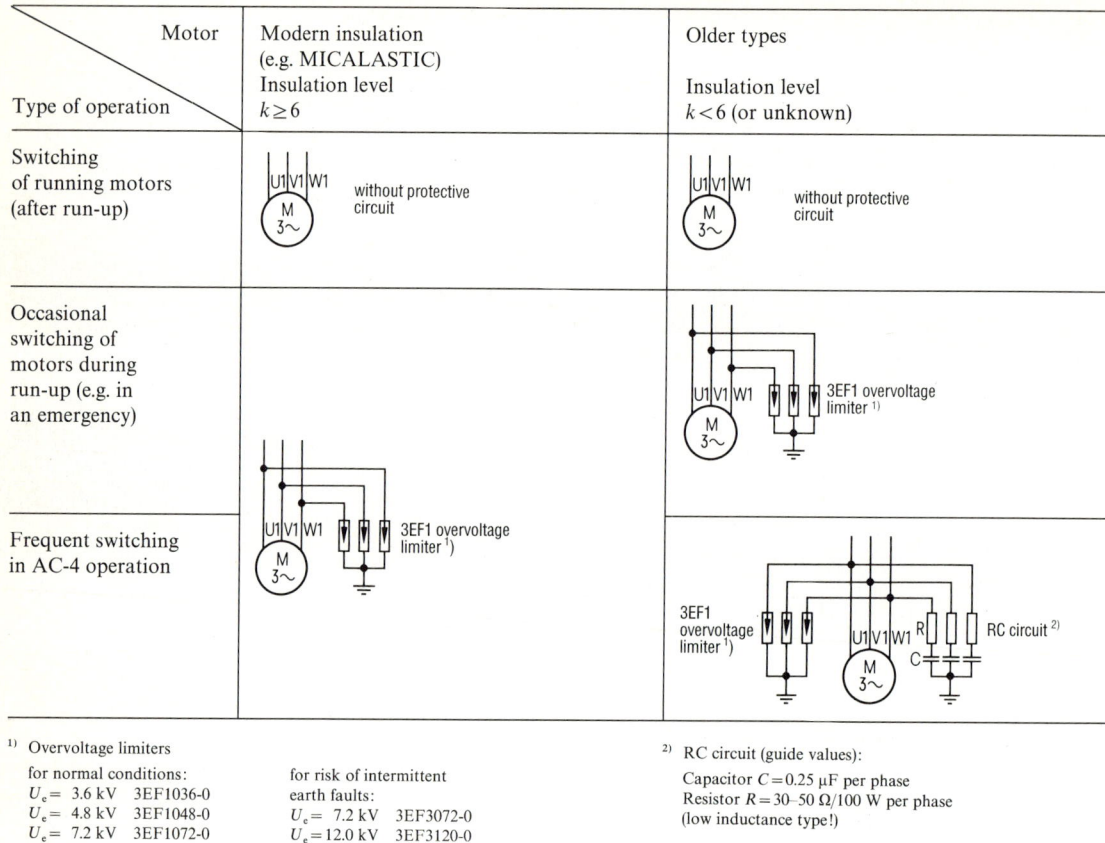

[1] Overvoltage limiters

for normal conditions:
$U_e = 3.6$ kV 3EF1036-0
$U_e = 4.8$ kV 3EF1048-0
$U_e = 7.2$ kV 3EF1072-0
$U_e = 12.0$ kV 3EF3120-0

for risk of intermittent earth faults:
$U_e = 7.2$ kV 3EF3072-0
$U_e = 12.0$ kV 3EF3120-0

[2] RC circuit (guide values):
Capacitor $C = 0.25$ µF per phase
Resistor $R = 30$–50 Ω/100 W per phase
(low inductance type!)

Figure 3.68
Examples of circuit diagrams for the overvoltage protection of motors with starting current up to 600 A at a rated voltage of 3.6 to 12 kV

motor winding becomes large), strong reflection of the travelling wave (factor ≤ 1.7) may necessitate the installation of a protection circuit closer to the motor terminals.

If all the necessary network data for a tuned overvoltage protection circuit (capacitance and inductance of network and load sides, cable type, cable lengths, number of parallel feeders on the network side and the operating duty of the motor) are known, then a circuit diagram of the protection system (including component ratings) can be supplied on request (Fig. 3.69).

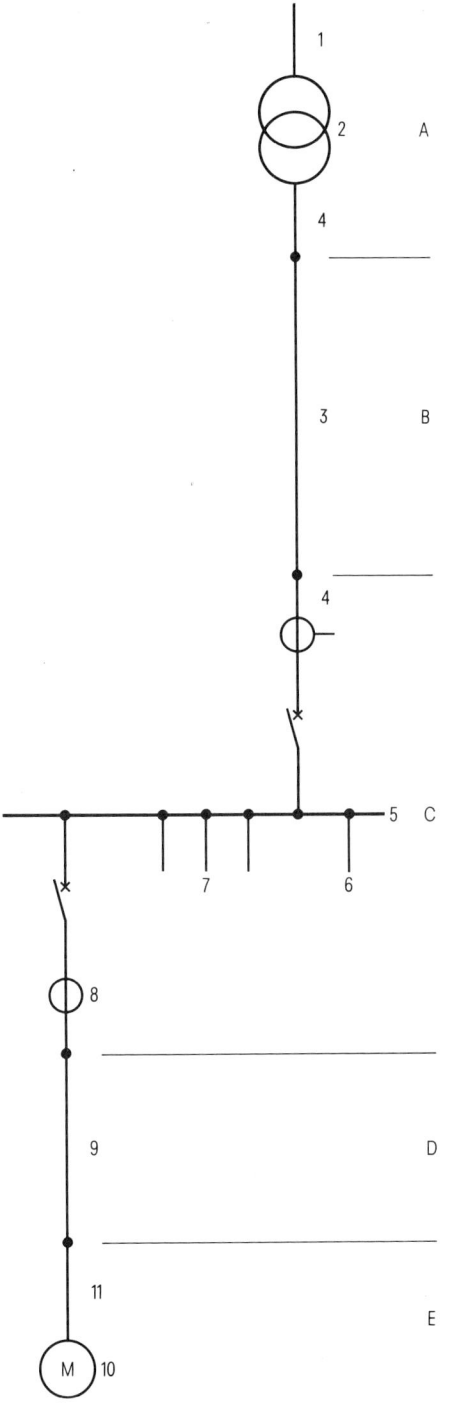

A Feeder transformer

1. Rated frequency 50 or 60 Hz,
   Operating voltage (maximum value),
   Output power rating,
   Short-circuit voltage

2. Low-voltage winding in star or delta connection?
   If star connection: Is the star point earthed, or not (floating)?

B Connection to the substation

3. Cable and/or open conductors in air?
   Length of cable and/or conductor, effective capacitance $\bar{C}$ in µF/km (only for cable), cable insulation (PE, PVC, paper), type designation of the cable

4. Is overvoltage protection installed on the cable at the transformer and substation ends?
   Voltage arrestor?
   Capacitor?

C Substation

5. Substation type
   Length of busbars between the incomer and motor switching cubicles

6. Other loads which are connected to the busbars at the same time
   Number and lengths of cables,
   type designation of cables
   effective capacitance $\bar{C}$ (in µF/km),
   cable insulation (PE, PVC, paper)

7. Are power factor correction capacitors present in the substation?
   Static reactive power compensator?
   Filter circuits?

8. Number of current transformers

D Connection to the drive

9. Busbar or cable connection?
   Length of busbar/cable,
   type designation of the cable
   effective capacitance $\bar{C}$ (in µF/km),
   cable insulation (PE, PVC, paper)

E Drive

10. Three-phase induction motor
    Rated voltage,
    Power rating,
    Insulation level (BIL),
    Motor type

11. Overvoltage protection fitted?
    (voltage arrestor, RC elements)

Figure 3.69
Required data for the design of a tuned overvoltage protection circuit for three-phase induction motors with rated voltage 3.6 to 12 kV

3.5.3 Vacuum contactors, type 3TF6, for the switching of three-phase induction motors with slip-ring or short-circuited rotors up to 1000 V

In general, the same principles apply for the Siemens 3TF6 units as for the high-voltage vacuum contactors type 3TL6.

The 3WS vacuum circuit-breakers possess a rated service and ultimative short-circuit breaking capacity of 50 kA at AC 500 V ($I_{cs} = I_{cu}$). They are suitable for use up to rated voltages of AC 1000 V, permit a large number of short-circuit interruptions and have an extremely high short-time withstand current capability. They incorporate the same electronic overcurrent tripping units as the 3WN air circuit-breakers.

Overvoltages and their effects

In low-voltage motors too, the dielectric strength of the line-end coil determines the permissible voltage stressing of the motor.

Generally, the technique of "mush winding" is used for these motors. This means that the first and the last conductor of a coil, and with lower probability the conductors of different coils of the same phase, may lie next to each other in the winding overhang. As a result, the dielectric strength of the wire insulation and the voltage drop on the first coil are of utmost importance.

In the case of new wire, the a.c. dielectric strength between two wires is typically 2 to 2.5 kV. It is however reduced through thermal ageing and can under certain circumstances drop to the level of the operational voltage.

Basic research tests have shown that overvoltages resulting from multiple restriking may reach peak values of up to 5 kV. These would therefor be sufficiently high to damage even new wire insulation. This means that other protective measures are required.

**Protective measures
for three-phase induction motors up to 1000 V**

The relationship between insulation levels in low-voltage motors and the overvoltage peak values resulting from multiple restriking led to the decision by Siemens to equip all 3TF6 vacuum contactors with integrated RC circuits. In contrast to conditions in and around high-voltage contactors, the smaller creapage distances and clearances in air permit the circuit components to be mounted directly into the housings of low-voltage contactors.

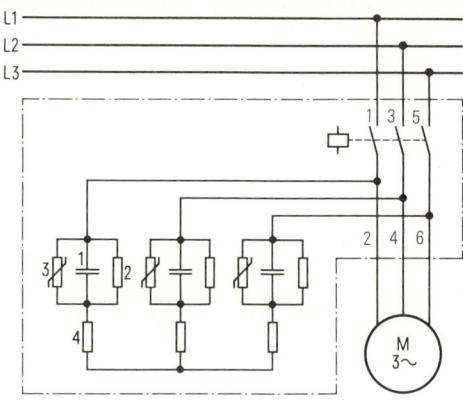

1 Capacitor 0.22 µF, 1000 V
2 Resistor 1.5 MΩ
3 Varistor 514 kΩ, 680 V
4 Resistor 100 Ω, 0.5 W

Figure 3.70
Internal wiring diagram of the Siemens 3TF6844-0C vacuum contactor

Figure 3.70 shows the circuit diagram. The varistor protects the capacitor from overvoltage spikes which do not originate from switching operations of the contactor itself. The connection of the RC circuit must always be made at the load side on the main current path assembly. They are factory-connected to the terminals 2, 4 and 6.

The advantage of this integrated RC circuit is that no special consideration need be given to the particular network conditions when selecting a Siemens 3TF6 vacuum contactor.

The 3WS vacuum circuit-breakers also incorporate an integrated RC circuit for the main current paths. It is connected across the load-side terminals and ensures that switching overvoltages are suppressed to uncritical levels. If the infeed is from the bottom (i.e. terminals 2, 4, 6) then the RC circuit must be connected across terminals 1, 3 and 5. This should be indicated during ordered.

Note: If high frequency oscillations (e.g. harmonics) are created by other equipment in the vacuum contactor or circuit-breaker circuit, e.g. by inverter load, then the integrated RC circuit may become overloaded. In such cases, please enquire as to whether the integrated RC circuit should be removed.

3.6 Leakage current and earth-fault protection

The use of leakage-current protection devices (differential-current protection) is becoming increasingly important for the protection of installations and personnel. The advantage of these devices lies in the versatility of their protective functions:

▷ Protection in the event of indirect contact,
▷ Additional protection in the event of single-pole direct contact (rated leakage current $I_{\Delta n} \leq 30$ mA),
▷ Earth-fault protection,
▷ Protection against fires resulting from earth-fault leakage currents.

3.6.1 Construction and method of operation

The construction of an earth-leakage protection device (Fig. 3.71) may be divided into three function groups:

1. Summation current transformer for the detection of the leakage current,
2. Release mechanism to convert the electrically measured value into a mechanical movement (e.g. holding magnet),
3. Switch mechanism and main contacts in the case of earth-leakage circuit-breakers (e.l.c.b.), alternatively switching mechanism and auxiliary contacts in the case of earth-leakage releases/tripping devices/relays.

The summation current transformer (1), a ring band core with high quality magnetic properties, surrounds and encloses all active conductors including the neutral conductor if present. In a fault-free installation, the vector (geometric) sum of the currents flowing to and from the load equals zero. In this case no voltage is induced in the secondary winding of the summation transformer (also known as a core-balance current transformer).

If, however, a leakage current flows to earth (protective conductor) downstream from the earth-leakage protection device, the sum of the currents will no longer be equal to zero (differential current). This leakage current causes a magnetic flux to circulate within the transformer core which in turn induces a voltage in the secondary winding. The release mechanism (2) is energized and the faulty circuit is switched off.

The rated earth-leakage current $I_{\Delta n}$ is a characteristic value of the device. The tripping current, however, usually lies between 50 and 100% of this earth-leakage current rating.

To test the functionality of the earth-leakage protection device, it is equipped with a test feature. Operation of the test button causes a test current to flow outside of the transformer core. This is interpreted as a leakage current and must cause the device to trip.

Figure 3.72 shows an application example of earth-leakage circuit-breakers in a TN network (system).

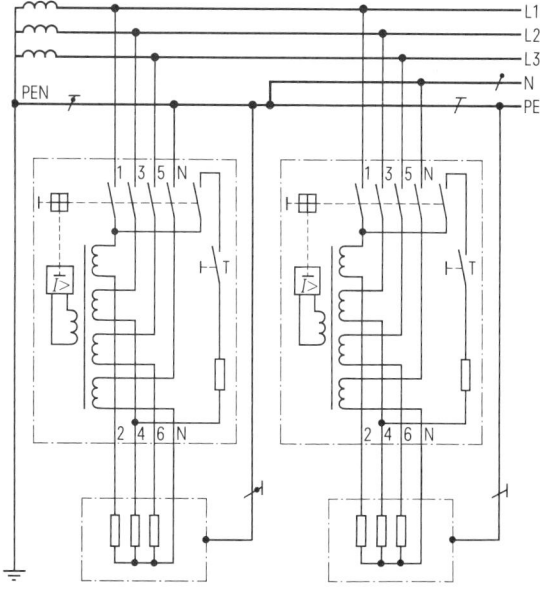

Figure 3.72
Application of earth-leakage circuit-breakers in TN networks (example)

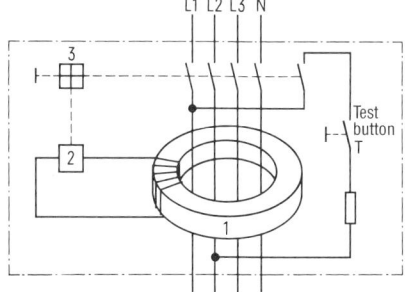

Figure 3.71
Four-pole earth-leakage circuit-breaker

Table 3.24 Tripping currents to DIN VDE 0664

Type of current	Symbol	Tripping current
A.C. leakage currents	∼	0.5 to $1 \cdot I_{\Delta n}$
Pulsing d.c. leakage currents (pos. and neg. half-waves), half-wave current	⌢⌢ ⌣⌣	0.35 to $1.4 \cdot I_{\Delta n}$
Chopped half-wave currents: phase angle 90° el 135° el	⌐⌐ ⌐⌐	0.25 to $1.4 \cdot I_{\Delta n}$ 0.11 to $1.4 \cdot I_{\Delta n}$
Half-wave current superimposed with smooth d.c. current of 6 mA	⌢⌢ ═══	max. $1.4 \cdot I_{\Delta n} + 6$ mA

3.6.2 Leakage-current circuit-breakers for a.c. and pulsing d.c. leakage currents

Owing to the increasing use of electronic components in equipment, leakage current is frequently of non-sinusoidal waveform. For the protection in the event of indirect contact to DIN VDE 0100 part 410, devices which are sensitive to pulse currents in accordance with DIN VDE 0664 must be used for the leakage-current protection measures (see Table 3.24). These devices also trip in the case of d.c. leakage currents which become zero or nearly zero within a period of the network frequency.

3.6.3 Discriminative leakage-current circuit-breakers

The application possibilities of leakage-current protective circuits are extensively increased if series connected leakage-current protection devices switch off in a discriminative manner. The earth leakage circuit-breakers types 5SZ and 5SV are available in discriminative versions. This also applies to the earth leakage protection device 3UL2.

3.6.4 Product range of Siemens leakage-current circuit-breakers and tripping devices

Siemens manufactures leakage-current protection devices with various current ratings and rated leakage currents. For use in domestic applications, the 5SZ earth-leakage circuit-breakers are available in two-pole versions up to $I_n = 40$ A ($I_{\Delta n} = 0.01$ A to 0.3 A) and four-pole versions up to $I_n = 63$ A ($I_{\Delta n} = 0.03$ A to 0.5 A). The next size of the 5SZ earth-leakage circuit-breaker has a rated current of 125 A ($I_{\Delta n} = 0.03$ to 1 A).

The 5SV circuit-breakers have a current rating of 160 and 224 A ($I_{\Delta n} = 0.3$ to 1 A).

In addition to these earth leakage circuit-breakers, the 3UL2 leakage-current protection unit offers leakage-current and earth-fault protection at even higher rated currents. It consists of two separate devices, viz. a summation current transformer and a tripping relay. The summation current transformer is offered in three sizes with inner diameter of 40 mm, 65 mm and 120 mm respectively. The 3UL2 may be used in conjunction with other switchgear. It may also be used for monitoring and alarm purposes.

3VF moulded-case circuit-breakers may be factory-fitted with differential-current tripping units. These offer adjustable tripping currents and delay times.

Further technical data is given in the Siemens low-voltage switchgear catalogues. Also refer to Section 8.8 on page 486.

3.7 Application of low-voltage switchgear in main circuits

3.7.1 Parallel and series connection of current path assemblies

Parallel connection

If the current path assemblies of a multi-pole switching device are connected in parallel, then the total current is divided between the individual poles in accordance with the respective ohmic resistances and the mutual inductive coupling between the current paths. The ohmic resistance of each pole, and thus current path, is mainly determined by the contact resistance. This changes with time owing to effects of switching arc erosion and oxidation. Thus, the current distribution between the poles is neither uniform nor constant with respect to time. Individual current paths may become overloaded and the overload releases or relays may operate prematurely (nuisance tripping).

If no other specific details are given in the catalogues, the following applies for the continuous loading of parallel connected poles:

▷ If three identical current path assemblies of a switching device are connected in parallel, the continuous current carrying capacity increases to 2.5 times that of the single current path. For two paths in parallel, the factor is 1.8. It should be noted, however, that the making and breaking capacities for capacitive load do not increase because the contacts do not close and open simultaneously and therefore the contacts of one pole must be able to make or break the total current. In the case of ohmic or motor load, the making capacity increases by a factor of approx. 1.5 for two current paths in parallel. The breaking capacity of the switching device is *not* increased by parallel connection of current paths.

▷ The connecting cables should be dimensioned and arranged in such a way that each individual current path is connected to the same length of cable. More uniform sharing of the total current can be achieved by making the parallel connection as far away as possible from the switching device.

▷ A short-circuit current will be divided between the current paths in accordance with their respective resistances. Thereby, it may be possible that the current is not high enough in any of the path assemblies to reach the set tripping value of the instantaneous electromagnetic releases.

The making and breaking capacities of contactors, with respect to the load currents for two or three poles connected in parallel, are illustrated in Table 3.25.

Table 3.25
Making and breaking capacities of contactors expressed as multiples of their rated current I_e for three-pole switching and parallel connection of two or three poles in terms of IEC 947-4; Rated voltages to IEC 947-4

	three-pole switching [1]	2 poles in parallel [1] $I'_e = 1.8 \cdot I_e$	3 poles in parallel [1] $I''_e = 2.5 \cdot I_e$
Making capacity	$12 \cdot I_e$ (utilization category AC-4)	$\dfrac{12 \cdot I'_e}{1.8} = 6.67 \cdot I'_e$	$\dfrac{12 \cdot I''_e}{2.5} = 4.8 \cdot I''_e$
Breaking capacity	$10 \cdot I_e$ (utilization category AC-4)	$\dfrac{10 \cdot I'_e}{1.8} = 5.55 \cdot I'_e$	$\dfrac{10 \cdot I''_e}{2.5} = 4 \cdot I''_e$

[1] Voltage across each switching gap $U = \dfrac{U_e}{\sqrt{3}}$

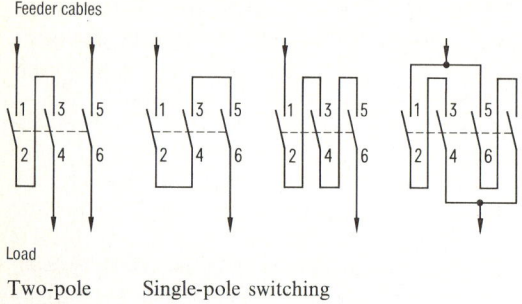

Two-pole switching   Single-pole switching

Figure 3.73 Series connection of current paths

Series connection

If the current path assemblies, or poles, of a switching device are connected in series, then it may be used to switch higher operating voltages. This characteristic is employed when triple-pole contactors or circuit-breakers are used to switch single-phase a.c. and, more significantly, d.c. load currents (Fig. 3.73). In some cases, the direction of energy flow must be taken into account.

Naturally, this higher operating voltage may not exceed the rated insulation voltage of the switching device.

Series connection of current paths in circuit-breakers and overload relays

Regardless of the number of poles connected in series, circuit-breakers retain the same rated breaking current, since the thermal rating of the breakers is fully utilized in all cases. However, when two or three poles are connected in series, the circuit-breakers may be used at higher system voltages than when only one pole is used.

When using triple-pole circuit-breakers for switching d.c., the ability to switch higher system voltages when several poles are connected in series, is used to great advantage.

For information on using a.c. switchgear in d.c. networks, see Section 3.7.4, on page 192.

The Siemens l.v. switchgear catalogues give details on the switching of d.c. loads for all triple-pole circuit-breakers, including cases where several poles are connected in series.

For single-phase a.c. or d.c. loads, the series connection of circuit-breaker poles for single or double-pole switching is recommended to ensure that the overload releases in all three poles are loaded uniformly (see Section 3.4.3.4, page 131).

The time-current tripping characteristics of the thermally delayed overcurrent relays and releases are based on uniform current loading of all three bimetal strips. The series connection of all three poles guarantees uniform current loading. In devices with phase-loss sensitivity, all the poles must be series connected in single-phase applications to ensure correct operation.

Series connection of contactor poles

For all Siemens contactors, the permissible operating currents and system voltages for the switching of d.c. loads using a single pole only or two/three poles in series, are given in the relevant l.v. switchgear catalogue.

If the triple-pole contactors are used with all three poles connected in series, then they may be used to their full rated operating current for switching d.c. loads at operating voltages up to 110 V. For voltages up to DC 220 V, this rule also applies for contactors from size 6 (3TF50) upwards. This means, that up to this operating voltage, the use of special d.c. contactors may be avoided.

3.7.2 Applications of four-pole switchgear

A number of DIN VDE specifications call for the use of four-pole switchgear:

▷ DIN VDE 0100 part 728, 03.90 "Standby generating plants".
  If stationary or mobile standby generating plants are used for the power supply of consumers during a failure of the distribution network, then *all the phase conductors and the non-earthed neutral conductor* of the distribution network as well as those of branch circuits which are not to receive power, must be isolated. Note: The PEN conductor must not be switched.

  If all of the connected loads are to be supplied from the standby generator, then the required four pole isolation can be accomplished by means of the main network circuit-breaker. If, however, only the selected important loads are to be fed from the standby generating plant in the event of a network failure, a bus-coupler circuit-breaker may be used for the isolation.

▷ DIN VDE 0100 part 530
  (identical to IEC 364-5-53).
  At present, the disconnection of the neutral con-

ductor is still required in leakage-current protection systems. In terms of the draft specification, a disconnection of the neutral conductor is not necessary in TN-S networks.

▷ DIN VDE 0100 part 430, 11.91
In IT networks, the neutral conductor (if present) must also be disconnected if an overcurrent protection system trips.

▷ DIN VDE 0100 part 410, 11.83, Section 7, as well as changes A1 to A3 (drafts)
If fault-voltage protection systems are used, the neutral conductor must also be disconnected.

In terms of French specifications:

▷ NFC 12-100 "Protection of employees in plants which use electric current", I. Decret I/Art. 2 and 8, IV. Fishe of Notes Techniques § 1, Comments according to PROMOTELEC, Exploitations Agricolas, Locaux Artisanaux, installation électrique:

At the power source of every three-phase circuit which has a neutral conductor, a disconnector which connects and disconnects the neutral conductor at the same time as the three phases, must be installed. The protection and disconnection function can be accomplished by the use of four pole switchgear such as the combination of fuses and disconnectors, circuit-breakers or contactors backed up by fused disconnectors.

▷ NCF 15-100 regulations for the erection and maintenance of low-voltage switchgear installations, clause 473.3.3–switching of neutral conductors:
If the switching of a neutral conductor is specified, then the connection and disconnection must be so arranged that the neutral conductor is not disconnected before the phase conductors and that it is connected either simultaneously or earlier than the phase conductors;

▷ NCF 15-100, chapter 53, appendix II and comments according to PROMETELEC: "Application examples for the disconnection of neutral conductors"

Note
In terms of the harmonization agreements, France is also obliged to waive the stipulation that neutral conductors in TN systems have to be switched (refer to the harmonizing document HD 384.4.46).

3.7.3 Influence of network frequency and harmonic currents on the operation of switching devices

If low-voltage switching devices designed for 50/60 Hz are to be used at other rated frequencies, the following characteristics must be examined closely:

▷ thermal loadability of the current path assemblies,
▷ switching capacity,
▷ service life of the contact system,
▷ tripping behaviour of the releases,
▷ behaviour of solenoid and motorized operating mechanisms.

3.7.3.1 Thermal load carrying capacity of the current path assemblies and conductors in dependence of the network frequency

In contrast to direct current, alternating current does not flow uniformly through the whole cross-section of a conductor. The current density is greater towards the surface, and this phenomenon increases with increase in frequency. At very high frequency the center of the conductor carries almost no current and the current only flows in a thin layer on the surface of the conductor. This is commonly known as "skin effect". The result of this "skin effect" is that the conductor cross-section is only partly used for carrying current and that the impedance of conductors increases linearly with increasing frequency. For example, in copper wire with a diameter of 10 mm, the useful cross-section is:

approx. 60% at 1 kHz, and
approx. 20% at 10 kHz.

A further phenomenon is also caused by alternating current at higher frequencies. A magnetic induction effect is found in neighbouring ferromagnetic materials. In conductive materials this magnetic induction causes power losses and produces heat. These so-called "hysteresis losses" increase rapidly with rising frequency.

The voltage induced in neighbouring metal materials and the resulting currents are proportional to the frequency of the phase current. These circulating currents also produce heat, known as "eddy-current losses".

Thus, an increase in frequency is associated with higher conductor impedance due to the skin effect, as well as higher hysteresis and eddy-current losses

Frequency f_1	Permissible rated operating current
100 Hz	$0.933 \cdot I_e$
200 Hz	$0.871 \cdot I_e$
300 Hz	$0.836 \cdot I_e$
400 Hz	$0.812 \cdot I_e$

Table 3.26 Permissible rated operating currents for contactors at frequencies up to 400 Hz

in the metallic materials, such as switching shafts, latching mechanisms, fixing brackets etc., situated near to the current paths.

The effect of the frequency on the operating current of the conductors themselves and on the neighbouring magnetic material, results in the following consequences for the load carrying capacity of the current path assemblies in the switching device:

Loadability of a.c. contactors

The following equation may be used to calculate the permissible operating current of Siemens contactors at a frequency f_1 in the range from 50 to 1000 Hz:

$$I_e(\text{AC-1}, f_1) = I_e(\text{AC-1, 50 Hz}) \cdot \sqrt[10]{\frac{50}{f_1}}.$$

The permissible operating currents for a number of frequencies below 1 kHz are given in Table 3.26.

For frequencies f_2 from 1 to 10 kHz, the following approximation applies:

$$I_e(\text{AC-1}, f_2) \approx I_e(\text{AC-1, 50 Hz}) \cdot 0.7 \cdot \sqrt[4]{\frac{1 \text{ (kHz)}}{f_2}}.$$

A special assessment is always required for particular applications. No generally valid statement can be made.

Loadability of circuit-breakers

Circuit-breakers designed for an a.c. voltage at 50/60 Hz can be used for at least the same rated currents at lower frequencies.

In contrast to this however, the permissible operating current may have to be reduced at higher frequencies, above approx. 100 Hz, to ensure that the specified temperature rise limits are not exceeded. For example, at 400 Hz, the permissible load carrying capacity of a circuit-breaker with a high proportion of metal parts, may be reduced to 80% or even 50% of that which applies at 50 Hz.

Loadability of busbars

The permissible operating currents at frequencies above 50 Hz, may be estimated by using the following equation:

$$I_e(f_x) = I_e(50 \text{ Hz}) \cdot \sqrt{\frac{50}{f_x}}.$$

For example, this means that the effective utilization of conductor cross-section is only approximately:

22% at 1 kHz, and
 7% at 10 kHz.

3.7.3.2 Switching capacity at network frequencies other than 50 Hz

In contrast to direct current, alternating current assists arc quenching in that the arc is quenched each time the current passes through zero and can only exist again if it is re-ignited during the next half wave. The arc is effectively quenched if restriking is prevented after the current zero. In a.c. arc quenching therefor, the charge-carriers produced by the arc (ionized gas) must be removed from between the contacts after they have separated, to restore the dielectric strength of the contact gap and thereby prevent the recovery voltage from restriking the arc once more.

At higher frequencies, the current zeros follow each other more rapidly. Thus the arc duration per half wave is shorter and the degree of ionization in switching chamber is lower than at 50 Hz.

Consequently, in terms of the influence of frequency on their switching characteristics, a.c. switching devices may generally also be used at frequencies higher than 50 Hz (up to approx. 500 Hz).

Motor switching capacity of contactors at frequencies above 50 Hz

The starting currents of motors for higher frequencies are mostly higher than those designed for use 50 Hz. In unfavourable instances, these may be as high as 15 times the rated current at 200 Hz and up to 20 times the rated current at 400 Hz. In addition, some of these motors may run at a power factor, or cos φ, of only 0.25.

In total, a reduction in the rated operating current I_e for Siemens contactors used at higher frequencies only results from the thermal considerations, and is given in Table 3.26.

Switching capacity of contactors at the network frequency $16^2/_3$ Hz

At $16^2/_3$ Hz and 400/380 V, the full value of rated operating current for three-pole operation at 50 Hz is permitted if two poles are connected in series. At operating voltages up to 500 V and $16^2/_3$ Hz, all three poles must be connected in series for use up to the full operating current.

Switching capacity of contactors at network frequencies lower than $16^2/_3$ Hz

For frequencies lower than $16^2/_3$ Hz, contactors must be selected in terms of their d.c. switching capacities. These are given in the relevant Siemens l.v. switchgear catalogue.

Switching capacity of circuit-breakers

Middle frequency generators can normally produce only relatively low values of short-circuit current. Therefore, a reduction of the 50/60 Hz breaking capacity of circuit-breakers, which may be necessary at frequencies higher than 400 Hz, does not generally lead to any difficulties in their application.

In single-phase networks and frequencies lower than 50 Hz, the rated breaking capacity of three-pole circuit-breakers as given in the catalogues is retained, if two poles are connected in series for a.c. operating voltages above 220 V and up to 400 V. At a.c. voltages over 400 V and up to 500 V, three poles must be connected in series in order to achieve these values. In this case, only single-pole switching is possible.

Irrespective of the above, it must always be ensured that the current flows through all three poles of the thermally delayed overload releases. In other words, one must select from one of the two possibilities for series connection (see Section 3.7.1, page 185).

3.7.3.3 Contact service life

A long contact service life is a specific important requirement for items of switchgear, such as contactors, which are to be used for frequent switching operations.

If the frequency deviates from 50/60 Hz, then a different rate of contact erosion owing to switching arcs, and thus a different contact service life will result. The actual values must be determined in each individual case.

3.7.3.4 Tripping response of releases and relays

Overcurrent releases

Mechanical overcurrent releases

Overload protection by means of thermal (current-dependent) delayed overload releases (a-releases) and relays

▷ Overload releases and relays

Thermally (current-dependent) delayed overload releases and relays usually employ bimetal strips. These are either heated directly by the total operational load current or by the output current of a current transformer.

Units which are not used in conjunction with an a.c. current transformer may therefor also be used with direct current. Their tripping characteristics remain unchanged.

Up to approximately 500 Hz, directly heated bimetal strips are heated principally by the thermal losses produced by the current. Any additional temperature rise owing to inductive effects is so small at these frequencies, that they can be ignored. The resulting tripping characteristics will be only slightly "quicker" than at 50 Hz. At frequencies above 500 Hz, however, inductive heating becomes more pronounced and this tends to produce a noticeably "quicker" tripping response.

In the case of bimetal relays connected to, or incorporating, current transformers with a high overcurrent factor, frequencies over 50 Hz and up to 400 Hz cause the tripping characteristics to become somewhat "quicker" than at 50 Hz.

In contrast, bimetal relays used in conjunction with a saturating current transformer for heavy duty starting with delayed tripping characteristic, have a much "quicker" tripping characteristic at frequencies up to 400 Hz.

▷ Short-circuit releases

Short-circuit protection by means of instantaneous electromagnetic overcurrent releases (n-releases).

The energizing of an electromagnetic overcurrent release depends not only on the magnitude of the operating or tripping current, but also on the duration for which it flows. At 50 Hz the armature of an electromagnetic overcurrent release pulls in at approximately the peak value of a single half wave. During the half wave the attractive force is sufficient to pull the armature completely to its end position. At higher frequencies the duration of a half wave may be too short. In such cases, the armature then pulls in at the effective, or r.m.s., value of the current. For example, for a change in frequency from $16^2/_3$ Hz to 500 Hz, the pull-in current increases approximately in the ratio of 1 to $\sqrt{2}$. Since, however, the construction and design of the release plays a major role, no general rules can be given.

Example

Power circuit-breaker, rated current 10 A; instantaneous electromagnetic overcurrent release set at 120 A ($12 \cdot I_e$).

At 200 Hz the response current can increase by approximately 30% to 155 A.

This increase in the response current of the electromagnetic overcurrent release may result in the inadequate short-circuit protection of the associated overload releases.

Electronic overcurrent releases

In the case of electronic overload releases, the current is measured via current transformers and is evaluated electronically. The following discussion is based on the Siemens 3WN power circuit-breakers which are equipped with a microprocessor-based overcurrent release system.

▷ Overload releases

Depending on the version of the overload release module, both the operating current (current setting I_r) and the time-lag behaviour of the release may be set. Thereby, the release may be matched e.g. to the run-up conditions of a motor load.

In contrast to thermally delayed overload releases, the tripping characteristic of an electronic overload release is not effected by prior loading within the operational current range.

The "cooling" behaviour after the overload has tripped can be adapted to the operating conditions by the switching in of a memory function (immediate or slow "cooling").

The electronic overload release modules and associated current transformers in the 3WN circuit-breakers have been so designed as to be virtually immune to the influences of harmonics (high frequency oscillations).

▷ Short-circuit releases

The response time of the electronic short-circuit release depends on whether or not a current was flowing through the circuit-breaker, and therefor through the current transformers of the release, before the instant of short-circuit. If not, the electronic overload release must first be "powered up" before it can respond, and so it may be a few milliseconds slower than if the short-circuit occurred after operational current had been flowing for some time.

To prevent the electronic overload release from responding to the short inrush current peaks which occur upon motor switch-on, the initial response time may be delayed by an additional few milliseconds instead of setting the response current to an extremely high value. In this way it is possible to set the short-circuit response current low enough for the electronic overload release to trip even at lower short-circuit currents.

The electronic overload release in 3WN circuit-breakers has an extremely fast response time. It may therefor respond to extremely short current peaks such as those which are typically caused by thyristor switching.

Within the range of 40 to 400 Hz, the response behaviour of the electronic overload release in 3WN circuit-breakers is only slightly effected by the frequency of the alternating current.

To improve discrimination with respect to upstream and downstream fuses, the short-time delayed short-circuit release may be switched from the usual current-independent delay to an I^2-dependent time lag. In this way, cables and conductors may also be protected more efficiently.

▷ Earth-fault protection release

The earth-fault protection prevents earth-fault currents, resulting from e.g. damaged cable insulation, from flowing for prolonged periods of time and possibly causing fires

The response current of the earth-fault protection release is lower than the set tripping current of the overload release, so that even relatively low earth fault currents cause the circuit-breaker to trip.

The earth fault protection release may also incorporate either a current-independent or I^2-dependent time delay.

Undervoltage releases and shunt trip units

Undervoltage releases and shunt trips function in a similar manner to other magnetic operating mechanisms and therefor have the same frequency-dependent performance. In terms of DIN VDE 0660 Part 1 these devices must operate within the following tolerances:

Undervoltage	0.7 to 0.35 · U_s	Circuit-breaker releases is tripped (opened),
	0.85 to 1.1 · U_s	Circuit-breaker can be switched on (closed).
Shunt trip	0.5 to 1.1 · U_s	Circuit-breaker is tripped (opened),
	0.7 to 1.2 · U_s	in accordance with DIN VDE 0660, part 102,
	0.7 to 1.1 · U_s	in accordance with DIN VDE 0660, part 101.

In order to maintain this working range during operational frequency fluctuations, the number of turns in the coil must be based on the relevant rated operating voltage and rated frequency.

3.7.3.5 The effect of harmonic currents on the tripping response of overload releases and relays

Harmonic currents are encountered if the network voltage contains harmonics, or loads are connected which operate in the saturation range, as is the case e.g. with shunt reactors of fluorescent lamps. Loads with phase-angle control also produce harmonics in the network. If harmonics are present in the network, then capacitors draw a higher current than would be the case in networks with pure sinusoidal current waveform and no harmonic content.

Thermally delayed overload relays with bimetal strips respond to the r.m.s. value of the current. This also applies to overload relays which are fed via current transformers with a high overcurrent factor.

In the case of overload relays or releases which are heated by the secondary current of saturation current transformers (i.e. current transformers with a low overcurrent factor n), currents with a high harmonic content cause the tripping characteristic to become substantially "quicker".

Electronic overcurrent relays and releases are generally designed for use with sinusoidal current at a frequency of 50 to 60 Hz.

In networks with a high harmonic content, the use of thermally delayed overload relays or thermistor motor protection devices is recommended, alternatively the manufacturer should be consulted prior to selection.

3.7.3.6 Electrical operating mechanisms for switchgear

According to DIN VDE 0660, the control circuits of electrical power operating mechanisms and all their functional elements must be dimensioned in such a way as to ensure their proper operation within the specified voltage tolerance range.

Either magnetic or motorized operating mechanisms are commonly used for the remote electrical operation of switching devices.

Magnetic operating mechanisms for a.c. excitation

The pull-in force of an a.c. electromagnet is proportional to the square of the flux density. The flux density, on the other hand, is inversely proportional to the frequency of the voltage applied across the terminals of the magnet coil. Therefore, as the frequency increases, the pull-in force of an electromagnetic system with a coil designed for a specific rated frequency, decreases. Conversely, as the frequency decreases the pull-in force increases. In order to attain the same pull-in force at a frequency other than, e.g. the rated 50 Hz, the operating voltage must be increased or reduced accordingly.

Example 1 (for contactors)

On the label of the coil, the rated values "220 V, 50 Hz" are printed. The contactor is to be used at a frequency of 60 Hz. This is possible at a control voltage of

$$U_s = \tfrac{60}{50} 220 \text{ V} = 264 \text{ V}.$$

The power consumption of the magnet coil in the closed position is increased by approximately the same relationship, i.e. 60:50 (factor 1.2). This conversion can only be used in the frequency range from 50 to 60 Hz!

The magnet systems of a.c. contactors are designed for control voltages with a frequency in the range 50 to 60 Hz. For frequencies outside this range, the control voltage must be rectified and contactors for d.c. excitation must be used. To ensure that the full rated voltage tolerance of the coil (0.8 to 1.1 · U_s) is retained after rectification, the rated control supply voltage of the magnet coil should be 0.9 times the a.c. control system voltage.

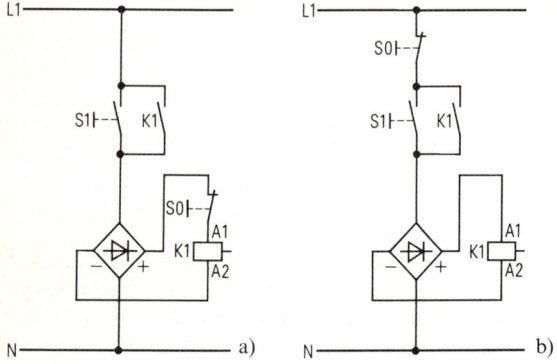

Figure 3.74
(a) Switching in the d.c. circuit (delayed drop-out only upon loss of control voltage), (b) Switching in the a.c. circuit (drop-out always delayed)

Example 2

The control system voltage is 220 V/120 Hz: contactors for d.c. excitation must be used.

In the case of full wave rectification, the coil voltage (rated control supply voltage) is calculated as: $0.9 \cdot 220 \text{ V} = \text{DC } 198 \text{ V}$.

Smoothing capacitors are not necessary and should not be used, since they would cause the d.c. voltage across the coil terminals to be higher!

To ensure a quick drop-out of the contactor upon switch-off, the coil must be switched in the d.c. circuit (Fig. 3.74a). If the switching is done in the a.c. circuit (Fig. 3.74b), the drop-out time will typically increase by approx. 9 times. Owing to the danger that the contactor may not drop out smoothly ("two-stage drop-out"), this configuration is only permissible in cases where the design of the magnet system allows the use of a freewheel diode across the coil terminals.

Magnetic operating mechanisms for d.c. excitation

For the excitation with direct current, contactors either incorporate a special d.c. magnet system (solid core) or utilize an a.c. magnet system (laminated core) with a so-called "economy switching circuit".

Alternating current magnet systems with economy switching circuits require a high closing, or pull-in, power (approx. 10 to 40 times the hold-in power). The operation of the magnet system is done either with a coil comprising separate pull-in and hold-in windings, or via a series resistance which is switched in upon closing to reduce the hold-in current. The closing times are approximately half as long as would be the case with a.c. excitation.

In the case of direct-current magnet systems, the pull-in power is the same as the continuous hold-in power. This means that e.g. control transformers and output modules of electronic control systems may be dimensioned according to the continuous operational current and do not have to be "overdimensioned" to cope with current surges. The switching times are somewhat longer than those of the a.c. magnet system. Thereby, a "softer" switching can be achieved, and this results in a longer mechanical service life in comparison to that of the a.c. magnet system with an economy switching circuit.

Motorized operating mechanisms

Siemens offers motorized operating mechanisms for several low-voltage switching devices. These operating mechanisms are generally designed for use in 50 and 60 Hz networks. The permissible operating voltage tolerance is from 0.85 to $1.1 \cdot U_s$; the frequency may deviate by $\pm 5\%$ from the rated value.

Since a.c./d.c. motors are used in most cases, the control voltage can be rectified for use at frequencies higher than 60 Hz. Motorized operating mechanisms which are fitted with rectifiers during manufacture, may be used with d.c. or a.c. control voltage (a.c. frequency ≥ 40 Hz).

3.7.4 Use of a.c. switchgear in d.c. networks

If low-voltage a.c. switching devices designed for use at 50/60 Hz are to be used in d.c. networks, the following characteristics must be examined closely:

▷ load carrying capacity of the current path assemblies,
▷ service life of the contacts,
▷ switching capacity,
▷ tripping behaviour of the releases,
▷ performance of motorized or magnetic operating mechanisms.

3.7.4.1 Load carrying capacity of the current path assemblies

In contrast to alternating current there is no "skin effect" to be considered for direct-current applications. The current flow is uniformly distributed through the cross-section of a straight conductor.

Apart from that, other phenomena associated with alternating current, such as hysteresis and eddy-current losses produced in neighbouring ferromagnetic materials (e.g. operating shafts, latching mechanisms, fixing brackets etc.) do not occur in d.c. systems. Consequently, switchgear designed for alternating current can conduct at least the same value of direct current.

3.7.4.2 Contact service life

The contact service life is almost exclusively determined by the breaking current together with the voltage and frequency of the network. For direct current a different type of arc erosion, and thus a different contact service life is to be expected than for switching with a.c. voltage at 50/60 Hz. The actual value of the contact service life which may be achieved by an item of switchgear in a d.c. circuit depends very much on the application, and it must therefor be determined separately for each individual case.

3.7.4.3 Direct-current switching capacity

In alternating-current circuits, arc quenching is assisted by the fact that the current passes through zero, and that the current can only continue to flow if the arc is re-struck across the open contacts during the following half wave. Direct current does not provide such assistance. In this case, a high arc voltage

Table 3.27 Application examples for a.c. circuit-breakers or contactors in d.c. networks

Rated d.c. voltage +10%	Required series connection of switching points	Possible circuits with	
		3-pole contactors or circuit-breakers	4-pole circuit-breakers
up to 300 V		Single-pole with two (left) or three (right) current paths in parallel	Double-pole Single-pole with two (left) or four (right) current paths in parallel
above 300 up to 600 V [1]		Double-pole only if one pole is earthed	Single-pole Double-pole with two current paths in parallel
above 600 up to 1000 V [1]		Single-pole	

[1] Only if fitted with arc chute extensions

3 Selection criteria for low-voltage switchgear in main circuits

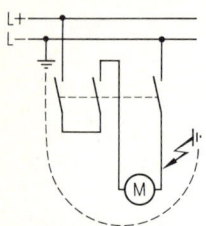

a) Earth fault with two-pole switching, earthed system

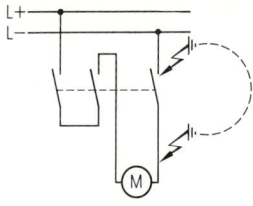

b) Double earth fault with two-pole switching, non-earthed system

Figure 3.75
Bridging of a switching point by an earth fault (a) or a double earth fault (b)

must be developed in order to stop the flow of the d.c. current. Therefore, the d.c. switching capacity depends on the arc quenching method employed by the switching device, on the network voltage and on the inductive reactance of the circuit.

Generally, a.c. switching devices are either single-pole (e.g. miniature circuit-breakers), triple-pole (e.g. contactors) or four-pole (e.g. circuit-breakers). In order to achieve a higher arc voltage, several switching points connected in series can be introduced into the current path (see Section 3.7.1, page 185). Table 3.27 (on page 193) shows a number of application examples in which a.c. circuit-breakers are used in d.c. networks. However, it must be taken into consideration that if the network is earthed and if an earth fault or double earth fault occurs, one of the switching points may be bridged out by the earth path (refer to Fig. 3.75).

This will remove its contribution to the total arc voltage and may reduce the effective breaking capacity to a dangerous level.

When using universal (a.c./d.c.) circuit-breakers in direct-current networks, particular care must be taken to ensure the correct polarity of the connections. The polarity is clearly labelled on the device in the vicinity of the main terminals. If a universal circuit-breaker is used in a d.c. application, the response value of its n-release will be 1.4 times ($\sqrt{2}$ times) higher than when it is used with alternating current.

Releases

Thermally delayed overload releases and relays

Overload releases with bimetal strips which are directly heated by the operating current may also be used in d.c. systems. Their tripping characteristics remain unchanged.

Overload relays and releases which operate via current transformers are not suitable for use in d.c. systems.

Electromagnetic overcurrent releases and relays

Electromagnetic overcurrent releases and relays may be used in d.c. systems if they are specially adjusted/calibrated for this purpose.

Undervoltage releases and shunt trip units

Undervoltage releases and shunt trip units have a similar operating principle to magnetic operating mechanisms. Unaltered, they generally cannot be used with direct current, since the voltage tolerances would no longer meet the specifications. Versions for use with d.c. auxiliary supply voltages are offered in the l.v. switchgear catalogues.

3.7.5 Use of a.c. contactors in networks with square-wave voltages

If a.c. contactors are used in networks which are supplied from batteries via a power inverter or chopper unit, special consideration must be given to their switching capacity and to the energy required by their magnet system.

Switching capacity

In contactors, effective quenching of the switching arc is determined by the amplitude and rate-of-rise of the current while the arc is burning, and by the

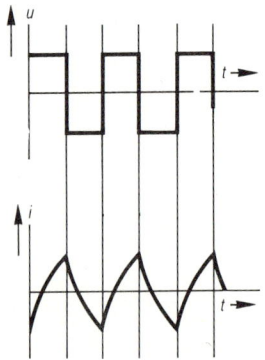

Figure 3.76
Square-wave a.c. driving voltage and the resulting wave form of the alternating current in the case of an inductive load

amplitude and rate-of-rise of the recovery voltage after the arc has been extinguished.

In the case of a pure ohmic load and a square-wave driving voltage, the current will also have a square waveform. If the load is inductive, the current will have a waveform as is shown in Figure 3.76.

If the rate-of-rise of the voltage at the point of zero crossing is less than 2 kV/ms (as is the case with Siemens inverters), then the switching capacities of the contactors, when used with square-wave a.c. voltage having the same r.m.s. values of current and voltage and at the same frequency, will be equivalent to the full a.c. switching capacities applicable for the respective utilization categories AC-1 to AC-4.

Magnetic operating mechanisms

The force with which the magnetic operating mechanism closes the contacts of the contactor is determined by the magnetic energy. The magnetic energy is directly proportional to the square of the effective current (r.m.s. value) flowing in the coil.

A contactor coil can therefor be operated with square-wave a.c. voltage if:

$I_{reff} = I_{eff}$ where

I_{reff} square-wave current (r.m.s. value)
I_{eff} rated control current at rated control supply voltage (r.m.s. value)

This condition is achieved if:

$U_{er} = 1.1 \cdot U_s$,
$f_c = f_{er}$,
$\cos \varphi_{on} \leq 0.75$.

U_{er} square-wave supply voltage (effective and maximum value)
U_s rated control supply voltage (sinusoidal)
f_c rated frequency at U_s
f_{er} rated frequency at U_{er}
$\cos \varphi_{on}$ power factor of the contactor coil at switch-on and at U_s.

Two examples

1. If it is used in a control circuit with a 50 Hz square-wave control voltage, a contactor having a 220 V/50 Hz coil must be energized with 242 V.
2. A contactor must be fitted with a 200 V/50 Hz coil if it is to be used in a control circuit with a 50 Hz square-wave control voltage of 220 V.

3.7.6 Switchgear for the switching of three-phase capacitors

3.7.6.1 Switching of capacitors with circuit-breakers

When switching capacitors it must be taken into account that the capacitors will draw harmonic currents according to the harmonic content of the network. The r.m.s. value of the fundamental current plus the harmonic currents must not exceed the rated operating current of the breaker. As a general rule, the current rating of the capacitors should not exceed 75% of the circuit-breaker rated operating current.

3.7.6.2 Switching of capacitors with contactors

If contactors are used for the switching of capacitors (load and pre-charge contactors as per Table 3.28), the capacitors must be discharged by means of resistors or discharge chokes prior to switch-on.

Depending on the phase angle at the instant of switch-on, severe charging current peaks can result if the capacitors are still partially charged. These peaks can reach values which are twice as high as would be the case if the capacitors are switched on when fully discharged, and could cause welding of the main contacts at the instant the contactor closes.

3.7.6.3 Switching on of single capacitors

The permissible loading of 3TF and 3TB contactors when used to switch low-inductance (MKV, MKK) three-phase capacitors, may be found in the Siemens l.v. switchgear catalogue for contactors. The given values apply for an estimated contact service life of 100 000 switching operations.

3.7.6.4 Switching of capacitor banks

If a capacitor is switched on in parallel to other capacitors already in the circuit, then these act as an additional energy source. The closing contactor is subjected to additional stresses by the transient currents which flow between this "new" capacitor and the parallel units already in circuit.

Protective measures

The effective capacitor-load switching capacities of a.c. contactors and their contact service life can be increased by the use of choke coils and/or pre-charge resistors.

3 Selection criteria for low-voltage switchgear in main circuits

Table 3.28 Contactors for the switching of three-phase capacitors used in individual or group power factor correction

Capacitor rating	Load contactor Operational voltage				Pre-charge contactor	Pre-charge resistor Operational voltage								Back-up fuses maximum [1]	
	230 V	400 V	500 V	690 V		230 V		400 V		500 V		690 V		For contactor type [2]	I_n
kvar						Ω	≤W	Ω	≤W	Ω	≤W	Ω	≤W		A
12.5	3TK44	3TK42	3TK42	3TK42	—	—	—	—	—	—	—	—	—	3TK42	35
25	3TF48	3TK44	3TK44	3TK44	3TF40	3×1	25	—	—	—	—	—	—	3TK44	100
30	3TF48	3TF46	3TF44	3TF44	3TF40	3×1	40	3×1.8	20	3×2	15	3×3	15	3TF44	63
40	3TF50	3TF46	3TF46	3TF46	3TF42	3×1	60	3×1.5	30	3×2	30	3×3	25	3TF46 3TF48	100 160
50	3TF52	3TF48	3TF46	3TF46	3TF43	3×0.68	70	3×1	45	3×1.5	30	3×2	30	3TF50	250
60	3TF52	3TF50	3TF48	3TF46	3TF43	3×0.68	90	3×1	60	3×1.5	55	3×2	35	3TF52	250
75	3TF54	3TF50	3TF50	3TF48	3TF44	3×0.5	110	3×0.68	70	3×1	60	3×1.5	50	3TF54	315
100	3TF56	3TF52	3TF50	3TF50	3TF44	3×0.3	130	3×0.5	90	3×0.68	75	3×1.5	60	3TF56	500

[1] Easily separable contact welding without further damage is permissible
[2] Refer to the technical data in the relevant Siemens l.v. switchgear catalogue for the capacitive load switching capacities of 3TF, 3TK and 3TB contactors

Choke coils

The transient condition which is caused when switching in on an existing capacitive load may be damped, if an inductance of at least 6 μH is connected in the parallel connecting leads between the capacitors.

Pre-charge resistors

For the switching of larger capacitor loads, a combination comprising two standard 3TF or 3TB contactors, used as load and pre-charge contactor respectively, and suitable pre-charge resistors, is recommended (see Table 3.28). The advantages of using these combinations include a reduced stressing of the network and an improved utilization of the thermal load capabilities of the contactors.

3.7.6.5 Switching of capacitors with lower power ratings

Siemens offers two special capacitor contactors, viz. types 3TK42 and 3TK44, for the switching of smaller capacitive loads. In these contactors, the capacitors are pre-charged via early-make contacts and integrated pre-charge resistors before the main contacts close.

The 3TK4 capacitor contactors and the contactor/pre-charge resistor combinations given, may be used with capacitors for p.f. correction of single loads or in centralized group power factor correction systems. The minimum inductivity normally required with capacitor banks may then be omitted.

Figures 3.77 and 3.78 show the internal wiring diagrams of the Siemens 3TK42 capacitor contactor and a combination comprising load contactor, pre-charge contactor and pre-charge resistors, respectively. For further information on power factor correction please refer to Section 7.4.8 on page 363.

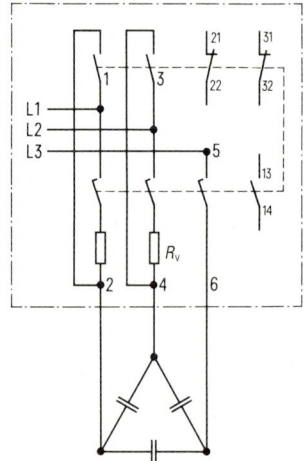

R_v Pre-charge resistors, each 1.6 Ω – 10%

Figure 3.77
Internal wiring diagram of the 3TK42 contactors for the switching of capacitors (pre-charging is accomplished with the early-make contacts of this 8 pole contactor; contactor and pre-charge resistors form a single unit

Switching of capacitors · Contact service life · Utilization categories

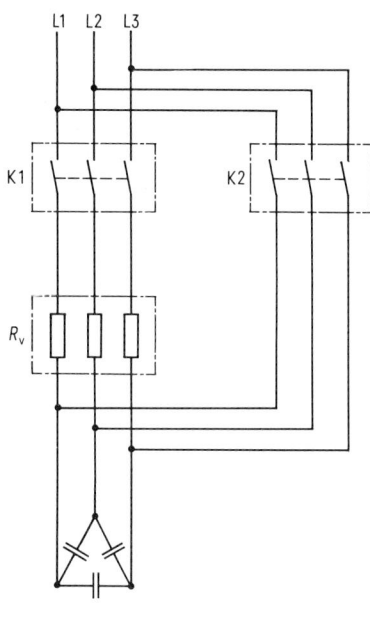

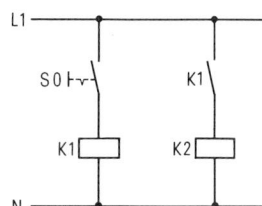

K1	Pre-charge contactor
K2	Load contactor
R_v	Pre-charge resistors

Figure 3.78
Combination comprising load contactor, pre-charge contactor and pre-charge resistors

3.7.7 Selection of 3TF and 3TB contactors in terms of contact service life and utilization category

Selection in terms of contact service life

Characteristic curves for the service life of the main contacts in 3TF, 3TB and 3TC contactors are given in the Siemens l.v. switchgear catalogue (Fig. 3.79). The dotted lines in Figure 3.79 overleaf indicate the maximum permissible motor rating for each 3TF contactor.

Examples

▷ Given, a squirrel-cage motor; 22 kW; cos φ = 0.83; 380 V; I_n = 45 A; starting current I_{an} = 6.4·I_n = 288 A. Select a contactor which will allow 10^6 switching operations without replacement of the contact pieces.

From the diagram in Figure 3.79: At $I_a = I_n = I_e$ = 45 A (I_a = breaking current), the contactor type 3TF46 is found to be suitable (contact service life > 10^6 switching operations).

▷ Given, a squirrel-cage motor; 200 kW; cos φ = 0.89; 380 V; I_n = 360 A; I_{an} = 6.8·I_n = 2456 A. Select a contactor which will allow 4·10^5 switching operations without replacement of contact pieces.

From the diagram in Figure 3.79, at P_n = 200 kW: The contactor type 3TF54 is found to offer approximately 7·10^5 switching operations at $I_a = I_n = I_e$ = 360 A.

Selection in terms of utilization category

AC-1 operation

As an example, select a suitable contactor for switching an electric heating device of 40 kW; cos φ = 1; U_e = 380 V; I_n = 61 A:

From the Siemens catalogue: "Technical data / Load ratings of a.c. contactors / AC-1 duty, switching resistive load / Ratings of three-phase loads p.f. ≥ 0.95/ at 380 V", the contactor type 3TF46 is found to be suitable for the switching of 52.5 kW. The contact service life from Figure 3.79 is found to be approximately 7·10^5 at I_a = 61 A.

For further information on utilization categories, refer to Section 3.3.4.1 on page 118.

AC-2 operation

The AC-2 utilization category is of particular importance in the selection of contactors for use with slipring motors in cranes.

The tables in DIN VDE 0660, part 102 (alt. IEC 947-4-1 appendix B) to be used as a basis in the definition of the electrical service life, stipulate a making and breaking current of 2.5 times the rated operational current for the utilization category AC-2. This is not always a realistic value, since in many instances the actual dynamic loading of the contactor may be substantially lower (e.g. switch-off while running at constant speed and/or partial load). The practical significance of this, is that the contacts may well have a longer service life than that which is indicated by the test conditions for electrical service life in terms of DIN VDE 0660, part 102. For further details, please refer to the Siemens H3 catalogue.

3 Selection criteria for low-voltage switchgear in main circuits

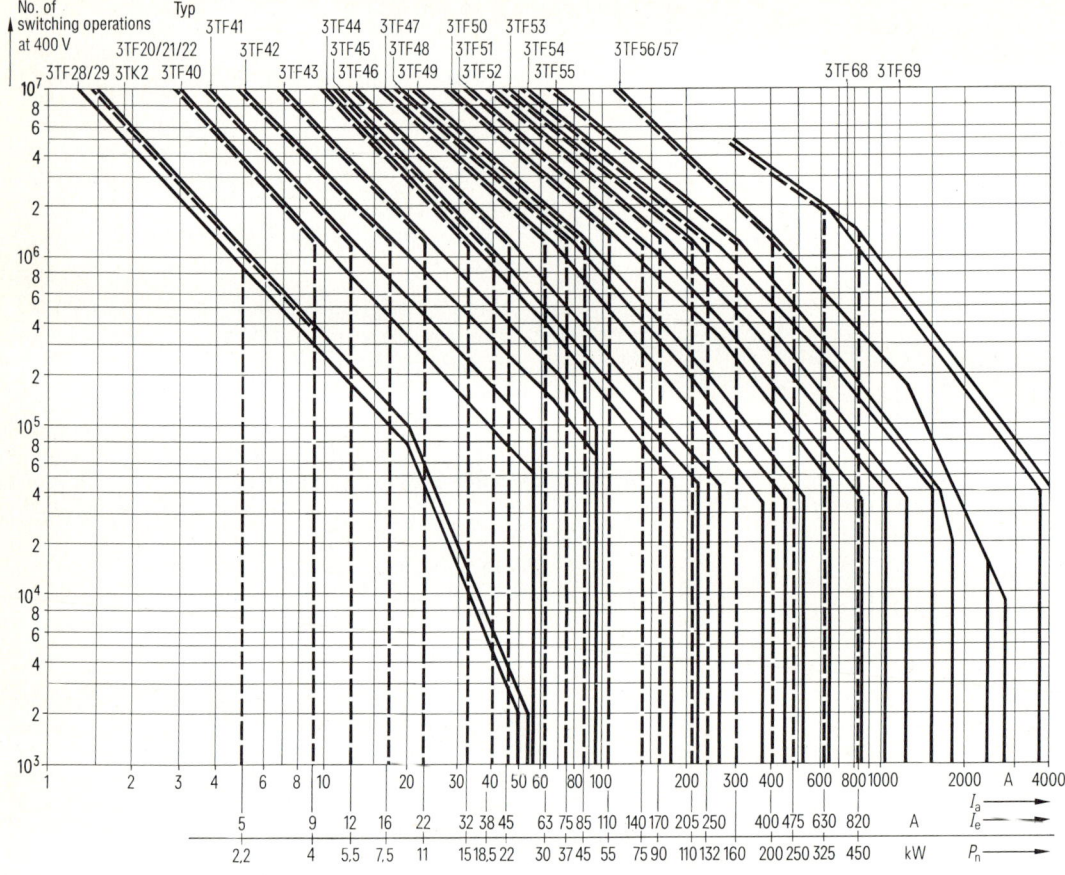

P_n Rated output power of three-phase induction motors with squirrel-cage rotors at AC 400 V, 50 Hz
I_a Breaking current
I_e Rated operational current

Figure 3.79
Contact service life (in no. of switching operations) of the 3TF contactors plotted as a function of breaking current at an operational voltage of AC 400 V, 50 Hz

AC-3 operation

As an example, a suitable contactor is to be selected for switching a three-phase asynchronous motor (squirrel-cage motor); 37 kW; $\cos \varphi = 0.85$; 380 V; $I_n = 72$ A; $I_{an} = 6.4 \cdot I_n = 460$ A:

From the Siemens catalogue, "Selection and ordering data", the contactor type 3TF48 which is capable of switching 37 kW at 380 V, is selected. The contact service life, as per Figure 3.79, for $I_n = I_a = 72$ A, is approx. $1.3 \cdot 10^6$ switching operations.

AC-4 operation

As an example, find a suitable contactor for switching the above motor in inching duty and intermittent periodic operation S4 (see pages 100 and 119):

From the Siemens catalogue, "Technical data/Load ratings of a.c. contactors/AC-4 duty" the contactor type 3TF52, which is capable of switching 40 kW at 400 V, is selected. Based on a breaking current I_a of 460 A, the contact service life will be approx. $2 \cdot 10^5$ switching operations (see Fig. 3.79).

If a shorter contact service life is acceptable, then

the rated operational current at AC-4 operation may be increased. If the switching frequency is very low, then a contactor for inching / jogging duty (AC-4), may also be selected according to the AC-3 selection and ordering data.

Mixed operation

If the operating duty of the contactor is such, that it is sometimes called upon to interrupt the load current I_e in terms of AC-3, and sometimes multiples of the load current in terms of AC-4 utilization category, then one speaks of "mixed operation". It is a mixture of normal switching operations and occasional "inching" or "jogging" duty. The contact service life of the contactor can be estimated by means of the following equation:

$$X = \frac{A}{1 + \frac{C}{100}\left(\frac{A}{B} - 1\right)} \qquad (3.1)$$

X Contact service life in no. of switching operations for mixed operation,
A Contact service life in no. of switching operations for normal operation ($I_a = I_n$),
B Contact service life in switching operations for inching operation (I_a = multiple of I_n),
C Proportion of inching operations as a percentage of the total no. of switching operations.

The contact service life for 3TF contactors with 10, 20, 50, or 100% inching duty when switching squirrel-cage motors with 6 times motor rated current, or slip-ring motors with 2.5 times motor rated current I_n when starting, may be estimated by using the diagrams in Figures 3.80 and 3.81 respectively.

Alternatively, the PC program "ELLE" (Order No. E86010-D 1802-A117-A2) may be used to calculate the electric service life of Siemens contactors under any given switching conditions.

Example for determining the contact service life of the 3TF48 contactor

▷ A three-phase motor 37 kW, $\cos \varphi = 0.85$, 380 V, $I_n = 72$ A, $I_{an} = 6.4 \cdot I_n = 460$ A, is to be switched with contactor type 3TB48 in mixed operation, where 30% of the switching operations are "inching" operations. The contact service life X, is to be found for this contactor.

From Figure 3.79:

$A = 1.2 \cdot 10^6$ operations (at $I_a = I_n = I_e = 72$ A),
$B = 5 \cdot 10^4$ operations (at $I_{an} = 6.4 \cdot I_n = I_e = 460$ A).
$C = 30\%$.

The contact service life of the contactor is calculated for this mixed operation from equation (3.1) as follows:

$$X = \frac{1.2 \cdot 10^6}{1 + \frac{30}{100}\left(\frac{1.2 \cdot 10^6}{5 \cdot 10^4} - 1\right)}$$

$= 15.2 \cdot 10^4$ switching operations.

Examples for the selection of contactors for a specified number of switching operations

▷ Given: Squirrel-cage motor, 30 kW $\cos \varphi = 0.83$, 400 V, $I_n = 60$ A, $I_{an} = 6.4 \cdot I_n = 384$ A, 10% inching duty.

Required: Contact service life of 10^6 switching operations.

To be found: Suitable contactor.

From Figure 3.80, interSection point 1, select contactor type 3TF50.

▷ Given: Squirrel-cage motor, 55 kW $\cos \varphi = 0.87$, 400 V, $I_n = 104$ A, $I_{an} = 6 \cdot I_n = 624$ A, 20% inching duty.

Required: Contact service life of 10^6 switching operations (in this case, the complete contactor must be selected for a service life of 1 million switching operations; i.e. replacement of contact sets is permissible).

To be found: Suitable contactor.

From Figure 3.80, the following possibilities may be considered:

Intersection point 2
Contactor type 3TF50, rated operating current 110 A, contact service life $0.21 \cdot 10^6$ switching operations,

Intersection point 3
Contactor type 3TF52, 170 A, $0.5 \cdot 10^6$ switching operations,

Intersection point 4
Contactor type 3TF54, 250 A, $1.1 \cdot 10^6$ switching operations.

In selecting the most suitable solution, it must be taken into consideration that although the 3TF54,

3 Selection criteria for low-voltage switchgear in main circuits

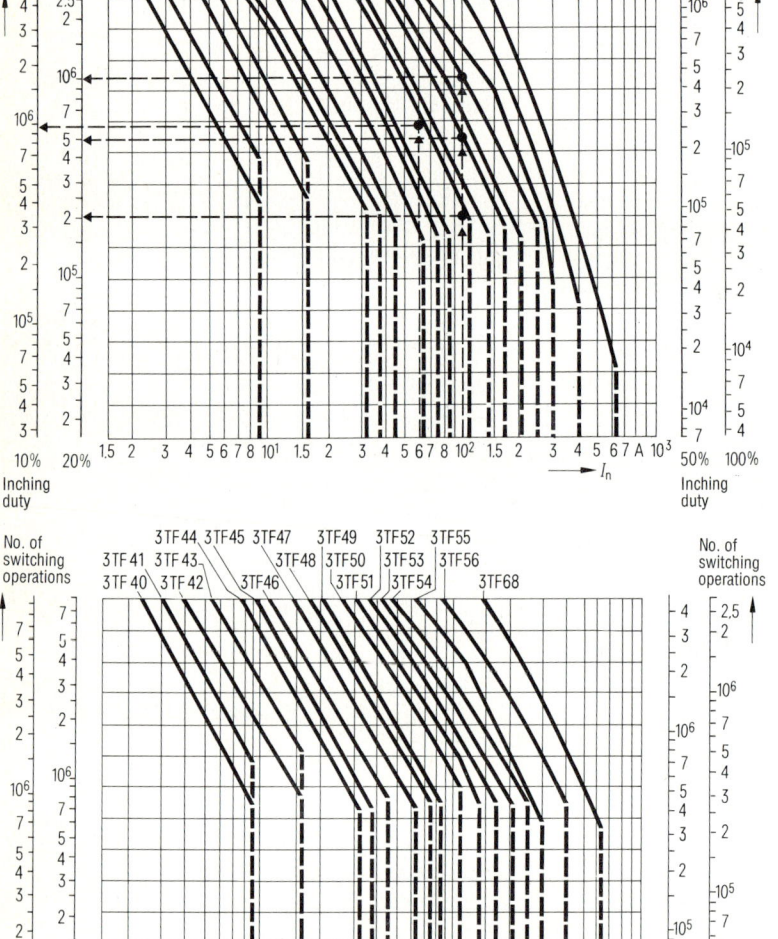

Figure 3.80
Contact service life of the 3TF contactors in mixed operation and *6 times* rated motor current I_n (squirrel-cage motors)

Intersection points 1 to 4 from the examples in the text

Figure 3.81
Contact service life of the 3TF contactors in mixed operation and *2.5 times* rated motor current I_n (slip-ring motors)

with its current rating of 250 A, is doubtlessly overdimensioned (cf. $I_n = 104$ A), and although it is more expensive and bigger than the 3TF50 contactor, no replacement of contact sets will be required.

Here, selection must be made according to the priorities set by the user.

Basically, one may conclude:

If it is not desirable that contacts be renewed during the operational life of a contactor, then an overdimensioned contactor may be selected. The cost for contactor and housing will be approximately double.

Selection of contactors using a personal computer (PC)

For the selection of contactors in terms of

▷ power,
▷ utilization category and
▷ electrical service life,

Siemens offers an applications software package known as "ELLE"
(Order No. E 86010-D1802-A117-A2)

Hardware requirements: Siemens, PCD-3T or any other AT-compatible PC (industrial standard).

Software requirements: Operating system MS DOS, version 3.0 or higher.

3.7.8 Selection of 3TF contactors for short-time and intermittent periodic duty

If contactors are to be used for short-time duty S2 (also known as temporary duty), or intermittent periodic duty S3, S4 or S5, then different selection criteria apply than for normal continuous duty S1 (see Section 2.2.2 and 2.2.3, from page 98 et seq.).

Guidelines for the selection of contactors for short-time duty S2

In dimensioning contactors for use in short-time duty applications, the permissible load currents can be determined by:

$I_{s2} = n \cdot I_{eAC-1}$

For Siemens contactors, the factor n may be read from Figure 3.82. It gives the values of n at an ambient temperature of 55 °C, expressed as a function of the on-load period t_{S2}.

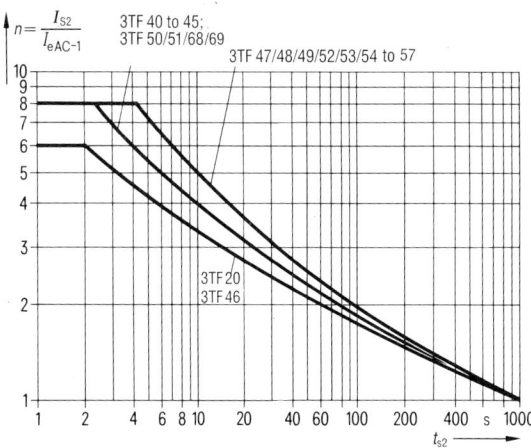

n	Multiples of the rated operational current to AC-1
t_{S2}	On-load period in seconds during short-time duty S2

Figure 3.82
Short-time duty S2. Permissible loading of contactors as a function of their rated operating current I_{eAC-1}

Table 3.29
Switching capacities (r.m.s. values) and minimum cooling times of contactors in short-time duty S2

Contactor Type	I_{eAC-1} A	Making capacity (r.m.s.) A	minimum cooling time t_0 min	Breaking capacity (r.m.s.) at $U_e =$ 400/380 V A
3TF2	16	100	5	72
3TF40	20	160	5	120
3TF41	20	200	5	120
3TF42	30	300	5	200
3TF43	30	300	5	200
3TF44/45	55	500	5	300
3TF46	80	640	5	500
3TF47	90	950	5	700
3TF48/49	100	1150	5	825
3TF50/51	160	1800	10	1200
3TF52/53	210/220	2500	10	1850
3TF54/55	300	4000	15	2750
3TF56	400	5000	15	4400
3TF68	630	7600	15	7200
3TF69	850	10000	15	7800

The off-load period t_0 during the short-time duty must be at least as long as the minimum cooling time of the contactor. The maximum permissible loading during S2 duty is determined by the switching capacity of the contactor. Appropriate values for both these criteria are given in Table 3.29.

Example

Short-time duty S2 as shown in Figure 3.83:

Motor starting current	$I_{an} = 2000$ A
Starting time	$t_{an} = 1$ s
Rated motor current	$I_n = 600$ A
On-load period	$t_{S2} = 15$ s
Off-load period	$t_0 = 14$ min
Load current	not constant;

Load steps as determined in Figure 3.84:

$I_1 = 2000$ A for 1 s and
$I_2 = 600$ A for 14 s.

The load current $I_q = I_{S2}$ is determined as being:

$$I_q = \sqrt{\left(\frac{I_1^2 \cdot t_1 + I_2^2 \cdot t_2}{t_1 + t_2}\right)} \qquad (3.2)$$

$$= \sqrt{\left(\frac{2000^2 \cdot 1 + 600^2 \cdot 14}{15}\right)} = 776.3 \text{ A}.$$

3 Selection criteria for low-voltage switchgear in main circuits

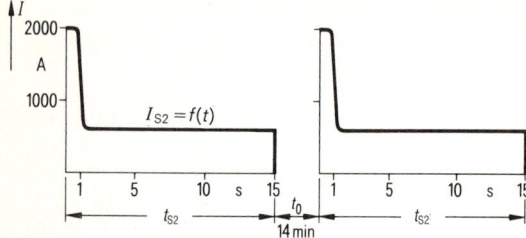

Figure 3.83
Current-time diagram for the operation of a three-phase induction motor (short-time duty S2)

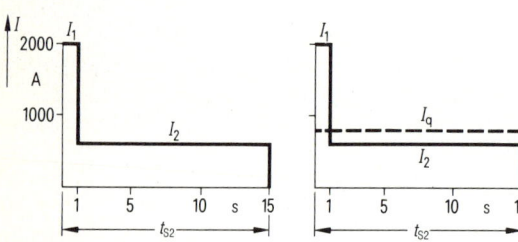

Figure 3.84
Determining the equivalent effective load current I_q (using load steps) of the three-phase induction motor during short-time duty as illustrated in Figure 3.83

Before the suitable contactor can be selected, the making/breaking capacities and the cooling time must be checked (see Table 3.29):

▷ for the making current $I_1 = 2000$ A: contactor 3TF52

▷ for the breaking current $I_2 = 600$ A: contactor 3TF47

Based on the necessary making current, the 3TF52 is favoured.

▷ For the 3TF52 contactor, and $I_q = I_{S2} = 776.3$; $I_{eAC-1} = 200$ A, the starting current ratio n is calculated as

$$n = \frac{I_{S2}}{I_{eAC-1}} = \frac{776.3}{200} = 3.88.$$

In terms of Figure 3.82, for a starting current ratio of $n = 3.88$, the permissible on-load period t_{S2} is 18 s. Since t_{S2} is greater than the required on-load period of 15 s, the 3TF52 contactor is found to be suitable for the required operating conditions.

Guidelines for the selection of contactors for intermittent periodic duty S3, S4, S5

Selection data is given in the Siemens l.v. switchgear catalogue for the stator and rotor contactors of *slip-ring motors* in intermittent periodic duty with operating cycle times of up to 10 min (refer to example 1, below). The permissible operating currents for intermittent periodic duty given in the catalogue, are based on the cyclic duration factors (c.d.f.[1]) 20, 40, 60 and 80% respectively. If, in practice, an on-load factor of a different value is calculated, then it is recommended that the contactor corresponding to the next higher factor given in the catalogue be chosen.

For the operation of *three-phase induction motors (squirrel-cage)* in intermittent periodic duty, the contactor is normally selected as for continuous duty. This is because the switching capacity of the contactor, and not its thermal rating, determines its suitability in this case.

If the motor has a smaller starting current than is given for utilization category AC-3 (e.g. 5 times I_n), then the contactor may be selected in the same way as a stator contactor for slip-ring motors in intermittent periodic duty. However, the switching capacity of the contactor chosen must be checked to be sufficiently high in comparison to the actual starting current (see examples 2 and 3 on page 203).

Example 1

Intermittent periodic duty of a three-phase asynchronous motor with slip-ring rotor as shown in Figure 3.85:
stator current $I_{es} = 45$ A,
operating cycle time $t_s < 10$ min,
shortest off-load interval = 1 min
(i.e. not short-time duty. Refer to the minimum cooling times in Table 3.29, page 201).

In terms of the equation, the cyclic duration factor

$$\text{c.d.f.} = \frac{\Sigma t_B}{\Sigma t_B + \Sigma t_{st}} \cdot 100\% \tag{3.3}$$

$$\text{c.d.f.} = \frac{t_1 + t_2}{t_1 + t_2 + t_{st1} + t_{st2}} \cdot 100\%$$

$$= \frac{1+2}{1+2+1+5} \cdot 100 = 33\%.$$

---
[1] Also known as the on load factor or ED from the German: Einschaltdauer

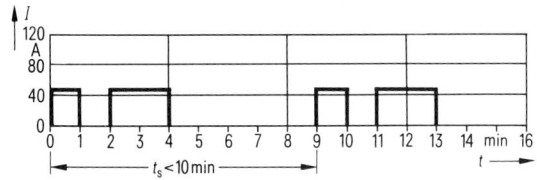

t_s Operating cycle time

Figure 3.85
Current-time diagram of the stator current during the operation of a three-phase asynchronous motor with slip-ring rotor (simplified illustration without switch-on current peaks)

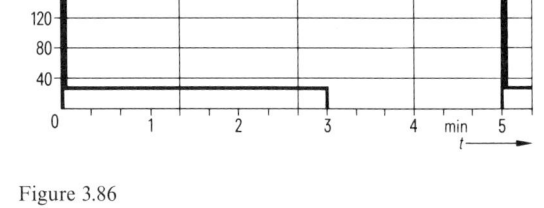

Figure 3.86
Simplified current-time diagram for the operation of a three-phase asynchronous motor with squirrel-cage rotor in intermittent periodic duty

For a stator current I_{es} of 45 A and an on-load factor of 33%, the contactor 3TF44 (alternatively 3TF34, or 3TB44) is selected from the Siemens catalogue (actual contactor rating: up to 55 A at on-load factor 40%).

Example 2

Operating conditions as shown in Figure 3.86:

Squirrel-cage motor with $I_n = 33$ A (AC-3), 400 V, starting current approx. $I_{an} = 5.5 \cdot I_n$, run-up time approx. 0.5 s, intermittent periodic duty with on-load factor c.d.f. = 60%, operating cycle time $t_s = 5$ min.

For squirrel-cage motors, the contactor can be selected as a stator contactor from the Siemens catalogue, provided that the starting current does not exceed that which is permissible for the contactor in question.

$I_{an} = 5.5 \cdot I_n = 5.5 \cdot 33$ A $= 181.5$ A.

The current flow during the intermittent periodic duty is shown in Figure 3.86.

The on-load period is determined as

On-load period $= \dfrac{t_B}{t_s}$; $t_B = \dfrac{\text{c.d.f.} \cdot t_s}{100} = \dfrac{60 \cdot 5}{100} = 3$ min.

In comparison with the on-load period of 3 minutes, the motor starting current flows for only a relatively short time (0.5 s) and has practically no effect on the heating up of the contactor during the on-load period.

A stator contactor for an on-load factor of 60%, and with a making capacity of at least $I_{an} = 181.5$ A, can be selected in the Siemens l.v. switchgear catalogue. The contactor type 3TF42 (alternatively 3TF32 or 3TB42) is found to be suitable.

Example 3

Operating conditions as in example 2, but with an operating cycle time $t_s = 20$ s.

Squirrel-cage motor with $I_n = 33$ A (AC-3), 400 V, starting current $I_{an} = 5.5 \cdot I_n$, run-up time 0.5 s, intermittent periodic duty with
on-load factor c.d.f. = 60%,
operating cycle time = 20 s.

The current flow during the intermittent periodic duty is shown in Figure 3.87.

The on-load period is determined as

On-load period $= \dfrac{t_B}{t_s}$; $t_B = \dfrac{\text{c.d.f.} \cdot t_s}{100} = \dfrac{60 \cdot 20}{100} = 12$ s.

The motor starting current, which flows for 0.5 s, will have an effect on the heating up of the contactor during the on-load period of 12 s.

The load current during the on-load period is not constant. The root mean square value I_q of the load current must be determined, using equation (3.4) below, as follows:

$$I_q = \sqrt{\dfrac{I_1^2 t_1 + I_2^2 t_2 + \cdots + I_n^2 t_n}{t_1 + t_2 + \cdots + t_n}}, \quad (3.4)$$

$$= \sqrt{\dfrac{181.2^2 \cdot 0.5 + 33^2 \cdot 11.5}{12}} = 49.2 \text{ A}.$$

3 Selection criteria for low-voltage switchgear in main circuits

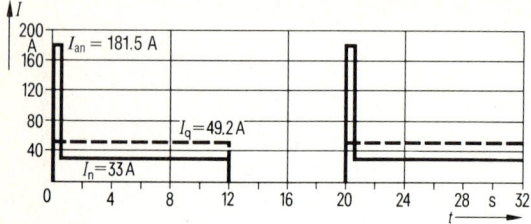

I_{an} Starting current
I_q Root mean square value of the load current
I_n Rated current of the motor

Figure 3.87
Simplified current-time diagram for the operation of a three-phase asynchronous motor with squirrel-cage rotor in intermittent periodic duty

In the Siemens l.v. switchgear catalogue, one finds that for a stator current of 49.2 A (in this case also applicable for squirrel-cage motors), and for an on-load factor of 60%, the contactor type 3TF46 (up to 87 A) appears to be the most suitable.

During this intermittent periodic duty, the starting current $I_{an} = 5.5 \cdot 33$ A $= 181.5$ A. Since the 3TF46 has a making capacity of 640 A and a breaking capacity of 500 A at $U_e = 380/400$ V, it does indeed meet the requirements.

3.7.9 Selection of contactors for three-phase pole-changing induction motors

Please refer to the Siemens M 11 catalogue for the various types and versions of three-phase pole-changing induction motors. For the basic circuit diagrams, see Section 8.2.5, from page 422 onwards.

Figure 3.88 shows the contactor power circuit diagram for the operation of three-phase pole-changing induction motors with a single winding and six terminals, in either Dahlander (4/2 and 8/4 pole) or PAM connection (6/4 and 8/6 pole).

The contactors are selected according to the respective rated currents I_n of the motor:

Contactor K1 for I_{n1} at low speed,
Contactor K2 for I_{n2} at high speed,
Contactor K3 (star contactor) for $I_{n3} = \dfrac{I_{n2}}{2}$.

For double star connection, the star contactor is always dimensioned for half the line current.

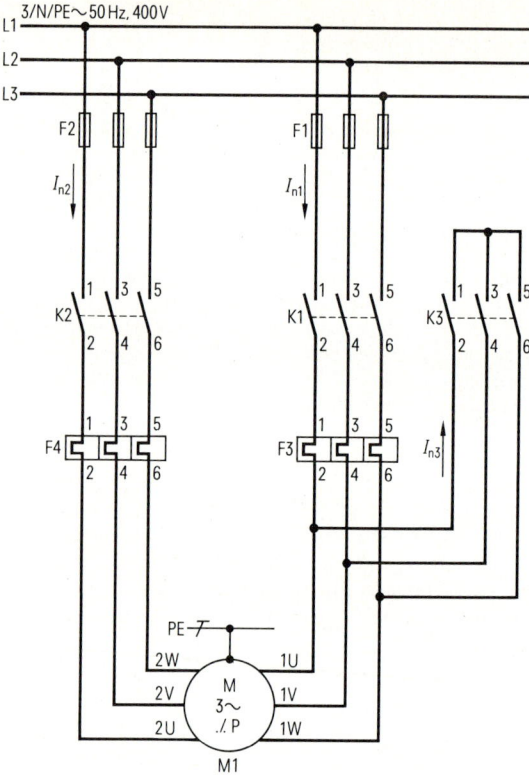

a) Circuit diagram

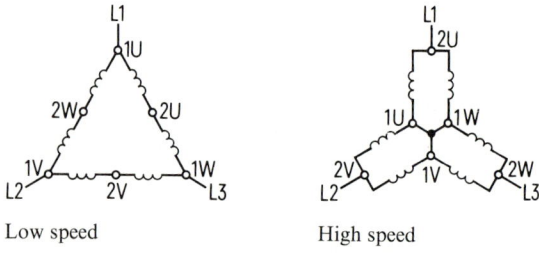

Low speed   High speed

b) Motor winding connections

Figure 3.88
Circuit for a three-phase pole-changing induction motor with a single winding and six terminals, in Dahlander connection

Figure 3.89 shows the contactor power circuit diagram for the operation of three-phase pole-changing induction motors with two separate windings brought out on six terminals, and two speed switching (6/4 and 8/6 pole).

Selection of contactors for pole-changing motors

Again, the selection of the contactors is made according to the motor rated currents I_n:

Contactor K1 for I_{n1} at low speed,
Contactor K2 for I_{n2} at high speed,

The contactor power circuit diagram for three speed switching (6/4/2, 8/4/2, 8/6/4/ or 8/6/2/ pole) of three-phase pole-changing induction motors having two separate windings brought out on nine terminals, is shown in Figure 3.90. The contactors are selected according to the motor rated currents I_n:

Contactor K1 for I_{n1} at low speed,
Contactor K2 for I_{n2} at medium speed
Contactor K3 for I_{n3} at high speed,
Contactor K4 (star contactor) for $I_{n4} = \dfrac{I_{n3}}{2}$.

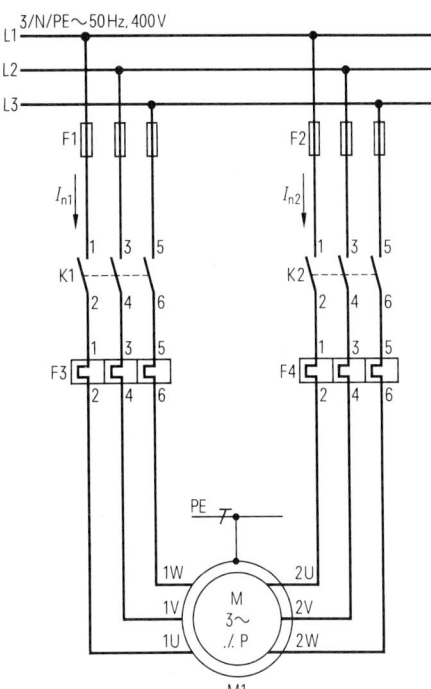

a) Circuit diagram

b) Motor winding connections

Figure 3.89
Circuit for a three-phase pole-changing induction motor with two separate windings, single direction of rotation and six terminals

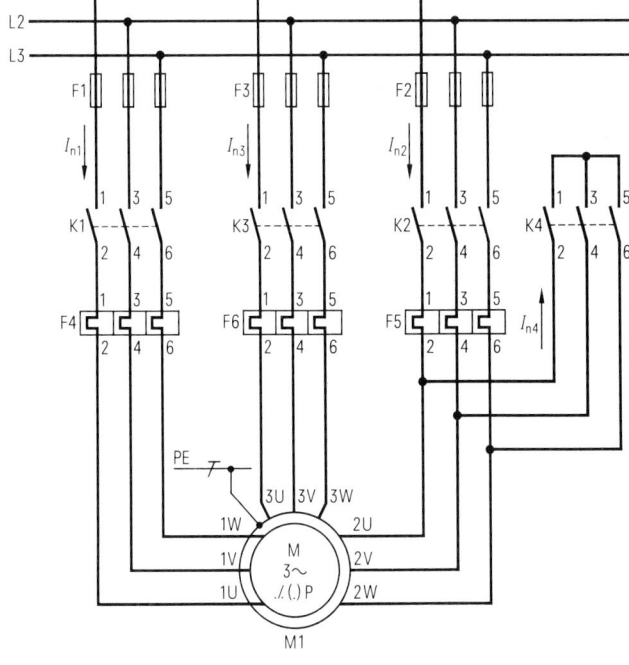

a) Circuit diagram

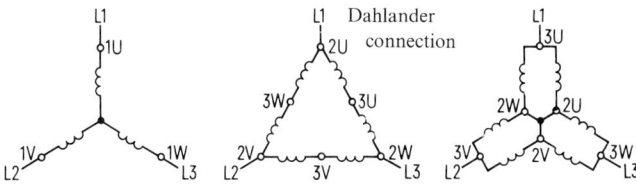

Two separate windings

b) Motor winding connections

Figure 3.90
Circuit for a three-phase pole-changing induction motor with two separate windings brought out on nine terminals, for three speeds of rotation

3.7.10 Selection of 3TF and 3TK contactors as well as circuit-breakers for the switching of lamps

In the selection of contactors and circuit-breakers for switching of lamps, the expected switch-on currents are of particular importance. Table 3.4 on page 116 lists the switch-on currents, the cos φ and the starting times of various lamp types, as well as the selection criteria for contactors in terms of the appropriate utilization category.

Selection guidelines for contactors used to switch fluorescent lamps are given in Table 3.30. Table 3.31 gives the data for high-pressure and halogen metal-vapour lamps, where-as Table 3.32 lists the data for incandescent lamps.

Table 3.30 Selection of 3TF and 3TK contactors for the switching of fluorescent lamps

	Lamp type	Capacitor value µF	Rated current per ball. unit A	Max. no. of ballast units per current path for contactors at 230 V, 50 Hz						
				3TF20, 3TK20	3TF40/ 3TF41	3TF42/ 3TF43	3TF44/ 3TF45	3TF46	3TF47	3TF48/ 3TF49
Luminaires with fluorescent lamps and compact fluorescent lamps (DULUX) with choke ballast	non-p.f. corrected (without capacitor)									
	S11W	–	0.16	100	125	187	281	500	500	625
	L18W	–	0.37	43	54	81	121	216	216	270
	L24W	–	0.34	47	58	88	132	235	235	294
	L36W	–	0.43	37	46	69	104	186	186	232
	L58W	–	0.67	23	29	44	67	119	119	149
	p.f. corrected (parallel capacitor)									
	S11W	4.5	0.07	22	29	54	78	111	155	185
	L18W	4.5	0.11	22	29	54	78	111	155	185
	L24W	4.5	0.13	22	29	54	78	111	155	185
	L36W	4.5	0.21	22	29	54	78	111	155	185
	L58W	7	0.32	14	19	34	50	71	99	119
	lead/lag circuit (DUO circuit)									
	S11W	–	0.07	228	284	428	642	1142	1142	1428
	L18W	–	0.11	144	180	272	408	726	726	908
	L24W	–	0.14	114	142	214	320	570	570	714
	L36W	–	0.21	76	94	142	214	380	380	476
	L58W	–	0.32	50	62	92	140	250	250	312
Luminaires with fluorescent lamps and electronic ballast units	AC operation									
	electronic ballast unit with single lamp									
	L18W	6.8	0.1	63	84	154	224	315	441	525
	L36W	6.8	0.18	35	46	85	124	175	245	291
	L58W	10	0.27	23	31	57	83	116	163	194
	electronic ballast unit with two lamps									
	L18W	10	0.18	35	46	85	124	175	245	291
	L36W	10	0.35	18	24	44	64	90	126	150
	L58W	22	0.52	12	16	29	43	60	84	101
	DC operation and three current paths in series									
	electronic ballast unit with single lamp									
	L18W	6.8	0.1	25	84	154	224	315	441	525
	L36W	6.8	0.18	13	46	85	124	175	245	291
	L58W	10	0.27	9	31	57	83	116	163	194
	electronic ballast unit with two lamps									
	L18W	10	0.18	13	46	85	124	175	245	291
	L36W	10	0.35	7	24	44	64	90	126	150
	L58W	22	0.52	4	16	29	43	60	84	101

Selection of contactors for the switching of lamps

Table 3.31 Selection of 3TF and 3TK contactors for the switching of high-pressure and halogen metal-vapour lamps

Lamp type		Power W	Capacitor value µF	Rated current per lamp A	Max. no. of lamps per current path for contactors at 230 V, 50 Hz						
					3TF20, 3TK20	3TF40/ 3TF41	3TF42/ 3TF43	3TF44/ 3TF45	3TF46	3TF47	3TF48/ 3TF49
High-pressure metal-vapour and mercury-vapour lamps (HQL)	non-p.f. corrected (without capacitor)										
		50		0.61	13	16	24	36	65	65	81
		80		0.8	10	12	18	28	50	50	62
		125		1.15	6	8	13	19	34	34	43
		250		2.15	3	4	6	10	18	18	23
		400		3.25	2	3	4	6	12	12	15
		700		5.4	1	1	2	4	7	7	9
		1000		7.5	1	1	2	3	5	5	6
	p.f. corrected (parallel capacitor)										
		50	7	0.27	14	19	34	50	71	99	119
		80	8	0.41	12	16	30	44	62	87	104
		125	10	0.62	10	13	24	35	50	64	80
		250	18	1.21	5	7	12	18	27	33	41
		400	25	1.93	4	5	7	11	20	20	25
		700	40	3.34	2	2	4	6	11	11	14
		1000	60	4.75	1	2	3	4	8	8	10
Metal-halide lamps (HQI)	non-p.f. corrected (without capacitor)										
		70		1.0	8	10	15	22	40	40	50
		150		1.8	4	5	8	12	22	22	27
		250		3.0	2	3	5	7	13	13	16
		400		3.5	2	2	4	6	11	11	14
		1000		9.5	0	1	2	2	4	4	5
	p.f. corrected (parallel capacitor)										
		70	12	0.38	8	11	20	29	41	58	69
		150	20	0.78	5	6	12	17	25	34	41
		250	32	1.25	3	4	7	11	15	21	26
		400	35	1.75	2	3	6	10	14	19	23
		1000	85	4.77	1	1	2	4	5	8	9
High-pressure sodium-vapour lamps (NAV)	non-p.f. corrected (without capacitor)										
		50		0.77	10	12	19	29	51	51	64
		70		1.0	8	10	15	22	40	40	50
		100		1.2	6	8	12	18	33	33	41
		150		1.8	4	5	8	12	22	22	27
		250		3.0	2	3	5	7	13	13	16
		400		4.4	1	2	3	5	9	9	11
		1000		10.3	0	1	1	2	3	3	4
	p.f. corrected (parallel capacitor)										
		50	8	0.28	12	16	30	44	62	87	104
		70	12	0.38	8	11	20	29	41	58	69
		100	12	0.53	8	11	20	29	41	58	69
		150	20	0.77	5	6	12	17	25	34	41
		250	36	1.25	2	3	6	9	13	19	23
		400	45	2.05	2	2	5	7	11	15	18
		1000	100	4.96	1	1	2	3	5	6	8

3 Selection criteria for low-voltage switchgear in main circuits

Table 3.32 Selection of 3TF and 3TK contactors for the switching of incandescent lamp.

		Max. rated power per current path for contactors at 230 V, 50 Hz						
		3TF20, 3TK20	3TF40/41	3TF42/43	3TF44/45	3TF46	3TF47	3TF48/49
Incandescent lamp	kW	1.7	2.3	4.2	6.1	8.6	12.0	14.3

Table 3.33
Maximum permissible lamp load (number of lamps) for a miniature circuit-breaker when switching lamps *without* power factor correction

Lamp data		Rated current per lamp	B/C10 [1]	B/C16 [1]	B/C20 [1]	B/C25 [1]
W	V	A				
Fluorescent lamps (electronic ballast, single lamps)						
L18/20	220	0.37	27	43	53	66
L36/40		0.43	23	37	46	58
L58/65		0.67	15	24	30	37
High-pressure mercury vapour lamps (HQL)						
50	220	0.6	10/8	15/13	18/16	23/20
80		0.8	6/6	9/10	11/12	14/15
125		1.15	4/4	6/7	7/9	9/10
250		2.15	2/2	3/3	3/4	4/5
400		3.25	1/1	2/2	2/3	2/3
700		5.4	–/1	1/1	1/1	1/2
1000		7.5	–/–	–/1	1/1	1/1
Metal-halide lamps (HQI) [2]						
150	220	1.8	4/2	6/4	7/5	9/6
250		3	2/1	3/2	4/3	5/4
400		3.5	2/1	3/2	4/2	5/3
1000		9.5	–/–	1/–	1/1	1/1
2000	D380	10.3	–/–	–	–	–
2000	N380	8.8	–/–	–	–	–
3500	380	18	–/–	–	–	–
High-pressure sodium-vapour lamps (NAV) [2]						
50	220	0.77	9/6	14/10	18/13	22/16
70		1.0	7/5	11/8	14/10	17/12
100		1.2	6/4	9/6	11/8	14/10
150		1.8	4/2	6/4	7/5	9/7
250		3.0	2/1	3/2	4/3	5/4
400		4.4	1/1	2/1	3/2	4/2
1000		10.3	1/–	1/–	1/1	2/1

[1] For fluorescent lamps the values only apply for miniature circuit-breakers with B characteristic. In the case of multi-pole m.c.b.'s, the permissible number of lamps is reduced by 20% respectively

[2] In the case of HQI and NAV lamps, switch-on current surges of up to max. $20 \cdot I_n$ can occur

Electronic ballast unit capacitances for the lamps in Table 3.33

Electronic ballast unit type 5LZ...-1	single lamp	double lamp
16 W, 32 W, 34 W	6.8 µF	10 µF
50 W	10 µF	22 µF

To determine the number of incandescent lamps which may be switched with 5SN and 5SX miniature circuit-breakers, the following rules may be applied:
— For miniature circuit-breakers with B-type characteristic, 0.5 times m.c.b. rated current I_n
— For miniature circuit-breakers with C-type characteristic, 1.0 times m.c.b. rated current I_n

Table 3.33 gives assistance in the selection of miniature circuit-breakers with type B or type C characteristics for the switching of lamps without power factor correction. If lamps with power factor correction are to be switched, see Table 3.34.

The lamp load values given for fluorescent lamps are based on the assumption of a conductor impedance equal to 800 mΩ.

This impedance corresponds to a 1.5 mm² feeder conductor (wire) of 15 m between the power distribution point and the first lamp, and a further 20 m of the same conductor to the middle of the load circuit.

In the case of 400 mΩ or 200 mΩ conductor impedance, the permissible values are reduced by 10% and 20% respectively.

Table 3.34
Maximum permissible lamp load (number of lamps) for a miniature circuit-breaker when switching lamps *with* power factor correction

| Lamp data | | Capacitor value µF | B/C10[1] | | B/C16[1] | | B/C20[1] | | B/C25[1] | |
W	V		KVG[2]	EVG[3]	KVG[2]	EVG[3]	KVG[2]	EVG[3]	KVG[2]	EVG[3]
Fluorescent lamps, p.f. corrected										
L18/20	220	4.5	32	18	51	26	64	33	82	42
L36/40		4.5	32	18	51	26	64	33	82	42
L58/65		7	20	8	33	12	41	15	53	19
High-pressure mercury vapour lamps (HQL)										
50	220	7	10/19		15/31		18/39		23/49	
80		8	6/12		9/19		11/24		14/30	
125		10	4/7		6/12		7/15		9/19	
250		18	2/4		3/6		3/7		4/9	
400		25	1/2		2/4		2/5		2/6	
700		40	–/1		1/2		1/2		1/3	
1000		60	–/1		–/1		1/2		1/2	
Metal-halide lamps (HQI)[4]										
150	220	20	7/5		11/8		14/10		17/12	
250		32	5/3		7/5		9/6		11/8	
400		35	3/2		5/4		7/5		8/6	
1000		85	1/–		1/1		3/1		3/2	
2000	D380	60	1/–		2/1		2/1		3/2	
2000	N380	37	–		1/–		1/1		2/1	
3500	380	100	–		–		–		–	
High-pressure sodium-vapour lamps (NAV)[4]										
50	220	10	16/11		24/17		31/22		38/27	
70		12	12/8		18/13		23/16		29/20	
100		12	10/7		16/11		20/14		25/17	
150		20	7/5		11/8		14/10		17/12	
250		36	5/3		7/5		9/6		11/8	
400		45	3/2		4/3		5/4		7/5	
1000		100	1/–		1/1		2/1		3/2	

[1] For fluorescent lamps the values only apply for miniature circuit-breakers with B characteristic; for miniature circuit-breakers with C characteristic particular attention must be paid to DIN VDE 0100 part 410. In the case of multi-pole m.c.b.'s, the permissible number of lamps is reduced by 20% respectively.
[2] Conventional choke ballast and single lamp type, switching on per group (KVG = *k*onventionelles *V*orschaltgerät; transl. conventional choke ballast)
[3] Electronic ballast units and double lamp types, no. of lamps to be switched simultaneously (EVG = *e*lektronisches *V*orschaltgerät; transl. electronic ballast unit)
[4] In the case of newer HQI and NAV lamps, switch-on current surges of up to max. $20 \cdot I_n$ can occur. If necessary, the values must be reduced accordingly.

3.7.11 Switching of three-phase transformers up to 1000 V with 3TF contactors

The extremely high switch-on current peaks associated with the switching of low-voltage three-phase transformers make particular demands on the making capacity of contactors. As mentioned in Section 3.2.2.6, on page 117, these rush currents are determined by a number of different factors.

In most cases, a switch-on rush current of 20 to 30 times rated operating current may be regarded as normal, and must be taken into account when contactors are selected (see Table 3.35).

In cases where the rush factor deviates from this value, the permissible apparent power of the low-voltage three-phase transformer to be switched by the contactor, must be calculated:

$$S_x = S_{nT} \cdot \frac{20}{x}.$$

S_{nT} Transformer apparent power rating assuming a switch-on rush of $20 \cdot I_n$

S_x Permissible transformer apparent power rating

x Actual switch-on rush factor referred to the rated current

According to IEC 947-4, the switching capabilities of contactors for the switching of three-phase transformers can be derived from the AC-3 switching capacity data. For a switch-on rush factor of up to 30: $I_{eT} = 0.45 \cdot I_{e\,AC-3}$ applies.

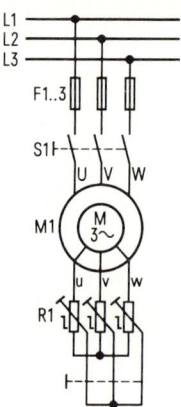

Figure 3.91
Three-phase induction motor with slip-ring rotor and variable starting resistors in the rotor circuit

3.7.12 Starting of three-phase induction motors with slip-ring rotors

Starters are switching devices or circuits which are used to bring electrical machines from standstill to their operating condition. Starters which limit and/or control the starting current and torque during the starting procedure (and, in some cases, during running and/or stopping) may be divided into two groups:

Rotor resistance starters for three-phase induction motors with slip-ring rotors (rotor starters), and

Reduced voltage starters for three-phase induction motors with squirrel-cage rotors (stator starters).

This section deals with the starting of the three-phase induction motors with slip-ring rotors. The starting

Table 3.35
Permissible apparent output power ratings S_{nT} of low-voltage three-phase transformers having a switch-on rush of $20 \cdot I_n$ to be switched by 3TF contactors

Size Order No.		00 3TF2	0 3TF40/41	1 3TF41/43	2 3TF44/45	3 3TF46	3 3TF47
230 V	kVA	2.0	4.6	4.7	9.4	13	19
400 V	kVA	3.5	7.9	8.0	16.2	23	33
500 V	kVA	4.6	10.4	10.6	21.3	30	43
690 V	kVA	6.0	13.7	14	28	40	57

Size Order No.		4 3TF48/49	6 3TF50/51	8 3TF52/53	10 3TF54/55	12 3TF56	14 3TF68
230 V	kVA	23	36	51	82	102	156
400 V	kVA	41	63	88	142	177	270
500 V	kVA	54	83	116	187	233	355
690 V	kVA	71	109	153	246	307	469
1000 V	kVA	108	166	231	374	465	710

of three-phase induction squirrel-cage motors by means of the SIKOSTART solid-state starter is discussed in Section 3.7.15, on page 223. KUSA-resistance starting is discussed in Section 3.7.13 on page 219.

The slip-ring starter has the task of maintaining the rotor current and the torque of the motor at a desired mean value during the starting sequence.

This mean value is achieved if the voltage on the rotor is gradually reduced. This can be done either by the introduction of variable resistances into the rotor circuit (Fig. 3.91), or by the application of a variable primary voltage. The first method is known as resistance starting, whereas the second is termed voltage starting.

Typical contactor-based starters for three-phase slip-ring induction motors are constructed as

▷ oil-cooled starters with control and resistance unit as a single device, or as

▷ starter switch with separate resistance unit.

The choice of starter version depends on the installation conditions, available space, the required degree of protection, the operational values and on economic considerations.

3.7.12.1 3PA3 Oil-cooled starters

Application

These starters with oil-cooled resistors are used for the starting of three-phase slip-ring motors with rated output up to 6400 kW and rotor standstill voltages up to 200 V.

Oil-cooled starters can absorb much more heat during the starting sequence than is the case with starters having air-cooled resistor units. The heat stored in the oil is dissipated to the environment relatively slowly. Particular attention must therefor be paid to the starting frequency h during operation up to maximum permitted oil temperature.

Twin starters are offered for the simultaneous starting of two motors or for motors of very high output rating. Starters equipped with additional water cooling permit a higher starting frequency h.

Starters without stator contactors are also suitable for the starting of high-voltage motors.

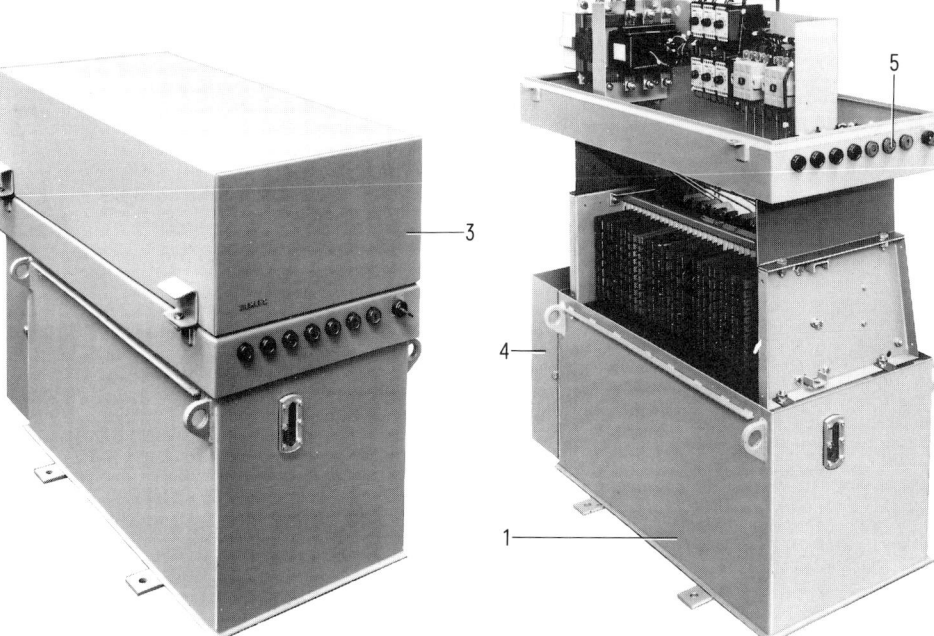

1	Oil tank
2	Equipment tray
3	Steel cover
4	Cable cover
5	Indicator lights

Figure 3.92   3PA oil-cooled starter shown closed (left), and opened (right)

3 Selection criteria for low-voltage switchgear in main circuits

Specifications

The 3PA3 oil-cooled starters comply with the specifications DIN VDE 0660 part 301, DIN 46062 and IEC 947-4-1.

Construction

As can be seen in Figure 3.92, the starters consist of:

1. the oil tank housing with the resistor unit (1),
2. the equipment tray which closes the oil tank and carries the controlgear (2),
3. the sheet-steel hood which covers the controlgear,
4. the cable cover (accessory, to be ordered separately).

The resistance unit of the smallest starter is made up of sheet-steel elements. All the other sizes use cast-iron elements.

The starting resistors are bridged out one after the other by the rotor-circuit contactors. Time relays determine the period between starting steps.

Each starting step is signalled by an indicator light (5).

Twin starters consist of two single starter units in which both sets of rotor circuit contactors are controlled by the same time relays (located in one of the units).

Starters with additional water cooling incorporate an integrated heat-exchanger unit.

The 3PA3 starters have a degree of protection IP 54. The environmental conditions given in the Siemens l.v. switchgear catalogue (place of installation, temperature, climatic conditions), are to be taken onto account.

Selection

Depending on whether normal or special starting conditions prevail, either the "simple selection" or the "calculation method" must be applied.

▷ Normal conditions prevail when the moment of inertia of the driven machine is not excessive ($J_G \leq 10 \cdot J_M$), and the required starting data corresponds with that given in the catalogue.

▷ Special starting conditions are e.g. an extremely short or long starting time, high starting frequency, heavy duty starting ($f > 2$) or a high moment of inertia of motor and driven machine (see examples).

The following terms and operational values are used for the calculation and the selection of starters:

P_n	Rated output power of the motor	kW
U_e	Rated operational voltage	V
u_{20}	Rotor standstill voltage (If the value is not given on the rating plate of the motor, the rotor standstill voltage must be determined by measurement)	V
I_m	Mean starting current	A
I_n	Rated current of the motor	A
I_e	Rated operational current	A
i_2	Rated rotor current (If the value is not given on the rating plate of the motor, the rated rotor current can be calculated by $i_2 \approx \frac{P_n \cdot 630}{u_{20}}$)	A
M_n	Rated motor torque	Nm
M	Motor torque (instantaneous value)	Nm
M_a	Locked rotor torque (breakaway torque)	Nm
M_L	Load torque	Nm
M_m	Mean starting torque	Nm
M_{Lm}	Mean load torque	Nm
M_B	Accelerating torque	Nm
n_n	Rated speed of the motor	min^{-1}
n	Motor speed (instantaneous value)	min^{-1}
n_S	Synchronous speed	min^{-1}
w	Angular velocity	s^{-1}
s	Speed reduction (slip)	%
J_M	Moment of inertia of the motor	kgm^2
J_A	Moment of inertia of the driven machine	kgm^2
J_G	Moment of inertia of the motor and driven machine	kgm^2
f	Starting load factor	–
t_a	Starting time (run-up time)	s
W	Starting energy	kJ
z	Number of starts	–
h	Starting frequency	h^{-1}
k_a	Starter characteristic value	Ω
k	Rotor characteristic value	Ω

Starting of three-phase induction motors using oil-cooled starters

Derived values and terms for starter selection

Starting load factor f

$$f = \frac{I_m}{I_n} \approx \frac{M_m}{M_n} \quad (3.6)$$

Rated torque of the motor M_n

$$M_n = \frac{P_n \cdot 9.55 \cdot 10^3}{n_n} \quad (3.7)$$

Preferred values for f are:

$f = 0.7$ for half-load starting,
$f = 1.0$ for fan starting,
$f = 1.4$ for full-load starting,
$f = 2.0$ for overload or heavy duty starting.

The diagrams in Figures 3.93 to 3.96 below, show the typical starting curves for the above-mentioned preferred values of f.

Starting time t_a

The starting time t_a is the time during which the starting resistor or parts thereof conduct current. It corresponds to the full duration of the starting procedure and is therefore the time taken for the motor to reach its rated speed from standstill. The following equation applies if the driven machine has a virtually constant load torque.

$$t_a = \frac{J_G \cdot w}{M_B} \text{ (in s), or } t_a = \frac{J_G \cdot n_n}{9.55 \cdot M_B} \text{ (in s)} \quad (3.8)$$

(Note that the shortest possible starting time is 10 s)

Where the accelerating torque

$M_B = M_m - M_L$ (in Nm)

If a transmission system is present between the motor and the driven machine, all moments of inertia are to be referred to the motor speed n_n, and (if necessary) converted accordingly.

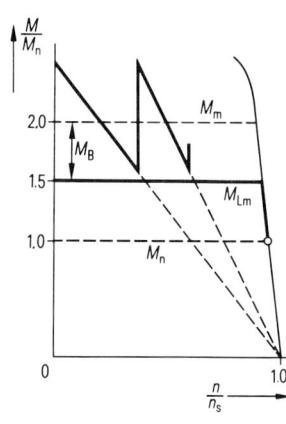

Figure 3.93
Overload or heavy duty starting: $f = 2.0$, e.g. dough mixing machines, crushers

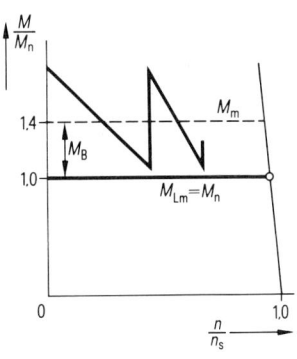

Figure 3.94
Full-load starting: $f = 1.4$, e.g. machine tools under load, milling machines

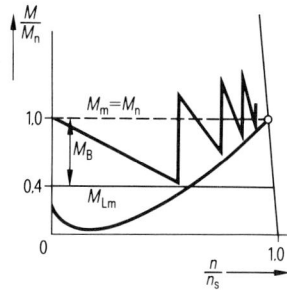

Figure 3.95
Fan starting: $f = 1.0$, e.g. fans, centrifugal pumps

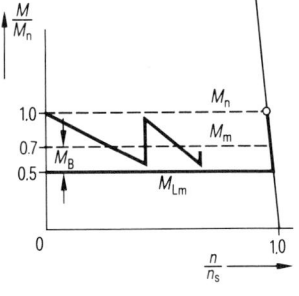

Figure 3.96
Half-load starting: $f = 0.7$, e.g. lathes, grinders, punching machines

The total moment of inertia J_G is found by adding that of the motor to the converted moment of inertia of the driven machine, i.e.

$$J_G = J_M + J_A \cdot \left(\frac{n_2}{n_n}\right)^2. \tag{3.9}$$

Starting energy W

The starting energy W is the energy to be absorbed by the starter during the starting sequence, and is expressed in kJ. It is calculated using the equation:

$$W = 0.5 \cdot P_n \cdot f \cdot t_a$$

The factor 0.5 indicates that half the starting energy drawn from the network is converted into heat in the resistance unit. The other half serves to accelerate the motor.

Number of starts z

Number of starts z, is the number of sequential starts the resistance unit may perform (beginning from the cold state) until its maximum temperature rise is reached under the specified conditions of starting time and mean starting current, and allowing a cooling period of twice the starting time between each starting operation.

The minimum number of starts is always selected as 2, so that a second starting operation is possible without a long waiting period being necessary. This may be particularly important if a repeat of the starting procedure is required during commissioning of a motor and starter.

The number of starts z can be calculated from:

$$z = \frac{\text{possible starting energy (see selection data)}}{\text{actual starting energy (calculated value)}}.$$

In terms of DIN 46062, starters for a motor rating of 100 kW or higher (full load starting) must be dimensioned for at least $z = 2$.

Starting frequency h

The starting frequency h of a starting device is the number of starting operations repeated at regular intervals which may be carried out per hour continuously, once operational temperature has been reached.

In the case of air-cooled resistors, one can estimate h from the number of starts z. For this, z must be ≥ 2.

The following relationships exists:

For cast-iron resistor elements $h \approx 5 \cdot z$,
For wire-wound resistors $h \approx 7.5 \cdot z$.

Relationship between starting energy W, number of starts z and the starting frequency h

In oil-cooled starters, a specific relationship exists between the starting energy W, the starting time z and the number of starts h. This can be determined from diagrams as is illustrated in Figures 3.97 to 3.100.

Starter characteristic value k_a

The starter characteristic value k_a for starters of three-phase slip-ring motors is derived from the rotor characteristic value of the motor and the starting load factor f:

$$k_a \approx 1.4 \cdot \frac{k}{f} \text{ (in } \Omega\text{)}. \tag{3.10}$$

Where the rotor characteristic value k of the motor is:

$$k = \frac{\text{Rotor standstill voltage}}{\sqrt{3} \cdot \text{rated rotor current}} = \frac{u_{20}}{\sqrt{3} \cdot i_2} \Omega. \tag{3.11}$$

Standard characteristic values k_a for three-phase starters:

| k_a | 0.4 | 0.5 | 0.63 | 0.8 | 1 | 1.25 | 1.6 | 2 | 2.5 |
| k_a | 3.2 | 4 | 5 | 6.3 | 8 | 10 | 12.5 | 16 | |

Smaller values tend to apply for heavy-duty starting, bigger values for half-load starting.

Three selection examples are given below. The order numbers of the respective starters can be found in the relevant Siemens low-voltage switchgear catalogue.

Example 1

Given:

- Slip-ring motor,
  Type 1RS6 314-6AA10 (Siemens) with
  Rated output power $P_n = 160$ kW,
  Rated operational voltage $U_e = $ AC 380 V,
  Rated operational current $I_e = 306$ A,
  Rotor standstill voltage $u_{20} = 285$ V,
  Rated rotor current $i_2 = 347$ A.
- Driven machine with
  Starting load factor $f = 1.4$ (Milling machine)
  Starting time $t_a = 16$ s,
  (derived from motor and machine data).

Required:

- Starter with stator contactor and safety tripping in the event of a locked rotor,
  Degree of protection IP 54
  Number of starts $z \geq 5$,
  Starting frequency $h \geq 1.5$.

– Determining of the starter type from the required starting energy W:

$$W = 0.5 \cdot P_n \cdot f \cdot t_a$$
$$= 0.5 \cdot 160 \text{ kW} \cdot 1.4 \cdot 16 \text{ s} = 1792 \text{ kJ}$$

The diagrams in Figures 3.97 to 3.100 can be used to show that the type 3PA3 02 starter will provide the necessary starting energy of 1792 kJ at the desired number of starts $z \geq 1.5$ and the starting frequency $h \geq 1.5$.

If one of the values is not achieved, the next larger starter must be selected.

The number of starts z may also be determined using

$$z = \frac{\text{possible starting energy}}{\text{actual starting energy}}$$

$$= \frac{9450 \text{ kJ}}{1792 \text{ kJ}} = 5.27 \approx 5$$

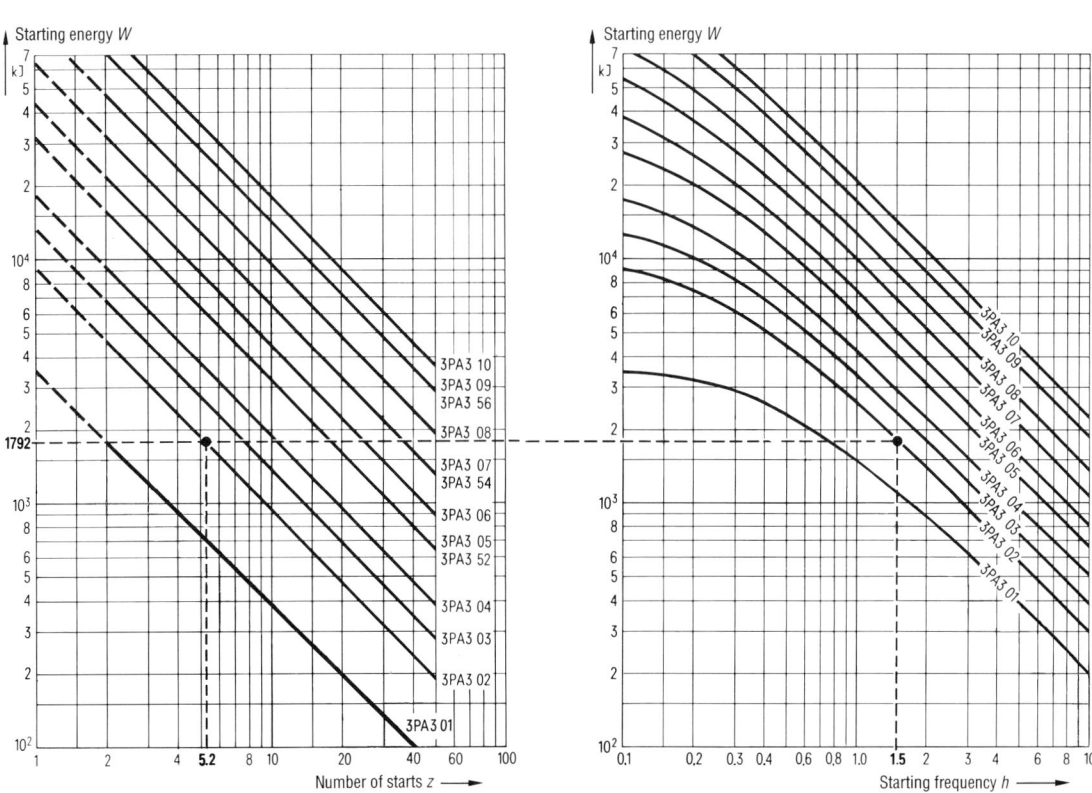

Figure 3.97
Diagram of starting energy as a function of the number of starts z for 3PA3 oil-cooled starters *without* additional water cooling

Figure 3.98
Diagram of starting energy as a function of the starting frequency h for 3PA3 oil-cooled starters *without* additional water cooling

3 Selection criteria for low-voltage switchgear in main circuits

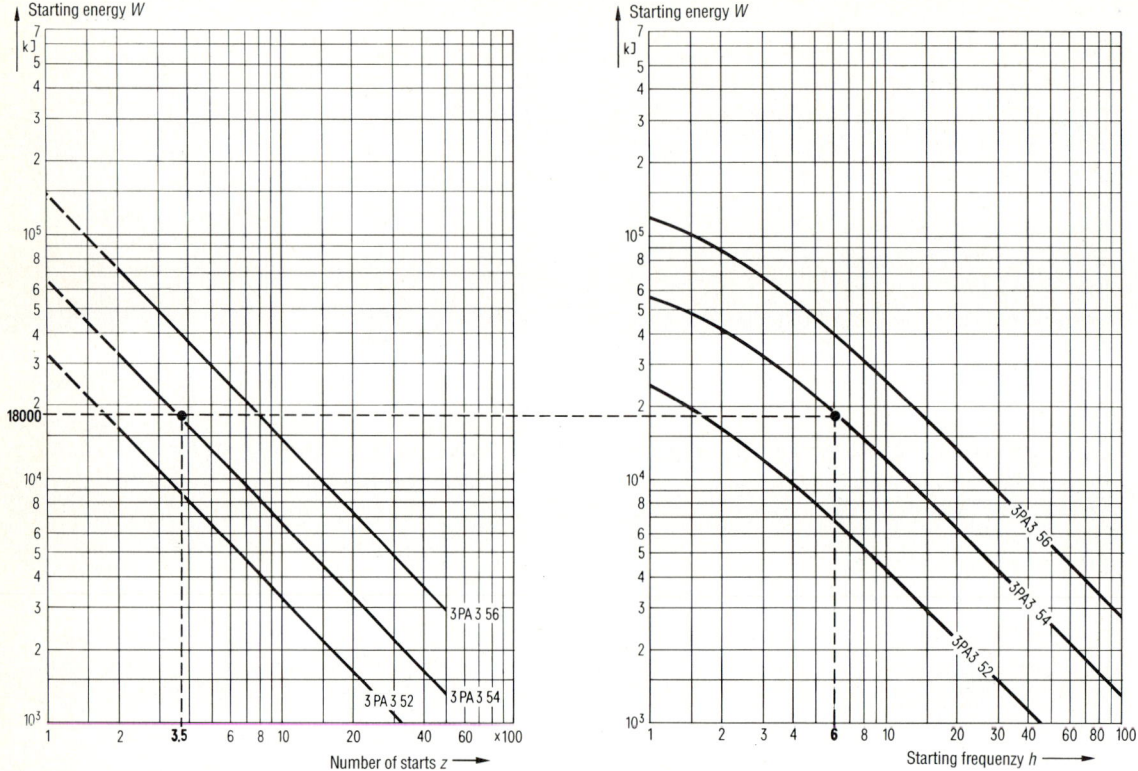

Figure 3.99
Diagram of starting energy as a function of the number of starts z for 3PA3 oil-cooled starters *with* additional water cooling

Figure 3.100
Diagram of starting energy as a function of the starting frequency h for 3PA3 oil-cooled starters *with* additional water cooling

The number of starts $z=5$ means, that beginning with a cold condition, five starting operations with a cooling period of double the starting time between them, may be repeated one after the other.

The starting frequency of $h=1.5$ means, that with the starter at operational temperature, starting operations may be performed at regular intervals of 40 min (60 min/1.5).

Thus, the 3PA3 02 starter permits five starts of the 160 kW motor from the cold state, with cooling periods of $2 \cdot 16 = 32$ s and a starting time $t_a = 16$ s. Once the device has reached operational temperature, the starting sequence may be repeated every 40 minutes.

To determine the rest of the order number for the oil-cooled starter, the factor k_a must be calculated

and the following points must be taken into account:

Starter characteristic value k_a

$$k_a \approx 1.4 \cdot \frac{k}{f} \text{ and}$$

$$k = \frac{u_{20}}{\sqrt{3} \cdot i_2}, \text{ therefore}$$

$$k_a = \frac{1.4 \cdot 285 \text{ V}}{1.4 \cdot \sqrt{3} \cdot 347 \text{ A}} = 0.47 \, \Omega,$$

– Control voltage,

– Rated rotor current i_2 (basic or heavy duty version of the oil-cooled starter),

– Rated current of the stator for the selection of the contactor and the setting range of the overload relay,

Table 3.36 Technical data of the 3PA3 oil-cooled starters

Permissible operating values									Starter data				
Motor output rating P_n				Rated rotor current i_2		Rotor stand-still voltage u_{20} V	Rated stator oper. current I_e (AC-2) A	Rated stator voltage U_e V	Order No.	Starting energy W kJ	Start-ing time t_a s	No. of starts z –	Starting fre-quency h starts/hour
Over-load starting $f=2.0$ kW	Full load starting $f=1.4$ kW	Fan starting $f=1$ kW	Half-load starting $f=0.7$ kW	Basic version A	Heavy duty version A								
Starters with stator contactor													
70	100	140	200	150	250	1320	180	660/690	3PA3 01	3830	18	3	1.4
155	225	315	450	250	450	1500	400	660/690	3PA3 02	9450	20	3	0.8
225	320	450	640	250	450	1500	630	660/690	3PA3 03	13440	20	3	0.75
Starters without stator contactor													
70	100	140	200	150	250	1320	–	–	3PA3 01	3830	18	3	1.4
155	225	315	450	250	450	1500	–	–	3PA3 02	9450	20	3	0.8
225	320	450	640	250	450	1500	–	–	3PA3 03	13440	20	3	0.75
315	450	630	900	450	630	1500	–	–	3PA3 04	18900	20	3	0.7
440	630	880	1260	450	630	1500	–	–	3PA3 05	30870	36	2	0.3
625	900	1250	1800	630	1100	2000	–	–	3PA3 06	44100	36	2	0.3
875	1250	1750	2500	630	1100	2000	–	–	3PA3 07	64750	36	2	0.25
1250	1800	2500	3600	1100	1600	2000	–	–	3PA3 08	93240	40	2	0.2
1750	2500	3500	5000	1100	1600	2000	–	–	3PA3 09	140000	40	2	0.2
2250	3200	4500	6400	1100	1600	2000	–	–	3PA3 10	179200	40	2	0.18
Twin starters without stator contactor													
2×1250	2×1800	2×2500	2×3600	2×1100	2×1600	2000	–	–	3PA3 40	2×93240	40	2	0.2
2×1750	2×2500	2×3500	2×5000	2×1100	2×1600	2000	–	–	3PA3 41	2×140000	40	2	0.2
2×2250	2×3200	2×4500	2×6400	2×1100	2×1600	2000	–	–	3PA3 42	2×179200	40	2	0.18
Starters with additional water cooling, without stator contactor													
440	630	88	1260	450	630	1500	–	–	3PA3 52	30870	36	2	2.2
875	1250	1750	2500	630	1100	2000	–	–	3PA3 54	64750	36	2	3
1750	2500	3500	5000	1100	1600	2000	–	–	3PA3 56	140000	40	2	2.8

– Starting time t_a,
– Special version of the oil-cooled starter which provides safety tripping in the event of a locked rotor.

Example 2

Given:

– Slip-ring motor,
  Rated output power $P_n = 5000$ kW,
  Rotor standstill voltage $u_{20} = 1990$ V,
  Rated rotor current $i_2 = 1580$ A,

– Driven machine,
  Starting load factor $f = 0.7$ (wood grinder)

Required:

– Starter without stator contactor,
  Degree of protection IP 54,
  Control voltage 240 V, 60 Hz.

– Determination of the starter type:

  From Table 3.36, the size 3PA3 09 in heavy-duty version ($i_2 = 1600$ A) is suitable for the starting of a 5000 kW motor with starting load factor $f = 0.7$ (half-load starting).

  The starting time $t_a = 40$, the number of starts $z = 2$, and the starting frequency $h = 0.2$ of the device must be taken into account.

For the completion of the order number the following is still required:

- Starter characteristic value k_a,

$$k_a = \frac{u_{20}}{\sqrt{3} \cdot i_2} \cdot \frac{1.4}{f}$$

$$= \frac{1990 \text{ V}}{\sqrt{3} \cdot 1580 \text{ A}} \cdot \frac{1.4}{0.7} = 1.45 \, \Omega,$$

- Control voltage (AC 60 Hz, 240 V):
- Rated rotor current i_2 (heavy-duty version),
- Starting time t_a (see Table 3.36).

Example 3

Given:

- Slip-ring rotor with

  Rated output power $P_n = 900$ kW,
  Rotor standstill voltage $u_{20} = 945$ V,
  Rated speed $n_n = 1480$ min^{-1}.

- Driven machine with large flywheel
  Starting load factor $f = 0.78$.

- Moment of inertia of motor and driven machine
  $J_G = 540$ kgm^2,
  Load torque $M_L = 0.5 \cdot M_n$.

Required:

- Starter without stator contactor,

  Degree of protection IP54,
  Control voltage AC 50 Hz, 220 V,
  Three successive starting operations, i.e. number of starts $z \geq 3$
  followed by further starting operations every 15 minutes, i.e. $h \geq 4$.

- Determination of the starter type from the starting energy W:

$W = 0.5 \cdot P_n \cdot f \cdot t_a$ (in kJ)

The starting time t_a is unknown, and must be calculated using:

$$t_a = \frac{J_G \cdot n_n}{9.55 \cdot M_B} \text{ (in seconds)}$$

In this equation, M_B is unknown. The mean accelerating torque M_B is calculated from:

$M_B = M_m - M_L$.

The load torque M_L is given as $0.5 \cdot M_n$.

The mean starting torque is:

$M_m \approx f \cdot M_n = 0.78 \cdot M_n$

The rated torque is calculated as being:

$$M_n = \frac{P_n \cdot 9550}{n_n} = \frac{900 \cdot 9550}{1480} = 5800 \text{ Nm},$$

(P_n in kW, n_n in min^{-1}).

Therefore, it follows:

$M_B = (0.78 - 0.5) \cdot M_n$
$ = 0.28 \cdot 5800 \text{ Nm} = 1620 \text{ Nm}$

$$t_a = \frac{J_G \cdot n_n}{9.55 \cdot M_B} = \frac{540 \cdot 1480}{9.55 \cdot 1630} = 52 \text{ seconds}$$

(J_G in kgm^2, n_n in min^{-1},

M_B in Nm = kgm^2/s^2).

$W = 0.5 \cdot P_n \cdot f \cdot t_a$
$ = 0.5 \cdot 900 \text{ kW} \cdot 0.78 \cdot 52 \text{ s} = 18250 \text{ kJ}$.

For the required starting energy of 18250 kJ, the diagrams in Figures 3.99 and 3.100 indicate that the specified number of starts ($z \geq 3$) and starting frequency ($h \geq 4$) can be achieved using the 3PA3 54 oil-cooled starter with additional water cooling.

In actual fact, beginning from a cold condition, the 3PA3 54 starter will permit 3.5 starting operations one after the other ($z = 3.5$), and a starting frequency of $h = 6$ (i.e. one start every 10 min) after the device has reached operational temperature.

To complete the selection and determine the full order number, the following must still be taken into account:

- Starter characteristic value k_a,
  which is calculated by

$$k_a \approx 1.4 \cdot \frac{k}{f} \text{ and}$$

$$k = \frac{u_{20}}{\sqrt{3} \cdot i_2}, \text{ therefor}$$

$$k_a = \frac{1.4 \cdot 945 \text{ V}}{0.78 \cdot \sqrt{3} \cdot 600 \text{ A}} = 1.63 \, \Omega,$$

- Control voltage,
- Rated rotor current i_2 (basic version),
- Starting time t_a (calculated value).

3.7.12.2 Starter and start-control switches

Starter and start-control switches 3PK4 are also used for starting of motors with slip-ring rotors. In the case of start-control switches, the slip may be adjusted during operation, i.e. the speed may be set.

The required starting/start-control resistors are mounted separately.

Selection

The selection (calculation) is done in accordance with the same principles as described for oil-cooled starters. In addition, the reduction in speed as well as the torque behaviour (constant or square function), must be considered in the case of start-control switches.

For technical data, selection and further descriptions, please refer to the corresponding Siemens l.v. switchgear catalogues.

3.7.13 Starting of three-phase induction motors with stator-resistance starters

Wherever possible, three-phase induction motors are started direct-on-line (full-voltage-across-the-line). If the starting current must be reduced owing to network or operating conditions, star-delta switching of the stator winding is mostly used for smaller motors.

The starting current may also be reduced by the introduction of resistors into the stator circuit. A reduction in the starting current causes in marked drop in the available motor torque, since although the current is directly proportional to the applied voltage, the torque is reduced by the square of the voltage on the motor terminals. If a resistance is introduced into only one phase of the stator supply, the starting torque of the motor is also reduced, the current is however only reduced in this line (KUSA circuit).

Single-pole stator-resistance starters (KUSA circuit)

KUSA is the abbreviation for the German "Kurzschlußläufer-Sanftanlauf", which means soft-starter for short-circuited (squirrel-cage) rotors. The KUSA starter circuit is used when a smooth starting is to be achieved, and the reduction of starting current is not required (Fig. 3.101). The value of the starting torque depends on the value of the resistance introduced into the stator feeder line.

This can be calculated as

$$R_{Ku} = \varrho \cdot \frac{U \cdot \sqrt{3}}{I_{an}} \qquad (3.12)$$

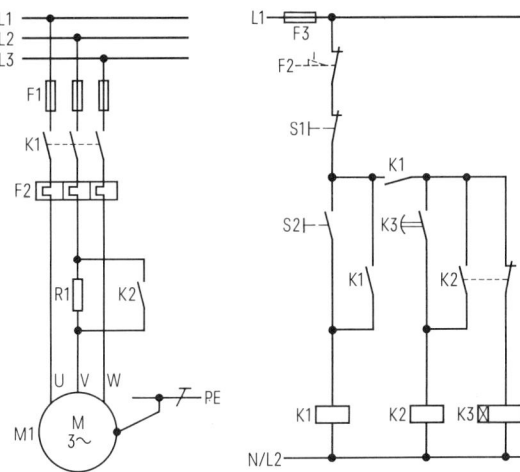

Figure 3.101
Power and control circuit for a single-pole stator-resistance starter (KUSA circuit)

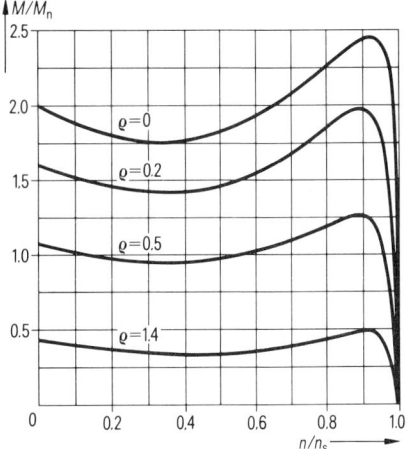

Figure 3.102
Torque-speed curves of a squirrel-cage motor with various values of KUSA resistance

The effect of the KUSA resistance on the torque-speed curve of the motor is illustrated in the diagram shown in Figure 3.102.

If a KUSA resistance is to be calculated for a specific starting torque M_{Ku}, the value of starting torque M_{an} and the starting current I_{an} of the motor in normal operation must first be known.

3 Selection criteria for low-voltage switchgear in main circuits

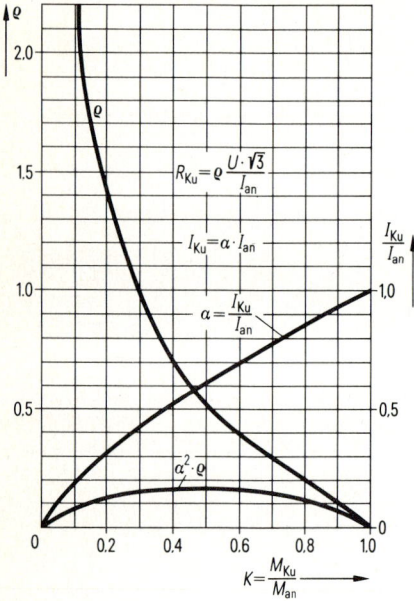

Figure 3.103
Diagram for determining the values of KUSA resistors

Using the equation

$$K = \frac{M_{Ku}}{M_{an}}, \quad (3.13)$$

the value of ϱ can be read from the diagram in Figure 3.103.

Associated with the relationship K is a specific factor

$$\alpha = \frac{I_{Ku}}{I_{an}}, \text{ which gives } I_{Ku} = \alpha \cdot I_{an}, \quad (3.14)$$

where I_{Ku} is the starting current with the KUSA resistance, and may also be found using Figure 3.103.

In order to select the size of the required resistance unit, the starting time and the starting frequency must also be known.

To facilitate the selection of the starter size and its price for specific output ratings, normal starting times and starting frequencies are given in the selection tables. For the starting of high-inertia or flywheel loads, the starting time must be calculated.

Circuit diagrams and a guide for the selection of the contactors are given in Section 8.3.6 on page 454.

*Example
for the calculation of a KUSA resistor*

Given:

Three-phase motor 1LA6 283-4AA70, 90 kW, 380 V, 50 Hz, 168 A
Starting torque $M_{an} = 2.5 \cdot M_n$
Starting current $I_{an} = 6.8 \cdot I_n$

Desired starting torque with KUSA resistor:
$M_{Ku} = 0.7 \cdot M_n$

Starting time 6 s, number of starts $z = 2$, starting frequency $h = 8$

Calculation:

$$K = \frac{M_{Ku}}{M_{an}} = \frac{0.7}{2.5} = 0.28.$$

From the diagram in Figure 3.103, the value for ϱ is found to be 1.08 at $K = 0.28$.

Then, the KUSA resistance is calculated as:

$$R_{Ku} = 1.08 \cdot \frac{380 \cdot \sqrt{3}}{6.8 \cdot 168} = 0.62 \; \Omega.$$

From the same diagram, at $K = 0.28$ find the value of $\alpha = 0.4$.

The current flowing in the KUSA resistor is calculated from

$$I_{Ku} = \alpha \cdot I_{an} = 0.4 \cdot 6.8 \cdot 168 = 457 \text{ A}.$$

Using the calculated resistance value R_{Ku} and the given data ($P_n = 90$ kW, $t_A = 6$ s, $z = 2$, $h = 8$), the resistance unit may be selected.

In this case, a cast-iron resistance unit type 3PR3 is required.

Three-pole stator-resistance starters

It must be noted that the starting torque is reduced by approximately the square of the reduction in starting current. Stator starting-resistances can therefore only be used for starting under low load conditions.

The selection tables apply for

$$\alpha = \frac{I_{RV}}{I_{an}} = 0.45 \text{ to } 0.9. \quad (3.15)$$

From this it follows, that

$$K = \frac{M_{RV}}{M_{an}} = 0.2 \text{ to } 0.8.$$

I_{an} Starting current
I_{RV} Starting current with stator resistor (R1)
M_{an} Starting torque
M_{RV} Starting torque with stator resistor (R1)

The diagrams in the Siemens catalogue may be used to determine the size of the devices if normal values for the starting time and starting frequency apply for the starting operation. For the starting of machines with a high inertia or flywheel loads, the starting time must by calculated.

The resistance value R_V (R1) for the desired starting torque may be determined by using the diagram in Figure 3.104. This diagram may also be used to determine the current flowing through the three-pole stator resistor.

Figure 3.105 shows the main and control circuit diagrams for a three-pole stator-resistance starter.

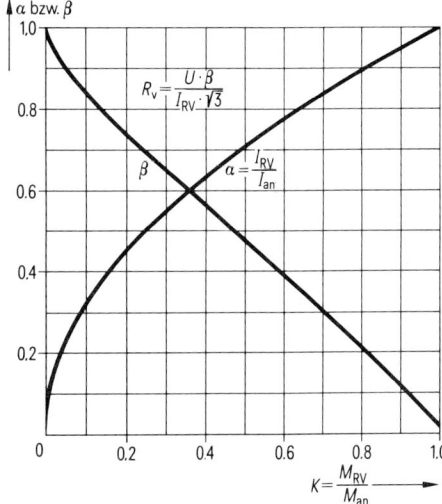

Figure 3.104
Diagram for determining the value of a three-pole starting resistance

Example for the calculation of a three-pole resistor for stator-resistance starting.

Given:

Three-phase motor	1LA6 283-4AA70, 90 kW, 400/380 V, 50 Hz, 168 A
Starting torque	$M_{an} = 2.5 \cdot M_n$
Starting current	$I_{an} = 6.8 \cdot I_n$

Desired starting torque with resistor:
$M_{RV} = 0.7 \cdot M_n$

Starting time 6 s, number of starts $z = 2$, starting frequency $h = 8$

Calculation:

$$\frac{M_{RV}}{M_{an}} = \frac{0.7}{2.5} = 0.28.$$

From the diagram in Figure 3.104,

$\beta = 0.66; \alpha = 0.53,$

Starting current $I_{RV} = \alpha \cdot I_{an} = 0.53 \cdot 6.8 \cdot 168 = 606$ A

Required three-pole starting-resistor:

$$R_V = \frac{U \cdot \beta}{\sqrt{3} \cdot I_{RV}} = \frac{380 \cdot 0.66}{\sqrt{3} \cdot 606} = 0.24 \, \Omega.$$

Using the calculated resistance value R_V and the giv-

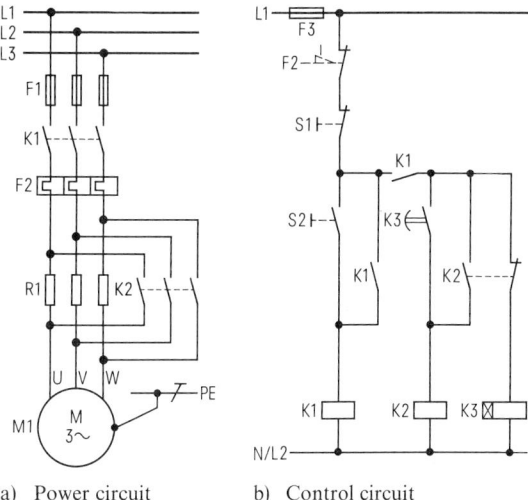

a) Power circuit   b) Control circuit

Figure 3.105
Power and control circuit for an automatic three-pole stator-resistance starter

en data ($P_n = 90$ kW, $t_A = 6$ s, $z = 2$, $h = 8$), the resistance unit may be selected.

In this case, a cast-iron resistance unit type 3PR3 is required.

For the selection and dimensioning of the contactors, see Section 8.3.7 on page 456.

3 Selection criteria for low-voltage switchgear in main circuits

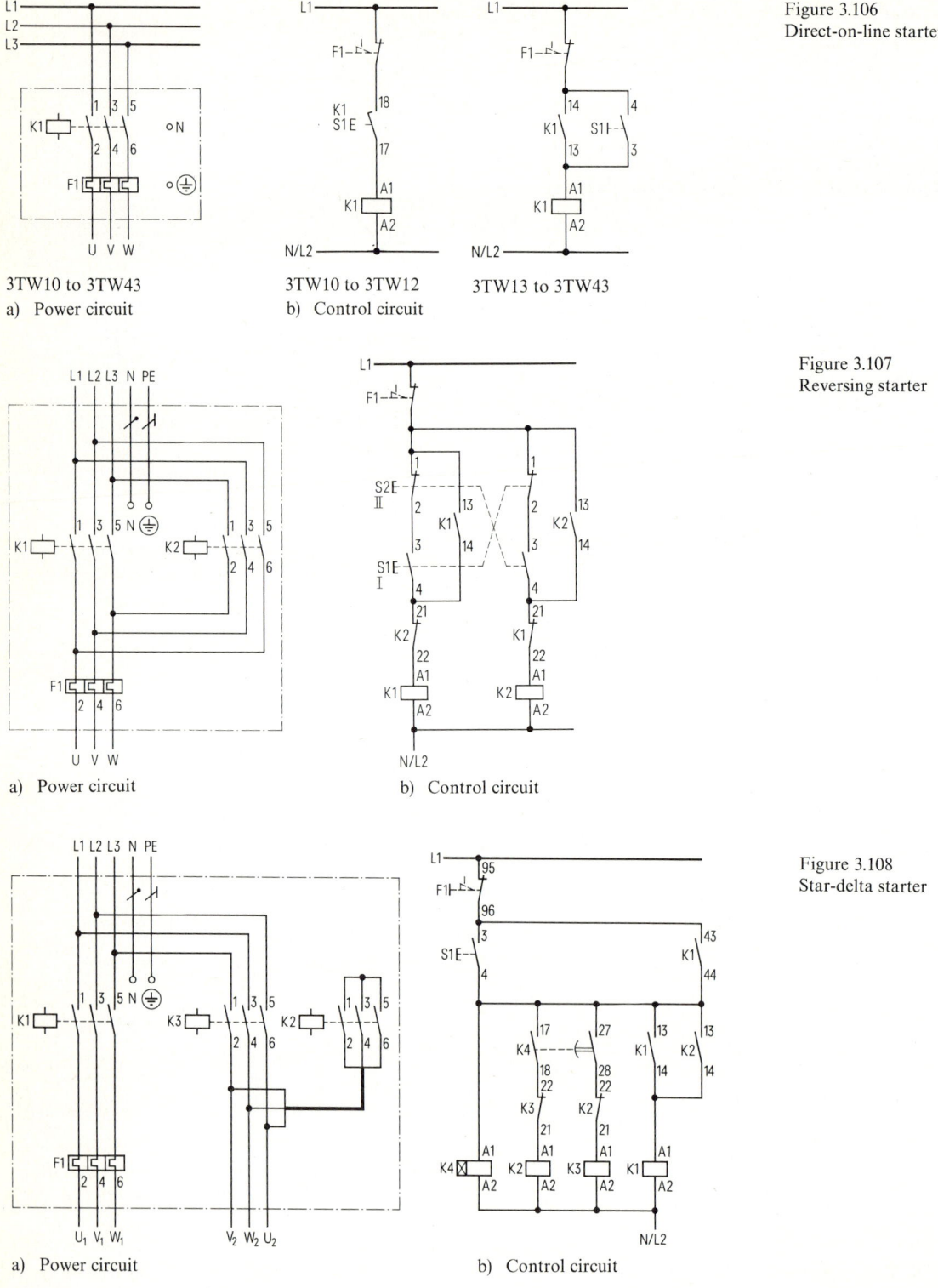

Figure 3.106
Direct-on-line starter

Figure 3.107
Reversing starter

Figure 3.108
Star-delta starter

3.7.14 Direct-on-line starting of three-phase induction motors with type 3TW motor starters

Application

These motor starters are mostly used if the drives of pumps, valves, mixers, conveyer belts, fans etc. are being switched from a remote point. The contactor in the motor starter also offers the possibility of remote switching via e.g. a float switch, position switch, sensor, or pushbutton. An automatic restarting of the drive, after an interruption in the power supply, is prevented since the contactor will remain in the dropped-out state when the voltage returns. A renewed start signal is required before the contactor will close again.

Construction

The motor starters are contained within an insulating protective moulded plastic housing with degree of protection IP 65. The pushbuttons for "On", "Off" and "Reset" are mounted in the lid of the housing.

The starters are supplied complete with a three-pole contactor and the control circuit wiring.

An overload relay, selected in terms of the motor current, must be retrofitted.

The following standard starter versions are offered.

▷ Direct-on-line starters (Fig. 3.106),
▷ Reversing starters (Fig. 3.107),
▷ Star-delta starters (Fig. 3.108).

For information on the short-circuit protection of the starters, please refer to relevant the Siemens low-voltage switchgear catalogue.

3.7.15 A.C. semiconductor motor controllers and starters, SIKOSTART 3RW22

Owing to its robust construction and its extensive range of applications, the three-phase squirrel-cage induction motor is undoubtedly the most widely used and cost-effective form of electric motor encountered in modern industrial machinery. Depending on the application and/or the available power at the point of installation, however, the high initial value of starting torque and starting current associated with direct-on-line switching of these motors, may require special consideration.

Since the initial torque (locked rotor torque) developed at the shaft of a three-phase induction motor

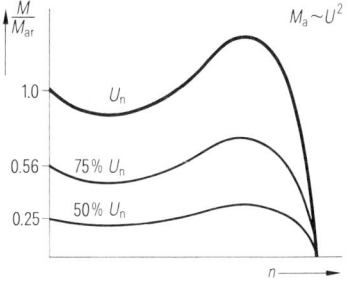

a)

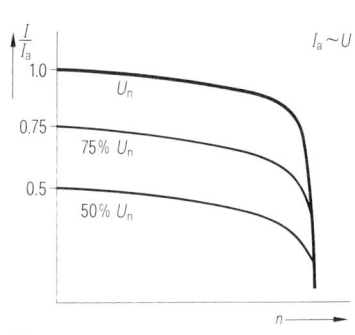

b)

Figure 3.109
(a) Torque-speed and (b) current-speed characteristic curves during direct-on-line starting of a three-phase squirrel-cage motor at various values of terminal voltage

is virtually proportional to the square of the applied terminal voltage, i.e.

$$M_a \sim U^2 \qquad (3.16)$$

and the starting current is almost directly proportional to the applied terminal voltage, i.e.

$$I_a \sim U \qquad (3.17)$$

both starting torque and starting current may be reduced by the initial application of a voltage which is lower than the rated terminal voltage U_n of the motor (Figure 3.109).

Examples of reduced-voltage starters described elsewhere in this handbook include star-delta starters, single-pole and three-pole stator-resistance starters and autotransformer starters (for fundemental circuit diagrams, refer to Section 8). Typical characteristics of star-delta starting are shown in Figure 3.110.

3 Selection criteria for low-voltage switchgear in main circuits

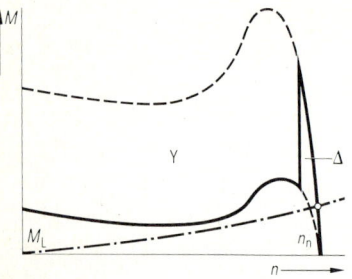

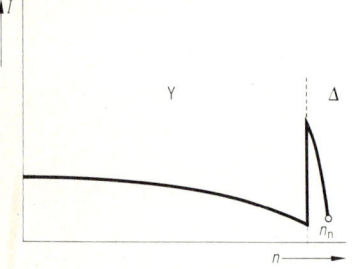

Figure 3.110
(a) Torque-speed and (b) current-speed characteristic curves during star-delta starting of a three-phase squirrel-cage motor.

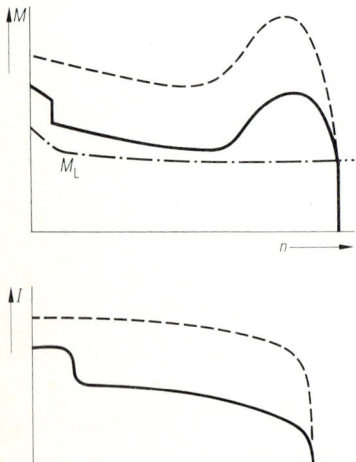

Figure 3.111
(a) Torque-speed and (b) current-speed characteristic curves during starting of e.g. a rolling mill using a three-phase squirrel-cage motor and a SIKOSTART solid-state motor starter

Solid-state reduced voltage starters are becoming increasingly popular for the "soft" starting of three-phase induction motors. Not only can they offer additional features such as optimized soft starting, controlled soft stopping, energy saving at partial load and d.c. braking, they generally provide a greater reduction in current surges and in mechanical stresses on drive couplings, belts, gears, pumps, valves, etc. (see Figure 3.111). This not only reduces the loading on the supply network and switchgear during starting, but also reduces mechanical wear and tear, and can thus increase the service life of the driven machine itself.

A distinction must be made between:

▷ A.C. semiconductor motor controllers and starters which regulate the value of the motor voltage and current at constant (system) frequency for relatively short periods of time e.g. during starting and stopping – also generally referred to as "solid-state soft starters" (IEC 947-.. specification in preparation), and

▷ variable speed drives (converters) which can alter both the voltage and frequency of the motor supply for the purposes of controlling torque and/or speed during various stages of operation.

The Siemens SIKOSTART 3RW22 solid-state motor controller and starter is offered in the power range from 3 to 710 kW (400 V, 40 °C). Its operation is based on phase-angle control of a 6 pulse thyristor connection (fully controlled bridge with 2 power thyristors per phase in antiparallel arrangement). The digital control is achieved by means of a microprocessor and software stored on an EPROM or EEPROM memory chip. During starting, the SIKOSTART unit increases the terminal voltage applied to the motor in accordance with the pre-programmed parameters and the starting function. The frequency of the supply is not altered.

The basic applications, features and selection criteria for solid state motor controllers and starters, with particular reference to the SIKOSTART 3RW22, are described below. Further details may be found in the special Siemens SIKOSTART Handbook (Order no. E20001-P285-A484-7600) and the computer program for 3RW22 selection (Order no. E20001-P285-A485-7600, incl. above-mentioned Handbook).

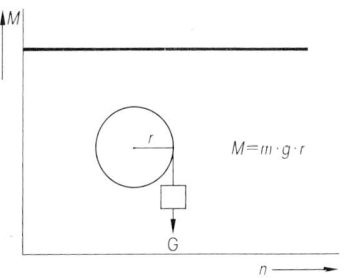

a) Constant load torque

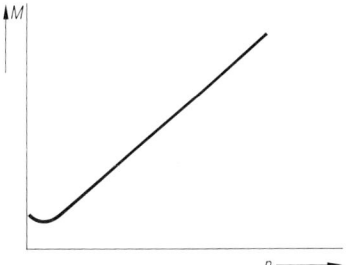

b) Load torque increases in direct proportion to the speed

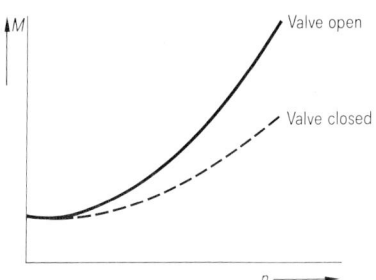

c) Load torque increases with the square of the speed

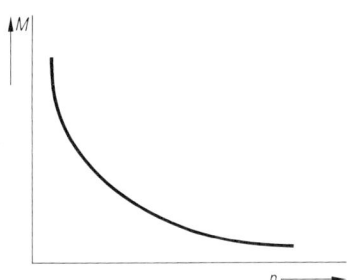

d) Load torque is inversely proportional to the speed

Figure 3.112 Simplified illustrations of the torque-speed behaviour of various driven loads

Typical applications, examples

Typical applications in which the use of an a.c. semiconductor motor controller and starter may be particularly advantageous include:

▷ drives for machinery in which sudden or jerking movements are to be avoided (smooth starting and stopping),

▷ drives for pumps (slow acceleration and deceleration of the pumped fluid),

▷ drives with long no-load and/or partial load periods (c.f. energy saving under partial or no-load conditions),

▷ machines incorporating gearboxes, chain drives, belt drives, drive-shaft couplings or other transmission systems,

▷ drives with a high moment of inertia, and

▷ applications in which the voltage drop in the supply lines to the motor would be too high to permit d.o.l. starting (e.g. long supply lines).

Examples of the above applications may be divided into 4 groups according to the torque-speed characteristics of the driven load.

a) Constant load torque
(see Fig. 3.112a)

$M_L < k$

The simplest example for this group is shown in Figure 3.112a. It illustrates a hoist which winds or unwinds a rope on a drum with radius r to raise or lower the mass m. If the mass of the hanging length of rope is ignored and the radius of the drum is assumed to remain constant (simplification), then

$M_L = m \cdot g \cdot r$,

where $g = 9{,}8$ ms^{-2} (acceleration due to gravity).

Further examples:

- cranes, hoists,
- piston-type pumps and compressors operating against a constant back-pressure,
- rolling mills,
- conveyer belts,
- mills without fan action,
- machine tools with constant cutting power (excluding flywheel-type),
- sometimes also applicable for shears, punching machines, presses, wood planers (depending on design).

b) Load torque increases in direct proportion to the speed
(see Fig. 3.112b)

$M_L \sim n$

Examples:

- calenders, mangles i.e. machines in which e.g. paper, cloth or board is glazed, smoothed or pressed by passing between rollers
- screw or worm conveyer systems (starting with empty conveyer troughs)

c) Load torque increases with the square of the speed
(see Fig. 3.112c)

$M_L \sim n^2$

Examples:

- centrifugal or turbine-type pumps, fans, or blowers feeding into an open pipe network (in the case of a closed valve or damper, the final torque value is about 50% of that applicable for a fully opened valve),
- centrifuge machines e.g. separators,
- screw drives of boats,
- stirring mechanisms, agitators, mixers.

d) Load torque inversely proportional to increase in speed
(see Fig. 3.112d)

$M_L \sim 1/n$

Examples:

- ball mills,
- turning machines, lathes, mills and similar machine tools (particularly common in automated manufacturing processes),
- reeling machines (take-up roller machines).

Operating modes and features of the SIKOSTART 3RW22

Soft starting

The standard form of soft start is achieved by means of a simple *linear voltage ramp*. The microprocessor control unit causes the motor terminal voltage to increase linearly from a preset initial value (adjustable from 20% supply voltage) to 100% of the supply voltage within the preselected ramp time (Fig. 3.113a). By means of the selectable feature "run-up detection" the terminal voltage may be increased to 100% supply voltage before the ramp time is over, if the SIKOSTART senses that the motor has already achieved its operational condition.

In addition to this linear voltage ramp, the SIKOSTART unit also offers a number of more complex starting techniques. For example, a *start impulse* (also known as "kick start" feature) may be activated to overcome initial static friction in the drive and/or the terminal voltage may be controlled in such a way as to achieve *limiting of the starting current* to a specific value before full supply voltage is applied to the motor (Fig. 3.113b). Alternatively, by means of *voltage limiting*, the terminal voltage may be held constant for a period of time during the run-up (e.g. for part of the linear voltage ramp) to achieve the desired load acceleration characteristic (Fig. 3.113c).

By means of the *emergency start* feature, the motor may be started in an emergency even if a single thyristor or a thyristor pair has become permanently conductive (i.e. failed).

Soft stopping

A motor may be switched of by the instantaneous interruption of its supply current (i.e. terminal voltage

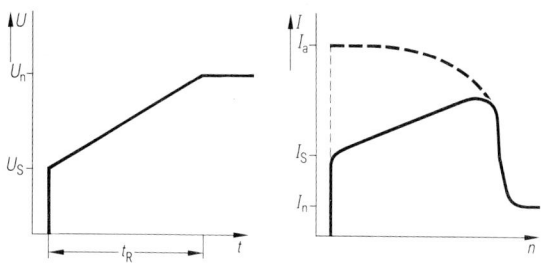

a) Soft start with a linear voltage ramp

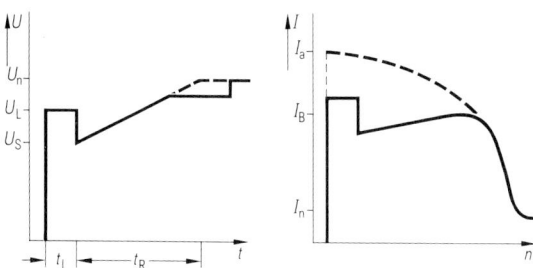

b) Soft start by means of a start impulse followed by current limiting

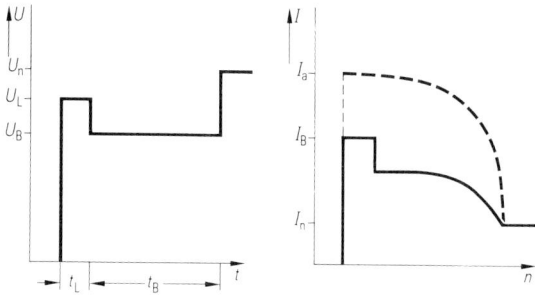

c) Soft start by means of a start impulse followed by voltage limiting

Figure 3.113
Illustration of some of the starting parameters which may be used to obtain optimum load acceleration throughout the starting time

stop feature, the motor terminal voltage is reduced gradually after the switch-off signal. Thereby, the initial value of the stop ramp U_B is set to 90% of the supply voltage and the final value U_{off} to 85% of the initial start voltage U_S (see Fig. 3.114).

A special type of soft stop, known as *pump control*, may be required for centrifugal pumps.

Owing to the generally low moment of inertia of these pump drives, it may happen that the fluid in the pipe system comes to an abrupt halt upon switch-off of the pump motor. The resulting "water hammer" caused by the back pressure of the pumped fluid, may be extremely loud and may damage non-return valves and other system components. To avoid this phenomenon, the terminal voltage must be reduced in such a way upon switch-off, that the delivery of the pump is reduced slowly. This, in turn, requires that the run-down torque-speed characteristics of the pump and motor, which will differ from one case to another, must be taken into account.

SIKOSTART 3RW22 offers a pump control feature based on "intelligent" software which recognises the behaviour of the run-down characteristic and adjusts the terminal voltage accordingly during the preset run-down time. During the run-down period, the motor will draw a current which is greater than its rated operational current. Since the run-down time may be as long as 90 seconds, care must be taken in the selection and adjustment of the overload protection devices. For this reason, pump control is only possi-

to zero). In this case, the motor will coast down as determined by the moment of inertia of the total drive system and the friction. If this moment of inertia is small, the drive may come to an abrupt standstill which is not alway desirable, e.g. in the case of conveyer belts, escalators or hoists. By means of the soft-

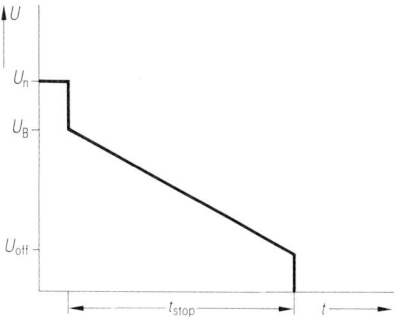

Figure 3.114
Voltage behaviour during run-down with the soft-stop feature

ble on SIKOSTART 3RW22 units which incorporate electronic device protection.

D.C. braking

In the case of a drive with a high moment of inertia, the d.c. braking feature may be activated to reduce the run-down time. A direct current is injected into the stator winding. Thereby, a stationary magnetic field is created and circulating currents are induced in the rotating short-circuited (squirrel cage) rotor. The resultant braking torque slows the motor down. In the 3RW22 starters, the d.c. excitation is created during the positive half-waves in phases T1 and T3. The braking torque may be increased by the inclusion of a free-wheeling contactor between T2 and T3. This braking contactor is energized by a command from the SIKOSTART control unit.

Energy saving at partial load

Under partial load and no-load conditions, the soft starter may be called upon to reduce the terminal voltage to such an extent, that the motor is supplied with just enough energy for it to maintain the minimum required torque and to ensure that possible load pulses do not lead to excessive speed reductions. Not only are the copper losses reduced by the improvement of the power factor and the resulting reduction in current, but the iron losses in the motor core drop to

$P_{Fe} = P_{Fen} \cdot (U_{nl}/U_n)^2$.

The extent of energy saving will always depend on the motor rating, its speed (no. of poles), design and the cyclic duration factor (c.d.f.) of the load (see Section 2.2.3, page 99).

Operation with a bypass contactor

Unless the operational duty calls for short repetitive on-load periods (e.g. rapid start / stop operation), it is recommended that the solid-state soft starter be bypassed by a contactor once the motor has reached full speed. On the one hand this saves the heat losses in the thyristors, and on the other it permits the power unit to cool down to ambient temperature quickly prior to the next start. This technique also allows one and the same starter to be used for several motors started one after the other.

Protection and control functions

A number of protection and control functions are permanently programmed into the control units of the 3RW22 SIKOSTART devices.

▷ Electronic device protection (option)

Based on a mathematical model of the thermal characteristics of the power electronics module, this feature continuously monitors the behaviour of the temperature in the device with respect to elapsed time and compares it to the reference values. It offers optimum overload protection for the thyristors while ensuring maximum utilization of the unit resources. Additional motor protection is only required if the motor has lower thermal reserves than the starter.

▷ Monitoring of the starting time during starting with current limiting

In units which do not incorporate electronic overload protection, the start is terminated if the motor has not reached operational speed within a fixed time period after the starting command.

▷ Temperature monitoring

The temperature of the thyristor heat sink is monitored by means of a thermistor. Starting of the motor is prevented if a the temperature limit is exceeded.

▷ Self testing in the full-on mode

After run-up of the motor, the thyristors are operated in the full-on mode i.e. they are continuously conductive. In this mode of operation, the power unit may be continuously overloaded at max. 115% of its rated current. The 3RW22 control unit continuously monitors the condition of the thyristors by means of random minimum phase-angle control operations. It detects malfunctions such as loss of current in one phase, missing voltage between phases, failure of a thyristor to fire, bypass contactor not closed, etc.

Selection

Normally, a 3RW22 SIKOSTART unit can be selected from the ordering data given in the corresponding Siemens l.v. switchgear catalogue. Selection is carried out in terms of the rated output power of the motor (or the rated operational current), the power circuit supply voltage and the ambient temperature at the point of installation. The rated operational current of the motor must be at least 20% of the rated operational current of the selected SIKO-START unit. 3RW22 devices incorporate an internal DC 24 V power supply. The auxiliary supply input voltage may be AC 100 to 200 V, AC 200 to 240 V or AC 380 to 415 V.

In addition to the basic versions, the range includes units which incorporate electronic device protection and a special pump control feature as well as devices with PC interface for communication and commissioning purposes.

The selection data given in the catalogues is based on the following assumptions:

▷ The torque class (also known as rotor class) of the motor is at least 10 (i.e. $KL \geq 10$).

The torque class KL10 means that, with direct-on-line starting, run-up will be possible (even at 5% undervoltage) against a load torque which is 100% the rated torque of the motor. A torque class of KL16 allows run-up at a load torque equal to 160% of the motor rated torque, etc. It is an indication of the rotor slot form.

▷ The breakaway torque of the load M_{LB} is less than or equal to the rated torque of the motor ($M_{LB} \leq M_n$).

▷ The moment of inertia of the complete drive J_{Tot} (i.e. motor plus load) is less than or equal to 10 times the moment of inertia of the motor itself. ($J_{Tot} \leq 10 \cdot J_M$), i.e. not heavy duty starting.

▷ The maximum number of starts per hour do not exceed the nominal values given.

It should be noted that the number of starts per hour depends on the starting current and starting time. A higher number of starts per hour may well be permissible with low inertia drives (e.g. centrifugal pumps).

If the actual parameters deviate from the above, or if a more exact knowledge of the voltage, current, torque and speed characteristics during starting is desired, than a calculation is necessary. One form of the calculation involves dividing the torque-speed curves of the motor and load into several portions and determining the accelerating torque for each portion. This procedure is described and illustrated in the SIKOSTART Handbook.

Alternatively, the calculation and selection of 3RW22 starter units may be carried out by means of a computer software package specially developed for this purpose (Order number, incl. Handbook: E20001-P285-A485-X-7600).

Hardware requirements:
– PC, IBM AT-compatible, graphics card (e.g. VGA), hard disk and floppy disk drive (3,5″),
– at least 422 kB of free memory,
– printer with graphics capabilities (IBM-graphics, HP Laserjet Series II or Epson FX80)

Software requirements:
– MS DOS from version 3.xx

The SIKOSTART 3RW22 program guides the user through the various decision and input steps required to specify the correct unit and allows the various operating modes and starting parameters to be simulated graphically. The following input data is called for:

▷ general project information, e.g. project title, system voltage, frequency, ambient temperature, etc.

▷ motor data – i.e. the rated output power, speed, current and torque, the torque class (or rotor class KL), the moment of inertia, a table of torque-speed values (in %) and a table of starting current-speed values (in %).

The program provides all the required standard data on Siemens 1LA5, 1LA6, 1LA8 and 1M motors. If one of these motors is to be used, only the type number needs to be entered – the program automatically loads the corresponding data. In the case of other motors which may deviate from the Siemens types, the corresponding data must be entered by hand.

▷ load data – i.e. type of load, power requirement or load torque, speed, moment of inertia and type of torque-speed characteristic.

The program suggests a set of torque-speed values (in %) for a constant, linear and square-law characteristic respectively (see application groups 1 to 3 above). Alternatively, the values may be entered or edited by hand.

▷ starting parameters e.g. start impulse (if required), initial voltage, ramp time, current limiting (if required).

After calculation, the program provides the following output data:

▷ the order number of the recommended 3RW22 SIKOSTART unit,

▷ the run-up time which would be achieved by direct-on-line and star-delta starting methods,

▷ the run-up time with the SIKOSTART unit,

▷ the maximum and the effective starting current during starting with SIKOSTART,

▷ the permissible starting frequency.

and the following graphical illustrations:

▷ the torque-speed characteristics of the motor showing both the starting curve under d.o.l. and soft starter conditions (soft starter parameters as entered) superimposed onto the torque-speed curve of the load.

By means of this set of curves, the acceleration torque, which should be greater or equal to 15% of the rated motor torque throughout the entire run-up ($M_B \geq 15\% \, M_n$) may be checked (Figure 3.111). This will ensure that the drive will accelerate smoothly and that the motor will not be heated unnecessarily.

▷ the current-speed characteristics of the motor showing the starting curve under d.o.l. and soft-start conditions (soft starter parameters as entered).

▷ curves of the terminal voltage, motor current and motor speed plotted against time.

If a suitable printer is available, the above-mentioned input data, calculation results and graphical illustrations may be printed out.

Integration of the SIKOSTART unit in the motor feeder circuit

The installation of a SIKOSTART unit in an existing or new motor control feeder circuit is normally very simple and may be likened to the use of a contactor. Figure 3.117 illustrates the simplest circuit configuration by means of a single-line diagram.

If power factor correction capacitors are present, these must be connected between the SIKOSTART and the incoming supply (never between the motor and the output terminals of the starter!).

Optimum protection for the complete motor feeder circuit comprises:

▷ HRC fuses (operation class gL/gG) for overload and short circuit protection of the supply cables and the switchgear,

▷ HRC semiconductor fuses (operation class aR) for the short-circuit protection of the power thyristors and

▷ an overload relay for protection of the motor

If applicable, the overload relay also protects the starter against overload (integrated electronic device protection permits full utilization of the thermal resources of the starter). A circuit-breaker may be used instead of the HRC fuses and the overload relay.

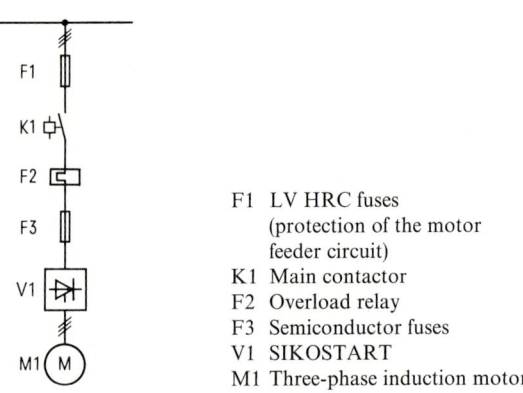

F1 LV HRC fuses (protection of the motor feeder circuit)
K1 Main contactor
F2 Overload relay
F3 Semiconductor fuses
V1 SIKOSTART
M1 Three-phase induction motor

Figure 3.117
Single-line diagram of a typical motor feeder circuit with a SIKOSTART solid-state starter

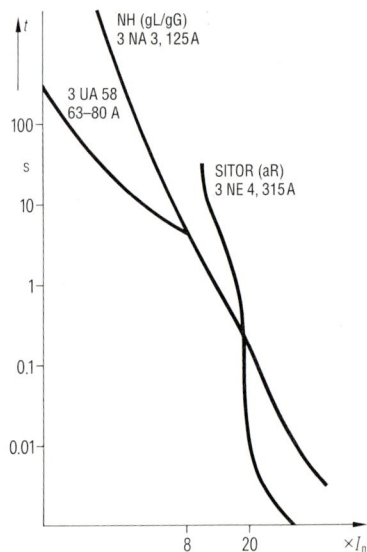

Figure 3.118
Illustration of the time-current characteristics of a 3UA overload relay, a 3NA HRC fuse and a 3NE SITOR semiconductor fuse as used in a motor feeder circuit with a solid-state motor starter

The suggested types and sizes of semiconductor fuses (e.g. Siemens SITOR fuses, type 3NE) are listed together with the technical data of the starters in the Siemens low-voltage catalogue.

By means of the respective time-current characteristic curves, Figure 3.118 shows that each protective device has its justification in a specific range of overload and/or short-circuit current. One school of thought suggests the omission of the semiconductor fuses, particularly if the risk of a short-circuit between the starter and the motor is small. It is, however, suggested that a detailed examination of the actual application (including fault levels, cable lengths, accessibility of spares, qualification of personell, the risks of a short-circuit and the cost of replacing a damaged thyristor) be carried out before such a decision is made.

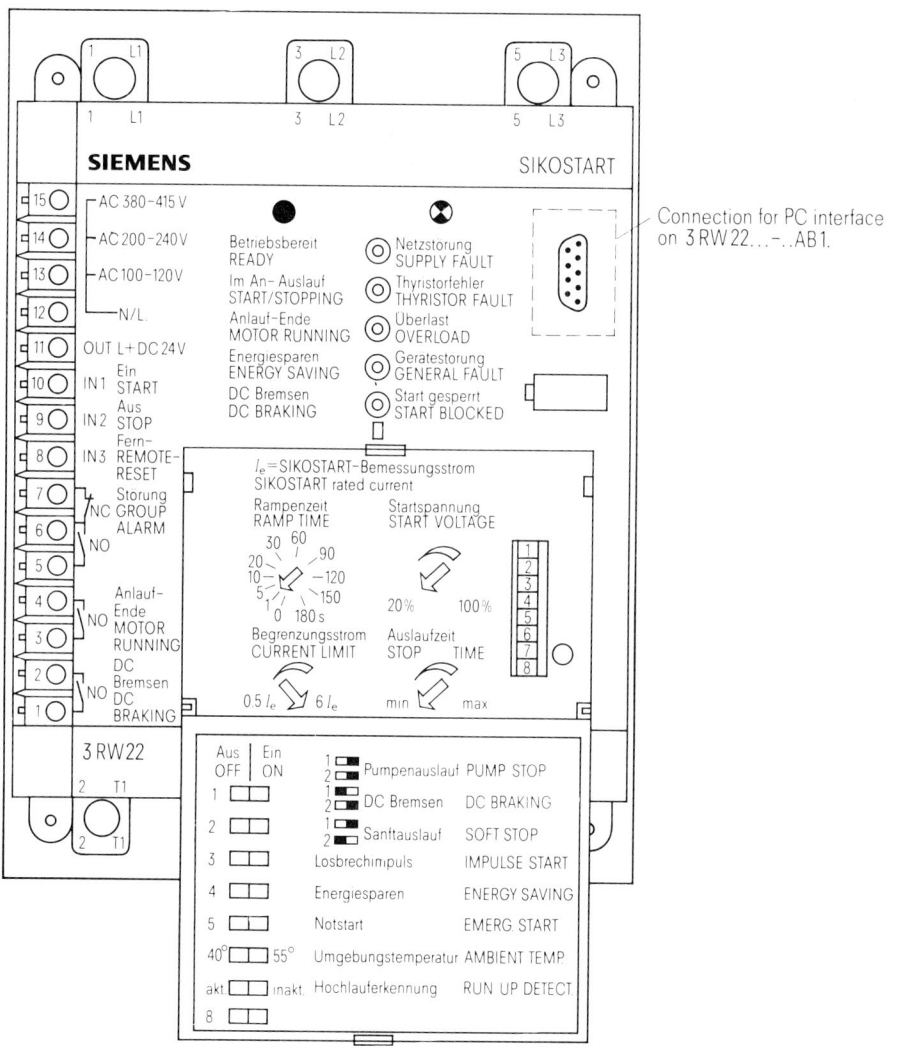

Figure 3.119
The control panel of a SIKOSTART a.c. semiconductor motor controller, type 3RW22 shown with the hinged cover flap opened to permit setting of the starting parameters

Several examples of circuit diagrams are given in the SIKOSTART Handbook. These include:

- direct in-line motor control (as with a contactor),
- circuits with a bypass contactor and/or a brake contactor to increase the effect of the d.c. braking feature,
- circuits for reversing operation,
- circuits for the soft starting of pole-changing motors,
- circuits for the sequential starting of several motors with one soft starter.

Commissioning

On the basic units, all the starting and stopping parameters are set up on potentiometers and DIL switches located under a protective flap on the front of the control panel (see Figure 3.118).

On units which incorporate a PC interface, even more extensive starting and stopping parameters may be entered by means of a personal computer and the COM SIKOSTART communications program (for ordering details, please refer to the SIKOSTART selection data in the corresponding Siemens l.v. catalogue).

This programm not only allows for the more accurate input of the parameters in numerical form and the indication of start parameters currently stored in the EEPROM, but also permits the data to be stored on disk for transfer to other similar installations.

Furthermore, the COM SIKOSTART program offers the facility of entering up to three independent sets of parameters. This is particularly useful for applications such as pole-changing and sequential starting of motors having different ratings, loads and/or torque-speed characteristics. COM SIKOSTART also allows a more extensive specification of the pump control soft stop parameters.

4 Selection criteria for low-voltage switchgear in auxiliary circuits

Switchgear in auxiliary circuits is used for control, indication and interlocking purposes. Primarily this includes command devices such as pushbuttons, control switches and position switches, contactor relays, auxiliary contacts of switchgear in main circuits, and time relays. Apart from the electrical demands made on switchgear for auxiliary circuits, a number of further requirements pertaining e.g. to operating mechanism variants, degrees of protection, switching elements for specific applications and aesthetic design must be met. In addition, it is expected that the devices are simple to install, have clear terminal markings and feature a time and cost effective termination technique. The Siemens control and signalling devices, proximity switches, position switches, contactor relays, time relays, control-circuit transformers and terminal blocks optimally fulfill all these requirements.

4.1 Operating voltages in auxiliary circuits

Standardized operating voltages, permissible voltage tolerances and voltage sources are given in Section 2.1.1 on page 76. Alternatively, they may be found in the relevant DIN VDE specifications and IEC publications.

4.1.1 Contact reliability in the case of low voltages

Among other factors, contact reliability depends on the condition of the point of contact and on the voltage level. In the case of low voltages and currents the self-cleaning effect of the point of contact by the switching arc is, in most cases, not evident. Contamination of the the contact point, such as a layer of foreign non-conducting substance on the contact surfaces, may adversely effect the quality of the contact or even prevent it. In most cases, the film of foreign substance which builds up on contact surfaces is so thin, that it will be punctured and broken down by a potential of only a few volts (this phenomena is known as "fritting"). Generally speaking, it therefor poses no real problem. In the case of multiple switching points in series, however, the risk of contact problems and unreliable signalling at low voltages increases.

High contact reliability is of particular importance in applications where mechanical switching devices are used in conjunction with electronic control systems. To ensure the contact reliability at the common control supply voltage of DC 24 V and currents of only a few milli-amps, special precautions are necessary:

▷ avoidance of series connected contacts,
▷ use of switchgear with double auxiliary contact bridges,
▷ selection of suitable contact material such as silver or silver compounds,
▷ high contact pressure by means of a suitable shape of the contact pieces,
▷ wiping action of the moving contacts,
▷ installation in dust-free environment or, if necessary, encapsulation.

Auxiliary contacts are constructed with these characteristics in mind to achieve the highest possible contact reliability. For this reason, the use of double contact bridges for auxiliary contacts is standard practice at Siemens. For example, they are used in all the poles of contactor relays, command devices, position switches as well as in the auxiliary contacts of power contactors and other switchgear for power circuits (also refer to Section 6.2.4 on page 293).

4.1.2 Voltage instability in auxiliary circuits

If the auxiliary supply voltage is sourced from the main power circuit, then e.g. short-circuit switching operations in the high voltage network or discriminative interruptions of short-circuits in the low-voltage network, or even the direct-on-line switching of large drives, can cause the control or operating voltage in the auxiliary circuit to be interrupted or to fall below the permissible tolerance level for a short period. To ensure the continued operation of voltage dependent switching devices (e.g. contactors), the following measures are suggested:

▷ Contactors with drop-out delay

Separately mounted drop-out delay units are offered for the Siemens contactor relays types 3TH2, 3TH8 and 3TH40 to 3TH43 as well as for the power contactors 3TF2, 3TB40 to 3TB43, 3TF30 to 3TF33 and 3TF40 to 3TF43. Depending on the unit version, the delay times at 110 V and 220 V, 50 to 60 Hz lie between 0.1 and 1.2 seconds. Exact values may be found in the Siemens l.v. switchgear catalogue.

▷ 3TH2 and 3TH83 contactor relays with mechanical latching

The contactor relays latch after being energized and remain in a closed or "on" state even after their coil supply voltage is removed. Switching on or off may either be done electrically (via the latch and delatch coils respectively) or manually. Owing to this "memory function" of the relays, a production process (for example) may be continued without reset delay once the supply voltage returns.

▷ Interface contactors for electronic controls

Specially designed contactors for motor power ratings of up to 11 kW at 380/400 V are available for use with electronic control systems. The particular properties of the interface contactors are:

– wide operating voltage tolerance of DC 17 to 30 V
– low (reduced) rated power consumption of the magnet coils
– auxiliary contacts with double moving contact bridges
– integrated suppression of the magnet coils to safeguard against switching overvoltages.

▷ Undervoltage release with delay unit

The undervoltage release of a circuit-breaker can be equipped with a delay unit (rc release). The delay is achieved by means of a capacitor delay element. The delay time is typically 1 second but it may be increased to 3 seconds by the insertion of a bridge.

4.2 Operating conditions

4.2.1 Utilization categories to DIN VDE and IEC

Utilization categories for auxiliary switches are specified in DIN VDE 0660 part 200 (07.92), EN 60947-5-1 and in IEC 947-5 (refer to Tables 4.1 and 4.2). In conjunction with the rated operating voltage U_e, the rated operating current I_e and the test cycle, these tables clearly specify the application purpose.

In DIN VDE 0660, part 200, 07.92 the previously used utilization categories AC-11 and DC-11 have been replaced by the utilization categories AC-12 to AC-15 and DC-12 to DC-14 respectively. For details, please refer to the specification.

Table 4.1
Utilization categories for the switching of auxiliary loads to DIN VDE 0660, part 200, 07.92

Current type	Utilization category	Typical application
Alternating current	AC-15	Control of electromagnetic load (>72 VA)
Direct current	DC-13	Control of electromagnets

4.2.2 Special considerations for the selection and the use of low-voltage switchgear in Canada and the USA

Generally speaking, auxiliary contacts are only classified by CSA and UL in terms of either "Heavy Duty" or "Normal Duty" as well as the maximum permissible voltage. This is indicated by the short form designations as are given in Table 4.3.

---

[1] The abnormal operating conditions are to be simulated by means of an electromagnet which is physically blocked in to open position (i.e. prevented from closing), as the load.
[2] If the breaking current differs from the making current, then the test duty cycle of 50 ms refers to the making current. There-after the current is reduced to the breaking current for a suitable duration, e.g. 0.05 s.
[3] The first 50 switching cycles must be performed at $U/U_e=1.1$, whereby the load is set up at U_e.
[4] The stated power factors (cos φ) are conventional values and apply for test circuits used to approximate the electrical reference data of coil circuits.
[5] In the case of solid-state switching elements, an overload protection device as specified by the manufacturer, should be used during the simulation of abnormal conditions.
[6] The value "$6 \cdot P$" is derived from an empirical equation which applies for most d.c. magnet loads up to $P=50$ W, whereby 6 (ms/W)·P(W)= 300 ms. Here it is assumed that individual loads with a rated power greater than 50 W do not occur and that for greater loads the actual load consists of smaller parallel connected sub-loads. Therefor, 300 ms is an upper limit, irrespective at the power. For solid-state switching elements, the maximum time constant must be 60 ms.

Table 4.2
Listing of the rated making and breaking capacities of switching elements to DIN VDE 0660, Part 200, 07.92.
Conditions for making and breaking in terms of the utilization categories AC-15 and DC-13

Current type	Utilization category	Normal operating conditions						Abnormal operating conditions [1]					
		Making			Breaking			Making			Breaking		
		I/I_e	U/U_e [3]	$\cos\varphi$	I/I_e	U/U_e [3]	$\cos\varphi$	I/I_e	U/U_e	$\cos\varphi$	I/I_e	U/U_e	$\cos\varphi$
Alternating current	AC-15 [2]	10	1	0.3 [4]	1	1	0.3 [4]	10	1.1	0.3 [4]	10	1.1	0.3 [4]
				$t_{0.95}$	I/I_e		$t_{0.95}$			$t_{0.95}$	I/I_e		$t_{0.95}$
Direct current	DC-13 [5]	1	1	$6 \cdot P$ [6]	1	1	$6 \cdot P$ [6]	1.1	1.1	$6 \cdot P$ [6]	1.1	1.1	$6 \cdot P$ [6]

I Making or breaking current
I_e Rated operational current
U Voltage before making
U_e Rated operational voltage
$t_{0.95}$ Time in milli-seconds until 95% of the steady-state current is reached
$P = U_e \cdot I_e$ (Rated power in W)

Table 4.3 Designation of auxiliary switches and contacts to CSA and UL

Classification in terms of duty type	Maximum ratings per pole				
	Voltage V	Switching capaciity "On" A	"Off" A	Continuous current A	Designation
Heavy Duty (Abbreviation: HD or HVY DTY)	AC 120	60	6	10	A 150
	AC 240	30	3		A 300
	AC 480	15	1.5		A 600
	AC 600	12	1.2		A 600
	DC 125	2.2	2.2		N 150
	DC 250	1.1	1.1		N 300
	DC 600	0.4	0.4		N 600
Standard Duty (Abbreviation: SD or STD DTY)	AC 120	30	3	5	B 150
	AC 240	15	1.5		B 300
	AC 480	7.5	0.75		B 600
	AC 600	6	0.6		B 600
	DC 125	1.1	1.1		P 150
	DC 250	0.55	0.55		P 300
	DC 600	0.2	0.2		P 600
–	AC 120	15	1.5	2.5	C 150
	AC 240	7.5	0.75		C 300
	AC 480	3.75	0.375		C 600
	AC 600	3	0.3		C 600
	DC 125	0.55	0.55		Q 150
	DC 250	0.27	0.27		Q 300
	DC 600	0.1	0.1		Q 600
–	AC 120	3.6	0.6	1	D 150
	AC 240	1.8	0.3		D 300
	DC 125	0.22	0.22		R 150
	DC 250	0.11	0.11		R 300

4 Selection criteria for low-voltage switchgear in auxiliary circuits

Example

A NEMA *A600* relay means: Contactor relay with 10 A continuous current rating; 600 V maximum permissible a.c. voltage (creapage distances and clearances for AC 600 V) and a making capacity of 7200 VA.

For a number of products in the Siemens catalogues, it is specified that the auxiliary switches or auxiliary contacts may only be used at *equal potential* above a certain voltage. This means that the incoming terminals of the contacts must be connected to the same pole of the control supply voltage. For example, "AC 600 V above AC 300 V same polarity" means that for voltages above AC 300 V and up to AC 600 V, the contact elements of adjacent auxiliary switches must be at the same potential.

Particularly in Canada and the USA, rubber insulated wires are routed within conduit piping. This applies particularly to the feeder conductors of manufacturing machines and machine tools. The conduit piping system must be completely sealed and electrically conductive along its entire length. This is achieved by the use of special couplings, bushings and elbow pieces. Since the conduit is also used for earthing purposes, metal-encapsulated or sheet-steel enclosed switchgear with cable entries for glands with Pg thread must be equipped with special adapters to accept the thread on the conduit (e.g. NPT – National Pipe Thread – in inches, or metric thread).

The necessary Pg adapters are listed as accessories to the switchgear in the Siemens catalogues. Unless otherwise indicated, they are to be ordered separately. Unused cable entries may be sealed by means of the usual plugs.

4.2.3 Short-circuit protection in auxiliary circuits

In terms of DIN VDE 0100, part 725, 09.91 not only the operating equipment, but also the cables and wires must be protected against the effects of short-circuits.

For the selection of the short-circuit protection apparatus, e.g. for the contacts, the relevant data given in the Siemens low-voltage switchgear catalogues should be taken into account.

For the selection of the overload protection apparatus for the cables and wires, DIN VDE 0100 part 430 applies (also refer to Section 9.4.2 on page 521).

Table 4.4
Short-circuit protection of an auxiliary circuit (example)

Circuit diagram	Auxiliary contacts	Maximum fuse rating acc. to manufacturers' data[1] A	Fuse rating to be selected A
L1 — F1 1.5 mm² Cu F2 S1 K2 Q1 K1 N/L2 —		–	6[1]
	Thermal overload relay	10	
	Command switch	16	
	Contactor K2 (e.g. interlocking)	10	
	Auxilary contact of the circuit-breaker	6	
	Contactor coil	16	

[1] The smallest single fuse rating specified determines the maximum fuse rating to be selected for the whole control circuit.

Table 4.4 illustrates the selection of the fuse rating for an auxiliary circuit with four series connected contact sets and a conductor cross-section of 1.5 mm² Cu (For cable protection circuit-breaker terminals refer to Section 8.7.2 on page 484).

The following applies according to DIN VDE 0113 part 1/02.86 paragraph B 2.3:

For low voltages and long conductors or conductors with small cross-sections as well as for circuits which are fed from transformers with low output power (e.g. control circuits) or which are connected via protective devices with a high impedance, the lowest possible value of prospective steady-state short-circuit current I_k must be calculated and taken into account when selecting the fuses.

4.2.4 Short-circuit and overload protection of control transformers

Depending on the phase angle at the instant of contact closing and on the load, high transient current peaks (closing inrush) can occur when control transformers are switched on. According to the size of the transformer, these peaks may attain values as high as 20 to 40 times the transformer rated current. This phenomena must be taken into account when selecting the overcurrent and short-circuit protection apparatus.

On the one hand, this switch-on inrush current must not cause tripping of the protection device on the primary side. On the other hand, the protection device must respond fast enough in the event of a short-circuit or overload to prevent thermal overloading of the transformer with resultant weakening of the insulation and danger to the operator.

Thus, the required protection equipment is determined by the design and size of the transformer. In addition to the possibility of protection by means of DIAZED fuse elements, the 4AM and 4AT control transformers are specifically designed to permit circuit configurations without fuses. In terms of their switch-on and short-circuit currents, they have been matched to the tripping characteristics of the 3VU and 3VF circuit-breakers (for the coordination of the protection devices, please refer to the relevant Siemens l.v. switchgear catalogues).

4.2.5 Thermistor protection in control transformers

Transformers may be protected from excessive temperature rise by means of thermistors. PTC thermistors are used as temperature sensors and are embedded in the winding on each limb of the transformer core. The rated operating temperature of the thermistors is selected to lie marginally above the temperature limit for continuous operation, alternatively for short-circuited output.

The connecting leads of the thermistors are brought out onto terminals. Separate (different) thermistors are used for warning and tripping respectively. In the case of the transformer core having several limbs, the thermistors are connected in series and the two ends are brought out onto terminals. Also refer to Section 3.4.3.5 on page 138.

4.3 Operating conditions for low-voltage switchgear in auxiliary circuits

4.3.1 Prevention of operational down-time in contactor control systems

Clear and unambiguous signals

Contactors require clear and unambiguous signals for switching on and off. This cannot always be guaranteed if the command device, such as e.g. a thermostat or monitor, has an unstable contact system, and particularly if this device is subject to vibration or mechanical shocks. In such cases flutter commands may cause a chattering of the contactor. The contactor contacts may weld as a result of the excessive switching frequency and thermal overloading.

If a flutter command by the command device cannot be avoided, it is recommended that a contactor relay with drop-out delay be used between the command device and the power contactor.

For a basic circuit diagram, see Section 8.1.5.1 on page 412.

Appropriate interlocking to avoid short-circuits

Electrical interlocking is usually required between the contactors in reverser and star-delta starter circuits to prevent short-circuits from being created when the load is switched from one contactor to another. It should not be neglected that mutual electrical interlocking may also be required between e.g. the respective command devices. If appropriate interlocking is not provided, a switching arc may still exist in the contactor which has been switched off at the instant when the next contactor closes. The resultant phase-to-phase short-circuit can cause the contacts of the closing contactor to weld.

Mechanical interlocking may also be required if e.g. the forward and reverse contactors in a reverser starter are subject to vibration or shock in the direction of their closing operation.

For reliable switching of contactor-based star-delta starters, refer to the tried and tested basic circuit diagrams in Sections 8.3 and 8.10.1 on page 444 et seq. and page 488 respectively.

Prevention of voltage dips in control circuits

In terms of DIN VDE 0660, part 102 contactors must still operate perfectly at $0.85 \cdot U_s$ (U_s Rated control supply voltage). Also refer to Section 5.3.2, page 279. If the voltage is lower than this undervoltage tolerance of 15% (20% for 3TF and 3TB contactors) at the instant of closing, e.g. because a large load is being switched on in a weak network and the voltage drops owing to the starting current the moment that the main contacts touch, then the contactor may "chatter". This chattering is caused when the magnetic force of the contactor coil system is insufficient to ensure proper pull-through of the moving contacts. During chattering, the contactor operates at a switching frequency up to twice the frequency of the supply voltage and is thermally overloaded by the switching arcs. It usually becomes jammed in a position in which the contacts touch and weld together.

Voltage recorders, which are often installed to monitor voltage fluctuations in electrical installations, are generally far too slow to register such voltage dips during operation. Even moving iron and moving coil instruments do not allow clear detection and measurement. These voltage dips can only be measured by means of an oscilloscope, oscillograph or other electronic measuring instrument.

It is therefor imperative that not only the continuous currents, but also the starting currents of loads which may be switched on simultaneously be taken into account during the design and engineering of electrical installations.

The voltage relationships in the whole installation, i.e. including network transformer, cables and conductors on the low-voltage side, control transformers and control conductors may be checked, e.g. during the design phase, using the principles outlined in the procedure below.

Voltage drop along the control circuit conductors at instant of switch-on

From the basic equation for the voltage

$$U = I \cdot R_{SL} \cdot 2 l_{SL} \cdot \cos \varphi_{on}$$

the voltage drop along the control circuit conductors can be calculated as

$$u_{SL} = \frac{S_{on} \cdot l_{SL}}{5 \cdot U_S^2} R_{SL} \cdot \cos \varphi_{on} \qquad (4.1)$$

where,

u_{SL} Voltage drop along the control circuit conductor in %

S_{on} Power consumption of the contactor coil during closing in VA (for values, see Siemens catalogue)

l_{SL} Single length of the control circuit conductor in m

U_S Rated control supply voltage in V

R_{SL} Ohmic resistance per conductor and km of the control circuit conductor in Ω/km (for values see Section 9.4.5, page 530)

$\cos \varphi_{on}$ Power factor of the contactor coil during switch-on (for values see Siemens catalogue)

Average voltage drop in the control transformer

This voltage drop u_{ST} is approx.

$$u_{ST} = 5\% \qquad (4.2)$$

provided that the transformer has been carefully and correctly selected (see Sections 2.1.1, page 76 and 4.3.5, page 257)

Voltage drop along the conductors in the main circuit, low-voltage side

The voltage drop along the conductors in the main circuit is calculated as:

$$u_L = \frac{\sqrt{3} \cdot I \cdot l_L}{10 \cdot U_n} (R_L \cos \varphi + X_L \sin \varphi). \qquad (4.3)$$

u_L Voltage drop along the main circuit conductor in %

I Maximum current which can arise upon switch-on of a load in A

l_L Single length of the main circuit conductor in m

U_n Rated voltage in V (e.g. no-load voltage on the low-voltage side of the transformer)

R_L Ohmic resistance per conductor and km of the main circuit conductor in Ω/km (for values see Section 9.4.5, page 530)

$\cos \varphi$ Power factor during motor starting

$\sin \varphi$ Apparent power factor, corresponding to $\cos \varphi$, see Table 4.5

X_L Inductive impedance per conductor and km of the main circuit conductor in Ω/km (for values see Section 9.4.5, page 530)

Table 4.5
Average values for $\cos \varphi$ and $\sin \varphi$ of three-phase induction motor starting currents

Motors with rated current	$\cos \varphi$	$\sin \varphi$
up to 16 A	0.65	0.76
above 16 A	0.35	0.94

Voltage drops in main and auxiliary circuits

Voltage drop along the busbars

For busbars with a length of up to 5 m, the voltage drop may be ignored. For longer busbars, the equation 4.3 may be used, with:

$X_L = 0.25\ \Omega/\text{km}$.

Voltage drop in the network transformer

The calculation of the regulation on the low-voltage side of the transformer resulting from internal voltage drop is done using:

$$u_\varphi = u'_\varphi + \frac{u''^2_\varphi}{200}. \qquad (4.4)$$

u'_φ and u''_φ are calculated as follows:

$$u'_\varphi = u_x \sin\varphi + u_R \cos\varphi,$$
$$u''_\varphi = u_x \cos\varphi - u_R \sin\varphi. \qquad (4.4a)$$

u_φ Regulation in % for most common transformers with $u_{kr} = 4\%$ (Figure 4.1a) or 6% (Figure 4.1b)
u_x Reactance voltage in % ($u_x = \sqrt{u_{kr}^2 - u_{Rr}^2}$)
u_{Rr} Rated value of the ohmic voltage drop in %
u_{kr} Rated value of the short-circuit voltage in %
φ Phase angle

At a given current loading I, the voltage drop is

$$u_T = u_\varphi \cdot \frac{I}{I_n} \qquad (4.5)$$

u_T Voltage drop in the transformer in %
u_φ Regulation in %
I Given current loading in A (e.g. short-time loading during starting of a motor)
I_n Rated current in A (at the transformer low-voltage side)

The current I is calculated as:

$$I = \sqrt{I_w^2 + I_q^2} \qquad (4.6)$$

I_w Effective current
I_q Reactive current

$$I_n = \frac{S_{nT}}{\sqrt{3} \cdot U}. \qquad (4.7)$$

I_n Rated current of the transformer in A
S_{nT} Power rating of the transformer
U Transformer secondary voltage

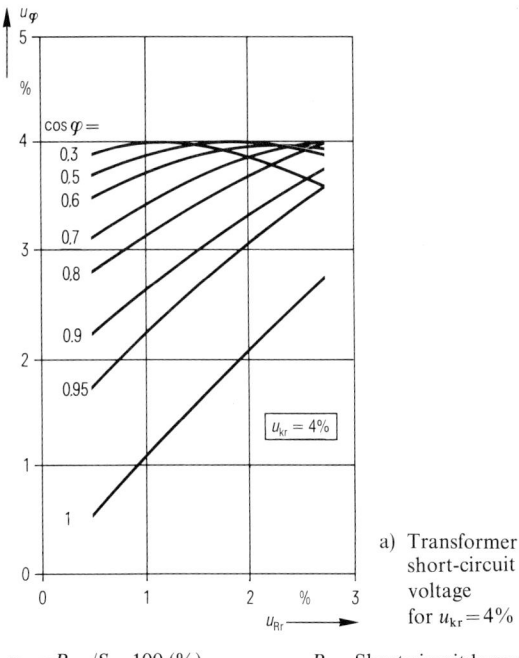

a) Transformer short-circuit voltage for $u_{kr} = 4\%$

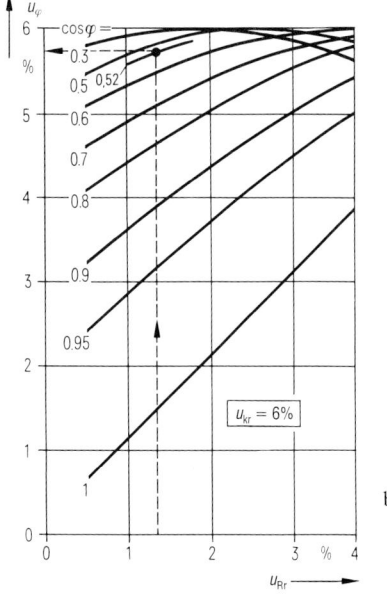

b) Transformer short-circuit voltage for $u_{kr} = 6\%$

$u_{Rr} = P_{krT}/S_{nT}\ 100\ (\%)$

P_{krT} Short circuit losses of the transformer in kW
S_{nT} Rated output power of the transformer in kVA

Figure 4.1   Regulation u_φ at various transformer short-circuit voltages

4 Selection criteria for low-voltage switchgear in auxiliary circuits

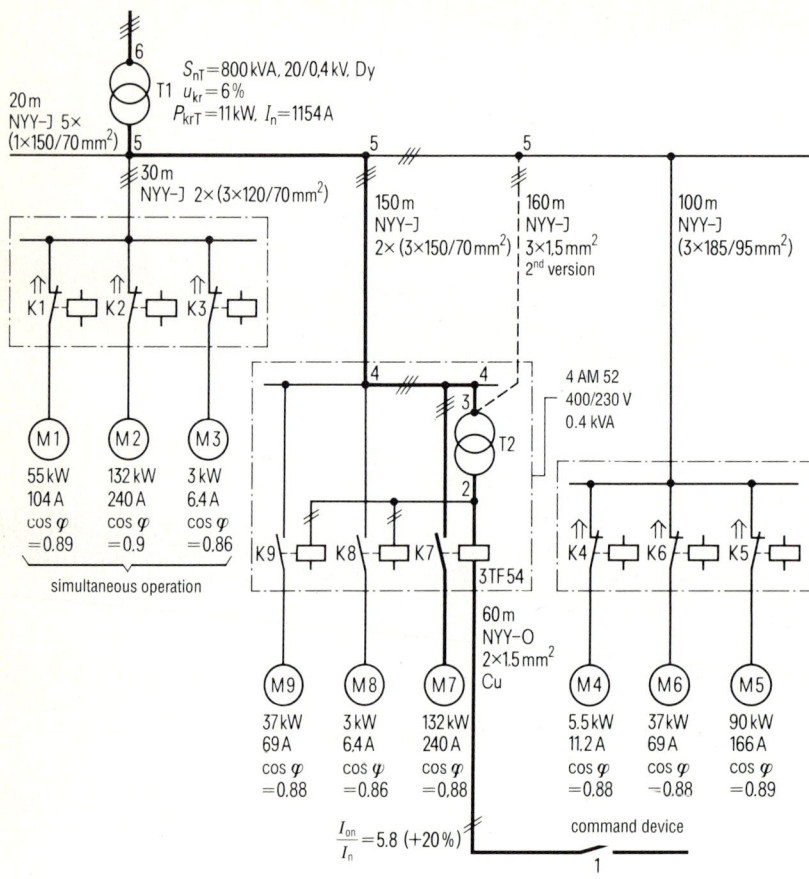

Figure 4.2
Example for the calculation of possible voltage dips in the control circuit of an installation. The N.O. contacts K1 to K6 are already closed when the motor M7 is switched on

T1	Network transformer (cast resin type)
T2	Control transformer
K1 to K9	Contactors
M1 to M9	Motors
1 to 6	Node points for the calculation of the individual voltage drops

Table 4.6
Total transformer loading as per Figure 4.2 at the instant when the motor M7 is switched on (Calculation using equation (4.6)

Motors already running	Motor currents I (from Fig. 4.2)		$\cos \varphi$ (from Fig. 4.2)	$\sin \varphi$	Effective current $I_w = I \cdot \cos \varphi$	Reactive current $I_q = I \cdot \sin \varphi$
	A				A	A
M1	104	(Rated current)	0.89	0.45	92.5	46.8
M2	240	(Rated current)	0.9	0.43	216	103.2
M3	6.4	(Rated current)	0.86	0.51	5.5	3.2
M4	11.2	(Rated current)	0.88	0.48	9.8	5.3
M5	166	(Rated current)	0.89	0.45	147.7	74.7
M6	69	(Rated current)	0.88	0.48	60.7	33.1
Additionally switched on: M7	1670	(Starting current)	0.35	0.94	588	1579.2
Total transformer loading when motor M7 is switched on:					1120.2	1845.5

Examples of voltage drop calculations

Voltage drop along the conductors in the main circuit, high-voltage side

Generally, this voltage drop is so small that it may be neglected. If necessary, it may be calculated by means of the equation 4.3.

Example (Figure 4.2)

Nine motors labelled M1 to M9 are fed from a common 800 kVA transformer. They are switched on in the same sequence as their label numbers (for technical data of the transformer, the motors and the contactor switching arrangement, see Figure 4.2). The contactor K7 switches the largest motor M7 (132 kW).

The total voltage drop comprises:

u_{SL} (between nodes 1 and 2),
u_{ST} (between nodes 2 and 3),
u_L (between nodes 3 and 5),
u_T (between nodes 5 and 6).

The total loading of the transformer at the instant when the motor M7 is switched on, is given in Table 4.6. The motors M8 and M9 are switched on at a later stage. Table 4.7 shows the known values and the interim calculations for determining the various voltage drops

Determining the voltage drops

▷ The voltage drop along the control circuit conductors using equation (4.1):

$$u_{SL} = \frac{S_{on} \cdot l_{SL}}{5 \cdot U_s^2} R_{SL} \cdot \cos \varphi_{on}$$

$$= \frac{1430 \cdot 60}{5 \cdot 230^2} \cdot 14.47 \cdot 0.34 = 1.7\%$$

▷ The voltage drop in the control transformer T2 using equation (4.2):

$u_{ST} = 5\%$

▷ The voltage drop along the conductors in the main circuit, low-voltage side, using the equation (4.3):

$$u_L = \frac{\sqrt{3} \cdot I \cdot l_L}{10 \cdot U_N} (R_L \cos \varphi + X_L \sin \varphi)$$

$$= \frac{\sqrt{3} \cdot 1670 \cdot 150}{10 \cdot 400} (0.075 \cdot 0.35 + 0.04 \cdot 0.94)$$

$$= 6.9\%$$

▷ The voltage drop on the busbars is negligible.

Table 4.7
Known values and interim calculations for determining the voltage drops

Parameter	Value	Derived from
For equation (4.1):		
Contactor	3TF54	Fig. 4.2
S_{on}	1430 VA	catalogue
$\cos \varphi_{on}$	0.34	catalogue
l_{SL}	60 m, $q = 1.5$ mm²	Fig. 4.2
U_S	230 V	Fig. 4.2
R_{SL}	14.47 Ω/km	Section 9.4.5, pg. 530
For equation (4.3):		
l_L	150 m, 2×150 mm²	Fig. 4.2
U_n	400 V	Fig. 4.2
R_L	$\frac{0.15}{2}$ Ω/km at 70 °C	Section 9.4.5, pg. 530
X_L	$\frac{0.08}{2}$ Ω/km	Section 9.4.5, pg. 530
Motor M7 on:		
$\cos \varphi$	0.35	Table 4.6
$\sin \varphi$	0.94	Table 4.6
$\frac{I_{on}}{I_n}$	$5.8 + 20\%$ [1]	Fig. 4.2
I_n	240 A	Fig. 4.2
$I_{on} = 5.8 \cdot 1.2 \cdot 240$	1670 A	
For equation (4.5):		
P_{krT}	11 kW	Fig. 4.2
S_{nT}	800 kVA	Fig. 4.2
$u_{Rr} = \frac{P_{krT}}{S_{nT}} \cdot 100$	$\frac{11}{800} \cdot 100$	
u_{kr}	6%	Fig. 4.2
u_φ	5.75%	Fig. 4.1a
U_0	400 V	Fig. 4.2
For equation (4.6):		
$I_w = I \cdot \cos \varphi$	1120.2	Table 4.6
$I_q = I \cdot \sin \varphi$	1845.5	Table 4.6
$I = \sqrt{I_w^2 + I_q^2}$	$\sqrt{1120^2 + 1845.5^2}$ = 2158.87	
$\cos \varphi = \frac{I_w}{I}$	$\frac{1120.2}{2158.87} = 0.5188$	
For equation (4.7):		
$I_n = \frac{S_{nT}}{\sqrt{3} \cdot U}$	$\frac{800\,000}{\sqrt{3} \cdot 400} = 1154$ A	–

[1] Permissible tolerance to DIN VDE.

▷ The voltage drop in the feeder transformer T1 using equation (4.5):

$$u_T = u_\varphi \cdot \frac{I}{I_n}$$

$$= 5.75 \cdot \frac{2158.87}{1154} = 10.8\%$$

▷ The voltage drop along the conductors in the main circuit, high-voltage side, is negligible.

Total voltage drop

The voltage drop on the primary side of the control transformer T2 (between nodes 3 and 6) is

$$u_L + u_T = u'_{3-6} = 6.9 + 10.8 = 17.7\%$$

i.e. on the primary side of the control transformer T2 (node 3), the voltage is

$$U_3 = 400 - \frac{17.7}{100} \cdot 400 = 329 \text{ V}.$$

Note: If the rated primary voltage of the control transformer had been e.g. 380 V, then the voltage drop must be referred to this value, i.e.

$$u'_{3-6} = \frac{380 - 329}{380} \cdot 100 = 13.4\%.$$

The voltage drop u_{1-6} from the transformer T1 (node 6) to the contactor coil K7 (node 2) is calculated as:

$$u_{1-6} = u_{SL} + u_{ST} + u'_{3-6} = 1.6\% + 5.0\% + 17.7\%$$
$$= 24.3\%.$$

Thus, it is shown that a proper functioning of the contactor K7 can no longer be guaranteed since $u_{1-6} > 20\%$ (for 3TF contactors, 20% is still permissible).

Changes to the installation

The following changes to the installation are possible to ensure the proper functioning of the contactor K7:

▷ use a larger feeder transformer T1,
▷ increase the cross-section of the cable between nodes 4 and 5,
▷ run a separate control conductor of $2 \times 1.5 \text{ mm}^2$ between the primary side of the control transformer T2 and the busbar node point 5 (shown in Figure 4.2 as a dotted line).

For the voltage drop along this control conductor between nodes 3 and 5 use the equation (4.1) as follows:

$$u'_{SL} = \frac{S_{on} \cdot l}{5 \cdot U_s^2} R_L \cdot \cos \varphi_{on},$$

with $l = 160$ m from Fig. 4.2

$U_S = 400$ V,

S_{on}, $\cos \varphi_{on}$, R_{SL} from Table 4.7,

$$u'_{SL} = \frac{1430 \cdot 160}{5 \cdot 400^2} \cdot 14.47 \cdot 0.34 = 1.41\%.$$

In comparison to u_{SL} a reduction in the voltage drop of

$$u_L = u'_{SL} = 6.9 - 1.41 \approx 5.49\%$$

is achieved. Thus, the total voltage drop

$$u'_{1-6} = u_{1-6} - 5.49 = 24.3 - 5.49 = 18.8\%.$$

A separate control conductor between the feeder transformer T1 and the control transformer T2 results in $u_{1-6} < 20\%$ and the contactor K7 will operate correctly even during the starting of motor M7.

4.3.2 Long control conductors – problems and solutions

If long control conductors (e.g. longer than 100 m at AC 230 V) are required in the auxiliary circuit of contactors, relays, undervoltage releases, etc., then the following points must be considered to ensure proper functioning of the devices:

▷ Maximum permissible voltage drop along the control circuit conductors
▷ Influence of conductor or line capacitance on the operation of a.c. operated contactors and relays
▷ Additional loading of contacts owing to capacitive transient currents (e.g. reed switches)

Voltage drop

Contactors still operate satisfactorily if the voltage on the coil is reduced to 0.85 times the rated operating supply voltage (in the case of Siemens contactors 3TC, 3TF, 3TH to 0.8 times the rated value). In general, the network voltage may fluctuate by $\pm 10\%$. Thus, the voltage drop on the control circuit conductors may not exceed 5% (for contactors 3TC, 3TF, 3TH up to 10%). In the case of special contactors with extended coil voltage tolerance, this fluctuation may be greater.

The permissible single length of control conductor l_{perm} can be estimated as follows:

Long control conductors · Effect of capacitance

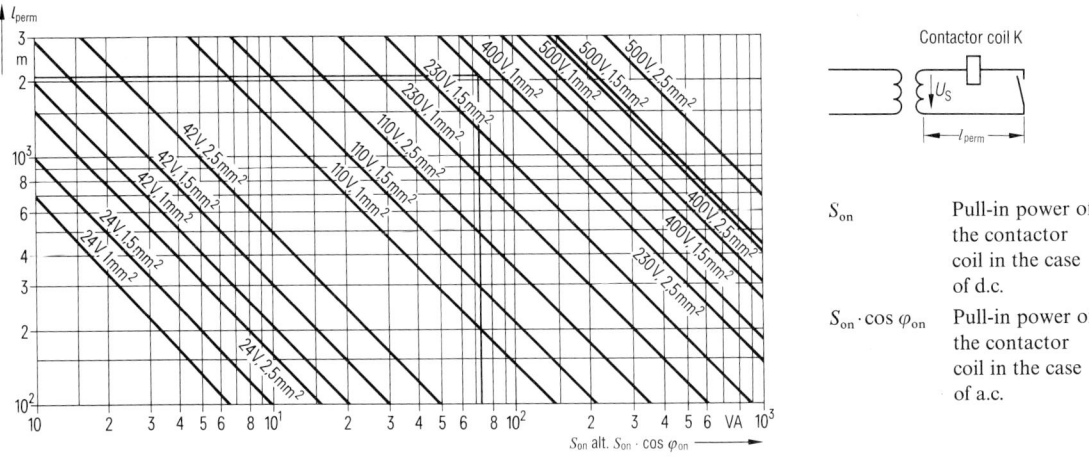

Figure 4.3
Permissible single length l_{perm} of control conductors (copper) as a function of the load for the switching on of contactor coils at various rated control supply voltages and conductor cross-sections (permissible voltage drop along the control conductor $u_{SL} = 5\%$)

for alternating current;

$$l_{perm} = \frac{5 \cdot U_S^2 \cdot u_{SL}}{R_{SL} \cdot S_{on} \cdot \cos\varphi_{on}} \quad \text{(in m)} \quad (4.8)$$

for direct current;

$$l_{perm} = \frac{5 \cdot U_S^2 \cdot u_{SL}}{R_{SL} \cdot S_{on}} \quad \text{(in m)}, \quad (4.9)$$

U_S Rated control supply voltage
u_{SL} Voltage drop along the control conductor in %
R_{SL} Ohmic resistance per conductor and km of the control conductor in Ω/km
S_{on} Power consumption of the contactor coil during closing in VA (for values, see Siemens catalogue)
$\cos\varphi_{on}$ Power factor of the contactor coil during switch-on (for values see Siemens catalogue)

Using the equations (4.8) alternatively (4.9), it follows that:

The higher the rated control supply voltage or the greater the cross-section of the control conductor is (R_{SL} becomes smaller), the longer the control conductor may be. If the rated control supply voltage of a.c. controlled contactors is to be increased however, the effect of the conductor capacitance must be considered.

Figure 4.3 illustrates the permissible lengths of control conductors with commonly used cross-sections for various rated control supply voltages U_s and a maximum permissible voltage drop of $u_{SL} = 5\%$.

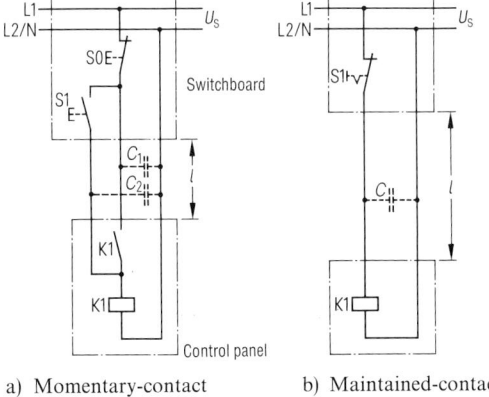

a) Momentary-contact control
b) Maintained-contact control

C, C_1, C_2 Control conductor capacitances
l Single length of control conductor (distance)

Figure 4.4
Connection of the control supply voltage via the command device. Command device located near to the incoming control voltage feed point. Contactor located some distance away

4 Selection criteria for low-voltage switchgear in auxiliary circuits

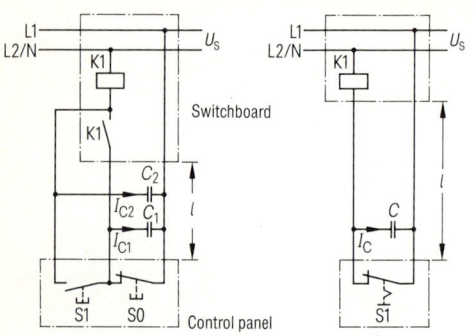

a) Momentary contact control

b) Maintained contact control

C, C_1, C_2 Control conductor capacitances
I_C, I_{C1}, I_{C2} Capacitive leakage currents
l Single length of control conductor (distance)

Figure 4.5
Usual connection of the control supply voltage to the contactor. Contactor installed in the switchboard. Command device located some distance away

Table 4.8
Critical conductor capacitances and corresponding conductor lengths at a rated control supply voltage of 230 V, 50 Hz, taking a 10% overvoltage into account and assuming a conductor capacitance of 0.3 µF/km

Contactor	Critical conductor capacitance	Critical conductor length for	
		momentary-contact control	maintained-contact control
Type	µF	m	m
3TF2, 3TH2	0.058	89	177
3TH30, −32, −33	0.085	130	260
3TH40, −42, −43	0.085	130	260
3TF30, −31, −32 −33, −34, −35	0.085	130	260
3TF40, −41, −42 −43, −44, −45	0.085	130	260
3TF46, 3TF47	0.145	221	442
3TF48, 3TF49	0.273	416	833
3TF50, 3TF51	0.435	644	1326
3TF52, 3TF53	0.495	754	1510
3TF54, 3TF55	0.717	1098	2185
3TF56	0.982	1500	2990
3TF68, 3TF69	1.200	1820	3640
3TC44	0.085	130	260
3TC48	0.222	338	677
3TC52	0.393	600	1145
3TC56	0.717	1093	2186
3TC78	1.366	2080	4165

Effect of the conductor (or line) capacitance on the operation of a.c. operated contactors

Basically, the conductor capacitance of 2 and 3 core control conductors, lines or cables have no detrimental influence on the operation of a contactor, if the command device is located close to the incoming control voltage feed (Fig. 4.4).

In general, however, the contactor is located in the vicinity of the incoming feed and the command device is located separately, some distance away e.g. in the control panel (Fig 4.5).

In this case, a potential difference exists between the conductors of the control cable after the contacts of the command device are opened.

The cable cores, which are at different potentials, behave like a capacitor whose capacitance increases with the length of the cable and the number of cores at different potentials. In control conductors or cables, the capacitance between 2 cores can be assumed to be approximately 0.3 µF/km.

When 3 cable cores are considered, as in Fig. 4.5 (momentary contact control of a contactor with self holding contact), where two cores are at the same potential and the third core is at different potential when switch S0 is opened, the capacitance will be approximately 0.6 µF/km.

Current flows through the contactor coil via the capacitance even after the contacts of the command device have separated.

If this current has a sufficiently large value, the contactor coil may remain energized and the contactor remain closed even after S0 has been opened.

Critical conductor capacitance and length

The critical conductor capacitance and conductor length are specified values which, if exceeded, may cause the contactor or relay to remain closed even after the contacts of the command device have been opened. In the case of very long control conductors, e.g. of a few hundred meters, it is advisable to determine whether the conductor lengths and the capacitances exceed the critical values for the respective contactors during the planning phase of the installation. If the calculated values exceed the critical values for conductor capacitance and/or conductor length, then suitable measures must be taken to prevent malfunction.

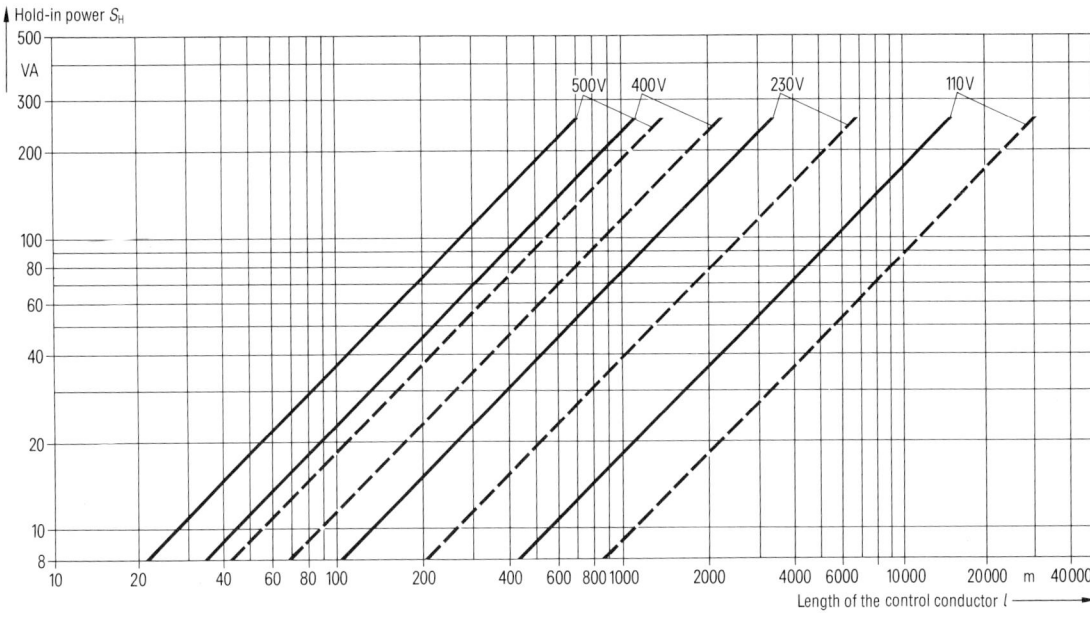

—— for momentary-contact control
--- for maintained-contact control

Figure 4.6
Guide values for permissible lengths of control conductors or cables l as a function of the rated control supply voltage U_S at 50 Hz allowing for a 10% overvoltage and based on the hold-in power rating S_H of the respective contactors. At 60 Hz, the values found are to be reduced by 20%

For critical conductor length:

in the case of momentary contact control

$$l_{\text{crit}} = \frac{500 \cdot S_H}{2 \cdot 0.3 \cdot U_S^2} \cdot 10^3 \quad \text{(in m)}, \qquad (4.10)$$

in the case of maintained contact control

$$l_{\text{crit}} = \frac{500 \cdot S_H}{0.3 \cdot U_S^2} \cdot 10^3 \quad \text{(in m)}. \qquad (4.11)$$

S_H Coil power rating when closed (holding)

For 60 Hz networks, all values obtained using the equations (4.10) and (4.11) must be reduced by 20%.

Refer to Table 4.8 and Fig. 4.6, which are based on the equations (4.10) and (4.11), for a quick estimation of the critical conductor capacitances and lengths for contactors in AC 230 V, 50 Hz networks. For AC 220 V 50 Hz, the values for critical conductor length in Table 4.8 may be increased by approx. 10%.

Determining the capacitive effect on contactor operation by measuring stray currents in existing contactor control systems

If long control conductors are run in close proximity to each other (see Fig. 4.7), stray capacitive currents which cannot be calculated in the planning stage, also flow between the adjacent circuits. In addition, changes may become necessary during the wiring of the installation. During the commissioning of a contactor control system, it may happen that the actual conductor capacitances become greater than the critical values for the contactors in question.

The actual conductor capacitances in the system may be determined by measuring the capacitive stray currents.

Instead of the contactor coil K1, a sensitive ammeter is connected into the circuit. To avoid a possible short-circuit via the ammeter, it is advisable to disconnect the cables on both sides (Fig. 4.7b). Next, all the appropriate cable cores are placed at operating potential (e.g. by bridging 1 to 2 as well as 3 to 4,

4 Selection criteria for low-voltage in auxiliary circuits

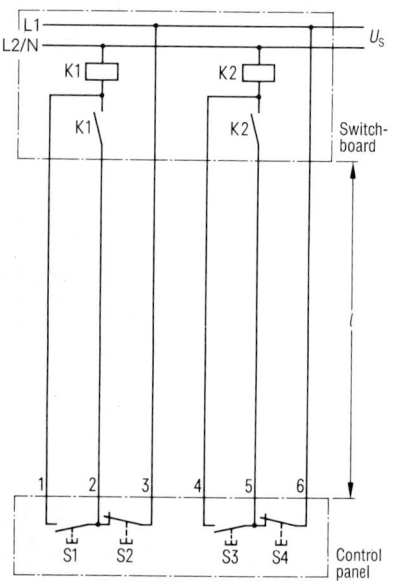

a) Control circuit diagram

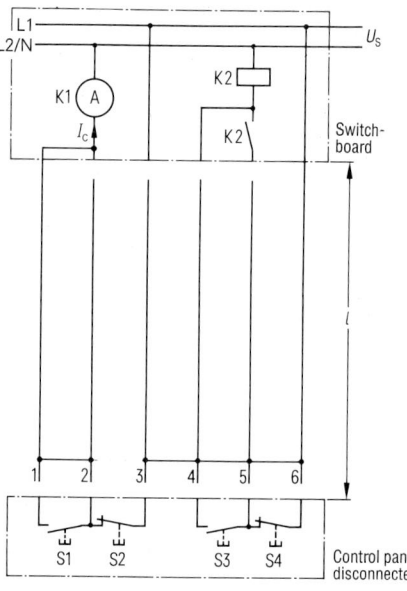

b) Circuit for measurement

Figure 4.7
Determining the effective conductor capacitance by measuring the capacitive stray current I_c between long control conductors

U_S Rated control supply voltage

4 to 5 and 5 to 6 as in Fig. 4.7b), to test for the correct drop-out of contactor K1 when contactor K2 is closed.

The effective conductor capacitance for the contactor is found by:

$$C_L = \frac{1000 \cdot I_c}{2 \pi f \cdot U_s} \quad \text{(in μF)} \qquad (4.12)$$

I_c Measured stray capacitive current in mA
f Rated frequency of the control circuit voltage in Hz
U_S Rated control supply voltage in V

For the normal control supply voltage of 230 V, 50 Hz,

$C_L = 0.0138 \cdot I_c$

For the most common rated control supply voltages in 50 Hz networks, the conductor capacitances as a function of the measured stray currents may be read from Fig. 4.8.

Measures to be taken to avoid disturbing effects caused by conductor capacitance

If, during the planning stage or the commissioning of an installation with long control conductors, the actual line lengths are found to be greater than the critical lengths (e.g. from Table 4.8 or Figure 4.6) and contactors do not drop out when an "Off" command is given, then the following measures may be taken to overcome the effects of the conductor capacitances.

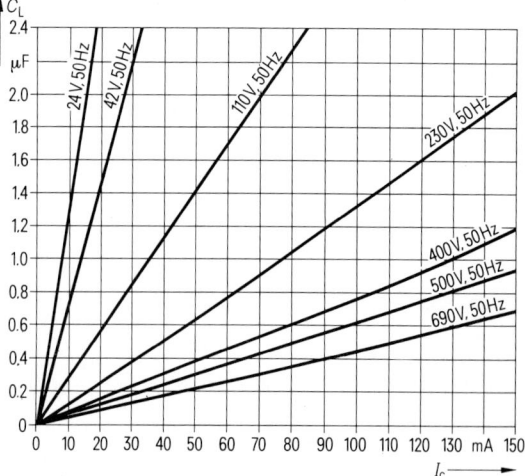

Figure 4.8
Conductor capacitance C_L as a function of the measured stray capacitive current I_c for various rated control supply voltages in 50 Hz networks (derived from equation (4.12))

Measures against effects of conductor capacitance

Use of a d.c. control supply voltage

If a d.c. control supply voltage is used, then the conductor capacitances are discharged via the contactor or relay coil within the total opening time when an "Off" command is given. A noticeable increase in the opening delay does not occur.

Use of a lower control supply voltage

At lower control voltages the critical conductor capacitance and conductor lengths are greater.

When using lower control voltages, however, special attention must be paid to the permissible voltage drop on the control conductors. Use of the equations (4.8) and (4.9) is recommended.

Lower control supply voltages are especially economical in larger contactor installations. The voltage must be selected in accordance with the contactor circuit having the largest conductor capacitance. The conductors are selected in terms of a calculation of the critical conductor capacitance C_{crit}.

Parallel connection of additional loads (shunt loads)

This is only economical in the case of control systems with smaller contactors (3TH contactor relays and 3TF contactors).

The voltage drop across the coil is reduced to a value lower than the hold-in voltage by an additional load connected in parallel to the contactor coil. Either resistive or capacitive shunt loads may be used.

Resistive shunt load

These are generally only recommended if the power loss is smaller than 10 W. The required resistance can be calculated by:

$$R_P = 1000 \frac{1}{C_L} \quad \text{(in } \Omega\text{).} \tag{4.13}$$

The power rating of the shunt resistance is given by:

$$P_P = \frac{U_S^2}{R_P} \quad \text{(in W).} \tag{4.14}$$

Where, should be <10 W, and:

C_L Conductor capacitance in µF
(from equation (4.12) or measured value)
U_S Rated control supply voltage in V
R_p Shunt resistance in Ω
P_p Power rating of the shunt resistance in W.

When installing the shunt resistance it must be borne in mind that the operating temperature of the resistor may be very high.

Capacitive shunt load

Capacitive shunt loads (RC elements) have lower losses. They are connected as shown in Figure 4.9.

Note:
In spite of the same circuit arrangement, capacitive shunt loads must not be mistaken for RC elements used for limiting overvoltages during contactor opening operations (see Section 4.3.3, page 248). Although they also provide a degree of overvoltage limitation, capacitive shunt loads are larger than RC elements for switching-spike suppression.

The size of the capacitive shunt load can be calculated as follows:

$$C_P = 4.5 \cdot C_L \quad \text{(in µF)} \tag{4.15}$$

The capacitor must be selected in accordance with the rated operating a.c. voltage. The resistor R_p should be approximately 100 Ω. The power rating of the resistor can be calculated by:

$$P_P = R_P \cdot \left(\frac{U_S \cdot 2\pi f \cdot C_P}{10^6}\right)^2 \quad \text{(in W)} \tag{4.16}$$

C_L Conductor capacitance in µF
(from equation (4.12) or measured value)
R_p Shunt resistance in the RC combination in Ω; 100 Ω recommended
U_S Rated control supply voltage in V
f Rated frequency in Hz
C_p Shunt capacitance in the RC combination in µF.

For $U_S = 230$ V, 50 Hz and $R_p = 100$ Ω, from equation (4.16) it follows that:

$$P_p = 0.52 \cdot C_p^2 \quad \text{(in W)}$$

If a capacitive shunt load is used, the opening time of the contactor could increase up to 50 ms.

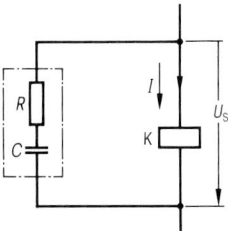

Figure 4.9
Connection of the capacitive shunt load to the contactor coil

Capacitive shunt loads are only economical for individual contactors with long control lines. Since a capacitive shunt load is required for each contactor coil, this solution will be relatively expensive for installations where several contactors need to be equipped.

Short-circuiting of the contactor coil

If the contactor coil is short-circuited during the "Off" command by an additional bridge connection, as is shown in Figure 4.10, then reliable drop-out of the contactor is guaranteed.

An additional control conductor is required between the command device and the contactor; the command device requires an additional normally-closed contact.

The opening time of contactor may increase to several 100 ms.

Use of a larger contactor

From Table 4.8 and Figure 4.6 it is evident that the critical conductor length increases for larger contactors, i.e. greater coil power ratings, and lower control supply voltage.

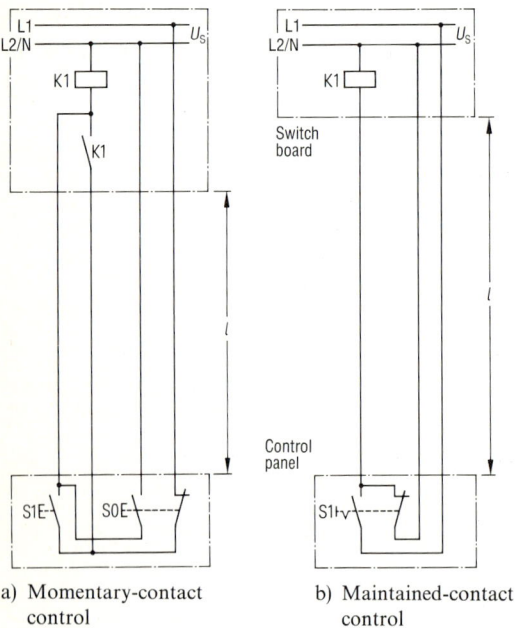

a) Momentary-contact control

b) Maintained-contact control

Figure 4.10
Control circuit with long control conductors ($l > l_{crit}$) and short-circuited contactor coil in the "Off" state

Additional loading of the contacts by capacitive stray currents

Long control conductors should also be avoided in the case of switching devices having low making capacities (e.g. reed switches). During closing, the contacts must conduct the relatively high conductor capacitance discharge current. Welding of light contacts may result.

(For further details and an example, refer to the Siemens l.v. switchgear catalogue, under "Technical Data" for the 3SE6 magnetically-operated position switch)

4.3.3 Limiting of overvoltage spikes caused by the switching off of contactors (overvoltage suppression)

For the reliable operation of an electrical control system, the *electromagnetic compatibility* (EMC) between the various components in the installation must be considered. Although the demands related to EMC in conventional control systems may be relatively low, special protective measures to guard against disturbing or even destructive influences of switching overvoltages are essential in systems employing electronic control elements, since the actual signal levels are generally low.

4.3.3.1 Causes of overvoltages

The most important cause of overvoltages are switching operations in inductive circuits. The output stages of most of the electronic control systems and logic controllers found on the market, are either protected against overvoltage spikes from directly-connected inductive loads, or switch in such a way that the spikes do not arise (a.c. output stages with switch-off at the natural zero crossing). There is, however, the danger of feed-back interference from neighbouring mechanically switched circuits on electronic input or output stages.

During switching off of an inductive load (e.g. contactor coil), the inductance "attempts" to oppose the interruption of the current flow, whereby the circuit exists through the presence of the inherent coil capacitance. In the case of sufficient dielectric strength in the circuit, decaying (transient) voltage and current oscillations are caused within this circuit.

Owing to the high resonance impedance of the open-circuited coil, the amplitude of the oscillations are calculated to be in a range of some 10 kV, whereas the rate-of-rise of the voltage spikes is typically

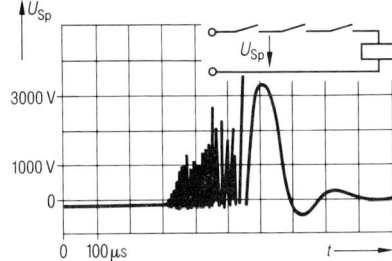

Figure 4.11
Switching overvoltages of an unsuppressed contactor relay coil (230 V, 50 Hz, 10 VA)

around 1 kV/μs. As a result, rapid restriking (or shower discharge) of the current initially occurs between the opening contacts, until the dielectric recovery voltage of the contact gap exceeds the decaying amplitude of the voltage oscillations.

Figure 4.11 shows a typical oscillograph of this rapid restriking. It shows the switching off of a contactor relay coil (230 V, 50 Hz, 10 VA) by means of six series connected mechanical contacts which open simultaneously. The rapid restriking period lasting approx. 250 μs is followed by a damped oscillation with peak amplitude of approx. 3.5 kV.

The rapid restriking can cause severe contact erosion. Furthermore, the steep wave flanks can cause severe disturbances in neighbouring systems owing to capacitive coupling effects. An appropriate suppression circuit for this source of "switching noise" may be necessary.

The overvoltages caused by the switching off of contactors and relays may not only endanger voltage-sensitive components connected in parallel to their coils, but may also cause severe disturbances in neighbouring electronic control systems owing to the capacitive coupling between control conductors. In practice, it is recommended that the overvoltages are suppressed directly at their source, i.e. at the contactor or relay coil, to ensure that the cross-coupling of the disturbing spikes into adjacent control conductors is minimized.

The following suppression elements are commonly used, and are connected parallel to the contactor coil:

– RC elements (resistor and capacitor in series)
– Freewheel diodes
– Varistors

4.3.3.2 Overvoltage suppression with RC elements

A.C. operated contactors

RC elements are mostly used with a.c. operated contactors. The effective increase in the capacitance of the coil circuit reduces the amplitude and rate-of-rise of the switch-off overvoltage spikes to such an extent, that no rapid restriking occurs. The RC elements are therefor particularly effective in the protection of du/dt-sensitive output stages (Triacs!) against sporadic signals. In cases of high coupling capacitance (range > 10 nF), there is a greater risk that input stages of the low-pass filter type may be disturbed, than if unsuppressed coils or varistor suppression elements are used. Optimal suppression requires the correct selection of the components in terms of the respective rated control supply voltage and frequency (Figure 4.12).

The components in the RC elements for Siemens contactors and contactor relays at the most common a.c. rated control supply voltages from 24 to 240 V, are given in Table 4.9. The overvoltage limiting factor n is approx. 2 to 3, and can be calculated from:

$$n = \frac{\text{max. overvoltage (peak value)}}{\sqrt{2} \cdot U_s} \quad (4.17)$$

U_S Rated control supply voltage (r.m.s. value)

For a.c. rated control supply voltages from 380 to 440 V, two RC elements for 230 V may be connected in series (Figure 4.13).

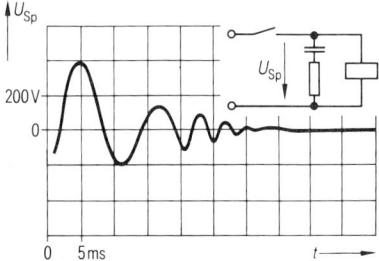

Figure 4.12
Switching overvoltages of a contactor relay coil (230 V, 50 Hz, 10 VA) suppressed by means of an RC element 110 Ω, 0.22 μF

4 Selection criteria for low-voltage switchgear in auxiliary circuits

Table 4.9 RC elements for a limiting factor $n=2$ to 3 with 3TH, 3TF, 3TK and 3TC a.c. operated contactors

Contactor Type	Rated control supply voltage at 50/60 Hz													Power rating of the resistor P_R
	24 V		42 to 48 V		60 V		110 to 127 V		220 V		230 to 240 V			
	R Ω	C μF	R Ω	C μF	R Ω	C μF	R Ω	C μF	R Ω	C μF	R Ω	C μF		W
3TH2	39	2.2	150	0.68	220	0.33	680	0.082	2200	0.02	2200	0.02		0.33
3TH3/3TH4	22	3.9	68	1.0	150	0.68	220	0.2	1000	0.068	1000	0.068		0.33
3TF2	39	2.2	150	0.68	220	0.33	680	0.082	2200	0.02	2200	0.02		0.33
3TF30–35 3TF40–45	22	3.9	68	1.0	150	0.68	220	0.2	1000	0.068	1000	0.068		0.33
3TF46/47	10	6.8	33	2.2	68	1.6	270	0.33	1000	0.1	1000	0.1		0.5
3TF48[1)]/49	4.7	15	15	4.7	27	2.7	120	0.68	390	0.22	390	0.22		0.5
3TF50[1)]/51	3.9	18	10	5.6	22	3.3	100	0.82	330	0.27	330	0.27		0.5
3TF52[1)]/53	2.7	22	10	6.8	15	3.9	68	1.0	270	0.33	270	0.33		1.0
3TF54[1)]/55	2.2	39	4.7	2.7	10	6.8	47	2.2	150	0.68	150	0.68		1.0
3TF56[1)]	2.2	47	4.7	15	10	8.2	47	2.7	150	0.82	150	0.82		1.5
3TF57/68/69	For up to $U_s=600$ V, varistors are factory fitted in the standard 3TF57, 3TF68 and 3TF69 contactors													
3TC44	68	3.9	68	1.0	150	0.68	220	0.2	1000	0.068	1000	0.068		0.33
3TC48	4.7	15	15	4.7	27	2.7	120	0.68	390	0.22	370	0.22		0.5
3TC52	2.7	22	10	6.8	15	3.9	68	1.0	270	0.33	270	0.33		1.0
3TC56	2.2	47	4.7	15	10	8.2	47	2.7	150	0.82	150	0.82		1.5

[1)] The values also apply for the corresponding AC-1 contactors, type 3TK4 and 3TK5.

D.C. operated contactors

In the case of d.c. operated contactors, the required capacitance C is determined by:

$$C = k \cdot \frac{\tau}{R} \qquad (4.18)$$

k Factor from Figure 4.14, page 251 (contains various parameters, e.g. required limitation of the overvoltage)

The resistance in the RC element should be selected to be approximately equal to the resistance of the coil:

$$R = \frac{U_s^2}{P_h} \approx R_c \qquad (4.19)$$

R Resistance of the contactor coil in Ω
R_c Resistance in the RC element in Ω

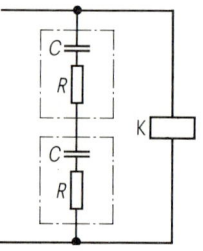

Figure 4.13
Series connection of RC elements parallel to the contactor coil

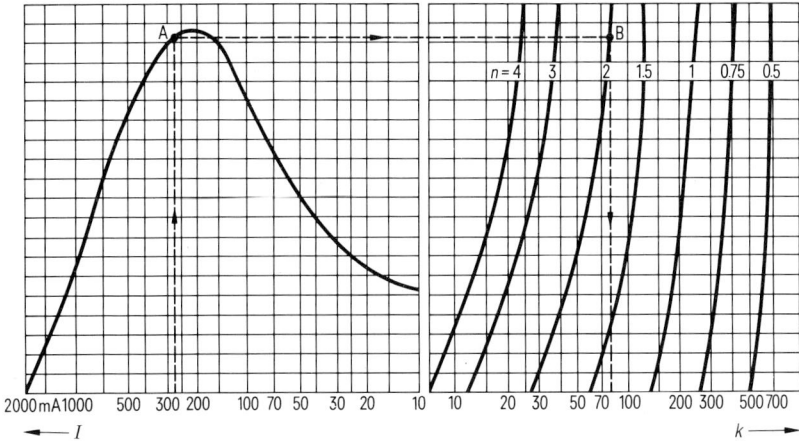

Figure 4.14
Determination of the factor k for equation (4.18), page 250 (Point A is found from the rated value of the coil current $I(I = P/U)$ in the diagram on the left. Point B is the intersection of a horizontal, drawn from A across to the right-hand diagram, and the curve corresponding to the required limitation value n. The factor k can then be read from the x-axis of the right-hand diagram)

In the case of direct current, no current flows through the RC element during normal operation. Therefore, the power rating of the resistor is based on the pull-in and drop-out power of the contactor, and can be calculated as follows:

$$P_R = 1.5 (\tau \cdot 10^{-3} \cdot P_H + 3 \cdot C \cdot 10^{-6} U_S^2) z \cdot \frac{1}{3600} \quad (4.20)$$

P_H Hold-in power of the coil in W
P_R Power rating of the resistor in the RC element in W
C Capacitance of the capacitor in µF
τ Time constant of the contactor in ms (Table 4.10)
U_S Rated control supply voltage in V
z Switching frequency per hour

Example 1:
Contactor with d.c. magnet system

The operating coil of a 3TB52 contactor has the following data:

Rated control supply voltage $U_S =$ DC 110 V,
Hold-in power rating $P_H = 30$ W,
Time-constant $\tau = \frac{L}{R} = 150$ ms,
Switching frequency $z = 600$ switching operations/hour.

Desired limitation of the overvoltage to maximum 220 V ($n = 2$). The capacitance must be found by using equation (4.18).

To determine the factor k from Figure 4.14, the rated value of the coil current I must first be found:

$$I = \frac{P_H}{U_S} = \frac{30}{110} \cdot 1000 = 273 \text{ mA}.$$

For $n = 2$, the factor from Figure 4.14 is found to be $k = 80$.

The resistance R of the coil, by equation (4.19) is

$$R = \frac{U_S^2}{P_H} = \frac{110^2}{30} = 404 \text{ } \Omega.$$

Therefore, the capacitance

$$C = k \cdot \frac{\tau}{R} = 80 \cdot \frac{150}{404} = 30 \text{ µF}.$$

The next larger preferred value of capacitance may be selected.

The resistance R_c is determined by using equation (4.19):

$$R_c \approx R = 404 \text{ } \Omega$$

The preferred value of 430 Ω is selected.

By equation (4.20), the power rating of the resistor is 4 W.

Thus, the RC element consists of $R = 430$ Ω, 4 W and $C = 33$ µF, DC 250 V.

Table 4.10
Values of the time-constant τ for a closed magnet system, and values of the hold-in power ratings P_H of the coils of d.c. operated contactors

a) D.C. magnet system

Contactor	Time constant τ	Hold-in power of the coil P_H
Type	ms	W
3TH2, 3TF2	18	3
3TH4, 3TH3	40	6.5
3TF40-3TF43	40	6.5
3TF30-3TF33		
3TB44	43	10
3TB46	73	15
3TB47, 3TB48	94	19
3TB50	130	25
3TB52	150	30
3TB54	140	60
3TB56	140	86
3TC44	43	10
3TC48	94	19
3TC52	150	30
3TC56	130	86
3TC74	67	46 [1]) ($U_n \leq 230$ V)
3TC78	67	92 [1]) ($U_n \leq 230$ V)

b) D.C. economy connection

	Time constant τ		Pull-in power [2])	Hold-in power
	Pull-in winding	Hold-in winding		
Type	ms	ms	W	W
3TF46/47	7	20	330	2.1
3TF48/49	9	16	400	2.7
3TF50/51	10	24	470	2.7
3TF52/53	12	23	820	4.8
3TF54/55	15	40	1050	7.1
3TF56	21	35	1560	7.7

[1]) Hold-in power: If applicable, including the series resistor alt. economy resistor (for non-standard coil voltages, the power consumption increases proportionally with the voltage).
[2]) Owing to the high pull-in power, an RC suppression of the pull-in winding requires the selection of correspondingly rated components.

Example 2:
Contactor with d.c. economy circuit

The hold-in coil of a 3TF56 contactor has the following data:

Rated control
supply voltage $\quad U_S =$ DC 220 V,
Hold-in
power rating $\quad P_H = 7.7$ W,
Time-constant $\quad \tau = L/R = 35$ ms,
Switching
frequency $\quad z = 1000$ switching operations/hour.

Desired limitation of the overvoltage to maximum DC 440 V ($n=2$). The capacitance must be found by using equation (4.18).

To determine the factor k from Figure 4.14, the rated value of the coil current I must first be found:

$$I = \frac{P_H}{U_S} = \frac{7.7}{220} \cdot 1000 = 35 \text{ mA}.$$

For $n=2$, the factor from Figure 4.14 is found to be $k=55$.

The resistance R of the coil, by equation (4.19) is

$$R = \frac{U_S^2}{P_H} = \frac{220^2}{7.7} = 6286 \, \Omega.$$

Therefore, the capacitance

$$C = k \cdot \frac{\tau}{R} = 55 \cdot \frac{35}{6286} = 0.31 \, \mu\text{F}.$$

The next larger preferred value of capacitance may be selected.

The resistance R_c is determined by using equation (4.19):

$$R_c \approx R = 6286 \, \Omega$$

The preferred value of 6200 Ω is selected.

The power rating of the resistor is found by means of equation (4.20):

$$P_R = 1.5 \cdot (\tau \cdot 10^{-3} \cdot P_H + 3 \cdot C \cdot 10^{-6} \cdot U_S^2) \cdot \frac{z}{3600}$$

$$P_R = 1.5 \cdot (36 \cdot 10^{-3} \cdot 7.7 + 3 \cdot 0.33 \cdot 10^{-6} \cdot 220^2) \cdot \frac{1000}{3600}$$

$$P_R = 0.13 \text{ W}.$$

A resistor with power rating 0.33 W is selected. This only applies for this specific rated control supply voltage.

Thus, the RC element consists of $R = 6200\ \Omega$, $1/3$ W and $C = 0.33\ \mu F$, DC 630 V.

4.3.3.3 Overvoltage suppression with diodes

The switching-off overvoltages of d.c. operated contactors and relays may be avoided completely by the use of "free-wheel" diodes (Fig. 4.15). Diodes connected parallel to the coil, however, cause the drop-out times of the contactors to increase by a factor of 6 to 9 times the normal values. This effect may be used to advantage if e.g. voltage dips of a few milliseconds are to be bridged. Owing to the risk of "two-stage drop-out", overvoltage suppression by means of diodes is only permissible for contactors up to size 2. Care must be taken to ensure the correct polarity of the free-wheel diode connection.

Suppression units containing diodes are offered in various forms for direct mounting onto the contactor housing. For switching devices which are specially designed for use with electronic control systems (e.g. interface contactors and relays, input/output units), the suppression elements are sometimes integrated within device itself (see Siemens low-voltage switchgear catalogue).

4.3.3.4 Overvoltage suppression with varistors

Varistors (voltage-dependent resistors), connected parallel to the coil, limit the maximum value of the switching overvoltage since they become highly conductive for voltages above the threshold value. Untill this threshold value is reached, shower discharge (rapid restriking), similar to the case of unsuppressed coils, occurs. The duration of this phenomeno is however shorter. Suppression with varistors is particularly suitable for the protection of input stages with low-pass filter characteristics. It is less suitable for protection of du/dt-sensitive output stages with insufficient inherent spike suppression. In contrast to RC elements, they do not reduce the rate-of-rise of the voltage spikes (Figure 4.16). They are suitable for use with alternating and direct current, and have a negligible effect on the switching (drop-out) times. They are particularly suitable for applications in which contactors are used in conjunction with industrial electronic control systems (e.g. SIMATIC). These systems contain only the basic elements required for electromagnetic compatibility (EMC).

In the Siemens l.v. switchgear catalogue, varistor-based overvoltage suppression units for the 3TC, 3TF, and 3TH contactors are offered to cover three ranges of rated control supply voltage.

Thereby, depending on the actual rated control supply voltage in question, the overvoltage limiting factor n ranges from 2.5 to 7.5. The continuous load ratings of the varistors have been selected in terms of the maximum switching frequencies of the contactors and contactor relays during AC-3 and AC-11 duty respectively.

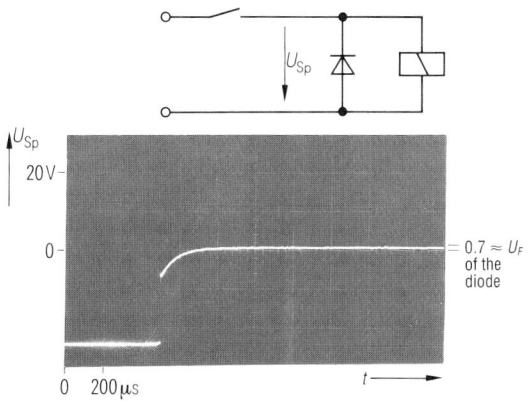

Figure 4.15
Switch-off of a contactor coil (DC 24 V, 3 W), suppressed by means of a freewheel diode

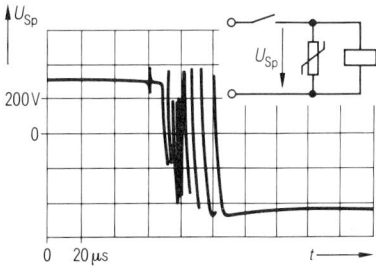

Figure 4.16
Switch-off overvoltage of a contactor coil (230 V, 50 Hz, 10 VA) suppressed by means of a 275 V varistor (initial transient; the voltage drops to zero after approx. 3 ms)

4 Selection criteria for low-voltage switchgear in auxiliary circuits

4.3.4 The use of contactor relays in safety circuits

If contactor relays are used in circuits intended to ensure the safety of operating personnel e.g. emergency-stop arrangements (safety circuits), then in terms of DIN VDE 0113 part 1/EN 60204 part 1, Paragraph 5.7.1 redundancy through the use of several contactor relays must be provided. The auxiliary contacts of the contactor relays must be so interconnected as to ensure that in the case of a fault in (or a failure of) one of the relays, the safety circuit must remain operational and reliable (so-called "contactor safety combinations").

The contactor relays in the safety combination must be checked automatically for correct opening and closing at least once during each on/off cycle. An on/off cycle is understood to means e.g. the switching on or off of the main switch, the disconnection or drop-out of the control supply voltage, or the operation of emergency-stop apparatus.

In terms of the accident prevention guidelines of the Swiss institute for accident insurance (SUVA) and the (German) safety regulations for controls of power operated presses in the metal industry (ZH1/457, edition 2, 1978), the contactor relay configuration must fulfill the conditions for "positively driven contacts". These conditions specify that the normally open and normally closed contacts of the switching device (or combination) may never be closed simultaneously.

Depending on the application, the contactor safety combination may comprise two contactor relays without positively driven contacts, or three contactor relays with positively driven contacts.

For the application in controls of power operated presses in the metal industry in accordance with ZH1/457 and ZH1/508, self-monitoring is not only required in combinations comprising relays and contactors, but also for circuits with position switches and pushbuttons. Appropriate examples of circuit diagrams are available from Siemens.

Contactor safety combinations may be built from 3TH4 contactor relays as per the circuit diagrams given below. Alternatively, factory assembled and tested combinations contained within a common housing (e.g. 3TK28), are offered. For selection and ordering data, please refer to the Siemens l.v. switchgear catalogue.

Contactor safety combinations consisting of two contactor relays, internal interlocking by means of overlapping contacts

A.C. excitation

Two identical 3TH4 contactor relays with overlapping contacts are used in the contactor safety combination (see Figure 4.17). The early-make normally-open contact and the late-break normally closed contact are required for the internal interlocking of the contactor safety combination.

The remaining "normal" contacts may be used for enabling, or alternatively, disconnection of the relevant circuit(s) being controlled.

This contactor safety combination comprising two contactor relays with overlapping auxiliary contacts does not fulfill the conditions for positively driven contacts (not essential for "normal" controls)

The enabling output consists of the series connection of two normally-open contacts - one from each contactor relay.

The number of possible enabling outputs is determined by contactor relays used.

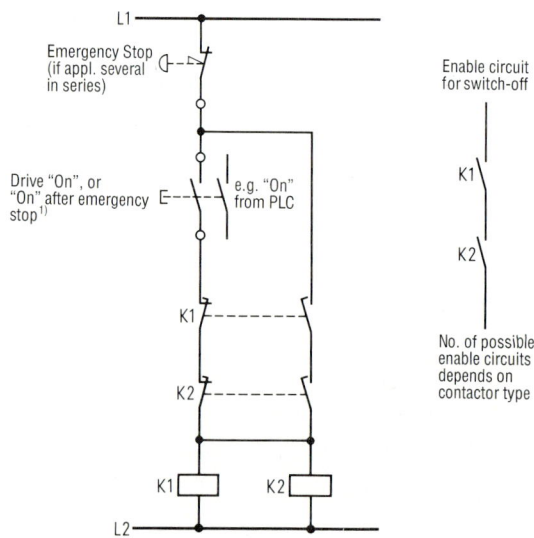

[1] The inclusion of the "On" button is necessary to avoid automatic restart after a delatching of the emergency stop facility

Figure 4.17
Contactor safety combination consisting of two identical contactor relays, internal interlocking by means of overlapping contacts, a.c. excitation

Contactor relays in safety combinations

If a contactor safety combination with more enabling outputs is required, then a further two contactor relays must be used. Thereby, their coils must be connected in parallel with the coils of the first two contactor relays. Their early-make and late-break contacts must be connected in series with the existing extended auxiliary contacts.

D.C. excitation

Owing to relatively wide tolerance bands, the closing times of contactor relays with d.c. magnet systems from different manufacturing batches may differ to a significant degree.

For this reason, occasional switching errors (no enabling output) may occur in contactor safety combinations consisting of two d.c. operated contactor relays with overlapping contacts. The integrity of the circuit, however, remains intact and is retained.

In the interest of a reliable operation, and owing to the fact that contactor relays with positively driven contacts *are* available, it is recommended that contactor safety combinations consisting of *three* contactor relays be used for both a.c. and d.c. applications (Figure 4.18).

Operation of the contactor safety combination shown in Figure 4.17

The coils of the contactor relays K1 and K2 are connected in parallel. If voltage is applied by an operation of e.g. "Drive On", then the contactor relays K1 and K2 close via the series connected late-break normally-closed contacts. The early-make contacts of K1 and K2 close, and the contactor relays remain energized as long as the supply voltage is maintained.

If the control supply voltage is interrupted by the operation of the emergency-stop facility, both contactors K1 and K2 drop out. If, for example, contactor relay K1 fails at this instant, i.e. does not drop-out, the circuit is nevertheless interrupted by the dropping out of contactor K2. Renewed energizing of the combination and resultant closing of the enable circuit is normally no longer possible, since although the late-break normally-closed contact of K2 and the early-make normally-open contact of K1 are closed, the late-break normally-closed contact of K1 and the early-make normally-open contact of K2 are both open.

Since the extended auxiliary contacts of contactors K1 and K2 are not positively driven, it may be possi-

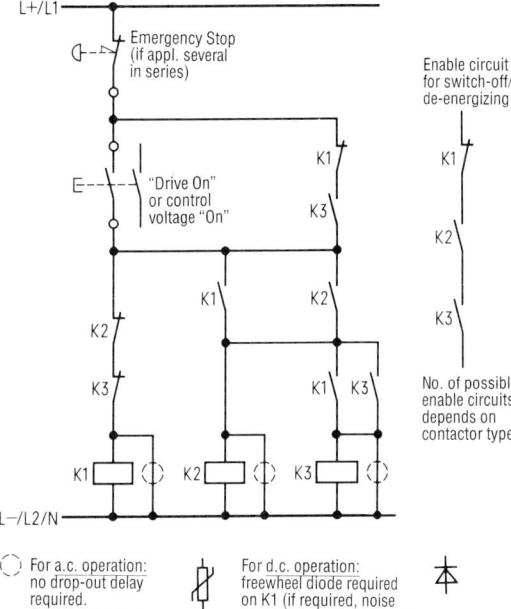

Figure 4.18
Contactor safety combinations comprising *three* contactor relays, internal interlocking and enabling outputs via positively driven contacts.
Suitable for a.c. and d.c. excitation

ble that under unfavourable conditions (e.g. partial drop-out of K1 due to light welding), the combination may re-energize in spite of a failure of one of the contactor relays.

Contactor safety combinations consisting of three contactor relays, internal interlocking and enabling outputs via positively driven contacts

Owing to the insufficient supervision of the function in the case of contactor relays with overlapping contacts, safety combinations comprising three contactor relays with positively driven contacts are becoming more prevalent.

The important advantage of this arrangement is that standard contactor relays may be used. The only precondition is that their auxiliary contacts are positively driven, since only positively driven contacts permit reliable supervision of the contact element operation and thus redundancy with supervision.

4 Selection criteria for low-voltage switchgear in auxiliary circuits

D.C. excitation

These contactor safety combinations consist of the three contactor relays K1, K2 and K3. The contactor relays are "normal" 3TH2 or 3TH4 types with d.c. excitation.

The internal interlocking of the contactor safety combination requires contactor relay K1 to have two normally-open and one normally-closed contacts, and contactor relays K2 and K3 each to have one normally-open and one normally-closed set of contacts.

The remaining auxiliary contacts may be used for enabling or disconnecting of the relevant circuit.

The enabling outputs with normally-open function for disconnection purposes, consist of the series connection of a normally-open contact from K1 and a normally-closed contact from each of K2 and K3.

The number of possible enabling outputs is determined by contactor relays used.

In the case of d.c. excitation, the contactor relay K1 must have a delayed drop-out to overcome possible differences in switching times owing to tolerance effects. To achieve this, its coil must be equipped with an external freewheel diode or a contactor relay with integrated bridge rectifier must be used.

The addition of freewheel diodes to K2 and K3 for overvoltage suppression has no adverse effect on the operation of the combination.

If a contactor safety combination with more enabling outputs is required, then a circuit comprising additional contactor relays Kx, Ky, etc. may be built. These contactor relays must be interconnected as is indicated in Figure 4.19.

A.C. or D.C. excitation

These contactor safety combinations consist of the contactor relays K1, K2 and K3 (see Figure 4.18)

Either a.c. or d.c. operated 3TH2 or 3TH4 contactor relays are recommended. In the case of d.c. excitation, the coil of contactor relay K1 must be equipped with a freewheel diode. Alternatively, contactor relays with integrated bridge rectifiers may be used (if desired, these may also be used for a.c. applications).

Operation

If the control voltage is applied to the combination via a command device, then the contactor relay K1

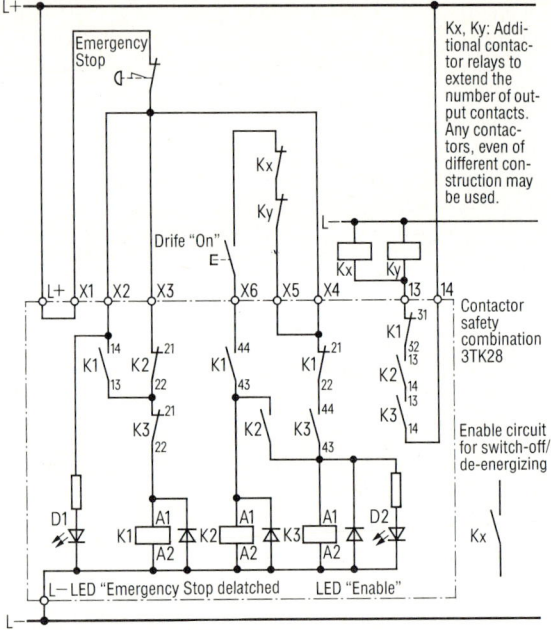

Figure 4.19
Example of extending the number of output contacts in a contactor safety combination

pulls in. The contactor relays K2 and K3 are energized via the normally-open contact of K1 and seal in via their own normally-open contacts as long as the "on button" or command device remains activated.

The normally-closed contacts of K2 and K3 cause K1 to drop out; contactor relays K2 and K3 remain sealed in via their own normally-open contacts even after the command device has been released provided that K1 has, in fact, dropped out.

The enabling output is switched on (closes the relevant circuit) once K1 has dropped out and both K2 and K3 have pulled in. If the contactor relay K1 should fail and not drop out, the enabling output cannot switch on. If K2 or K3 fails after the on command has been completed, the switching off of the relevant circuit is still guaranteed, since the normally-open contacts of K1 and K2 are connected in series. A renewed switch-on is not possible since the normally-closed contacts of K2 and K3 are series connected on the supply side of K1.

Selection criteria for control transformers

Circuit with increased reliability

The circuit diagrams shown in Figures 4.18 and 4.19 differ from those previously given in the Siemens NS 2 catalogue (up to 1989), and also differ from those given in the VDE documentation series 26. In the circuit diagrams shown in Figures 4.18 and 4.19, the drop-out of the contactor relay K1 is also monitored. Thereby, it is ensured that an unwanted switch-on of the enabling circuit is also prevented in the event of a temporary failure.

Circuit diagrams for two channel emergency-stop contactor safety combinations (consisting of three contactor relays), circuit diagrams for the supervision of protection systems by means of contactor safety combinations (consisting of three contactor relays) as well as methods for extending the number of enabling outputs of contactor safety combinations comprising three contactor relays by connecting additional contactor relays, are given in the Handbook "Switching off dangers with the greatest safety – the 3TK28 back-up combination unit" (Order No. E20001-P285-A452-X-7600).

Reverse-current braking (plugging),
Enabling circuit with a PLC

In braking by plugging, the relevant main contactors may only remain connected to the supply voltage

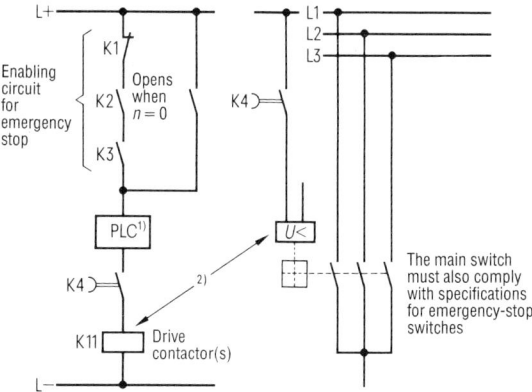

[1] PLC = Programmable Logic Controller

[2] Alternatively, the relevant contactor(s) or main switch is (are) switched off after a period which is longer than the max. braking time.

Figure 4.20
Redundancy in circuits for reverse-current braking (plugging) using a drop-out delayed time relay
Conditions for redundancy are only fulfilled if both the speed and the drop-out delayed time relay are monitored, e.g. by the inclusion in a contactor safety combination (in series with the normally-closed auxiliary contacts of K2 and K3)

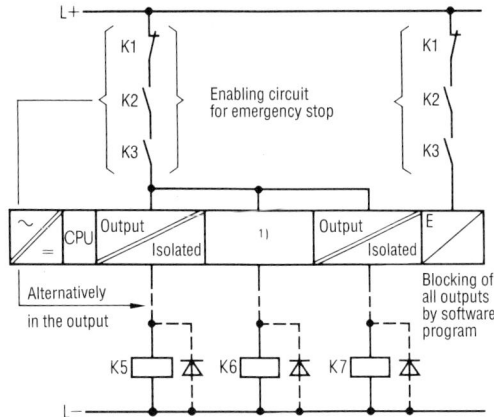

[1] In the case of non-isolated outputs, the switching/enabling must be done in the output

K5, K6, K7 Contactors of drives which must be switched off in the event of an emergency stop

Figure 4.21
Connection of the enabling circuits via the load current path of an electronic control system

until the drive has reached zero speed. To comply with the stipulations of DIN VDE 0113 part 1/EN 60204 part 1 Paragraph 5.7.1, a redundancy is required. This may be achieved e.g. by connecting an additional drop-out delayed time relay (K4) in parallel to K3 in the case of Figure 4.18. Refer to Figure 4.20.

Figure 4.21 shows how the enabling circuit is to be connected in an electronic control system (PLC) to ensure safe and reliable switch-off.

4.3.5 Selection criteria for low-voltage control transformers

In addition to meeting the general operational reliability and safety demands, low-voltage transformers must also meet the demands made by a wide variety of applications and must fulfill the relevant specifications. Thus, in the selection of transformers it is important to consider the prevailing environmental and operating conditions as well as the standards and specifications applicable to the field of use.

4.3.5.1 Operating conditions

Rated power

The rated apparent power P_S of a transformer is based on fixed reference conditions under which loading at the given rated output is permissible.

4 Selection criteria for low-voltage switchgear in auxiliary circuits

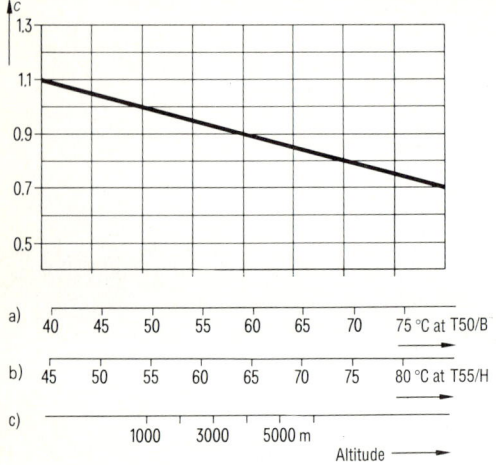

a) Thermal stability T50/B [1)]
b) Thermal stability T55/H
c) Installation altitude in m above sea level

[1)] Insulation class B for 50 °C ambient temperature

Figure 4.22
Load characteristic curve of a transformer as a function of the ambient temperature and the installation altitude above sea level

For example, in the case of the 4AM and 4AP transformers, these are:

▷ Continuous operation duty S1,
▷ Ambient temperature +50°C,
▷ Installation altitude up to 1000 m above sea level,
▷ Frequency 50/60 Hz,
▷ Degree of protection IP 00.

Operating and environmental conditions which differ from these influence the permissible loading by the consumer P_{SCons} in relation to the given rated apparent power P_S of the transformer (Figure 4.22).

$$c = \frac{P_{SCons}}{P_S}$$

P_{SCons} Permissible consumer load power
P_S Transformer rated output power
c Factor

Factors which reduce the permissible consumer power load ($P_{SCons} < P_S$) are:

▷ Increased ambient temperature,
▷ Increased installation altitude,
▷ Restriction of cooling (e.g. through increased degree of protection)

Factors which permit an increase in the consumer power load ($P_{SCons} > P_S$) are:

▷ Low ambient temperature,
▷ Short-time duty.

Required design rating P_{Sreq}

For similar consumer loads P_{SCons}, the applicable operating and environmental conditions may dictate different required design ratings P_{Sreq}.

Fixed reference conditions have been laid down for the design ratings (type ratings) of the transformers listed in the selection tables of the Siemens l.v. switchgear catalogue. These include:

▷ Type of duty,
▷ Frequency,
▷ Installation altitude,
▷ Ambient temperature,
▷ Degree of protection,
▷ Electrical ratings.

In terms of these conditions, the transformers may be loaded up to their output ratings ($P_{Sreq} = P_{SCons}$).

Factors which necessitate the selection of a transformer with a larger design rating ($P_{Sreq} > P_{SCons}$) include

▷ a higher ambient temperature,
▷ restricted cooling,
▷ a higher degree of protection,
▷ installation altitudes higher than 1000 m above sea level,
▷ frequency < 50 Hz,
▷ lower voltage drop,
▷ reduced losses,
▷ lower switch-on current,
▷ lower no-load current,
▷ tappings for constant power,
▷ several windings which may not be loaded simultaneously.

Factors which permit the selection of a transformer with a lower design rating ($P_{Sreq} < P_{SCons}$) include

▷ short-time duty,
▷ frequency > 50 Hz,
▷ lower ambient temperature,
▷ switching of economy circuits.

4.3.5.2 Duty types

Small transformers may not be overloaded continuously. Duty types which differ from continuous operation influence $P_{S\,req}$ and therefore the transformer size. In the case of short-time loading, the actual load current value, its duration and the frequency at which this load is repeated determine the thermal stressing of the transformer (see Section 2.2.4, page 102).

In general, the effective mean load current is calculated as being:

$$I_{mean} = \sqrt{\frac{\Sigma I^2 \cdot t}{\Sigma t_{load} + \Sigma t_{pause}}}.$$

The effective mean current as a function of time is (Figure 4.23):

$$I_{mean} = \sqrt{\frac{I_1^2 \cdot t_1 + I_2^2 \cdot t_2 + I_3^2 \cdot t_3}{t_1 + t_{01} + t_2 + t_{02} + t_3 + t_{03}}}.$$

The cyclic duration factor (c.d.f.) is:

$$\text{c.d.f.} = \frac{t_{load}}{t_{cycle}} \cdot 100 \text{ (in \%)}$$

$$t_{cycle} =ت_{load} + t_{pause}$$

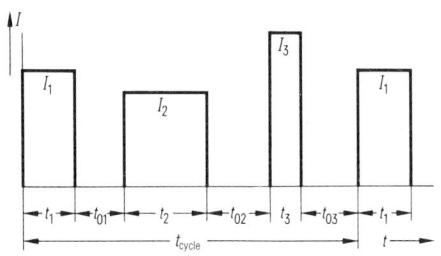

Figure 4.23   Load current as a function of time

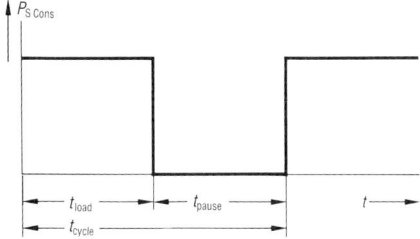

Figure 4.24   Load as a function of time

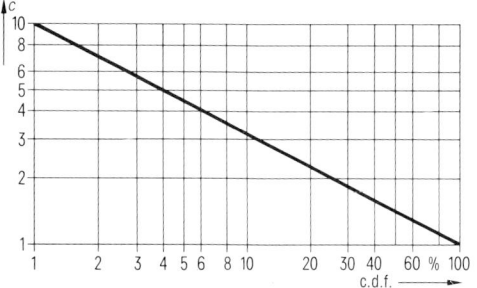

Figure 4.25
Load curve as a function of the cyclic duration factor in percent (c.d.f.)

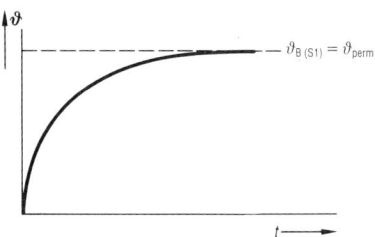

Figure 4.26
Characteristic curve of continuous operation duty (S1)
(Legend s. Fig. 4.29)

The cycle duration is limited to 10 minutes. $t_{cycle} > 10$ min corresponds to transformers in continuous operation duty (Figure 4.24).

$$P_{S\,req} = P_{S\,Cons} \sqrt{\frac{\text{c.d.f.}}{100}} \text{ (in \%) or}$$

$$P_{S\,req} = \frac{P_{S\,Cons}}{c} \quad \text{(Factor } c \text{ from Fig. 4.25)}$$

The most important duty types for low-voltage control transformers are continuous operation (S1), short-time operation (S2), intermittent periodic operation (S3) and continuous operation with intermittent loading (S6).

Continuous operation duty (S1)

The operating period lasts at least long enough for the thermal equilibrium temperature $\vartheta_{B(S1)}$ to be reached (Figure 4.26).

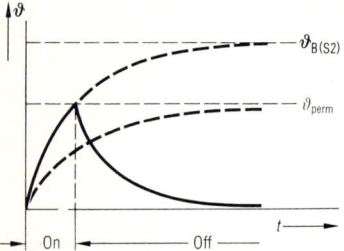

Figure 4.27
Characteristic curve of short-time operation duty (S2)
(Legend s. Fig. 4.29)

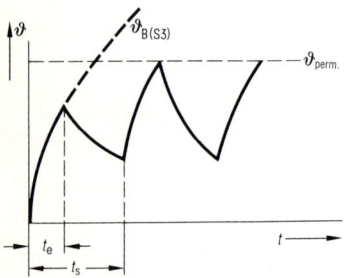

Figure 4.28
Characteristic curve of intermittent periodic operation duty (S3)
(Legend s. Fig. 4.29)

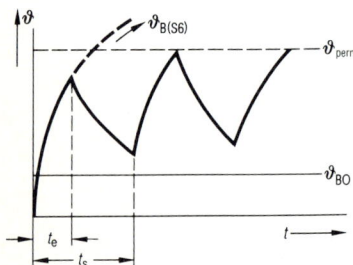

$\vartheta_{perm.}$ Permissible temperature rise
ϑ_{B0} Equilibrium temperature at no-load
$\vartheta_{B(S1)}$ Equilibrium temperature at rated load
$\vartheta_{B(S2)}$ Equilibrium temperature in short-time duty
$\vartheta_{B(S3)}$ Equilibrium temperature in intermittent periodic duty
$\vartheta_{B(S6)}$ Equilibrium temperature in continuous duty with intermittent loading
t_e Load duration
t_s Cycle time

Figure 4.29
Characteristic curve of continuous operation duty with intermittent loading (S6)

Short-time operation duty (S2)

The specified permissible operating period is so short, that the thermal equilibrium temperature $\vartheta_{B(S2)}$ is not reached. The following load interval, during which the transformer primary is disconnected from the network, is however long enough to allow cooling to the ambient temperature (Figure 4.27).

Intermittent periodic operation duty (S3)

The specified permissible operating period is so short, that the thermal equilibrium temperature $\vartheta_{B(S3)}$ is not reached. The following load interval, during which the transformer primary is disconnected from the network, is however not long enough to allow cooling to the ambient temperature (Figure 4.28).

Continuous operation duty with intermittent loading (S6)

The primary of the transformer is continuously connected to the network, and the specified permissible operating period is so short, that the thermal equilibrium temperature $\vartheta_{B(S6)}$ is not reached. The following load interval, during which the transformer primary remains connected to the network (no-load current in the primary), is however not long enough to allow cooling to the no-load equilibrium temperature (Figure 4.29).

4.3.5.3 Transformer types

Transformers must be suited to the widest variety of application conditions. Special transformer versions are required to meet the particular demands of the various applications and to fulfill the applicable specifications.

Control transformers

Application

The use of transformers for the power supply of control and indication circuits is laid down in DIN VDE 0113, part 1/EN 60204, part 1 "Electrical equipment of Industrial Machines". Also refer to page 79. For example, these specifications stipulate that control transformers must be used if

▷ machines contain more than five electromagnetic operating coils (e.g. in contactors, relays or valves),

▷ control and/or indicating devices are located outside of the control cabinet or

▷ electronic control and indication circuits are to be supplied with power.

Control transformers must have separate primary and secondary windings and must be installed downstream from the main switch. They should preferably be connected between two phases.

Specifications and designation labels

For control transformers, the DIN VDE 0550 specifications, parts 1 and 3, apply.

Control transformers are single-phase transformers and must carry the label ⏂. They are intended for the power range ≤ 16 kVA and for output voltages ≤ 250 V. The preferred output voltages are 230 V (220 V), 110 V, 42 V and 24 V.

The short-time rating of a control transformer should be as high as possible to ensure the reliable switching of magnet coils while simultaneously permitting the selection of the smallest possible output rating for the application in question.

In addition to the rated output power for continuous operation duty S1, the short-time rating and $\cos \varphi$ of the load must be stated on the rating plate. At these rated loads, the output voltage may not drop more than 5% from the rated output voltage.

Determining the control transformer size

In the selection of a control transformer, both the maximum expected load in continuous operation (S1) and the maximum expected load in continuous operation with intermittent loading (S6), must be considered. The load on a control transformer typically comprises the vector sum of the hold-in power of already energized contactors, the ohmic load of operating indication lights, the pull-in power of contactors and contactor relays as they are energized, as well as conductor losses.

The required transformer size is determined by two calculations, viz. calculation of the short-time load for continuous operation with intermittent loading (S6), and calculation of the load in continuous operation (S1).

- Calculation of the short-time load for continuous operation with intermittent loading (S6):

Firstly, the transformer size for duty type S6 is determined by the max. permissible voltage drop of $5\% \cdot U_N$ in accordance with DIN VDE 0550, part 3 to ensure reliable operation of contactors. The actual maximum apparent power P_S during S6 and S1 which results in the corresponding voltage drop can be calculated from:

$$P_S = \sqrt{P_p^2 + P_q^2}. \qquad (4.21)$$

P_p Active power
P_q Reactive power

The power factor (only required for the calculation of the short-time load for continuous operation with intermittent loading) is determined by:

$$\cos \varphi = \frac{P_{p(S1)} + P_{p(S6)}}{P_S} \qquad (4.22)$$

The selection of the transformer size from the type/load table (Table 4.11) is based on these values of $\cos \varphi$ and P_S.

- Calculation of the load for continuous operation (S1):

In the case of inductive loading (e.g. only contactor coils), the calculation of the maximum short-time loading usually determines the selection of the control transformer. For duty type S1, the thermal load capacity of the control transformer must be taken into account.

For additional ohmic loading, e.g. by indicator lights and by additional energized contactor magnet systems, the required transformer output power is calculated as:

$$P_{S\,req} = \sqrt{P_{p(S1)}^2 + P_{q(S1)}^2} \qquad (4.23)$$

The $\cos \varphi$ is not taken into account for the calculation of the load in S1.

If the required transformer rating, in terms of equation (4.23), is found to be greater than that selected from Table 4.11 in terms of power factor and apparent power, then the selection must be made according to the continuous load requirements.

Example

Loading of the control transformer:

Existing load:

8 indicator lights
1 contactor type 3TF47 (size 3).

To be switched in:

3 contactors type 3TF52 (size 8)

4 Selection criteria for low-voltage switchgear in auxiliary circuits

Table 4.11 Short-time power ratings of Siemens control transformers $P_{S(S6)} = f(\cos\varphi)$ for $U_2 = 0.95 \cdot U_{2n}$

Transformer Type	Rated output power $P_{S(S1)}$ at IP 00 kVA	Apparent power consumption P_S of all consumers in continuous duty with intermittent loading S6								
		$\cos\varphi = 0.2$ kVA	$\cos\varphi = 0.3$ kVA	$\cos\varphi = 0.4$ kVA	$\cos\varphi = 0.5$ kVA	$\cos\varphi = 0.6$ kVA	$\cos\varphi = 0.7$ kVA	$\cos\varphi = 0.8$ kVA	$\cos\varphi = 0.9$ kVA	$\cos\varphi = 1$ kVA
4AM804	0.063	0.33	0.253	0.206	0.175	0.152	0.135	0.122	0.11	0.104
4AM344	0.1	0.62	0.46	0.37	0.31	0.26	0.23	0.21	0.19	0.17
4AM384	0.16	0.98	0.73	0.58	0.49	0.42	0.37	0.33	0.3	0.28
4AM404	0.25	1.62	1.24	1.0	0.85	0.74	0.66	0.59	0.54	0.51
4AM814	0.315	2.15	1.63	1.33	1.12	0.97	0.86	0.77	0.71	0.67
4AM464	0.4	2.53	2.0	1.67	1.44	1.26	1.13	1.0	0.95	0.92
4AM484	0.5	3.75	2.9	2.4	2.0	1.75	1.55	1.4	1.3	1.25
4AM524	0.63	3.85	3.15	2.7	2.35	2.1	1.9	1.75	1.65	1.6
4AM554	0.8	5.8	4.65	3.9	3.4	3.0	2.7	2.5	2.3	2.25
4AM574	1.0	8.85	7.0	5.85	5.0	4.4	3.95	3.6	3.3	3.2
4AM494	1.6	11.6	9.7	8.3	7.4	6.6	6.0	5.6	5.3	5.3
4AM994	2.5	24.4	18.3	14.6	12.2	10.5	9.3	8.3	7.5	7.1
4AT303	4.0	32.6	25.4	20.9	17.8	15.5	13.8	12.5	11.5	11.0
4AT361	5.0	36.7	27.9	22.6	19.0	16.5	14.6	13.1	12.0	11.2
4AT363	6.3	42.1	33.8	28.4	24.5	21.7	19.5	17.8	16.5	16.1
4AT391	8.0	53.6	43.0	36	31.1	27.5	24.8	22.6	21.0	20.4
4AT393	10.0	53.3	45.8	40.5	36.4	33.3	30.9	29.1	27.9	29.3
4AT430	11.2	85.8	67.8	56.3	48.3	42.4	37.9	34.5	31.9	30.7
4AT431	12.5	89.5	72.9	61.8	53.8	47.9	43.3	39.8	37.2	36.7
4AT432	14.0	90.0	75.9	66.0	58.7	53.1	48.8	45.5	43.2	44.2
4AT450	16.0	140.0	112.0	94.0	81.2	71.7	64.5	59.0	54.7	53.4

▷ Calculation of the short-time load for continuous operation with intermittent loading S6 (for individual load ratings, refer to Table 4.12):

$$P_p = P_{p(S6)} + P_{p(S1)}$$
$$= 3 \cdot 346\,W + 8 \cdot 3\,W + 5\,W$$
$$= 1067\,W$$

$$P_q = P_{q(S6)} + P_{p(S1)}$$
$$= 3 \cdot 842\,var + 16\,var$$
$$= 2542\,var$$

$$P_S = \sqrt{1067^2 + 2542^2}\,VA$$
$$= 2756\,VA$$

$$\cos\varphi = \frac{1067}{2756} = 0.39$$

From Table 4.11, via the column $\cos\varphi = 0.4$ and with $P_{S(S6)} = 3.9$ kVA, the required transformer rating is found to be 0.8 kVA ($\hat{=}$ Type 4AM554).

Table 4.12
Electrical values of the contactors 3TF47/3TF52 and the 3SB indicator lights (see example)

Consumer	3TF47 contactor size 3	3TF52 contactor size 8	3SB indicator lights
Hold-in power (S1)			
$P_{S(S1)}$	17 VA	58 VA	
$\cos\varphi$	0.29	0.26	
$P_{p(S1)}$	5 W [1]	15 W [1]	3 W
$P_{q(S1)}$	16 var [1]	56 var [1]	
Pull-in power (S6)			
$P_{S(S6)}$	183 VA	910 VA	
$\cos\varphi$	0.6	0.38	–
$P_{p(S6)}$	110 W [1]	346 W [1]	
$P_{q(S6)}$	146 var [1]	842 var [1]	

[1] Values calculated from data given in the Siemens l.v. switchgear catalogue

Control transformers for indicator lights

▷ Calculation of the load for continuous operation (S1):

$$P_{P(S1)} = 8 \cdot 3\,W + 3 \cdot 15\,W + 5\,W$$
$$= 74\,W$$

$$P_{Q(S1)} = 16\,var + 3 \cdot 56\,var$$
$$= 184\,var$$

$$P_S = \sqrt{74^2 + 184^2}\,VA$$
$$= 198\,VA$$

From Table 4.11, a transformer rating of 0.25 kVA is found to be suitable ($\triangleq$ 4AM404).

The actual selection of the transformer must be based on the maximum required rating. In this example, therefore, the calculation of the rating for S6 duty determines the selection, and the control transformer 4AM554 with a rated output power of 0.8 kVA is chosen.

The selection of Siemens' control transformers is facilitated by a PC software package known as "ASIST" (Order No. E86010-D1802-A127-A2). Also refer to Section 8.13, page 496.

Transformers for indicator lights

Transformers with low output power ratings tend to have a relatively large increase in secondary voltage during partial loading. If indicator lights are connected to such a control transformer, then this voltage increase under low transformer load conditions may require special attention (Figure 4.30). Overvoltage significantly reduces the service life of incandescent lamps. Both the service life and the luminous flux output of an indicator lamp are functions of the operating voltage (Figure 4.31). If the reduced service life of indicator lamps is unacceptable, then a separate transformer is required for the indicator light load. Thereby, the secondary of the transformer must be such, that even under partial load consisting of only one indicator light, its rated operational voltage is not exceeded:

$$U_2 < U_{nLamp}.$$

The secondary transformer voltage under rated load conditions is selected so that the rated luminance of the lamps is achieved. Transformer secondary tappings may be used if switch-over to threshold luminance is required, e.g. in control room applications.

The threshold luminance is that brightness at which the human eye will just perceive that the lamp is switched on.

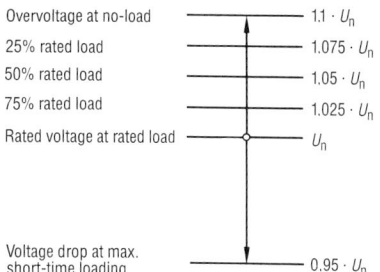

Figure 4.30
Voltage rise at the secondary of a 4AM344 (P_S = 0.1 kVA) control transformer with indicator light load

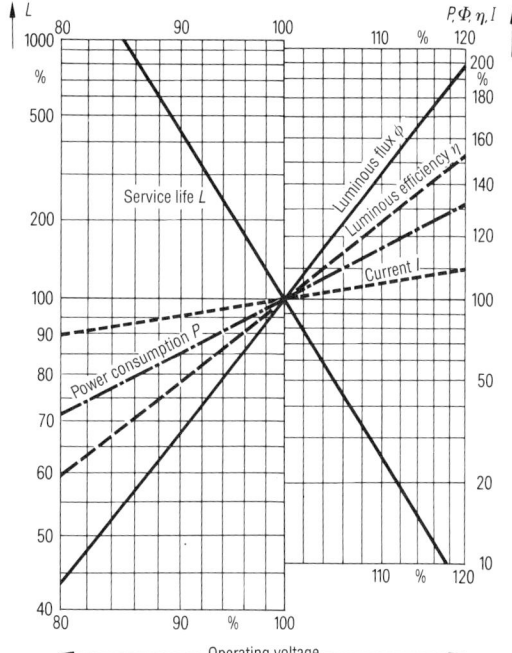

Figure 4.31
Characteristics of incandescent lamps (Values as % of the rated data)

Transformers for rectifier operation

Provided that no particular demands concerning the stability and a.c. content of the supply voltage is made by the d.c. load, the simplest form of rectifier circuits may be used in conjunction with transformers for d.c. power supplies. The basic calculation parameters

Table 4.13
Basic calculation parameters for single-phase (B2) and three-phase (B6) bridge rectifier circuits

Circuit type		
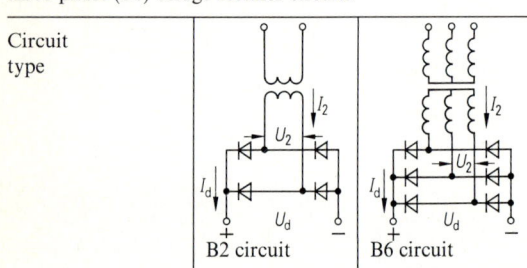 B2 circuit / B6 circuit		
For ohmic load		
$U_{2n} = 2 \cdot U_F + U_d \cdot F$	$F = 1.11$	$F = 0.74$
$I_2 = I_d \cdot F$	$F = 1.11$	$F = 0.82$
$P_s = U_{2N} \cdot I_2 \cdot F$	$F = 1$	$F = 1.73$
w	48.5%	4.2%
f_w	100	300
With smoothing capacitor for $w \leq 5\%$		
$U_{2n} = 2 \cdot U_F + U_d \cdot F$	$F = 0.79$	
$I_2 = I_d \cdot F$	$F = 1.57$	
$P_s = U_{2n} \cdot I_2 \cdot F$	1	

U_{2n} Rated outut voltage (r.m.s. value)
I_2 Output current of the transformer
P_s Rated apparent power of the transformer
U_d D.C. output voltage
I_d D.C. output current
f_w Frequency of the superimposed a.c. voltage (in Hz)
U_F Conducting-state voltage of the diode
U_w Superimposed a.c. voltage (r.m.s. value)
w A.C. content of the D.C. output voltage (in %)

$$w = \frac{U_w}{U_d} \cdot 100$$

for single-phase and three-phase bridge rectifier circuits are given in Table 4.13.

It is recommended that transformers with suitable voltage tappings be used to achieve the required d.c. voltage.

Rectifier units with DC 24 V for the power supply of electronic controls

To ensure the operational reliability of electronic control systems, specific demands are made on their DC 24 V power supplies with regard to the voltage tolerance and the ripple content. The following voltage limits for DC 24 V at a ripple content of 5% are laid down in DIN 19240 ("Peripheral interfaces of electronic controls – power supply and binary interfaces"):

Upper limit 30.2 V as peak value
Lower limit 20.4 V as arithmetic mean value
alternatively
18.5 V as peak value

The Siemens rectifier units 4AV2 and 4AV3 comply with these voltage limits, and are also suitable for use with SIMATIC S5 logic controllers. The units comprise single-phase or three-phase transformers with secondary-connected bridge rectifiers and capacitive smoothing/filtering. They provide an unregulated d.c. voltage of DC 24 V with a ripple content of less than 5%. Electronic control systems are supplied with the permissible DC 24 V in accordance with DIN 19240 irrespective of the load (from no-load to rated current) and fluctuations in the network voltage (from -10% to $+6\%$).

Isolating transformers

Application

Isolating transformers are used to provide galvanic separation between circuits and thereby to limit the danger if earthed and live parts are touched at the same time. These live parts may also be e.g. metal parts which carry voltage as a result of insulation failure.

Amongst other applications, DIN VDE 0100 part 410 Paragraph 6.5 and DIN VDE 0100 part 706 specify the use of isolating transformers for the operation of electrical tools which are held or partially enclosed by the hand(s) during working. Isolating transformers offer increased protection in areas of particular risk, e.g. metallic conductive working areas inside tanks, on metal structures or walkways, in the hull of a ship, etc. as well as with wet polishing and grinding machines, angle grinders, hand-held pounding machines, drills and vibrators.

In the case of metallic conductive working areas, the metal housing of the load (e.g. power tool) must be connected to the surrounding metal (e.g. tank, steel structure) by means of a special, visible, exposed or insulated protective strap or cable of at least 4 mm² copper. This measure also serves to maintain equal potential in the case of a fault.

In the case of work carried out in tanks, the isolating transformer must be installed outside of the working area. Cable joints in the output circuit are only permitted if each conductor has a minimum length, e.g.

Table 4.14
Scope of DIN VDE 0551 part 1. 09/89, EN 60742, IEC 742 for isolating transformers

Trans-former	Maximum rated values for			
	Input voltage V	Frequency Hz	Output voltage (no-load and loaded) V	Power kVA
single-phase	1000	500	1000	25
three-phase	1000	500	1000	40

Table 4.15
Scope of DIN VDE 0551 part 1. 09/89 for safety isolating transformers

Trans-former	Maximum rated values for			
	Input voltage V	Frequency Hz	Output voltage (no-load and loaded) V	Power kVA
single-phase	1000	500	50	10
Three-phase	1000	500	50	16

longer than 10 m. The joints must have increased mechanical strength in terms of DIN VDE 0620. Switches may not be connected in these output cables.

Specifications and designation labels

For isolating transformers the DIN VDE 0551 specifications (part 1 of 09.89) apply. See Table 4.14. To distinguish isolating transformers from transformers for other applications, isolating transformers must carry the label ⊖ .

Isolating transformers may be used to fulfill the conditions for "protection by means of safety separation", i.e. the operating equipment to be connected is safely isolated from the supply network. In this case, neither the body or a conductor (phase) of the equipment may be earthed or connected to the protective conductor (exception: equipotential strap as described above).

Safety isolating transformers

Safety isolating transformers are intended to supply a distribution system, device or other equipment with voltage up to AC 50 V, alternatively via rectifiers up to DC 120 V (under no-load and full-load conditions), with the purpose of preventing dangerous and excessively high touch voltages from arising in the event of a fault.

Furthermore, safety isolating transformers are used where the load conditions are particularly arduous, i.e. in cases where the usual protective measures such as "protection by disconnection or indication" are insufficient. For example, safety isolating transformers are demanded in DIN VDE 0100 parts 701 and 702 (for explosion protected equipment in Zone 0, however, only values smaller than AC 50 V) as well as in part 706.

Specifications and designation labels

For safety isolating transformers, part 1 of the DIN VDE 0551 specifications apply. The corresponding values are given in Table 4.15.

To distinguish safety isolating transformers from transformers for other applications, safety isolating transformers must carry the label ⊖ . Safety isolating transformers may be used to fulfill the conditions for "protection by means of safety extra-low voltage (SELV), i.e. floating output voltage is so low that it itself poses no danger.

Transformers for IT Systems (networks) in medical environments

Application

The applicable DIN VDE specifications, DIN VDE 0107 (11.89), require that IT systems (networks) be used for the supply of life-supporting medical equipment in hospitals, clinics or medical practices. The reason is that a fault e.g. "fault to frame" or "earth fault" in a consumer circuit, should not cause disconnection of the supply with consequential danger to the life of a patient. The fault should, however, be recognized and indicated so that it may be repaired. Transformers are specified for the current source of these IT systems (networks).

Specifications

These transformers must be constructed as isolating transformers for fixed installation (non-transportable) to DIN VDE 0551 part 1, and must have double or reinforced insulation. A passive screening (screen winding) with an accessible insulated terminal for connection of the protective conductor, is required between the input and output sides.

Single-phase transformers must have a center tapping for the connection of insulation monitoring apparatus. The rated voltage on the secondary may not exceed 230 V (between phases in the case of three-phase transformers).

The limit values of no-load current I_a at 3%, short-circuit voltage u_{kr} at 3% and switch-on current at eight times rated current, must be adhered to.

Overload protection causing a switch-off of the transformer is not permitted. A supervision of the load (by means of temperature or current monitoring) with optical or acoustic indication must be provided.

The rated output of the transformers should be within the range of 3.15 to 8 kVA.

Autotransformers

Application

If merely a voltage adaptation is required without the need for galvanic separation from the network, then autotransformers offer an economic alternative.

For the same output power to the consumer (known as "transformer throughput rating"), and depending on the transformation ratio, autotransformers offer a significantly smaller volume and weight in comparison to transformers with separate windings.

The required transformer design rating P_{Sreq} can be calculated for single-phase and three-phase autotransformers as follows:

For step-up transformers ($r > 1$)

$$P_{Sreq} = P_{SCons} \frac{U_2 - U_1}{U_2},$$

For step-down transformers ($r < 1$)

$$P_{Sreq} = P_{SCons} \frac{U_1 - U_2}{U_1}.$$

4.3.6 Application and selection of position switches, type 3SE

Position switches (also known as limit switches) are used to determine the position of moving machine parts, doors or objects and to convert these positions into electrical signals for further processing in the control circuit.

Siemens offers three ranges of position switches. These not only differ in their physical dimensions and technical data, but are each suited for use in different applications.

The criteria for their selection and most important data are shown in Table 4.16 (For further technical data, refer to the Siemens l.v. switchgear catalogue):

▷ 3SE3 020 and 3SE3 023 open-type position switches,

▷ 3SE3 200 moulded-plastic enclosed position switches which comply inter alia with DIN EN 50047 "Auxiliary switches, position switches 30 × 55",

▷ 3SE3 210 and 3SE3 220 moulded-plastic enclosed position switches in wide housings with two cable entries,

▷ 3SE3 120 metal-enclosed position switches which comply inter alia with DIN EN 50 041 "Auxiliary switches, position switches 42.5 × 80",

▷ 3SE3 100, 3SE3 303 and 3SE3 404 metal-enclosed position switches in wide housings with three cable entries,

▷ 3SE3 100 and 3SE3 120 metal-enclosed position switches with incorporated indicator light (neon or LED) for switch state indication (optical indication). For basic circuit diagrams refer to Section 8.6.2, page 480.

▷ 3SE4 miniature position switches,

▷ 3SE6 magnetically operated proximity-type position switches.

Environmental conditions

The position switch version, i.e. open type, moulded-plastic or metal-enclosed with a suitable degree of protection (e.g. IP 20, IP 65 or IP 67) must be selected according to the conditions existing at the point of installation.

The moulded-plastic enclosed devices fulfil the requirements of total insulation, and are resistant to many aggressive chemicals and fluids.

The build up of condensation experienced in humid environments with severe temperature fluctuations is significantly reduced with moulded-plastic housings.

The metal-enclosed position switches withstand high mechanical stressing and are also suitable for installa-

Table 4.16 Selection criteria for position switches

Construction	3SE3						3SE4	3SE6
	Open type	Moulded-plastic enclosed			Metal-enclosed		Miniature pos. switches	Proximity-type (switching magnet)
				Safety pos. switch				
To DIN EN	–	50047	–	–	50041	–	–	–
Rated a.c. voltage up to AC	500 V	500 V	500 V	500 V	500 V	500 V	380 V	250 V
Rated operational current I_e/AC-15, at 230 V	6 A	6 A	6 A	6 A	6 A	6 A	6(3) A	0.5 A
Degree of protection IP 00	–	–	–	–	–	–	×	–
IP 20	×	–	–	–	–	–	–	–
IP 65	–	×	×	×	–	–	–	–
IP 67	–	–	–	–	×	×	–	×
Actuator								
Simple plunger	×	–	–	–	×	×	×	–
Overtravel plunger	–	×	×	–	×	×	–	–
Roller plunger	–	×	×	–	×	×	–	–
Angular roller lever	–	×	×	–	×	×	–	–
Roller crank	–	×	×	–	×	×	–	–
Roller crank with adjustable length	–	×	×	–	×	×	–	–
Rod	–	×	×	–	×	×	–	–
Spring rod	–	×	×	–	×	×	–	–
Fork lever	–	–	–	–	×	×	–	–
Overtravel plunger with central fixing M 18	–	×	×	–	–	–	–	–
Roller plunger with central fixing M 18	–	×	×	–	–	–	–	–
Switching magnet	–	–	–	–	–	–	–	×
Separate actuator	–	–	–	×	–	–	–	–
Switching elements								
snap-action contacts	×	×	×	–	×	×	×	–
creep-action contacts	×	×	×	×	×	×	×	–
galvanically separated contacts with double contact bridges	×	×	×	×	×	×	–	–
reed contacts	–	–	–	–	–	–	–	×
Forced opening	×	×[1]	×[1]	×	×[1]	×[1]	×[2]	–
Repeat accuracy of the switching point in mm	±0.05	±0.05	±0.05	±0.05	±0.05	±0.05	±0.05	±1
Rated mechanical service life in million switching operations	30	10	10	1	30	30	1	30
Electrical service life of the contacts in million switching operations. Load = 3TH4 or 3TF40 to 3TF44	30	10	10	1	30	30	1	30
Max. switching frequency in switching operations per min. at a.c., 50 Hz	100	100	100	100	100	100	160	1200
Protection measures								
Totally insulated	–	×	×	×	–	–	–	×
Earth terminal	–	–	–	–	×	×	–	–

[1] With the corresponding actuators [2] Only with creep-action contacts

Table 4.17 The range of position switch actuators

Actuator	Advantages	Comments
Simple plunger / Overtravel plunger / Roller plunger	Approach possibilities in the direction of plunger stroke or with an actuating bar at right angles to the plunger axis from any direction; Possibility of overtravel in the case of the overtravel and roller plungers, and therefor a longer actuating travel.	The position switch must not be used as an end stop; Approach and trailing angles should be equal; Correct matching of materials and regular lubrication in the case of lateral actuation with an actuating bar or cam disc; Overtravel and roller plungers require higher actuating force.
Roller lever / Angular roller lever	For high approach speeds; very high mechanical service life.	–
Roller crank / Rod atuator	For high approach speeds; many approach possibilities; insensitive to environmental influences such as oil, dirt, sawdust, ice.	Also available with overtravel absorber for special applications.
Spring rod	Suitable for undefined actuation and changing approach conditions; approach from any direction.	Only suitable for use with snap-action contacts.
Overtravel plunger with central fixing, M 18 × 1	For rapid mounting and simple adjustment; approach from any direction.	–
Roller plunger with central fixing, M 18 × 1	For rapid mounting and simple adjustment.	–
Fork lever	Latching actuator.	–
Separate actuator	Versions with actuation from the side and end-on actuation (axial) respectively; cannot easily be manipulated (tamper-proof).	The actuator cannot be exchanged for a standard type.
Actuation with switching magnet (3SE6)	For high approach speeds and switching frequency; proximity-type actuation, insensitive to environmental influences such as oil, dirt, dust.	Protect against strong vibrations and mechanical shocks.

tion in areas in which they are exposed e.g. to flying hot shavings and/or sparks. They are insensitive to most industrial cleaning agents and solvents.

The available installation space determines the dimensions of the position switch to be selected.

Actuation

The construction and method of operation of the machine or plant to be controlled determines the nature of the actuation. Position switches are offered with a wide range of different actuators, each suited for different approach travel or angles and operating speeds (Table 4.17).

Actuating elements

The external actuating element can be in form of a switching bar, a cam disc or shaft, a door, a moving object, etc. Switching bars and cams should be appropriately shaped to ensure that their leading and trailing angles do not cause undue mechanical stressing of the position switch. The selection of the correct actuating element/actuator combination will ensure that a long mechanical service life is achieved, even if only a minimum of maintenance (greasing of the actuating element for plain and overtravel plungers, oiling of the roller in the case of the roller plunger) is carried out.

Overtravel absorber

For special applications, position switches with long roller crank and rod actuators can also be supplied with an overtravel absorber (Figure 4.32).

The purpose of the overtravel absorber is to prevent spurious switching which could otherwise be caused during rebound (swinging back) of the lever after it has been operated. Position switches with an overtravel absorber should be used in situations where the roller crank or rod actuator deflects by more than 45 ° and swings back freely from this position.

Operating or actuation speed

The permissible speed of the actuating element in relation to the direction of actuation is given in the Siemens technical l.v. switchgear catalogue. The minimum length of an actuating bar is dependent on the operating speed, i.e. the contacts in the position switch must remain operated for a sufficient length of time.

Switching elements (contact blocks)

Position switches are available with one or more switching elements which feature different contact arrangements for various control applications (Figures 4.33 and 4.34).

Snap-action contacts

These are specially suitable for:

▷ almost simultaneous switching of the normally-open and normally-closed contacts (changeover time approximately 3 ms) irrespective of the actuating speed,

▷ full contact opening distance even in the case of machines in which the moving parts stop immediately (i.e. without any overrun), after the position switch has given the off command (owing to the contact travel difference or hysteresis between the NC and NO contacts, vibrations do not cause spurious signals),

▷ low operating speeds

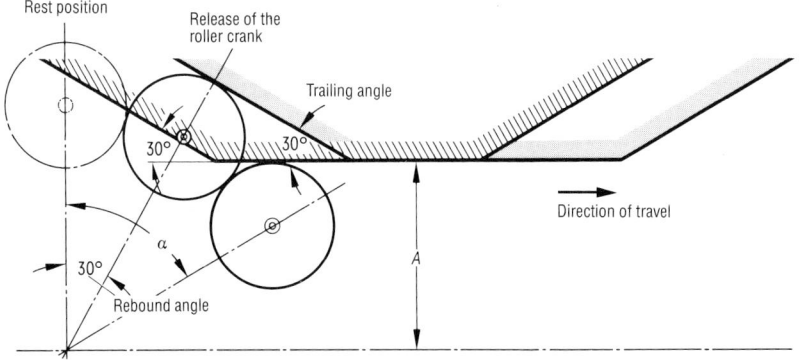

α Angle of deflection of the roller crank in relation to distance A

Figure 4.32
Rebound of the roller crank actuator

4 Selection criteria for low-voltage switchgear in auxiliary circuits

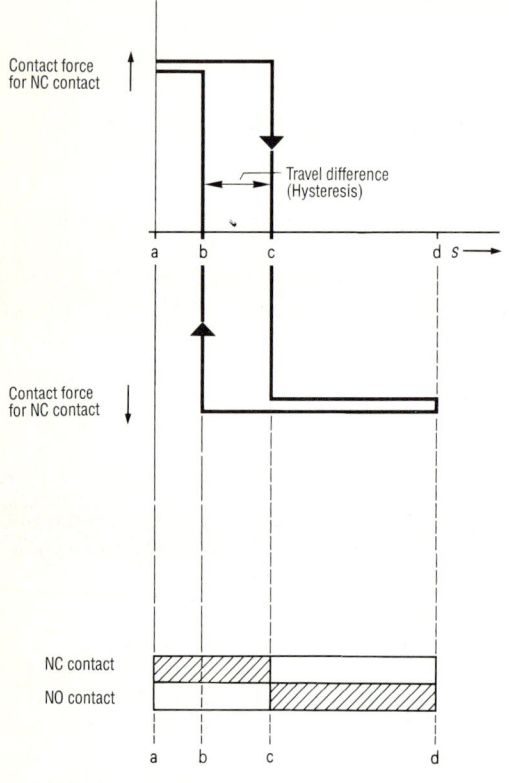

F   Contact force
s   Actuator travel
a   Start of plunger movement
b   Switch-back point
c   Switching point
    (NC opens, NO closes)
a–d Maximum actuator travel

Figure 4.33
Snap-action contacts. Contact force/travel diagram (above) and switch/travel diagram (below).

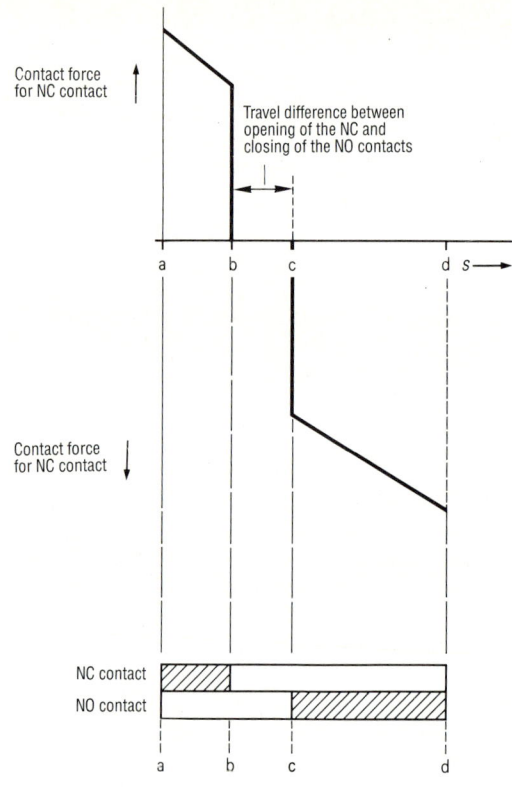

F   Contact force
s   Actuator travel
a   Start of plunger movement
b   NC contact opens
c   NO contact closes
a–d Maximum actuator travel

Figure 4.34
Creep-action contacts. Contact force/travel diagram (above) and switch/travel diagram (below).

Creep-action contacts

These are specially suitable for:

▷ a definite time interval between the switching of the normally-closed and the normally-open contacts as a function of the operating speed,
▷ make-before-break contacts and staggered switching operations,
▷ applications in which it is important that switching of the contacts occurs at exactly the same point during operation and release of the actuator (no contact travel difference or hysteresis).

The switch/travel diagram in Figure 4.34 shows after which distance the normally-closed contacts open and the normally-open contacts close, as well as the remaining overtravel. The overtravel allows for mounting and installation tolerances. The maximum actuating distance must not be exceeded, since this would destroy the position switch mechanically.

Forced opening of the normally-closed contacts in accordance with DIN VDE 0113

Both the creep-action and the snap-action contacts (excluding those of the 3SE4 miniature position switches and the 3SE6 magnetically-operated position switch) possess "forced opening". This means that the opening contact is mechanically driven by the switch plunger to ensure reliable opening (Figure 4.35). This important property is expressly demanded by the specifications for the electrical equipment of machine tools and processing machines, i.e. DIN VDE 0113 and the IEC Publication 204-1, for position switches used in safety circuits.

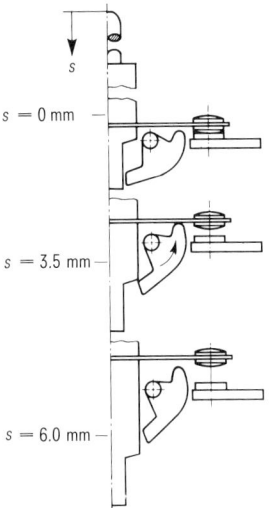

s Plunger travel (stroke)

Figure 4.35
Operating principle of stroke-dependent forced opening of the normally-closed contact in the snap-action contact block

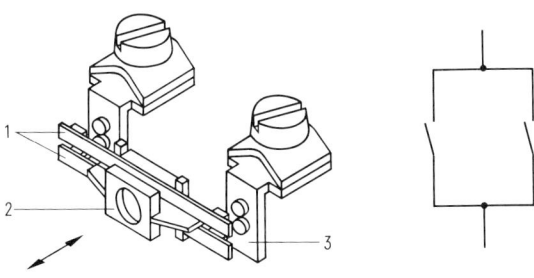

1 Double moving contact bridge
2 Spring retainer (carries the double moving contact pieces without rigid mechanical connection)
3 Fixed contact piece and connection terminal

Figure 4.36
Construction and switching principle (circuit) of contacts with a double moving contact bridge

Galvanic separation of the contacts

Galvanic separation permits the switching of voltages at differing potentials in the case of contact blocks with several contact elements. Thereby one may, for example, control a device with AC 230 V and indicate the operation with DC 60 V.

If one allows for a reduction in the switching capacity, this is sometimes possible with contact blocks which employ a common moving contact bridge for their normally-closed and normally-open contacts.

Galvanically separated contacts with double moving contact bridges are specially suitable for:

▷ extremely high contact reliability, even when used at low currents and voltages,
▷ use with electronic as well as conventional electro-mechanical control systems,
▷ the switching of voltages at differing potentials.

4.3.6.1 Position switches with safety function

Position switches used in safety applications must guarantee the safe operational state of the machine by providing a positive and forced opening of the relevant circuit. It must not be possible for the integrity of the safety system to be jeopardized by simple manipulation. This may be ensured by the appropriate inaccessible positioning of category 1 position switches. Position switches are of category 1 if the switching and actuation elements form one functional unit (e.g. roller crank or angular roller lever position switches).

Position switches of category 2 are devices in which the switching and actuation elements are separate units and they are operated when these are brought together or pulled apart (Figure 4.37). Position switches with separate actuator are used particularly in applications where, for safety reasons, the position of sliding doors, hinged doors or covers are to be monitored. An opening or removal of the protective cover causes the dangerous moving parts of the machine to be brought to an immediate standstill. Thereby, it must also be ensured that the moving parts do not coast to a stop.

An appropriate shape or coding within the actuator head provides protection against manipulation. The switch is operated when the actuator is inserted to its full depth into the actuator head. It is switched off the moment the actuator is withdrawn. Both the normally-closed and normally-open contacts have forced positive opening.

4 Selection criteria for low-voltage switchgear in auxiliary circuits

| Actuation from the side, moulded plastic housing | End-on actuation (axial), moulded plastic housing | Actuation from the side, metal housing |

Figure 4.37 Position switch with separate actuator

The following points should be borne in mind when position switches are used in safety applications:

▷ Compliance with the given minimum travel distances ensures positive separation of closed contacts.

▷ Only the matching (supplied) actuators guarantee positive switching.

▷ The position switches must be securely fixed in their mounting positions. In the case of the metal-enclosed position switches, both 5.3 mm round screw holes must be used. For the secure fixing of moulded-plastic enclosed position switches, use the L-shaped holes at 20 mm (40 mm apart in the case of wide housings). In cases of severe vibration rivetting pins may be necessary.

▷ Short-circuit protection must be provided. Use max. 10 A (DIAZED flink) or 6 A (DIAZED träg) fuse, or alternatively a max. 10 A miniature circuit-breaker (Characteristic C).

▷ Siemens moulded-plastic enclosed position switches have degree of protection IP 65. The metal-enclosed units have IP 67.

▷ The mounting position is not restricted.

▷ The actuation and approach specifications must be adhered to.

▷ In the case of position switches with separate actuator, it is advantageous to provide accurate guiding of the actuator and to minimize the mechanical play of the door, cover or screen.

The position switches listed in Table 4.18 all have positive opening and may be used in safety applications in compliance with IEC 947-5,-1,-3. They may therefor carry the international symbol: ⊖→

Table 4.18
Siemens position switches in compliance with IEC 947-5-1, section 3

Actuator	Housing type	Snap-action contacts	Creep-action contacts
Simple plunger	Open type	3SE3 020-1A	3SE3 020-0A
	Metal	3SE3 120-1B	3SE3 120-0B
Overtravel plunger	Metal	3SE3 120-1C	3SE3 120-0C
	Moulded plastic	3SE3 200-1C 210-1C 220-1C	3SE3 200-0C 210-0C 220-0C
Roller plunger	Metal	3SE3 120-1D	3SE3 120-0D
	Moulded plastic	3SE3 200-1D 210-1D 220-1D	3SE3 200-0D 210-0D 220-0D
Roller lever	Metal	3SE3 120-1E	3SE3 120-0E
	Moulded plastic	3SE3 200-1E 210-1E 220-1E	3SE3 200-0E 210-0E 220-0E
Angular roller lever	Metal	3SE3 120-1F	3SE3 120-0F
	Moulded plastic	3SE3 200-1F 210-1F 220-1F	3SE3 200-0F 210-0F 220-0F
Roller crank, 4 positions at 90°	Metal	3SE3 120-1J	3SE3 120-0J
	Moulded plastic	3SE3 200-1G 210-1G 220-1G	3SE3 200-0G 210-0G 220-0G
Overtravel plunger, roller plunger with central fixing	Moulded plastic	3SE3 200-1L 210-1L 220-1L 3SE3 200-1M 210-1M 220-1M	3SE3 200-0L 210-0L 220-0L 3SE3 200-0M 210-0M 220-0M
With separate actuator from the side end-on side	Moulded plastic	–	3SE3 200-0XB 3SE3 200-0XD
	Metal	–	3SE3 120-0XB

Position switches with safety function in terms of DIN VDE 0660, part 206 may carry the symbol: ⊙↔

This symbol is, however, being superseded by the international IEC symbol.

5 Installation, operation and maintenance of low-voltage switchgear

5.1 Installation

5.1.1 Mounting aids

A number of mounting aids are offered which greatly facilitate both the mounting and fixing of the switching devices:

▷ *Snap-mounting*, which permits the rapid and simple mounting of the devices on mounting rails (Figure 5.1a), ensures secure and reliable fixing

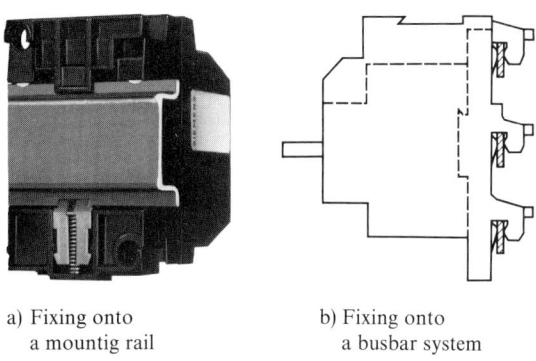

a) Fixing onto a mountig rail

b) Fixing onto a busbar system

Figure 5.1   Switching devices for snap-mounting

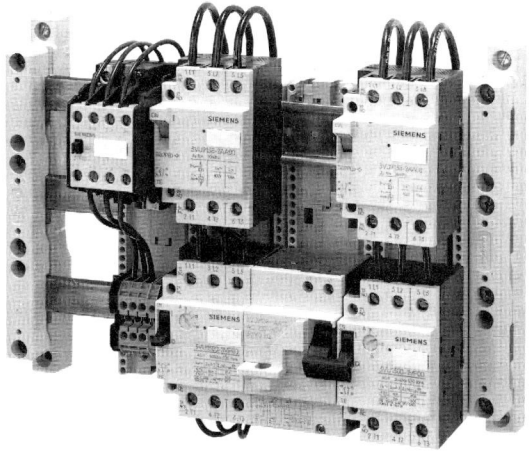

even for long periods of operation (for smaller devices: 35 mm wide mounting rails to DIN EN 50 022; for larger devices 75 mm wide mounting rails to DIN EN 50 023).

▷ *Adapters for busbar systems* permit items of switchgear to be mounted directly onto busbars and allow optimal space-saving (Figure 5.1b). Thereby, the mechanical fixing of the device simultaneously provides the electrical connection. The adapters are either designed to carry specific switching devices (e.g. adapters for 3VU13 and 3VE3 circuit-breakers) or are integrated in the product design (e.g. 3NP4 fuse switch disconnectors).

Alternatively, adapters are offered as empty carriers complete with one or two 35 mm mounting rails for the mounting of other standard equipment. The adapters are particularly suited for motor feeder circuits without fuses and are then typically equipped with a circuit-breaker and a contactor. The electrical connection between the busbars and the switchgear is accomplished by means of appropriately prepared connecting leads (Figure 5.2). Further details may be found in the Siemens l.v. switchgear catalogues.

▷ *Plug-in systems.* Here, the plug-in base or socket is mounted and wired in the switchboard. The circuit-breaker (e.g. MCCB types 3 V.5 to 3 V.7), equipped with appropriate blade connectors, may then be inserted from the front. A tripping pin protruding from the back of the breaker ensures that the unit can only be inserted or withdrawn in the off or tripped state (Figure 5.3). If an attempt is made to withdraw the breaker while it is switched on, the release of the spring-charged tripping pin will cause the breaker to trip before the blade connectors leave their sockets.

Figure 5.2
Mounting of switchgear on busbars by means of an adapter:
11 kW motor feeder without fuses, comprising 3VU13 circuit-breaker with remote control unit and limiter

▷ *Withdrawable design* for 3WN and 3WE air circuit-breakers.

Withdrawable air circuit-breakers (e.g. 3WN1 in Figure 5.4) comprise the circuit-breaker itself and a guide frame or cradle.

The guide frame contains socket contacts for both the main and auxiliary connections, the spindle drive for moving the breaker back and forth and a number of position switches for indicating the position of the breaker within the frame.

The withdrawable air circuit-breaker itself is equipped with blade contacts for the connection and disconnection of the main current paths and with automatically locating plug connectors for the auxiliary circuits.

▷ Mechanical latching prevents the circuit-breaker from being moved within the guide frame while the main contacts are closed.

The 3WN withdrawable air circuit-breaker may be moved to three distinct positions within the frame:

– Connected, or operating position
(main plug-in contacts connected, auxiliary plug-in contacts connected)

– Test position
(main plug-in contacts disconnected, auxiliary plug-in contacts connected)

– Disconnected, or isolated position
(main and auxiliary plug-in contacts disconnected)

A fourth position is available for inspection and maintenance purposes. In this position, the circuit-breaker is withdrawn completely from the frame and rests on supporting cantilevered rails outside the front of the switchboard cubicle.

5.1.2 Mounting position

During installation of switchgear, the permissible mounting position must be taken into account. The permissible angular deviation from the perpendicular to the vertical mounting surface is given in the catalogues (Figure 5.5). Larger deviations can cause a reduction in the rated operating current, the voltage

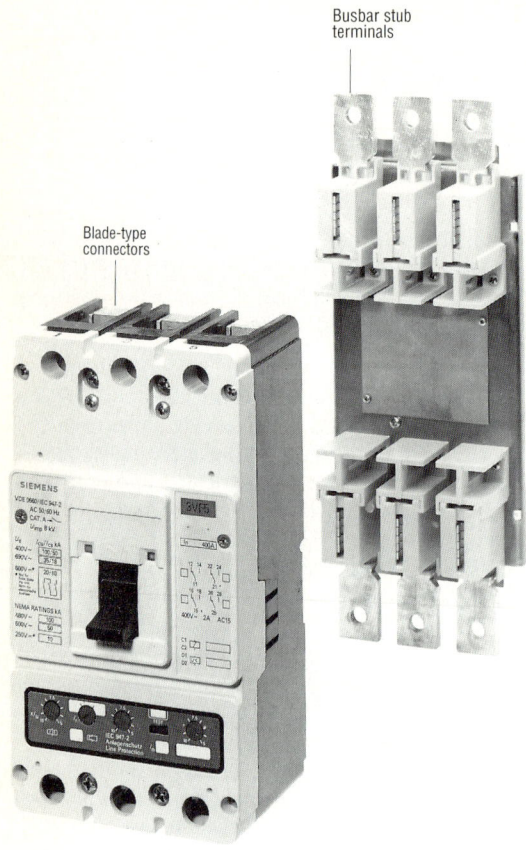

Figure 5.3 Plug-in circuit-breaker with socket

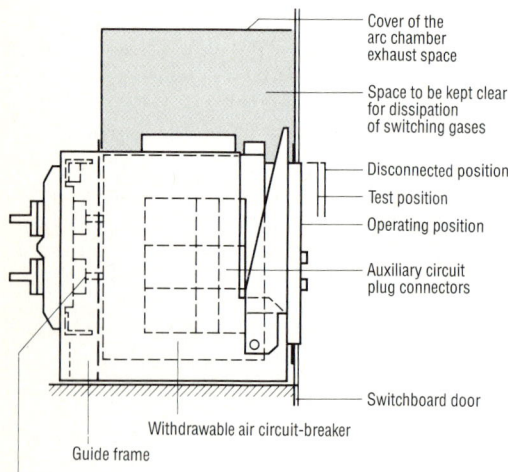

Figure 5.4
3WN1 air circuit-breaker in withdrawable design

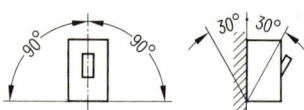

Figure 5.5
An example of the permissible mounting position for a circuit-breaker

tolerance of magnet coils, the switching capacity, the electrical and mechanical service life and/or a change in the tripping characteristics.

5.1.3 Clearance for switching-arc gases

Switching arcs are created whenever electrical current is interrupted. Their intensity is dependent on the switching power, i.e. the voltage and the current values. These switching arcs are quenched and extinguished within the arc chamber of the switching device. The ionized gases, which are released during this quenching procedure, are expelled with a certain pressure from the arc chamber and switching device.

To prevent these gases from causing flash-over, there must be a minimum clearance between the exhaust openings of the device and current carrying or earthed parts. Albeit smaller, a minimum clearance to insulated parts, e.g. insulating barriers, must also be provided to allow the unrestricted dispersion of the ionized gases.

The minimum dimensions of the space required for the safe dissipation of switching-arc gases are given in the respective Siemens low-voltage switchgear catalogues (see Figures 5.3 and 5.6). In some cases, this clearance may depend on the rated operational voltage.

If necessary, additional free space may be required for the unrestricted removal of the arc chambers once the device has been installed.

If busbars or items of switchgear are located above or behind the switching device, and if these are live after a current interruption, then the arc chamber exhaust space should be enclosed to prevent the escape of ionized gases to these areas (see Figure 5.4).

5.2 Termination

5.2.1 SIGUT termination technique

The Siemens SIGUT termination technique offers a number of advantages for the wiring of low-voltage switchgear in auxiliary circuits as well as for the termination of power circuit conductors in the case of switching devices with lower current ratings (Figure 5.7).

The SIGUT termination technique offers:

▷ Reliable mechanical and electrical connection through the use of terminal screws with *crimping washers* or *box terminals*. The functional reliability is increased by ridges on the surface of the curved elastic crimp washers, and by corrugations inside the box terminals which fix the conductor securely within the terminal. The conductor may be solid, stranded or finely stranded and may be with or without end sleeve to DIN 46228.

▷ *Pre-opened terminals.* The wire end need only be inserted and the terminal screw tightened.

▷ *Captive screws.* Although this may appear to be a minor detail, it saves unnecessary irritation. Also, the screws of unused terminals do not have to be tightened to prevent them from falling out at a later stage.

▷ *Funnel-shaped wire entries.* Clearly partitioned terminals, wire guides and end-stops ensure easy and problem-free insertion of the wire into the terminal to the correct depth.

▷ *Screwdriver guides* permit the use of power screwdrivers. These not only offer a degree of working comfort, but also ensure that all terminal screws

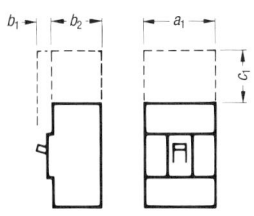

Type	a_1	b_1	b_2	c_1
3VF3	105	15	86	86
3VF4	105	17	104	127
3VF5	140	18	104	164
3VF6	210	33	104	156

Minimum distances to neighbouring earthed and non-insulated live parts. Bare conductors must be insulated if they extend to within the arc chamber exhaust space.

Figure 5.6
The minimum clearance for switching-arc gases (arc space) required for a circuit-breaker
(Example: 3VF circuit-breakers)

Figure 5.7
Switching device showing the SIGUT termination technique

are tightened to the same torque, even after several consecutive tightening operations.

▷ *Unambiguous terminal and other markings* contribute significantly to the prevention of wiring errors. All the terminals are provided with clearly visible position and function numbers which ensure positive identification and correlation to wiring diagrams. The devices themselves are equipped with snap-on device labels.

▷ *Protection against accidental touching of live parts.* The terminals and terminal screws are recessed within the contour of the device housing. Thus, the devices fulfill the conditions for finger-touch and back-of-hand protection in terms of DIN VDE 0106 part 100 and the additional demands of the trade associations for electrical installations and equipment (VBG4) in the simplest way (also refer to Section 1.4.1.2 on page 34).

Finger-touch protection

Electrical equipment must be "finger-touch protected" (Figure 5.8) if located in the immediate vicinity of an actuating device or element (30 mm distance). It must not be possible to bring a straight test finger into contact with parts which may be dangerous to touch (also see Section 1.4.1.2, page 34).

Back-of-hand protection

In addition to the finger-touch protection, devices located at a further distance from an actuating device (30 to 100 mm distance) must fulfil the conditions for "back-of-hand protection". It must not be possible to bring a sphere of 50 mm diameter into contact with parts which may be dangerous to touch (Figure 5.9). This also applies to devices which are mounted in the door of a switchboard or control cubicle (also see Section 1.4.1.2, page 34).

Protection area

In terms of DIN VDE 0106 part 100, the protection area is that area into which the operator must reach in order to operate an actuating device. This area must be kept free of parts which may be dangerous to touch. If such parts extend to within this protection area, then they must at least fulfil the conditions for back-of-hand protection. The shape of this protection area was dictated by human ergonomics, and is dependent on the stance and position of the operator (standing or kneeling), as well as the position, shape and size of the actuating device (Figure 5.10).

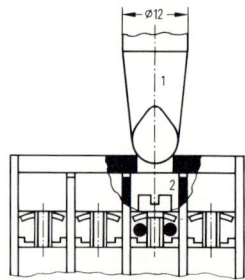

1 Test finger to DIN VDE 0470 part 1
2 Part which may be dangerous to touch

Figure 5.8 Finger-touch protection

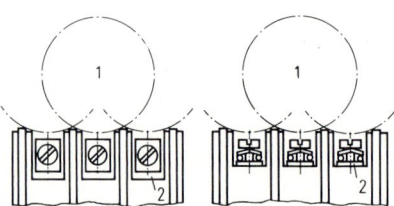

1 Sphere with 50 mm diameter
2 Part which may be dangerous to touch

Figure 5.9 Back-of-hand protection

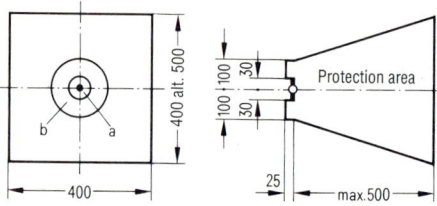

Figure 5.10
Finger (a) and back-of-hand (b) protection as well as the protection area for push actuation (dimensions in mm)

Switchgear with SIGUT terminals, and those with touch-proof box terminals may be installed within the protection area without the need for additional terminal covers or shrouds.

5.2.2 Tab connectors

Pre-cut and prepared wires with tab connectors may be used for control conductors and power conductors of smaller current ratings. These facilitate rapid termination without the need for special tools (Figure 5.11). Depending on the device design, each terminal can usually accept either one B6.3 or two B2.8 tab connectors to DIN 46 247 respectively.

Figure 5.11
3TH2 Contactor relay with 2.8 mm × 0.8 mm tab connectors

Guides for the B2.8 tab connectors simplify plugging-in of the wires and ensure a strong and durable bond between the plug and the sleeve.

Connectors which are accessible from the front permit a smaller spacing between rows of switchgear, e.g. between rows of contactor relays.

All potentially live parts of the devices are protected against accidental touching. The tab connections are recessed in the device housing. If insulating sleeves are used, the devices are finger-touch protected to DIN VDE 0106 part 100.

5.2.3 Box terminals

In the case of Siemens switchgear for higher rated operational currents, box terminals offer the same ease of termination and touch protection as the SIGUT termination technique – without the need for additional covers or shrouds.

In comparison to busbar connectors, the use of box terminals means a saving of cable lugs and termination time. As an example, the box terminals for the main power connections on the 3TF44 to 3TF50 contactors are each suitable for the termination of two bare copper wire ends which may also be of differing cross-section.

The 3V. circuit-breakers too offer the possibility of connecting bare copper wires without the need for crimping or soldering. In addition, this range offers the option of either busbar connector pieces or rear-mounting stud terminals.

5.3 Operation

The operation of switching devices can be divided into two groups:

▷ Manual operation,
▷ Electrical or powered operation
e.g. magnetic operation in the case of contactors; motorized, motorized stored energy and magnetic operation in the case of circuit-breakers

5.3.1 Manual operation

This group includes pushbuttons, toggle levers, twist levers or handles and rotary operating handle mechanisms. For larger isolators and knife switches, type 3K., spade handles and linkage lever or switch stick levers are offered.

The direction of actuation, grouping of the effects of actuation and the arrangement of actuating parts to IEC 337−2 and DIN 30 600 are illustrated in the Tables 5.1 to 5.4.

Rotary operating handle mechanisms

Rotary operating handle mechanisms provide

▷ a detachable connection to the switching device (door coupling operating mechanisms)
▷ a locking facility, switch position indication and colour combinations to DIN VDE 0113 part 1,
▷ the compliance with special specifications for motor control center (MCC) operating handles.

For further details on rotary operating handle mechanisms, please refer to the Siemens l.v. switchgear catalogues.

Direction of operation and arrangement of actuators to IEC and DIN

The operation of an actuator on an item of switchgear should cause the desired action quickly and unambiguously. For this purpose, the IEC Publications 447, 1974 and DIN 43 602, July 1975 specify:

▷ a number of effect groups to categorize the required final result of the actuation (Table 5.1),
▷ the direction of actuation and the effect produced by the actuation (Table 5.2),
▷ the arrangement of two actuators (e.g. pushbuttons) one above the other or side by side (Table 5.3),
▷ the location of the stop position in the case of several actuators arranged one above the other or side by side (Table 5.4).

Table 5.1 Effect groups

Effect Group 1 (increase effect)	Effect Group 2 (decrease effect)
Change of condition, e.g. activate start accelerate close an electric circuit ignite	Change of condition, e.g. deactivate stop brake open an electric circuit extinguish
Movement of an object or vehicle, e.g. upwards, to the right, forwards	Movement of an object or vehicle, e.g. downwards, to the left, backwards

Table 5.3 Arrangement of two actuators (e.g. pushbuttons)[1]

Arrangement of the actuators	Effect Group 1 increase effect		Effect Group 2 decrease effect	
One above the other	● ○	Actuation of the upper device	○ ●	Actuation of the lower device
Side by side	○ ●	Actuation of the right-hand device	● ○	Actuation of the right-hand device

[1] Also refer to the colour codes for pushbuttons in Section 1.5.5, page 62

Table 5.2 Direction of rotary actuation and linear movements

Type of move-ment	Type of actu-ator	Direction of actuation for			
		Effect Group 1 (switch-on, increase effect, etc.)		Effect Group 2 (switch-off, decrease effect, etc.)	
Turning movement	hand wheel, twist knob, rotary handle	↷	clock-wise	↶	anti-clock-wise
Vertical movement	lever, handle	↑	upwards	↓	down-wards
Horizontal movement	lever, handle	↗	away from the operator (push or draw)	↙	towards the oper-ator (pull or draw)
		→	to the right	←	to the left

Table 5.4 Arrangement of three actuators (e.g. pushbuttons)[1]

Effect	One above the other		Side by side
One direction of move-ment	○ ○ ○	Start II (e.g. fast) Start I (e.g. slow) Stop	○ ○ ○ Stop Start I Start II
Two directions of move-ment	○ ○ ○	Raise Stop Lower	○ ○ ○ Left Stop Right

[1] Also refer to the colour codes for pushbuttons in Section 1.5.5, page 62

5.3.2 Powered operation

Magnetic operation of contactors

Reliable voltage levels for operation

In terms of DIN VDE 0660 part 102, contactors must switch reliably in the following ranges:

▷ Closing (pull-in) between 0.85 and $1.1 \cdot U_s$ [1]
▷ Opening (drop-out) between 0.10 and $0.75 \cdot U_s$ [1]

Siemens 3TF, 3TK and 3TB contactors (and 3TH contactor relays) offer a substantially greater independence from voltage variations. The coil voltage tolerances for these contactors are given below:

Contactor size	Magnet system	A.C.	D.C.
00–14	Closes	0.8 to $1.1 \cdot U_s$	0.8 to $1.1 \cdot U_s$
00	Opens	0.2 to $0.5 \cdot U_s$	0.13 to $0.3 \cdot U_s$
0–14	Opens	0.4 to $0.65 \cdot U_s$	0.15 to $0.35 \cdot U_s$

Snap-action behaviour

The specification NFC 63–110[2] requires that the functional limits be documented. Among other stipulations, these requirements include the following:

▷ It must be shown that the contactor pulls in reliably if a voltage which increases steadily at $0.2 \cdot U_S/s$ is applied to the coil terminals. In this test, there may be no noticeable hesitation or delay at the instant the main contacts close.

▷ In addition, it must be shown that the contactor drops out reliably and fully if the voltage applied to the coil terminals is decreased steadily at $0.2 \cdot U_S/s$. In this test, the contact pressure between the main contacts may not drop significantly before the contactor drops out rapidly and completely.

Both these tests are to be done under current-free conditions.

The Siemens contactors fulfill these conditions, and therefor possess a snap-action operation characteristic.

Motorized, motorized stored energy and magnetic operation

Motorized and magnetic operating mechanisms are offered for remote switching purposes. Motorized operating mechanisms with stored energy serve for remote switching and synchronization.

---
[1] U_S Rated control supply voltage
[2] NFC 63-110: Normes francaises, appareillage industriel à basse tension

▷ *Operating mechanisms without stored energy:*
Motorized operating mechanisms for switching on and off (typically for 3 V. circuit-breakers (MCCB types) as well as for 3KE disconnectors and 3KV/3KW knife switches).
Motorized operating mechanisms for switching on only (typically for 3WE, 3WF and 3WV air circuit-breakers). Switching off is accomplished by means of a shunt or undervoltage release.
Magnetic operating mechanisms for switching on and off (typically for 3VU13 and 3VF circuit-breakers). Switching off may also be accomplished by means of a shunt or undervoltage release (e.g. with 3V. circuit-breakers).

▷ *Operating mechanisms with stored energy and energy release unit (3WN, 3WE, 3WS):*
– motor for charging the closing springs. Switching on from remote station via energy release solenoid,
– switching off by means of a shunt or undervoltage release,
Circuit-breakers with motorized stored energy operating mechanisms and energy release units may be used for synchronization of two supply voltages (high speed).

These operating mechanisms are all also provided with a manual means of operation for use in cases of auxiliary power loss. Examples include switching handles, manual charging pump or crank handles for closing springs, manual release units, etc.

5.4 Measures to facilitate the checking/replacement of parts and maintenance work

The maintenance of switchgear may be kept to a minimum by the correct selection of the devices in terms of the mechanical and electrical service life demanded by the operating conditions.

In general, the electrical service life will be shorter than the mechanical service life. This is particularly true for contactors and circuit-breakers. When economically justifiable, the design of the switching device therefor must permit the replacement of contact pieces and arc chambers. Hereby, emphasis is placed on quick and simple replacement. Often, it is merely a question of economics as to whether a contactor should be selected to have an expected electrical service life corresponding to its rated mechanical service life, or whether repeated replacement of the contacts as they become worn is not a more viable solution (see Sections 3.3.3, 3.7.3 and 3.7.7 on pages 118, 187 and 197 resp.).

In the case of 3WN and 3WE air circuit-breakers it is recommended that the condition of the contacts and the arc chambers (arc chutes) be inspected after every short-circuit interruption to determine their continued serviceability. The necessary guidelines are given in the operating instructions.

The following measures are examples of design features which greatly facilitate checking and/or replacement of parts as well as other maintenance work:

▷ Integrated snap-mounting facilities;
▷ Snap-on auxiliary contact blocks in the case of contactors and contactor relays;
▷ Plug-in versions;
▷ Device labels which permit the clear identification of the devices and e.g. their association with specific circuits, particularly in the case of complicated control systems;
▷ Test button on electronic overload relays and thermistor tripping units;
▷ Withdrawable versions with test, disconnected and maintenance positions;
▷ Functional testing of the electronic overload release module in 3WN air circuit-breakers by means of a test unit without the need for removing the breaker from the switchboard;
▷ Indicating the condition of HRC fuses installed in 3NP fuse switch disconnectors by means integrated fuse monitoring.

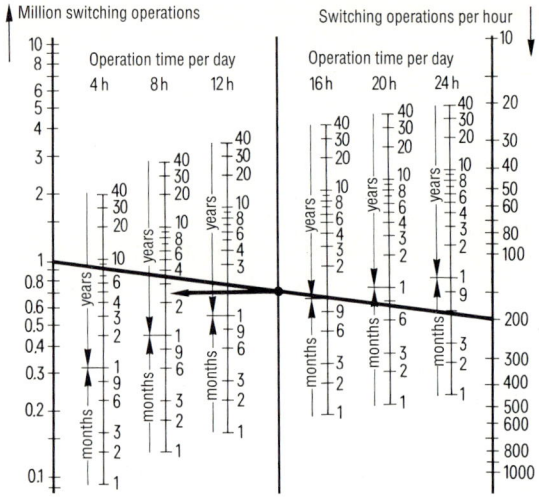

Example

Operating requirements:	Service life of 1 million operations, 200 switching operations per hour, 8 hours of operation per day
Procedure:	Draw a line from 1 million (switching operations) on the left-hand scale to 200 (switching operations per hour) on the right-hand scale. Now draw a horizontal line from the point at which the first line intersects the center scale towards the left to the 8 hour axis.
Result:	Total operating time approx. $2^1/_2$ years

Figure 5.12
Nomogram for determining the service life of contacts with an example based on 250 working days with operation for 4, 8, 12, 16, 20 and 24 hours (also refer to Section 9.8). If the no. of working days is 365 instead of 250, then the total operating time found from the nomogram must be multiplied by 0.68.

5.5 Checking the condition of contact pieces in 3TF contactors; assessment criteria

Based on the prevailing operating conditions, the expected contact service life may be estimated in months and years by using the chart in Figure 5.12. Alternatively, Siemens offers a software package (known as "ELLE") for AT-compatible personal computers by means of which a more accurate calculation of the contact service life and the selection of the contactor may be done.

The program ELLE permits the selection of contactors in terms of the power to be switched, the utilization category and the service life. The user is guided through the program by a series of dialog screens so that a particular knowledge of computing is not required. Before data is entered, the user is presented with the option of printing a data-entry form in which the program sequence and the required operating data is listed.

The program produces a list of suitable contactors for the particular application. For each contactor, it states the calculated electrical service life in no. of operations, years and months, as well as the number of contact sets required. This list may be labelled with a reference or key word and the date before it is printed out.

The program allows the simple and quick selection of contactors for the widest range of applications. (ELLE, Version 2 – Order No. E86010–D1802–A117–A2, multilingual)

From time to time, routine inspections should be carried out since a number of factors could cause the service life achieved in practice to be longer or shorter than the value determined by theoretical calculation.

These factors which influence the contact service life do not only include tolerance differences, but may also be due to operating conditions which could not be predicted prior to the theoretical calculation.

Inspection may be carried out during shut-down periods. It significantly contributes to the operational reliability of the installation as a whole and can prevent expensive forced shut-downs.

Visual inspection is also necessary after the occurrence of a fault, e.g. a short-circuit. Here, attention is drawn to the fact that DIN VDE 0660 and IEC 947-4-1 permits welding of contactor contacts during short-circuit conditions. See, for example, Section 3.4.4.1 on page 141.

The condition of the contacts in non-energized (and isolated!) Siemens contactors (from size 2 upwards) may be inspected very simply. The arc chamber is removed by means of operating two quick-release fasteners. In the case of 3TF contactors from size 3 upwards, removal of the arc chamber causes the contactor to latch mechanically in the open position as a safety precaution. The disconnection of cable terminals is not required. Once the arc chamber has been removed, the contacts are clearly visible so that it may be easily determined whether their replacement is necessary (Figure 5.13).

Figure 5.13
Simple replacement of the contacts in a 3TF contactor

Maintenance personnel should be in a position to judge the condition of the contacts and to decide whether they should be replaced or not. It would, for example, be incorrect to replace contacts because the contact pieces have merely become rough and discoloured due to the effect of switching arcs. This is an absolutely normal occurrence and replacement of such fully functional contacts would only cause unnecessary expense.

Whether a contact piece is still good for further service or not, practically only depends on the volume of remaining contact material.

6 Transducing sensors and signal processing systems

Numerous sensing, evaluation and recognition tasks which are carried out in modern fully automatic manufacturing and processing lines require suitable transducers and signal processing systems to provide the control system with the necessary signals and data. Mechanically operated position switches (limit switches) can only be used if the object to be sensed is brought into physical contact with the actuator (see Section 4.3.6, on page 266).

A great number of manufacturing processes, conditions and sequences however, prohibit the use of sensors or position switches which require physical contact with the object or material to be monitored. This may, for example, be due to the consistency or nature of the object to be sensed, its distance from the sensing device, its high velocity or the lack of sufficient actuating force. In some cases, it may even be necessary to distinguish between and identify materials and

Table 6.1 Overview of transducing sensors or sensor systems for the detection and evaluation of objects and procedures

Transducing sensor or sensor system	Operating principle	Sensing of objects			Signal	Signal processing with		
		type of material	maximum distance	operating method		contactor control system	electronic control system	process computer
Position switch	mechanical actuation	solid	—	direct contact	binary	×	×	—
Inductive BERO proximity switch	change of an electromagnetic field	metallic	60 mm with steel	by proximity	binary	×	×	—
Capacitive BERO proximity switch	change of an electrostatic field	any	35 mm with earthed steel	by proximity	binary	×	×	—
Opto-BERO photoelectric proximity switch	change between light and dark (light intensity)	any, provided it blocks or reflects light	15 m	by proximity	binary	×	×	—
Sonar-BERO ultrasonic proximity switch	echo time measurement of pulsed ultra-sound waves	any, provided sufficiently sound reflectant	8 m	by proximity	binary, digital or analog	×	×	—
OPTOMAT	anamorphotic imaging (simplified image capturing with photo-electric diodes)	solid	0,5 m	by proximity	digital	—	×	×
VIDEOMAT	picture capturing with TV camera	solid	20 m	by proximity	digital	—	—	×

shapes of objects. In such cases, sensing and/or measurement must be done without the need for contact with the object being sensed.

The sensors, detection systems and proximity switches used to meet these requirements vary significantly in their operating principles. Table 6.1 gives a general overview of some of these transducing sensors and detection systems.

For simple tasks, mechanical position switches and electronic proximity switches are predominantly used. Owing to their more complex operating principles, systems such as OPTOMAT and VIDEOMAT are generally reserved for more demanding applications.

The VIDEOMAT system, for example, which uses a television camera for picture registration is intended for the capturing and processing of large amounts of data. A detailed description of its operation and application falls outside the scope of this book.

6.1 Selection criteria for BERO proximity switches

Position switches which operate without touching the object to be sensed, and which do not have mechanical contacts are termed "proximity switches" in the relevant standards and specifications. The Siemens BERO[1] proximity switches are available in inductive, capacitive, photoelectric and ultrasonic versions. The applicable definitions, classifications, characteristics, performance requirements and verification tests for all these types of proximity switches are laid down in the IEC 947-5-2 Standard, first edition 08.1992 (DIN VDE 0660 part 208).

6.1.1 Inductive and capacitive proximity switches for operating distances up to 65 mm

Inductive and capacitive BERO proximity switches differ in their operating principles. Inductive proximity switches are used to sense metal objects, whereas capacitive proximity switches may be used to sense any materials in solid, liquid or powder form.

A BERO proximity switch operates (change of signal at its output) when the object (or target) is brought to within a specific distance from its active surface, or sensing face. Refer to the explanation of "operating distances" in the Appendix, page 607.

The beneficial properties of BERO proximity switches include:

▷ actuation without physical contact,
▷ a high switching frequency,
▷ extended operating distances,
▷ consistent repeat accuracy
▷ a practically unlimited service life which is independent of the switching frequency,
▷ high operating voltage tolerances,
▷ versions for a.c./d.c. operation,
▷ high output ratings,
▷ integrated protection features,
▷ fully suited to industrial use (e.g. high level of EMC),
▷ state of the art semiconductor technology (hybrid technology).

The Siemens direct-current and a.c./d.c. inductive BERO proximity switches can be divided into six groups:

BERO for normal requirements
– operating voltage range DC 10 to 30 V,
– 3 or 4 wire versions,
– pnp or npn output.

BERO optimized for PLC use
– operating voltage range DC 10 to 30 V,
– 2 wire version,
– output specially suited for connection to PLC's.

BERO for more severe operating conditions
– extended operating voltage range DC 10 to 65 V,
– 2 wire a.c./d.c. version for up to AC 265 V alt. DC 320 V,
– extremely high switching frequency (up to 10 kHz).

BERO for extremely harsh environmental conditions (IP 68)
– 2 wire a.c./d.c. version, UC 20 to 320 V,
– 3 or 4 wire version, DC 10 to 30 V,
– 3 wire version with extended operating voltage range DC 10 to 65 V.

BERO with extended operating distance
– permits the use of smaller sizes (e.g. M18 replaces M12).

NAMUR[2] BERO
– intrinsically safe to DIN 19 234 [EEx ia] IIC,
– for use in hazardous areas (potentially explosive atmospheres).

---

[1] BERO is the German abbreviation for "Berührungsloser Endtaster mit rückgekoppeltem Oszillator", which means "Proximity position switch with feedback oscillator". This is a Siemens registered trade name.
[2] Normenausschuß für Meß- und Regelungstechnik (≈ Standards committee for measurement and control)

6 Transducing sensors and signal processing systems

Table 6.2 Operating distances and switching frequencies of inductive and capacitive BERO proximity switches

Model (examples)		Maximum operating distance mm	Maximum switching frequency	
			d.c. version Hz	a.c. version Hz

Inductive BERO

	Cylindrical with fixing thread	35	10 000	25
	Cylindrical, smooth, with small dimensions	3	2 000	–
	Cubic, plug-in type, active face adjustable to 5 directions	40	1 000	25
	Cubic	65	100	25
	Cylindrical, smooth	25	50	3
	Mini cylinder (or "button")	5	100	–
	Cubic, with central fixing	7	5 000	25
	Micro-switch	2	500	–

Capacitive BERO

	Cylindrical with fixing thread	25	20	3
	Cylindrical, smooth	35	20	3

Selection in terms of operating distance and switching frequency

Table 6.2 provides an overview of the construction forms, operating distances and switching frequencies. In comparison to the BERO switches for direct-current applications, the a.c. versions operate at a lower operating frequency which corresponds to the permissible operating frequency of contactor relays.

The highest switching frequency is achieved if the surface area of the target (of ST37 steel) corresponds in size and shape to the surface area of the active surface (or sensing face) of the BERO, and the actua-

Table 6.3 Task and application examples for inductive and capacitive BERO proximity switches

Application examples	Task	Advantageous properties
Sorting machines Transport systems	Positioning	High repeat accuracy (e.g. 0.01 mm in the case of M 8 flush-mounting version)
	Counting	High switching frequency (to 10000 Hz)

Machine parts onto which (e.g. for technical reasons) special actuating elements (cams, switching bars) cannot be fitted
Machine parts in which the distance to the BERO may vary within certain limits
  Automatic packaging machines,
  Conveyer systems,
  Doors, gates, screens,
  Actuator drives,
  Palletizing machines,
  Hoists and elevators
	Proximity sensing	Actuation without touching

Digital counting applications Stand-still monitoring Monitoring direction of rotation (two BERO's required)	Pulse generation	Short switching times (0.03 to 0.9 ms)
Scanning of pointer positions in instruments which do not produce sufficient force for actuation by mechanical means Scanning of stations or units on guided conveyer systems (coding systems)	Scanning	Actuation without touching
Feed of tape or strip material into automatic manufacturing machines or presses Feeding of parts into assembly machines	Supervision	High traverse speeds up to 150 m/s

6 Transducing sensors and signal processing systems

tion distance is equal to $0.5 \cdot s_n$[1] (to DIN EN 50010; also refer to the Siemens l.v. switchgear catalogue).

Selection in terms of the functional task and application

Table 6.3 illustrates a number of task and application examples for inductive and capacitive BERO proximity switches.

Owing to the extremely large number of application constraints and parameters to be considered in the selection of inductive and capacitive proximity switches, such as

- availabe space, degree of protection required,
- surrounding material, ambient temperature, possible presence of aggressive chemicals, extent and nature of shock and/or vibrations,
- material, shape, size and approach of the target,
- function and position in the control circuit,
- danger of electromagnetic interference (e.g. welding machines, h.v. motors),
- connection method (e.g. terminals or plug-in),
- nature of the load to be switched, etc.,

a selection cannot always be made in terms of the operating distance, switching frequency and available auxiliary supply voltage only. Extensive technical data is given in the selection tables of the Siemens low-voltage switchgear catalogues. Several of the terms used are defined in the Appendix, see Section 9.12 from page 556 onward.

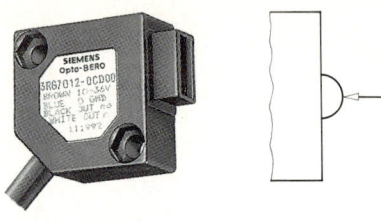

a) Type T: Through-beam

 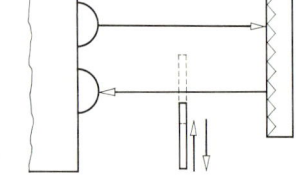

b) Type R: Retroreflective

 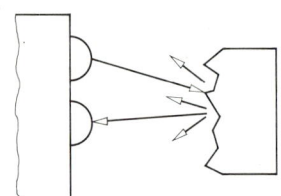

c) Type D: Diffuse-reflective

Figure 6.1 Types of photoelectric proximity switches

6.1.2 Opto-BERO photoelectric proximity switches

These proximity switches sense the presence of an object which either reflects or interrupts visible or invisible (infra-red) light. They incorporate a semiconductor output switching element.

One can distinguish between three types of photoelectric proximity switches (Fig. 6.1):

In types T and R photoelectric proximity switches, the emitter and receiver are usually contained within the same housing. Units with so-called "fibre-optic amplifiers" (Fig. 6.2) may be used to sense objects in confined spaces, in potential-free areas (e.g. potentially explosive atmosphere) or at points in extremely strong electromagnetic fields (e.g. welding machines)

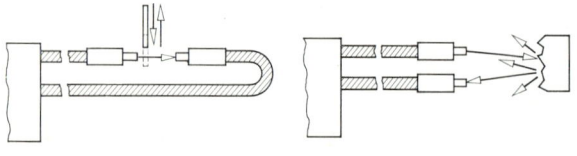

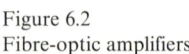

Figure 6.2
Fibre-optic amplifiers

---

[1] s_n Rated operating distance

Dirt or dust build-up on the lenses of the emitter and/or receiver elements can eventually cause malfunction as the intensity of the light to be sensed decreases. Units with the so-called "surplus detection indication" can signal the deterioration of the surplus light impinging on the receiver and thereby give advance warning so that appropriate preventative cleaning or maintenance work may be carried out.

Extensive technical data is given in the Siemens low-voltage switchgear catalogue to assist with the selection of the correct Opto-BERO for the application at hand. Several of the terms used are defined in the Appendix, see Section 9.12 from page 556 onward.

6.1.3 Sonar-BERO ultrasonic proximity switches

The heart of the Sonar-BERO ultrasonic proximity switch is a piezoceramic transducer which sends out ultrasonic impulses. If these are reflected by an object, the BERO receives the echo and can convert it into an electric signal. In this way, BERO ultrasonic proximity switches can sense objects at distances of 6 to 800 cm.

The object to be sensed may be solid, liquid, granular or in powder form. The material may be transparent, of any shape and colour and may have a polished or matt surface. The only criteria is that the material must be able to reflect sound waves to a sufficient degree.

Among others, the devices offer the following advantages:

▷ sensing of even small objects at large distances,
▷ selective sensing of objects by means of adjustable sensing ranges is possible,
▷ selectable background suppression,
▷ simple handling and commissioning,
▷ maintenance-free operation.

The Sonar-BERO switches are available in two modular versions, viz. Modular Range I and II, and in three compact versions known as Compact Ranges I, II and III respectively.

Modular Ranges I and II

The modular Sonar-BERO proximity switches are extremely versatile in their application possibilities and may even be used for complex sensing tasks. To facilitate adaptation to the particular task, these devices are offered in two versions:

Modular Range I

Sonar-BERO proximity switches of the Modular Range I comprise the sonar sensor, the signal evaluation unit and, if required, a number of additional optional modules.

Sonar sensors

The heart of the sonar sensor is the piezoceramic transducer which sends out ultrasonic impulses and receives echoes from the target object.

Three versions are offered (Figure 6.3):

Sensors with a 400 kHz transducer for distances from 6 to 30 cm,

Sensors with a 200 kHz transducer for distances from 20 to 99 cm,

Sensors with an 80 kHz transducer for distances from 80 to 600 cm.

The sensors for distances from 6 to 99 cm have a resolution of 1 cm. The resolution of the sensors for distances from 80 to 600 cm is 10 cm. Figure 6.4 illustrates the definition of the operating distances.

a) Sensing range 6 to 30 cm

b) Sensing range 20 to 99 cm

c) Sensing range 80 to 600 cm

Figure 6.3
Sonar sensors of the Modular Range I Sonar-BERO ultrasonic proximity switches for three sensing distances from 6 to 600 cm

6 Transducing sensors and signal processing systems

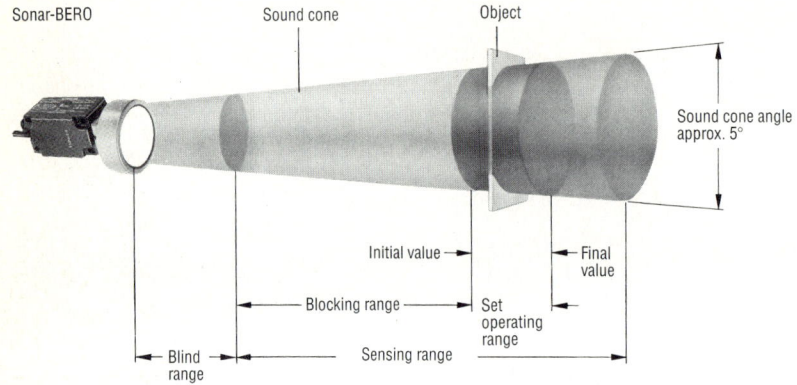

Figure 6.4
Definition of the operating distances of the Modular Range I Sonar-BERO ultrasonic proximity switches

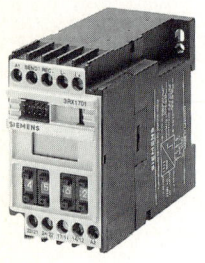   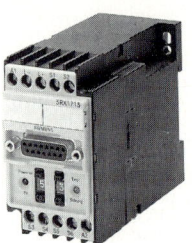

a) Signal evaluation unit   b) Digital output unit   c) Analog output unit   d) Multiplexer unit

Figure 6.5
Signal evaluation unit and additional modules of the Modular Range I Sonar-BERO ultrasonic proximity switches

Signal evaluation unit

The function of this unit is to evaluate and process the signals received from the sensor. If the presence of an object in the selected range is sensed, then a signal is produced (contacts change state) at the output of the signal evaluation unit.

The signal evaluation unit is the same for all three sonar sensors and adapts itself automatically to the particular sensor connected (Figure 6.5a).

Additional modules

The functions of the Sonar-BERO may be extended by the addition of the following modules to the basic configuration:

▷ Digital output: This unit provides the actual distance of the sensed object from the active face of the Sonar-BERO in binary or BCD format for further processing by the PLC or NC system (Figure 6.5b).

▷ Analog output: This unit provides the actual distance of the sensed object from the active face of the Sonar-BERO in analog format (4 to 20 mA) for further processing in an analog control system (Figure 6.5c).

▷ Multiplexer: By means of this unit, 2 to 6 sensors are scanned cyclically. It is particularly used if:
 – the outputs from several sensors are to be combined into a single output signal
 – an interference between sensors located closely together is to be excluded

▷ Monitoring module: This watchdog unit serves to monitor and supervise the correct operation of the Modular Range Sonar-BERO ultrasonic proximity switch circuit.

▷ Network filter module: This unit is included if the supply voltage has an extremely high noise content.

Modular Range II

In addition to the integration of several features of the Modular Range I units, these devices offer a number of additional functions as well as improved properties.

Sonar sensors

The new sonar sensors are supplied with a plug-in temperature probe. By means of a connecting lead, the probe may also be installed separately at some distance from the sensor.

Three groups of sensors are available for the Modular Range II:

– cylindrical (similar to those of the Compact Range, see Fig. 6.7)

– cubic (see Fig. 6.6a)

– spherical (see Fig. 6.6b)

Signal evaluation unit (Figure 6.6c)

The signal evaluation unit contains two switching outputs, an analog output, temperature compensation and a connection for a reference sensor to take other environmental influences on the velocity of sound waves into account. Furthermore, the unit offers the possibility of programming various measurement parameters for the evaluation.

a) Cubic sensor

b) Spherical sensor (0.8 to 8 m) for connection to the Modular Range II signal evaluator unit

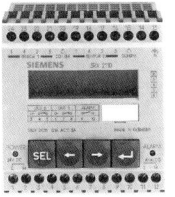
c) Signal evaluator unit

Figure 6.6
Modular Range II
Sonar-BERO ultrasonic proximity switches

a) Sensing range
6 to 30 cm

b) Sensing range
20 to 130 cm

c) Sensing range
80 to 600 cm

Figure 6.7
Compact Sonar-BERO ultrasonic proximity switches with integrated distance evaluation for 3 sensing distances from 6 to 600 cm

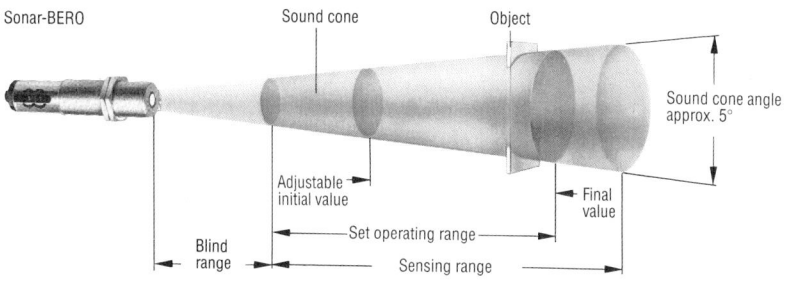

Figure 6.8
Definition of the operating distances of the Compact Range Sonar-BERO ultrasonic proximity switches

6 Transducing sensors and signal processing systems

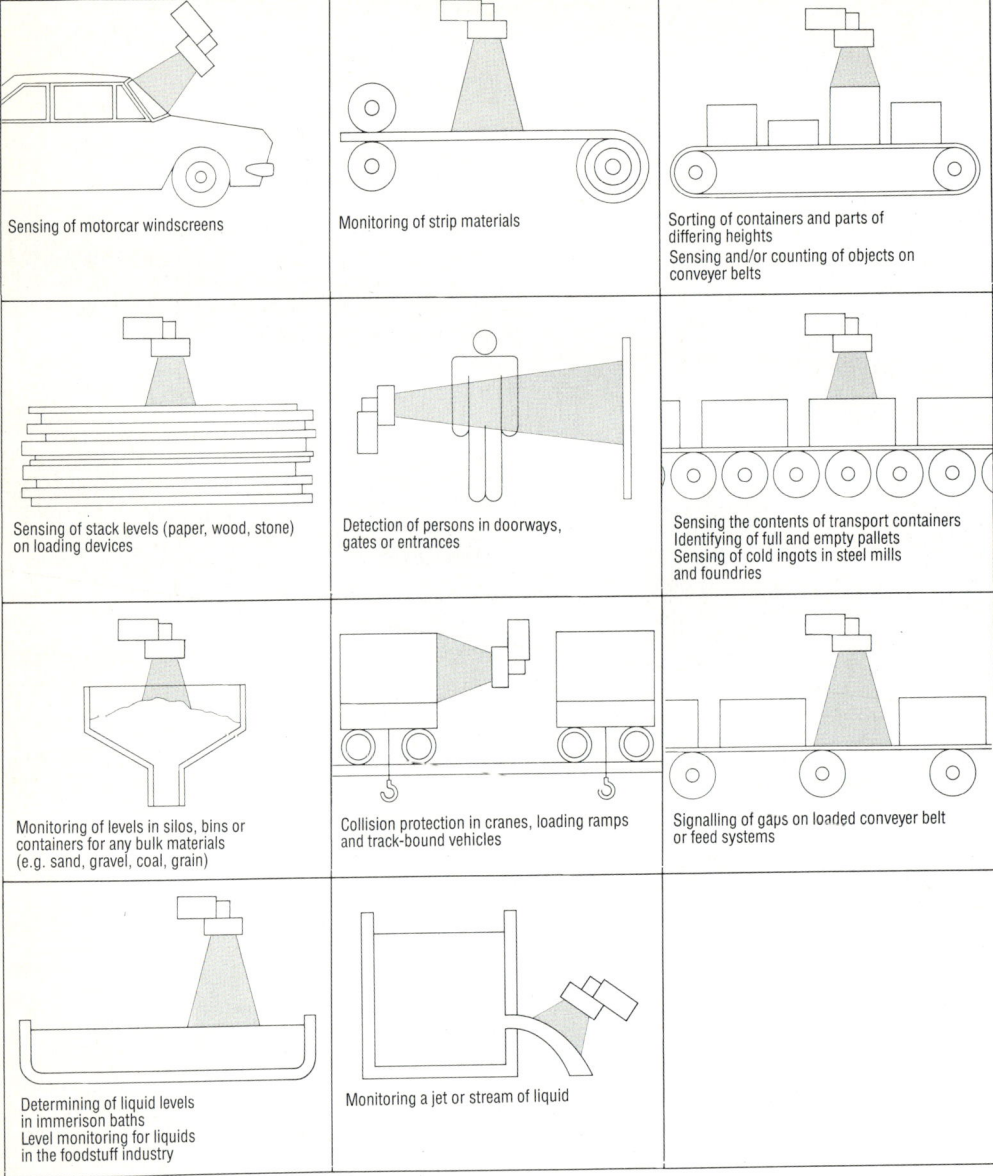

Figure 6.9 Application examples for the Sonar-BERO ultrasonic proximity switches

The resolution has been improved by a factor of 10 in comparison with Modular Range I, and is 1 mm for the two sensors with shorter sensing distance and 1 cm for the sensor with sensing distance up to 8 m.

The modular units are characterized by the following features:

▷ The operating range of the sensing distance in which the object is to be detected, can be selected and adjusted digitally,

▷ The output function is programmable (normally-open or normally-closed)

▷ Operational and fault indication is displayed,

▷ The distance of the object from the active face of the BERO is given in analog or digital format,

▷ Multiplexed operation is possible,

▷ The operation of the sensors can be monitored,

▷ The units may be used in a.c. or d.c. applications.

Compact Ranges I, II and III

These Sonar-BERO ultrasonic proximity switches are complete, compact devices and are simple to install (see Figures 6.7 and 6.8).

The compact ranges are characterized by the following features:

▷ Both the sensor and the evaluating unit are contained within the same cylindrical housing. (The Compact Range I also offers a version with a separate transducer head for installation in confined spaces).

▷ The units are designed for connection to a DC 24 V supply voltage.

▷ The output state is indicated by means of an LED.

▷ The operating range in which the object is sensed can be adjusted by means of two potentiometers (initial and final values). In the case of Compact Range II and III, several parameters may be set by means of the SONPROG PC-program and the 3RX3000 interface unit.

▷ The output is suitable for connection to all common programmable logic control systems.

▷ Sensing ranges up to 600 cm

▷ Multiplexing capabilities

To assist the user with selection of the devices and with the design of his system, Siemens offers a comprehensive manual on 3RG60 and 3RG61 Sonar-BERO ultrasonic proximity switches (Order No. E20001-P285-A358-X-7600).

Figure 6.9 shows a number of application examples for Sonar-BERO ultrasonic proximity switches.

6.2 Electronically compatible control and signalling with low-voltage switchgear

Auxiliary and control circuits of low-voltage switchgear often include electronic devices. As opposed to circuits containing only electromechanical switching devices, such applications place additional demands on the switchgear:

▷ reliable operation of the signal recipient by means of an output from an electronic control system,

▷ adaptation of the rated voltage and operating voltage tolerances of the signal recipient to the system values,

▷ suppression of damaging overvoltages which may have their origin inside or outside of the control system,

▷ high contact reliability of the control and signalling devices (e.g. auxiliary contacts) at low voltages and currents, while maintaining the full switching capacity at higher voltages.

6.2.1 Reliable operation by electronic output stages

In principle, there are three possibilities for operating contactors from an electronic control system:

1. via semiconductor outputs with transistors for the usual system voltage of DC 24 V and for voltages up to a maximum of DC 60 V,
2. via semiconductor outputs with triacs or thyristors for voltages up to AC 230 V and
3. via relay outputs for voltages up to AC or DC 250 V.

In practice, the use of DC 24 V is preferred for reasons based mainly on safety and cost. The signal recipients (e.g. contactors or relays) should therefor be equipped with d.c. magnet systems. In comparison to other methods (a.c. operation, d.c. economy connections), the pure d.c. magnet system has the following advantages:

▷ extremely high mechanical service life ($15 \cdot 10^6$ to $30 \cdot 10^6$ switching operations),

▷ extremely low pull-in current with low rate-of-rise,

▷ pull-in current equal to the hold-in current and therefor no change-over switching required,

▷ low losses (e.g. 1.2 W in interface contactor relays for use with electronic control, which corresponds to 50 mA at DC 24 V),

▷ no humming and low stray magnetic fields.

6 Transducing sensors and signal processing systems

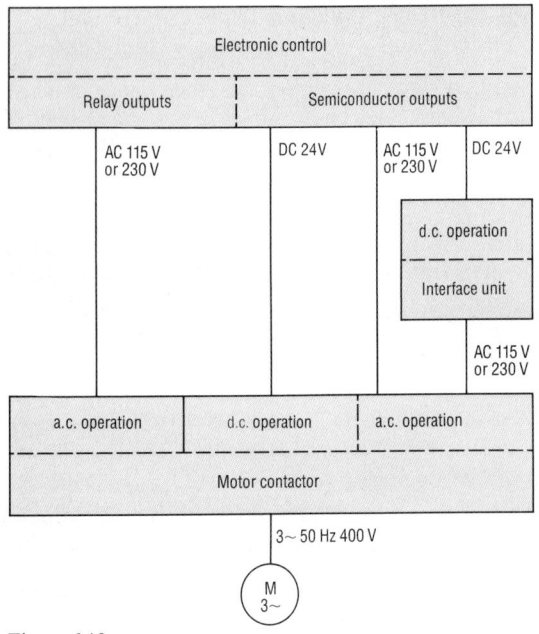

Figure 6.10
Possibilities for the operation of motor contactors in an electronic control system

Operating mechanisms with a power consumption too high for direct connection to electronic output modules may be connected via an interface contactor relay (Figure 6.10).

Newly developed switching devices operate with a wider control supply voltage tolerance and at reduced power consumption, e.g.

▷ 3TX4180-3C: 0.5 W at DC 17 to 30 V,
▷ 3TX70..-.AB00   0.5 W at DC 19 to 30 V,
▷ 3TH20..-0FB4   1.2 W at DC 19 to 30 V.

Larger motor contactors with a.c. magnet systems may be switched via these interface units.

For further technical details, please refer to the Siemens low voltage switchgear catalogues.

6.2.2 Adaptation of the operating voltage tolerances

Figure 6.11 illustrates the usual voltage tolerances for electronic control systems and contactor operation respectively. In terms of DIN 19240, 07.1985, the DC 24 V rated voltage for the power supply of electronic control systems may range from 20.4 to 28.8 V. If allowance is made for an additional voltage drop of up to 3 V within the output stage itself, then it follows that the magnet system of the contactor must operate reliably for voltages ranging from 17.4 to 28.8 V.

Siemens contactors and interface contactor relays for use with electronic controls operate reliably from 17 to 30 V, which corresponds to a coil voltage tolerance of 0.7 to $1.25 \cdot U_s{}^{1)}$. This represents an extremely wide permissible operating voltage range when compared to the tolerance of 0.85 to $1.1 \cdot U_s{}^{1)}$ for contactors as is specified in DIN VDE 0660, part 102.

In the desgin of the magnet and coil systems, both the minimum required excitation and the maximum permissible temperature rise, have been taken into account.

6.2.3 Overvoltage suppression

Electrical control systems only operate consistantly and reliably when a sufficient degree of "*electromagnetic compatibility*" (EMC) exists between the various components. Although the demands related to EMC in conventional control systems may be rela-

---
[1] U_S Rated control supply voltage

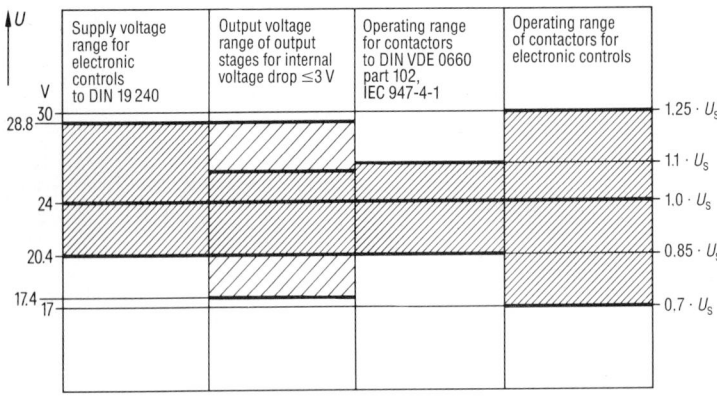

Figure 6.11
Voltage tolerances for electronic control systems and contactor magnet systems with a rated control supply voltage $U_S = $ DC 24 V

tively low, special protective measures to guard against disturbing or even destructive influences of switching overvoltages are unavoidable in systems employing electronic control elements, since the signal levels themselves are generally low.

The following suppression elements are commonly used, and are connected in parallel with the contactor coil:

▷ RC elements,
▷ RC-V elements with integrated varistor,
▷ Diodes,
▷ Varistors.

For the effect and operation of overvoltage suppression elements, refer to Section 4.3.3, on page 248.

Suitable suppression units are offered in various forms. For switching devices which are specially designed for use with electronic control systems (e.g. interface contactors and relays, input/output units), the suppression elements are sometimes integrated within device itself.

6.2.4 Contact reliability

Conventional electromechanical switching and signalling devices must be able to communicate not only with each other, but also within electronic systems. Thereby, a number of different demands are made on the devices. Figure 6.12 illustrates the typical signal levels found in conventional electromechanical and electronic systems, respectively.

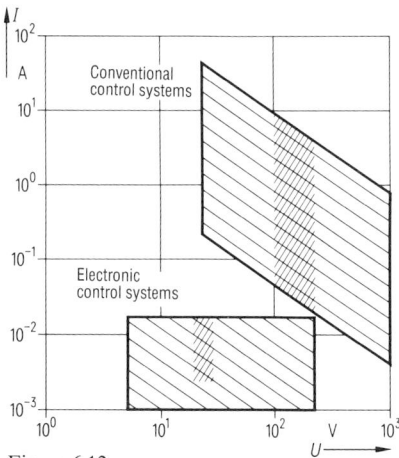

Figure 6.12
Ranges of signal levels for transducing sensors, signalling devices and signal recievers in conventional and electronic control systems (cross-hatched areas show preferred ranges)

Conventional control systems are characterized by:

▷ switching of relatively large ohmic and inductive currents and the production of switching arcs (magnet coils),
▷ predominant use of AC 230 V as the control voltage.

Thereby, the construction and nature of the contact elements are mainly influenced by factors related to electrical service life and switching capacity.

On the other hand, the input circuits of electronic controls are characterized by:

▷ switching of relatively low ohmic or ohmic-capacitive circuits without the presence of switching arcs (input filters)
▷ predominant use of DC 24 V as the control voltage.

For the mechanical switching devices used here, factors related to the contact reliability dominate.

Minimum current and voltage

The determination of the minimum current and voltage at which switching devices with mechanical contact elements will operate reliably, is extremely difficult. Although it is possible to derive such limit values from the physical properties and nature of a set of contacts, they tend to lure the user into a sense of false security since the published values cannot take all the operating and environmental conditions into account.

Since no generally recognized test procedure exists at present, the following method was developed by Siemens to judge the suitability of contacts for their use in electronic control systems.

Contact reliability is defined as the probability of a contact condition which leads to malfunction of a

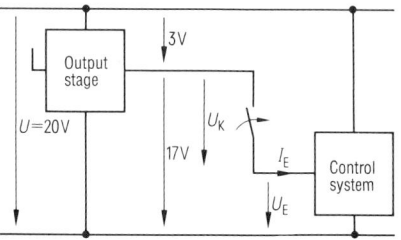

Figure 6.13
Voltage distribution and input current in a control system for the determination of the permissible voltage drop (contact resistance)

control circuit. Thereby, the contact condition is described by the permissible contact resistance R_K at a specified current value.

In terms of Figure 6.13, contact failure is defined as the condition when the voltage drop U_K across the contacts is so great, that the control input voltage U_E drops to below the permissible level:

$U_E = U - U_K < U_{E,min}$

Test circuit for contacts in electronic controls

DIN 19 240 of July 1985 specifies the following electrical limit values for DC 24 V in an electronic control system:

Supply voltage $U = 20.4$ to 28.8 V,
H signal $\quad U_E = 13$ to 30.2 V, $I_E \leq 15$ mA,
L signal $\quad U_E = -3$ to $+5$ V.

Typical operational values for SIMATIC S5 programmable logic controllers are

$U_E = 24$ V,
$I_E = 8.5$ mA.

The voltage drop within an output stage can be 2 to 3 V.

Since the interlocking functions are programmed within the electronic systems (PLC), there is normally only one contact point connected between an electronic output stage and the next input. Thus, under unfavorable conditions, the constellation illustrated in Figure 6.13 is possible.

Based on the typical data mentioned above, the following operational values for the point of contact may be expected:

Input current $I_E = \dfrac{13\text{ V}}{24\text{ V}} \cdot 8.5$ mA,

$\qquad\qquad\quad = 4.6$ mA,

permissible contact resistance

$R_K = \dfrac{17\text{ V} - 13\text{ V}}{4.6\text{ mA}}$

$\quad\;\; = 870\ \Omega.$

Therefor, the following values (rounded) have been laid down for the test circuit:

$U_{test} \quad = 17$ V,
$I_{test} \quad = 5$ mA,
$R_{K,perm} = 800\ \Omega$

Evaluation of the test results

The contact failure rate H_F is defined as the number of contact failures occurring within a specified number of switching operations.

Hereby, a large quantity of current paths are tested so that the published results represent statistic mean values.

If, for example, a contact failure rate of $H_F = 2 \cdot 10^{-8}$ is stated, this means that the permissible contact resistance was only exceeded twice in a large batch of test samples subjected to a total of 10^8 (100 million) switching operations. The practical meaning of this result can be formulated as follows:

No contact failure is expected over a long period of operation (independent of the number of switching operations). Contact failures, however, do not occur in terms of pure probability function with respect to unit time; they may e.g. be more frequent at either the beginning or the end of the service period. In addition, certain individual unit tolerances may have to be taken into account.

Achieving a high contact reliability

The goal of all the constructional design aspects for a contact set must be to retain conductivity in spite of the effects of contamination and dust.

Some salient design possibilities to improve contact reliability of universal mechanical switching devices include:

▷ high contact forces (pressure),
▷ the shape of the contact pieces (pressure per unit area),
▷ wiping action of the moving contact piece (This, however, causes increased mechanical wear and contact erosion.),
▷ single instead of double-break contacts,
▷ current paths in parallel (double moving-contact bridges, redundancy),
▷ encapsulation of the switching chambers.

Design features

Figure 6.14 illustrates three design examples for the arrangement of contact elements in the current path of a switching device.

Contact reliability

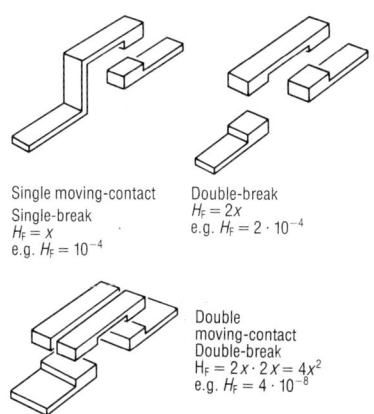

Single moving-contact
Single-break
$H_F = x$
e.g. $H_F = 10^{-4}$

Double-break
$H_F = 2x$
e.g. $H_F = 2 \cdot 10^{-4}$

Double moving-contact
Double-break
$H_F = 2x \cdot 2x = 4x^2$
e.g. $H_F = 4 \cdot 10^{-8}$

H_F Contact failure rate

Figure 6.14
Various configurations of contact elements

Let us assume the contact failure rate H_F of the single contact with the single break point to be x. It follows that the single contact with double break will have $H_F = 2x$, since two contact points are effectively connected in series.

If two single contacts with double break are connected in parallel, then the resulting contact failure rate for this double contact will be $H_F = 2x \cdot 2x$, i.e. $H_F = 4x^2$.

These results apply for ideal conditions (e.g. equal contact forces on all the respective contact points, unhindered movement of the contact pieces). Which contact arrangement will produce the highest contact reliability in a particular item of switchgear, will depend e.g. on the service life, the switching frequency and the switching capacity required.

All three variants are used successfully in the auxiliary and control circuits of Siemens low-voltage switchgear, although the use of double moving-contact bridges is more predominant.

Contact reliability
in electronic control applications

3TH4 and SIMICONT 3TH2
contactor relays

In these devices, the moving contacts consist of two parallel current paths. These can assume their positions relative to the fixed contacts independently of each other, i.e. there is no rigid bond between them (double moving-contact bridges with double break).

The excellent results of research using relay chain circuits were achieved under the test conditions for control and signalling devices of electronic control systems (DC 17 V, 5 mA ≙ 85 mW). The contact failure rate was found to be $H_F < 10^{-8}$.

For the purpose of comparison, Figure 6.15 illustrates the mean contact failure rate H_F of eight current paths in two 3TH contactor relays, equipped with single and double moving-contact bridges respectively. Double moving-contact bridges provide the following advantages:

▷ a significant reduction in the contact failure rate and

▷ a drastic reduction in the number of early failures.

In the device with single moving-contact bridges, contact failure occurred in seven current paths, whereas only one current path in the device with double moving-contact bridges showed signs of excessive contact resistance during the test.

Auxiliary contacts of a.c. power contactors

Up to the size 1 contactors, auxiliary contacts are integrated within the contactor housing. They are equipped with double moving-contact bridges.

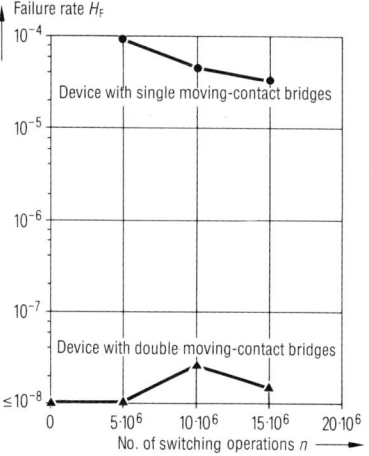

Figure 6.15
Mean contact failure rate H_F of the eight current paths in 3TH contactor relays equipped with single and double moving-contact bridges respectively, as a function of the number of switching operations n (test circuit: DC 17 V, 5 mA)

6 Transducing sensors and signal processing systems

For practical orientation purposes, research was carried out with 3TF42 contactors in an environment excessively contaminated with cement dust (approx. 1.3 g/l). The results showed that devices with double moving-contact bridges exhibited contact failure frequencies which were lower by a factor of 10^{-2} to 10^{-4} compared to those of devices fitted with single moving-contact bridges.

The snap-on auxiliary contact blocks of the 3TF3 contactors (and 3TH3 contactor relays), and the side-mounted auxiliary contact blocks of the Siemens contactors from size 2 to size 14, are also equipped with double moving-contact bridges.

For these units, the contact failure frequencies lie between 10^{-6} and 10^{-8}.

Auxiliary contacts of 3VU13, 3VU16 and 3VE3 circuit-breakers for motor protection

Circuit-breakers for motor protection are not intended for a particularly high number of switching operations or high switching frequency. Consequently, contamination resulting from the wear of moving parts within the devices plays an insignificant role. For this reason, well encapsulated auxiliary current paths with single-break contacts are used.

During each switching operation, the contact points are subjected to a sliding or wiping movement. This wiping action ensures a high contact reliability even in the case of low voltages and currents found in electronic control systems.

Auxiliary contacts of 3V. circuit-breakers

These auxiliary contacts too are used for a relatively low number of switching operations and at relatively low switching frequencies. They employ double-break single moving-contact bridges. An encapsulating housing prevents the ingress of contamination. The contact failure rate has been found to be $H_F < 2.7 \cdot 10^{-8}$.

3SB Pushbutton switching elements

All the current paths are equipped with double moving-contact bridges (similar to those of the 3TH2 contactor relays). These devices too have well encapsulated switching chambers. During the research testing of these devices, no contact failures were noted. The contact failure rate is therefor $H_F < 10^{-8}$.

3SB pushbuttons are also frequently used as signalling and command devices in electronic controls with

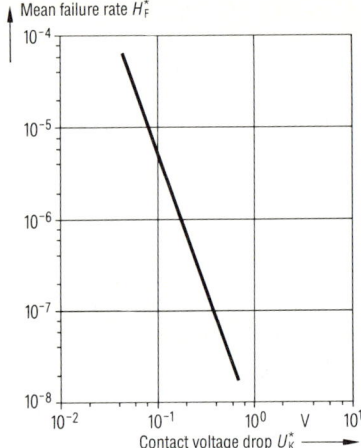

Figure 6.16
Mean contact failure rate H_F^* for 3SB1 pushbuttons as a function of the permissible contact voltage drop U_K^* (test circuit: DC 5 V, 1 mA)

5 V signal levels. For this reason, additional tests have been carried out to measure the voltage drop across the contacts in test circuits with DC 5 V, 1 mA ($\cong 5$ mW). Normally, the voltage and current used in such test circuits is DC 17 V, 5 mA (see page 293). From these tests one can derive a new mean contact failure rate H_F^*, in which contact failure is defined as the occurrence of a voltage drop U_K^* across the contacts, which causes the calculated permissible voltage drop for the input stage of the electronic control system, to be exceeded. Figure 6.16 shows the typical relationship between the mean contact failure rate H_F^* and the voltage drop U_K^* across the contacts. For $U_K^* \geq 0.8$ V, the mean contact failure rate H_F^* is less than 10^{-8}. The increase in H_F^* for smaller permissible contact voltage drops indicates the increasing effect of contamination on the contact surfaces.

BERO electronic proximity switches

Electronic proximity switches do not have mechanical contacts. The electronic switching circuits used have an extremely high reliability and service life expectancy. If at all, difficulties only arise when the specified maximum ambient conditions (e.g. humidity) and/or the electrical limit values (e.g. permissible cable lengths in a.c. operation) are exceeded (refer to the Siemens l.v. switchgear catalogue).

6.3 Assessment criteria for electromechanical and electronic controls

Automation of industrial processes requires the sensible use of the available control devices and systems. According to the task at hand, it must be decided whether e.g. contactor relays, time relays and/or hard-wired or programmable electronic controllers offer the most economical solution.

The diversification of automation in industry ranges from the simplest machine control to complete process control, management and optimization systems. Under consideration of the economic aspects, it is at present not possible to fulfill the requirements of this vast range of applications with one technology.

Owing to the continued technical development and sinking prices in the field of semiconductors, the use of programmable logic controllers such as SIMATIC S5, has increased substantially over the past years.

In the past, logic operations had to be achieved by a multiple of interconnected (hard-wired) auxiliary contacts and magnet coils. Nowadays, these connections and logic operations are often done in the software of programmable logic controllers (PLC's). The possibilities of simple modification and the reduced installation costs are only two of the well-known benefits of this technology.

A wide range of applications, however, particularly for simpler and smaller tasks, remains for hard-wired solutions with relays, contactor relays and time relays and for electronic logic modules such as SIMATIC C1, C2 and C3.

Of particular interest, is the Siemens 7PT SICONTROL manual logic controller which offers the flexibility of programming, but employs the basic elements of discreet hard-wired technology.

The selection of the optimal control equipment is extremely complex. For each application, the advantages and disadvantages related to planning, engineering, construction, installation, commissioning and operation of the control system, as well as the initial, maintenance and running costs of each possible solution must be compared and assessed.

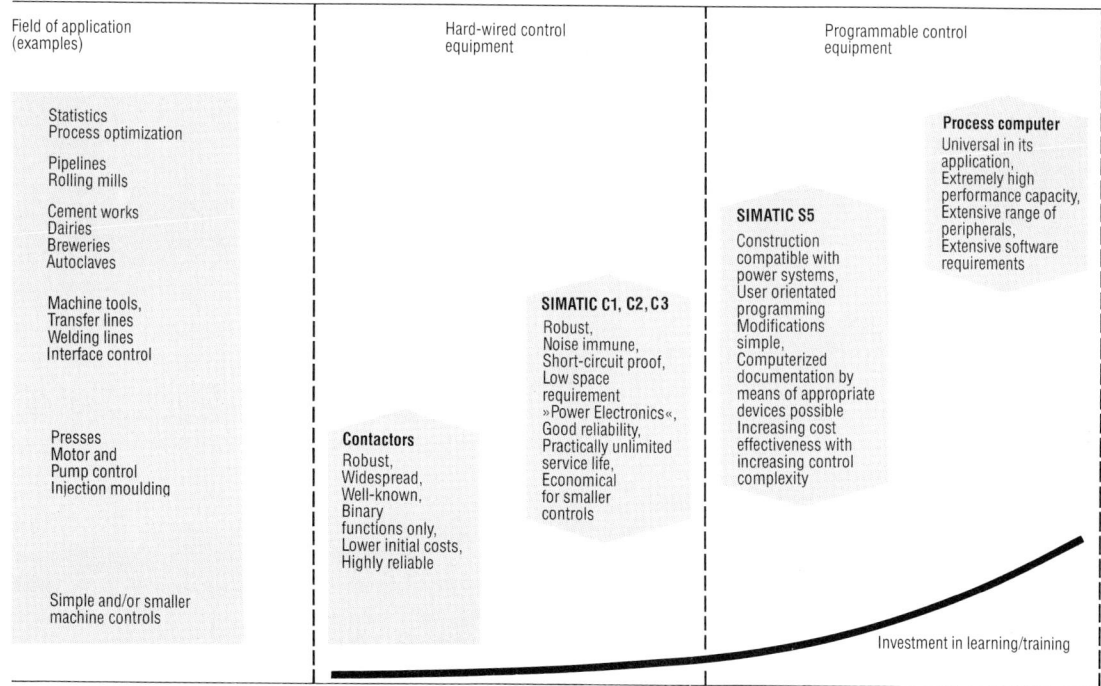

Figure 6.17 An appropriate technology is offered for each control application

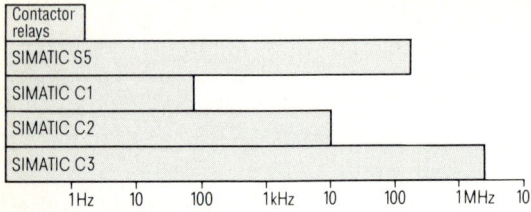

Figure 6.18
Comparison of signal processing frequency of contactor relay and SIMATIC based systems

In all these considerations, the abilities and technical qualifications of the manufacturing, operating and maintenance personnel should also be taken into account.

The following aspects may provide a degree of orientation and inspiration for selection of the control equipment optimally suited for a particular task.

Increasing the degree of automation

An increase in the degree of automation means that more functions are to be performed by the control system. If, in addition, shorter process times are to be achieved, then the use of electronic systems becomes inevitable (Figure 6.18).

Size and scope of the plant

Owing to lower initial costs, the use of SIMATIC C1 and C2 controls has increased in fields of application previously dominated by contactor and time relays. SIMATIC S5 control systems offer a viable alternative if the advantages of programmable logic controllers, such as "quick and simple modification", "duplication of control systems" and/or "computerized program and process documentation" overshadow the somewhat higher initial costs. Further information may be found in the Siemens ST50 to ST59 catalogues.

7 Type-tested switchgear assemblies (TTA)

7.1 General

Low-voltage switchboards and distribution boards may be regarded as the link between the apparatus for generation (generators, alternators), transmission (cables or overhead lines) and transformation (transformers) of electrical power on the one hand, and the consumer or load equipment such as motors, solenoid valves, heating, lighting and air-conditioning plant, on the other (Refer to Section 1.4 for corresponding specifications and regulations).

Figure 7.1 gives a summary of the design features of low-voltage switchboards, distribution boards and control panels presently offered by Siemens.

A new generation of low-voltage switchboards, known as SIVACON, is presently being launched. The SIVACON switchboards will incorporate at least all the features and options listed in the first three columns of Figure 7.1 and offer a maximum main busbar current rating of up to 6300 A. The type-tested assemblies employ a particularly universal modular construction. Their design may be regarded as having its origins in the well-tried and trusted 8PU range which has enjoyed world-wide acceptance and recognition for many years and which is described in more detail later in this chapter. If further information on SIVACON is required at this stage, kindly contact your Siemens branch office for assistance.

7.1.1 Versions and designs

The appropriate type of low-voltage switch and/or distribution board is determined by a number of specific criteria.

Type of distribution:
▷ Point (or radial) distribution,
▷ Linear distribution (busbar trunking systems),
▷ Installation distribution.

Currents:
▷ Rated currents of busbars,
▷ Rated currents of incoming feeders,
▷ Rated currents of outgoing feeders,
▷ Short-circuit withstand capability of the busbars.

Nature of protection and installation:
▷ Degree of protection,
▷ Protective measures,
▷ Method of installation (against a wall, free-standing),
▷ Number of sides which must be accessible for operation,
▷ Enclosure material,
▷ Ambient conditions.

Equipment mounting/installation:
▷ Snap or screw mounting,
▷ Fixed, i.e. non-withdrawable,
▷ Removable sub-assemblies,
▷ Withdrawable.

Application:
▷ Switching, disconnection, distribution and control of loads (i.e. consumers of electrical energy).

In most cases, the switch and distribution boards are labelled in terms of their application or function in the plant (e.g. "Main Pump Control"). As a rule, one distinguishes between "Main Switchboards" or "Main Distribution Boards", and "Sub-Distribution Boards" (Fig. 7.2).

Usually, each busbar section of the main switchboard is connected directly to an incoming transformer. Downstream motor control centres, distribution boards for lighting, heating, air-conditioning, etc. which are fed from the main switchboard or distribution level, all fall under the term "Sub-Distribution". Main and sub-distribution systems may also be characterized by the type of distribution viz. "point or radial distribution" and "busbar trunking distribution".

Point (or radial) distribution boards

The term "point distribution" (Fig. 7.3) is used to describe all switchboards and distribution boards which distribute electrical power radially from a "point source" via cables and/or overhead lines to remotely located consumers (loads). The necessary

7 Type-tested switchgear assemblies (TTA)

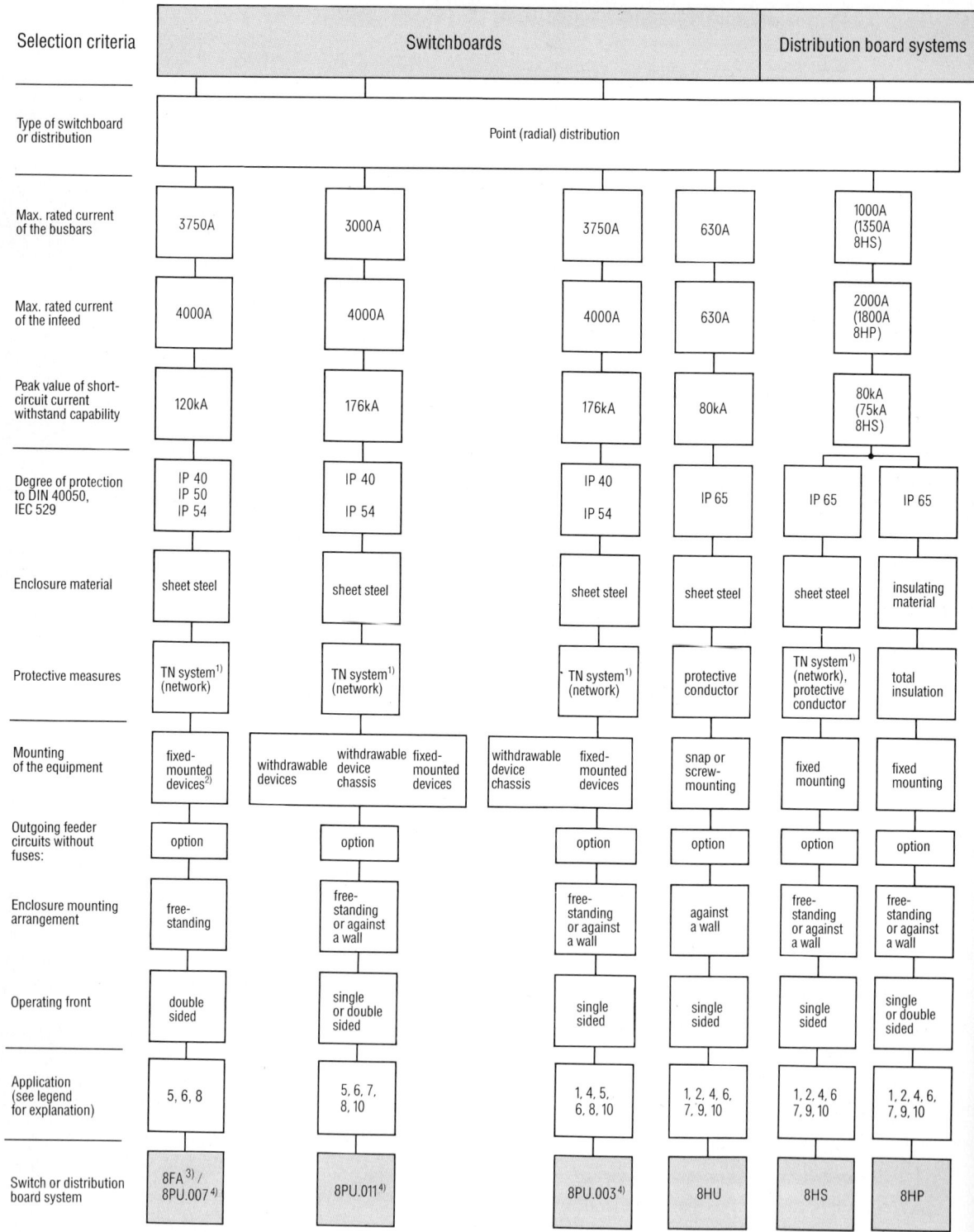

300

Versions and types of construction of l.v. switchboards, distribution boards and control systems

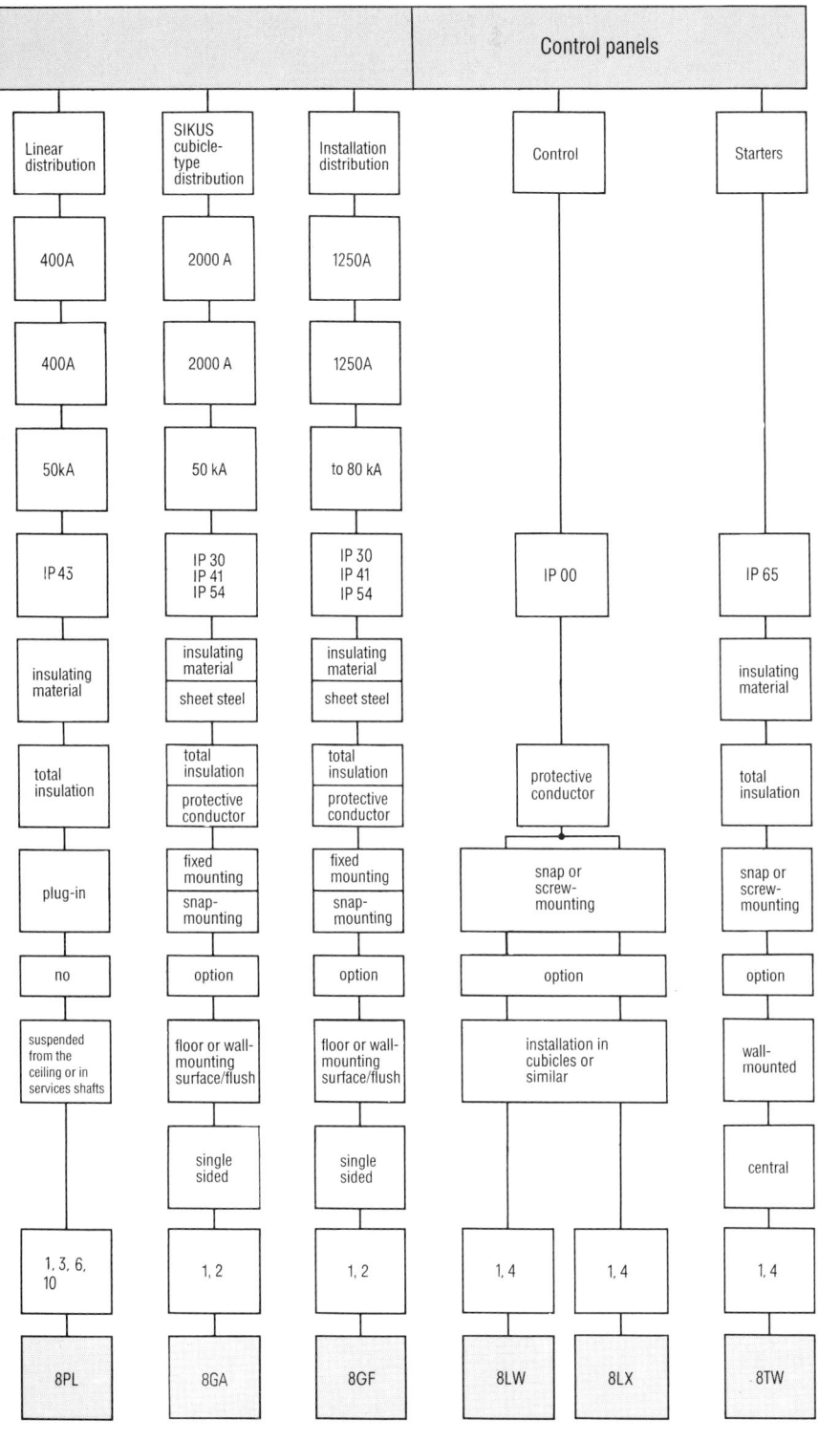

Figure 7.1
Selection criteria for low-voltage switchboards, distribution boards and control systems

Application:
1. Distribution for lighting and power circuits
2. Installation (domestic) distribution
3. Linear distribution
4. Control systems
5. Power factor correction
6. Industrial distribution
7. Motor control centres
8. Main switchboards
9. Main distribution boards
10. Sub-distribution

[1] TN system (network):
Direct earthing of a network point. Body connected directly via the protective and/or PEN conductor to the earthed star point.
[2] Alternatively, withdrawable circuit-breakers.
[3] Transformer rating 400 to 1250 kVA (in case of AF cooling, 40% higher).
[4] Please refer to the comment on the new generation SIVACON switchboards on page 299.

7 Type-tested switchgear assemblies (TTA)

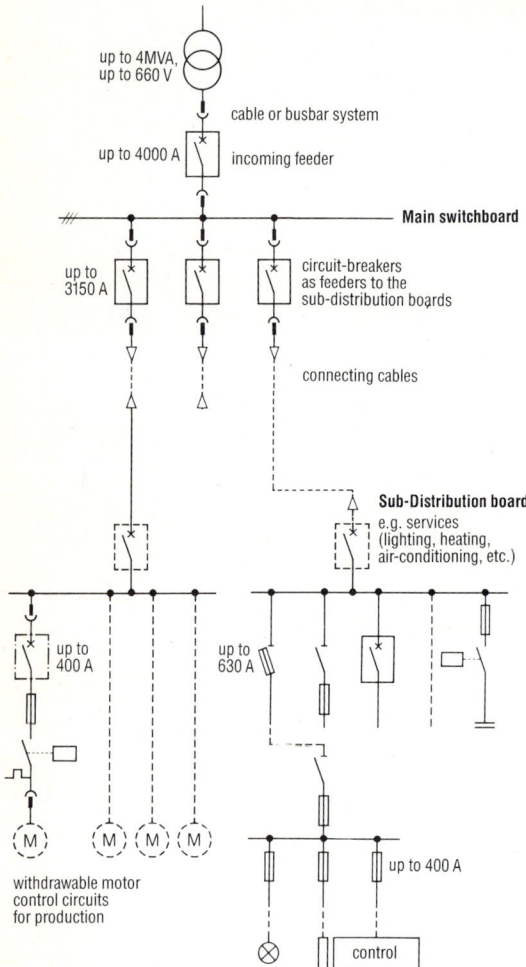

Figure 7.2
Main switchboard, sub-distribution and control systems in an industrial low-voltage network

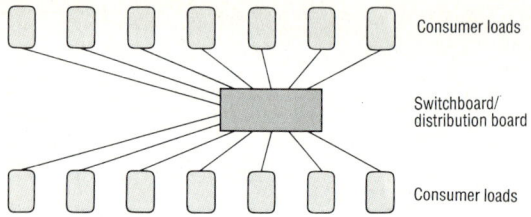

Figure 7.3 Point distribution system (principle)

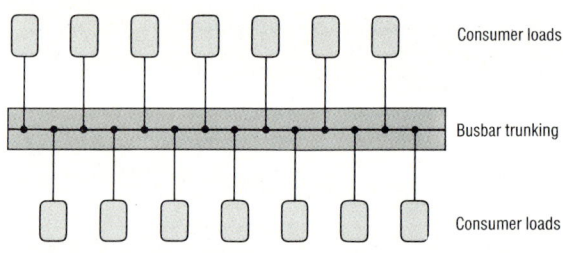

Figure 7.4 Busbar trunking system (principle)

switching, protective and measuring devices are grouped centrally in the switch or distribution board.

Linear distribution (busbar trunking systems)

In the case of linear distribution systems (Fig. 7.4), the power is distributed through a single run of relatively long enclosed busbars to the various points of installation of consumer equipment. The loads located along the length of the busbar run are connected to the busbars via short stub lines or cables and through tap-off boxes with HRC fuses. Busbar trunking systems (with tap-off units of various sizes located at appropriate positions) are typically used to distribute power in spatially extended factories, assembly halls and workshops. Since tap-off units can be provided (or relocated) at virtually any point on the busbar run, busbar trunking systems are especially suitable in cases where consumers (e.g. machine tools) are added to the supply or may have to be moved from one location to another (e.g. factory reorganization).

Control systems

While point and linear distribution systems are exclusively employed for the distribution of electrical energy, control boards (or panels) are part of the machine tool or manufacturing/processing machine whose operation they control.

7.1.2 Types of construction

On a regional level (Europe) the basis of the design, manufacture and testing of switchboards and distribution boards produced at the responsibility of the manufacturer is EN 60439 part 1, alternatively VDE 0660 part 500 (specifications for low-voltage switchgear assemblies as type-tested assemblies TTA and partially type-tested assemblies PTTA with rated voltages up to AC 1000 V or DC 1500 V). This specification describes various types of construction and contains guidelines for the installation of the switchboards (also refer to Section 1.3.1, page 26).

Types of construction

Open-frame construction

In the open-frame type of construction (Fig. 7.5) which normally does not employ covers or shrouding of any kind, parts that may be live during operation, such as main busbars, busbars droppers, items of equipment, terminals and conductors are accessible from all sides. Installation of open-frame equipment is permitted only in closed electrical operating areas. Owing to the restrictions and conditions imposed by the trade associations on the operators of electrical plant and equipment (e.g. Safety Regulations 4, VBG 4 – also refer to pages 34 and 37), this type of construction is only rarely used today.

Panel construction

Unlike the open-frame design, the panel-type of construction provides protection against contact with live parts on the operating side, but is not enclosed on the other sides, which permit access to the equipment. This type of switchboard, therefore, may also only be installed in closed electrical operating areas. For the reasons mentioned above, the use of the panel construction is very rarely considered.

Cubicle or enclosed construction

The cubicle-type of construction (Fig. 7.6) is enclosed on all sides so that contact with live parts during operation is prevented (minimum degree of protection IP2X). Installation is permissible in generally accessible operating areas. In most cases the enclosed construction has a height greater than 1000 mm (standard height 2200 mm), and is made up of a number of cubicles (individual chassis frames) bolted together side-by-side. A group of up to four cubicles (known as a section) normally constitutes a transport unit.

Cubicle construction is the most widely used nowadays. Of all the possible designs it represents the optimum for the user with regard to the protection of personnel and plant.

In most applications, the doors on these switchboards are of the same height as the cubicles (full-length). Individual compartment doors are used if the switchgear is withdrawable or of the plug-in type, or if the entire compartment assembly (also known as a "bucket" or "draw-out unit") can be removed and replaced.

Figure 7.5 Switchgear assembly with open-frame type of construction

7 Type-tested switchgear assemblies (TTA)

In particular cases, individual compartment doors are also used with fixed-mounted equipment.

Withdrawable construction

This is a special form of the enclosed or cubicle-type construction, in which the cubicles are divided into a number of compartments, each with its own door (Fig. 7.7). Each compartment contains a "draw-out unit". These consist of an assembly or chassis-frame in which a number of items of equipment and switchgear are grouped and interconnected to form a functional entity (e.g. motor feeder or starter, incomer, bus-coupler). The units may be withdrawn from the switchboard and simply replaced in the event of a failure or for maintenance purposes. As they usually plug onto the busbar droppers, they often also permit disconnection from the main supply (as required for isolators according to DIN VDE 0660 part 107) while remaining mechanically attached to the switchboard (e.g. by means of an intermediate position on draw-out slide rails).

This type of construction offers the highest degree of operational reliability and safety for personnel.

Box-type, or modular construction

Box-type distribution boards (Fig. 7.8), typically made of insulating material or sheet steel, consist of a number of boxes (or enclosures) containing items of equipment such as busbars, fuses, switches and contactors securely assembled together to form a unit. Contact with parts that may be live during operation is prevented by means of the complete encapsulation of each section. Distribution boards of this design can therefore be installed in generally accessible operating areas.

If a protective cowl or roof is attached, and provided that the boxes have the appropriate degree of protection (minimum IP 55), this type of distribution board, unlike those described earlier, may also be installed outdoors.

Figure 7.6
Switchgear assembly of the cubicle or enclosed construction type

Types of construction

Figure 7.7
Replaceable draw-out unit
(3WN air circuit-breaker)

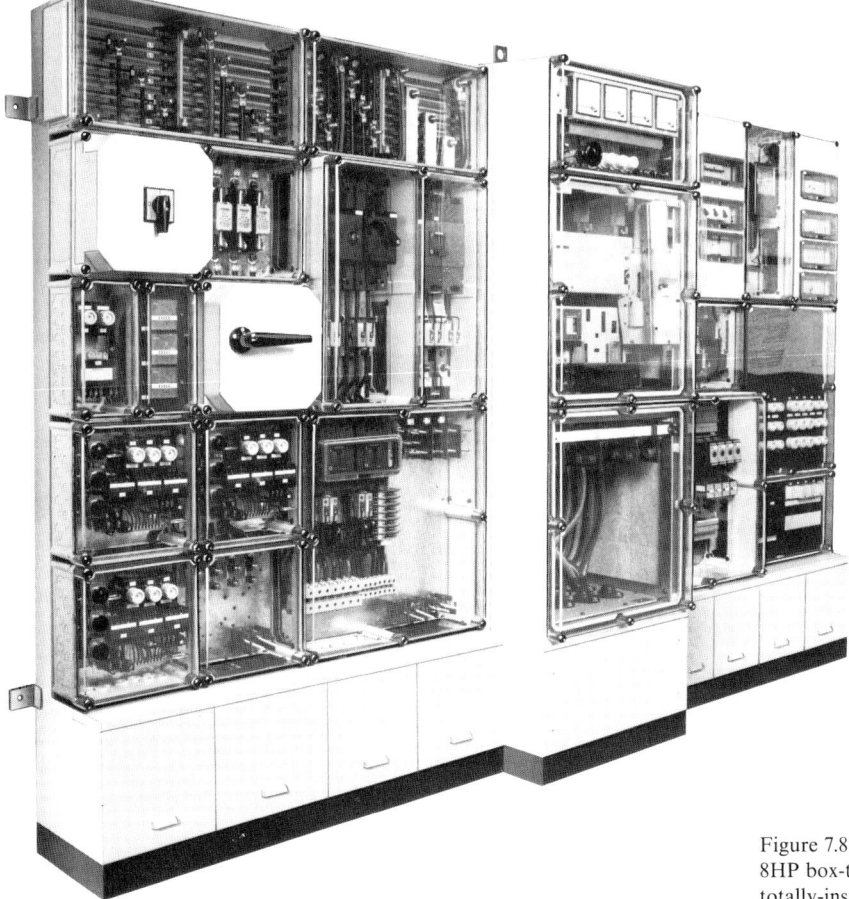

Figure 7.8
8HP box-type (modular)
totally-insulated distribution board

7.1.3 Selection criteria

The type of construction and the size of the switchgear assembly, or switchboard, is mainly determined by the rated currents of the incomers, feeders and busbars.

In terms of Figure 7.1 a distinction is made between "switchboards", "distribution board systems" and control panels.

The new modular generation of low-voltage switchboards presently being launched by Siemens (known as SIVACON) offers solutions for all these applications.

Switchboards

Switchboards are characterized primarily by:

▷ high current ratings of the equipment, up to 6300 A,
▷ high short-circuit withstand capability, up to 220 kA peak value
▷ degree of protection of the enclosure normally IP 40; up to IP 54 on request,
▷ sheet steel as the enclosure material,
▷ a standard height of 2200 mm,
▷ mounting methods for the equipment: fixed, or as removable or withdrawable units.

When considering switchboards, the type 8FA/8PU.007 shown in Figure 7.1 may be of special interest. Whereas all the other switchboards and distribution systems comprise only the low-voltage switchgear and equipment, the 8PU.007 switchboard may be regarded as the low-voltage section of a transformer load centre type 8FA. This transformer load centre may contain a 12 or 24 kV high-voltage section with switch disconnectors for ring or spur feed, and a transformer with a rating up to 1250 kVA.

Distribution systems

Distribution systems, on the other hand, generally have the following features:

▷ rated currents of the incorporated equipment up to about 2000 A,
▷ short-circuit withstand capability up to 80 kA peak value,
▷ degree of protection up to IP 65,
▷ various enclosure materials (insulating material or sheet steel),
▷ height of the individual enclosures normally < 1000 mm,
▷ mounting methods for the equipment: generally fixed.

In the selection of a switchboard or a type of distribution board best suited to the requirements of a particular application, use of the "selection criteria" presented in Figure 7.1 is recommended. If the various stages are worked through from top to bottom, the table will almost automatically lead to the most suitable system.

Short-circuit withstand capability

The prospective short-circuit current (I_k'' = initial symmetrical short-circuit current, I_p = peak asymmetrical short-circuit current) at the point of installation of the switchboard or distribution board – that is, between the terminals of the infeed transformer on the one side and the cable-connected loads on the other side – may not exceed the short-circuit withstand capability quoted for the switchboard by the manufacturer.

In terms of DIN VDE 0660 part 500, the manufacture must state the short-circuit withstand capability in the following manner:

▷ by stating the maximum permissible prospective short-circuit current at the incoming line terminals of the switchboard which incorporates short-circuit protection in its incomer section
▷ by stating the rated short-time withstand current and the rated peak withstand current of the complete switchboard
▷ by stating the conditional rated short-circuit current and giving the corresponding data of the current limiting switching device.

In the case of a switchboard with several incomers which are not simultaneously in operation, the short-circuit withstand capability may be stated for each of the incomers separately.

For a switchboard having several incomers which may be in operation simultaneously, as well as for a switchboard with one incomer and one or several outgoing feeders for rotating machinery of larger rating which could contribute to the short-circuit current, the short-circuit withstand capability of the switchboard must be determined and stated in accordance with the information regarding the individual continuous short-circuit currents, as supplied by the end-customer.

If this information is not available, then a short-circuit calculation according to the guidelines for the calculation of short-circuit currents as set out in DIN VDE 0102, 01.90, must be carried out. The calculation may be simplified by using an appropriate calculation sheet, or a software package (PC program) such as KUBS.

For further details on this subject, refer to Sections 2.1.3 and 2.1.5. (pages 81 and 85 et.seq.).

Degree of protection

Depending on the installation location and the surrounding conditions, a switchboard or distribution board design should be chosen that provides the necessary degree of protection against contact with live parts and against the ingress of foreign bodies and water (DIN 40050 and IEC 529). A list of protection methods and requirements for the enclosure of an item of equipment is given in Section 1.5.6, page 63. According to DIN VDE 0100, switchboards and distribution boards with a degree of protection up to IP2X may only be installed in "electrical operating areas" or in "closed electrical operating areas".

In rooms that are accessible to anyone, switchboards and distribution boards must be so designed that protection is provided both against accidental contact and against contact with parts which may be live during operation: that is, the degree of protection must be at least IP2X. This requirement is met in all the constructional designs described.

Enclosure and shrouding material

In certain distribution system designs (up to busbar current ratings of 1250 A) there is a choice between metal and insulating material (e.g. glass-fibre reinforced polyester resin) for the enclosure.

▷ *Enclosure of insulating material*

The insulated enclosure offers wide ranging protection against corrosion as well as a high degree of protection against contact with live parts. Furthermore these distribution systems (8HP, 8GF and 8PL) comply with the additional protective measure "total insulation" in accordance with DIN VDE 0660 part 500 and DIN VDE 0100 part 410.

▷ *Metal enclosures*

In all types of switchboard construction and corresponding distribution systems, the enclosures, shrouds and parts of the mounting structure made of metal are protected by a high-quality surface finish and, except in case of damage, need no further treatment or maintenance. Most of the switchboard enclosure surfaces are provided with a durable baked epoxy powder coating which has extremely good mechanical properties.

Protective measures

Switchboards must be built in such a way that, if need be, they provide the necessary protection against contact with live parts and, if applicable, they may be incorporated into the protection measures against excessively high contact voltage to DIN VDE 0100.

For the required application of the protective measures, please refer to Section 1.4, page 33.

In order that the most economic constructional design for a switchboard or distribution board can be chosen for a particular application, the characteristic features of the various systems should be compared, and the space, erection and access requirements on site should be analyzed by the user before any manufacture is undertaken or any building and installation site alterations are carried out.

Erection, access, delivery

Erection site

▷ generally accessible operating areas,
▷ electrical operating areas,
▷ closed (locked) electrical operating areas.

Installation nature

▷ on the floor, against a wall,
▷ on the floor, free-standing in the room,
▷ fixed to a wall, in a stairwell, alcove or a recess,
▷ suspended from the ceiling,
▷ mounted on a rack.

Nature of access

▷ on one side or on two sides for operation,
▷ front or back access for cable connections and alterations,
▷ top or back access for modifications to, or installation of busbars.

Quoted installation dimension

▷ height, width and depth.

Delivery conditions

▷ height and width of doors,
▷ lift/elevator dimensions,
▷ where necessary, lifting capability of cranes/hoists.

Ambient conditions

The ambient temperature and climatic conditions at the installation site must be taken into consideration during the design and selection phases, since these parameters determine the load carrying capacity of the electrical equipment. Ambient temperatures and humidity values which vary to a great extent, as well as outdoor installation have a strong influence on the type of switchboard to be selected (e.g. degree of protection, protective roof, anti-condensation breather valves, heaters, heat exchangers).

Equipment mounting methods

Switch and controlgear as well as measuring instruments may be fixed-mounted onto removable chassis or withdrawable units in a switchboard. Hereby, the possibilities for

▷ the replacement of gear in the event of a fault, and

▷ the modification of, or maintenance to a switchboard without general shutdown

are enhanced.

Special requirements

Possible special requirements such as, for example, explosion protection, protection against aggressive atmospheres, designs to withstand induced shock (e.g. earthquake or similar vibrational loads) require additional agreement between the manufacturer and the user.

7.2 Switchboards in standardized design

7.2.1 Introduction

Application

Switchboards for rated currents up to about 6000 A are mostly used as main switchboards in industrial plants and large commercial establishments, in power stations and refineries, and in high-rise buildings and large hospitals; i.e. wherever a large power consumption has to be catered for.

Network (system) configuration

A logical division of an industrial power supply system into a "main switchboard" (or "power center") with incomers, feeders and bus-couplers which consist only of circuit-breakers and supplied directly from the transformer(s) on the one hand, and the various "sub-distribution boards" on the other, has the following advantages:

▷ The main switchboard is installed in the immediate vicinity of the infeed transformers. The cable or busbar connections to the transformers are short.

▷ The incomer, outgoing feeder and busbar coupler circuit-breakers incorporated in the power centre are all of the same constructional design with only five different current rating frame sizes (1000 to 6000 A).

▷ If these circuit-breakers are of the withdrawable type, there is no need for an additional disconnector upstream of each breaker, as would be the case for fixed-mounted circuit-breakers.

▷ A double infeed in conjunction with a bus-section switch (bus-coupler) – open in normal operation – offers the possibility of retaining power supply to all the sub-distribution level in the event of the failure of one infeed. If this should occur, the faulty incomer is disconnected and the bus coupler closed. The arrangement reduces downtime and increases the supply availability for all the loads.

If necessary, depending on the degree of utilization of the transformer capacity, load shedding may be necessary during operation with one incomer only and a closed bus-coupler.

▷ The sub-distribution boards are placed centrally in relation to their respective loads. This results in shorter feeder cables.

▷ As a result of the relatively long cable or busbar runs between the main switchboard and the sub-distribution boards, the sub-distribution boards may incorporate busbars and components of lower short-circuit withstand capabilities.

Busbar and cable incomers (infeeders)

Switchboards in the immediate vicinity of the transformers have the advantage that they may be fed directly from the transformer terminals via shrouded busbars. Entry into the switchboard is then possible either as with cables from below or, by means of additional panel sections, from above. When considering connection by means of cables, it should be borne in mind that the labour involved in connecting the transformer to the main switchboard may be significantly more extensive and costly than would be the case with busbar connection. If, for example, a 2 MVA transformer with a rated current of about 3000 A at 400 V must be connected, up to 14 parallel cables each with a cross-section of 240 mm² have to be secured and terminated within the incomer panel.

Outgoing feeders, cable runs

As a rule, cables are used for outgoing feeders. Depending on the distance to the load and the permissible voltage drop at the load current, parallel connected cables are used above about 250 A.

Typically, the load cables are fed out from underneath the switchboard into a cable trench or through the floor. In the latter case, cable trays are suspended from the ceiling of the storey below to support the cable runs. In ground-floor factory areas, where no cellar is available, false floors allowing about half a meter of height are often used, so that the cables can be laid with relative ease and remain accessible.

In the design and assembly of cable racks, galleries trenches and false floors, care must be taken to ensure that the permissible bending radii of the cables can be maintained.

Operation and maintenance isles

The operation and maintenance isles must be dimensioned in accordance with DIN VDE 0100 part 729 (Fig. 7.33, page 334). In the case of switchboards with withdrawable equipment – especially if this equipment is of high current rating, and therefore large – it should be borne in mind that sufficient space must be left on the operating side between the switchboard and the opposite surface (e.g. wall, next switchboard, machine or other obstruction) to allow access for personnel and to permit withdrawal of the equipment and possible positioning of a transport trolley (also refer to Section 1.4.4, page 54).

Possibilities for disconnection (isolation) of the incoming supply

When several transformers feed onto a common set of busbars, or onto a number of busbar sections (linked by means of bus-coupler circuit-breakers), provision must be made for an additional means of isolation between each incomer circuit-breaker and the busbars, so that it may be disconnected for the purposes of inspection, maintenance, or in the event of a fault.

This can be achieved by means of:

▷ withdrawable circuit-breakers (up to 6000 A),
▷ a disconnector or on-load switch disconnector (up to 4000 A),
▷ LV HRC fuse bases with isolating links (up to 1250 A),
▷ LV HRC fuse switch-disconnectors with isolating links (up to 630 A).

The first-mentioned possibility not only saves costs, but also space in the switchboard, and is therefore the most frequently adopted means of isolation. The same applies for the sectionalizing of the busbars by means of bus-coupler circuit-breakers between two busbar sections.

Busbar sectionalizing

In the case of two or more infeeding transformers, the purpose of busbar sectionalizing is to limit the fault level requirements of the individual switchboard sections to that of a single transformer. A basic requirement is of course the normal operation of the switchboard with the bus-coupler circuit-breaker in the open position i.e. each section fed by its own separate transformer.

In addition, busbar sectionalization is used to ensure that in the event of failure of one of the infeeding transformers, the power supply to its load circuits – albeit at reduced capacity – can be maintained via the closed bus-coupler circuit-breaker.

Cross-coupling

Switch-disconnectors or circuit-breakers can similarly be used to interconnect two switchboards installed back-to-back, each with its own busbar run. Such an interconnection is referred to as a cross-coupling.

Isolation or disconnection of outgoing feeders

Provision must also be made for isolating the individual outgoing feeder units from the main busbars or section dropper busbars in a simple manner to cater e.g. for the replacement of draw-out units or for modification and maintenance purposes. In most cases the LV HRC fuse links in fuse bases or the LV HRC fuse switch-disconnectors that are provided for the short-circuit protection of the outgoing feeder circuit can be used for this purpose. The highest degree of safety and convenience for maintenance and repair is provided when the equipment can simply be withdrawn from its compartment in the switchboard as a complete unit. This facility is afforded by switchboards of the plug-in or withdrawable construction types.

Type of busbar system

Since unbalanced loads such as heating and lighting circuits and single-phase motors are rarely connected to main switchboards, they normally incorporate a four-conductor busbar system, viz. the three main phase (or line) conductors L1, L2, L3, and the PE or PEN busbar rail which is usually located in the lower part of the switchboard. The main phase or line conductors are either fitted in the upper part of the switchboard and run across the top of each cubicle in a separate busbar chamber, or they are located across the back or roughly at half cubicle depth and about eye-level. Vertical dropper busbars, or a plug-in busbar trunking system (in the case of withdrawable or plug-in construction) down the back of each cubicle are terminated on the main busbars and distribute the power to the individual outgoing feeders, control circuits or compartments (draw-out). If necessary, an additional fifth insulated neutral conductor rail can be incorporated in either or both of the horizontal or vertical busbar systems.

Busbar material

Nowadays it is common practice to use electrolytic copper of strength F30 for the busbar material. Aluminium and copper-clad aluminium are only used to a minor extent in switchboards although this material is still found in switchboards originating from and in the former German Democratic Republic and a number of east European countries.

Marking and identification of busbars

For the identification of busbars, heat-resistant adhesive tapes are typically used. The tapes are wrapped around the individual busbars, at least at one point within each cubicle or section. They are lettered L1, L2 and L3 on the phase or line conductors and are coloured green/yellow on the PE or PEN protective conductor.

Current-carrying capacity of the busbars

In the case of switchboards manufactured in accordance with DIN VDE 0660 part 500, only the data for the particular product, as stated by the manufacturer for the load capacity of the busbars in terms of their rated current, the rated operational current and the short-circuit withstand capability (from the thermal and dynamic points of view) are relevant. The values given in catalogues, data sheets or other publications have been determined by actual testing at the values stated or by extrapolation of test results obtained in accordance with VDE-prescribed type testing procedures.

The rated and short-circuit currents of the busbars have different values according to the type of switchboard in which they are incorporated, and depend on:

▷ the mounting position within the cubicles,
▷ the physical arrangement of the conductors with respect to each other,
▷ the cross-sectional area/thickness of the conductor material,
▷ the strength and pliability of the conductor material,
▷ the distance between the busbar supports,
▷ possible heating effects from other components.

Cross-section of the PE conductor

The PE conductor bar which is electrically connected to the framework of the switchboard must be adequately dimensioned to carry any short-circuit current that may arise.

The cross-sectional areas of protective conductors must at least comply with the values given in Table 1.16, page 43. For "internal" connections of protective conductors (protective conductors to bodies which cannot be incorporated in the overall protective measures by suitable mounting), DIN VDE 0660 part 12 applies.

Cross-section of the PEN conductor

In principle, the same conditions apply for the PEN busbar as for the protective conductor rail, although higher prospective asymmetrical currents may have to be taken into account.

A reduction of the protective conductor cross-section is not permitted if rectifiers with current limitation are used in earthed networks. In this case, the protective conductor must be dimensioned in the same way as a secondary side phase (line) busbar.

Position of the incomer

An important aspect of the specification of the busbars is the physical position of the incomer or infeed within a switchboard. Figure 7.9 illustrates possible arrangements of incomers and outgoing feeders and the resulting stresses. In these cases the busbars are not divided by bus couplers even when there are two or more transformer infeeds. This would, however,

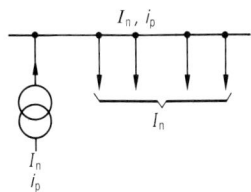

a) A single infeed from the side

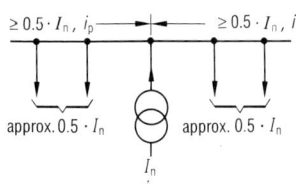

b) A central infeed with an almost symmetrical distribution of the consumer loads

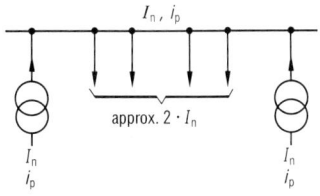

c) Two infeeds from the sides

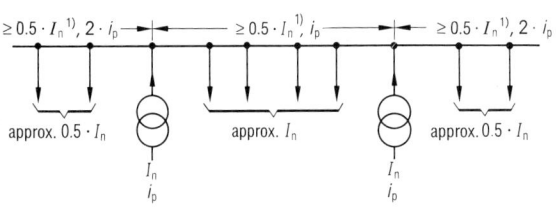

d) Two central infeeds with nearly symmetrical arrangement of the consumer loads

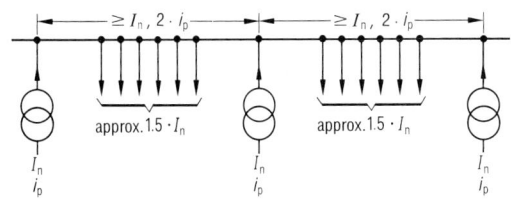

e) One central infeed and two infeeds from the sides

I_n Rated current of the transformer
i_p Asymmetric short-circuit current (peak value) through the transformer in the case of connection to a network with unlimited short-circuit power

[1] When the transformers are in operation simultaneously and the busbars are not sectionalized

Figure 7.9 Distribution of busbar loading according to the positions of the incomers

be advisable on the grounds of cost saving in the procurement of the switchboard (equipment with lower breaking capacity and lower short-circuit withstand capability) and on the grounds of operational reliability (limited damage in case of a fault).

Motor loads –
increase in the prospective short-circuit current

If large motor loads are connected to the outgoing feeders of a switchboard, they will contribute to in-

Table 7.1
Rated operational currents and short-circuit currents of three-phase transformers with ratings from 50 to 3150 kVA and rated voltages of 400 V and 525 V, 50 Hz

Transformer rated power S_{nT}	Rated voltage AC 400 V			Rated voltage AC 525 V		
	Rated current I_n	Rated value of the short-circuit voltage: $u_{kr}=4\%$ [1] Short-circuit current I_k [3]	$u_{kr}=6\%$ [2]	Rated current I_n	Rated value of the short-circuit voltage: $u_{kr}=4\%$ [1] Short-circuit current I_k [3]	$u_{kr}=6\%$ [2]
kVA	A	A (r.m.s.)	A (r.m.s.)	A	A (r.m.s.)	A (r.m.s.)
50	72	1 805	1 203	55	1 375	916
100	144	3 610	2 406	110	2 750	1 833
200	288	7 220	4 812	220	5 500	3 667
250	360	9 025	6 015	275	6 875	4 583
315	455	11 375	7 583	346	8 660	5 775
400	578	14 450	9 630	440	11 000	7 333
500	722	18 050	12 030	550	13 750	9 166
630	910	22 750	15 160	693	17 320	11 550
800	1 156	28 900	19 260	880	22 000	14 666
1 000	1 444	36 100	24 060	1 100	27 500	18 330
1 250	1 805	45 125	30 080	1 375	34 375	22 916
1 600	2 312	57 800	38 530	1 760	44 000	29 333
2 000	2 890	72 250	48 170	2 205	55 075	36 660
2 500	3 613	90 325	60 210	2 750	68 825	45 880
3 150	4 550	113 800	75 870	3 470	86 725	57 820

[1] To DIN 42500 and 42503 for power ratings from 50 to 630 kVA
[2] To DIN 42511 for power ratings from 100 to 1600 kVA
[3] For high-voltage side connection of the transformer to a rigid network (infinite bus) $I_k \approx I_k''$, see page 80.

crease the short-circuit current during a fault, and this should be taken into account in dimensioning the busbars. It is recommended that the short-circuit currents should be calculated with the aid of the standard form in Section 2.1.5 on page 92.

Rated and short-circuit currents of transformers

For the purposes of dimensioning switchboards, Table 7.1 offers selected typical data on the rated operational currents and short-circuit currents of three-phase transformers with ratings from 50 to 3150 kVA, rated impedance voltages u_{kr} of 4% and 6%, and rated voltages at the transformer terminals of 400 V and 525 V.

To determine the values at 690 V, the figures quoted for rated current and short-circuit current at 400 V for the same transformer power rating should be divided by 1.73 ($\sqrt{3}$).

In the case of switchboards built in accordance with VDE 0660 part 500, and in the absence of a contrary specification by the customer, the r.m.s. value of the short-circuit current I_k can be converted to the maximum asymmetrical short-circuit current i_p (peak value) by application of the factor n. Multiplication of the r.m.s. value by the factor n (Table 7.2) gives

Table 7.2
Preferred values of the factor n for determining the maximum asymmetrical short-circuit current at various r.m.s values of steady-state short-circuit current

Effective (r.m.s.) value of the steady-state short-circuit current	Power factor	Time constant	Factor n
kA	$\cos \varphi$	ms	
> 4.5 to 6	0.7	5	1.53
> 6 to 10	0.5	5	1.7
> 10 to 20	0.3	10	2.0
> 20 to 50	0.25	15	2.1
> 50	0.2	15	2.2

the maximum asymmetrical short-circuit current i_p, (peak value of the first half-cycle of the short-circuit current, including the d.c. component), which determines the degree of dynamic stress on the busbar and supporting hardware during the fault (also refer to Section 2.1.2, page 79).

The correlation between the power factor (cos φ) and the current values apply for most conditions found in practice. If, in particular cases, e.g. in the vicinity of large transformers, more severe factors apply, then special agreements should be made between manufacturer and customer.

7.2.2 Standard switchboards, type 8PU.

8PU. is the generic designation for a Siemens range of switchboards offering the following types of construction:

▷ Fixed-mounted switchgear
▷ Removable modular assemblies
▷ Withdrawable design.

The 8PU. standard switchboards qualify as "type-tested switchgear assemblies" (TTA) in terms of DIN VDE 0660 part 500.

Versions which deviate from the standard design, and which have not specifically been re-subjected to testing for proof of compliance with temperature rise limits and short-circuit withstand capability are regarded as "partially type-tested switchgear assemblies" (PTTA) to DIN VDE 0660 part 500. A summary of the required proof by testing for compliance with TTA or alternatively by testing and calculation for PTTA is contained in DIN VDE 0660 part 500 (Section 8.2.1.1, Table VII).

Constructional designs

The complete range of 8PU. standard switchboards, developed for both domestic and overseas manufacture, caters for the most diverse applications in three constructional designs having the following features and characteristics:

8PU.003: Fixed-mounted switchgear; also able to incorporate withdrawable-type circuit-breakers (e.g. incomers); modular groups with LV HRC fuse switch-disconnectors as outgoing feeders; main busbars located at the rear (full-depth) of the cubicles; switchboard can be installed against a wall.

8PU.007: Fixed-mounted switchgear; also able to incorporate withdrawable-type circuit-breakers (e.g. incomers); modular groups with LV HRC fuse switch-disconnectors as outgoing feeders; main busbars located centrally (at mid-depth) in the cubicles; double-fronted; free-standing; typical low-voltage switchboard of the 8FA transformer (S-type) load centre. For a detailed description of the 8PU.007, please refer to Section 7.2.3, pages 320 and 325).

8PU.011: Fixed-mounted switchgear, removable modular assemblies or fully withdrawable design for incomers and/or outgoing feeders; main busbars at the top of the cubicles in a separate compartment; for installation against a wall, free-standing or back-to-back depending on the section type.

The essential features of the constructional designs described above are listed in Table 7.3. Table 7.4, gives a summary of data and characteristics common to all three designs.

Cubicle types

Every switchboard is made up of one or more cubicles which accommodate the items of switchgear and other equipment in the required type of mounting arrangement. Various types of cubicle have been developed to facilitate planning, design, commissioning, maintenance, and also manufacture. Each cubicle type contains specific equipment, or switchgear groups/modules with clearly defined characteristics and function. These modules generally only differ in their current ratings, and are mounted in one of the three arrangements mentioned, viz. fixed-mounted individual items, fixed-mounted modular assemblies or withdrawable.

Fixed-mounted design

In the case of "fixed-mounted individual items" the outgoing feeder groups are permanently connected to the vertical cubicle busbars (droppers) by means of cables or bars. To exchange, modify or maintain the equipment, it is necessary to disconnect and isolate the main busbars. The cable to the motor and the auxiliary cables for control and measurement, are

Table 7.3
8PU constructional designs with indication of equipment mounting method, main busbar location, cubicle type, cubicle dimensions and installation possibilities

Design	Mounting of equipment			Busbar position			Cubicle type	Cubicle dimensions			Installation		
	fixed	modular assemblies	withdrawable	top	back	centre		width W mm	height H mm	depth D mm	against a wall	free-standing	back-to-back
8PU.011 [1]	×	×	×	×			L1, S, K, T1, T3, F, C, Z	600 800	2200	400	×	×	×
								600 800		600	×	×	
								600 800		800		×	
8PU.003 [2]	×	×	× [4]		×		L1, L2, T1, T4, F, C, Z	600 800 900		600	×	×	×
8PU.007 [3]	×	×	× [4]			×	L1, L2, T1, T4, C, Z	800		800		×	

[1] manufactured in the factory
[2] manufactured in the factory/workshop
[3] manufactured in the workshop
[4] only with 3WN withdrawable air circuit-breakers

connected directly to the switchgear terminals (e.g. overcurrent relays) or to terminal blocks, and must also be disconnected prior to removing a piece of equipment.

Removable modular assemblies; semi-withdrawable design

In contrast to the fixed-mounted design, removable modular assemblies or chassis frames represent a considerable step towards convenience when it comes to exchanging pieces of equipment.

The input power for each module is derived from the cubicle busbars (droppers) via the isolating plug-in facility on a load switch disconnector with LV HRC fuses. After the load switch disconnector with fuses has been switched off, and the motor and control cables have been disconnected, the unit can be removed from the cubicle by undoing two or four screws. This may be done while the rest of the cubicle is still live, provided the appropriate precautions, as outlined in DIN VDE 0105, are taken. Figure 7.10 illustrates a semi-withdrawable or removable unit of a type S cubicle for a motor feeder.

Fully withdrawable design

The fully withdrawable arrangement offers the operator the highest degree of convenience for service and maintenance work, and also offers the highest degree of personal safety and operational security.

Figure 7.10
ES-type withdrawable unit for a motor feeder in an S-type cubicle of the 8PU. range

Table 7.4
Data and characteristics common to all three 8PU. constructional designs

Specifications and/or Publications	DIN VDE 0660 part 500 and IEC 439-1
Rated voltages	up to 690 V, 40 to 60 Hz
Rated currents for busbars	up to 3000 A / 3750 A (depending on design)
Material for main, dropper and stub busbars	F30 electrolytic (standard) copper
Peak short-circuit current capability of the busbars	up to 176 kA (peak value)
Infeed currents	up to 4000 A
Degree of protection to DIN 40050, alt. IEC-Publ. 529 and IEC 947-1	IP 40 (IP 54)
Chassis construction	self-supporting individual cubicles
Outer cladding	sheet-steel, colour RAL 7032, pebble grey
Height of the cubicles	2200 mm
Depth of the cubicles	400 mm, 600 mm or 800 mm (depending on design)
Width of the cubicles	600 mm, 800 mm or 900 mm (depending on design)
Selection of the internal equipment	for an ambient temperature of 35 °C (24-h mean value) around the switchboard. In the case of higher ambient temperature, the rated operational current of the equipment is reduced.
Altitude of installation[1]	up to 2000 m above sea level

[1] If the installation site is at an altitude higher than 2000 m, the manufacturer should be asked to supply the permissible (derated) load values for the switchgear assemblies.

Figure 7.11
VS-type withdrawable unit for a motor feeder in an S-type cubicle of the 8PU. range

By this form of construction, the requirements of the so-called VBG 4 specification, as drawn up by the trade association and brought into force on 1.4.79 to ensure operator safety, are met most effectively. Both the input and the output of an outgoing feeder unit (e.g. motor starter) in this arrangement are plug-connected; i.e. both line and load side incorporate three-pole isolating contacts. Unbolting or unscrewing of the feeder cable terminals is not necessary. The auxiliary circuits (up to 30 poles) are also automatically disconnected and connected by means of a multi-pin plug-in arrangement (Fig. 7.11). VS-type withdrawable units for controlling motor loads are exclusively installed in plug-bar (S) cubicles.

Motor Control Centers (MCC)

As opposed to the common practice in Europe, which simply includes a position switch in the holding circuit of the main contactor coil for the electro-mechanical interlocking of a withdrawable unit to prevent it from being drawn while it is switched on, the American Motor Control Center (MCC) arrangement according to NEMA ICS 2-322 requires an additional door-interlocked main switch. This main switch in every motor feeder unit must have a motor-current switching capacity (6 to $8 \cdot I_n$), and ensures that the compartment door in front of the unit can only be opened e.g. during the motor run-up period (possibly with locked rotor, or with welded contactor and associated current) once it has been switched off and the main power to the unit has been safely disconnected. After the main switch has been opened the unit is completely isolated, so that e.g. the changing of a fuse or the setting of an overcurrent relay in the compartment can be carried out safely. The operating mechanism of the main switch can be locked in the off position by up to a maximum of three padlocks to prevent unauthorized operation.

Figure 7.12 illustrates an MCC cubicle (S-type) which accommodates withdrawable units in individual compartments. Figure 7.13 shows a schematic single-line diagram of an MCC withdrawable unit.

Table 7.5 provides an overview of the various cubicle types and their most important features and characteristics.

Clearances and creepage distances

In terms of DIN VDE 0110 parts 1 and 2, the following points must be taken into account in the dimensioning of clearances and creepage distances of horizontal and vertical busbar systems in low-voltage switchboards:

– overvoltage category
– degree of contamination
– tracking resistance of the insulating materials,
– rated voltage and rated impulse voltage

To permit a worldwide manufacture of the 8PU switchboard range in compliance with other international standards and specifications, the development of the design was based on values which exceed those stipulated in DIN VDE 0110. The values for the clearances and creepage distances are based on the standard which applies for MCC's in America, viz. NEMA ICS 2-322. They are:

– 1 inch (25.4 mm) for clearances between busbar parts (L1, L2, L3) and for clearances and creepage distances between busbar parts (L1, L2, L3) and earthed parts (structure, covers, etc.)

– 2 inches (50.8 mm) for creepage distances between busbar parts (L1, L2, L3)

Internal subdivision

Owing to the completely segregated installation of the items of equipment within the cubicle, the constructional design of the 8PU.011, in particular, offers an extremely high degree of safety and operational security. The internal subdivision of this type of cubicle encompasses

▷ the main busbar chamber,
▷ the equipment area,
▷ the cable connection chamber,
▷ the cubicle busbar or plug-in busbar (dropper) chamber.

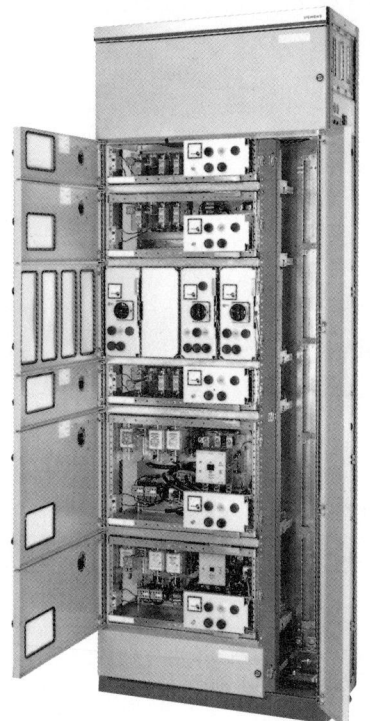

Figure 7.12
S-type cubicle with VS and KS-type withdrawable units; compartment doors opened

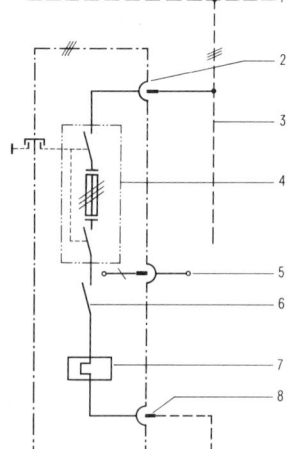

1 Main busbars
2 Incoming plug-connector
3 Dropper busbars
4 Main switch with rear-mounted 3-pole busbar plug, door interlocking handle mechanism and LV HRC fuses (load disconnector with fuses)
5 10, 20 or 30 pole plug arrangement for auxiliary circuits
6 Contactor
7 Thermal overload relay
8 Outgoing plug-connector

Figure 7.13
Schematic single-line diagram of an MCC withdrawable unit with main switch, contactor and overload relay

Table 7.5 Cubicle types, range of equipment and applications

Cubicle type	Device type	Mounting	Typical applications, features
L1	Circuit-breakers 3WN	with-drawable	Infeed from transformers up to 4 MVA at AC 690 V, Busbar coupling/sectionalizing or cross-coupling inside the switchboard up to 3000 A, Feeders to sub-distribution up to 3000 A
L2	Circuit-breakers 3WN	fixed-mounted	Infeed from transformers up to 3.15 MVA at AC 690 V, Busbar coupling inside the switchboard up to 3000 A, Feeders to sub-distribution up to 3000 A
S	Contactors and contactor combinations 3TF, Load switch disconnectors with fuses 3KL/3KM, Moulded case circuit-breakers 3VF	with-drawable	Motor feeder circuits (MCC) up to 400 A, with or without fuses, Cable feeders to sub-distribution with load switches or moulded case circuit-breakers up to 400 A, up to 20 draw-out units per cubicle
K	Contactors 3TF, Contactor reverser combinations 3TD	with-drawable	Small draw-out units for actuators, servo-drives and electro-magnetic valves up to 32 A, Motor feeder circuits up to 32 A, up to 30 draw-out units per cubicle
T1	Fuse switch disconnectors 3NP5	modular assemblies	Cable branch circuits of 160 A to 630 A as infeeds for sub-distribution boards, max. 12 of 160 A feeders or 6 of 250 A to 630 A per cubicle
T3	Load switch disconnectors with fuses 3KL	modular assemblies	Cable branch circuits of 250 A to 630 A as infeeds for sub-distribution boards, Installation of equipment in individual compartments, 4 feeders per cubicle
T4	Load switches with fuses 3NJ6	modular assemblies	Cable branch circuits of 250 A to 630 A as infeeds for sub-distribution boards
F	Contactors 3TF, Contactor combinations 3TF/3TK, Load switch disconnectors with fuses 3KL, Load switch disconnectors 3KE, Fuse switch disconnectors 3NP, Circuit-breakers 3VF, Auxiliary contactor relays 3TH	modular assemblies	Cable branch circuits up to 800 A as infeeds for sub-distribution boards, Motor feeder circuits up to 630 A, Control systems
C	Capacitors with contactor(s), LV HRC fuses or fuse switch disconnector, Power factor correction controller	modular assemblies	Power factor correction, capacitor banks of up to 100 kVAr, total power per cubicle up to 450 kVAr, only one power factor correction controller required per switchboard
Z	As dictated by the application, e.g. DIAZED fuses, LV HRC fuses, Rectifier units	fixed mounted	Empty cubicle for the installation of devices or configurations not covered by the other cubicle types (PTTA)

Elastic (spring-loaded) door fastenings

Spring-loaded door fastenings and locks are provided to prevent an excessive air pressure build-up resulting e.g. from an arcing fault within a compartment or elsewhere in the cubicle. Thereby, it is ensured that the doors or cover plates are not blown off from the frame during a fault. The direction of pressure relief is controlled.

Protection from fault arcs

The 8PU. switchboards are so constructed, that no fault arc can arise under normal operating conditions. Since, however, the possibility of the occurrence of fault arcs (e.g. owing to external influences) cannot be ruled out completely, and the operating personnel must be afforded the maximum safety, 8PU. switchboards were subjected to fault arc testing.

The tests were based on DIN VDE 0679 part 601, 09.84 and the PEHLA recommendations No. 4/12.84, since there are no official national or international specifications or standards which prescribe the fault arc testing of low-voltage switchboards (see page 326).

In accordance with DIN VDE 0679 part 601, the condition of the switchboard after fault arc testing is assessed in terms of the criteria numbers 1 to 6, as follows:

1. No doors opened.
2. No parts, which could cause danger, were blown from the switchboard.
3. No holes in the outer cladding or walls were caused by burning or tearing.
4. No vertical indicators were ignited.
5. No horizontal indicators were ignited.
6. The quality of the earth connections was not reduced.

The type 8PU.011 and type 8PU.003 switchboards (for 8PU.007 refer to Section 7.2.3, page 320) were subjected to tests in terms of the above-mentioned specifications and withstood these successfully. The fault arcs were contained within the cubicles in which they were initiated for the entire specified test duration.

Features which significantly contributed to this success include:
▷ the strict separation of the various functional chambers and compartments as well as the constructional design of the cubicles themselves,
▷ the busbar supports, which are made of a fabric-based laminate bonded in synthetic-resin, act as flash barriers between the cubicles and prevent the fault arc from passing from one cubicle to the next, and
▷ the elastic (spring loaded) door catches (8PU.011).

Automatic shutters

Shutters incorporated in the shrouding around the vertical busbar droppers in each S-type cubicle, provide an automatically operating protection against contact with live parts. The shutter mechanism is operated when the withdrawable unit is removed from or inserting into the compartment. The shutters prevent access to the live busbars when the withdrawable unit has been taken out of its compartment (Fig. 7.14).

If required, similar automatically operating shutters can be provided for the line and load side terminals of withdrawable circuit-breakers installed in L1-type cubicles (Fig. 7.15).

Moving/withdrawing of circuit-breakers

Withdrawable circuit-breakers are physically moved within their guide frames by means of a crank handle which operates the guide frame spindle. The breaker

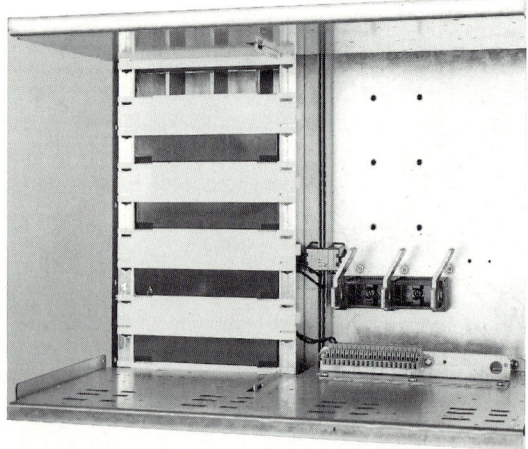

Figure 7.14
VS-type of withdrawable unit chamber showing the automatic shutters in the dropper busbar chamber (The top shutter has been held open for the purposes of illustration)

Standard switchboards, type 8PU · Fault arcs

Figure 7.15
Guide frame of a withdrawable 3WN air circuit-breaker showing the automatic shutters which provide protection against contact with live parts after the breaker has been removed

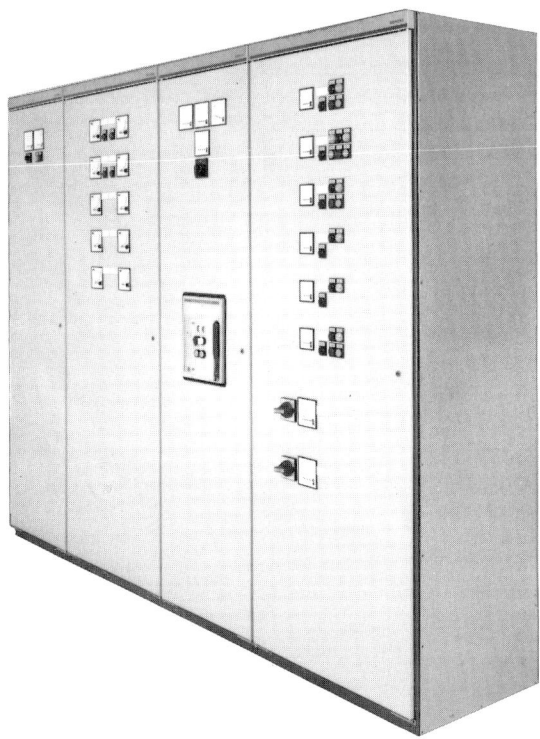

can be moved from the operational position to the test and disconnected positions and vice versa without the compartment door having to be opened. In fact, a mechanical interlock can be provided to prevent opening of the compartment door if the breaker is in the connected or operational position.

Arrangement of the various cubicle types

Irrespective of cubicle width and depth, the various cubicle types may be joined together in any order provided that the main busbars are located in the same position.

Figure 7.16 shows an 8PU.003 switchboard consisting of various cubicle types.

Extensions

To facilitate the extension of existing switchboards by the addition of new cubicles, all the cubicles are produced with the necessary fixing holes in their structural frames and in the busbars. All that has to be done on site is to unscrew and remove the side covers of the existing switchboard before bolting and connecting the new unit in place.

Installation possibilities

The 8PU. switchboards may be installed

▷ against a wall,

▷ free-standing (one operating front), or

▷ free-standing back-to-back (two operating fronts).

Cable entry, busbar trunking infeed

All types are basically designed for input and output cable entries from the bottom (cable trench, through the floor, etc). In sections with a minimum cubicle depth of 600 mm and the main busbars located at the top, however, cable entry from the top is also possible. To this end, the top cover plates of the cubicles can be unscrewed, and must be fitted with the appropriate cable glands on site.

Provided that, e.g. the cubicle dimensions and main busbar configuration and position permit it, an infeed

Figure 7.16
8PU.003 low-voltage switchboard comprising various cubicle types (viz. a C, T1, L1 and T4-type respectively)

by means of busbar trunking from nearby transformers is possible. The cubicle types L1 and L2 may be considered for this arrangement.

Cable termination

Depending on the type of cubicle and the rated current of the equipment, a number of various cable connection methods are offered (Table 7.6).

Double busbars

All the various cubicles are fitted with a main busbar system in which there are two bars in parallel for each phase. This has the advantage that no holes need to be provided in the busbars for the connection of vertical dropper bars or cable lugs. They are connected by means of a special bolted clamp.

In the case of cubicles arranged in a back-to-back formation, and each having a depth of 400 mm, a second main busbar system can be fitted across the top of the cubicles. Depending on the application, the two busbar systems may be connected together by means of a cross-coupling circuit-breaker.

7.2.3 Transformer load-centre (S) substations for up to 24 kV and 1250 kVA

Application

Typically, a compact 8FA substation incorporating a GEAFOL cast resin transformer is used when, for economic reasons, the high voltage must be brought to close proximity of the load centre. Examples include the decentralized power distribution in large buildings, warehouses, industrial plants, workshops and other manufacturing premises.

Advantages

▷ Installation in generally-accessible (indoor) working areas, in accordance with DIN VDE 0100, DIN VDE 0101 and DIN VDE 0660 part 500 is possible by virtue of the appropriate enclosure of the electrical equipment, the use of GEAFOL cast-resin transformers as well as additional measures incorporated in the components of the S-type transformer substation to ensure the safety of personnel. These measures not only permit safe operation and other work on the substation, but also limit the direct effects of an arcing fault when the doors are closed so that direct risks to persons in the vicinity of the substation are avoided.

▷ Environmental acceptability and adaptability is ensured through the use of GEAFOL cast-resin transformers. These will withstand a continuous overload of up to 140% of the rated power with forced air cooling, affording a power reserve to meet extensive load surges or emergency operation without the need for over-rating the transformers in their initial selection.

▷ The space requirement is small by virtue of the compact and flexible construction.

▷ Project planning is simplified through the use of standardized and type-tested modules.

Construction

An S-type transformer substation consists of the following standardized and type-tested modules:

▷ HV (high-voltage) module,

▷ GEAFOL cast-resin transformer in the transformer cubicle,

▷ LV switchboard type 8PU.007.

The HV infeed can alternatively be arranged without the need of the HV module. In this case an earthing switch is provided directly on the HV terminals of the transformer.

Table 7.6
Cable termination possibilities in the various cubicle types

Cubicle type	Termination possibilities
L1, L2	cable termination busbar stubs for 3 to 14 parallel cables; auxiliary wires up to 2.5 mm² on terminal blocks
S	cable up to 50 mm²: busbar-mounting terminals for termination without cable lug; >50 up to 240 mm²: terminal lugs; auxiliary wires up to 2.5 mm² on screwless plug-in contacts
K	up to 35 mm² on terminals; auxiliary wires either soldered, plug-in or up to 2.5 mm² on terminal blocks
T1, T4	directly onto the switching devices
T3	terminal lugs with holes for cable lugs
F	directly onto the switching devices; auxiliary wires either on terminal blocks or plug-in terminals

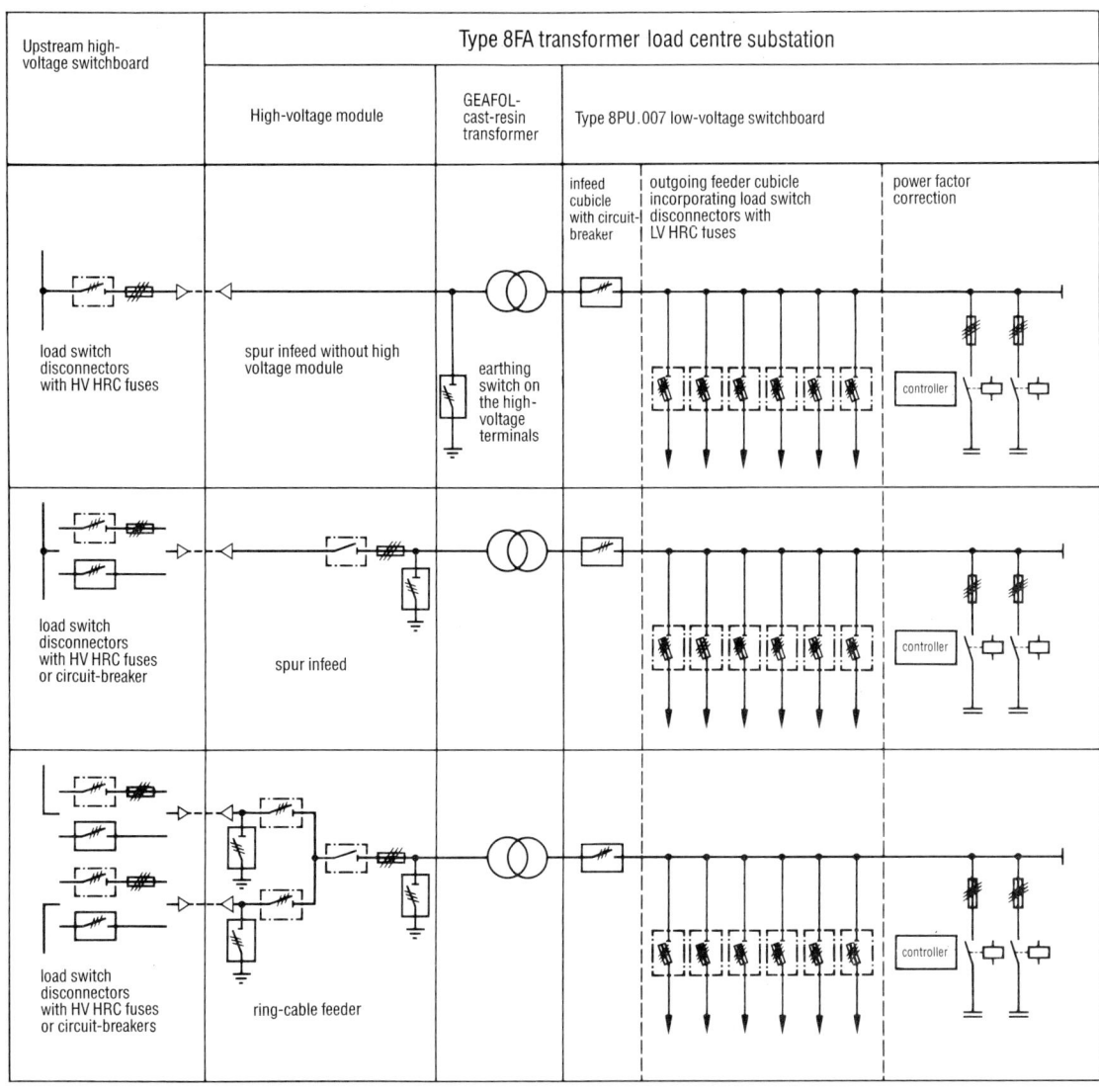

Figure 7.17   Basic circuit diagrams of type 8FA load-centre transformer substations

Various configurations of the type 8FA load centre transformer substations are illustrated by the single-line circuit diagrams shown in Figure 7.17. These differ essentially in the nature of the HV infeed. The low-voltage switchboards can be arranged in any way and can be of any size depending on the transformer power rating.

HV module

The HV switchgear section is a sheet-steel cubicle to accommodate either a spur feed (radial system) or a ring main feed.

In the case of spur feed it contains the switchgear combination comprising a transformer on-load switch disconnector with HV HRC fuses for short-circuit protection. For a ring main feed two additional switch disconnectors are incorporated (refer to Fig. 7.18).

Earthing arrangement

For earthing and short-circuiting of the primary incomer (high-voltage feeder cable) the transformer and cable disconnectors are equipped with earthing switch contacts.

Operational safety

The switch-disconnectors can be fitted with auxiliary contacts for the purpose of interlocking as a protection against incorrect switching operations (Fig. 7.18).

Measures taken against the effects and dangers of fault arcing

A pressure relief duct is built into the back of the HV module. In the event of a fault, ionized gases are discharged upwards along this duct.

Fault arc testing

For the verification of fault arc withstand capability, the HV module has been tested in accordance with the PEHLA Recommendations No. 2 of January 1974. The test criteria numbers 1 to 5 of these recommendations were fulfilled (see page 318).

Cable connections

The high-voltage cables are fed into the switchgear section from below.

Connection of the GEAFOL cast-resin transformer

The connections between the transformer switch disconnector and the GEAFOL cast-resin transformer are made with preformed single-core cables.

The dimensions of the HV module for radial spur or ring main feed incomers are given in Figure 7.19, below.

Technical data

Rated voltage	up to 12 kV, List 2
	up to 24 kV, List 1
Rated operational current	Load switch disconnector up to 630 A
Short-time current (1 sec.)	up to 20 kA
Degree of protection to DIN 40050 alternatively IEC Publ. 529	IP 40
Colour of paint finish	RAL 7032 (pebble grey)

The HV module conforms to VDE 0101.

Figure 7.18
HV module for voltages up to 12 kV; on-load cable disconnector switch and earthing switch

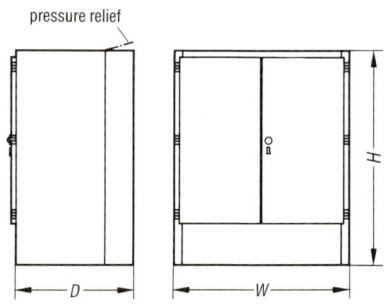

Figure 7.19
Dimensions of the high-voltage module

High voltage	Width (W) mm	Height (H) mm	Depth (D) mm
up to 12 kV, List 2	1470	2200	960
up to 24 kV, List 1	1735	2200	1130

GEAFOL cast-resin transformers

The GEAFOL cast-resin transformer is a self-cooling dry-type transformer. Its insulation consists of a mixture of epoxy resin and quartz powder which is classified as an environmentally acceptable material and which renders the windings maintenance-free, moisture-proof and suitable for use in tropical environments, virtually non-inflammable and self-extinguishing. Even under the influence of fault arcs, no toxic gases are produced.

These cast-resin transformers comply with DIN VDE 0532 part 1 and the IEC Publication 76-1, as well as with DIN 42 523.

Power ratings and technical data

Cast-resin transformers with power ratings from 400 kVA to 1250 kVA may be installed in the transformer load-centre substations.

When the cast-resin transformer is installed in the transformer cubicle, the rated power in *AN operation* (natural air cooling) must be reduced by 10%.

In the case of *AF operation* (forced air cooling), i.e. with the roof fans switched on, the transformer can be operated at 140% of its rated power. This means that the transformer may be rated for the normal load, while peak loads or emergency operation can be covered by *AF operation*.

Further technical details of the cast-resin transformers used in transformer load-centre substations are:

Rated (primary) high voltage to DIN VDE 0111	> 3 up to 12 kV, List 2 >12 up to 24 kV, List 1
Rated (secondary) low voltage	400 V
Rated frequency	50 Hz
Rated value of the impedance voltage	6%
Taps	±5%
Vector group	Dyn 5

The GEAFOL transformer cubicles are available in two sizes. Table 7.7 shows the relationship between cubicle size and transformer rating.

Earthing switch

In the case of spur infeed to the load-centre substation without the use of an HV module, an earthing switch is fitted directly to the high-voltage terminals of the transformer (Fig. 7.20). This switch can be

Figure 7.20
Type 8FA transformer load-centre substation incorporating a 630 kVA, 10/0.4 kV GEAFOL cast-resin transformer; spur infeed from the top with earthing switch on the high voltage terminals of the transformer

Table 7.7
Relationship between cubicle size and transformer power rating

Size of transformer housing	Installation of cast resin transformer with	
	Rated high voltage	Rated power
1	up to 12 kV, List 2 up to 24 kV, List 1	400 kVA to 800 kVA 400 kVA to 630 kVA
2	up to 12 kV, List 2 up to 24 kV, List 1	1000 kVA and 1250 kVA 800 kVA to 1250 kVA

operated through the closed door by means of a lockable drive mechanism. It has a making capacity corresponding to the full system fault level (earthing and short-circuiting take place simultaneously) to ensure safety in the event of an inadvertent switching operation.

By means of an early-make auxiliary contact on the operating mechanism of the earthing switch, a trip

Figure 7.21
Size 2 transformer cubicle for AN operation

command can be given to the load disconnector switch of the high-voltage transformer and/or to the low-voltage incomer circuit-breaker.

For this type of spur infeed, an upstream high-voltage switchboard incorporating an HV load switch disconnector and HV fuses is required.

Transformer cubicle

The GEAFOL cast-resin transformer is housed in a transformer cubicle, which prevents both direct and indirect contact with the transformer and other live parts. The degree of protection according to DIN 40050 and IEC Publ. 529 is:

IP 20 for the roof, and
IP 40 for the side walls.

All parts of the enclosure are finished with a durable synthetic resin lacquer in RAL 7032 (pebble grey).

Variants, methods of cooling

Depending on the rated power of the transformer and the level of high voltage, two sizes of transformer cubicle are offered. In addition, there are two further cubicle variants depending on the type of cooling required (in accordance with VDE 0532 part 1):

a) transformer cubicle for natural air cooling, *AN operation* (Fig. 7.21),
b) transformer cubicle for forced air cooling, *AF operation* (Fig. 7.22).

Figure 7.22
Type 8FA transformer load-centre substation incorporating a 630 kVA, 10/0.4 kV GEAFOL cast-resin transformer, a low-voltage switchboard type 8PU.007 and a transformer cubicle for AF operation

In both cases the cooling air enters through prod-proof ventilation slots in the lower part of the cubicle on three sides.

AN operation

In the case of AN operation the transformer cubicle is fitted with two roof cowls to allow the heat produced by the transformer losses to escape (Fig. 7.21).

A transformer monitoring system, actuated by PTC thermistor temperature sensors embedded in the low-voltage windings, protects the transformer winding by disconnecting the load in the event of excessive heating.

AF operation

In the case of AF operation, two heat extraction fans are fitted in the roof of the transformer cubicle (Fig. 7.22). The fans are switched on and off by a fan control system which is controlled by PTC thermistor temperature sensors embedded in the low-voltage windings of the transformer.

In the event of a fan failure, or insufficient cooling, the fan control system sends an opening command to the high-voltage transformer switch disconnector and/or to the low-voltage incomer circuit-breaker.

Dimensions

The dimensions of the two transformer cubicles are shown in Table 7.8.

Protection against fault arcs

The GEAFOL cast-resin transformer in its cubicle is protected by HV HRC fuses which, by virtue of their current-limiting characteristics and short clearing times, prevent the generation of any serious overpressure. This has been verified by arcing tests in accordance with PEHLA Recommendations No. 2.

Cable entry

In the case of spur cable infeed (radial feed) without the use of an HV module, the cable entry can be from the top or the bottom.

Connection of the transformer to the low-voltage switchboard

The transformer is connected to the low-voltage switchboard by means of cables.

Low-voltage switchboard type 8PU.007

The type 8PU.007 low-voltage switchboard is classified as a type-tested switchgear assembly (TTA) in accordance with DIN VDE 0660 part 500.

The low-voltage switchboards may be assembled to the specified requirements from individual sheet steel cubicles. These contain the actual switchgear and other equipment which may be fitted from either the front or the back. Examples:

▷ incomer or outgoing feeder cubicle, containing withdrawable or fixed-mounted circuit-breakers (L1 or L2-type cubicle – Fig. 7.23),

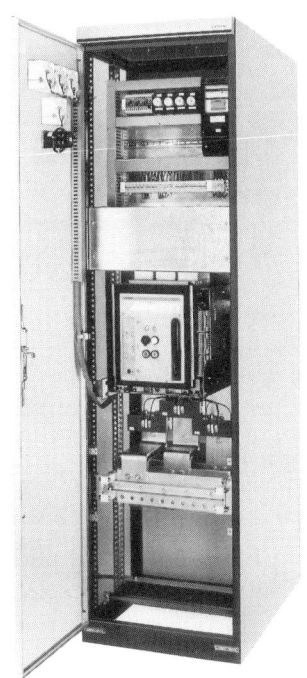

Figure 7.23
Type L1 cubicle containing a 3WN withdrawable air circuit-breaker as incomer

Table 7.8
Dimensions of the transformer cubicles

Transformer housing	Dimensions		
	Width mm	Height mm	Depth mm
AN operation			
Size 1	1800	2370	1200
Size 2	2200	2370	1400
AF operation			
Size 1	1800	2530	1200
Size 2	2200	2560	1400

7 Type-tested switchgear assemblies (TTA)

Figure 7.24
Type T4 cubicle containing in-line fuse switch disconnector units.
Above: 7 × 250 A, below 3 × 630 A

▷ outgoing feeder cubicle, containing mounting chassis with fuse switch disconnectors (T1-type) or in-line fuse switch disconnector units (T4-type cubicle – Fig. 7.24),

▷ cubicle with switched capacitor banks and electronic controller for central power-factor correction mounted on removable chassis-plates (C-type cubicle).

Operational safety

The circuit-breakers with motorized operating mechanisms can be switched while the cubicle doors are shut. This ensures the highest degree of safety for personnel.

In the case of outgoing feeder cubicles, appropriate shrouding and barriers provide protection against contact with live parts for fuse switch disconnectors which have to be operated with the cubicle doors open. The vertical busbar droppers, the connecting bars and the equipment connections are provided with protection against contact (touching of live parts) in accordance with DIN VDE 0106 part 100.

The busbars running across the middle of the cubicles are similarly provided with protection against touching.

Pressure relief

The door mountings and fastenings are suitably strengthened, and the top cover is so designed as to provide pressure relief in the event of a fault.

Fault arc testing

Arc testing is not called for in the constructional specifications for type tested switchgear assemblies (TTA) EN 60439 part 1/DIN VDE 0660 part 500, or in the corresponding IEC publications.

However, in order to ensure the maximum safety for operating personnel, as well as for anyone who may chance to be in the vicinity of the substation at the time of a fault, arcing tests have been carried out in accordance with PEHLA Recommendation No. 2 of January 1974. The test criteria numbers 1 to 5 of this recommendation were fulfilled (refer to page 318).

Cable entries

Cable entry for the outgoing feeders can be at the top or the bottom of the cubicle.

Technical data

For technical data of the 8PU.007 low-voltage switchboard, please refer to Section 7.2.2 (Table 7.4, page 315).

Arrangement of the modules and installation sites for the transformer load-centre substations

Arrangement of the modules

The modules may be arranged to suit the dimensions and shape of the installation site. Figure 7.25 illustrates the various arrangement possibilities.

The low-voltage switchboards may also be

▷ fed from two GEAFOL cast-resin transformers, or be

▷ installed separately from the GEAFOL cast-resin transformer.

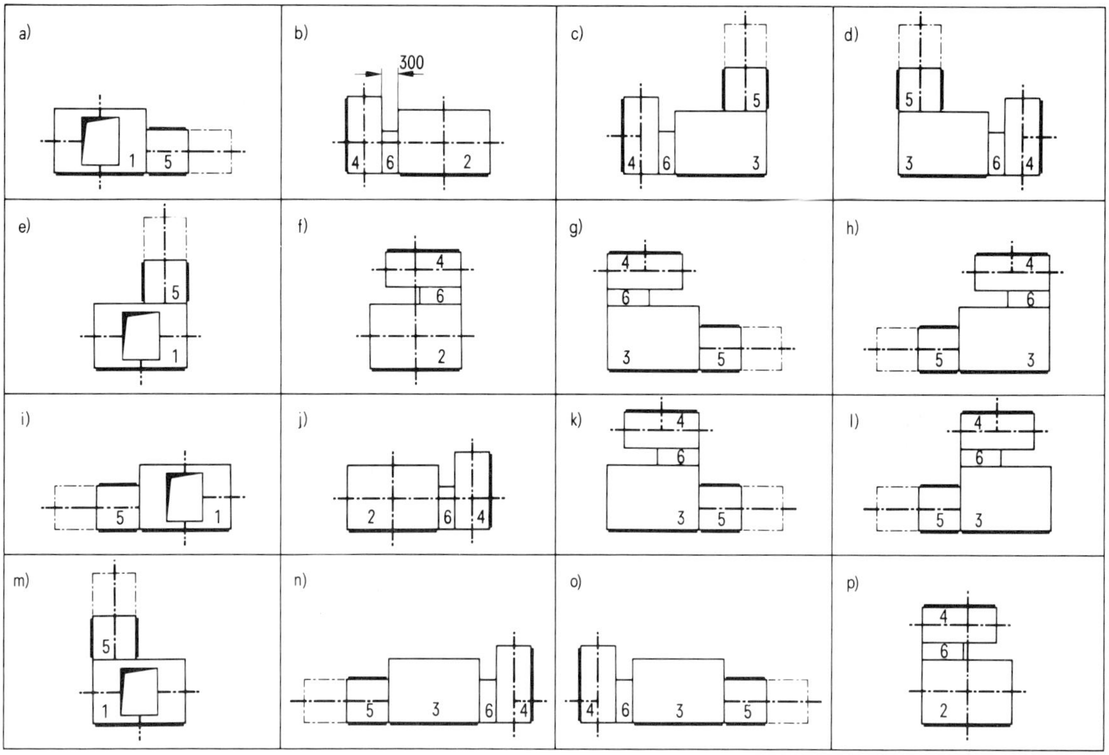

1 Transformer housing over a cable trench
2 Transformer housing over a cable canal
3 Transformer housing
4 High-voltage module
5 Low-voltage switchboard
6 Pressure-relief channel

----- possible cable route to be determined by the site conditions
——— operating side

Figure 7.25 Possible arrangement of the modules comprising a transformer load-centre substation

Installation site

As mentioned previously, it is not necessary to provide a "closed electrical operating area" in accordance with DIN VDE 0100 and DIN VDE 0101 to accommodate these transformer load-centre substations.

Specific minimum distances between the transformer cubicle and walls or neighbouring equipment must be provided (Fig. 7.26).

The transformer load-centre substations must not necessarily be installed on the ground, but may also be erected on mezzanine levels or suspended structures as is illustrated in Figure 7.27.

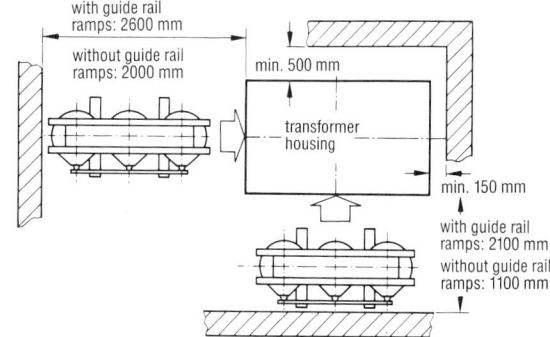

Figure 7.26
Minimum distances for the installation of the transformer cubicle. The distances (with and without guide rails ramps) apply for a transformer rating of 1250 kVA

7 Type-tested switchgear assemblies (TTA)

Figure 7.27
Transformer load-centre substation with a GEAFOL cast-resin transformer installed on a mezzanine level in a factory

7.3 Distribution board systems

Application

Main and sub-distribution boards typically have busbar current ratings of 250 to 1350 A. These boards can be built up from the elements of a distribution board system.

Designs

Generally speaking, distribution boards consist of a number of modular, factory-finished cabinets or enclosures which can be combined to make up a distribution configuration of any size or function. The basic elements of a distribution board system are modular enclosures of differing sizes but with standardized dimensions to facilitate their combination, or bolting together, in the workshop. Such "building block" systems permit the flexible construction of custom-designed distribution boards which are tailored to the particular application. Nowadays, the enclosures are generally made of either:

▷ sheet steel or
▷ synthetic insulating material (e.g. polycarbonate, glass-fibre reinforced polyester resin, etc.).

7.3.1 8HS sheet-steel enclosed distribution boards

Application

The 8HS sheet-steel enclosed distribution boards are suitable for main and sub-distribution, as well as for the enclosure of control systems in power stations, industrial plants, workshops, etc.

Design

The 8HS distribution board system comprises four enclosure sizes with a common depth of 380 mm, whereby the size 4 enclosure is also available in a deeper versions to permit the installation of larger circuit-breakers. Two door versions are offered, viz. in solid sheet steel or with a transparent panel. These distribution boards (Fig. 7.28), which have the degree

8HU (sheet steel) and 8HP (fully insulated) distribution boards

Figure 7.28 8HS sheet-steel enclosed distribution board

of protection IP65, can be installed as self-supporting units on a baseframe or plinth.

The main characteristics of the 8HS distribution board system are:

▷ high operational reliability by virtue of the high degree of protection, robust construction and well-organized lay-out of the switchgear and other equipment;

▷ ease of wiring in the workshop by virtue of spacious terminal compartments, adequate wiring space;

▷ convenient installation by virtue of simple self-supporting structure;

▷ ease of extension by virtue of standardized system elements;

▷ simple installation and termination of switchgear after board installation through the use of clip-on busbar mounting clamps and terminals;

▷ low maintenance by virtue of the durable synthetic-resin coating of enclosure parts.

7.3.2 8HU sheet-steel enclosures

Application

The 8HU sheet-steel enclosures may either be used as individual cubicles, or may be combined to form distribution board units. By using switchgear mounting trays and other modular elements of the 8PU switch and distribution board system, separate mounting of individual items on chassis plates as well as busbars rated up to 630 A, the 8HU sheet-steel enclosures may be used to build main, sub-distribution and control boards. They are particularly suitable for use under extremely harsh environmental conditions, such as in heavy industry, steel works, etc.

Design

The basic enclosures, or boxes, are offered in six different sizes. The 2.5 mm welded sheet-steel design offers an extremely robust encapsulation. The standard version enclosures are closed on all sides. Knock-out panels in the side walls are provided to permit the boxes to be bolted together to form larger enclosures or to attach flanges for cable entry. Boxes which are open on one or more sides, can also be supplied.

The particular characteristics and features of these enclosures are:

▷ strong and robust construction,
▷ high degree of protection,
▷ weatherproof baked epoxy powder coating,
▷ easily extendable,
▷ attachable to the discontinued 8HE distribution board system (U-system)

7.3.3 8HP insulated distribution board system

Application

The 8HP fully insulated distribution board system is used to construct main, sub-distribution and control boards for use in power stations, in a great variety of industrial and trade applications, in offices, schools and residential buildings. By means of a protective roof, outdoor installation is also possible. The system is used particularly in applications, including those in aggressive environments, where "protective insulation" or "total insulation" in accordance with DIN VDE 0100, is demanded.

Design

The basic elements or "building blocks" of the 8HP distribution board system are five enclosures (or boxes) which are sized in terms of their base surface area in the ratio $0.5:1:1.5:2:4$.

Equipped with low-voltage switchgear devices, the enclosures may be used individually, e.g. as starter boxes, or they may be joined together to form a complete switch or distribution board (Fig. 7.29).

7 Type-tested switchgear assemblies (TTA)

The handling and use of the various standardized system elements including the basic building blocks, equipment mounting modules, trays and other hardware, busbar and termination systems and an extensive range of accessories is facilitated by a number of constructive measures and techniques.

Particular characteristics and features of the 8HP insulated distribution board system include:

▷ simple project planning,
▷ simple, economical stockholding,
▷ simple and universal handling of only a few parts,
▷ protective insulation
  (also known as total insulation),
▷ high degree of protection (IP65),
▷ corrosion proof,
▷ resistant to thermal and climatic influences,
▷ high mechanical strength,
▷ simple assembly,
▷ maintenance-free,
▷ flame-retardant (self-extinguishing),
▷ halogen-free (thereby, even no consequential damage resulting from external fire)
▷ low weight of the system components.

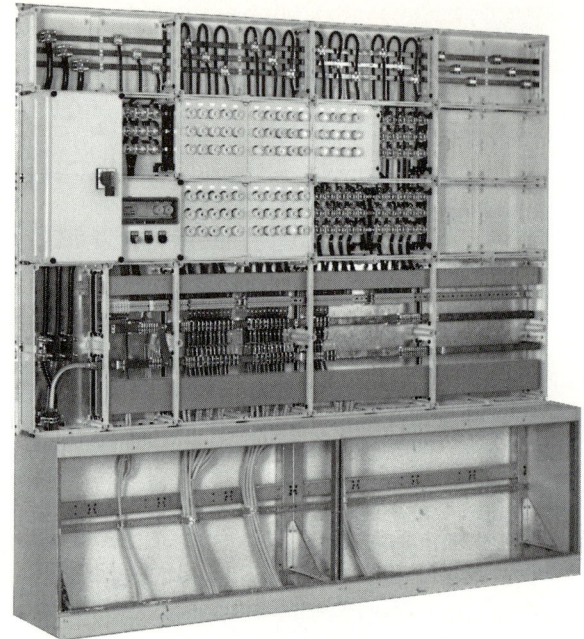

Figure 7.29 An 8HP totally insulated distribution board

7.3.4 8PL insulated busbar trunking system (L-system)

Application

These busbar distribution, or busbar trunking systems are basically lengths of copper busbar sets encapsulated in an insulating material and are used either horizontally or vertically. In the horizontal configuration, they are typically used to distribute electrical energy in production halls and factories of the machine and automotive industry, in manufacturing plants of the wood, paper and textile industries as well as in manufacture and processing areas of the food, and beverage industries. Furthermore, they are extensively used in workshops, machine shops, test stations, laboratories, exhibition stands and trade fair centres (Fig. 7.30).

In the vertical arrangement, they are typically used as the main power distributor, or "rising mains", in high-rise residential, office block or utility buildings to distribute power to the distribution and meter boards on the various floors.

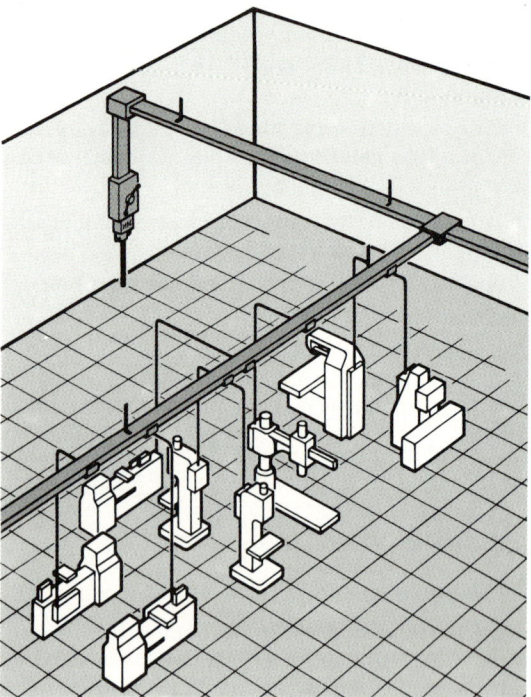

Figure 7.30
Insulation encapsulated busbar trunking installed in a machine shop

Machines, motors and other items of electrical equipment which need to be relocated or regrouped from time-to-time in terms of a varying production programme or owing to expansion or modification of the plant, are connected to the busbar trunking via plug-in tap-off units.

Alternatively, busbar trunking may also be used for the direct infeed of domestic and other sub-distribution boards.

Design

The busbar trunking is manufactured as "type-tested switchgear assemblies" (TTA) and "partially type-tested switchgear assemblies" (PTTA) in accordance with DIN VDE 0660 parts 5 and 500. In particular, it must comply with the specifications of the IEC 439-2 publication, and must be installed in accordance with DIN VDE 0100 and the manufacturer's instructions.

The busbar trunking consists of lengths of copper busbar, encapsulated in glass-fibre-reinforced polyester material with several openings along its length for the attachment of tap-off units.

The system elements include type-tested busbar trunking sections, straight, corner and knee trunking connector sections, end caps, infeed, transfer, thermal expansion and branch units as well as tap-off units in various designs and versions. Accessories include luminaire attachments and a host of mounting hardware. Specially designed and type-tested fire barriers compliment the system.

The busbar trunking is delivered in kit form for simple and quick assembly and installation on site. It is of particular advantage, that the insulating encapsulation can accommodate copper busbar sets of three different current ratings viz. 160, 250 and 400 A. This enhances the flexibility of the system, as only a single range of tap-off units which fits on any trunking section irrespective of its current rating, is required.

Busbar trunking systems offer the following benefits:

▷ simplified location and relocation of machines
▷ long-lasting operational and functional reliability by virtue of the high quality, complete insulation as well as sealed and interlocked tap-off units for voltage-free replacement of fuses,
▷ clear and visible overview of the reticulation
▷ small dimensions,
▷ low weight,
▷ no maintenance in normal environments,
▷ well sealed enclosures,
▷ quick, safe and reliable attachment and removal of tap-off units with one hand,
▷ a single range of tap-off units for all three busbar trunking current ratings.

7.3.5 8L. mounting and wiring systems for control circuits

Application

The mounting systems 8L. were developed to rationalize the manufacture of conventional control systems (e.g for machine tools, processing machinery, conveyer belt systems) and for the mounting of contactor banks typically found in the basic materials and processing industries as well as in power stations.

Version with rapid wiring from the front

The 8LW and 8LV mounting and wiring systems are both designed to offer a quick and simple method of point-to-point interconnection of equipment from the front of the switchboard. Through the use of pre-cut and finished wires, the workshop costs per connection are reduced significantly in comparison with wiring in cable channels (ducting).

The basic elements of the system are mounting trays which are attached to a chassis or to the framework of the switchboard by means of special spacing brackets. These spacing brackets provide room for the wires between the back of the mounting trays and the back wall of the switchboard cubicle. The flexible wires are held in place by supports and guides clipped onto to the 35 mm mounting rails (DIN EN 50022), or they are pressed between the tines of comb-like supports which are fitted to the edges of the mounting trays (Fig. 7.31).

The characteristic features of the 8LW and 8LV mounting and wiring systems include:

▷ all wiring can be done from the front,
▷ simple assembly,
▷ low space requirement,
▷ devices are screwed to the equipment trays or are clipped onto 35 mm mounting rails,
▷ use of pre-finished flexible wire lengths,

7 Type-tested switchgear assemblies (TTA)

Figure 7.31
Illustration of switchgear in a control panel using the 8LW mounting and wiring system

▷ simple fixing and arranging of the wires by the use of comb-like plates and strips of pliable plastic material,

▷ standardized dimensions of the elements match the modular sizes of the 8MF cubicle range.

Version with direct point-to-point wiring at the back of the panel

The 8LX mounting system (X system) is designed to allow wiring from behind the equipment trays. The basic elements of the system are equipment mounting trays of various widths and heights. These are attached, one above the other, on a framework to form a panel. Special PVC hole-plates are fitted in the spaces between the individual equipment mounting trays. The control wires are fed through these hole-plates to the back of the panel, where they take the most direct and convenient path between the individual pieces of equipment.

The 8LX mounting and wiring system has the following characteristic features:

▷ wiring from the front and the back of the control panel,

▷ fast and simple assembly, mounting of equipment and wiring,

▷ clear and space-saving arrangement of the devices,

▷ devices are screwed to the equipment trays or are clipped onto 35 mm mounting rails,

▷ rational and unrestricted wiring,

▷ fixing of wires by means of PVC hole-plates.

7.3.6 8MF cubicle system for switch, distribution and control boards

Application

The 8MF cubicles are suitable for the installation of switchgear devices and other equipment used in conventional or electronic control systems including automation, measurement, process control, protection and communication. They are also suitable for use as low-voltage switchboards (Fig. 7.32).

For example, the construction permits the simple and cost-saving installation of the industrially common 19 inch electronic card rack systems, as well as outgoing feeder modules for energy distribution.

The 8MF cubicles are also offered in EMC-versions (*e*lectro*m*agnetic *c*ompatibility), as well as in versions which incorporate special measures taken against induced vibration (e.g. owing to earthquake).

Figure 7.32
8MF system cubicle

Design

The structure is based on two angle-welded side frames and four identical profiles for the top and bottom bracing.

Locking brackets ensure that the cubicles are always assembled at right-angles and that the dimensions are always within design tolerances.

Since no assembly jigs or measuring instruments are required, the cubicles may easily be assembled on site. The structure is designed to carry the total weight capacity of the cubicles. The total weight may be up to approx. 500 kg crane-hook weight, and an additional 200 kg of material may be added once the cubicle has been placed in its final position. Crane-hook rings can be attached to the top of the cubicles for transport purposes. Two holes of 14 mm diameter are drilled into each bottom brace to permit the cubicles to be bolted to the floor.

Two 90° angle brackets with regular holes at 20 mm distance along their lengths are fitted inside both side wall frames of the cubicles. The top and bottom bracing also have strips of mounting holes. Thus, installation of equipment is possible in all three axes within the cubicle. A series of holes (100 mm centre-to-centre) in the side frames serve to attach the side covers, or to join cubicles to each other. The door hinges are equipped with a quick release mechanism which allows them to be attached or removed with the greatest of ease. In spite of the fact that the hinges are recessed and covered, the doors may be opened through an angle of about 180° in the case of single cubicles (approx. 150° in the case of two or more cubicles joined together).

The 8MF switchboard cubicle system has the following characteristic features:

▷ degree of protection IP40 or IP54 (it is also possible to upgrade from IP40 to IP54 after delivery),

▷ available as single cubicle or as series connected units,

▷ available assembled or in knocked-down kit form,

▷ low stocking space required in the case of knocked down cubicles,

▷ large range of versions for individual applications.

7.4 Guidelines for project planning of low-voltage switch, distribution and control boards or systems

7.4.1 General

In order to supply electrical power to all consumer loads of e.g. a home, an apartment block, an administrative building or a factory reliably, the appropriate low-voltage switch and distribution boards or even control systems at each point in the network must be so planned and constructed that they comply with the specifications and conditions prevailing at their respective points of installation, and that they afford the necessary switching functions, protection and monitoring for both the connected loads and the cables.

The end user of the system – if necessary in conjunction with the manufacturer of the switchboards, distribution boards and control systems or a consultant – specifies the local operational and environmental conditions, and supplies the manufacturer with all the electrical data pertaining to the switchboards and distribution boards and to the supply system at the point of installation, so that the technically most suitable and most economical installation can be planned for each point in the network.

Environmental and installation site conditions

The following points are of particular importance to the planning of electrical installations:

▷ mechanical stresses,

▷ degree of protection to DIN 40050, IEC 529, IEC 947-1, protection against contact with live parts, protection against ingress of dust and water,

▷ ambient temperature and climatic conditions,

▷ corrosive influences,

▷ type of installation and mounting (e.g. wall-mounting, free-standing),

▷ cubicle, frame or box-type distribution board,

▷ covers or doors transparent or not,

▷ maximum permissible dimensions of switchboard, distribution board or control system for transport and installation,

▷ cable duct (existing or non-existent), cubicle base cladding if required,

▷ cable entry and exit (top or bottom),

▷ nature of cable laying (cable ducting, cable shafts or trenches, cable trays, etc.),

7 Type-tested switchgear assemblies (TTA)

▷ equipment mounting (non-withdrawable, alternatively, removable e.g. withdrawable for quick and simple replacement),

▷ overall dimensions; if necessary the maximum permissible outer dimensions of the distribution board, switchboard or control panel must be taken into account,

▷ accessibility of the equipment and switchgear devices. Devices which need to be operated under normal running conditions (e.g. fuses and fused switches, miniature circuit-breakers, MCCB's, etc.) should be grouped together within the switchboard and so arranged as to provide the necessary specific and easy access (e.g. under a quick-release cover). In other words, contactors and associated fuse-gear may have to be mounted in separate compartments.

Protection against condensation

If distribution boards with a degree of protection IP 55 or IP 65 (e.g. 8HP or 8HS systems) are installed outside, and are subject to significant simultaneous changes in humidity and ambient temperature, then it is recommended that breathers be fitted to the bottom of the cubicles to prevent condensation inside the enclosures.

Breather glands should always be fitted to the underside of the cubicles in such a way as to prevent the ingress of water (e.g. water spray or jet) through the breathers themselves. This can also be achieved by a shield plate or cover strategically fitted in front of the breather opening.

In addition to these measures, the use of a continuous heater of low power consumption (e.g. heater resistance or element) inside the enclosure, can prevent the inside temperature from dropping below the condensation point.

Nature of the installation; accessibility

To ensure that the most suitable and economical construction is selected for a particular application, the various characteristic features and constructional details of the different switch and distribution board types should be compared early in the planning stages. Examples of such features include:

▷ open or closed construction (nature of the working area),

▷ self-supporting construction: free-standing, against a wall or in an alcove,

▷ non self-supporting construction: for fixing to a wall, on a supporting structure, or in an alcove,

▷ accessibility, e.g. for mounting, maintenance and operation,

▷ dimensions (height, depth, width),

▷ constructional differences.

The minimum isle widths are shown in Figure 7.33. If doors which only open against the direction of the escape route are fitted to the switchboard, then the distances given in Figure 7.33 apply from the opened door to the outer wall or, alternatively, to the next switchboard (refer to Section 1.4.4, page 54).

Open-air installation

Distribution boards which are installed outside (open-air), must be protected from direct sunlight by means of a protective roof or shield. All elements and components, unless inherently resistant to corrosion, must have a high quality surface treatment (e.g.

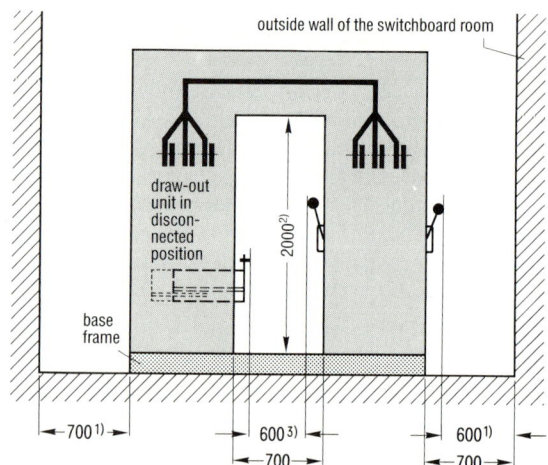

[1] Minimum dimensions to be adhered to for the installation of the switchboard assemblies (in keeping with the terms of DIN VDE 0101, a mounting isle width of 500 mm behind closed back walls of switchgear assemblies, may be acceptable).

[2] Minimum passageway under obstacles, e.g. cladding with at least IP 2X degree of protection.

[3] Only in the vicinity of operating elements for electrical equipment or draw-out units in the disconnected position.

Figure 7.33
Minimum dimensions of isles and passageways within a switchboard installation delivered as a complete unit and with cladding offering a minimum degree of protection IP2X to DIN 40 050 against direct contact with live parts

galvanizing) to render them suitable for exposure to industrial atmospheres.

For further details, please refer to Section 7.4.7 "Degrees of protection, climatic and other ambient conditions" on page 360.

Electrical conditions and data

In addition to having a function plan, e.g. block diagram, and a circuit diagram in the case of control boards, the planning engineer must have access to the following electrical conditions and data:

▷ applicable DIN VDE specifications, regional or international specifications, norms and standards,

▷ required protective measures (refer to Section 1.4, page 33), such as protective insulation, protection by means of switch-off or indication, etc.,

▷ network type (system),

▷ rated network supply voltage and frequency,

▷ rated currents, number of conductors (3, 4, 5 line) in busbar systems,

▷ prospective short-circuit current at the point of installation,

▷ location of the feeder cable (from the top, bottom or side),

▷ nature (alternatively type and cross-section) and number of cables and/or conductors,

▷ location of outgoing feeders (to the top, bottom or side), cross-section and number of cores,

▷ number of outgoing feeder circuits and specification of their configuration (contactors, circuit-breakers, fuses etc.), rated power, rated or operational current, setting range of current-dependent delayed overload releases, etc.,

▷ rated auxiliary and control supply voltages and frequency in the case of control devices as well as position, colour and nature of command and indication devices,

▷ if applicable, duty factor or indication of loads and feeder circuits which can or cannot be operated simultaneously.

The clarity, detail and completeness of these specifications contribute significantly to cutting down the planning and design time, and ultimately contribute to the degree of success in finding specific solutions for the individual application at hand.

If the above-mentioned details are available, the project manager will be in a position to select the most technically and economically suitable type of switch and distribution boards. For example, he will be able to use a decision matrix such as is shown in Figure 7.1 (page 301) which employs a number of selection criteria and is designed to guide the user to the most suitable switchboard type for his application. This particular decision matrix is based on the range of switch and distribution boards described in Sections 7.2 and 7.3.

Short-circuit strength

The initial a.c. short-circuit current I_k'' at the point of installation may not exceed the value given for the short-circuit strength of the switch or distribution board. This condition may also be fulfilled by means of an upstream current-limiting protection device.

Guidelines for the calculation of the short-circuit current at the point of installation are given in Section 2.1.3, page 81.

Short-circuit protection devices for feeder cables

Generally speaking, it makes sense to install the short-circuit protection apparatus required for the switch and/or distribution board itself at the start of the infeed cables, so that these are included in the scope of the protection. This means, that the infeed circuit within the switch or distribution board itself normally need only incorporate a means of isolation (e.g. on-load disconnector, isolation links). In the case of control panels complying with DIN VDE 0113, the use of a main switch is stipulated; in all other applications, the type of disconnector is determined by the specific technical, safety and economical aspects.

In view of the fact that the shortest possible fault switch-off time is normally required in the low-voltage network, it is not recommended that this switch-off is provided by e.g. circuit-breakers or HV fuses on the high-voltage side of the feeder transformer alone. Fuses or circuit-breakers on the low-voltage side always provide a higher level of protection through quicker rupturing or switch-off times in the event of a fault.

Figure 7.34 illustrates the location of short-circuit protection devices for the protection of feeder cables and downstream distribution boards, as well as the arrangement of disconnecting devices within the distribution boards themselves.

7 Type-tested switchgear assemblies (TTA)

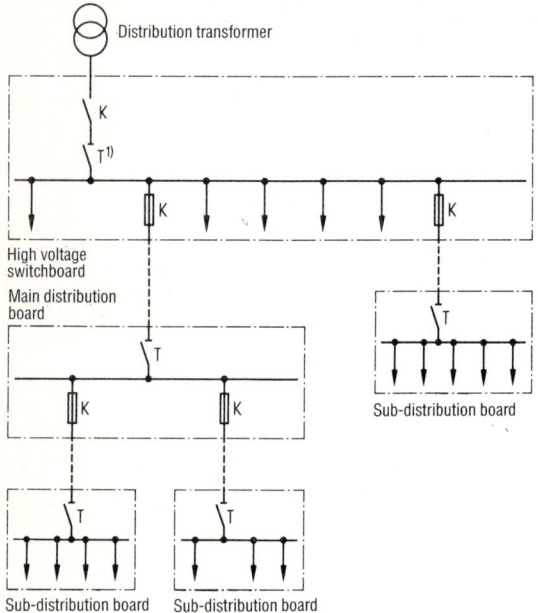

1) Only required if reverse voltage (back-e.m.f.) is possible

K Short-circuit protection for the feeder cables and downstream distribution boards
T Disconnection device in the infeed

Figure 7.34
Location of short-circuit protection and disconnection (isolation) devices in switchboard and distribution systems

Short-circuit and overload protection for outgoing cables, wires and switchgear incorporated within the switch or distribution board

In addition to the aspects mentioned above, the following points should be considered in the case of switch and distribution boards:

▷ short-circuit and overload protection of outgoing cables and wires for which (in terms of DIN VDE 0100 part 430) the fuse ratings must be selected in accordance with the conductor cross-sectional area (refer to Section 9.4, page 516),

▷ short-circuit protection of switchgear incorporated in the switch or distribution board. This protection must ensure that in the event of a fault, the short-circuit current must not exceed any of the following rated values of the switchgear:
 – switching capacity (e.g. circuit-breakers for motor protection),

 – dynamic stressing (e.g. in the case of short-circuit currents exceeding 50 kA r.m.s.),
 – resistance to contact welding (e.g. in the case of contactors, relays, etc.),
 – thermal capacity (e.g. in the case of overload relays).

Protection against direct contact with live parts

Protection against direct contact with live parts can e.g. be achieved by one or more of the following measures (refer to DIN VDE 0100 part 410 and DIN VDE 0660 part 500):

▷ insulation of active parts,
▷ covering or shrouding (degree of protection at least IP 2X, top side of horizontal surfaces IP 4X) or
▷ installation of open and unprotected switchboard versions (IP 00, IP 10) in closed and locked electrical rooms (or cabinets) in terms of the corresponding specifications in DIN VDE 0100.

In terms of DIN VDE 0100, covers and shrouding which serve to protect against contact with live parts in switch and distribution boards installed in generally accessible working areas, may only be removable by means of a tool (including a key). For further information, please refer to Section 1.4.1, page 33.

Particular guidelines for equipment which needs to be accessible for operation

Unless a switchboard is installed in closed electrical working areas, "quick-release covers", which can be removed without the need of a specific tool or key, may only be used on compartments of the distribution board (e.g. fuse box) if these are equipped with an (additional) internal insulating shrouding or barrier to provide the necessary protection against accidental contact with live parts when the cover is opened. Since, for example, fuses may have to be replaced occasionally under operational conditions, it makes sense e.g. to install contactors and their associated fuses in separate compartments. The fuse compartment only may then be equipped with the necessary internal shrouding and quick-release cover.

Protection against indirect contact with live parts

The term "protection against indirect contact with live parts" refers to measures taken in addition to normal operational insulation. In the event of insulation failure, these measures are intended to prevent personnel or working animals from coming into contact with an excessively high contact voltage on any part of the equipment, and to prevent the prolonged existence of such contact voltages (also refer to Section 1.4.2, page 40).

Total insulation

If totally insulated switch and distribution boards are used, care must be taken to ensure that the total insulation provided by the design and material of the units is not jeopardized by the installation of metal parts such as switch operating shafts, metal cable glands, screws, etc.

Inactive metal parts within the totally insulated enclosure, such as base plates and switchgear housings, may under no circumstances be connected to the PE or PEN conductor (earth), even if they are equipped with a special terminal for earth connection.

If covers or doors can be opened without the use of a key or a specific tool, then all accessible conductive parts located behind such covers or doors must be equipped with insulating barriers which provide degree of protection IP 2X or higher. These barriers may only be removable through the use of a specific tool.

Totally insulated enclosures provide a high degree of protection against corrosion and generally do not require any form of maintenance (refer to Section 1.4.3, page 53).

Metal enclosures

Generally, enclosures and metal parts of the supporting structure are protected against corrosion by means of suitable surface treatment. All metal parts of the switch and distribution boards, excluding the case of total insulation, are to be incorporated into the protection by PE conductor (also refer to Section 1.4.3, page 53).

Space requirements of installed equipment

Particularly in the case of enclosed switch and distribution boards, consideration must not only be given to the space requirements of the actual installed switchgear and devices themselves, but also to the additional space required for:

▷ clearances between live parts and the enclosure walls,

▷ heat dissipation and coolant air flow between the devices,

▷ possible arcing space in the case of switching devices,

▷ wiring and copperwork between the devices,

▷ termination of incoming and outgoing feeders (connecting chambers)

▷ labeling of the devices. Clear and unambiguous identification of the devices and their relation to one another in both the documentation and the finished switchboard is imperative. It is particularly important, for example, to establish and indicate a clear relationship between fuses and their associated circuits (also refer to Section 8.1.3, page 408).

Thus, devices should never be mounted as closely together as theoretically permitted by a mere comparison of their outer physical dimensions and the inner dimensions of the enclosure.

If possible, meters and other indicating devices should be mounted at eye-level. All hand-operated devices should be located within easy reach (approx. at a height between 0.6 and 1.8 m).

Under specific circumstances, it may be necessary to take certain restrictions and/or deratings into account when using a piece of equipment in a confined space. Where applicable, details of such restrictions pertaining e.g. to switching capacity or current ratings are given in the manufacturers' technical catalogues.

Temperature rise

In the case of an open construction, the heat losses produced in items of electrical equipment is carried away directly by the unrestricted flow of air past the devices and conductors. In the case of encapsulated switchgear, on the other hand, the heat transfer takes place between the surface area of the encapsulation and the surrounding air. Therefor, particularly in the allocation of space for typical heat-producing devices such as thermal overload relays, rectifiers, transformers, fuses, etc. careful consideration must be given to adequate heat dissipation within the enclosure. If the devices are mounted too closely together, and the enclosure is too small (surface area), then insufficient heat dissipation could lead to thermal overloading and ultimate destruction of the devices. This explains the derating factors which sometimes have to be applied to the rated current of several switching devices if these are to be installed in small enclosures.

In general, Siemens switchboards, distribution boards and control panels may be installed and operated in rooms with an ambient temperature of up to 40 °C without the need to reduce the rated values. At higher ambient temperatures it may be necessary to reduce the permissible continuous current rating. In these cases, the applicable derating factors found in the Siemens l.v. switchgear catalogues are to be applied.

As a rule of thumb, one can say that for every 1 °C rise in ambient temperature above 40 °C, the continuous current rating should be reduced by 1%. Owing particularly to the temperature rise limits of their magnet coils, special attention should be given to the use of contactors at temperatures exceeding 40 °C. Please refer to the appropriate data in the Siemens catalogue.

Further details on this subject may be found in Section 7.4.6.1 – "Heat extraction from cubicles with ventilation openings" and from sealed cubicles (page 353).

Cable termination (connection) chambers

The space available inside (and outside) the switchboard for the termination of incoming and outgoing cables, wires and other conductors contributes significantly to the ease, duration and efficiency of the installation and erection procedures.

The purchase price of a particularly small and compact enclosure may, at first, appear to be very economical and attractive. Owing to restricted space for the termination of incoming and outgoing feeder cables, it may however cause exorbitant initial installation costs and may even restrict the possibilities of future modification or extension. Unless complete and ready-wired units are chosen from the catalogues, special care must be taken in the selection of encapsulated switch and distribution boards and the equipment or kits to be fitted in them (e.g. 8HP or 8HS distribution board systems), as well as in the engineering and lay-out of distribution and control panels to ensure adequate space for the termination of incoming and outgoing cables and wires. In the case of cable with larger cross-section, allowance must be made for the bending radius and adequate space must also be provided for the spreading of the cable cores. As a guide, one can estimate that the unrestricted length of the terminating chamber should be at least four to six times the outer diameter of each single core.

Special requirements

Special requirements such as explosion protection, resistance to strong vibration, protection against aggressive atmospheric conditions or e.g. the exclusive use of a particular PE conductor must be taken into account in terms of the appropriate specifications or within the framework of additional agreements.

7.4.2 8PU low-voltage switchboards

Various positions of the main busbars

Owing to the different positions of the main busbars, the height of the equipment installation area in the various versions of the 8PU switchboards, differ from one another.

Main busbar position

The main busbars are located

▷ across the top of the cubicles in the 8PU.011 version,

▷ across the back of the cubicles in the 8PU.003 version and

▷ through the middle of the cubicles in the 8PU.007 version, thus facilitating the installation of devices from both front and back (these sections are always free-standing)

Separate busbar compartment

The 8PU.011 incorporates a busbar compartment which shrouds and separates the main busbars from the other electrical equipment and switchgear.

Nature of equipment installation

In terms of the equipment mounting technique employed, the following possibilities are offered:

▷ Fixed mounted / removable modular assemblies in the 8PU.003 and 8PU.007 designs as well as in some of the cubicles of the 8PU.011 design,

▷ simple withdrawable arrangement (ES) with incoming isolating contacts which plug onto busbar droppers – only in the S-type cubicle of the 8PU.011 design,

▷ fully withdrawable arrangement (VS, KS) only in the S-type cubicle, and small withdrawable units in the K-type cubicle of the 8PU.011 design,

▷ withdrawable arrangement in the L1-type cubicles of all three designs.

Project planning aids

The following list includes a number of aids which facilitate the project planning:

▷ Pre-printed drawing patterns by means of which the dimensions of the switchboard may be determined and the lay-out of the cubicles including the location of the individual compartments may be planned,

▷ Pre-printed data recording sheets for the registration of the electrical data, dimensions, cables to be connected, site and environmental conditions and even the agreed hand-over date in the factory if required,

▷ standard circuit diagrams; these may be included with the offer to illustrate its scope and to facilitate the passing on of clear and unambiguous instructions to the workshop in the event of an order being placed,

Among others, standard circuit diagrams are offered for the following applications:

▷ incoming feeder, bus-coupler and outgoing feeder withdrawable circuit-breakers in an L1 circuit-breaker cubicle

▷ motor control circuit e.g. for direct-on-line starting (Fig. 7.35), reversing, pole-changing, star-delta starting, Dahlander starting as well as feeder circuits for lighting and other power loads.

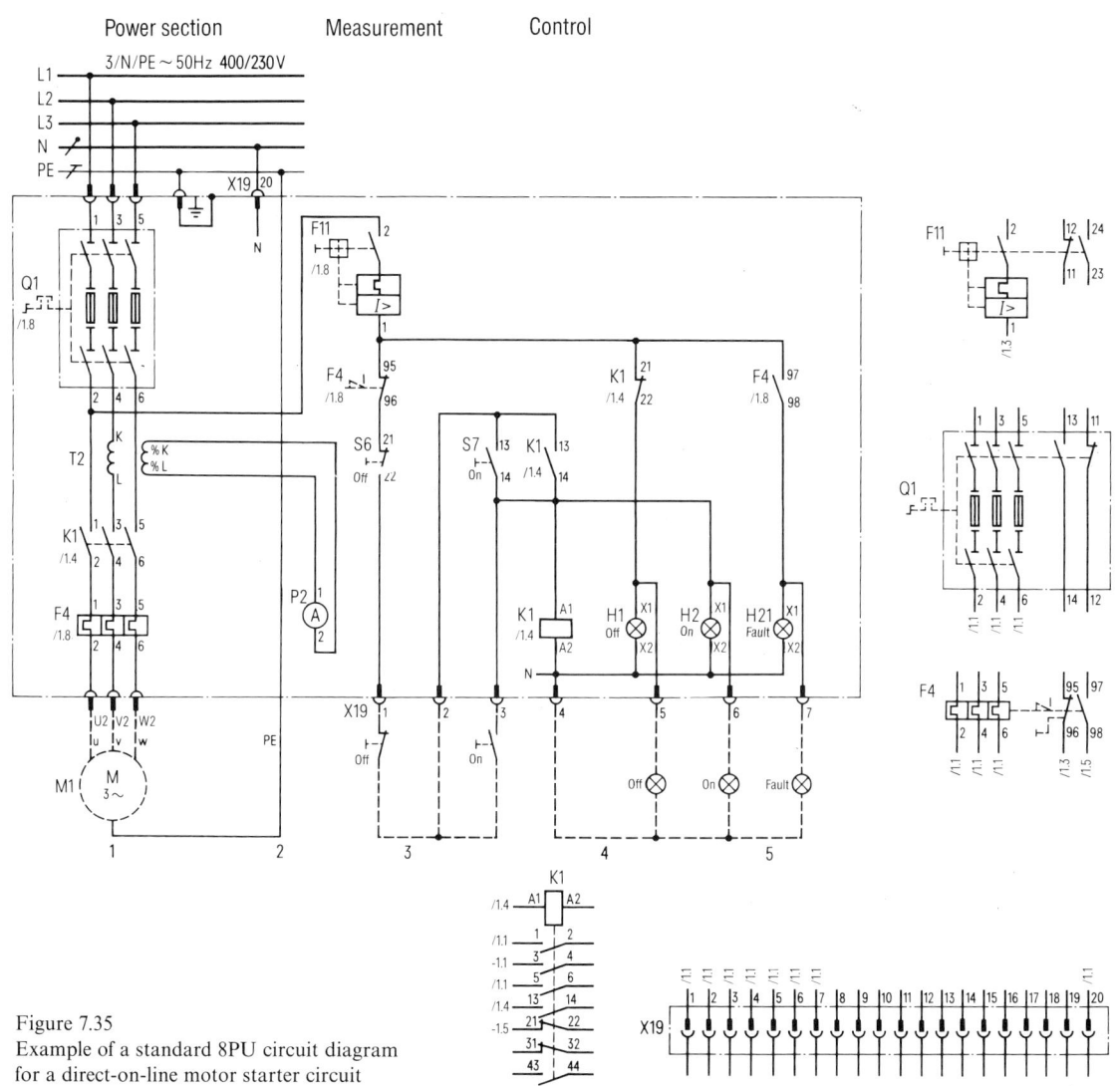

Figure 7.35
Example of a standard 8PU circuit diagram for a direct-on-line motor starter circuit

The circuit diagrams are created on a computer aided design system. The diagrams can easily and quickly be modified and adapted to the requirements of the specific task at hand; either manually, or by means of a computer-aided inter-active graphic process.

MCC technology

Owing to their compartmental design and the technology employed in the S and/or K-type cubicles of the 8PU.011 design, they are ideally suited to accommodate the control and feeder / starter circuits of a **M**otor **C**ontrol **C**entre (MCC). Each contactor / overload combination with ancillary equipment is mounted in its own withdrawable chassis, and is supplied via a main switch (with motor switching capacity). This main switch (3KM) not only carries the LV HRC fuses for the short-circuit protection of the starter and feeder cables, but also provides the necessary compartment door interlocking by means of its operating handle mechanism.

7.4.3 8HS, 8HP and 8HU distribution systems

The switchgear and/or the distribution assemblies for the incoming feeder are selected in accordance with the function diagram (single line diagram). If possible, the incoming feeder should be positioned in the middle of the distribution board. The incoming feeder terminals are to be positioned in such a way that the busbars are not loaded with more than their rated current at any point; even in the case of multiple infeed.

The switchgear and/or the distribution assemblies for the distribution feeder circuits are also selected in terms of the single line diagram. These should be located to the left and right of the incoming feeder in such a way that outgoing feeders for greater loads are closest to the incoming supply.

In the case of box-type distribution boards, the size and position of the busbar enclosure is determined by the size and position of the incoming and outgoing feeders as well as the ambient temperature (see Section 7.4.6.6, page 358).

Boxes for fuses and miniature circuit-breakers may, for example, be located above the horizontal busbar run if they are accessible for operation in this position. For feeder circuits with fuses, the back-up fuses should preferably be installed in the busbar enclosures (direct-mounting onto the busbars), if space permits. If the complete distribution assemblies or kits offered are not suitable, then the required switchgear (or multiple kits) may be installed in empty boxes. For further details of assembly and equipment mounting possibilities, please refer to the relevant Siemens l.v. switchgear and switchboard catalogues.

Cable entries, cable glands or rubber grommets are selected according to the cables and wires to be connected.

If possible, the different box sizes should be assembled in such a way as to produce a complete solid structure without any gaps. A system planning template (M 1:10) may be used to make constructional sketches of the distribution board and to determine the most efficient arrangement of the elements (see Section 8.13, page 496).

7.4.4 8L. mounting systems for control circuits

The switchgear and other components to be mounted are selected in terms of the circuit diagrams and parts lists.

The component lay-out must be drawn to scale. This process is at least partially facilitated by the use of plastic templates containing the outline of a large number of devices as well as the use of pre-printed planning sheets (refer to Section 8.13, page 496). Such a sketch may be used as a workshop drawing since it indicates the exact positioning and space requirements of the components.

If possible, one should not only position the components in terms of their electrical relation to one another and their direct interconnection in the lay-out plan, but should also separate the power and control functions.

Heavy devices should be positioned on the lowest mounting plate or rail, and in the case of wide switchboards, they should be mounted as close to the supporting framework as possible.

Devices which are known to produce heat, such as rectifiers, transformers, etc. should be mounted as high as possible so that the heat they produce does not adversely effect other devices above them.

7.4.5 Domestic and utility distribution boards

These can be divided into categories "fixed" and "transportable".

The category "fixed" includes domestic distribution boards, wall mounted and free-standing units as well as cable distribution boards, meter boxes, etc. Typical of the category "transportable" are building site, exhibition hall and camping site distribution boards.

According to their construction as main or sub-distribution boards, they are used for the supply of electrical power to loads in residential and office buildings, trade and industrial premises, markets and exhibition areas as well as building, construction and camping sites.

Distribution systems should be planned and installed by adequately qualified personnel. It is not sufficient that the application, purpose and site conditions of the distribution board be known. In addition, a single-line function diagram (and in the case of controls, a circuit diagram) should be placed at the disposal of the contractor.

Furthermore, the following specifications and DIN standard are of importance:

▷ DIN VDE 0603 Distribution and meter boxes with up to 250 V to earth,

▷ DIN VDE 0660 part 504, 04.92 – low-voltage switchgear combinations, EN 60439-3; domestic distribution boards, e.g. also 8 HP,

▷ DIN VDE 0660 part 501, 04.92, section 4 – particular requirements for building site distribution boards,

▷ EN 60439 part 1 / DIN VDE 0660 part 500 Low-voltage switchgear assemblies,

▷ DIN VDE 0660 part 503 Cable distribution boards,

▷ DIN VDE 0110 Erection of power installations,

▷ DIN VDE 0110 Isolation coordination for electrical equipment in low-voltage installations,

▷ EN 60204 part 1 / DIN VDE 0113 part 1 Specifications for the electrical equipment of industrial machines – General requirements, as well as

▷ the applicable national, regional or international specifications such as e.g. the technical connection specifications of the local electricity supply authority, CENELEC and IEC specifications,

▷ DIN 43871 Small consumer distribution boards for installation equipment up to 63 A,

▷ DIN 43870 Meter panels,

▷ DIN 43880 Panel-mounted installation devices.

Table 7.9 provides an overview of the various domestic distribution boards and their technical data.

A host of planning aids are offered to help manufacturers of low-voltage switch and distribution boards, consultants, project planners, end users and other customers to specify and plan quickly and effectively. The following list gives a number of examples of such planning aids:

▷ Computer software packages for the planning and calculation as well as for the preparation of lay-out and construction drawings and circuit diagrams (SIPUK 8 HP and SIPUK-CAD),

▷ Computer software packages to assist in the selection of circuit-breakers based on short-circuit calculations, back-up protection and discrimination consideration (KUBS),

▷ Magnetic lay-out planning sets,

▷ Self-adhesive symbols and drawing elements,

▷ Drawing and planning templates,

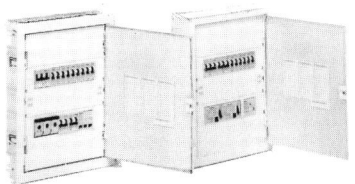

Figure 7.36
Various designs of domestic distribution boards

7 Type-tested switchgear assemblies (TTA)

Table 7.9 Overview and technical data of the Siemens domestic power distribution systems

	Fixed distribution boards					Movable distribution boards	
Designation/Type	N-type and small distribution boards 8GB	STAB wall distribution boards 8GD	SIKUS free-standing distribution boards 8GA	SIPRO universal system meter cubicles/ distribution boards	Cable distribution cubicles 8MB	Building-site distribution boards 8MM2	Campsite distribution boards 8MM6
Rated current	63 A	400 A	2000 A	400 A	1000 A	630 A/ 390 kW	40 A
DIN VDE specification	0603	0660 part 504 and part 500	0660 part 504 and part 500	0603	0660 part 500	0612	–
Dimensions standard	DIN 43871	DIN 43870	DIN 43870	DIN 43870	DIN 43629	DIN 43868	DIN 43629
Equipment standard	DIN 43880	DIN 43880	DIN 43880	DIN 46857	DIN 43623	–	DIN 46278
Depth in the design: flush	70/85 mm	130 and 220 mm	–	210 mm	–	–	–
surface	73/85 mm	160 and 220 mm	250/480/ 600 mm	210 mm	320/335 mm	360/480 mm	330 mm
canopy	74/85 mm	160 and 220 mm	–	–	–	–	–
hollow wall	71 mm	–	–	–	–	–	–
Floor mounting	–	–	yes	yes	yes	yes	–
Degree of protection to DIN 40050 alt. IEC-Publ. 529	IP 30	IP 30, 41, 54	IP 30, 41, 54	IP 31, 54	IP 43, 54	IP 43, 54	IP 43
Protection class	II	I and II	I and II	II	I and II	I and II	I and II
Size	–	–	–	–	0 to 3	–	–
Cubicle (section) width multiples of 250 mm	1	1 to 5	2 to 5	1 to 5	–	–	–
Rows of devices (mounting rails to DIN EN 50022) per cubicle (section) width	1 to 4	3, 6 and 9	15	as distribution board 5 bis 9	–	–	–
No. of dimension modules (18 mm) per cubicle (section) width	12 to 48	12 to 108	180	108	–	–	–

▷ Pads with pre-printed calculation and planning sheets,
▷ Standard drawings and patterns,
▷ Catalogues and equipment lists.

7.4.5.1 8GB Small distribution boards

Small distribution boards up to 63 A are suitable for virtually all domestic sub-distribution applications. Owing to their particularly shallow and compact construction, they are predominantly installed directly in the proximity of the main load points in domestic and trade premises. They comprise the distribution level directly following the meter panel or incoming supply; alternatively the main floor distribution point in the case of high-rise buildings.

These boards are usually equipped with N-type installation devices to DIN 43880. These devices have a snap-on mounting facility for $35 \times 7,5$ mounting rails to DIN 50022. The distance between the mounting rails in the small distribution boards is 125 mm.

The width of an N-type installation device is always expressed as a multiple of the modular width 17.5 mm. Each mounting rail in the 8GB small distribution boards can accommodate devices up to a total width of 12 modules.

If the cut-out in the insulating cover is enlarged, it is possible to extend the mounting width to 14 modules. The spacious design permits an unrestricted combination of all the installation devices, e.g. earth leakage circuit-breakers (e.l.c.b.) with miniature circuit-breakers (m.c.b.).

The N-type distribution boards are of a particularly shallow design and have a mounting depth of only 70 mm. They are intended for flush mounting in walls having only a single brick thickness to DIN 105, and typically accommodate N-type installation devices such as N-miniature circuit-breakers, N-earth leakage circuit-breakers and N-NEOZED fuses.

Depending on the application, one can select from the following versions, all having protection class 2 (total insulation):

– flush-mounting, surface-mounting and canopied versions with one to four rows of mounting rails for solid brick walls, and
– cavity wall versions with two to three rows of mounting rails for hollow walls (see Fig. 7.37).

All the versions have large lead termination and wiring spaces in accordance with DIN 43871. Some ver-

Figure 7.37
The range of 8GB small distribution boards

sions even have enough space behind the 35 mm mounting rails to allow routing of wires through the distribution board.

Large heat dissipation surfaces ensure adequate cooling of the installed devices. A neutral and earth terminal (N/PE) are provided for each circuit.

The housing material is a high impact cream-white (RAL 9001) polystyrene. The housings pass the glow-wire test with test temperatures of 650 °C for the flush-mounting, and 750 °C for the surface-mounting versions respectively. The housings of the cavity wall versions, on the other hand, are of a flame-retardant polystyrene and therefor withstand a test temperature of 960 °C.

Polystyrene wall boxes designed to have a thermal stability up to 100 °C are offered to facilitate the fitting of the cavity wall distribution boards into prefabricated hollow walls.

In order to increase the mechanical protection of the fitted devices, all the small distribution boards, with the exception of the canopied versions, are fitted with stove-enamelled sheet-steel doors in such a way that the characteristic total insulation is retained. Every small distribution board version from Siemens is tested to DIN VDE 0603 and carries the VDE mark. The flush-mounting version consists of three parts viz. the wall box with N/PE terminals, the N-device mounting tray with the insulation cover and the masking frame with the sheet-steel door. The moulded door handle is generously dimensioned, and the door is held closed by means of a magnetic latch.

If the wall box has not been fitted into the wall at right angles, this can be compensated for by means of the adjustable screw-slot connection between the device mounting tray and the masking frame with the sheet-steel door.

Junction boxes permit several flush-mounted and cavity wall distribution boards to be joined together in the vertical or horizontal direction.

The N-device mounting tray can also be installed into the 80 mm deep flush-mounting distribution board. In this way, a deeper wiring space can be created behind the 35 mm mounting rails. Further details may be found in the relevant Siemens catalogue.

7.4.5.2 8GD and 8GA STAB wall-mounting / SIKUS floor-standing distribution board system

The STAB wall-mounting and SIKUS floor-standing distribution boards are used for main and sub-distribution up to 400 A and 2000 A respectively. The two types of board with common kits of pre-assembled switch and fusegear combinations and mounting chassis can be regarded as parts of a complete distribution board system. Their modular design permits the construction of type-tested (TTA) and partially type-tested (PTTA) switchgear assemblies in accordance with DIN VDE 0660 part 500.

They carry the VDE mark and, depending on the type of distribution board, comprise a main housing, a wiring compartment, a base frame, kits of pre-assembled switch and fusegear combinations, mounting chassis and accessories. The system components are also supplied as separate individual items for assembly by switch and distribution board manufacturers. The flexibility of the system means that only a relatively low number of elements needs to be stocked to permit the construction of several different configurations and designs.

The range of boards that can be assembled include flush-mounting, surface-mounting and canopy-type distribution boards, as well as a number of floor-standing versions in various sizes and degrees of protection (Fig. 7.38).

The inner dimensions of the housings are based on one to five modular widths each of 250 mm, and one to five modular heights each of 375 mm. All housing sizes can be manufactured to these basic grid dimen-

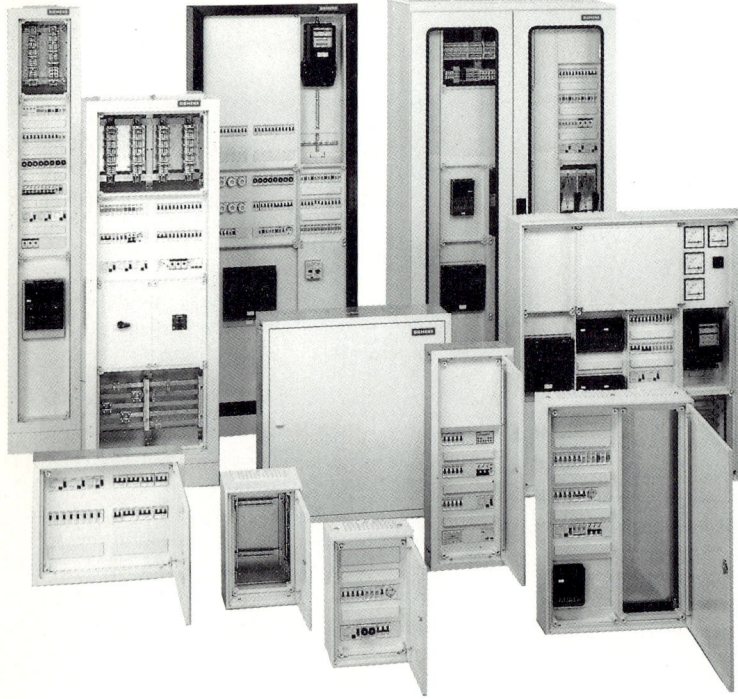

Figure 7.38
The range of SIKUS system free-standing distribution board cubicles and housings

Table 7.10
Cabinet sizes and dimensions of sheet steel STAB wall-mounted distribution boards with protection class I and II

Flush-mounting distribution boards				with wall boxes	without wall boxes	Canopied and surface-mounting distribution boards	
nominal sizes (inner dimensions) $H \times W$ mm	mounting depth mm	enclosure size		external dimensions $H \times W \times D$ mm	recess dimensions $H \times W \times D$ mm	recess dimensions $H \times W \times D$ mm	external dimensions[1] $H \times W \times D$ mm	mounting depth mm
375 × 250	130 / 220	H1 B1		461 × 336 × 136/226	475 × 350 × 140/230	445 × 320 × 140/230	451 × 300	160 / 220
375 × 500	130 / 220	H1 B2		461 × 586 × 136/226	475 × 600 × 140/230	445 × 570 × 140/230	451 × 550	160 / 220
375 × 750	130 / 220	H1 B3		461 × 836 × 136/226	475 × 850 × 140/230	475 × 820 × 140/230	451 × 800	160 / 220
375 × 1000	130 / 220	H1 B4		461 × 1086 × 136/226	475 × 1100 × 140/230	445 × 1070 × 140/230	451 × 1050	160 / 220
375 × 1250	130 / 220	H1 B5		461 × 1336 × 136/226	475 × 1336 × 140/230	445 × 1320 × 140/230	451 × 1300	160 / 220
750 × 250	130 / 220	H2 B1		836 × 336 × 136/226	850 × 1350 × 140/230	820 × 320 × 140/230	826 × 300	160 / 220
750 × 500	130 / 220	H2 B2		836 × 586 × 136/226	850 × 600 × 140/230	820 × 570 × 140/230	826 × 550	160 / 220
750 × 750	130 / 220	H2 B3		836 × 836 × 136/226	850 × 850 × 140/230	820 × 820 × 140/230	826 × 800	160 / 220
750 × 1000	130 / 220	H2 B4		836 × 1036 × 136/226	850 × 1100 × 140/230	820 × 1070 × 140/230	826 × 1050	160 / 220
750 × 1250	130 / 220	H2 B5		836 × 1336 × 136/226	850 × 1350 × 140/230	820 × 1320 × 140/230	826 × 1300	160 / 220
1125 × 250	220	H3 B1		1211 × 336 × 226	1225 × 350 × 230	1195 × 320 × 230	1201 × 300	220
1125 × 500	220	H3 B2		1211 × 586 × 226	1225 × 600 × 230	1195 × 570 × 230	1201 × 550	220
1125 × 750	220	H3 B3		1211 × 836 × 226	1225 × 850 × 230	1195 × 820 × 230	1201 × 800	220
1125 × 1000	220	H3 B4		1211 × 1086 × 226	1225 × 1100 × 230	1195 × 1070 × 230	1201 × 1050	220
1125 × 1250	220	H3 B5		1211 × 1336 × 226	1225 × 1350 × 230	1195 × 1320 × 230	1201 × 1300	220

[1] including cable flange plates

sions. The housing sections are designed to accommodate any of the pre-assembled switch and fusegear combinations, mounting chassis and busbar systems. The modular pre-assembled switch and fusegear combinations are fixed in place by means of rapid twist fasteners, and the standardized mounting rails to DIN 50022 have a row spacing of either 125 or 150 mm. In addition, individual elements such as terminal carriers, extra-deep mounting brackets, mounting trays and chassis, section covers and barriers for specific switchgear may be fitted. The individual elements are prepared in such a way, that virtually only a screwdriver is needed for the complete assembly.

The housings are offered in various degrees of protection (IP 30, IP 41, and IP 54) and with various protection classes (protection class I – protective-conductor terminal and protection class II – total insulation).

All the distribution cabinets with protection class I are made of galvanized and painted sheet steel (colour RAL 7035 – light grey). The class II totally insulated STAB wall-mounted housings have an additional cladding made of plastic insulating material sections, whereas the SIKUS free-standing cubicles are made completely of halogen-free insulating material (expanded Polyurethane, colour RAL 9011 – graphite black). All the housing versions (with the exception of the canopy types) have sheet steel doors and, depending on the door size, are fitted with twist locks or espagnolette door latches with twin blade locks and three-point latching. Instead of the standard door latches and locks, special versions can easily be fitted.

The section covers are made of a halogen-free glass-fibre reinforced polyester, and are fastened by means of 90° screw-twist catches. They are available in two mounting depths viz. 35 and 70 mm, and are reversible so that by means of only two different section covers, four different mounting depths may be achieved.

A large assortment of flanges and gland plates are offered to close the sides of the housings.

8GD STAB wall-mounted distribution board system

STAB wall-mounted distribution board housings, which can accommodate installation equipment and kits up to 400 A, are offered in a number of different sizes (see Table 7.10, page 345).

The 130 mm deep flush-mounting version and the 160 mm deep surface-mounting version have the same simple angle bracket fixing principle. They are both designed for the simple installation of standard pre-assembled switch and fusegear sets, mounting chassis and section cover plates. The inside of the 220 mm deep flush-mounting and surface-mounting versions, on the other hand, contain a system of transom support braces and brackets for the flexible installation of standard and/or custom-built switch and fusegear sets of the STAB wall-mounted distribution board system.

All the flush-mounted versions are equipped with adjustable trim frames to provide a neat finish and to cover untidy gaps in the plasterwork. They are all supplied with strong wall clamps to ensure a secure fixing inside the niche, and can be supplied with or without galvanized steel wall boxes with protection class I or II.

All the distribution board cabinets, with the exception of the canopy versions, have galvanized sheet steel doors with a gray lacquer finish and twist-lock latches, as well as internal hole strips to facilitate the possible fitting of indicator lights.

7.4.5.3 8GA SIKUS free-standing distribution board cubicles

The SIKUS free-standing distribution board system offers the electrical contractor and switchboard manufacturer a range of spacious cubicles for incoming supplies of up to 2000 A and maximum short-circuit impulse strength ($i_p < 80$ kA). These cubicles are not only easy to assemble, but offer a high degree of flexibility and ease of equipping and wiring.

Owing to the well designed modular system, the cubicles can be offered in a number of versions and sizes.

Cubicles are offered:

▷ in completely knocked-down kit form for space-saving storage,

▷ as assembled empty cabinets,

▷ partially equipped for completion and wiring in the workshop, or

▷ fully equipped and pre-wired as complete factory-built distribution boards.

The free-standing cubicles are available in the degrees of protection IP 30 (without door), IP 41 (with door) and IP 54 (with sealing door).

Owing to the modular grid dimension system, a large number of different cabinet sizes can be built. Particularly for free-standing applications, the system includes matching base frames (100 mm high) or wiring

Guidelines for project planning: SIKUS distribution board cubicles

plinths (275 mm high) as well as wiring compartments (475 mm high) which can be mounted on top of the main cabinet (Fig. 7.39). Special split versions of the wiring compartments even permit uncut wires and cables to be routed through the distribution board.

A removable back panel greatly contributes to the accessibility and the ease of wiring. Depending on the depth of the installation equipment to be accommodated, the mounting depth of the cubicle can be either 250, 480 or 600 mm (see Fig. 7.40). Standard factory-assembled and type-tested switch and fuse-gear kits, as well as busbar systems are fastened to the rigid upright/transom structure simply and quickly by means of solid brackets. In this a way the actual walls of the cabinets do not carry the weight of the equipment. The use of these type-tested kits and the standardized fastening methods not only offer the highest degree of electrical reliability, but also contribute to the static strength and ensure a solid and well-tried construction.

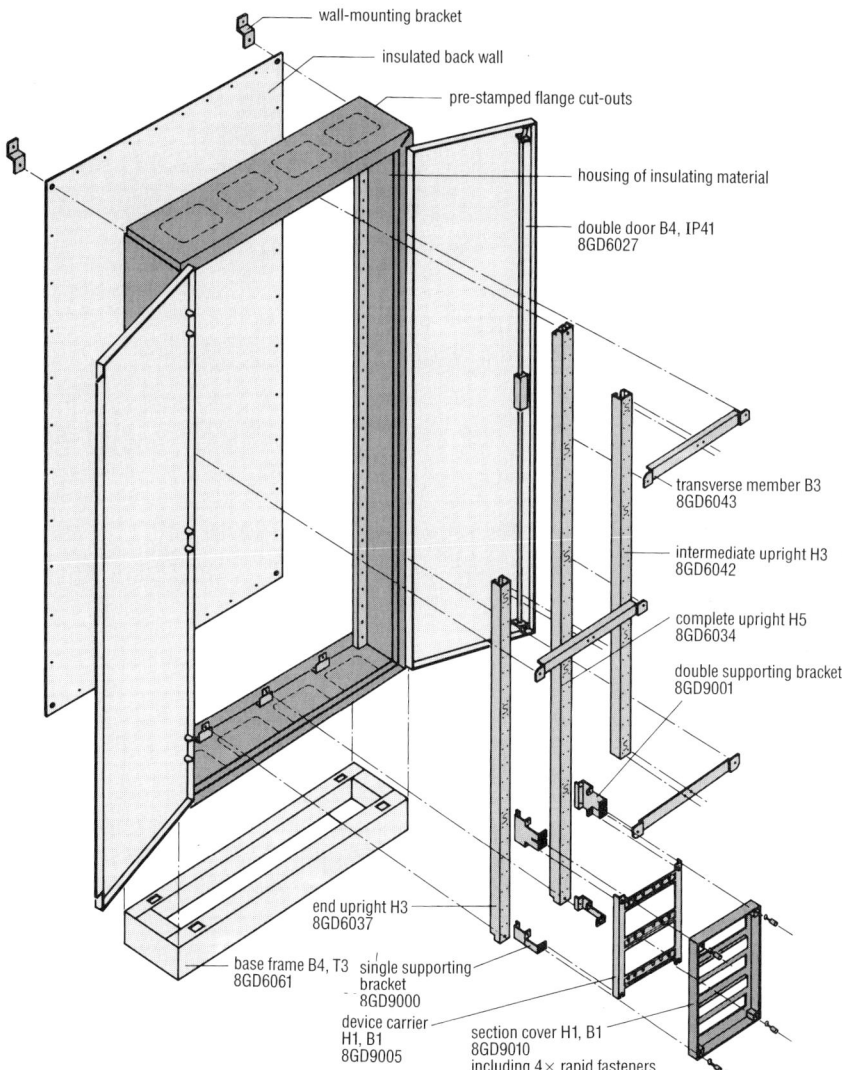

Figure 7.39
The basic elements and base frame of the SIKUS free-standing distribution board system

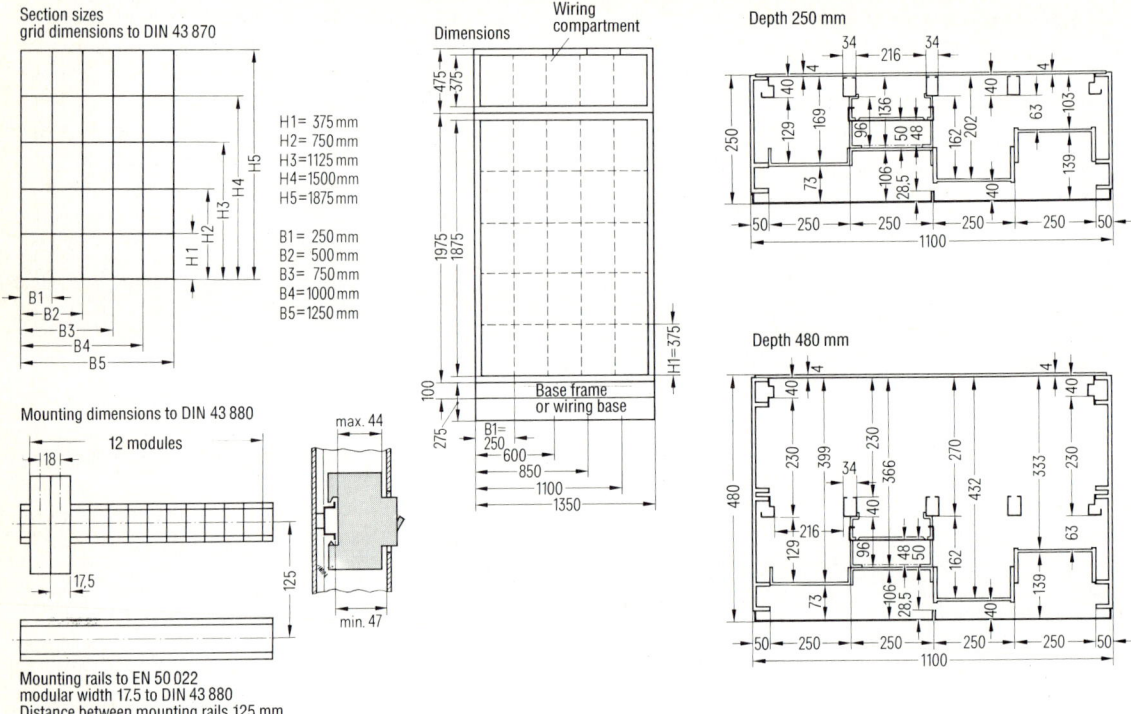

Figure 7.40
Grid dimensions, mounting dimensions, cabinet sizes and mounting depths of the SIKUS free-standing distribution board system

7.4.5.4 SIPRO universal system meter cabinets, meter/distribution cubicles, distribution cabinets, free-standing distribution boards

The energy supply authorities offer electricity for the power supply of consumer loads in domestic, agricultural, trade, industrial and a number of other applications. Before a customer may draw current from the electricity supply network via the local service box, a meter cabinet with an approved electricity meter must be installed by a recognized electrical contractor, in order to fulfill the particular regional connection specifications. In addition, the various applicable laws and by-laws pertaining to the supply of electricity to tariff customers, must be taken into account.

To avoid possible delays and approval problems, it is advisable to co-ordinate the selection of the meter cabinet version (including the proposed electricity meter and other internal equipment) with the supply authorities or local inspectorate, before the planning and erection phases begin.

Furthermore, for billing purposes, the customer is obliged to inform the supply authorities of any alterations or extension to the original installation. It is advisable, to include reserve space in the meter cabinet for later extension.

The SIPRO universal system from Siemens offers an extensive range of meter cabinets, meter/distribution cabinets and distribution panels, as well as complete free-standing cubicles for local distribution boards.

The SIPRO universal system is offered in five different delivery forms in accordance with the regional technical connection specifications (German *TAB*) and the market requirements:

1. Empty cabinets with separately supplied quick-mounting kits and installation elements for retro-fitting;
2. Empty cabinets with separately supplied and ready-equipped quick-mounting kits;

3. Empty cabinets with separately supplied, ready-equipped and pre-wired quick-mounting kits;
4. Partially equipped cabinets (cabinets with fitted meter section quick-mounting kit and prepared reserve space for additional meter or distribution sections or other equipment) with separate supply authority equipment packages;
5. Complete cabinets.

Construction

The SIPRO meter cabinets are constructed and erected in accordance with the applicable specifications DIN VDE 0603 and DIN VDE 0660 part 500 as well as DIN VDE 0100. They carry the VDE mark which is characteristic of the safety standards applied. There is only *one* cabinet version for flush-mounting, surface-mounting and partially recessed mounting, and it is offered in various sizes:

4 heights viz. 900 mm, 1050 mm, 1200 mm, and 1350 mm and up to
5 modular widths each of 250 mm.

All the cabinets have the protection class II (complete insulation). The degree of protection is IP 31 in the case of the simple door (protection against ingress of foreign particles up to 2.5 mm diameter and drip-proof protection against ingress of water) and IP 54 in the case of the sealing door (protection against ingress of dust and splashed water). The cabinets are constructed from metal reinforced plastic profiles which are joined together by means of corner braces and the solid rear panel. The sheet-steel doors are hung on rapid-fixing hinges.

Braces in the doors increase the mechanical strength and provide reliable protection for the installed equipment. The doors have a stove-enamelled finish (RAL 7030, light grey) and provide the cabinets with a neutral appearance. Inside the cabinets, the supporting profiles have two parallel-running slots for the attaching of the quick-mounting kits and the equipment covers of the upper and lower connection areas. The fitting of the quick-mounting kits can be accomplished very swiftly by inserting the catches at the base of the kit into the lower slots, swinging the kit up into the cabinet and then tightening the two captive screws. The equipment covers of the upper and lower connection areas are fitted in a similar way, except that only one screw is used in each case.

The lower connection area may be extended to accommodate cables and wires of larger cross-section by the flanging on of a cable connection chamber which can be sealed to prevent unauthorized opening. The profiles on the sides of the cable connection chamber also permit the fitting of busbar systems, HRC fuse bases and corresponding terminals, as may be required. Extension cabinets not only offer an even more spacious connection area, but also offer the possibility of installing additional equipment on quick-mounting kits. They are mounted in conjunction with the base frame to the underside of the main cabinet in such a way, that a free-standing unit is created out of the wall-mounting one.

The well-proven construction principle of the basic SIPRO cabinet which is divided into lower connection chamber, meter section and upper connection chamber, in conjunction with the custom-designed quick-mounting kits for the fitting of electrical equipment, meets all the demands of the various versions required by the electricity supply authorities.

The quick-mounting kits themselves have a modular construction. Each quick-mounting kit consists of two U-shaped profile uprights, to which an equipment carrier and the meter mounting tray are attached above the N/PE terminal rail. A large variety of switchgear and other electrical components may be mounted between these profile uprights which contain a series of threaded holes along their length. For example, busbar system kits, main terminals, HRC fused disconnectors, HRC fuse bases, base-mounting switches, back-up miniature circuit-breakers, etc. may be built into the lower connection chamber as may be required by the electricity supply authorities. In the middle section, electricity meters or tariff switching devices are mounted by means of standardized meter carrier trays to DIN 43870. In the upper connection chamber, all the installation switchgear and devices approved by the electricity supply authorities (e.g. N-type earth leakage circuit-breakers, power circuit-breakers, NEOZED fuse bases and N-type miniature circuit-breakers) as well as the N/PE sets of terminals are mounted on standard 35 mm DIN EN 50022 rails (Fig. 7.41).

Professional electrical contractors prefer to order the partially equipped cabinets which are supplied with completely pre-wired quick-mounting kits and separate supply authority equipment packages. These merely need to be erected on the building site and connected to the incoming supply.

Quick-mounting kits and other accessories such as equipment trays, profile uprights and transoms as well as the corresponding cover plates are offered

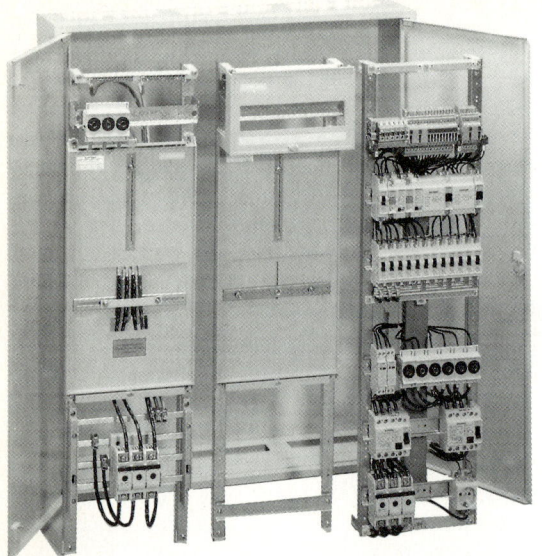

Figure 7.41
SIPRO metering cubicle with quick-mounting kits (completely equipped, partially equipped and distribution section kits)

7.4.5.5 8MB, 8MM and 8GR cable, distribution and meter cubicles for outdoor use

In cases where the installation, erection, accessibility and/or the ability to read the meter of an indoor meter box or distribution board is difficult or even impossible, outdoor distribution and meter cubicles are used. Such cases include e.g. remotely situated houses, holiday and recreation areas as well as outdoor public lighting installations, traffic signalling and railway installations, etc. These cubicles for outdoor use are made of press-moulded glass-fibre reinforced polyester, of type 833.5 to DIN 16913, by means of the SMC process (SMC = *S*heet-*M*oulding-*C*ompound). They conform to DIN VDE 0660 part 503, are type-tested and carry the VDE mark.

Several designs and versions are offered ranging from the complete cubicle (cabinet and pedestal form a single unit) to completely knocked down modular kits (cabinet and pedestal are assembled from component parts) (Refer to Fig. 7.42). In the design of the cubicles, specific emphasis has been placed on achieving sufficient ventilation between the pedestal and the cabinet, as well as below the roof. All ventilation channels are of the labyrinth type to prevent the ingress of foreign bodies. The front cover plates are equipped with rapid-release catches and thereby permit a simple replacement of damaged cabinet parts. All the press-moulded parts are uniformly coloured through their entire cross-section, and the roofs have an additional coating of special UV-resistant two-component polyurethane lacquer.

to permit the flexible construction of general and specialized cabinets containing e.g. current transformers, telemetering equipment and diverse low-voltage switchgear. Here too, the principle of pre-assembly and wiring of quick-mounting kits for subsequent installation into the cabinets, is retained.

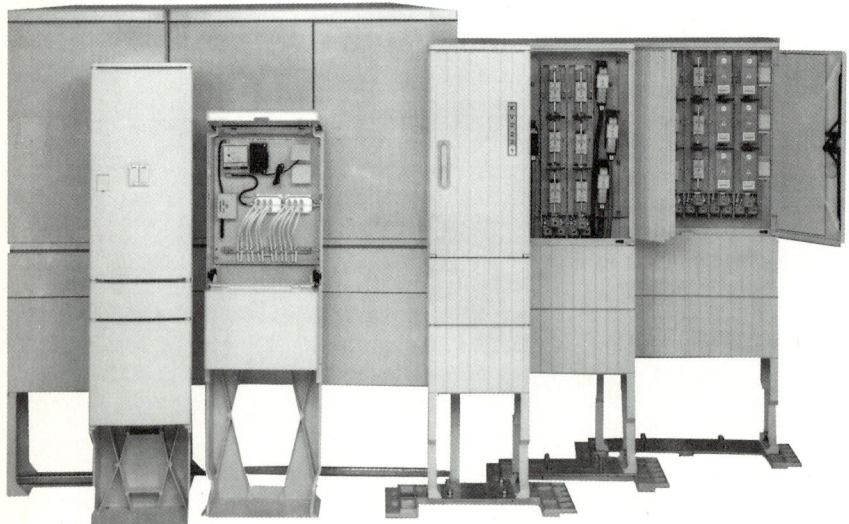

Figure 7.42
"Family picture" of the cable distribution cabinets

Thus, these cubicles may be installed in areas which are subject to extremely harsh environmental conditions. For their use in desert areas, on the coast or in high mountains, a complete coating of the cubicles is offered.

The infeed is accomplished via the service entrance box with HRC fuses to DIN 43627 part 2, and has a degree of protection IP 54. The meters, time switches and the audio-frequency remote control receivers (ripple control) are mounted in plastic housings with transparent covers. These too have a degree of protection IP 54. Furthermore, space is provided for additional switchgear upstream of the meter or for an additional consumer feeder circuit which may be accomplished by means of DIAZED or NEOZED fuses up to 63 A, an N-type m.c.b., or similar device.

In the case of extreme ground humidity, the foundation and pedestal are filled with hygroscopic airated clay marbles (extensively used in hydroculture) or dry sand up to the cable-fastening rail in the cubicle to prevent condensation.

The overall degree of protection is IP 44 in accordance with the VDEW guidelines 5.88. The doors are attached to the side walls by means of strong pin hinges, and have a robust three-point latching system which is designed for the retrofitting of a double-locking unit.

7.4.6 Air-conditioning of installations, switchboards and cubicles

7.4.6.1 General

The surrounding air temperature, humidity as well as the nature and extent of the air pollution determine the climatic environment around an electrical installation. These ambient conditions as well as the actual installation site, nature of the operation and the extent of effective heat losses influence the project planning in various ways:

▷ the installation may have to be overdimensioned,
▷ the failure rate of electronic components may increase,
▷ corrosion effects may be more pronounced,
▷ plastics may age more quickly,
▷ condensation may occur,
▷ the formation of mould may become a problem.

These could lead to the following consequences:

▷ the installation or cubicles must be cooled,
▷ the installation or cubicles must be heated,
▷ the air in the installation or cubicles must be purified and/or dried.

The methods and equipment which could be used to achieve this are determined by a number of specific constraints:

▷ degree of protection of the installation,
▷ permissible difference $\Delta\vartheta$ between the outer and inner temperature,
▷ permissible humidity,
▷ available space,
▷ effective heat losses,
▷ extent and cost of maintenance,
▷ purchase price, running costs.

Possible solutions to the problems include:

▷ thermal conduction and natural convection (although these only provide a limited degree of cabinet cooling),
▷ filtered fan cooling in the case of unproblematic ambient conditions,
▷ heat exchangers for increased cooling in the case of IP 54 degree of protection and problematic ambient conditions,
▷ refrigeration units which simultaneously dry the air in the installation and cool it to a temperature below that of the ambient,
▷ heating units to warm the air and to reduce the relative humidity.

An overview of the heat extraction methods commonly used, is given in Table 7.11.

In order to calculate the heat extraction measures necessary for switchboard cooling, the following parameters must be known:

1. the total value of *power losses to be dissipated* i.e. losses generated by the equipment in the cubicle P_{V1} (in W),
2. the permissible *temperature inside* the cubicle ϑ_i (in K),
3. the *ambient temperature* around the cubicle ϑ_u (in K),
4. the permissible *temperature difference* in K $\Delta\vartheta = \vartheta_i - \vartheta_u$,
5. the cooling *surface area* of the cubicle A_e (in mm²),

Table 7.11 Overview of the various heat extraction techniques

	Open cubicles	Closed cubicles				
Version	with ventilation slots	sealed on all sides	sealed on all sides with internal fan assembly	heat exchanger with separate inner and outer air-flow circuits	through-draught ventilation with filtered fans	refrigeration unit with separate inner and outer air-flow circuits
Illustration (The picture is not binding for the particular version) Cubicle dimensions: 2200 mm × 600 mm × 600 mm						
Principle form of heat withdrawal or dissipation (excluding radiation)	natural convection aided by ventilation slots	heat conduction	heat conduction (forced circulation prevents areas of heat concentration)	heat dissipation through heat exchange between internally heated air and cooler outside air; forced circulation of internal air	heat dissipation through forced expulsion of internally heated air and drawing-in of cooler outside air by means of fan	cooling of the internal air by refrigeration unit and forced circulation of internal air
Heat losses in cubicle P_{V1}	temperature rise in the top of the cubicle: 20 K					temperature in the cubicle: 10 K below ambient temperature
Power dissipated from free-standing single cubicles (P_2)	up to approx. 700 W	up to approx. 260 W	up to approx. 360 W	up to approx. 1700 W	up to approx. 2700 W[1]	up to approx. 5000 W
Reduction of the power dissipated from cubicles joined together in rows	up to 10%	up to 20%	up to 20%	up to 10%	up to 10%	up to 5%
Application in aggressive atmosphere	no	yes	yes	yes	no	yes
in dusty environment	no	yes	yes	yes	only if the filter matting is monitored regularly	
in humid environment	no	yes	yes	yes	no	yes
Degree of protection	up to IP 20	up to IP 54	up to IP 54	IP 54	up to IP 54	IP 54
Interior switchboard temperature	>ambient temperature	>ambient temperature	>ambient temperature	>ambient temperature	>ambient temperature	interior temperature regulated; adjustable to a value between + 25 °C and 40 °C

[1] up to approx. 1400 W with extra fine filter

6. the *heat transfer coefficient k*:

$$k = 2 \text{ to } 4 \, \frac{W}{m^2 \, K}.$$

The heat transfer coefficient k is extremely difficult to determine, and its value tends to vary greatly from one set of circumstances to the next. It depends, among others, on factors such as the actual temperature difference between the cubicle and the surrounding air (as opposed to some theoretical average), as well as the nature of the heat dissipation or extraction (e.g. fan type and rate of air flow). Thus, in fact, one cannot depend on a constant value of k. For the sake of simplicity, and based on practical experience, a constant value has been assumed for the purposes of project planning:

$k = 5 \, \frac{W}{m^2 \, K}$ in the case of cooling with heat exchangers,

$k = 2 \, \frac{W}{m^2 \, K}$ in the case of cooling with fans,

$k = 5 \, \frac{W}{m^2 \, K}$ in the case of cooling with refrigeration units.

In the case of non-forced ventilation, and for both sealed cubicles and cubicles with ventilation openings, the calculation of the effective cooling surface of the housing, of the dissipated power losses and of the air temperature rise inside the housing, is done in accordance with DIN VDE 0660 part 507.

At Siemens, these thermal calculations are done by means of special PC programs.

Heat extraction from cubicles with ventilation openings

Natural convection with through-draught ventilation

Cooling is mainly accomplished by the natural thermal convection whereby the outside air flows through the cubicle.

To achieve this natural cooling, the roof of the cubicle must be perforated or be of a special design to allow the rising hot air to escape. If the cubicle is closed at the bottom (floor plate), then the door and/or back wall must be equipped with rows of ventilation slots.

Heat extraction from sealed cubicles

Natural thermal convection

The heat dissipation can only take place via the exposed cubicle surfaces, and is therefore correspondingly low. To improve the cooling, the cubicles may be moved away from the wall (degree of protection: IP40 or IP54), provided the location permits this.

Forced air circulation by means of internal fan assemblies

The technique of internal forced air circulation by means of fan assemblies (see Siemens ET 4 catalogue), prevents heat accumulation in the top of the housing and achieves direct cooling of specific components. Localized "hot spots", or areas of pronounced heat accumulation, may also be prevented by this technique. Furthermore, since the warm air is circulated passed the walls of the cubicle, the overall heat dissipation via the housing surface to the outside air, is also improved.

The calculation method described in DIN VDE 0660 part 507 may be used to determine the power losses to be dissipated P_{v1}.

Heat exchangers

Heat exchangers operate with two separate air circuits. The warm air inside the cubicle is circulated by means of a fan from the top to the bottom, and passes over a large-surface louvred membrane which separates the inner and outer air-flow circuits. On the other side of the separating membrane, the outer air is circulated in the opposite direction. The temperature difference across the membrane causes a thermal transfer from the inside of the cubicle to the outside. In this way, heat accumulation in cubicles with an IP54 degree of protection can be avoided.

To calculate the power loss which can be extracted in this way through the heat exchanger, the following approximation may be used:

$$P_{v2} = \Delta \vartheta (q + k \cdot A_2). \tag{7.1}$$

q specific power of the heat exchanger in W/K

A_2 effective cooling surface of the cubicle A_e in m^2 reduced by the mounting surface of the heat exchanger

Through-draught ventilation using filter/fan assemblies

Electric fans draw in cool air through filters from the surroundings and force it through the cubicle. The major part of the dust particles is filtered out of the air stream. Exhaust port filters are used if a raised top plate or perforations in the roof do not afford the necessary protection (degree of protection:

IP 20 to IP 54). Standard filter matting is used for dust particles > 5 μm; extra-fine filter matting can be used for dust particles > 0.5 μm.

To calculate the power loss P_{V2} which can be extracted in this way, the following approximation may be used:

$$P_{V2} = \frac{V \cdot \Delta \vartheta}{3.1 \cdot 1.2} + k \cdot A_e \cdot \Delta \vartheta \quad \text{(in W)} \qquad 7.2$$

V air flow rate of the fan in m³/h

If extra-fine filter matting is used, the values given in Table 7.11 for the power loss P_{V2} which can be extracted, is reduced by about 50%.

Refrigeration units

These units operate on the same principle as domestic refrigerators and regulate the temperature inside the switchboard housing typically to within the range of 25 to 40 °C by means of separate inner and outer air-flow circuits. The forced air flow over the coolant circuit by means of the fan, simultaneously produces a forced circulation of the air in the switchboard cubicle.

To calculate the power loss which can be extracted by means of refrigeration units, the following approximation may be used:

$$P_{V2} = Q + k \cdot A_2 \cdot \Delta \vartheta \quad \text{(in W)} \qquad (7.3)$$

Q cooling power (capacity) of the refrigeration unit in W

In the case of refrigeration units, as opposed to the other methods described, the air inside the switchboard cubicle can be cooled to a temperature below that of the ambient. Thus, negative values of $\Delta \vartheta$ can result.

In order to determine the most suitable method of cooling, it may be necessary to calculate the thermal equilibrium. For this, the heat losses of all the installed equipment (devices, busbars, conductors, wires, etc.) must be determined and summated. Naturally, duty factors and rated loading factors must be taken into account. In addition, data concerning the ambient temperature, permissible temperature inside the cubicle and the exposed surface area of the housing must also be known. Once this information is known, the most suitable device for the heat extraction may be selected using the guidelines given in Table 7.11, or the project planning data given part 2 of the Siemens NV 21 catalogue.

7.4.6.2 8ME78 heat exchanger

Application

The Siemens 8ME78 heat exchangers (Fig. 7.43) are used for the heat extraction from electrical installations and switchboards while maintaining a degree of protection IP 54.

Through the use of two completely separate air flow circuits, the cubicles are protected from the outside climate. The heat exchangers are particularly well suited for use in harsh environments.

Principle of operation

The fan for the inner circuit draws in the heated cubicle air at the top and forces it through the cooling ducts of the heat exchanger (Fig. 7.44). The cooled air is returned to the cubicle at the bottom. The outer air-flow circuit operates in the opposite direction. Owing to the temperature difference between the two air streams passing over each side of the large separating membrane, heat is transferred from the inner to the outer air circuit.

Characteristic features

The particular characteristics of the heat exchangers include:

▷ two separate air circuits operating on the counter flow principle
▷ degree of protection IP 54

Figure 7.43
Heat exchangers fitted to the inside of 8MF cubicle doors

Guidelines for project planning: heat exchangers

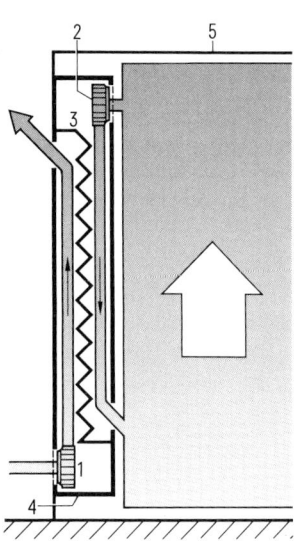

1 Fan for outside air
2 Fan for inside air
3 Separating membrane
4 Heat exchanger
5 Switchboard cubicle

P_{v1} installed power loss in W
$\Delta\vartheta$ permissible temperature difference between the air inside and outside the cubicle in K

$$k = 5 \frac{W}{m^2 K}$$

A_2 effective cooling surface of the cubicle A_e reduced by the mounting surface of the heat exchanger

7.4.6.3 8MR11 filtered fans

Application

8MR11.. filtered fans are used for the heat extraction from electrical installations and cubicles having degrees of protection IP 20 to IP 54 (Fig. 7.45). The protection against ingress of water (2nd digit), however, is only afforded with the fans at standstill. These fans provide for low-cost cooling under normal and unproblematic environmental conditions.

Principle of operation

The fans (Fig. 7.46, left) draw in cool air from the surroundings and force it through the cubicle. The major part of the dust and dirt particles is filtered out of the air stream. The temperature inside the cubi-

Figure 7.44
Operating principle of a heat exchanger

▷ mounting on the inside or outside of a door, rear or side wall
▷ shallow mounting depth of only 110 mm
▷ low weight; not more than 16 kg
▷ cooling ducts accessible over their entire length for cleaning
▷ optimal air-flow ducting
▷ air discharge angle 45°
▷ plastic membrane, self-extinguishing, flame retardant, resistant to weak acids, alkalines, alcohol, fatty and oily substances, sulphur dioxide, hydrogen sulfide, nitric oxides, ammonia
▷ temperature range -20 to $+65\,°C$
▷ connecting lead can be run as desired
▷ only one electrical connection required
▷ operating voltage AC 115 V, alternatively AC 220–240 V single-phase, 50/60 Hz as well as DC 24 V.

Planning guidelines

The required specific power q of the heat exchanger can be calculated using equation 7.1:

$$q = \frac{P_{V1}}{\Delta\vartheta} - k \cdot A_e \quad \text{(in W/K)}$$

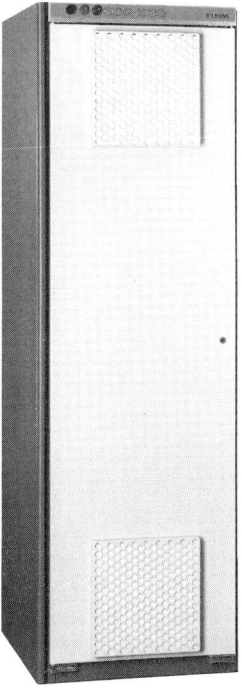

Figure 7.45
8MF cubicle with filtered fan and filtered exhaust port

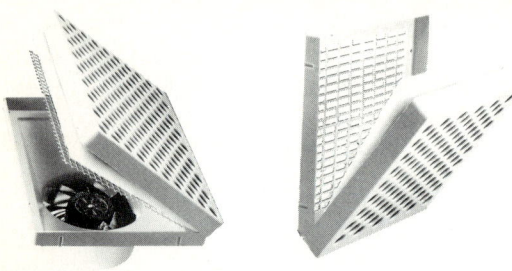

Figure 7.46
Opened fan assembly and exhaust port filter

cle depends on the magnitude of the power loss and the cooling-air flow rate. Exhaust port filters (Fig. 7.46, right) are used if a raised top plate or perforations in the roof do not afford the necessary degree of protection.

Characteristic features

The particular characteristics of the filtered fans include:

▷ housing of self-extinguishing thermoplastic material
▷ protective grille in front of fans for filter change during operation
▷ temperature range -10 to $+55\,°C$
▷ standard filter matting for dust particles $>5\,\mu m$ diameter (efficiency 92%)
▷ combination of extra-fine and standard filters for dust particles $>0.5\,\mu m$ diameter (efficiency 96%)
▷ operating voltage AC 115 V single-phase, AC 230 V single-phase, AC 400 V three-phase, 50/60 Hz etc.

Planning guidelines

The required specific power q of the fan can be calculated using equation 7.1:

$$q = \frac{P_{V1}}{\Delta\vartheta} - k \cdot A_e \quad \text{(in W/K)}.$$

P_{V1} installed power loss in W
$\Delta\vartheta$ permissible temperature difference between the air inside and outside the cubicle in K
$k = 2\,\dfrac{W}{m^2\,K}$
A_e effective cooling surface of the cubicle in m^2

7.4.6.4 8MR17 refrigeration units

Application

8MR17 refrigeration units (Fig. 7.47) are used for the cooling of electrical installations and cubicles if the inside temperature has to be cooled down to or below the ambient temperature. A secondary effect of this is that the air in the cubicle is dried.

Principle of operation

The 8MR17 refrigeration units operate on the same well-known principle as domestic refrigerators. Figure 7.48 shows a schematic diagram of their construction.

Characteristic features

The particular characteristics of the refrigeration units include:

▷ inside temperature of the cubicle is regulated to an adjustable value between $+25$ and $40\,°C$ (standard $+35\,°C$)
▷ two separate air-flow circuits
▷ degree of protection IP 54

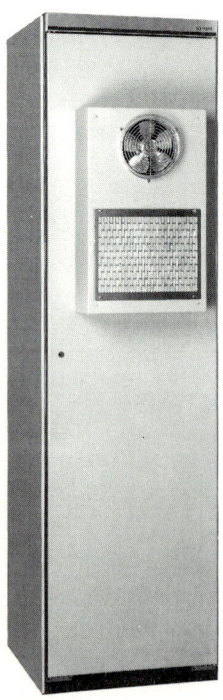

Figure 7.47
8MR17 refrigeration unit mounted on an 8MF cubicle

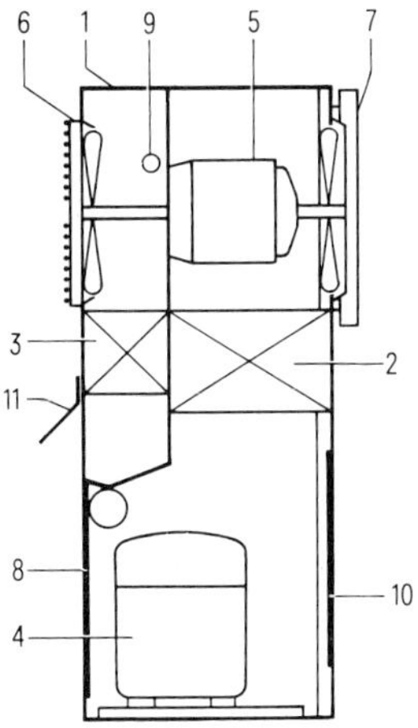

Figure 7.48
Construction of a refrigeration unit

1 Housing
2 Condenser
3 Evaporator
4 Compressor
5 Fan motor
6 Wire grid screen
7 Wire grid screen
8 Insulation
9 Thermostat
10 Filter matting
11 Baffle plate

▷ fault indication if the inside temperature should rise above 45 °C
▷ external heat-exchange circuit protected from contamination by a filter
▷ operating voltage AC 115 V single-phase, AC 230 V single-phase, AC 400 V three-phase, 50/60 Hz etc.
▷ 3 m connecting cable

Planning guidelines

The required cooling power Q of the refrigeration unit can be calculated using equation 7.3:

$Q = P_{V1} - k \cdot A_2 \cdot \Delta\vartheta$ (in W).

P_{V1} installed power loss in W

$k = 4 \, \dfrac{W}{m^2} K$

A_2 effective cooling surface of the cubicle A_e in m² reduced by the mounting surface of the refrigeration unit

$\Delta\vartheta$ permissible temperature difference between the air inside and outside the cubicle in K

7.4.6.5 8MR21 heating units

Application

8MR21 heating units (Fig. 7.49) are used to raise the temperature of the air inside a switchboard cubicle. Thereby, they reduce the relative humidity, guard against condensation and ensure that a minimum inside temperature is maintained.

Principle of operation

The heating elements are ceramic PTC thermistors which, owing to their positive resistance characteristic, have a variable power output as a function of the ambient temperature. The temperature in the cubicle is thus regulated automatically to some extent.

Characteristic features

The particular characteristics of the heating units include:

▷ operating temperature −20 to +60 °C, in the case of the unit with fan −10 to +50 °C
▷ radio interference level K, unit with fan: level N
▷ silicone connecting cable 3 × 0.75 mm², 0.3 m long
▷ operating voltage AC 115 to 265 V single-phase, 50/60 Hz (units with fan AC 230 V single-phase, 50/60 Hz)

Planning guidelines

To prevent condensation, the temperature inside the cubicle must be about 10° higher than the ambient temperature.

The required heating power P_H, for anti-condensation (operation standstill) heating can be calculated using the equation 7.4:

$$P_H = k \cdot A_e \cdot \Delta\vartheta \quad \text{(in W)}. \tag{7.4}$$

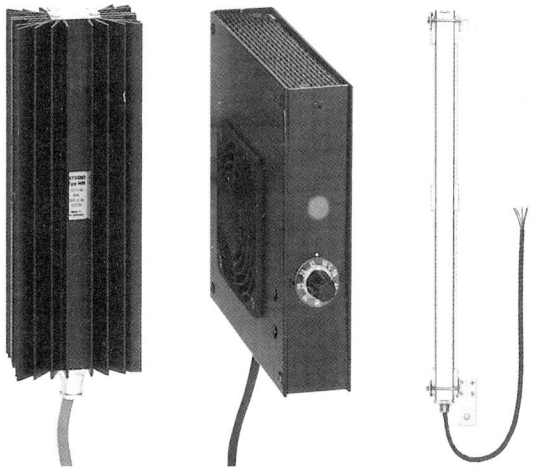

Figure 7.49
8MR21 heating units. Left: basic unit,
Centre: version with fan, Right: door-frame heater

For service heating:

$P_H = k \cdot A_e \cdot \Delta\vartheta - P_V$ (in W)

where

A_e effective cooling surface of the cubicle in m²

$k = 3 \; \dfrac{W}{m^2 \, K}$

$\Delta\vartheta$ permissible temperature difference between the air inside and outside the cubicle in K

P_V effective heat loss in W

7.4.6.6 Temperature rise inside insulation material and sheet steel enclosures

One of the type tests prescribed by DIN VDE 0660 part 500 for type tested switchgear assemblies (TTA), is a heat-rise test. In terms of this test, the permissible temperature rise of installed equipment, including wires, busbars, operating elements and exposed outer surfaces, is specified. This mainly applies to encapsulated TTA's.

Heat rise experiments and type tests have been carried out on a number of standard Siemens TTA's to ensure that the permissible temperatures are not exceeded on any parts. The results of these test are documented in the corresponding test reports.

In practice, however, it is more common that specific switchgear assemblies designed around a particular application, need to be encapsulated or installed into an enclosure. By means of the characteristic curve sets in Figures 7.50 and 7.51, the permissible heat loss inside the enclosures of the 8HS and 8HP systems may be estimated. In this way, the design engineer may save a great deal of time and money which would otherwise have been spent on thermal calculations or heat-rise tests. The values given are all based on practical tests, extrapolation and/or calculation, and should be regarded as guidelines only.

The average temperature inside the enclosure is determined as a function of:

▷ the enclosure size,
▷ the ambient temperature and
▷ the equivalent heat loss of the installed equipment.

By means of these sets of curves, the maximum power loss which can be dissipated as heat via the respective surface areas of the various enclosure sizes, may be estimated.

When using these curve sets (see Figs. 7.50 and 7.51), the following aspects should be taken into account:

The *average air temperature* is the temperature measured at the centre of the enclosure or encapsulation. The temperature in the top of the enclosure will be higher, and in the bottom lower than this value. In critical cases, therefor, equipment with greater heat loss should be mounted in the lower half of the enclosure! The (average) air temperature inside the enclosure is influenced by:

▷ the ambient temperature outside the enclosure
▷ the heat losses of the installed devices and conductors.

To ensure the reliable operation of the installed equipment, the (average) air temperature inside the enclosure must not exceed a specified value.

During the calculation of the resultant average air temperature inside the enclosure during operation, the project planner must take the corresponding data of the switchgear (as may be found in the catalogues) as well as that of the wires and other conductors into account.

For example, the permissible temperature (maximum average air temperature inside the enclosure) is 55 °C for

– 3V. circuit-breakers, 3KE on-load disconnectors, contactors and PVC insulated wires.

The term *ambient temperature* is understood to mean the temperature, measured under specified conditions, of the air surrounding an item of switchgear, a switchboard or a distribution board (refer to Section 1.6.1.1, page 65).

The given *maximum power loss* which can be dissipated as heat via the surface areas of the various enclosures apply for individual (separate) wall-mounted enclosures. In the case of enclosures which are joined together, e.g. to form a larger distribution board, the power losses are to be reduced by 20 to 30%.

The total installed power loss of the switchgear assembly must be worked out. To do this, the individual power losses of the switchgear components must be determined and added together, taking any applicable duty factors into account. Furthermore, the power losses from the cables and wires, which could amount to 30 or even 50% of the equipment power losses, must be added.

Thus, taking the ambient temperature and the permissible average air temperature in the enclosure for the device or devices into consideration, the maxi-

Guidelines for project planning: temperature rise

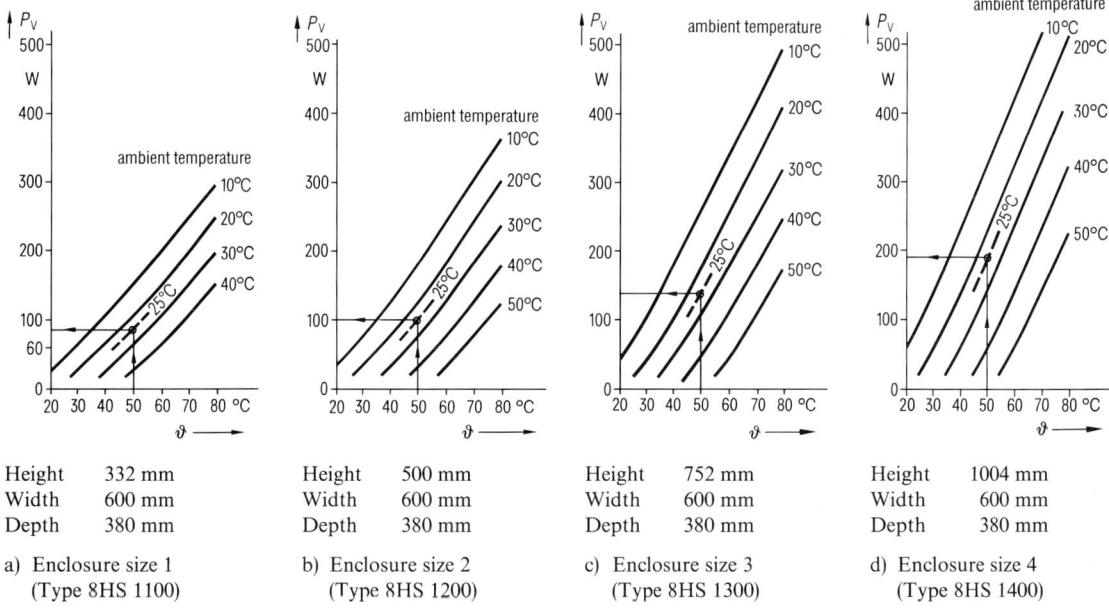

Figure 7.50
Estimation of the permissible installed power loss P_V as a function of the average air temperature ϑ inside the enclosure, in order to determine the required enclosure size from the 8HS sheet steel distribution board system (individual enclosures for wall-mounting)

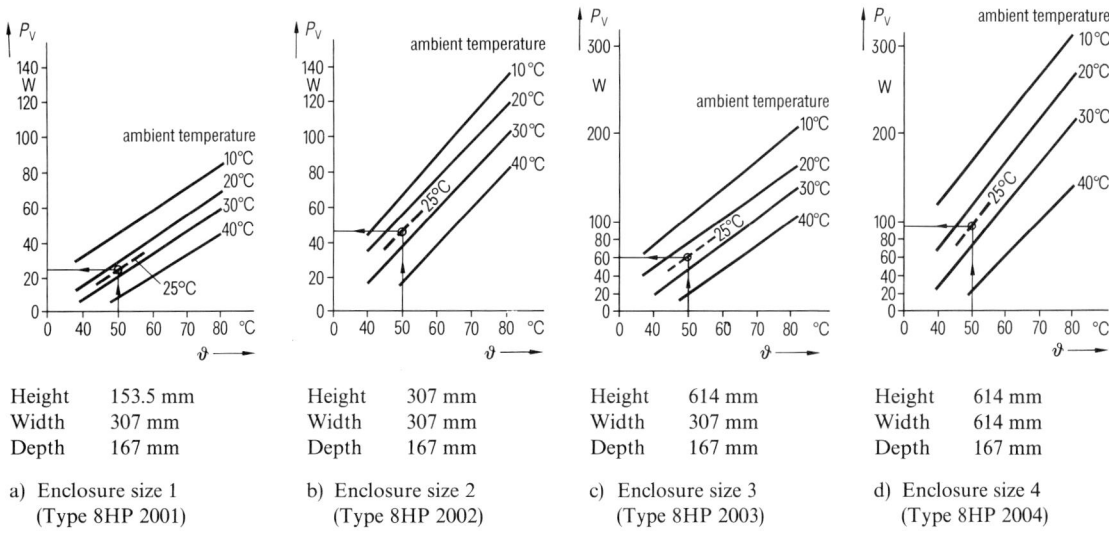

Figure 7.51
Estimation of the permissible installed power loss P_V as a function of the average air temperature ϑ inside the enclosure, in order to determine the required enclosure size from the 8HP fully insulated distribution board system

359

mum power loss per enclosure size may be determined. If necessary, the next bigger enclosure size may have to be selected.

Example

Given:

The total power losses of a switchgear assembly including wire and cable losses is 60 W,

the ambient temperature is 25 °C and

the permissible temperature around the equipment is 50 °C (max. average air temperature inside the enclosure) for a free-standing installation.

To be found:

The enclosure size from the 8HS and 8HP ranges respectively.

Solution:

From Figure 7.50a, the enclosure size 1, type 8HS 1100 is selected; the permissible power loss is approximately 85 W.

From Figure 7.51c, the enclosure size 3, type 8HP 2003 is selected; the permissible power loss is approximately 60 W.

In borderline cases, e.g. if the power losses of the installed equipment is equal or almost equal to the maximum permissible power losses of the enclosure size, and/or if the maximum ambient temperature is not known exactly, or if equipment must be installed in the upper half of the enclosure (where the average air temperature will be exceeded), then the next larger size of enclosure should be selected.

7.4.7 Degrees of protection, climatic and other ambient conditions

The ambient influences on a switchboard at a particular installation site may include those of dust, water and climatic conditions (indoor, outdoor, open-air climate), as well as a number of conceivably aggravating operational and local factors.

There is a common misconception among project planners and end users, that switch and distribution boards with a higher degree of protection, e.g. IP 65 "protection against ingress of dust and water jets from any direction", may be installed out in the open without the need for further protective measures. Great is the surprise, when regular failure of the installed equipment is experienced, the enclosures show

[1] For actual short-form and full definitions, please refer to the specifications mentioned

Table 7.12
Short description of the IP degrees of protection in line with DIN 40050 and IEC 529[1]

Code digit	First code digit: Degree of protection against contact and ingress of foreign bodies	Second code digit: Degree of protection against ingress of water
0	No particular protection	No particular protection
1	No protection against intentional access, however no access to large body surfaces; protection against ingress of foreign bodies with diameter greater than 50 mm	Protection against vertically falling water drops
2L	No access for fingers or similar items, protection against ingress of foreign bodies with diameter greater than 12.5 mm	Protection against water drops falling at an angle of 15°
3 (3L)	No access of wires or similar items with a diameter greater than 2.5 mm, protection against ingress of foreign bodies with diameter greater than 2.5 mm	Protection against water spray falling at an angle of 60°
4	No access of wires or similar items with a diameter greater than 1 mm, protection against ingress of foreign bodies with diameter greater than 1 mm	Protection against water splashed from any direction
5	Complete protection against contact with live parts; protection against harmful dust deposits	Protection against water jets from any direction
6	Complete protection against contact with live parts; protection against ingress of dust	Protection against heavy seas, or strong water jets
7	–	Protection against harmful ingress of water during submersion
8	–	Protection against ingress of water during submersion

excessive signs of wear, or if water flows out of the enclosure when the door is opened or the cover removed! What then, one might ask, is the meaning of the IP degree of protection, and what relevance does it have for the outdoor erection of switch and distribution boards?

Degrees of protection

The degrees of protection, as defined in DIN 40050 and IEC 529, are described in Table 7.12 (also refer to Section 1.5.6.1 on page 63). These IP degrees of protection may be regarded as "operative" classifications. On the one hand, they should indicate the protection afforded to operating personnel against contact with live parts, while on the other hand indicating the protection of the installed equipment against contact with dust and water where relevant. These specifications do not, however, include influences and stresses which may result from exposure to an open-air environment. Effects caused by outside installation of a piece of equipment are not taken into account in the tests for "protection against ingress of water" as outlined in DIN 40053, DIN VDE 0660 part 100 or IEC 947-1 app. C. These tests merely call for an indoor installation at relatively constant climatic conditions. This means, for example, that a distribution board erected out in the open, may cause serious failure of installed equipment owing to insufficient ventilation in spite of (or even because of) a high degree of protection.

Climatic conditions

In addition to the impact of factors directly related to and resulting from the technological processes, and depending on their location, switchboards, distribution boards and control panels are exposed to a number of climatic influences. If necessary, these should be taken into account separately during the planning, manufacturing and erection phases of the project (refer to Fig. 7.52). Typical climatic influences are:

▷ sunlight radiation,
▷ high temperatures,
▷ low temperatures,
▷ high relative humidity,
▷ rain, snow, wind.

Of particular importance are the influences of high ambient temperatures (also caused by direct or indirect sunlight) and the combination of a high relative

Figure 7.52
Sheet-metal encapsulated distribution board (degree of protection IP65) in a potash mine. No special protective measures are required here, since the humidity is low, and the temperature is relatively constant

humidity with varying temperature on switchboards (also refer to page 65).

DIN 50010 part 1 defines the following climates (among others):

An *indoor climate* is the climate in rooms which are such that objects inside them are protected to a large extent from the direct influences of an open-air climate

Note:

An indoor climate may be determined to a large extent by local heat or moisture sources, dust, chemical influences, e.g. by people, animals or plants and by industrial processes, heating or air-conditioning.

An *outdoor climate* is the climate in rooms which are such that objects inside them are protected from direct exposure to sunlight and precipitation as well as from wind, if applicable. In all other respects, however, the objects are exposed to the influences of an open-air climate.

As the term suggests, an *open-air climate* is that natural climate to which objects are exposed in the open.

If no specific data to the contrary is given in the Siemens catalogues for low-voltage switchboards, distribution boards and control panels, then these are suitable for use in areas with indoor climate.

Sunlight radiation

One can assume that switchboards, distribution boards and control panels, which are installed out in the open, will be exposed to direct sunlight. This can cause excessive heat rise in the cubicle and consequent overheating of installed equipment.

It is therefor generally recommended, that switchboards and other electrical cubicles and enclosures be protected from direct sunlight by means of awnings or roofs. In any event, breather glands should be fitted to the bottom of the cubicles (their function is described below).

High relative humidity and strong fluctuation of the ambient temperature

If cubicles and enclosures with IP 65 degree of protection are installed out in the open, where they are exposed to a high relative humidity and simultaneous strong fluctuations in ambient temperature, then with rising temperature the air inside the encapsulation is naturally also heated. It expands, and escapes to the outside past the door seals and other gaskets. When the ambient temperature drops again, a partial vacuum, or negative pressure, is caused inside the enclosure. This, in turn, draws in (damp) air from outside, and condensation forms on the inner walls. If this cycle is repeated a number of times, a considerable amount of water may collect inside the enclosure.

The following measures are recommended to guard against this phenomenon:

▷ breather glands
▷ anti-condenstion heaters

High relative humidity and constant ambient temperature

If the relative humidity at the installation site is high, then in the case of intermittent periodic duty, condensation can result even if the ambient temperature remains approximately constant. This is caused by the temperature fluctuation inside the cubicle owing to the periodic heating up and cooling down of the installed equipment (e.g. heat losses of general switchgear, overload, control and indication devices). A typical example of this, may be found in distribution boards and motor control centres for water-supply pump schemes (intermittent periodic duty). These are often installed directly inside the pump-house where the relative humidity is high and the air temperature virtually constant. For this type of application, breather glands and anti-condensation heaters must be fitted to ensure a free exchange of air between the inside and outside of the enclosures, and to achieve a constant temperature inside the switchboards.

Measures to be taken against climatic influences

Protected installation

Even in the case of a high degree of protection, switchboards, distribution boards and control cubicles should not be installed out in the open without some form of protection against the weather. Additional measures are always necessary to protect them from direct sunlight, precipitation (e.g. rain, snow) and, if applicable, strong winds. This is typically achieved by means of additional protective enclosures or extended roofs and awnings (see Fig. 7.53). Protective roofs and/or sunshades must be big enough to provide the desired extent of protection; if necessary, side and back walls may need to be erected.

Breather glands

The use of breather glands is generally recommended in cases of extreme climatic conditions (Fig. 7.54). These devices are similar to cable compression glands

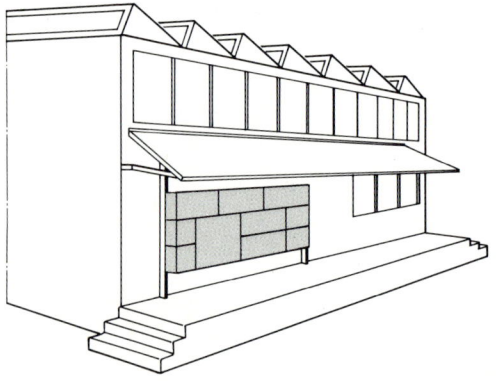

Figure 7.53
Example of a distribution board installed under an existing awning

Climatic influences · Power factor correction

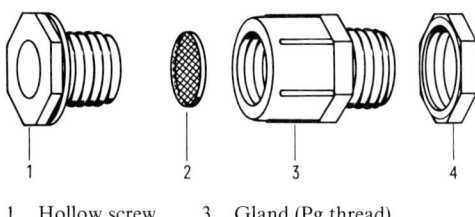

1 Hollow screw
2 Fine mesh
3 Gland (Pg thread)
4 Locknut

Figure 7.54
Breather glands: Construction and component parts

with either a Pg or metric thread. Instead of the rubber seal, however, they are fitted with a gauze or mesh insert to allow the flow of air between the inside and outside of the enclosure. In addition, since they are fitted to the underside of the cubicles, they act as drain outlets for any condensation water which may collect in the bottom of the enclosures.

Anti-condensation heating

In the case of strongly fluctuating ambient temperature and a high relative humidity (tropics), the use of anti-condensation heaters inside the switchboard cubicles is strongly recommended. These heaters prevent condensation by keeping the temperature inside the enclosures at a near constant value. Simple heating elements or standard switchboard anti-condensation heaters may be used. Typically, such heating devices may have a power consumption ranging from 5 to 150 W.

Protection from corrosion

If the installation site is in an aggressive environment, e.g. coastal sea-air climate or chemically polluted atmosphere, then it is recommended that the manufacturer of the switchboard be approached with the view to incorporating additional protective measures, such as e.g. special corrosion-resistant paint finish (also refer to Section 1.6.1.5, page 68).

7.4.8 Power factor correction in networks with or without harmonics

An overview of the selection criteria of capacitors and control panels for power factor correction in supply networks with or without harmonic content, is given in Figure 7.56.

7.4.8.1 Basic principles

The ratio between actual (working) power P and apparent power S gives the cosine of the angle φ; otherwise known as the power factor (refer to the vector diagram in Fig. 7.55):

$$\cos \varphi = \frac{P}{S}.$$

The angle φ is identical to the angular phase-displacement between the current and the voltage.

The (inductive) reactive power Q, to be compensated, can be calculated as:

$$Q = \sqrt{S^2 - P^2} \quad \text{(in kvar).} \tag{7.5}$$

A capacitor having the same value of (capacitive) reactive power Q_c would fully compensate this (inductive) reactive power and bring the power factor ($\cos \varphi$) up to a value of 1.

In practice, it is not usual for the power factor to be corrected fully to 1 by means of step-switched capacitor loading, since a fluctuation in the load and (an unavoidable) delay in the reaction of the controller, would lead to an over-compensation of the load. Typically, the electricity supply authorities prescribe to which value the power factor may be corrected.

The capacitor load Q_c required for correction to a specified power factor, may be calculated using the

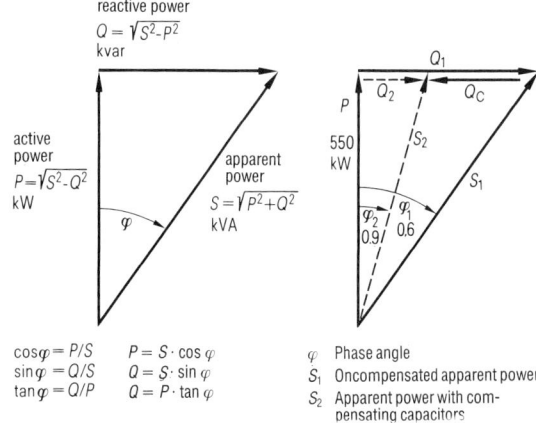

Figure 7.55
Power factor as the relationship between actual and apparent power

7 Type-tested switchgear assemblies (TTA)

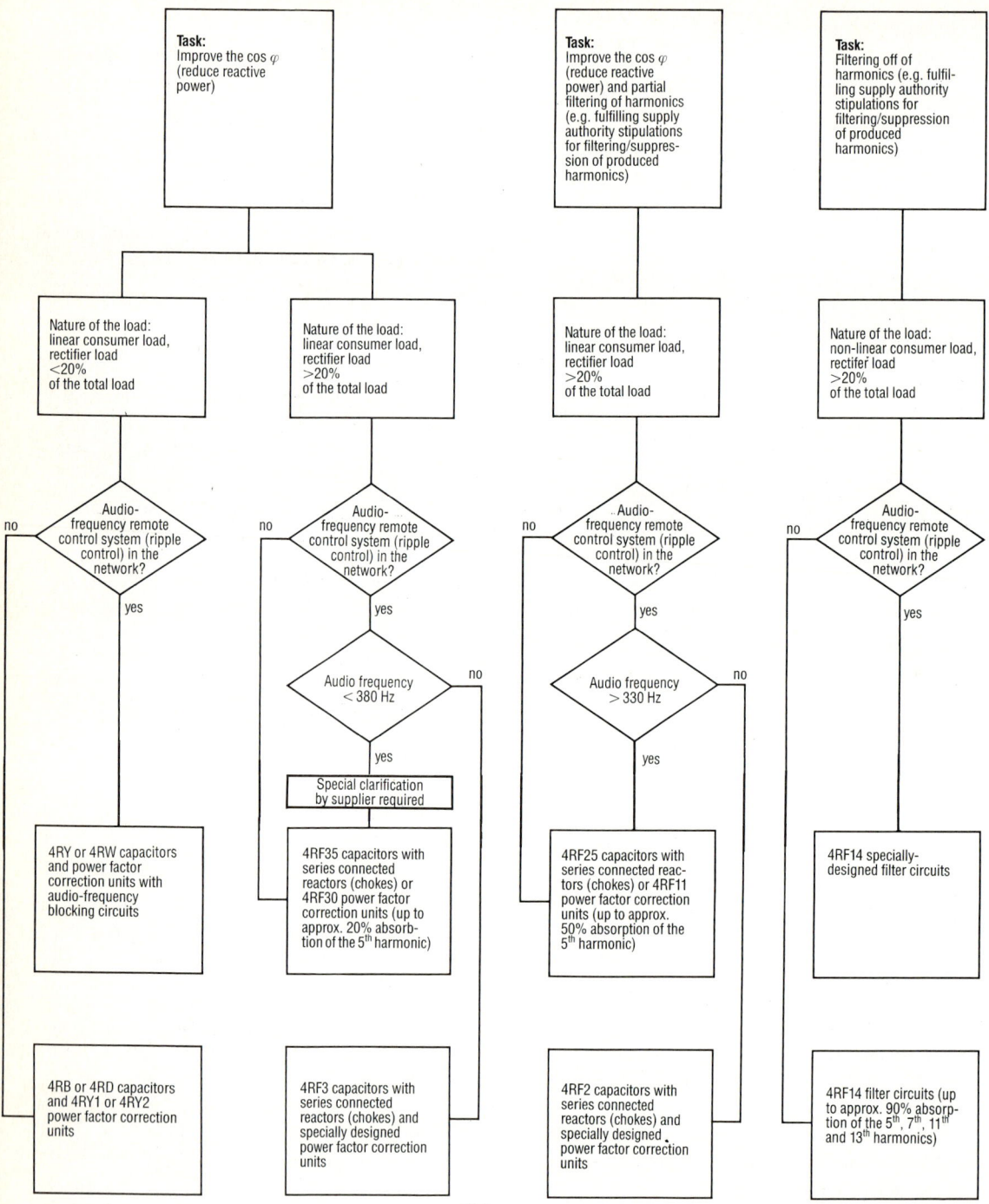

Figure 7.56
Selection diagram for capacitors and power factor correction control units

Note
Power factor correction units without series reactors (chokes) may not be used in parallel with choked units or filter circuits in the same network section. Power factor correction units with series reactors (choked units), may be used in parallel with filter circuits.

tangents of the phase-displacement angles, as follows:

$$Q_c = P \cdot (\tan \varphi_1 - \tan \varphi_2) \qquad (7.6)$$

(also refer to Section 7.4.8.4 Project planning, page 369).

For the power factor correction of a d.c. drive with constant torque:

$$Q_c = \frac{n_1}{n_n} \cdot \frac{1}{\eta} \cdot P_{dn} (\tan \varphi_1 - \tan \varphi_2) \qquad (7.7)$$

(also refer to page 376).

7.4.8.2 Types of power factor correction (reactive power compensation)

In electrical networks in which inductive loads (e.g. motors) are constantly being switched on and off, the power factor of the total load changes with every switching operation. Since, in an industrial environment, electricity charges are levied on the apparent power (e.g. maximum demand measurement), a bad power factor (or $\cos \varphi$) can be a major (and often avoidable) cost factor. There are basically three methods which are commonly used to compensate for the inductive reactive power drawn by a load or an electrical installation, and thereby improve its power factor.

Individual compensation

In the case of individual compensation, the capacitors are connected directly to the terminals of the individual loads and are switched in and out of circuit by the same common switching device (see Fig. 7.57).

Group compensation

In this case, the power factor correction equipment provides a fixed value compensation for a number of motors or e.g. fluorescent lamps which are switched as a group by means of a common contactor or circuit-breaker (see Fig. 7.58).

Centralized compensation

In most cases power factor correction units, which automatically switch power capacitors into or from the load circuit, are used for centralized power factor correction. They are usually directly associated with, and installed centrally at, a particular switchboard, distribution board or incoming supply (Fig. 7.59).

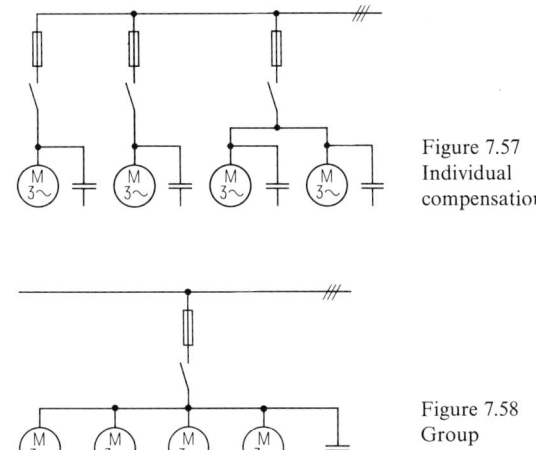

Figure 7.57 Individual compensation

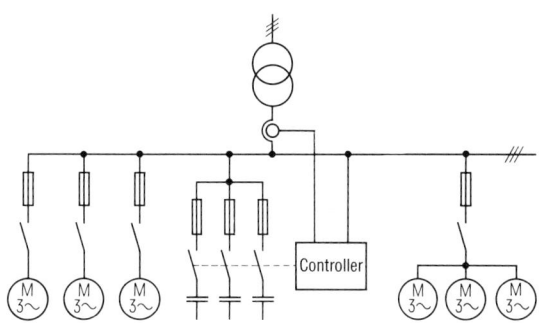

Figure 7.58 Group compensation

Figure 7.59 Centralized compensation

Selection of the most suitable type of compensation

A number of technical and economical aspects need to be considered before the decision is made as to whether fixed capacitor banks or automatic central power factor correction is the most suitable method of compensation. Generally speaking, central power factor correction units have a higher price per installed kvar. If one considers, however, that in most cases all the loads of e.g. a factory are never switched on at the same time, then the total installed reactive power (kvar) may be considerably lower in the case of centralized power factor correction, than if all the loads were equipped with their own fixed compensating capacitors. As a guideline, it has been found that the cost of the two methods is comparable for a diversity factor (or load factor) as high as 0.7 to 0.8. The lower the diversity factor, the more attractive the centralized method of compensation becomes.

Individual compensation can therefor only be economical in the cases of

▷ large single loads with
▷ constant power consumption and
▷ high duty factor (long running times).

Individual compensation also reduces the current in the cables feeding the load, but continuous adaptation of the required capacitive power is not possible.

Centralized compensation is particularly advantageous where

▷ a large number of smaller loads with
▷ varying power requirements and
▷ differing duty factors

are connected to the supply network.

The capacitive power is continuously matched to the reactive power requirements of the plant. Subsequent installation or later extension is relatively simple. Checking and monitoring of the power factor correction equipment is facilitated by its central location.

Centralized power factor correction units incorporate a control device which measures the instantaneous value of $\cos \varphi$, and automatically regulates the amount of compensating capacitive load connected to the load. By means of switching signals from the output relays on this controller, banks of capacitors are switched on or off via contactors until the preset target value of $\cos \varphi$ is achieved.

To avoid switching spikes in the network, large capacitor banks should not be switched on as single loads. In Siemens power factor correction units with capacitor banks greater than 25 kvar, the banks are divided into smaller loads (e.g. 25 kvar resp.) which are each equipped with its own capacitor contactor.

Only the first capacitor contactor of a bank receives its switching signal directly from the controller; all the others are switched via the auxiliary contact of their predecessor. For example, a 100 kvar capacitor bank is not connected as one load. Instead, 4×25 kvar are switched on in quick succession which results in a "softer" and less abrupt capacitive loading of the network (refer to Fig. 7.60).

After switch-off, the capacitors are discharged by means of resistors or discharge chokes in less than 10 seconds to ensure that no partially discharged capacitors are switched into the network.

The switching in of partially discharged capacitors can result in high transient currents which stress both the capacitors and the capacitor contactors if this should e.g. occur at the instant at which the phases are in opposition.

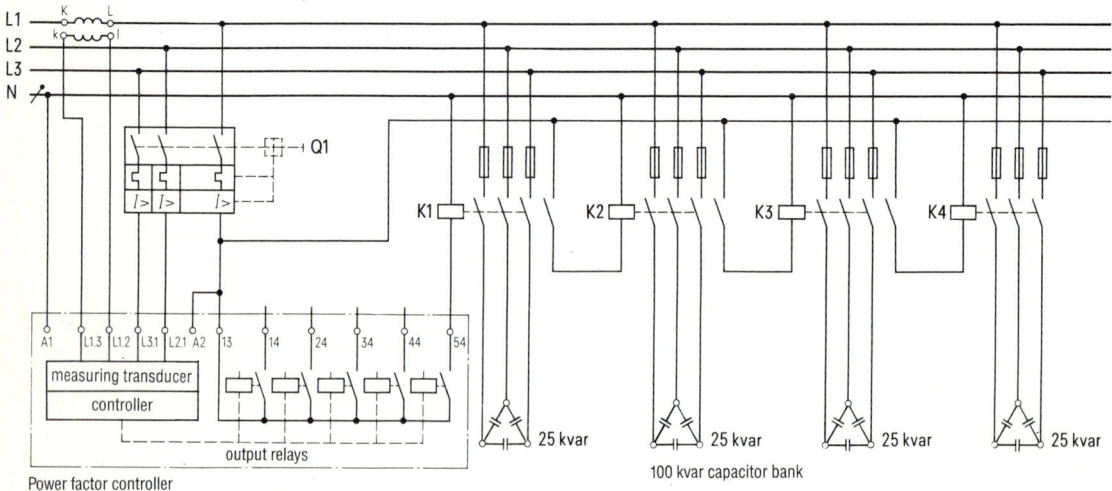

Figure 7.60 Example of the circuit for a power factor correction unit greater than 25 kvar

7.4.8.3 Power factor correction of three-phase induction motors and transformers

During the planning of electrical installations, one can usually assume that inductive loads operate with an average power factor of $\cos\varphi = 0.7$. For compensation to a power factor of $\cos\varphi = 0.9$, a reactive power Q_c equal to about 50% of the active power P, is required:

$$Q_c = 0.5 \cdot P.$$

The capacitive power which will be required to correct the power factor from an existing to a new value, may be found by means of Table 7.13.

Compensation of three-phase induction motors

In the case of three-phase induction motors, the compensating capacitive load Q_c may not exceed 90% of the no-load reactive power of the motor, otherwise self-excitation of the motor may occur during rundown. This causes an extreme overvoltage on the motor terminals.

As a practical guide:

$$Q_c \approx 0.35 \cdot P_{nM}$$

P_{nM} rated output power of the motor

Under less favourable circumstances, e.g. regular starting, prolonged running under no-load or partial-load conditions, the compensation determined in this way will not correct the power factor to above 0.9. In these cases, additional power factor correction, e.g. centralized compensation, may be advantageous.

For further details, please refer to Section 3.7.6, page 195 and the Siemens K catalogue.

Table 7.13
Required capacitor power for power factor correction

Present power factor $\cos\varphi_1$	Capacitor power in kvar per kW of active power to achieve required power factor $\cos\varphi_2$ of				
	0.8	0.85	0.9	0.95	1
0.4	1.54	1.67	1.81	1.96	2.29
0.42	1.41	1.54	1.68	1.83	2.16
0.44	1.29	1.42	1.56	1.71	2.04
0.46	1.18	1.31	1.45	1.6	1.93
0.48	1.08	1.21	1.34	1.5	1.83
0.5	0.98	1.11	1.25	1.4	1.73
0.52	0.89	1.02	1.16	1.31	1.64
0.54	0.81	0.94	1.08	1.23	1.56
0.56	0.73	0.86	1	1.15	1.48
0.58	0.66	0.78	0.92	1.08	1.41
0.6	0.58	0.71	0.85	1	1.33
0.62	0.52	0.65	0.78	0.94	1.27
0.64	0.45	0.58	0.72	0.87	1.2
0.66	0.39	0.52	0.66	0.81	1.14
0.68	0.33	0.46	0.59	0.75	1.08
0.7	0.27	0.4	0.54	0.69	1.02
0.72	0.21	0.34	0.48	0.64	0.96
0.74	0.16	0.29	0.43	0.58	0.91
0.76	0.11	0.24	0.37	0.53	0.86
0.78	0.05	0.18	0.32	0.47	0.8
0.8	–	0.13	0.27	0.42	0.75
0.82	–	0.08	0.21	0.37	0.7
0.84	–	0.03	0.16	0.32	0.65
0.86	–	–	0.11	0.26	0.59
0.88	–	–	0.06	0.21	0.54
0.9	–	–	–	0.15	0.48

Example

An existing power factor of $\cos\varphi_1 = 0.76$ must be corrected to $\cos\varphi_2 = 0.9$.

According to Table 7.13, 0.37 kvar of capacitive power would be required per kW of active power.

Thus, for an active power of 140 kW (e.g. as per electricity meter reading), the required capacitive power amounts to:

$0.37 \cdot 140 = 51.8$ kvar; 50 kvar is selected.

Compensation of three-phase induction motors with star-delta starting

In the case of direct-on-line starting, the individual three-phase power factor correction capacitors are directly connected to the motor terminals U, V and W. The motor and the capacitors are switched together by the same switching device.

If, in star-delta starting, the capacitors would be connected to U, V and W of the motor in the same way, a dangerous self-excitation of the motor would result if it were disconnected from the supply by means of its main contactor. The motor would operate as a generator receiving its excitation from the discharging capacitors. A potential of up to twice the network supply voltage (in the case of the star-connected winding) could be generated at the motor terminals while the inertia of the driven machine maintains the rotation of the shaft. In addition, the transient currents could severely stress the motor, switchgear and capacitors.

During the change-over period between the star and delta stages, in which the motor is momentarily disconnected from the supply, the capacitor would retain its charged state. It could happen, that the phases are in opposition at the instant the motor is reconnected to the supply.

The ensuing transient current surges not only stress the motor and the capacitors, but cause particularly severe erosion of the contact material in the contactors of the star-delta starter combination.

Therefore, if standard star-delta starters are to be used in conjunction with compensating capacitors for power factor correction, the capacitors must be connected via their own separate contactor (see Fig. 7.61).

This contactor (K5) should preferably be switched via a NO auxiliary contact on the main contactor (K1) of the star-delta combination. In this way, the full power of the capacitor is already effective during the run-up of the motor (in star connection). A discharge choke (or resistor combination) must be provided to discharge the capacitor between switching operations.

Compensation of transformers (no-load)

For the compensation of the no-load reactive power of transformers, the capacitor power is selected in terms of the transformer's reactive power consumption:

$$Q_0 \approx S_0 = \frac{i_0 \cdot S_{nT}}{100}. \tag{7.8}$$

Q_0 no-load reactive power in kvar
S_0 no-load apparent power in kVA
i_0 no-load current in % of the transformer rated current
S_{nT} rated apparent power of the transformer in kVA

As can be seen from Table 7.14, the values differ for oil and GEAFOL (cast-resin) transformers for no-load power losses to the DIN standard, whereby the loss values of the cast-resin transformers are roughly 50% of those given for the oil-filled types.

In the case of transformers with reduced no-load losses, the apparent power values of both transformer types are the same.

The given values only contain the magnetizing reactive power of the unloaded transformer. The use of a fixed-value, individual capacitor for the power factor correction would be appropriate in this case. However, the extent to which the installation of an individual capacitor would make economical sense, particularly in view of transformers with reduced no-load losses, would depend on the application at hand.

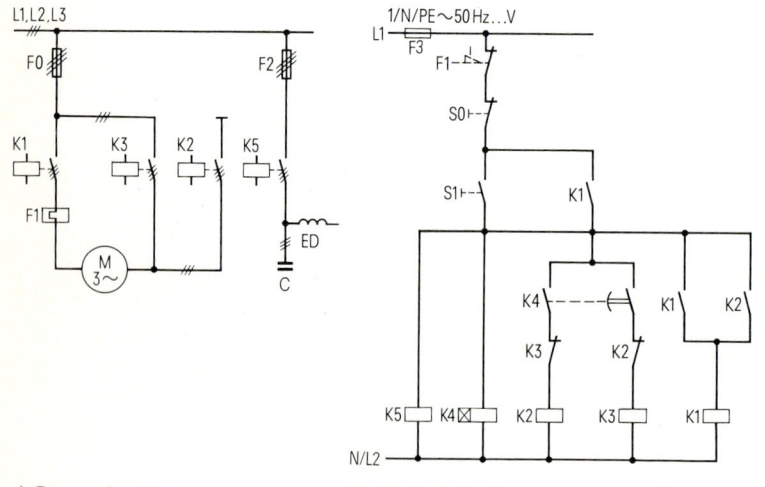

a) Power circuit   b) Control circuit

K1   Main contactor
K2   Star contactor
K3   Delta contactor
K4   Time relay
K5   Contactor for the capacitor
C   Capacitor
ED   Discharge choke (reactance)
F1   Overload relay
F0, F2, F3   Fuses

Figure 7.61
Preferred connection of a power factor correction capacitor in a star-delta starter

Table 7.14
Required capacitive power in dependance of the transformer reactive power for the fixed compensation of unloaded three-phase distribution transformers up to 20 kV HV, 400 V LV, $u_{kr} = 4\%/6\%$

Transformer rated power S_{nT}	Oil-filled and cast-resin transformers with no-load losses in accordance with the DIN standard				Oil-filled and cast-resin transformers with reduced no-load losses	
	Oil-filled transformers		GEAFOL (cast-resin) transformers			
	no-load apparent power of the transformer	capacitor power	no-load apparent power of the transformer	capacitor power	no-load apparent power of the transformer	capacitor power
kVA	kVA	kvar	kVA	kvar	kVA	kvar
250	4.5	5	2.8	3	1.6	2
315	7.9	7.5	3.6	4	1.9	2
400	8.2	7.5	4.4	4	2.2	2
500	10.6	10	4.7	5	2.5	3
630	13.2	12.5	5.7	5	3.2	3
800	15.5	15	6.5	7.5	3.6	4
1000	19.0	20	7.5	7.5	4.0	4
1250	20.6	20	10.1	10	5.0	5
1600	23.2	25	10.6	10	5.3	5
2000	27.0	25	–	–	7.0	7.5

7.4.8.4 Project planning

There are several different methods of determining the required capacitive power Q_c. One method is by using the corresponding tangents (Table 7.15) and the equation (7.6)

$$Q_c = P \cdot (\tan \varphi_1 - \tan \varphi_2). \quad (7.6)$$

Automatic central power factor correction

In installations or electric plants in which various loads are continuously switched on and off during operation, the cos φ changes constantly. Centralized compensation of the inductive reactive power by means of a power factor correction unit which measures and monitors the instantaneous value of cos φ and automatically switches capacitor banks in and out of the network as required, is often the most appropriate solution. Depending on the actual nature of the plant, a mixture of individual compensation of selected larger loads having a high duty factor, and centralized compensation of the rest of the load, may prove to be even more suitable.

To determine the "correct" power rating of the capacitive load, the load values and duty times should be known as accurately as possible.

In the case of installations which are still in the planning phase, one can assume that most industrial loads which are of an inductive nature, will have an average power factor of cos $\varphi_1 \approx 0.7$. For compensation up to cos $\varphi_2 \approx 0.9$, Table 7.15 gives a value of 0.54 for $\tan \varphi_1 - \tan \varphi_2$. Thus, a value for the capacitive power equal to approx. 50% the active power may be used as an orientation:

$$Q_c = P \cdot 0.5.$$

In installations which are already in operation, the required values may be determined by measurement. This is greatly simplified if electricity meters for both active and reactive power are installed. Since $\tan \varphi = Q/P$ (see Fig. 7.62 on page 371), the value of cos φ may easily be calculated or read from Table 7.15. It should be borne in mind, that electricity is often billed in terms of the "maximum demand" (kVA) drawn over a specific integral period. It is therefor important that the power factor and power consumption to be used as the basis for determining the capacitive power requirements, are measured under the appropriate load conditions. A PC programme, known as SIEBKO, is available from Siemens to assist with the project planning and preparation of quotations for power factor correction units (Order No. E 50001-V213-Y27-X-7600).

Table 7.15 Determination of the capacitive load required to correct the power factor from $\cos \varphi_1$ to $\cos \varphi_2$

Present power factor			Required power factor $\cos \varphi_2$									
			1.00	0.98	0.96	0.94	0.92	0.90	0.85	0.80	0.75	0.70
$\cos \varphi_1$	$\sin \varphi$	$\tan \varphi$	$\tan \varphi_1 - \tan \varphi_2$									
0.40	0.92	2.29	2.29	2.09	2.00	1.93	1.86	1.81	1.67	1.54	1.41	1.27
0.45	0.89	1.99	1.99	1.79	1.70	1.63	1.56	1.51	1.37	1.24	1.11	0.97
0.50	0.87	1.73	1.73	1.53	1.44	1.37	1.30	1.25	1.11	0.98	0.85	0.71
0.55	0.83	1.52	1.52	1.32	1.23	1.16	1.09	1.04	0.90	0.77	0.64	0.50
0.60	0.80	1.33	1.33	1.13	1.04	0.97	0.90	0.85	0.71	0.58	0.45	0.31
0.65	0.76	1.17	1.17	0.97	0.88	0.81	0.74	0.69	0.55	0.42	0.29	0.15
0.70	0.71	1.02	1.02	0.82	0.73	0.66	0.59	0.54	0.40	0.27	0.14	–
0.75	0.66	0.88	0.88	0.68	0.59	0.52	0.45	0.40	0.26	0.13	–	–
0.80	0.60	0.75	0.75	0.55	0.46	0.39	0.32	0.27	0.13	–	–	–
0.85	0.53	0.62	0.62	0.42	0.33	0.26	0.19	0.14	–	–	–	–
0.90	0.44	0.48	0.48	0.28	0.19	0.12	0.05	–	–	–	–	–

Example

$$\frac{\text{reactive power consumption (reactive work)}}{\text{active power consumption (active work)}} = \frac{131670 \, \text{kvar(h)}}{99000 \, \text{kW(h)}} = 1.33 = \tan \varphi$$

By means of Table 7.15, the corresponding value of $\cos \varphi$ is 0.6 This means, that the average power factor during the metering period was 0.6.

In order to correct the power factor to 0.9, the equation $Q_c = P \cdot (\tan \varphi_1 - \tan \varphi_2)$ is used to find the necessary capacitive power to be installed. P can be calculated from the measured active power consumption and the operating time (= measurement period)

e.g. $\dfrac{99000 \, \text{kWh}}{180 \, \text{h}} = 550 \, \text{kW}.$

The value for $\tan \varphi_1 - \tan \varphi_2$ from Table 7.15 ($\cos \varphi_1 = 0.6$; $\cos \varphi_2 = 0.9$) is found to be 0.85. Therefore, the require capacitive power is

$550 \cdot 0.85 = 467.5 \, \text{kvar}.$

The following equation leads to the same result:

$$Q_c = \frac{\text{reactive work} - (\text{active work} \cdot \tan \varphi_2)}{\text{measurement period}}$$

$$= \frac{131670 - (99000 \cdot 0.48)}{180}$$

$$= \frac{131670 - 47520}{180} = 467.5 \, \text{kvar}.$$

This method of calculation, however, is only accurate enough if the load remains relatively constant. In the cases of strongly fluctuating loads, e.g. several motors during normal production (inductive load) but only heating and incandescent lighting (ohmic load) during the night, the capacitor power determined from by the average load method will not be sufficient to compensate the total peak inductive load (maximum demand!). Here, it is recommended that the meter readings be recorded over a period of, say, one hour during peak inductive load conditions. Alternatively, the exact instantaneous values of the peak load current, voltage and $\cos \varphi$ may be measured by means of the appropriate instruments.

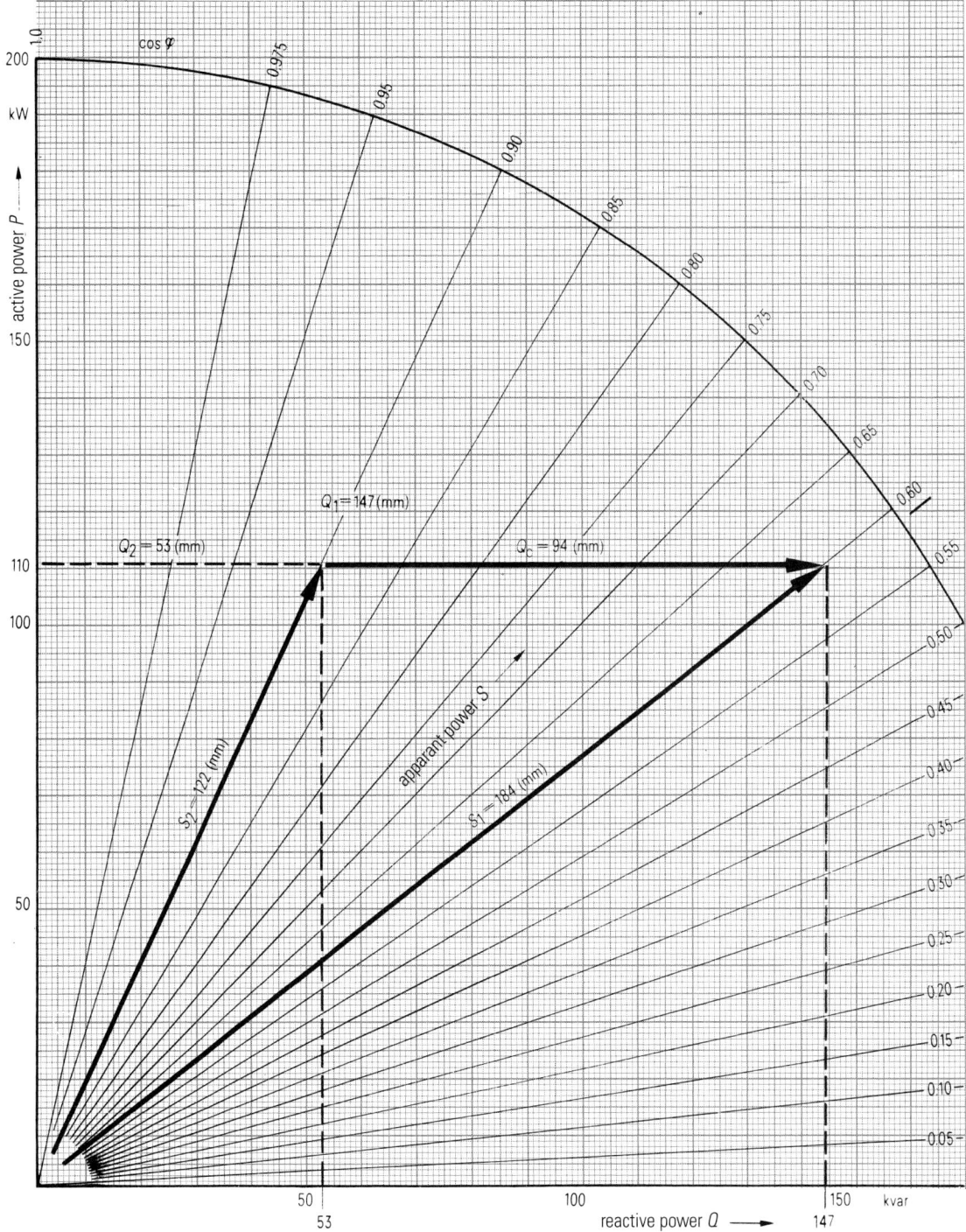

Figure 7.62
Determination of the required capacitive power as a function of the power factor and the active power using scaled vectors on special graph paper

Example

voltage $U = 400$ V
(apparent) current $I = 1330$ A
$\cos \varphi_1 = 0.6$

The product of the voltage and (apparent) current gives the apparent power:

$$S = \frac{U \cdot I \cdot \sqrt{3}}{1000} = \frac{400 \cdot 1330 \cdot 1.73}{1000} = 920 \text{ kVA};$$

the active power is:

$$P = S \cdot \cos \varphi_1 = 920 \cdot 0.6 \approx 550 \text{ kW}$$

the reactive power is:

$$Q_1 = S \cdot \sin \varphi_1 = 920 \cdot 0.8 \approx 736 \text{ kvar}.$$

A capacitive power of 736 kvar (power factor correction unit with 750 or 700 kvar) would enable compensation to $\cos \varphi_2 \approx 1$.

At compensation up to $\cos \varphi_2 = 0.9$ the resultant apparent power would be

$$S_2 = \frac{P}{\cos \varphi_2} = \frac{550}{0.9} = 610 \text{ kVA}.$$

The required capacitive power would be

$Q_c = Q_1 - Q_2$,

$Q_2 = S_2 \cdot \sin \varphi_2 = 610 \cdot 0.44 \approx 268$ kvar

or

$Q_2 = P \cdot \tan \varphi = 550 \cdot 0.48 \approx 264$ kvar [1],

$Q_c = Q_1 - Q_2 = 736 - 266 = 470$ kvar.

Calculation by using vectors

In order to save the above calculation procedures, the required capacitive power may also be found with the aid of a diagram (Fig. 7.62).

A mm scale results in a sufficient degree of accuracy of the results (1 mm = 1 kvar).

If larger units need to be calculated, the appropriate scale multipliers may be used.

---

[1] Difference owing to rounded φ values have a negligible effect in practice. We have assumed an average of 266 kvar

Example

Active power $P = 550$ kW $= 110 \cdot 5$ kW;

existing $\cos \varphi_1 = 0.6$, desired $\cos \varphi_2 = 0.9$.

Result:

Apparent power 1
$S_1 = 184 \cdot 5 = 920$ kVA,

Apparent power 2
$S_2 = 122 \cdot 5 = 610$ kVA.

Reactive power 1
$Q_1 = 147 \cdot 5 = 735$ kvar,

Reactive power 2
$Q_2 = 53 \cdot 5 = 265$ kvar,

Capacitive power
$Q_c = Q_1 - Q_2 = 94 \cdot 5 = 470$ kvar.

7.4.8.5 Voltage rise caused by capacitors

If a capacitive current flows through an inductive resistance, then the "output" voltage will be higher than the "input" voltage by an amount equal to the voltage drop across the inductance.

This could happen in practice if, for example, a number of capacitor banks remain connected downstream of a transformer after the inductive load has decreased (leading power factor). The voltage rise which will result depends on the relationship between the connected capacitive powers and the transformer power output. The following equation provides an indication of the expected voltage rise as a percentage with sufficient accuracy for most practical purposes:

$$\Delta U = u_{kr} \cdot \frac{Q}{S_{nT}} \text{ (in \%)} \qquad (7.9)$$

ΔU voltage rise in %
u_{kr} rated value of the transformer short-circuit voltage in %
Q capacitive power in kvar
S_{nT} rated apparent power of the transformer in kVA

In practice, noticeable increases in the voltage have become very rare, since the connected capacitive power is nowadays almost invariably regulated e.g. by means of automatic power factor correction units.

The vector diagrams in Figures 7.63 and 7.64 which illustrate lagging (inductive) and leading (capacitive) loads, help to explain this phenomenon.

Power factor correction: networks with harmonics

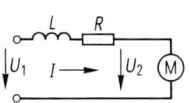

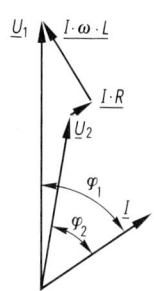

Figure 7.63
Equivalent circuit and vector diagram in the case of an inductive load (lagging p.f.)

U_1 No-load voltage
U_2 Network voltage on the motor terminals

$U_2 < U_1$

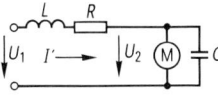

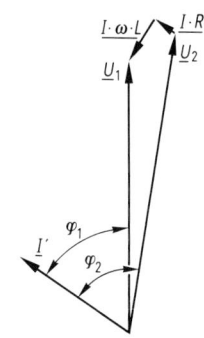

Figure 7.64
Equivalent circuit and vector diagram in the case of a capacitive load (leading p.f.)

U_1 No-load voltage
U_2 Network voltage on the motor terminals

$U_2 > U_1$

Since, however, they draw a non-sinusoidal current of which a significant proportion tends to be reactive, rectifiers (Figure 7.65) often distort the three-phase supply network and may need special attention as far as power factor correction is concerned.

Causes and effects of harmonics

The rectifier current waveform is composed of a number of sinusoidal currents; one fundamental waveform at supply frequency and a series of harmonics at integer multiples of the supply frequency. The harmonic currents are superimposed on the three-phase supply. Thereby, harmonic voltages are created across the network impedances and a distortion of the network voltage waveform results. This can cause disturbances in the network and even lead to failure of other consumer loads.

The frequencies of the harmonics depend on the pulse number of the rectifier circuit. The pulse number indicates the number of times rectification takes place within one period of the fundamental waveform. In a two-way rectification circuit (Graetz connection) the pulse number is two, where-as a full three-phase bridge rectifier has the pulse number 6.

The higher the pulse number, the less pronounced the "pulsing" of the rectified direct current will be.

The harmonic number v indicates at which multiples of the network frequency harmonics will occur:

$$v = p \cdot k \pm 1 \qquad (7.10)$$

p pulse number
k 1, 2, 3 etc.

Thus, a six-pulse rectification circuit can produce the 5th, 7th, 11th, 13th, etc harmonic. Figure 7.66 illustrates the idealized component waveforms of such a six-

Example

Transformer: rated output power 630 kVA, $u_{kr} = 6\%$;
Capacitive load: 200 kvar,

$$\Delta U = 6 \cdot \frac{200}{630} = 1.9\%.$$

In other words, the voltage U_2 is increased by 1.9% with respect to the no-load voltage U_1.

7.4.8.6 Compensation in networks with harmonics

Owing to developments in the field of power electronics, the number of rectifier and converter loads (e.g static convertors, variable speed drives) in industrial networks has increased significantly.

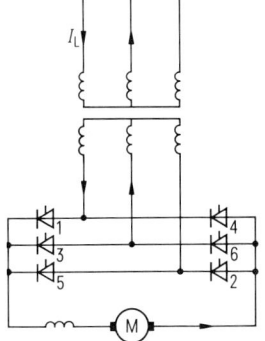

Figure 7.65
Three-phase bridge-rectifier circuit

7 Type-tested switchgear assemblies (TTA)

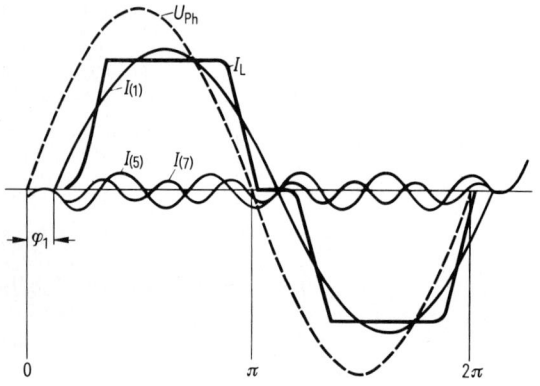

I_L Total load current
$I_{(1)}$ Fundamental waveform
$I_{(5)}$ 5th harmonic
$I_{(7)}$ 7th harmonic
U_{Ph} Phase voltage
φ_1 Angular phase shift between U_{Ph} and I_L

Figure 7.66
Rectifier current resolved into the fundamental waveform and a number harmonics

pulse rectifier current and shows both the 5th and 7th harmonics. Note that the amplitude of the harmonic currents are significantly lower than that of the fundamental. In theory, their values can be calculated using the equation

$$I_\nu = \frac{1}{\nu} \cdot I_1 \qquad (7.11)$$

This equation, however, only applies in ideal conditions and completely rectified direct current. In practice, the following values may be expected:

$I_{(5)} = 0.25 \cdot I_1$,
$I_{(7)} = 0.13 \cdot I_1$,
$I_{(11)} = 0.09 \cdot I_1$,
$I_{(13)} = 0.07 \cdot I_1$.

Resonance effects resulting from the use of power capacitors

More and more often, power factor correction needs to done in industrial networks containing a significant proportion of harmonic-producing load (e.g. rectifiers, static converters, variable speed drives). Great caution should be exercised in the use of power capacitors in such networks, however, since there could be a real danger that undesirable resonance effects may result.

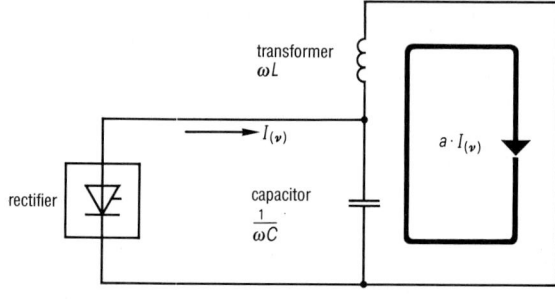

a amplification factor due to resonance

Figure 7.67
Resonance effects caused by the use of power capacitors

The reason for this is that, seen from the low-voltage side, the installed capacitors together with the reactive impedance of the incoming supply transformer and the rest of the network inductances, form an LC oscillating circuit. If the natural frequency of this oscillating circuit happens to coincide with that of one of the harmonic currents, then resonance will occur in the LC circuit. Extreme overcurrents result which lead to overloading of the installation and tripping of protection equipment (refer to Fig. 7.67).

Power factor correction using choked capacitors
(Detuned resonance circuits)

If power factor correction has to be carried out in networks containing a significant proportion (20%) of harmonic-producing load (e.g. rectifiers, static converters, variable speed drives), then chokes or inductors (reactor coils) should be connected in series with the capacitors to avoid the possibility of resonance at harmonic frequencies (Figs. 7.68 and 7.69).

The series connection of the capacitors and inductors form series oscillating circuits. The natural, or resonance frequency of these oscillating circuits must be lower than the frequency of the first expected harmonic. For example, if the first expected harmonic is the 5th (250 Hz), then the resonance frequency of the capacitor/choke combinations should be in the range of 200 to 220 Hz. In this way, the power factor correction unit appears to be inductive for all frequencies (harmonics) ≥ 250 Hz and resonance effects are no longer possible (refer to Figs. 7.69 and 7.70).

In the Siemens K catalogue, two versions of power factor correction units with the designation 4RF11 and 4RF30 respectively, are offered. These two units differ from each other in the type of choking they employ.

Harmonics · Resonance effects

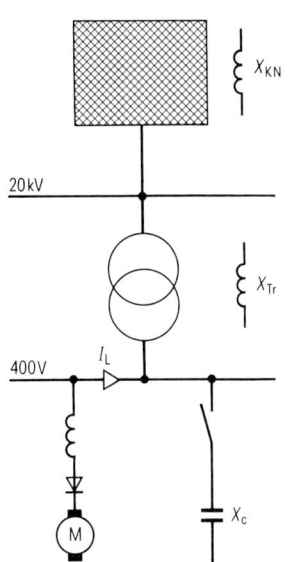

Figure 7.68
Network configuration showing the possibility of parallel resonance owing to rectifier loading

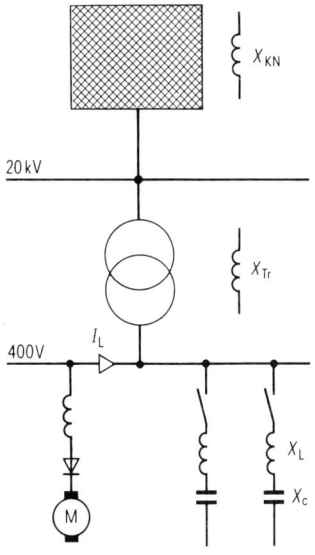

Figure 7.69
Choked capacitors (capacitors with series connected reactors) in a power factor correction application

4RF30 power factor correction units

In the case of the 4RF30 units, the resonance frequency of the series connected reactor and capacitor combination is 189 Hz (7% choking). Resonance is avoided and, in addition, approximately 20% of the 5th harmonic is suppressed (filtered out). The major portion of the harmonics are superimposed on the feeder network.

4RF11 power factor correction units

In the case of the 4RF11 units, the resonance frequency of the series connected reactor and capacitor combination is 210 Hz (5.6% choking).

The inductive impedance of this combination is lower at 250 Hz than that of the aforementioned power factor correction unit with a resonance frequency of 189 Hz. This means, that the 4RF11 units suppress (filter out) about 50% of the 5th harmonic.

In practice, it is recommended that choked power factor correction units should always be used in cases where the proportion of harmonic producing load is greater than 20% of the total load. The size of power factor correction unit is merely selected in terms of the reactive power (inductive) to be compensated, as is described above. The harmonics need not be considered in the planning or selection of the equipment, since possible overloading as a result of resonance is prevented by the reactance/capacitor series combinations.

Filter circuit
(Tuned resonance circuits)

If it is not possible to limit the harmonic content to an acceptable level by the selection of a suitable rectifier type or circuit, then the harmonics may have to be filtered from the network by means of specifically tuned oscillating circuits.

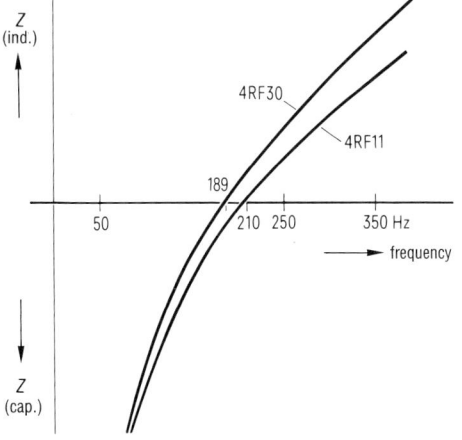

Figure 7.70
Behaviour of the impedance Z of series-connected reactors and capacitors in choked power factor correction units, as a function of the frequency

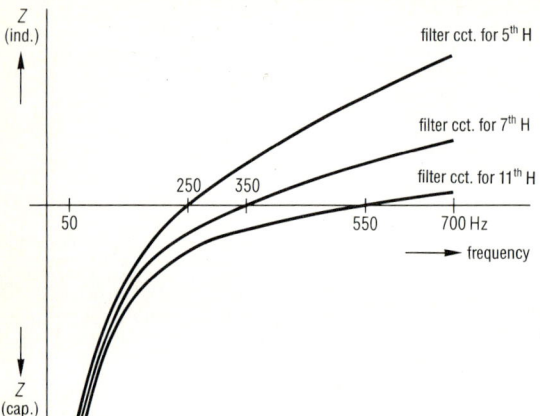

Figure 7.71
Behaviour of the impedance Z of series-connected reactors and capacitors in filter circuits tuned for the 5th, 7th and 11th harmonic (H) as a function of the frequency

These filters are also series oscillating circuits consisting of reactors and capacitors, but in contrast to the choked capacitor technique described above, they are specifically tuned to resonate at the frequency of a selected harmonic. At this frequency, the impedance of the particular filter is virtually zero (see Fig. 7.71).

By tuning the respective filter circuits to selected harmonic frequencies, e.g.

250 Hz   for the 5th harmonic,
350 Hz   for the 7th harmonic, and
550 Hz   for the 11th harmonic,

they will resonate at these harmonics and filter up to 90% of the respective waveform from the network.

Naturally, the filter circuits must be able to carry the harmonic currents produced by the rectifier load under all operation conditions, without becoming overloaded themselves. Therefor, the power capacity of the filter circuits must be designed in terms of the expected harmonic currents.

As far as current at the fundamental frequency is concerned, the filter circuits are always capacitive. Since, as a first approximation, the harmonic currents are proportional to the reactive power, the full filtering capacity must also be available at small phase-control delay angles of e.g. a static converter load. In this operational condition, however, care must also be taken to ensure that the fundamental reactive load is not overcompensated.

An overcompensation during reduced load periods can be avoided by the use of a load current dependent control system which regulates the connected capacitive power in a single step for small installations, or in several stages for larger applications.

If the feeder network (e.g. HV supply) has a high harmonic content, then this can cause additional loading of the filter circuits. The extent of this loading depends on the feeder network data and its configuration.

In practice, filter circuits are mostly installed for the 5th, 7th, 11th and 13th harmonics, whereby the 11th and 13th share a common filter circuit (see Figs. 7.72 and 7.73).

The efficiency of the filter system is always dependent on the short-circuit reactance of the feeder transformer.

If the reactive power drawn by the loads is not sufficiently compensated by the filter circuits, then an addition power factor correction unit with choked capacitors may be installed.

Note

As a matter of principle, unchoked p.f.c. units must not be operated on the same busbars in parallel with choked p.f.c. units or harmonic filter circuits. This operational condition could cause a parallel reso-

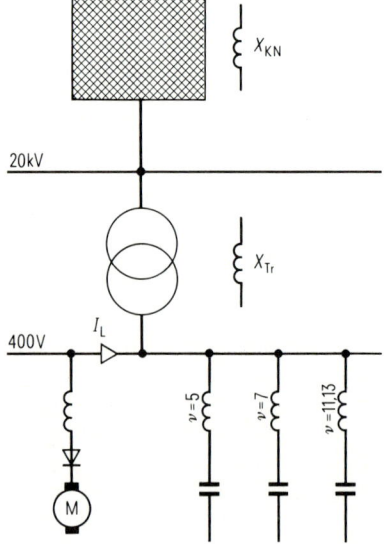

Figure 7.72
Tuned filter circuits to remove harmonic currents from the network

Harmonic filters · AF ripple control

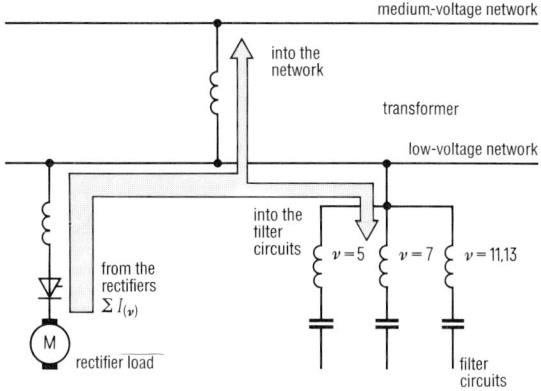

Figure 7.73
The drawing off of harmonic currents by means of tuned filter circuits

nance circuit with a fundamental frequency within the critical range of one of the expected harmonic currents.

Calculation of capacitive power for the compensation of a constant torque DC drive

Rated data of the drive:

output power P_{dn},
rated speed n_n,
efficiency η.

The following assumptions apply as approximations for the most common rectifier circuits:

$\eta = 0.95$,
$\cos \varphi_n = 0.88$.

Rectifier data on the three-phase side at rated operation ($n = n_n$):

active power P_n,
apparent power S_n,
displacement power factor $\cos \varphi_n$.

Operating condition 1 ($n = n_1$):

$S_1 = S_n$,

$P_1 = \dfrac{n_1}{n_n} \cdot P_n$,

$\cos \varphi_1 = \dfrac{n_1}{n_n} = \cos \varphi_n$.

The capacitive power necessary to compensate the lagging reactive power under operating condition 1 and achieve the desired power factor $\cos \varphi_2$, can be calculated by equation (7.6) as follows:

$Q_c = P_1 \cdot (\tan \varphi_1 - \tan \varphi_2)$,

$= \dfrac{n_1}{n_n} \cdot \dfrac{1}{\eta} \cdot P_{dn} \cdot (\tan \varphi_1 - \tan \varphi_2)$.

Q_c required capacitive power

Example

output power $\quad P_{dn} = 250$ kW,
rated speed $\quad n_n = 1200$ min^{-1}.

The drive operates at constant torque. At 50% rated speed, the power factor must be corrected to $\cos \varphi = 0.9$ by means of compensating capacitors.

Determining the necessary capacitive power:

$\cos \varphi_n = 0.88 \quad$ (assumption, see above),

$\cos \varphi_1 = \dfrac{600}{1200} \cdot 0.88 = 0.44$

$\tan \varphi_1 = 2.0$,

$\tan \varphi_2 = 0.48$,

$\eta = 0.95 \quad$ (assumption, see above),

$Q_c = \dfrac{600}{1200} \cdot \dfrac{1}{0.95} \cdot 250 \text{ kW} \cdot (2.0 - 0.48)$

$= 200$ kvar.

A capacitive power of 200 kvar is required if the desired power factor is to be achieved.

In these applications, either choked capacitor power factor correction units or tuned filter circuits need to be used.

Drives which do not operate at constant torque

The description given above does not apply in the case of drives which operate with a torque-speed curve which can be expressed by either a square or cubic function, e.g. fans or pumps.

The required data for these drives, is to be taken from the power diagrams of the particular drive in question.

7.4.8.7 The use of audio-frequency remote control systems

In networks which employ audio-frequeny remote control systems (AF ripple control), audio-frequency impulses are superimposed on the network supply to switch remote receiver relays on and off.

7 Type-tested switchgear assemblies (TTA)

If power capacitors or power factor correction units are installed in such networks, then the following points should be given serious attention:

Even if they are of a relatively low voltage, high frequency signals can cause large audio-frequency currents to flow in the network, since the impedance of capacitors is inversely proportional to frequency. This not only leads to additional loading of the audio-frequency signal generator, but may also cause the voltage of the switching signal to drop below the value necessary to operate the AF receiver relays in the vicinity of the capacitors. A remedy for this problem, is the fitting of audio-frequency blocking circuits (or AF traps) to the capacitive load.

One can distinguish between two versions of audio-frequency blocking circuits.

Version 1

These are almost invariably parallel resonance blocking circuits (low-pass filters) which are connected into the supply line upstream of the capacitors. Parallel resonance blocking circuits are the parallel connection of an inductance L and a capacitance C_p. In low-voltage networks, these are transformer-coupled into the line for economic reasons (Fig. 7.74).

The resonance frequency f_0 of the parallel blocking circuit is determined by the relationship between L and C_p. It is chosen to be equal to the audio-frequency switching signal of the ripple control system. At this frequency, the parallel resonance blocking circuit represents an extremely high impedance in the line (see Fig. 7.76). The responsible electricity supply authority in the area can advise which audio frequency needs to be blocked.

One must be extremely careful in networks with rectifier loads. If the rectifier loads represent more than 20% of the total load, then choked capacitors or tuned filter circuits should be used for power factor correction.

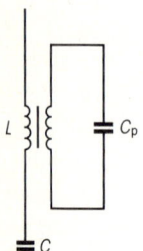

Figure 7.74
Transformer-coupling of a parallel resonance blocking circuit

Version 2

Some electricity supply authorities (e.g. Berlin) require that the method of "partial choking" be used in their areas of responsibility (audio frequency usually 750 Hz).

In the case of partially choked capacitors, the capacitor load to be blocked is itself part of the blocking circuit. The capacitive load is divided into C_1 and C_2. A reactor is connected upstream of the capacitance C_1 (see Fig. 7.75).

The capacitances C_1 and C_2 of the power capacitors and the inductance of the reactor L are selected in such a way, that their combination represents a virtually infinitely high impedance Z at the audio fre-

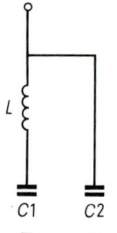

Figure 7.75
Partially choked capacitors

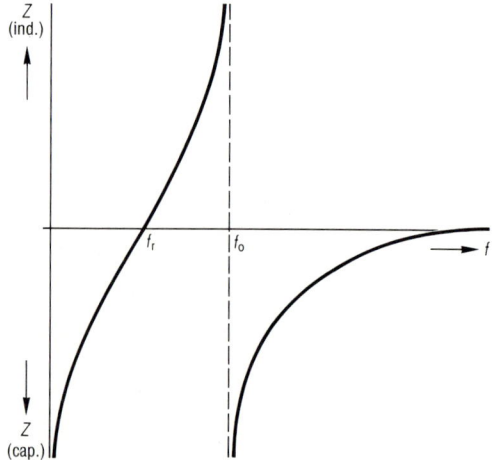

f_r Series resonance frequency
f_0 Parallel resonance frequency (audio frequency)

Figure 7.76
Behaviour of the impedance Z in relation to the frequency f

Audio frequency blocking circuits

quency f_0, e.g. 750 Hz. The series resonance frequency f_r is determined by the values of L and C_1 (refer to Fig. 7.76).

4RW power factor correction units with partially choked capacitors based on the above guidelines, are offered as complete cubicles ready for installation. They are used for centralized compensation of the reactive power in networks with predominantly linear loads (proportion of rectifier and other harmonic-producing loads < 20%).

Blocking factor a_f

The effectivity of capacitors and power factor correction units equipped with audio-frequency blocking circuits, is expressed in terms of the blocking factor a_f. It is defined as follows:

The blocking factor a_f of an audio-frequency blocking circuit is the ratio between the audio-frequency impedance Z_f of a capacitor blocked by this circuit, and the reactance X_c of the capacitor at sinusoidal voltage with the fundamental frequency f_1 (50 Hz).

$$a_f = \frac{Z_f}{C_{cf_1}} \qquad (7.12)$$

(VDEW[1] recommendation of 1958).

The VDEW[1] stipulates specific values for the blocking factor as a function of the audio frequency (see Figure 7.77).

---
[1] VDEW: Vereinigung Deutscher Elektrizitätswerke e.V. (transl. Association of German power stations)

The use of choked power factor correction units in networks with audio-frequency remote control systems

Choked power factor correction units can be used in networks with audio-frequency remote control systems (AF ripple control) *and* a rectifier load which exceeds 20% of the total load, provided that the following points are taken into consideration:

▷ Above an audio frequency of about 330 to 380 Hz, the choked power factor correction units achieve a sufficiently high blocking factor; additional measures for the blocking of the audio-frequency switching signal are not required.

▷ At audio frequencies lower than the above-mentioned values, the blocking factor drops to below 0.3, which is the minimum value prescribed by VDEW[1] for audio frequencies below 475 Hz. In this case, a special arrangement may need to be made with the responsible electricity supply authority.

▷ At audio frequencies in the range of the tuned frequency of the choked capacitors and below (mainly in the case of audio frequencies below 250 Hz), sufficient blocking of the audio frequency is no longer possible. Here, special precautions and measures may need to be taken.

▷ In principle, it should be noted that audio-frequency blocking circuits used in conjunction with choked power factor correction units require special attention with regard to their design, and that they can only be supplied subject to prior request and checking of all the relevant parameters.

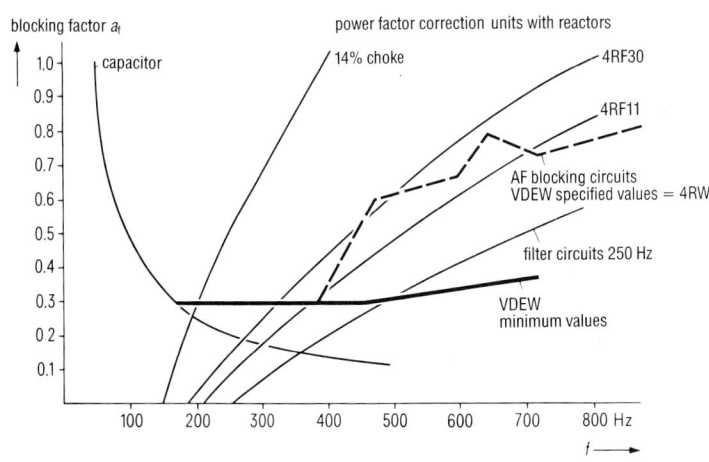

Figure 7.77
Comparison of the blocking factors achieved by the various techniques

7 Type-tested switchgear assemblies (TTA)

The use of audio-frequency blocking circuits in networks with a rectifier load up to 20% of the total load

In networks with a harmonic content and an audio-frequency switching signal in the range of 283 to 425 Hz, partial resonance may occur at the 5th and 7th harmonic of the rectifier load.

Audio-frequency blocking circuits for this frequency range therefor have a stronger design.

7.4.8.8 Range of products for power factor correction

The basic p.f.c. units are suitable for use in networks without or with rectifier loads adding up to max. 20% of the total load. Table 7.16 gives an overview of the range of Siemens power factor correction units. Extension units, having the same power ratings as those listed for the basic units, may be used for larger installations.

Table 7.17 illustrates the product range of the Siemens power factor correction units with choked capacitors.

In principle, all the units may be equipped with audio-frequency blocking circuits. Figures 7.78 and 7.79 show some examples of the products used for individual compensation. Figures 7.80 and 7.81 show p.f.c. units in cabinet and cubicle construction respectively. These incorporate power factor controllers with automatic C/k value setting.

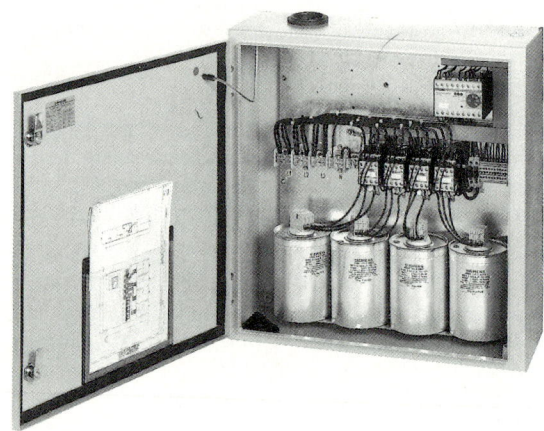

Figure 7.80
Sheet steel cabinet, controller with automatic C/k value setting

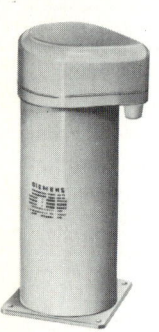

Figure 7.78
4RB3 capacitor,
12.5 kvar, 400 V, 50 Hz, for individual compensation

Figure 7.79
4RD3 capacitor bank, 100 kvar, 400 V, 50 Hz, for individual compensation

Figure 7.81
Power factor correction unit in cubicle construction

Table 7.16
Overview of the range of basic power factor correction units

	Rated voltage (at 50 Hz) V	Power kvar
Fixed compensation		
4RB	400, 440, 525, 690	2, 3, 5, 7.5, 10, 12.5
4RD	400, 525, 690	20, 25, 30, 35, 40, 50, 60, 75, 90, 100
Automatic power factor correction units		
4RY	400, 440, 525, 690	15, 20, 25, 30, 40, 50, 75, 100, 125, 150, 200, 250, 300, 350, 400

Table 7.17
Overview of the range of power factor correction units with choked capacitors for the compensation of reactive power in low-voltage networks which incorporate a significant proportion of rectifier load

	Rated voltage (at 50 Hz) V	Power kvar
Fixed compensation		
4RF35 [1]	400	5, 10, 20, 40
4RF25 [2]	400 525 690	25, 50
Automatic power factor correction units		
4RF30 [1]	400	20, 25, 30, 40, 50, 60, 80, 100, 120, 160, 200
4RF11 [2]	400, 525, 690	100, 150, 200, 250, 300, 350, 400

[1] Absorbtion of approx 20% of the 5th harmonic
[2] Absorbtion of approx 50% of the 5th harmonic

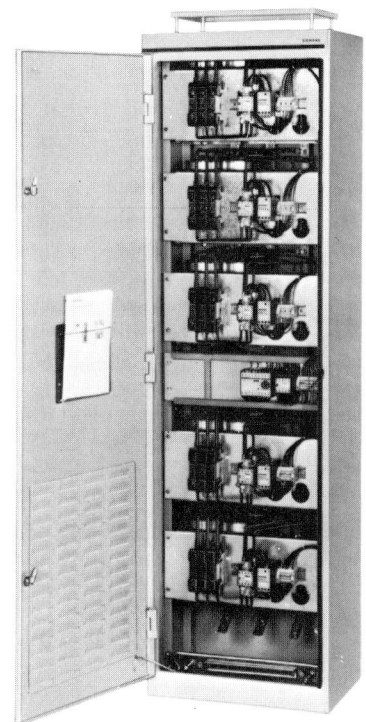

Figure 7.82
Power factor correction unit with choked capacitors

Extension units, having the same power ratings as those listed for the basic units and a similar construction, may be used for larger installations.

Figure 7.82 shows a unit with choked capacitors.

Since filter circuits need to be matched and tuned to the supply network and the connected loads, no standard filter units can be offered in the catalogues. An enquiry for a custom-built unit needs to be placed on the manufacturer.

7.5 Charging units for stationary standby battery installations

In the event an interruption in the primary electricity supply, e.g. owing to a fault in the supply network, a standby battery installation can take over the continued power supply of the most important or critical operational equipment. In these stationary battery installations, the charging apparatus is permanently connected to both the battery and the consumer load. The rated and operational data of the battery, charger and load circuits must be carefully matched to ensure the reliable and efficient functioning of the installation as a whole.

In the parallel stand-by mode, the charging unit not only charges the battery (or maintains its charge level), but also supplies power to the consumer loads. The required power output of the battery charger depends on the capacity of the battery and on the total power consumption of the consumer loads. For it to be determined, the load voltage and the rated value of the charging current must be known.

The total required charging unit current consists of the maximum consumer current I_v which can be drawn, plus a so-called reserve charging current I_{LR} for the charging of a battery which was discharged during a main power supply failure. As a rule of thumb, a minimum reserve charging current of 10 A for every 100 Ah of installed battery may be assumed.

The rated voltage of the charger must be selected to be the same as the consumer load voltage.

The Siemens 6RC55 battery charger product spectrum covers a power range of 0.36 to 44 kW. Standard units are offered for up to 500 A direct current and for DC 12, 24, 48, 60, 110 and 220 V (Fig. 7.83).

Automatic rapid recharging

During normal operation, the battery charger simply maintains the charge on the battery by supplying a charging current (trickle charge) which is just sufficient to compensate for the natural discharge process (floating operation).

The voltage required to maintain the fully charged state (trickle charging) is 2.23 V per cell in the case of lead (Pb) batteries, and 1.4 V per cell for nickel-cadmium (NiCd) batteries, respectively.

As an optional accessory, an automatic boost-charge controller may be installed which recharges the batteries at an increased voltage (2.35 V per cell for Pb batteries, and 1.55 V per cell for NiCd batteries) after each power supply interruption (boost charging). After a period of time, which can be adjusted to between 0.5 and 6 hours, the charger is switched back to the trickle-charging, or floating voltage again.

The guide values for recharging time given in Table 7.18, are based on the assumption that a battery charging current of 10 A per 100 Ah of installed battery capacity is available.

In the case of Pb batteries, an automatic charge controller is usually not required, since they are charged to over 90% within a short period of time by the trickle-charging voltage. The advantage of this is, that no special measures need be taken to limit the consumer d.c. voltage during active charging.

In the case of NiCd batteries, the trickle-charge voltage is not sufficient to achieve a full charge, and therefore an automatic boost-charge controller is indispensable.

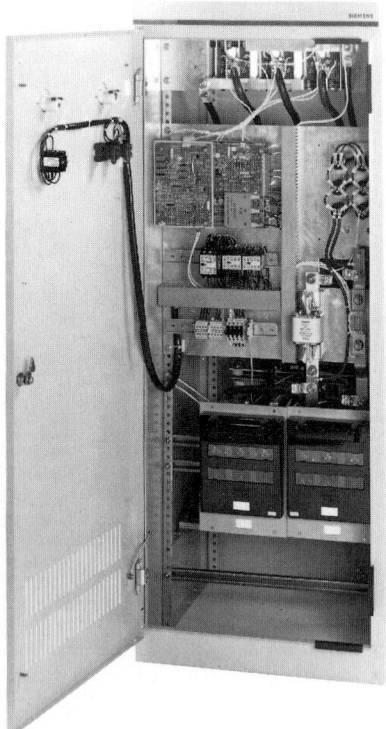

Figure 7.83
A Siemens 6RC55 regulated charger for stationary stand-by battery installations

Table 7.18
Recharging time and degree of charge of a battery as a function of the charging voltage, degree of prior discharge and battery type

Battery type	Charging voltage per cell	Degree of re-charg-ing	Re-charge time	Degree of dis-charge
Pb	2.23 V	93%	3 to 6 h	partially discharged
			20 h	fully discharged
		complete charging possible over several days depending on the degree of discharge		
	2.35 V	97%	3 to 6 h	partially discharged
		99%	20 h	fully discharged
NiCd	1.4 V	complete charging not possible		
	1.55 V	98%	10 h	partially discharged
			20 h	fully discharged

Limiting of the consumer d.c. voltage when using NiCd batteries

As far as the connected consumer loads are concerned, the battery represents a strongly fluctuating source of d.c. voltage. The behaviour of the battery voltage over a period of time which includes the typical operational conditions charging, trickle charging and discharging, is illustrated in Figure 7.84. It can be seen clearly that the maximum permissible consumer voltage tolerance of -15% to $+10\%$ (to DIN VDE 0160) is exceeded in some ranges.

As already mentioned above, Pb batteries do not necessarily require a charging voltage greater than 2.23 V per cell. The 11.1% overvoltage which occurs during trickle charging can be compensated by a corresponding reduction in the number of cells connected to the load, so that the consumer voltage remains within the permissible $+10\%$ of the rated value.

In the case of NiCd batteries, the resulting voltage range is definitely too wide, and special measures need to be taken to limit the consumer voltage to

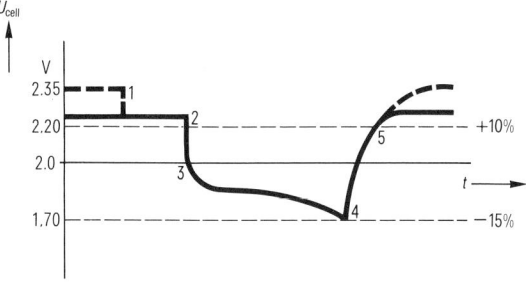

a) Pb batteries

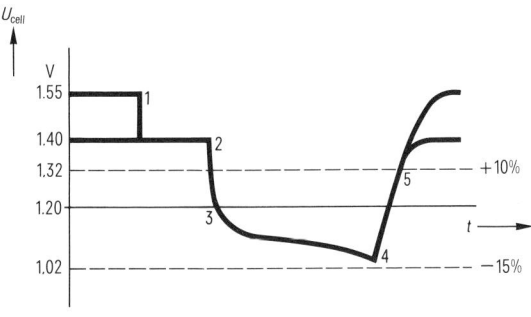

b) NiCd batteries

1 Charging with 2.35 V per cell (Pb) or 1.55 V per cell (NiCd)
2 Maintenance-charging (trickle-charging) with 2.23 V per cell (Pb) or 1.4 V per cell (NiCd)
3 Voltage at the start of the discharge period
4 Voltage at the end of the discharge period
5 Re-charging

Figure 7.84
Charging and discharging characteristics of Pb batteries (a) and of NiCd batteries (b)

an acceptable level. To achieve this, either *counter cells* or *split batteries* are used.

Counter cells

Owing to the losses produced by these cells, they are usually only used at lower currents up to 100 A. They consist of a string of series-connected silicon diodes (Fig. 7.85) and are connected between the battery and the consumer load. The voltage drop across this chain of diodes is independent of the load current. As the voltage of the battery drops during discharge, these counter cells are automatically bridged out (in one or two stages) so that the consumer voltage is remains constant within a narrow tolerance band.

7 Type-tested switchgear assemblies (TTA)

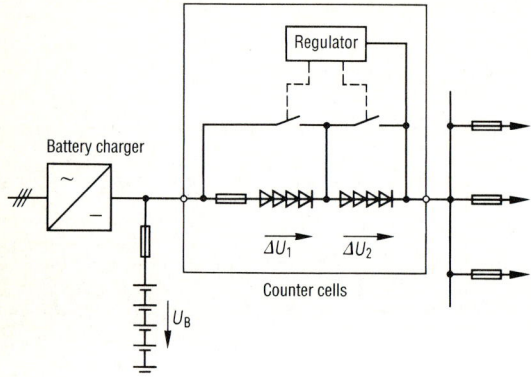

Figure 7.85
Schematic diagram of a two-stage counter cell

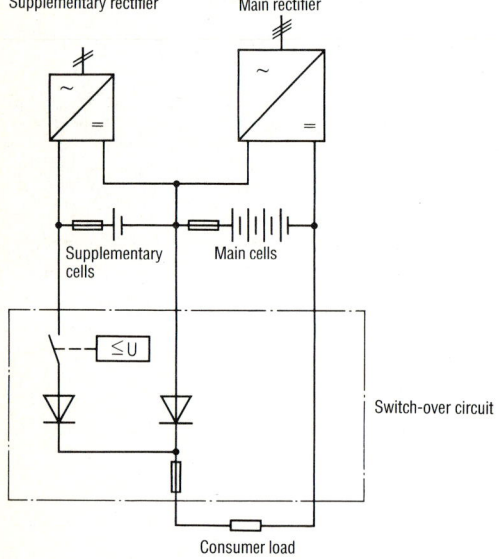

Figure 7.86 Schematic diagram of a split battery bank

The counter cells are selected according to

▷ the maximum load current which will flow through them and

▷ the required voltage drops during "trickle charging (ΔU_1)" and "boost charging (ΔU_2)"

Split batteries

For economic reasons, the split battery method is only employed at higher load currents where counter cells can no longer be used. In this method, the battery is split by means of a tapping into main and supplementary cells (Fig. 7.86).

In addition to the main rectifier, a supplementary charger (supplementary rectifier) is required to charge the supplementary cells. In the event of a power supply interruption and depending on the consumer voltage level, the supplementary cells are added to the main cells by means of a switch-over arrangement.

The following points need to be considered in the planning phase:

▷ The number of main cells must be selected in such a way, that only the maximum permissible limit of the consumer voltage tolerance will be reached during boost loading in the parallel stand-by mode.

▷ The switch-over arrangement must be selected for the maximum consumer load current.

Example

A suitable battery charger is to be selected for a stationary stand-by battery installation with a rated voltage of 60 V. In addition, the appropriate measures to ensure restriction of the consumer voltage to within the given tolerance limits, is to be specified. The following technical data is provided:

maximum consumer voltage $U_{v\,max} = 66$ V,
minimum consumer voltage $U_{v\,min} = 51$ V,
maximum consumer load current $I_{v\,max} = 40$ A,
battery type: NiCd,
battery capacity $K = 150$ Ah,
number of cells $n = 50$.

Specification of the battery charger:

The rated current of the charger I_{nL} is made up of the maximum consumer load current $I_{v\,max}$ and the reserve charging current I_{LR}:

$I_{nL} = 40$ A $+ 15$ A $= 55$ A.

The rated voltage of the charger U_{nL} is 60 V.

An automatic boost-charge controller is required to ensure that the NiCd batteries can be charged to their full capacity.

– Measures to limit the consumer voltage:

Since the load current is smaller than 100 A, counter cells may be used. The maximum consumer load current $I_{v\,max} = 40$ A is used as the basis for determining the current rating I_{nG} of the counter cells.

A two-stage counter cell arrangement is required in systems using NiCd batteries. The voltage drop across the first counter-cell stage for the trickle-charging mode is calculated as follows:

$\Delta U_1 \geq n \cdot 1.4\,V - U_{v\,max}$

$\Delta U_1 \geq 70\,V - 66\,V \geq 4\,V.$

The voltage drop across the second additional counter-cell stage for the boost-charging mode is calculated as follows:

$\Delta U_2 \geq n \cdot 2.55\,V - n \cdot 1.4\,V$

$\Delta U_2 \geq 77.5\,V - 70\,V \geq 7.5\,V.$

Thus, the required charger must have a rated output current of at least 55 A and a rated output voltage of 60 V. It must incorporate two stages of counter cells; the first for a 4 V and the second for a 7.5 V voltage drop. The counter cells must have a rated current of at least 40 A. This information is sufficient for selection of a suitable charger (as well as the optional accessories) from the Siemens product range.

7.6 Current transformers

Current transformers convert larger currents to values which are easier to measure and/or use in monitoring or protection devices. Owing to the characteristics of the current transfer, they also protect measuring instruments from short-circuit currents and overvoltages.

7.6.1 Basic designs

The basic designs of current transformers may be divided into two groups, viz. wound-primary type and window-type current transformers.

Wound-primary current transformers

As the name suggests, these current transformers incorporate a primary winding with a number of turns. They have fixed primary and secondary terminals (Fig. 7.87a).

Window-type current transformers

Window-type current transformers (also known as bar-type) have no primary winding and have fixed secondary terminals. They are slipped over the conductor, i.e. the conductor itself is the primary winding as it passes through the "window" of the c.t. (Fig. 7.87b).

These c.t.'s may also be installed as "thread-through" or "pin-wound" current transformers (Figure 7.88).

Earthing of current transformers

Current transformers for use at operational voltages up to 1000 V do not have to be earthed.

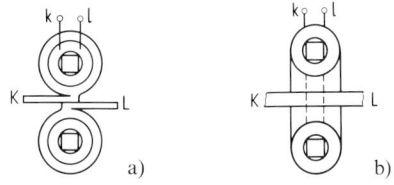

K, L   Primary terminals
k, l   Secondary terminals

Figure 7.87
Schematic representation of a wound-primary (a) and a window-type current transformer (b)

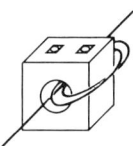

Figure 7.88
The use of a window-type current transformer as a thread-through c.t.

7.6.2 Current transformers for specific applications

7.6.2.1 Interposing current transformers

Interposing current transformers are used to change the value of the secondary current (of the main c.t.). In addition they can reduce the overcurrent or saturation factor of main current transformers and thus serve to protect equipment which is particularly sensitive to overcurrent. Wound-primary current transformers are used for interposing c.t's. Their own power consumption should be relatively low so that the main current transformer is not excessively loaded.

The required output power of the main current transformer (Figure 7.89) can be determined from:

$$S_{MCT} = S_{ICT} + S_C + S_L \qquad (7.13)$$

S_{MCT} required output power of the main current transformer in VA

S_{ICT} secondary output power of the interposing transformer, including wire losses in VA

S_C power consumption of the interposing current transformer in VA at the primary rated current

S_L wire losses between the main and the interposing transformers in VA at the primary rated current

7.6.2.2 Summation current transformers

Generally, summation current transformers are used to measure the sum of the currents (being at the same frequency and at the same phase displacement, but possibly at differing power factors) in two or more incoming or outgoing feeder circuits. If the currents in all three phases (i.e. L1, L2 and L3 respectively) of two feeder circuits are to be summated, then three summation c.t.'s and six main c.t.'s will be required (Figure 7.90a). Current differences may also be measured by means of the reversed connection of the main current transformer secondary conductors.

The number of primary windings required is determined by the number of component currents to be summated. Summation current transformers can have up to 10 primary windings. In order to design the correct primary windings, the transformation ratios of the main current transformers are required.

The scale or display of the summation instruments must be selected according to the sum of the rated primary currents of the main current transformers.

Any measurement errors (tolerances) in the main current transformers and the summation current transformers are added together vectorially. If primary terminals provided as reserves for later extensions are not used, then these circuits must remain open. If one of the main current transformers is not loaded on its primary side, then an additional error is produced with respect to the summation of the other component currents.

The connected burden (impedance of the secondary circuit) and the losses of the summation current transformers themselves should be as small as possible, so that the output power requirements of the main current transformers can be kept low. The output power S_n of the summation transformer, including its own power consumption, is shared between the main transformers in the ratio of their rated primary currents.

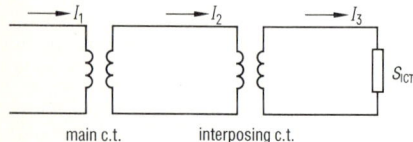

a) Schematic circuit diagram

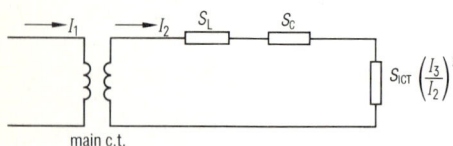

b) Equivalent circuit

Figure 7.89
Main current transformer and interposing current transformer

a) Circuit diagram for three summation c.t.'s and six main current transformers (infeed via two network branch circuits)

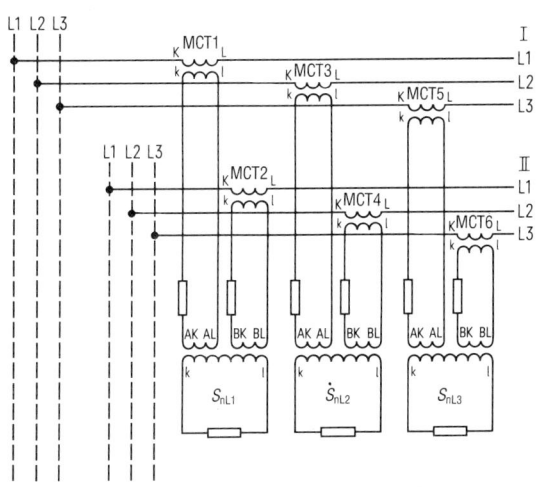

b) Summation of three load currents on one phase

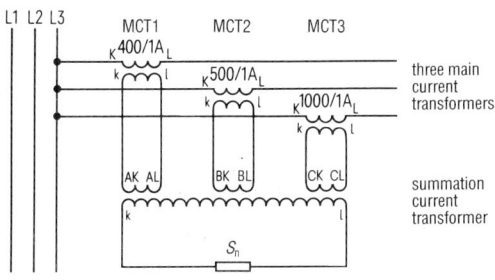

Figure 7.90
Summation of component currents using summation current transformers

Example – refer to Figure 7.90 b

The rated value of the total transformation ratio is:

400/1 A + 600/1 A + 1000/1 A = 2000/1 A.

Calculation of the respective power load S_T on each 3 main current transformers by means of the equation (7.14):

$$S_T = S_n \frac{I_{1nT}}{I_{1nS}} + S_L; \qquad (7.14)$$

which gives,

for the 1st main current transformer MCT1, 400/1 A

$$S_T = 20 \text{ VA} \frac{400 \text{ A}}{2000 \text{ A}} = 4 \text{ VA},$$

for the 2nd main current transformer MCT2, 600/1 A

$$S_T = 20 \text{ VA} \frac{600 \text{ A}}{2000 \text{ A}} = 6 \text{ VA},$$

for the 3rd main current transformer MCT3, 1000/1 A

$$S_T = 20 \text{ VA} \frac{1000 \text{ A}}{2000 \text{ A}} = 10 \text{ VA},$$

S_T power load on the individual main current transformers in VA,

S_n total power of the summation current transformer including its own consumption in VA, e.g. 20 VA,

I_{1nT} rated primary current of an individual main current transformer in A,

I_{1nS} sum of rated primary currents in A

S_L wire losses between the main current transformers and the summation current transformer in VA (negligible for short cable lengths and for secondary currents of 1 A)

7.6.2.3 Thread-through (or pin-wound) current transformers

Window-type current transformers can be used as thread-through transformers up to an operational current of approx. 200 A. For higher currents, this technique is impractical, since the wires required have cross-sections greater than approx. 35 mm² and can no longer be "threaded" through the transformer window. By using the thread-through method, different transformation ratios are achieved for a current transformer at the same output power rating.

Thread-through current transformers are primarily used in cases where low primary currents are to be measured.

Example

A current transformer with accuracy class 1, a rated overcurrent factor M5 and a rated output of 10 VA is required to measure an operating current of 50 A. A window-type current transformer with this technical data and a transformation ratio 250/1 may be selected. The primary conductor must then be threaded through the c.t. window five times.

7.6.2.4 Cable or busbar current transformers (split-core c.t.'s)

Cable or busbar current transformers, also known as split-core current transformers, (Fig. 7.91) can be retrofitted onto bare busbars up to an operational voltage of 660 V or onto fully insulated cables of any voltage rating. They consist of two U-shaped halves (hence, split-core) which are placed on either side of the cable or copper conductor and then screwed together. Cable or busbar current transformers are used extensively in the field of earth-fault and earth-leakage detection (core-balance application – refer to Section 3.6, page 183).

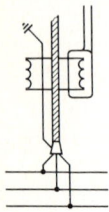

Figure 7.91
Cable current transformer

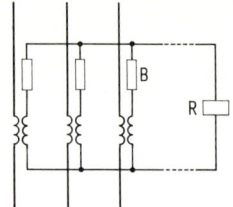

Figure 7.92
Three current transformers in the Holmgreen connection

B Burden
R Relay

In contrast to the method of using three separate c.t.'s in the Holmgreen connection (refer to Fig. 7.92), one does not have to contend with an error owing to differences between three separate c.t.'s when a cable current transformer is used. This is because the single core of the cable current transformer encloses all three phases and thus the vectorial sum of the currents is transformed to the single secondary winding.

7.6.2.5 Cast resin current transformers

The cores and windings of these transformers are embedded in cast resin so that they are protected against corrosion and the effects of tropical climates. The high dielectric strength and resistance to tracking of the cast resin make it possible to keep the dimensions of the current transformers to a minimum. Cast resin transformers are particularly resistant to impact and mechanical shock.

7.6.2.6 Explosion-protected current transformers and current transformers for mining applications

Wound-primary, window-type and summation current transformers which comply with the specifications for explosion-proof and mining industry apparatus are suitable for use in mines, parts of the chemical industry and refineries.

Explosion-proof current transformers must be approved and certified by the PTB (**P**hysikalisch-**T**echnischen **B**undesanstalt, Braunschweig), and must be built in accordance with DIN VDE 0170/0171 part 6 as well as DIN EN 80019. The certification numbers, approval numbers and test certificate numbers for the 4NB and 4ND current transformers are given in the relevant Siemens l.v. switchgear catalogue.

If these current transformers are used in mining applications in the Federal Republic of Germany, then they must also be approved and certified by the Bergwerksschaftlichen Versuchsstrecke Dortmund-Derne, and entered into the official records of the Oberbergamt in Dortmund.

7.6.2.7 Current transformers for protection purposes

Current transformers for protection purposes differ from those for measurement (i.e. instrument current transformers) in terms of their behaviour under overload (overcurrent, or saturation) conditions.

In the case of instrument current transformers with a rated overcurrent factor M5 or M10, a total transfer error of at least -15% is prescribed at 5 or 10 times the rated current respectively to protect connected measuring instruments from overload.

Figure 7.93 compares the overcurrent behaviour of an M5 instrument current transformer with that of a 5P10 current transformer for protection purposes.

Only a limited transfer error is permitted in the case of current transformers for protection purposes; even in the event of an overcurrent. For this reason, the protection classes specify that the secondary current must increase in direct proportion to the primary current up to a stipulated overcurrent.

At the accuracy limit primary current, and with the current transformer loaded at the rated burden, the total error may be -5% (5P) or -10% (10P) respectively.

7.6.2.8 Current transformers for power factor correction controllers

The standard design window-type or wound-primary current transformers may be used for p.f. correction control. When selecting the current transformer, the following points should be considered:

▷ The primary current of the c.t. should be

$$I_{1n} \geq \frac{S_{max}}{U_n \cdot \sqrt{3}} \qquad (7.15)$$

I_{1n} rated primary current of the current transformer, alternatively the rating of the incoming feeder

S_{max} maximum permissible power consumption of the installation (incoming transformer size)

U_n rated voltage

Example

$S_{max} = 600$ kVA,
$U_n = 400$ V

from the above-mentioned equation (7.15):

$$I_{1n} = \frac{600\,000}{400 \cdot \sqrt{3}} = 866 \text{ A}$$

i.e., a current transformer with a rated primary current of $I_{1n} = 1000$ A is selected.

▷ Accuracy class 1 is adequate for power factor correction control.

▷ Overcurrent factor M5 is adequate if an overcurrent protection device is fitted into the power factor correction unit.

▷ The power rating S_n of the current transformer must be sufficient. Its can be calculated from

$$S_n \geq S_R + S_M + S_L \qquad (7.16)$$

S_R power consumption of the controller
S_M power consumption of the measuring instruments
S_L power losses in the connecting wires

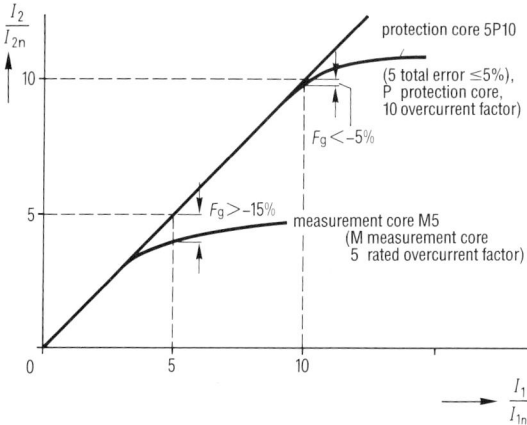

I_1 Primary current I_{1n} Rated primary current
I_2 Secondary current I_{2n} Rated secondary current
F_g Total error

Figure 7.93
Overcurrent behaviour of current transformers loaded with the rated burden

▷ The installation and wiring instructions must be adhered to, i.e. for Siemens current transformers and for the power factor controller respectively. The primary designations K and L on the current transformer must correspond to the direction of power flow from the incoming supply (i.e. K towards the incomer power transformer, L towards the load). The secondary terminals k and l of the current transformer must be connected to the controller terminals k and l respectively. Since most power factor controllers measure $\cos\varphi$ in one phase only, care must be taken to ensure that the current transformer is installed in the correct phase with respect to the voltage measured by the controller.

7.6.3 Accuracy classes of current transformers

Certification

Before a current transformer of accuracy class 0.5 to DIN VDE 0414 may be used in the Federal Republic of Germany for billing purposes, it must be officially certified in a testing station, i.e. it must have a certifying mark (stamp or label) and a security seal (lead seal). The certification is valid for an unlimited period. Details of whether a particular current transformer can be officially certified, are given in the Siemens catalogue.

Approval

In Germany, current transformers to be certified must first be type-approved by the PTB in Braunschweig. The Siemens catalogue gives details of which current transformers are suitable for approval (or, alternatively, which c.t.'s can be supplied as approved units).

The approval designation reflects the generic number as well as a design serial number.

Example

| 311 | generic number |
| 521 | design serial number |

Measuring ranges

Normal range

Current transformers shown in the Siemens low-voltage switchgear catalogue can be loaded continuously with up to 1.2 times their rated current. For the operational purposes of accuracy class 1, they even operate with sufficient accuracy at below 0.1 times the rated current I_n. The current errors at such low loading are still roughly within the limits (approx. $\leq 2\%$ referred to $0.1 \cdot I_n$).

Wide range

Wide range current transformers can be loaded continuously with up to twice their rated current. They are mainly used in high voltage networks. In type-tested switchgear assemblies (TTA) the cross-sections of conductors are dimensioned in terms of the rated operating current. Therefore, it does not make sense to install current transformers which can be loaded with twice their rated current.

Two range (extended range)

Two-range current transformers operate accurately from 0.01 to $1.2 \cdot I_n$.

Table 7.19   Power losses in connecting wires between the current transformer and the measuring equipment (wire losses)

Wire cross-section q_n	Power losses in the outward and return wires, referred to a 1 A secondary current (bracketed values apply to 5A), and a length of:									
	1 m	2 m	3 m	4 m	5 m	6 m	7 m	8 m	9 m	10 m
mm²	VA	VA	VA	VA	VA	VA	VA	VA	VA	VA
1.5	0.054	0.107	0.160	0.214	0.281	0.321	0.380	0.429	0.482	0.536
2.5	0.014 (0.36)	0.028 (0.71)	0.042 (1.07)	0.057 (1.43)	0.071 (1.78)	0.085 (2.14)	0.1 (2.5)	0.114 (2.86)	0.128 (3.21)	0.142 (3.57)
4	0.008 (0.22)	0.018 (0.45)	0.026 (0.67)	0.035 (0.89)	0.044 (1.12)	0.053 (1.34)	0.062 (1.56)	0.071 (1.79)	0.08 (2.01)	0.089 (2.24)

However, since the normal-range current transformers are usually accurate enough and are approved for billing purposes, special two-range current transformers do not have general application.

7.6.4 Secondary currents of current transformers

Standard design current transformers have a 1 A secondary current. The low secondary current offers the advantage that the transformer output power is virtually determined only by the power consumption of the connected equipment. Losses in wires of 2.5 mm² and up to 10 m in length can be neglected; only longer wires need be taken into account when determining the current transformer output (Table 7.19).

Wires with a cross-section as small as 1 or 1.5 mm² may be used between current transformers with 1 A secondaries and their associated measuring equipment. In the case of current transformers with a 5 A secondary current, this would lead to excessive losses (up to 25 times).

7.6.5 Rated output power and overcurrent factor of current transformers

The rated overcurrent factor is that factor, with which the upper limit of the rated primary current must be multiplied, to obtain accuracy limit primary current, e.g. M 5 (also refer to Table 7.21, page 395).

The overcurrent factor in terms of DIN VDE 0414 only equals the rated overcurrent factor if the current transformer is loaded at its rated output power.

If the output differs from the rated output, then the overcurrent factor varies approx. in inverse proportion.

Calculation of the overcurrent factor:

$$n_B = n \frac{S_n + S_i}{S_B + S_i} \qquad (7.17)$$

- n rated overcurrent factor (rated value at the rated output)
- n_B overcurrent factor at the operating burden (the burden of a current transformer is the impedance of the secondary circuit)
- S_n rated output of the core at the rated burden
- S_B output power at operating burden and rated current
- S_i power consumption (losses) of the core itself

Example
for an underburdened current transformer

Current transformer data:

rated overcurrent factor M5 (e.g. $n = 5$)

rated output power of the current transformer $S_n = 15$ VA

power consumption (losses) of the current transformer $S_i = 2$ VA

power consumption of the connected load $S_B = 10$ VA

The overcurrent factor at the operating burden can be calculated from equation (7.17):

$$n_B = n \frac{S_n + S_i}{S_B + S_i} = 5 \cdot \frac{15 + 2}{10 + 2} = 5 \cdot \frac{17}{12} = 7.08 > 5.$$

Result:

Since $n_B > n$, the overcurrent range is displaced and the transformer becomes saturated at a higher primary current (Figure 7.94).

Measuring instruments which are connected to underburdened current transformers are adequately protected during normal operation, although this may not be the case in the event of a short-circuit.

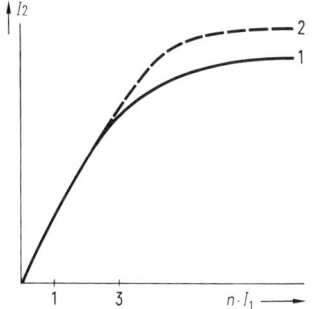

1   Rated operation
2   Underburdened current transformer
I_1 Primary current
I_2 Secondary current

Figure 7.94
Secondary current I_2 of an underburdened current transformer as a function of the primary current I_1

Example
for an overburdened current transformer

Current transformer data:

rated overcurrent factor M 5 (e.g. $n = 5$),

rated output power of the current transformer
$S_n = 15$ VA,

power consumption (losses) of the current transformer
$S_i = 2$ VA,

power consumption of the connected load
$S_B = 20$ VA.

The overcurrent factor at the operating burden can be calculated from equation (7.17):

$$n_B = n \cdot \frac{S_n + S_i}{S_B + S_i}$$

$$= 1.5 \cdot \frac{15 + 2}{20 + 2} = 1.5 \cdot \frac{17}{22} = 1.16 < 1.5.$$

Measuring instruments connected to overburdened current transformers are adequately protected in normal operation and also in the case of short-circuit. It can, however, happen that the current transformer does not meet up to the limits of the stated accuracy class, as the transformer becomes saturated at a lower value of the primary current.

7.6.6 Voltages across the secondary terminals of a current transformer

The voltage across the secondary terminals of a current transformer can be calculated from (7.18–7.20):

during normal operation:

$$U_{2\,\text{eff}} = \frac{S}{I_2} \qquad (7.18)$$

in the event of a short-circuit:

$$\hat{U}_2 \approx 50 \cdot U_{2\,\text{eff}} \qquad (7.19)$$

with the secondary terminals open-circuited:

$$\hat{U}_2 = U_S \cdot A_{\text{Fe}} \cdot W_2 \qquad (7.20)$$

$U_{2\,\text{eff}}$ effective (r.m.s.) value of the secondary voltage
$\hat{U}_2$ peak value of the secondary voltage
U_S specific magnetizing voltage which depends on the core material and the field strength
S power consumption of the connected load
I_2 secondary current
A_{Fe} cross-sectional area of the core (iron core)
W_2 number of secondary turns

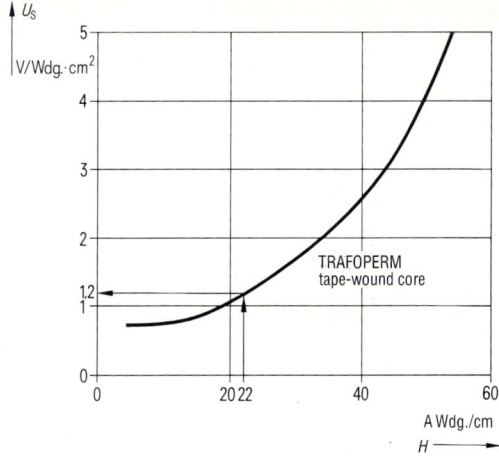

Figure 7.95
Magnetizing voltage U_S as a function of the magnetic field strength H

Table 7.20
Examples for the calculation of voltage levels across the secondary terminals of a current transformer, in normal operation and under short-circuit conditions (Values from Tables 7.19 and 7.22)

	1 Amp. c.t.	5 Amp. c.t.
required power for		
ampere meter	0.2 VA	0.2 VA
power factor meter	3.5 VA	3.5 VA
electricity meter	0.5 VA	1.0 VA
50 m wire, 2.5 mm²	0.71 VA	17.85 VA
total power S	4.91 VA	22.55 VA
selected c.t. power rating	5 VA	25 VA
voltage on the secondary terminals		
under normal operation $U_{2\,\text{eff}} = \frac{S}{I_2}$ (equation (7.18))	$\frac{4.91}{1} = 4.91$ V	$\frac{22.55}{5} = 4.51$ V
	no dangerous touch voltage	
under short-circuit conditions $\hat{U}_2 \approx 50 \cdot U_{2\,\text{eff}}$ (equation (7.19))	$4.91 \cdot 50 \approx 246$ V	$4.51 \cdot 50 \approx 226$ V
	dangerous touch voltage!	

Figure 7.95 shows a magnetizing characteristic curve by means of which the peak value of the secondary open-circuit voltage $\hat{U}_2$ may be determined.

Normal operation

Normal operation is with 0.1 to 1.2 times rated primary current and with the rated burden connected. In the case of 1 A current transformers and small rated output powers, no dangerous touch voltages (over 50 V) occur across the secondary terminals. An example is given in Table 7.20.

Operation in the event of a short-circuit

Whereas $U_{2\,\mathrm{eff}}$ increases only slightly in the saturation range of the current transformer, the peak value $\hat{U}_2$ can increase substantially, i.e. the limiting effect of eddy currents has little effect on it. Thus, the peak of the secondary voltage is not merely 2 times the effective value, but substantially higher. It can be typically approx. 50 times the effective value.

$\hat{U}_2$ generally attains values above 50 V in the event of a short-circuit in the primary (dangerous touch voltage).

An example is given at the end of Table 7.20.

Operation with open circuited secondary winding

If a current transformer is operated with open-circuited secondary terminals, it behaves as if an infinitely large burden is connected. The total primary current I_1 is used for magnetization (Fig. 7.96).

Owing to the saturation of the core, the effective (r.m.s.) value of the magnetic flux Φ does not increase appreciably. In the case of a current transformer with an open circuited secondary, however, the rate of rise through the current zero is considerably greater and causes high voltage peaks of approx. 0.2 to 0.7 ms duration since $\hat{U}_2$ is a function of $d\Phi/dt$. Even with small primary currents a current transformer is magnetized to saturation during each current halfwave, i.e. the magnetic flux in the transformer core is reversed twice in each cycle (Fig. 7.97).

If a current transformer is operated with open-circuited secondary terminals, the peak value of the voltage across the secondary terminals ($\hat{U}_2$) can be as high as a few kilovolts. This can damage the insulation of the secondary winding and the iron core can become too hot. The possibility of flashovers must be regarded as a fire hazard.

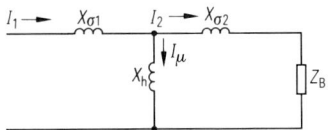

I_1	Primary current
I_2	Secondary current
I_μ	Magnetizing current
$X_{\sigma 1}$	Stray inductive reactance on the primary side
$X_{\sigma 2}$	Stray inductive reactance on the secondary side
X_h	Main inductive reactance
Z_B	Burden connected to the secondary side

Figure 7.96
Equivalent circuit of a current transformer with connected burden

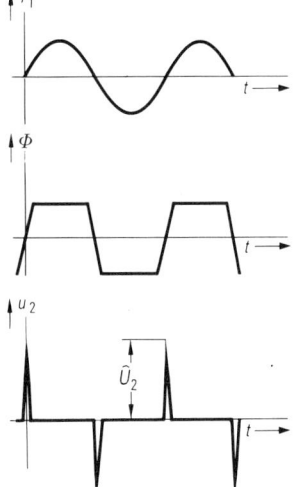

Figure 7.97
Behaviour of the primary current (i_1), the magnetizing flux (Φ) and the secondary voltage (u_2) with respect to time in a current transformer with open-circuited secondary terminals

Example

Operating data of the current transformer:

$I_1 = 200$ A,
$W_1 = 1$,
$W_2 = 200$,
$A_{Fe} = 25$ cm² $= 22$ A · Wdg./cm.

From equation (7.20),

$$\hat{U}_2 = U_S \cdot A_{Fe} \cdot W_2$$

and from Figure 7.95:

$$\hat{U}_2 = 1.2 \frac{V}{Wdg.cm^2} \cdot 25 \text{ cm}^2 \cdot 200 \text{ Wdg.} = 6 \text{ kV}.$$

Thus, an extremely dangerous touch voltage will exist across the open-circuited secondary terminals.

7.6.7 Selection criteria for current transformers

Table 7.21 provides an overview of the most important selection criteria for current transformers to DIN VDE 0414 (10 sections of specifications for instrument current transformers).

7.6.8 Power consumption and losses in current transformer secondary circuits

Losses in copper wires

The following equation may be used to calculate the power losses S_L in the wires of the secondary circuit:

$$S_L = \frac{I^2 \cdot 2 \cdot l}{q \cdot 56} \qquad (7.21)$$

I load current passing through the connecting wires
l single length in m
q cross-section in mm²

Further values may be read from Table 7.19 on page 390.

Power consumption in measuring instruments

The power consumption of various measuring instruments is given in Table 7.22 (page 396).

Table 7.21 Selection criteria for current transformers in terms of DIN VDE 0414

1. Application	Wound-primary c.t.'s from 1 to 100 A or window-type c.t.'s from 50 to 10 000 A	
2. Highest continuously permissible operating voltage	This is the r.m.s. value of the highest voltage between the conductors (e.g. busbars) of a network. The insulation of the c.t. must be rated for this voltage under normal operating conditions. Apart from the explosion-proof and mining-purpose c.t.'s, the transformers offered are all suited for an operational voltage of 800 V. Current transformers with 1200 V rated voltage are offered as special design only.	
3. Accuracy class	Class 0.5: For billing purposes, accurate measurement C.t.'s of this class may be certified and granted approvals.	Current error $\pm 0.5\%$ at $1 \cdot I_n$ and $1.2 \cdot I_n$
	Class 1: Operational measurement, internal metering	Current error $\pm 1\%$ at $1 \cdot I_n$ and $1.2 \cdot I_n$
	Class 3: Rough measurement	Current error $\pm 3\%$ at $1 \cdot I_n$ and $1.2 \cdot I_n$
	Class 5: Although this class is included in the standard, it is not commonly used in practice. Rough measurement!	Current error $\pm 5\%$ at $1 \cdot I_n$ and $1.2 \cdot I_n$
	Class 5P… and Class 10P… (please refer to Section 7.6.2.7 "Current transformers for protection purposes", page 389).	
4. Rated primary current	Unless otherwise specified, current transformers and measuring instruments can be continuously loaded with 1.2 times the rated primary current (I_n). Rated primary currents in A: 5, 10, 12.5, 15, 20, 25, 30, 40, 50, 60, 75 and the decadal multiples thereof. In view of the window size, the rated primary current of the current transformer should correspond to the rated current of the associated switching device(s). Taking into account that c.t.'s may be continuously overloaded by a factor 1.2, the rated currents of switching devices correspond to the standard rated primary currents of current transformers. The measuring range scales of instruments with c.t. connections also correspond to these standard rated primary currents.	
5. Secondary current	1 A (standard version) 5 A (special version) The standard current transformer may also be used with long connecting cables or wires.	
6. Rated power	The rated power of a current transformer is expressed in VA. The actual load, or burden, should be nearly equal to the rated power, but never greater if the operating current is 1.2 times the rated primary current of the c.t. At a frequency of 60 Hz, the rated power increases by the factor 1.2. At $16^2/_3$ Hz, the power decreases to $^1/_3$ of the 50 Hz rated value.	
7. Rated overcurrent factor	The overcurrent factor indicates that value of primary current, in multiples of the rated value, which will cause the core of a measurement c.t. to become saturated, or alternatively the minimum value at which the core of a c.t. for protection purposes may not saturate. It is, however, dependent on the connected burden.	
	Measurement cores: The rated overcurrent factor is indicated by the letter M and a number, e.g. M 5 or M 10 └──┴── Overcurrent factor └──── Measurement core i.e. at 5 or 10 times the rated current I_n respectively, the total error must be at least -15% (protection for connected measuring instruments).	Protection cores: The rated overcurrent factor is indicated by means of a supplement to the class designation, Class 5P… (Rated over- └── e.g. 10 current factor) └─── Protection └── Max. rated total error, i.e. at $10 \cdot I_n$ the total error may not be more than -5%
8. Rated thermal short-time current I_{th}	The thermal short-time current rating I_{th} is that maximum r.m.s. value of primary current which can flow for one second duration, while the secondary winding of the c.t. is short-circuited, without causing thermal damage to the current transformer (The values are given in corresponding tables). In the case of 4NA and 4NC current transformers, $I_{th} = 60 \cdot I_n$.	
9. Rated dynamic current I_{dyn}	The rated dynamic current I_{dyn} is the highest instantaneous value of the current after onset of the short-circuit, the dynamic force of which the current transformer can withstand without being damaged. It is quoted as a peak value. In the case of 4NA and 4NC current transformers, $i_{dyn} = 2.5 \cdot I_{th} = 150 \cdot I_n$.	

Table 7.22 Power consumption of measuring instruments in current transformer circuits

Type of measuring instrument	Current path, 50 Hz		Voltage path, 50 Hz	
	with quadrant, horizontal or sector scale	with circular scale	with quadrant, horizontal or sector scale	with circular scale
Indication measuring instruments				
Moving-iron measuring instruments				
for current (1 A c.t. connection)	0.2 VA	approx. 1 VA	–	–
for voltage	–	–	0.9 to 5 VA	5 to 9 VA
Moving-coil measuring instruments with built-in rectifier				
for current	approx. 0.25 VA	–	–	–
for voltage	–	–	current drawn approx. 10 mA	–
Bimetallic ampere meters	3 alt. 5 VA	–	–	–
Power meters with electrodynamic measuring element	approx. 1 VA	–	current drawn approx. 10 mA per voltage path	–
Power factor meters	approx. 3.5 VA	–	10 to 15 mA at 100 V per voltage path	–
Vibrating reed frequency meters	–	–	1 to 3 VA	–
Pointer-type frequency meters	–	–	1.2 VA (at 110 V)	–
	–	–	2.3 VA (at 230 V)	–
	–	–	3.7 VA (at 400 V)	–
Recording-type measuring instruments				
Recording instruments (e.g. plotters) with moving-coil element(s) and built-in rectifier(s)				
voltage recorder (per measuring element), current recorder (per measuring element)	approx. 1.2 to 2 VA	–	current drawn approx. 10 mA	–
recorder with electrodynamic measuring element as power meter	approx. 5 VA	–	approx. 1 VA	–
recorder with moving-coil element(s) for frequency recording	–	–	approx. 10 VA	–
Electricity meters (see Siemens Z catalogue)				
single phase	1 to 2 VA	–	2 to 4 VA	–
three-phase	0.5 to 1.5 VA	–	4 VA	–
Switching measuring instruments				
Network master relays	approx. 0.6 VA	–	approx. 13.5 VA	–

8 Fundamental circuit diagrams

8.1 General information

8.1.1 Terminal designations

The designation of terminals in circuit diagrams is based on the terminal markings on the items of electrical equipment themselves.

The terminal markings for switchgear are laid down in DIN EN 50005 of July 1977, and in DIN 40719, Part 2, of June 1978. DIN EN 50005 applies to industrial low-voltage switchgear for rated voltages of up to AC 1000 V and DC 1200 V. The standard is based on the generally applicable designation system as outlined in IEC publication 445 (1973) [1].

If standardized terminal markings are not available, it is recommended that the same designation system be used for the indication of terminals as for similar equipment used in low voltage switchgear combinations.

If, for example, a colour code according to DIN 41788 is used for an equipment terminal, the assignment to the corresponding graphic symbol or pictorial marking or to the alphanumeric code applicable to this standard, should be clear from the circuit documentation.

The DIN EN 50005 standard contains general stipulations. The following DIN EN standards were published for specific items such as contactor relays, auxiliary contacts of a.c. contactors and command devices:

DIN EN 50011, DIN EN 50012, DIN EN 50013.

The terminal designations apply to units in their state of delivery.

The general stipulations in DIN EN 50005 state that:

▷ The terminal markings on a piece of equipment must be unique, i.e., each marking may only be used once.

▷ The markings of the various terminals of a circuit element in any one conducting path must clearly indicate that the terminals belong to the same device.

▷ The terminal markings of impedances (e.g. coils) are always alphanumeric. Upper case is generally preferred. Should lower case be used, the characters are taken to have the same meaning as the corresponding upper case characters (Fig. 8.1 to 8.3).

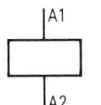

a) Coil with one winding, e.g. contactor coil

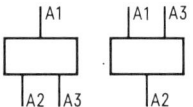

b) Coil with tappings

The tap terminals are marked in sequence, e.g. A1, A3, etc.

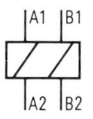

c) Coil with two windings

The terminals of the first winding are marked as A1, A2, the terminals of the second winding as B1, B2

Figure 8.1
Terminal designations for operating mechanisms

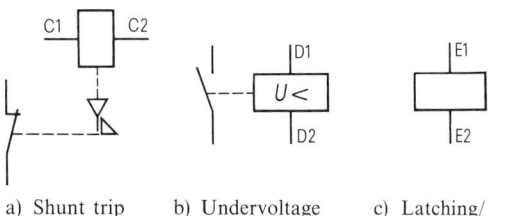

a) Shunt trip release   b) Undervoltage release   c) Latching/delatching magnetic drive

Figure 8.2
Terminal designations for auxiliary releases and latching/delatching magnetic drives

---

[1] Reference to the equivalent national standard may be included, if it has been harmonized according to CENELEC HD 241. DIN 42400 and IEC 445 contain the same specications.

397

8 Fundamental circuit diagrams

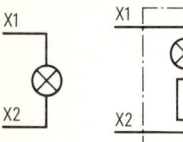
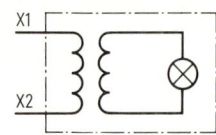
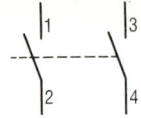

Figure 8.3
Terminal designations for indicator lights

Note:

The term "indicator light" includes an integrated resistor or transformer, if applicable.

▷ The terminal markings of a contact element are always numeric. A terminal is designated by an odd number, the other terminals of the same contact by the immediately following even number (see Fig. 8.4 to 8.9).

▷ If a distinction must be made between the input and output terminals of a circuit element, then the smaller number is assigned to the input (e.g. input 11 and output 12, input A1 and output A2).

Note:
The position of the terminals indicated in the designation examples bears no relation to the location of the terminals or markings on the item itself.

Main circuit contacts

Contacts in main circuits are designated by a single integer (Fig. 8.4). For each terminal marked with an odd number, the associated other terminal is marked with the even number immediately following.

Auxiliary circuit contacts

Contacts for auxiliary circuits are designated with double integer numbers, consisting of the function number preceded by the sequence number (Fig. 8.5).

Auxiliary contacts for special purposes, for example time-delayed auxiliary contacts, are assigned the function numbers 5 and 6 for NC contacts and 7 and 8 for NO contacts (Fig. 8.6).

Terminals of the same contact element are marked (designated) with the same sequence number (Fig. 8.7):

▷ Terminal marking/designation for overload protection equipment (Fig. 8.8).

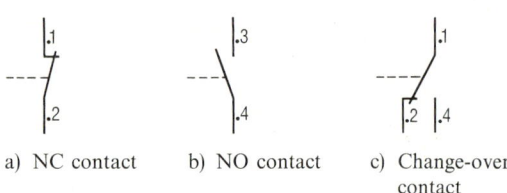

Figure 8.4
Designation of the contacts, e.g., for two- to five-pole main circuits. (If one switch is equipped with more than five main contacts, the alphanumeric marking according to the IEC publication 455 or to the harmonized national standard has to be applied).

a) NC contact   b) NO contact   c) Change-over contact

Figure 8.5
Designation of auxiliary circuit contacts with function number (the sequence number is indicated by a point. NC contacts are assigned the function numbers 1 and 2, NO contacts the function numbers 3 and 4. (NC contacts, NO contacts defined as in IEC publication 50/441).

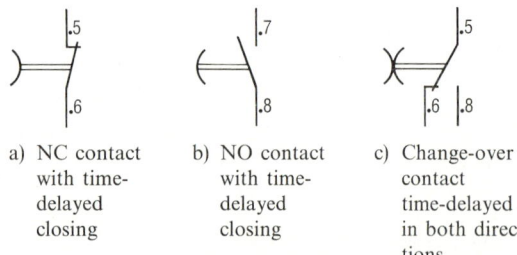

a) NC contact with time-delayed closing
b) NO contact with time-delayed closing
c) Change-over contact time-delayed in both directions

Figure 8.6
Designation of auxiliary contacts for special purposes. (Auxiliary contacts for special purposes, for example time-delayed auxiliary contacts, are assigned to the function numbers 5 and 6 for NC contacts, and 7 and 8 for NO contacts)

Terminal markings

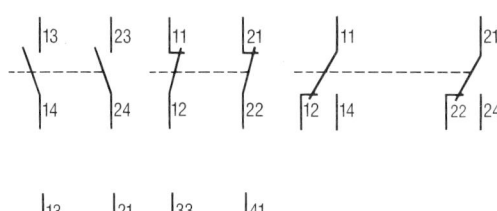

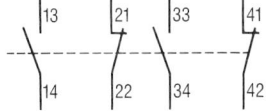

Figure 8.7
Examples of terminal designations with sequence and function numbers

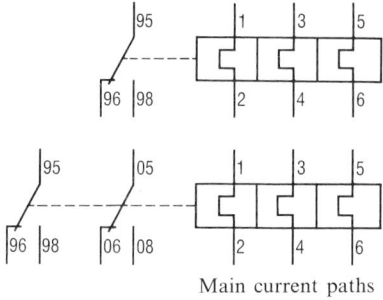

Figure 8.8
Examples of terminal designations for overload protection equipment. If a second sequence number is required, 0 should be used.

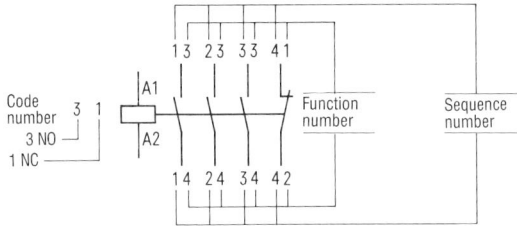

Figure 8.9
Example of the designation of terminals in the case of a contactor relay with the code number 31

▷ Code numbers for switchgear. Switchgear with a fixed number of auxiliary contacts (NO or NC contacts) may be assigned a double integer code number. The first integer indicates the number of NO contacts, the second the number of NC contacts while a third can be used for indicating change-over contacts, if applicable.

▷ Example for the designation/marking of a switchgear item. Figure 8.9 illustrates an example of the terminal designations of a contactor relay with three NO contacts and one NC contact.

The DIN EN 50005 standard of May 1978 for the terminal markings and code numbers/letters of specific contactor relays which have a denite fixed arrangement of contacts, requires that the following additional demands are met:

▷ The contacts of a unit are numbered consecutively from the left to the right (In the case of units with contacts on several levels, numbering starts with the level which is closest to the mounting surface).

The number, type and position of the contacts has to be indicated by a code number followed by a code letter.

Code number:
In accordance with DIN EN 50005 (as described above).

Code letter:
The code letter gives information on the position of the contacts in relation to each other, and on their terminal markings.

The versions of the contactor relays having the code letter E (Table 8.1) are defined in this standard.

Permitted deviations are indicated by the code letters X, Y, or Z.

The terminal markings and code numbers for auxiliary contacts of specific contactors are given in DIN EN 50012 of May 1978.

▷ The terminal markings of the contactors and the code numbers of the auxiliary contacts are the same as the terminal markings of corresponding contactor relays with the code letter E (DIN EN 50011).

▷ The number and type of the auxiliary contacts are defined in the standard, however their physical location on the contactor is *not*.

DIN EN 50013 of May 1978 applies for command devices.

This standard applies to command devices with two defined switching positions in accordance with IEC publication 337-1. These include pushbuttons, position switches and similar items, regardless of their design. The terminal markings of the contacts con-

399

8 Fundamental circuit diagrams

Table 8.1
Terminal markings of the contacts of specific contactor relays in accordance with DIN EN 50011 of May 1978 (extract)

Code number	Contacts	Code number	Contacts	Code number	Contacts
10E	(13-14)	01E	(21-22)	—	—
20E	(13-14, 23-24)	11E	(13-14, 21-22)	02E	(11-12, 21-22)
30E	(13-14, 23-24, 33-34)	21E	(13-14, 21-22, 33-34)	12E	(13-14, 21-22, 31-32)
03E	(11-12, 21-22, 31-32)	—	—	—	—
40E	(13-14, 23-24, 33-34, 43-44)	31E	(13-14, 21-22, 33-34, 43-44)	22E	(13-14, 21-22, 31-32, 43-44)
13E	(13-14, 21-22, 31-32, 41-42)	04E	(11-12, 21-22, 31-32, 41-42)	—	—
50E	(13-14, 23-24, 33-34, 43-44, 53-54)	41E	(13-14, 21-22, 33-34, 43-44, 53-54)	32E	(13-14, 21-22, 31-32, 43-44, 53-54)
23E	(13-14, 21-22, 31-32, 41-42, 53-54)	14E	(13-14, 21-22, 31-32, 41-42, 51-52)	05E	(11-12, 21-22, 31-32, 41-42, 51-52)
60E	(13-14, 23-24, 33-34, 43-44, 53-54, 63-64)	51E	(13-14, 21-22, 33-34, 43-44, 53-54, 61-62)	42E	(13-14, 21-22, 31-32, 43-44, 51-52, 61-62)
33E	(13-14, 23-24, 33-34, 41-42, 51-52, 61-62)	—	—	—	—
80E	(13-14, 23-24, 33-34, 43-44, 53-54, 63-64, 73-74, 83-84)	71E	(13-14, 23-24, 33-34, 43-44, 53-54, 61-62, 73-74, 83-84)	62E	(13-14, 23-24, 33-34, 43-44, 53-54, 61-62, 71-72, 83-84)
53E	(13-14, 23-24, 33-34, 43-44, 53-54, 61-62, 71-72, 81-82)	44E	(13-14, 23-24, 33-34, 43-44, 51-52, 61-62, 71-72, 81-82)	—	—

form to the terminal markings of corresponding contactor relays which have the code letter E in accordance with DIN EN 50 011.

▷ The location of the contacts on the command device does not have to correspond to the location given in the table.

For the drafting of specifications, it is recommended that standards DIN EN 50 011, 50012 and 50013 for the respective switching devices in question, be applied.

8.1.2 Graphic symbols according to DIN, ANSI, BS and IEC

Tables 8.2 to 8.4 illustrate a selection of graphic symbols in accordance with DIN 40900 Parts 1 to 13. The graphic symbols are practically identical to those of the IEC, ANSI and BS standards. Where applicable, deviations are indicated.

Guidelines for circuit diagrams and documentation are given in the publication:

▷ How to Read German Schematic Diagrams of Industrial Equipment (Publ. Siemens AG, D-91050 Erlangen, Germany, order number E10582-W346-U11)

Table 8.2
Graphic symbols and representation of voltage, current and frequency, according to DIN, ANSI, BS and IEC (examples)

Designation	Graphic symbol
Direct current	⎯⎯ or ≡≡≡
Alternating current	∼
Direct or alternating current	≂ [1)]
Rectified current with a.c. content	(rectified symbol)
Alternating square wave pulse	(square wave symbol) [1)]
Square wave pulse, positive, represented with a pulse duration of 2 µs and at a pulse frequency of 10 kHz	2 µs 10 kHz [1)]
Triangular pulse	(triangular pulse symbol) [1)]
Single-phase alternating current, e.g. $16^2/_3$ Hz	$1 \sim 16^2/_3$ Hz
Three-phase alternating current (three-phase system)	$3 \sim 50$ Hz 400 V
same, with neutral (four-wire system)	$3/N \sim 50$ Hz 400/230 V [2)]
same, with neutral and protective conductor	$3/N/PE \sim 50$ Hz 400/230 V [2)]
Direct current two-wire system	2 – 24 V
Direct current three-wire system with ground conductor	2/M – 24 V [2)]
Direct current three-wire system with protective conductor	2/PE – 220 V [2)]

[1)] only to DIN
[2)] In the case of IEC, ANSI and BS, the oblique (/) can be omitted when indicating the conductor, e.g. 3NPE ∼ 50 Hz 400/230 V

Note:

In the case of IEC, ANSI and BS, the previously used graphical symbols are also still partly in use (see also DIN).

Table 8.3
Graphic symbols for conductors and connections according to DIN, ANSI, BS and IEC

Designation	Graphic symbol
Conductor, line, wire	⎯⎯
Protective conductor (PE)	(symbol)
Neutral conductor (N)	(symbol)
Protective neutral conductor (PEN)	(symbol)
Line marked to denote the number of conductors, 3 conductors	(symbol) or (symbol with 3)
Double branching or junction of conductors, single-phase or three-phase	(symbols)
Connection or junction	●
Connecting point (e.g. terminals)	○ or preferably ●
Terminal strip with terminal markings	1 2 3 4
Line-up terminals, terminal strip, represented as terminal strip with isolating terminals	1 2 3 4 5 6 [1)]

[1)] only to DIN

These guidelines for circuit diagrams and documentation are based on the following (previously used) DIN standards with regard to graphic symbols and terminal markings:

DIN 40 700 [3)], DIN 40 714 [3)],
DIN 40 703 [4)], DIN 40 715,
DIN 40 708 [4)], DIN 40 719
DIN 40 713 [3)],

The guidelines for circuit diagrams and markings to DIN 40719 are still valid.

[3)] DIN standard still partially valid
[4)] The listed DIN standards are no longer valid and have been replaced by DIN 40900 Parts 1 to 13. A direct correlation between the respective parts of the old and new standards is not possible, since the new standards are subdivided in a different way.

8 Fundamental circuit diagrams

Table 8.4 A selection of frequently used graphic symbols according to DIN, ANSI, BS and IEC

Designation	Graphic symbol	Designation	Graphic symbol
Resistor – general	preferred alternative	Leading normally open (NO) contact early make (of a contact group)	
– variable		Normally closed (NC) contact	
– with tappings (two tappings represented)		Leading normally closed (NC) contact early break (of a contact group)	
Heating element			
Inductor (winding, reactor, coil)	preferred alternative	Change-over contact break-before-make	
– with fixed tappings (two tappings represented)		Change-over contact make-before-break, fleeting contact element	or
– with magnetic core			
Capacitor – general		Single-throw switch, NO maintained contact; manually operated	[2)]
– variable		Single-throw switch, NC maintained contact; manually operated	[2)]
– adjustable			
Piezoelectric crystal with two electrodes		NO contact with automatic return to the rest position, shown in the operated position Note: The double arrow indicates a state of operation which differs from the normal state, e.g. closed-circuit relay	[2)]
Earth connection, general			
Frame or chassis connection	or	Time delayed contacts – Make contact (NO), delayed make	or
Variability – non-inherent	non-linear	– Break contact (NC), delayed break	
– inherent (data on variable quantities may be added)	non-linear	– Make contact (NO), delayed break	
Adjustability	stepwise continuously	– Break contact (NC), delayed make	or
Switching function, general normally open (NO) contact	[1)] or	– Set of contacts: Instantaneous NO, after drop-out delayed opening of 2nd NO, delayed NC contact	

[1)] Preferred representation
[2)] DIN only

Graphic symbols for switchgear

Table 8.4 A selection of frequently used graphic symbols according to DIN, ANSI, BS and IEC *(continued)*

Designation	Graphic symbol	Designation	Graphic symbol
Power contactor with thermal overload relay and auxiliary contacts		Fuse – general	
		– marking of supply-side connection	
		Screw-in fuse, triple-pole representation: 10 A, Type DII	DII 1) / 10 A
Triple-pole circuit-breaker with latching mechanism – with thermal overload release, – with electromagnetic short-circuit release		Isolating link	or
		Plug and socket device	or
Circuit-breaker		Hand-operated mechanism, general	
Disconnector Off-load switch		Foot-operated mechanism	
		Actuation on approach	
Isolating circuit-breaker, derived from DIN/IEC		Emergency switch	
Triple-pole load-break switch		Actuated by electromagnetic operating mechanism	
Triple-pole fuse switch disconnector		Actuated by electromagnetic overcurrent protection	
Circuit-breaker with latching mechanism for motor protection, triple-pole representation with – three thermal overload releases, – three electromagnetic short-circuit releases, – undervoltage release and – auxiliary contacts		Actuated by thermal operating mechanism, e.g. bimetal relay, thermal overload relay	
		Cam-operated mechanism	
		Actuated by pneumatic or hydraulic means in direction of arrow	
Three-pole circuit-breaker for motor protection, with thermal and electromagnetic releases (single-pole representation)		Power-operated mechanism, general	
		Motor-operated mechanism	M

1) In this configuration to DIN only
2) Preferred representation of circuit-breakers with primary connections

403

8 Fundamental circuit diagrams

Table 8.4 A selection of frequently used graphic symbols according to DIN, ANSI, BS and IEC *(continued)*

Designation	Graphic symbol	Designation	Graphic symbol
Unidirectional latching device, unlatched		Operating mechanism with two separate coil windings – combined representation	
Bidirectional latching device	1)	– separated representation	
Notch, latched		Measuring relay with data of measured quantity, e.g. undervoltage	$U<$
Delayed operation	or	Electromechanical operating mechanism with – drop-out delay	
Mechanical coupling, disengaged		– pick-up delay	
engaged		– pick-up and drop-out delay	
Hand-operated momentary contact switch, general		Electromechanical operating mechanism of a polarized relay	
Hand-operated momentary contact switch, press action (pushbutton)	E	Electromechanical operating mechanism of a remanent relay	
Hand-operated maintained-contact switch	1)	Stepping relay, current impulse relay	1)
Multi-position switch, triple-pole, with 3 switch positions	1 0 2	Audio frequency ripple control relay	$\approx$ 1)
Maintained-contact switch, momentary-contact switch, control-discrepancy switch, triple-pole representation, hand operated by turning (latching in positions 2 and 3)	1) 1 2 3 4 2/3	Flasher relay, representation with a flashing frequency of 5/min	1) 5/min
Electromechanical operating mechanism, relay coil, general		Undervoltage relay: operating range 50 to 80 V, reset at 130%	$U<$ 50...80V 130%

1) to DIN only

Table 8.4 A selection of frequently used graphic symbols according to DIN, ANSI, BS and IEC *(continued)*

Designation	Graphic symbol	Designation	Graphic symbol
Relay for sensing turn-to-turn faults	$N<$	Surge voltage protector in a gas discharge tube	
Overload relay with two outputs, one is effective at five times the setting value, the other is effective with inverse time delay	$I>$ $5x$	Starter, general	
Fault voltage to chassis	U	Starter for star-delta connection	
Fault current to ground	I	Automatic starter	1)
Current between two neutral conductors	I_{N-N}		
Surge voltage protector, Lightning arrestor		Starter with thermal and magnetic releases	1)
Proximity sensor		Starter combination represented with – contactor stator starter for two directions of rotation	1)
Touch sensor		– three-phase slip-ring motor	M 3~
Positive pulse		– automatic rotor resistor starter	
Negative pulse			
AC pulse			
Sawtooth		Buchholz protection device	
Spark gap			

[1] to DIN only

Table 8.4 A selection of frequently used graphic symbols according to DIN, ANSI, BS and IEC *(continued)*

Designation	Graphic symbol	Designation	Graphic symbol
Transformer with two windings	[symbol] or [symbol]	Machine, general One of the following letters is inserted in the place of the asterisk (*): C Converter G Generator GS Synchronous generator M Motor MG Machine used either as generator or motor MS Synchronous motor Note: The graphic symbols 02-02-01 and 02-04-04 [2] may be added. See Main Sections 5 to 8, in part 6 of DIN 40900.	[symbol]
Single-phase transformer with two windings and screen	[symbol] or [symbol]		
Autotransformer	[symbol] or [symbol]		
Autotransformer, single-phase, with variable voltage control (Variac)	[symbol] or [symbol]	Three-phase induction motor with squirrel-cage rotor	[symbol M 3~]
Reactor, choke	[symbol] or [symbol]	Three-phase induction motor with slip-ring rotor	[symbol M 3~]
Current transformer	[symbol] or [symbol]	Three-phase induction series-wound motor	[symbol M 3~]
Voltage transformer	[symbol] or [symbol]	Three-phase induction motor in star connection with starter winding in the rotor	[symbol M Y]

[1] For single-pole representation, e.g. in single-line diagrams
[2] see DIN 40 900 Part 2

Miscellaneous graphic symbols

Table 8.4 A selection of frequently used graphic symbols according to DIN, ANSI, BS and IEC *(continued)*

Designation	Graphic symbol	Designation	Graphic symbol
Semiconductor diode, general		Cable sealing end, representation with a three-core cable	
Limiting diode – Z-diode, Zener diode, Esaki-diode		Tap joint, representation with three-leads, T-shaped tap joint, for all poles	
– Breakdown-diode		Horn, hooter	
Rectifier		Bell	
Bridge-connected rectifier		Siren	
Inverter		Whistle, electrically operated	
Rectifier/Inverter (reversable)		Lamp, general, indicator light	
Voltage stabilizer	[1)]	Indicator light, flashing	
Primary cell, accumulator cell		Sensor, proximity switch – mode of operation inductive	[2)]
Battery of primary cells		– mode of operation capacitive	[2)]
Transfer, energy flow, signal flow in one direction		– mode of operation ultrasonic	[2)]
Energy flow away from the busbar			

[1)] to DIN only
[2)] not contained in DIN but combined from graphic symbol elements and codes based on No. 07-19-02 of DIN 40900

8.1.3 Designation of equipment, conductors and general functions

Designation of equipment

Each device, and its constituent components, i.e. all items of equipment, have to be identified according to DIN 40719 part 2 of June 1978 (see Tables 8.6 and 8.7).

The simplest form of designation is:

— type/sequence number/function,

preceded by a minus sign (−), e.g. −K2A.

K code letter to identify the type of equipment (Table 8.6, page 409), e.g. K for contactor,

2 sequence number for differentiating between equipment of same type and/or function,

A code letter denoting the function of the equipment (see Table 8.7, Page 410), e.g. A for "Off".

The sign and/or the marking of the type and/or function may be omitted if there is no ambiguity. However, only the following combinations are permitted:

−K2A complete designation,

K2A complete designation without sign,

2 sequence number only,

−2 sequence number with sign

K2 identication of equipment without code letter for the function and without sign,

−K2 identication of equipment without code letter for the function and with sign.

Designation of conductors

The designation of some particular conductors and their terminals according to DIN and IEC are listed in Table 8.8, Page 410.

Designation of the terminals on current and voltage transformers

Table 8.5 illustrates the markings/designations of the primary and secondary terminals of current and voltage transformers respectively in terms of the specifications IEC, AS (Australia), BS (Great Britain), NBN (Belgium), NEN (Norway), SEN (Sweden), CSA (Canada), ANSI (USA) and DIN VDE.

Table 8.5
Terminal markings/designations of current and voltage transformers

Specifications	Primary		Secondary	
Current transformers				
			1 winding, several tappings	several windings
IEC 185; AS C388; BS 3938; NBN 134; NEN 3184; SEN 2703	P1	P2	S1 S2...Sn	1S1 1S2... nS1 nS2
CSA C13; ANSI C57.13	H1	H2	X1 X2...Xn	V1V2, W1W2... Z1Z2
DIN VDE 0414	K	L	kl1...$l$$n$	1k 1l... nk $n$$l$
Voltage transformers				
IEC 186, BS 3941, SEN 270821	A or A	N B	a1a2...n a1a2...b	1a1n,2a2n, or dadn —
AS 1243	A1	A2	a1...an	1a1 1a2, 2a1 2a2,...
CSA 13, ANSI C57.13	H1	H2	x1...xn	x1x2, y1y2,...
DIN VDE 0414	U or U	V X	u1u2...v —	— 1u1x... nunx or ux and en

General logic functions

Input and output control signals from/to contactor relays, sensors, actuators, time relays (SICONTROL), etc., as well as from/to programmable logic controllers may be combined via AND and/or OR gates in accordance with the rules of Boolean logic (algebra). Some of the symbols used are:

A = AND $\bar{A}$ = NAND ("not AND")
O = OR $\bar{O}$ = NOR ("not OR")

Example

1 A 0 = 0;   1 $\bar{A}$ 0 = 1;
1 O 0 = 1;   1 $\bar{O}$ 0 = 0;

Table 8.6
Code letters for the designation of the kind of equipment in accordance with DIN 40719 Part 2, Table 1, June 1978

Code letter	Kind of equipment	Examples
A	Assemblies, sub-assemblies	Amplifier, magnetic amplifier, laser, maser, switchgear assembly
B	Transducers, from non-electrical quantity to electrical quantity and vice versa	Transducer, thermoelectric sensor, thermo cell, photoelectric cell, dynamometer, crystal transducer, microphone, pick-up, loudspeaker, phase-sequence indicator
C	Capacitors	–
D	Binary elements, delay devices, storage devices	Delay line, bistable element, monostable element, core storage, register, disk recorder, magnetic tape recorder
E	Miscellaneous	Lighting device, heating device, devices not specified elsewhere in this table
F	Protective devices	Fuse, overvoltage suppressor, protective relay, release, miniature circuit-breaker
G	Generators, power supplies	Rotating generator, rotating frequency converter, battery, power supply device, oscillator
H	Signalling devices	Optical and acoustic indicators
J	-	not used
K	Relays, contactors	Power contactor, contactor relay, flasher relay, time relay, reed relay
L	Inductors	Induction coil, choke, reactance
M	Motors	–
N	Amplifier, regulator	Integrated circuit, impedance converter, operational amplifier
P	Measuring equipment, testing equipment	Indicating, recording, counting and measuring devices, signal generator, clock
Q	Mechanical switching devices for power circuits	Circuit-breaker, disconnector, fused disconnector
R	Resistors	Adjustable resistor, potentiometer, rheostat, shunt, thermistor
S	Switches, selector switches	Pushbutton, limit switch, control switch, selector switch, rotary switch, selector, thumb-wheel selector switch, command device
T	Transformers	Voltage transformer, current transformer
U	Modulators, changers	Discriminator, demodulator, frequency converter, encoder, inverter, converter
V	Tubes, semiconductors	Electronic tube, gas-discharge tube, diode, transistor, thyristor
W	Transmission paths, waveguides, aerials	Jumper wire, cable, busbar, waveguide, waveguide directional coupler, dipole, parabolic antenna
X	Terminals, plugs, sockets	Disconnecting plug and socket, test jack, terminal block or rail, soldering terminal strip, line-up terminal
Y	Electrically operated mechanical devices	Brake, clutch, valve
Z	Termination systems, filters, equalizers, limiters	Cable balancing network, compandor, crystal filter

8 Fundamental circuit diagrams

Table 8.7
Code letters for the designation of general functions to DIN 40 719 Part 2, Table 1, June 1978

Code letter	General function
A	Auxiliary function, function "Off"
B	Direction of movement (forward, backward, hoist, lower, clockwise, anti-clockwise)
C	Counting
D	Differentiating
E	Function "On"
F	Protecting
G	Testing
H	Signalling
J	Integrating
K	Jogging, inching
L	Conductor marking/identification
M	Main function
P	Proportional
Q	State (start, stop, limit)
R	Resetting, erasing, cancelling
S	Storing, recording
T	Timing, delaying
U	–
V	Speed (accelerating, braking)
W	Adding
X	Multiplying
Y	Analog
Z	Digital

Table 8.8
Alphanumeric codes of some particular conductors and their connections according to DIN 42400, Sept. 1983, DIN 40705, Feb. 1980, alt. IEC 445/446; designation by graphic symbols according to DIN 30600, Nov. 1988, DIN 40900, March 1988 alt. IEC 417 and 617

Type of conductor	Code alpha-numerical	graphic symbol
Alternating current network		
Phase conductor 1	L1[1]	
Phase conductor 2	L2[1]	
Phase conductor 3	L3[1]	∼
Neutral conductor	N	
Direct current network		
Positive	L+	
Negative	L−	or ═══
Middle (ground) conductor	M	
Protective conductor	PE	⏚
Neutral conductor with protective function	PEN	–
Protective conductor, not earthed	PU	–
Earth, ground	E	⏚
Low-noise earth	TE	
Frame, chassis	MM	⏚ or ⏚
Equipotential	CC	▽

[1] Alternatively, the motor terminal markings T4, T5 T6 may be used on the load side of switching or protection devices connected to motors.

8.1.4 Circuit diagrams

8.1.4.1 Types of circuit diagrams

In general, the graphic symbols in circuit diagrams are represented in a de-energized and mechanically non-operated state. Deviations from this rule must be clearly indicated in the circuit diagrams (see DIN 40900 Part 7, March 1988).

The different types of diagrams can be classified as follows:

▷ Survey diagram,
▷ Circuit diagram,
▷ Equivalent circuit diagram.

Survey diagram

A survey diagram (also known as an overview or block diagram) is a simplified representation of a circuit and shows only the most essential parts. It shows the principle of operation and the structure of the electrical installation (Fig. 8.10).

Circuit diagram

A circuit diagram is a detailed representation of a circuit with its individual components. It illustrates the principle of operation of the electrical equipment.

Nowadays, the circuit diagram is the most frequently

used representation of a circuit in the field of electrical engineering.

A circuit diagram is broken down into the main circuit and the auxiliary circuit (control circuit and signalling circuit). The various circuits are represented separately from left to right in this sequence (Fig. 8.11).

The equipment designations are not only used to identify the complete devices themselves, but also to identify all the used contacts and terminals of each device in the circuit. Thus, for example, all the contacts of a particular item of switchgear used in the various current paths are labelled with the same equipment designation.

Note:

It has been proposed to use uniform line widths (preferably 0.2 mm) for graphic symbols, electrical connections and linkage lines, when creating circuit diagrams using graphics software (e.g. CAD). To a large extent, this proposal has been adopted for main and auxiliary circuits in the following sections of this chapter.

Equivalent circuit diagram

An equivalent circuit diagram is a special version of an explanatory circuit diagram and serves the analysis and calculation of circuit characteristics.

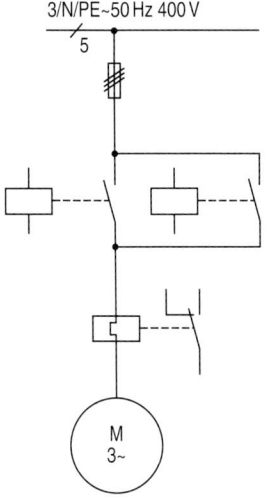

Figure 8.10 Survey diagram

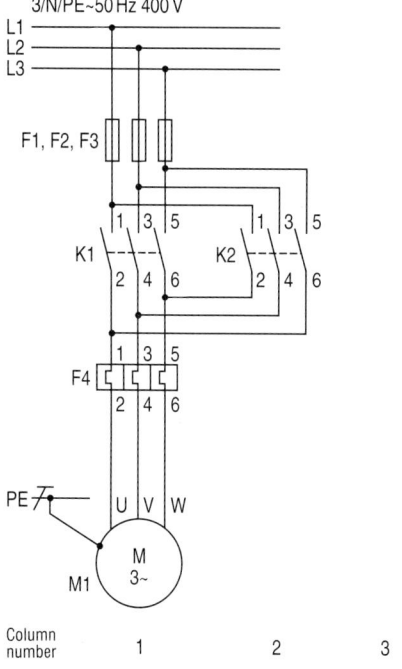

a) Main circuit

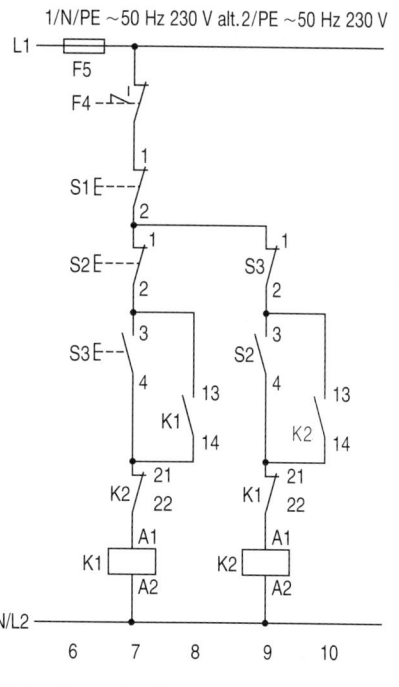

b) Auxiliary circuit

Figure 8.11 Circuit diagram with main and auxiliary circuit

8.1.4.2 Making use of the fundamental circuit diagrams or modifying them

The following fundamental circuit diagrams illustrate the most commonly used main and auxiliary circuits. They may be considered as generic diagrams and it should be noted that they give no indication of the actual physical location of the devices in the installation.

The main circuit diagrams generally include fuses for short-circuit protection and overload relays or releases for overload protection.

The representation of the auxiliary circuits was based on the following considerations:

A fuse is connected to the line side phase for short-circuit protection of the auxiliary circuit.

The auxiliary contacts of an overload relay protecting a particular motor are connected in the common supply circuit for the magnet coils of all the contactors associated with the motor. This prevents the motor from being switched on too soon via another contactor after the overload relay has tripped.

8.1.5 Switching via contactors

The contactor is one of the most important switching devices in industrial circuits. It comprises all the functions which are virtually always required in control systems:

▷ remote operation,
▷ high switching capacity and continuous rating,
▷ long mechanical service life,
▷ low space requirements,
▷ reliable contact making,
▷ complete maintenance-free operation

In order to fulfill these tasks properly, the contactor requires:

▷ clear command signals and
▷ observation of the coil voltage tolerances.

The contactor coils are connected either to the voltage between the phase and neutral conductors, or they are connected between two phases at the secondary side of a control voltage transformer in which case the second phase is preferably linked to the protective earth conductor. Control circuits connected via a control voltage transformer may also be operated without earth connection if they are provided with an insulation monitoring device.

The contactor coils in the circuit diagrams are so arranged, that on the one side they are connected directly to the neutral conductor (or to the earthed phase on the secondary side of a control voltage transformer), and on the other side to the main phase or line conductor via the various control circuit contacts. This prevents the spurious operation of the contactors in the event of an earth fault in the control circuits.

Contactors which must not be energized simultaneously are interlocked via their auxiliary NC contacts.

The control switches are interlocked via NC contacts to enable the immediate change-over from one switching state to another without actuation of the OFF pushbutton. Moreover, this interlocking arrangement prevents energizing commands from being given to several contactors at the same time (double commands).

The graphic symbols used in the fundamental circuit diagrams are derived from DIN 40900 Parts 1 to 12.

8.1.5.1 Contactors with drop-out delay unit for fluttering command signals

Fluttering command signals may lead to chattering of the contactor and may cause the destruction of the magnet system or may lead to welding of the contacts when high currents are being switched.

When using Siemens contactors, a voltage tolerance of 0.8 to $1.1 \cdot U_s$ must be observed. If the rated control supply voltage is too high, the service life will be reduced and it may even lead to burning-out of the coil. If the rated control voltage is too low, the contact pressure may be too low, which can cause contact welding. Furthermore, this too leads to overheating and possible coil burning-out and will certainly reduce the service life of the magnet system.

Contactors equipped with a drop-out delay (off-delay) unit prevent the damaging effects due to ambiguous and/or fluttering command signals.

Figure 8.12 illustrates an auxiliary circuit diagram of a contactor control with a drop-out delay unit.

Switching via contactors · Drop-out delay unit

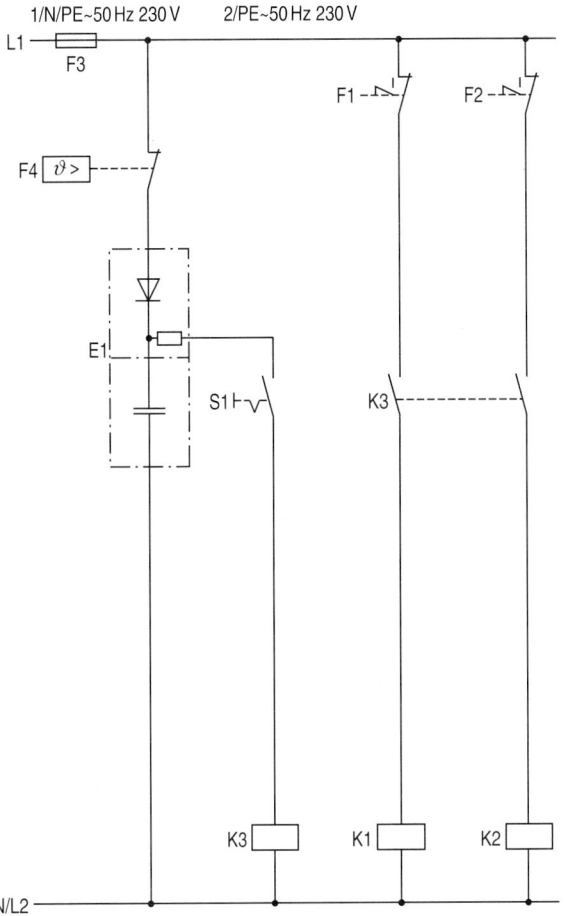

The maintained-contact switch S1 energizes the drop-out-delayed contactor relay K3 since the NC contact of the thermostat F4 is closed.

Contactors K1 and K2 are energized via the NO contacts of K3.

If the command signal applied flutters because, for example, the NC contact of F4 is subjected to vibrations, then contactors K1 and K2 remain reliably closed. The capacitor of E1 supplies the energy for contactor coil K3 if the NC contact of F4 opens for a short time.

For further details, refer to the Siemens low-voltage switchgear catalogue.

Required auxiliary contacts:

Contactor K3 with 2NO and d.c. economy circuit,
Maintained-contact switch S1 with 1NO,
Thermostat F4 with 1NC,
Overload relays F1, F2 each with 1NC

Figure 8.12
Circuit diagram for auxiliary circuit using a drop-out delay unit

8.1.5.2 Extended (early-make/late-break) auxiliary contacts in contactors (mainly for d.c. excitation of the coils)

Figure 8.13 illustrates the circuit diagrams for a contactor control with an early-make/late-break auxiliary contact.

Contactor K1 is directly energized by the actuation of maintained-contact switch S1 via the extended NC contact. The coil of contactor K1 is over-excited until the extended NC contact of K1 opens and the economy resistor R1 is connected in series with the coil. The motor is started.

This configuration is commonly known as a "d.c. economy circuit".

8.1.5.3 Drop-out delay units for contactors

For opening delay devices (drop-out delay units) for contactors, please refer to the Siemens l.v. switchgear catalogue.

8.1.5.4 Contactor safety combinations

For contactor safety combinations, please refer to the Siemens l.v. switchgear catalogue and to Section 4.3.4, page 254.

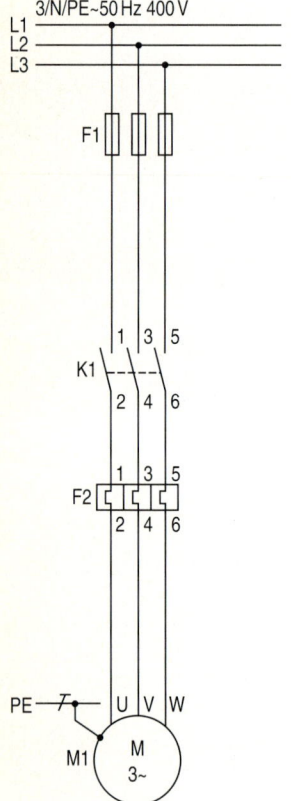

a) Main circuit

b) Auxiliary circuit

Required auxiliary contacts:

Contactor K1 with 1 extended NC,
Maintained-contact switch S1 with 1NO,
Overload relay F2 with 1NC

Figure 8.13
Circuit diagrams of a contactor control incorporating an extended auxiliary contact, showing the equipment designation and sequence numbers

8.2 Direct switching of three-phase induction motors

8.2.1 Switching on and off of three-phase induction motors

Figure 8.14 illustrates the circuit diagrams for the direct switching on and off of three-phase induction motors.

Pushbutton control / momentary contact (Fig. 8.14b)

Switching on: Contactor K1 is energized by the actuation of momentary-contact switch S1 (e.g. pushbutton). The sealing-in contact K1 closes, and the motor is switched on.

Switching off: Contactor K1 is de-energized by the actuation of momentary-contact switch S0 (e.g. pushbutton). The sealing-in contact K1 opens and the motor is switched off.

Switch control/maintained contact (Fig. 8.14c)

Contactor K1 is energized and de-energized via the maintained-contact switch S1. The motor is switched on and off by the main contacts of contactor K1.

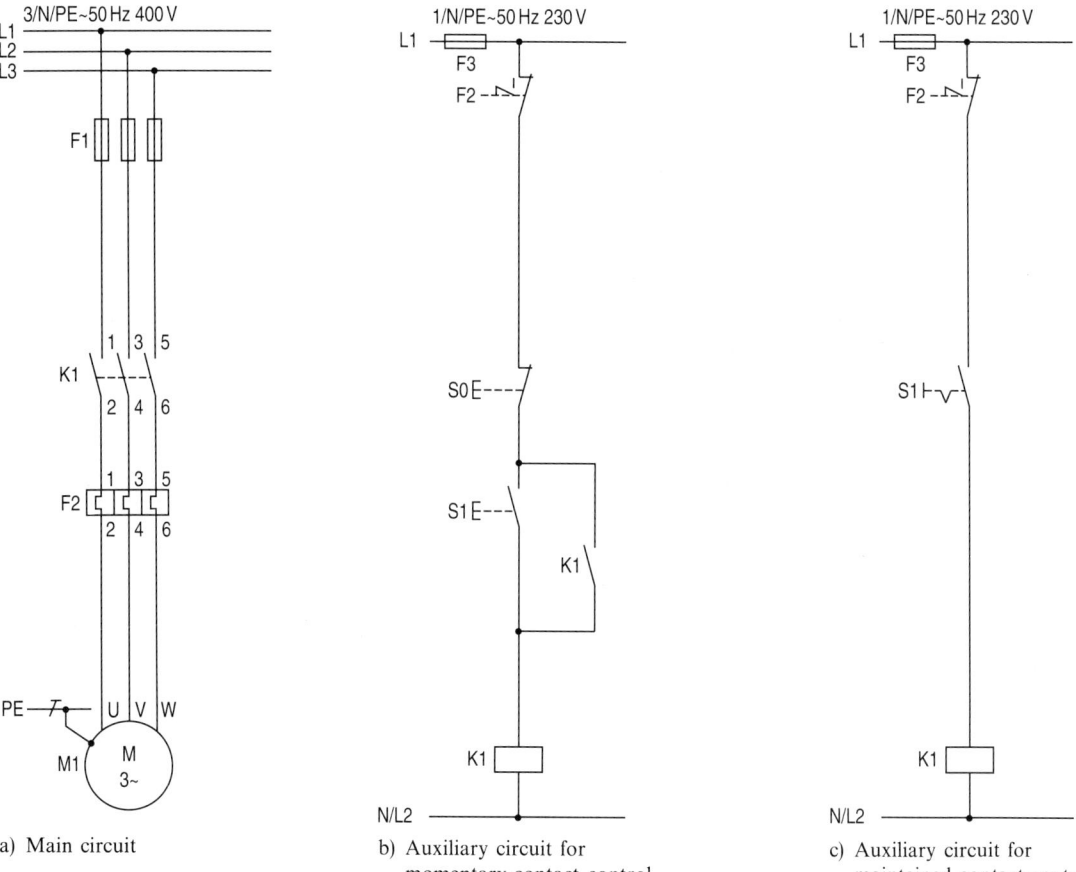

a) Main circuit

b) Auxiliary circuit for momentary-contact control

Required auxiliary contacts:

Contactor K1 with 1NO,
Momentary-contact switch S0 with 1NC,
Momentary-contact switch S1 with 1NO,
Overload relay F2 with 1NC

c) Auxiliary circuit for maintained-contact control

Required auxiliary contacts:

Maintained-contact switch S1 with 1NO,
Overload relay F2 with 1NC

Figure 8.14
Circuit diagrams for the direct switching on and off of a three-phase induction motor

8 Fundamental circuit diagrams

8.2.2 Switch-over of a three-phase induction motor from one supply network to another

Figure 8.15 illustrates the circuit diagrams for the switch-over of a three-phase induction motor from one supply network to another (also known as load transfer).

Pushbutton control/momentary contact (Fig. 8.15b)

Switching on: Contactor K1 is energized by the actuation of momentary-contact switch S1. The sealing-in contact K1 closes, and the motor is connected to Network 1.

Change-over: Contactor K1 is de-energized by the actuation of momentary-contact switch S2 and the opening of its NC contact. Contactor K2 is energized by the closing of the NO contact S2. This energizing command on K2 only becomes effective once the NC contact of K1 has closed (i.e. once K1 has dropped out). The sealing-in contact K2 closes, and the motor is connected to Network II.

The change-over of the supply from Network II to Network I is accomplished in the same way (reverse order).

Switching off: Contactor K1 or K2 is de-energized by the actuation of momentary-contact switch S0. The sealing-in contact K1 (alt. K2) opens and the motor is switched off.

Switch control/maintained contact (Fig. 8.15c)

Contactor K1 or K2 is energized or de-energized via the maintained-contact switch S1 according to its switch position indication. Thereby, the motor is connected to either Network I or Network II.

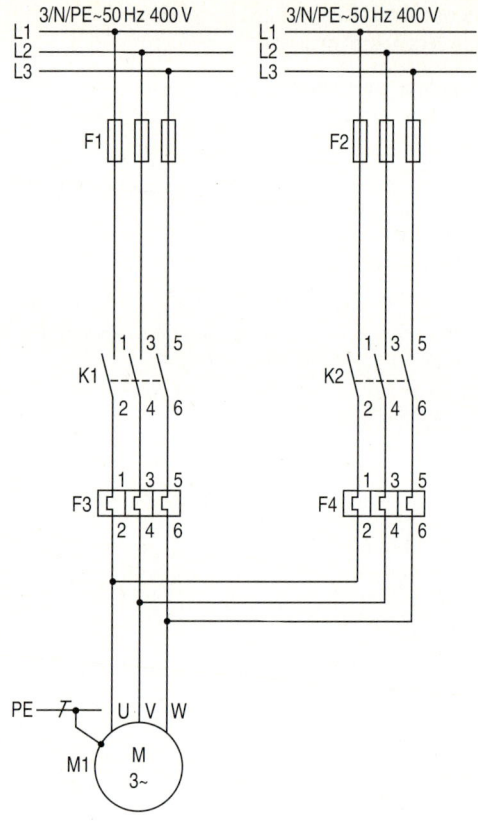

a) Main circuit

Figure 8.15
Circuit diagrams for the switch-over of a three-phase induction motor from one supply network to another

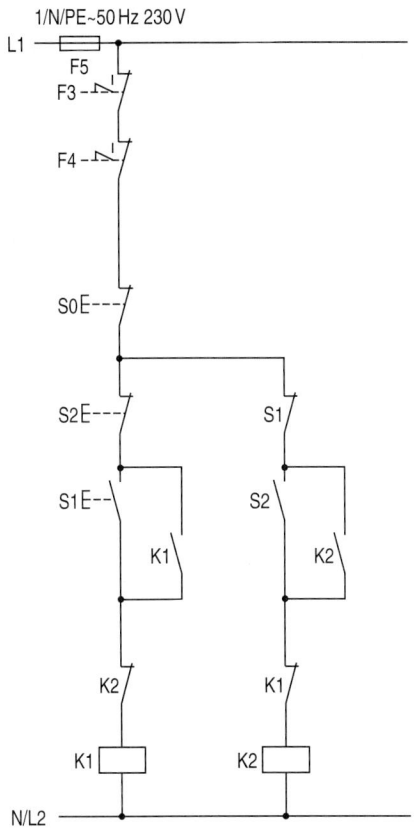
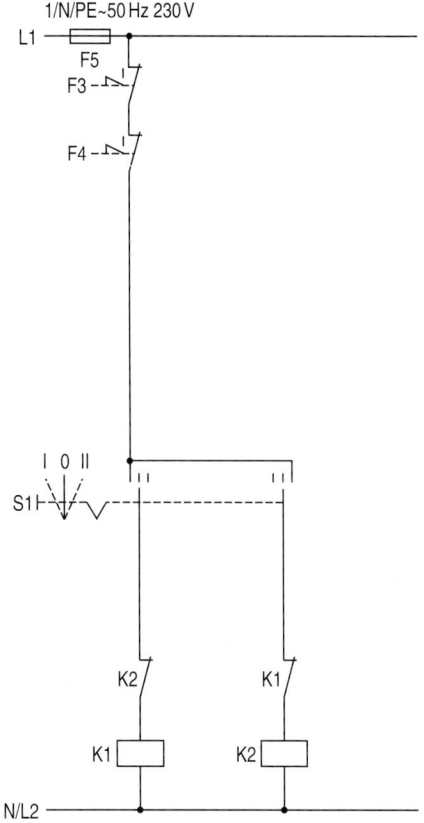

b) Auxiliary circuit for momentary-contact control

Required auxiliary contacts:

Contactor K1 with 1NO+1NC,
Contactor K2 with 1NO+1NC,
Momentary-contact switch S0 with 1NC,
Momentary-contact switch S1 with 1NO+1NC,
Momentary-contact switch S2 with 1NO+1NC,
Overload relays F3, F4 each with 1NC

c) Auxiliary circuit for maintained-contact control

Required auxiliary contacts:

Contactor K1 with 1NC,
Contactor K2 with 1NC,
Maintained-contact switch S1 with 3 switch positions,
Overload relays F3, F4 each with 1NC

Note:
The control supply voltage could preferably be derived from a battery or uninterruptible power supply system (UPS). If necessary, the auxiliary supply voltage must also be switched over from one network to the other (all poles must be transferred).

8 Fundamental circuit diagrams

8.2.3 Automatic sequential starting of three-phase induction motors

Figure 8.16 illustrates the circuit diagrams for the automatic switching-on of several three-phase motors in succession (sequential starting).

Momentary-contact control with a contactor relay (Fig. 8.16b)

Switching on: Contactor relay K5 is energized by the actuation of momentary-contact switch S1. The sealing-in contact K5 closes. Contactor K1 is energized via the NO contact of K5. The NO contact of K1 energizes contactor K2, etc. to K4, as the NO contacts of contactor relay K5 have closed. The motors are switched on in succession via contactors K1 to K4. The current peaks also occur in succession – the load on the network is relieved. (This circuitry is suitable for motors with extremely short run-up times or for low motor outputs. For longer run-up times, it is recommended to delay the individual motor circuits with respect to one another using time relays, in order to ensure a peak load reduction in the network).

Switching off: Contactors K1 to K4 and K5 are de-energized by the actuation of momentary-contact switch S0. All the motors are switched off simultaneously.

Momentary-contact control without a contactor relay (Fig. 8.16c)

Switching on: Contactor K1 is immediately energized by the actuation of momentary-contact switch S1. The sealing-in contact and the NO contact of K1 close. Contactor K2 is energized via the NO contact of K1, K3 is energized via the NO contact of K2, etc. Thus, the motors are switched on in succession (see momentary-contact control with a contactor relay).

Switching off: Contactors K1 to K4 are de-energized by the actuation of momentary-control switch S0. All the motors are switched off simultaneously.

Switch control/maintained contact (Fig. 8.16d)

Contactor K1 is energized by the closing of the maintained-contact switch S1. The NO contact of K1 closes. Contactors K2 to K4 are energized in succession via the NO contact of the preceding contactor (for other actions, see momentary-contact control without a contactor relay).

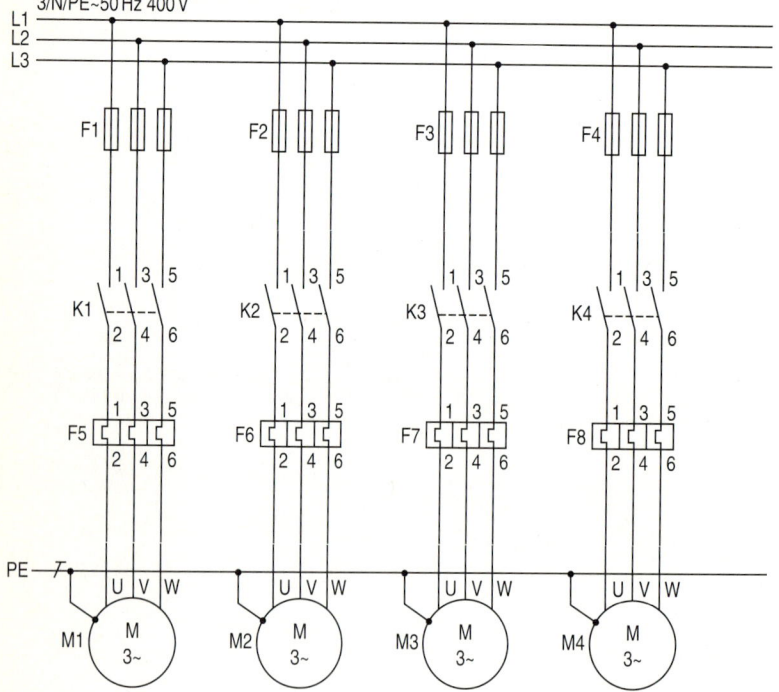

a) Main circuit

Figure 8.16
Circuit diagrams for automatic sequential starting of several three-phase induction motors

Automatic sequential starting of motors

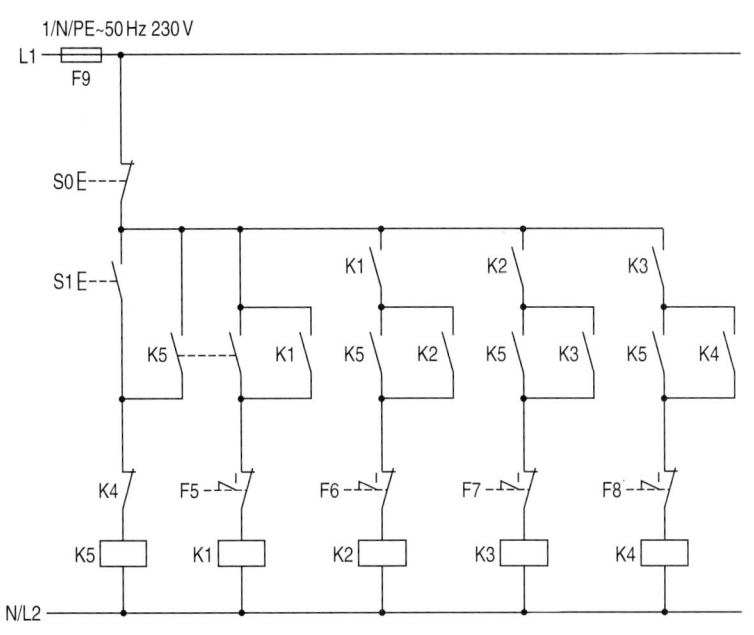

b) Auxiliary circuit for momentary-contact control with a contactor relay, overload relay *with* or *without* a lockout device[1]

Required auxiliary contacts:

Contactors K1, K2, K3 each with 2NO,
Contactor K4 with 1NO + 1NC,
Contactor relay K5 with 5NO,
Momentary-contact switch S0 with 1NC,
Momentary-contact switch S1 with 1NO

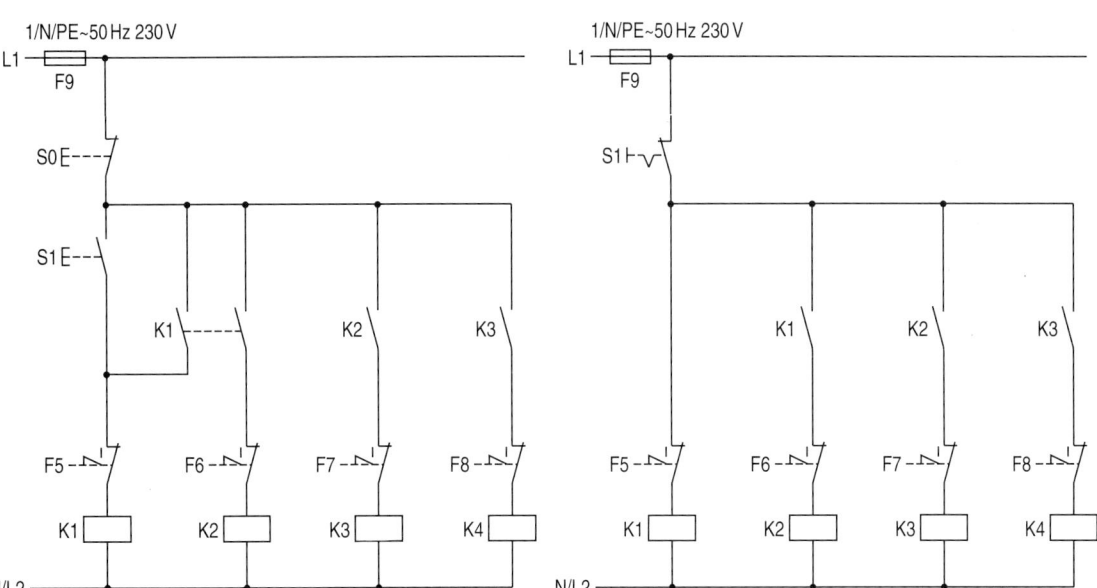

c) Auxiliary circuit for momentary-contact control without a contactor relay, overload relay *with* a lockout device[1]

Required auxiliary contacts:

Contactor K1, with 2NO,
Contactors K2, K3 each with 1NO,
Momentary-contact switch S0 with 1NC
Momentary-contact switch S1 with 1NO

d) Auxiliary circuit for maintained-contact control, overload relay *with* lockout device[1]

Required auxiliary contacts:

Contactors K1, K2, K3 each with 1NO,
Maintained-contact switch S1 with 1NO

---

[1] If an overload relay trips, the lock-out device ensures that the overload does not reset automatically. In circuits c) and d), the automatic resetting of an overload relay could cause unexpected (dangerous) restarting of a motor.

8.2.4 Reversing the direction of rotation of three-phase induction motors (reversing starters)

Figure 8.17 illustrates the circuit diagrams for reversing the direction of rotation of a three-phase induction motor.

For mechanically interlocked and factory wired reversing starters (TTA) as well as for the required individual devices and components for do-it-yourself assembly, please refer to the relevant Siemens low-voltage switchgear catalogue.

Pushbutton control/momentary contact (Fig. 8.17b)

<u>Switching on</u>: Contactor K1 is energized by the actuation of momentary-contact switch S1, the sealing-in contact K1 closes and the motor is started, e.g. in clockwise direction of rotation.

<u>Change-over</u>: Contactor K1 is de-energized by the actuation of momentary-contact switch S2 and the opening of its NC contact. Contactor K2 is energized by the closing of the NO contact S2. This energizing command on K2 only becomes effective once the NC contact of K1 has closed. The motor is slowed down (braked by the phase-reversal) and starts-up again in the anti-clockwise direction of rotation.

<u>Switching off</u>: Contactor K1 or K2 is de-energized by the actuation of momentary-contact switch S0. The motor is switched off.

Switch control/maintained contact (Fig. 8.17c)

The maintained-contact switch S1 energizes and de-energizes contactor K1 or K2, and changes over from contactor K1 to K2 via the OFF position and vice versa. The other actions are the same as with pushbutton control.

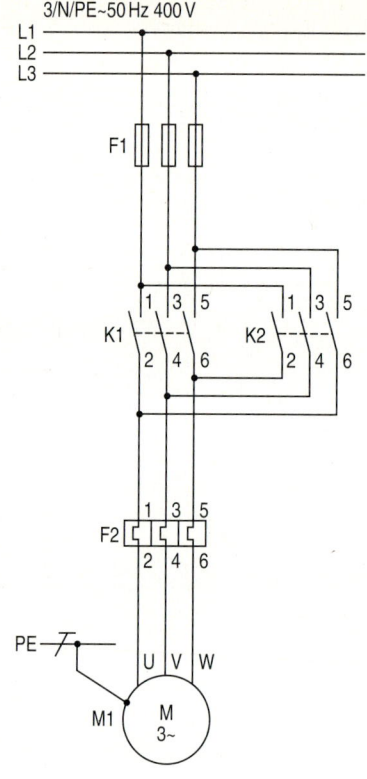

a) Main circuit

Figure 8.17
Circuit diagrams for reversing the direction rotation of a three-phase induction motor

Reversing of three-phase induction motors

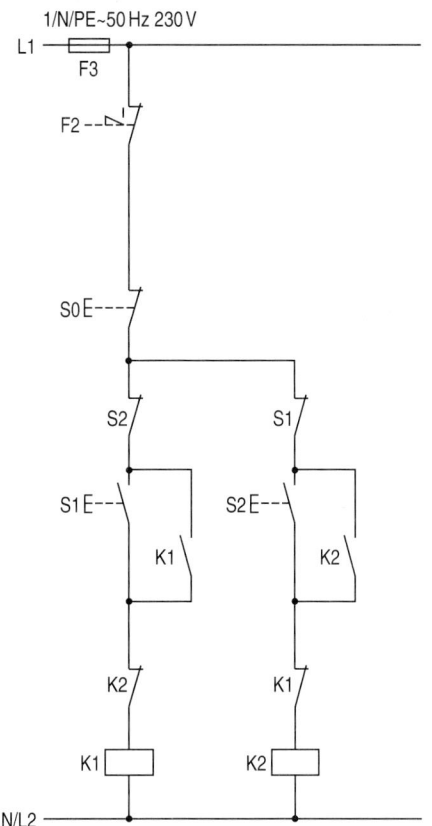

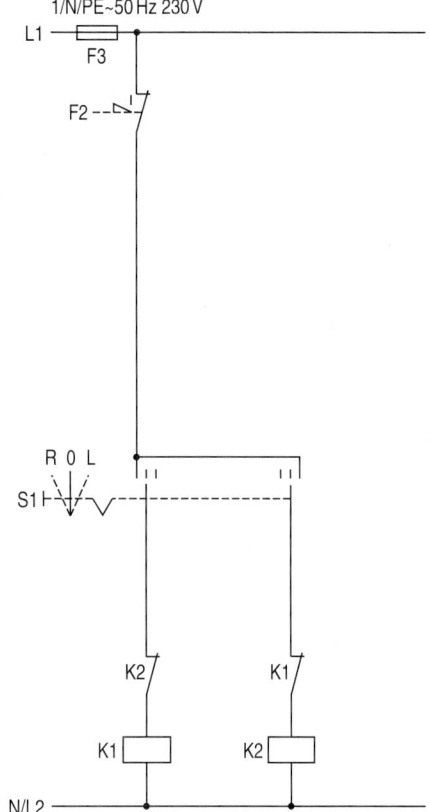

b) Auxiliary circuit for momentary-contact control

Required auxiliary contacts:

Contactor K1, K2 each with 1 NO+1NC,
Momentary-contact switch S0 with 1NC,
Momentary-contact switch S1, S2 each with 1NO+1NC,
Overload relay F2 with 1NC

c) Auxiliary circuit for maintained-contact control

Required auxiliary contacts:

Contactor K1, K2 each with 1NO+1NC,
Maintained-contact switch S1 with 3 switch positions,
Overload relay F2 with 1NC

8 Fundamental circuit diagrams

8.2.5. Switching of pole-changing three-phase induction motors

8.2.5.1 Pole-changing three-phase induction motor with one winding (Dahlander connection), two speeds, one direction of rotation

The use of a Dahlander connection for pole-changing three-phase induction motors with two speeds in the ratio of 1:2 is more effective than a pole-changing control with two separate windings, since the complete winding is used in each speed. The winding consists of two coil groups per phase.

The pole changing is achieved by the change-over and the reversal of the current in the corresponding coil groups. There are various Dahlander connections to optimize adaptation to the load torque.

The most frequently used connections are:

▷ Delta/star-star
   for drives with constant torque,
   output ratio $P_1/P_2 = 1:1.4$

▷ Star-star/delta
   for drives with constant output power,
   output ratio $P_1/P_2 = 1:1$

▷ Star/star-star
   for drives with square-law
   load torque (e.g. fan drives),
   output ratio $P_1/P_2 = 1:4$ to 8

Further advantages are:

Only six terminals are required. The speed can be changed by change-over and the creation of various star point connections.

In the case of the delta connection, star-delta starting is possible, where-as with the star-star connection, starting via a single-star conguration with current and torque ratio 1:4, is possible.

Figure 8.18 illustrates the circuit diagrams for a pole-changing three-phase induction motor with a single winding (Dahlander connection) for two speeds and one direction of rotation.

Pushbutton control/momentary contact (Fig. 8.18b)

<u>Switching on</u>: Contactor K1 is energized by the actuation of momentary-contact switch S1. Sealing-in contact K1 closes and the motor starts up at low speed.

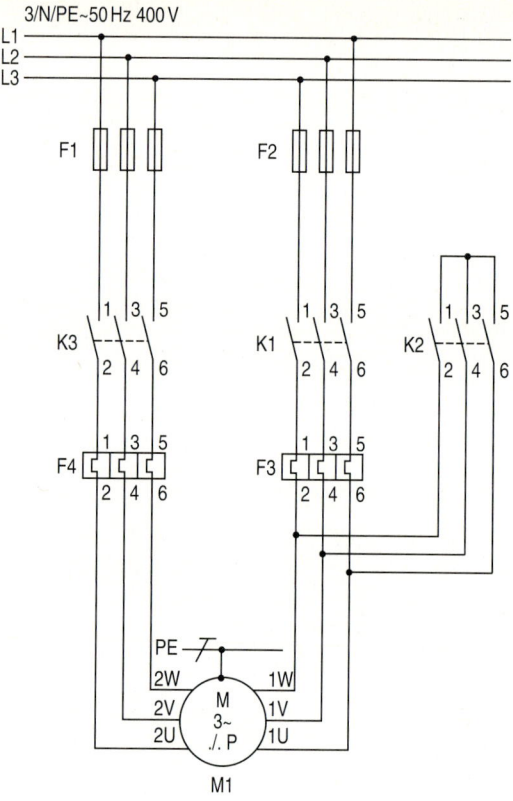

a) Main circuit

Figure 8.18
Circuit diagrams for a pole-changing three-phase induction motor with one winding (Dahlander connection) for two speeds and one direction of rotation

<u>Change-over</u>: Contactor K1 is de-energized by the actuation of momentary-contact switch S2 and the opening of its NC contact. Contactor K2 is energized by the closing of the NO contact of S2. This energizing command on K2 only becomes effective once the NC contact of K1 has closed. Sealing-in contact K2 closes. The line contactor K3 is energized via the NO contact of K2. The motor runs at high speed.

The change-over to a lower speed is effected in the reverse order.

<u>Switching off</u>: Contactors K1 or K2 and K3 are de-energized by the actuation of momentary-contact switch S0. The motor is switched off.

Switching of pole-changing three-phase induction motors

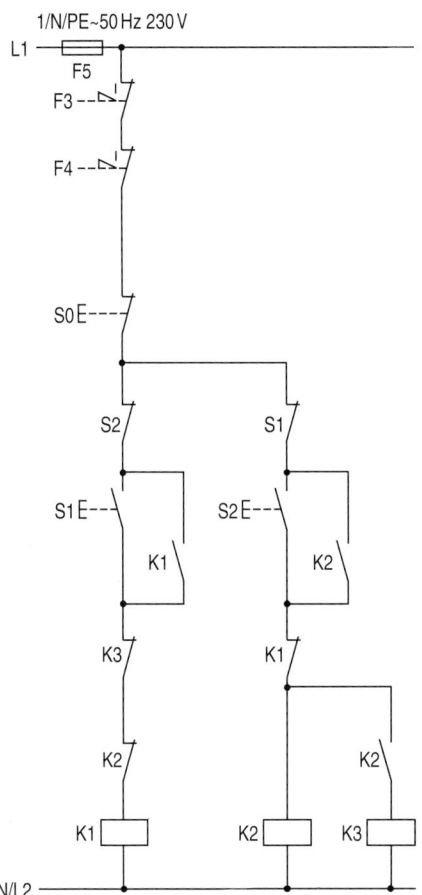

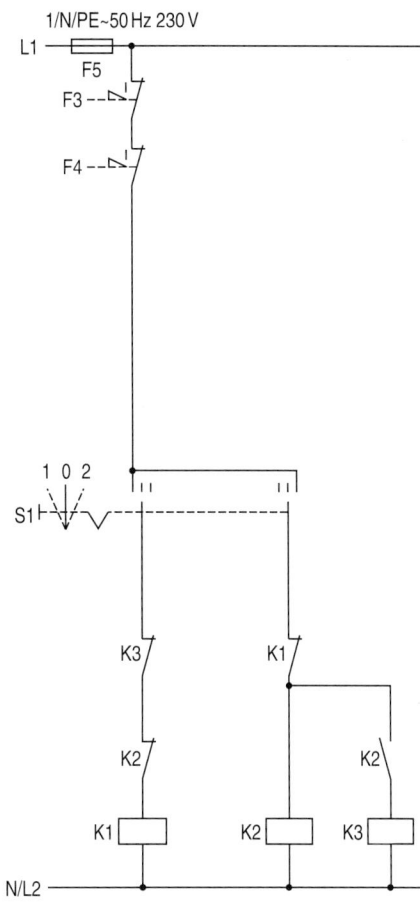

b) Auxiliary circuit for momentary-contact control

Required auxiliary contacts:

Contactor K1 with 1NO+1NC,
Contactor K2 with 2NO+1NC,
Contactor K3 with 1NC,
Momentary-contact switch S0 with 1NC,
Momentary-contact switches S1 and S2 each with 1NO+1NC,
Overload relays F3 and F4 each with 1NC

c) Auxiliary circuit for maintained-contact control

Required auxiliary contacts:

Contactors K1, K3 each with 1NC
Contactor K2 with 1NO+1NC
Maintained-contact switch S1 with three switch positions
Overload relays F3 and F4 each with 1NC

Switch control/maintained contact (Fig. 8.18 c)

Contactors K1 or K2 and K3 are energized, de-energized or changed over by the actuation of maintained-contact switch S1 as indicated by the switch position indicator; for other actions, see pushbutton control.

Note on the main circuit:

Fuses F1 and F2 have to be selected according to the motor rated currents at both speeds. A common fuse protection of motor contactors K1 and K3 is permissible if the motor rated currents vary only slightly from each other and if the overload relays F3 and F4 can be used with the same setting ranges.

8.2.5.2 Pole-changing three-phase induction motor with <u>one</u> winding (Dahlander connection), two speeds, <u>two</u> directions of rotation

Figure 8.19 illustrates the circuit diagrams for a pole-changing three-phase induction motor with one winding (Dahlander connection), two speeds, and two directions of rotation.

Pushbutton control/momentary contact (Fig. 8.19b)

Switching on the motor e.g., for clockwise rotation: Contactor K1 is energized by the actuation of momentary-contact switch S1. The sealing-in contact K1 closes. The motor is switched on at low speed.

Change-over to high speed: Contactor K1 is de-energized by the actuation of momentary-contact switch S3 and the opening of its NC contact. Star contactor K5 and contactor K3 are energized by the closing of the NO contact of S3. The energizing command on K5 only becomes effective once the NC contact of K1 has closed. The energizing command on contactor K3 only becomes effective once the NO contact of contactor K5 has closed. The sealing-in contacts of contactors K5 and K3 close and the motor runs at high speed.

Change-over to low speed: The change-over to low speed is effected in the reverse order.

Change-over to anti-clockwise rotation (high speed): Contactor K3 is de-energized by the actuation of momentary-contact switch S4 and the opening of its NC contact. Contactor K4 is energized by the closing of NO contact S4. This energizing command on K4 only becomes effective once the NC contact of K3 has closed. The motor runs at high speed in anti-clockwise direction.

Change-over to anti-clockwise rotation (low speed): The change-over at low speed is accomplished in the same way as with high speed, but by the actuation of momentary-contact switch S2.

Switching off: By the actuation of momentary-contact switch S0 all the contactors in the circuit are de-energized and thus the motor is switched off.

Switch control/maintained contact (Fig. 8.19c)

Contactor K1 or K2 is energized for low speed in clockwise or anti-clockwise rotation respectively, depending on which switch position of the maintained-contact switch S1 has been selected. This energizing command only becomes effective if the NC contact of K1, alternatively K2, is closed. When changing over to high speed in clockwise or anti-clockwise rotation, the star contactor K5 and contactor K3 or K4 are energized by the closing of the NO contact of K5. This energizing command on the star contactor K5 only becomes effective once the NC contacts of K1 and K2 have closed again.

Note:

For fuse protection of the main circuits, see the "Note on the main circuit", Figure 8.18 on page 423.

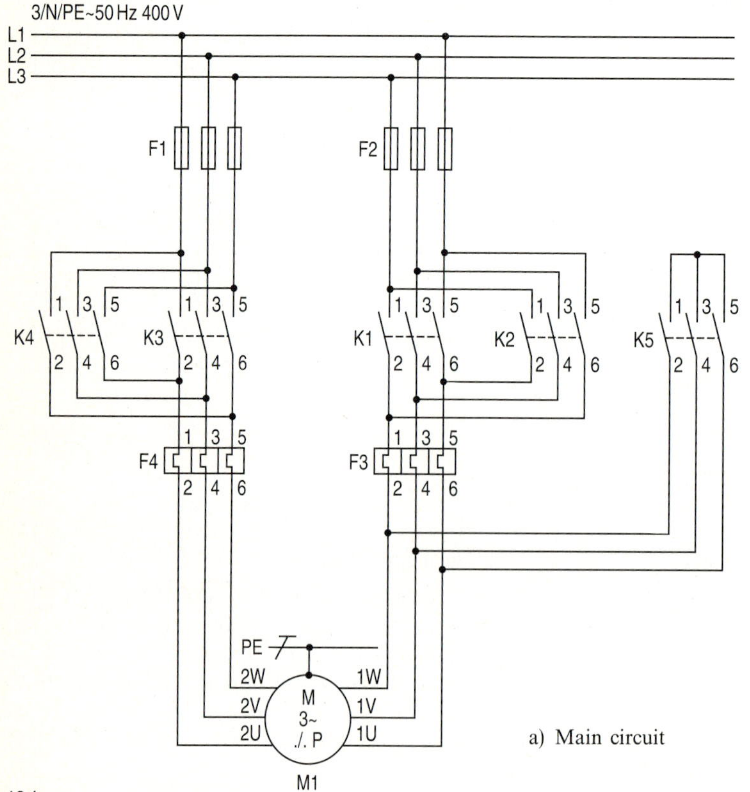

a) Main circuit

Figure 8.19
Circuit diagrams for a pole-changing three-phase induction motor with one winding (Dahlander connection) for two speeds and two directions of rotation

Switching of pole-changing three-phase induction motors

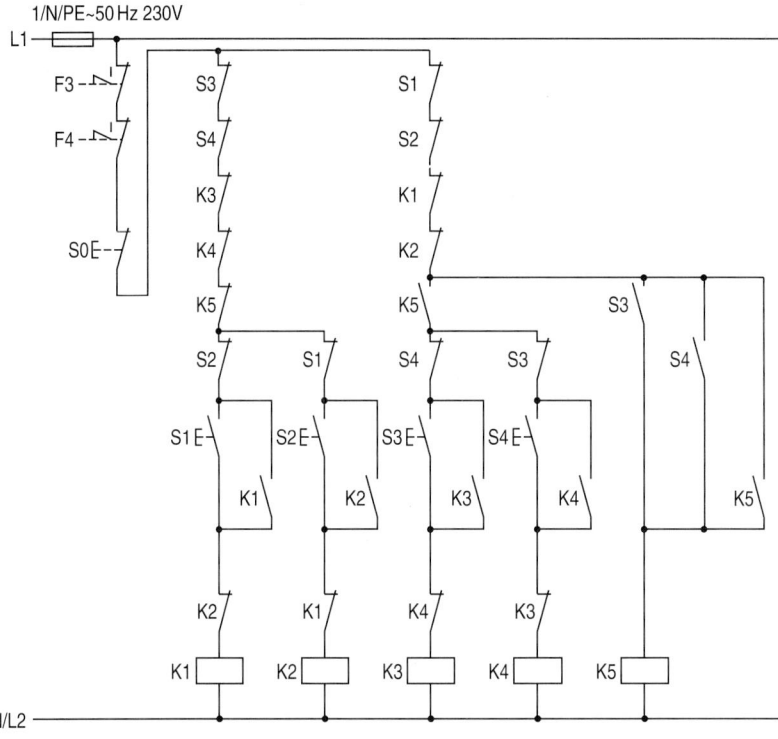

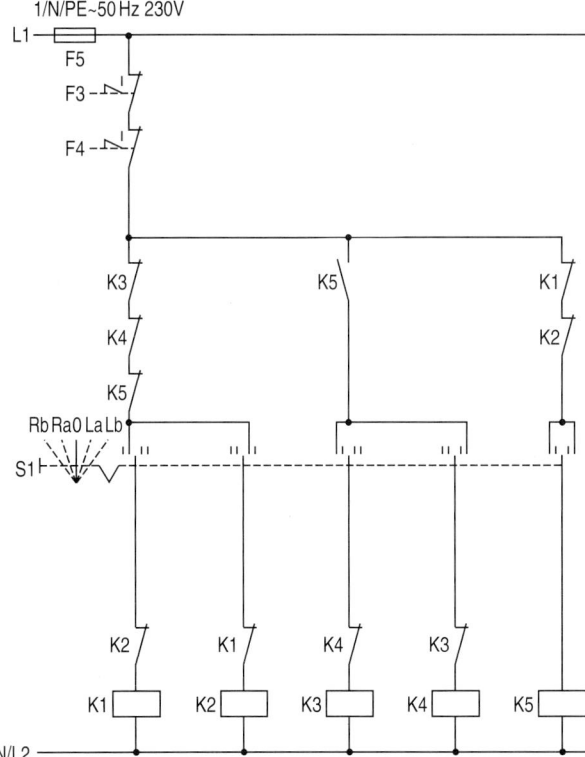

b) Auxiliary circuit for momentary-contact control

Required auxiliary contacts:
Contactors K1, K2, K3, K4 each with 1NO+2NC,
Contactor K5 with 2NO+1NC,
Momentary-contact switch S0 with 1NC,
Momentary-contact switches S1, S2 each with 1NO+2NC,
Momentary-contact switches S3, S4 each with 2NO+2NC,
Overload relays F3 and F4 each with 1NC

c) Auxiliary circuit for maintained-contact control

Required auxiliary contacts:
Contactors K1, K2, K3, K4 each with 2NC,
Contactor K5 with 1NO+1NC,
Maintained-contact switch S1 with 5 switch positions,
Overload relays F3 and F4 each with 1NC

8 Fundamental circuit diagrams

8.2.5.3 Pole-changing three-phase induction motor with two separate windings, two speeds, one direction of rotation

Figure 8.20 illustrates the circuit diagrams for a pole-changing three-phase induction motor with two separate windings, two speeds and two directions of rotation.

Pushbutton control/momentary contact (Fig. 8.20b)

Switching on: Contactor K1 is energized by the actuation of momentary-contact switch S1, and the sealing-in contact K1 closes. The motor starts up, for example, at low speed.

Change-over: Contactor K1 is de-energized by the actuation of momentary-contact switch S2 and the opening of its NC contact. Contactor K2 is energized via the NO contact of momentary-contact switch S2. This energizing command on K2 only becomes effective once the NC contact of contactor K1 has closed. The motor runs at high speed.

Switching off: Contactor K1 or K2 is de-energized by the actuation of momentary-contact switch S0. The motor is switched off.

Switch control/maintained contact (Fig. 8.20c)

Contactors K1 or K2 are energized, de-energized or changed over by the actuation of maintained-contact switch S1 (as with pushbutton control).

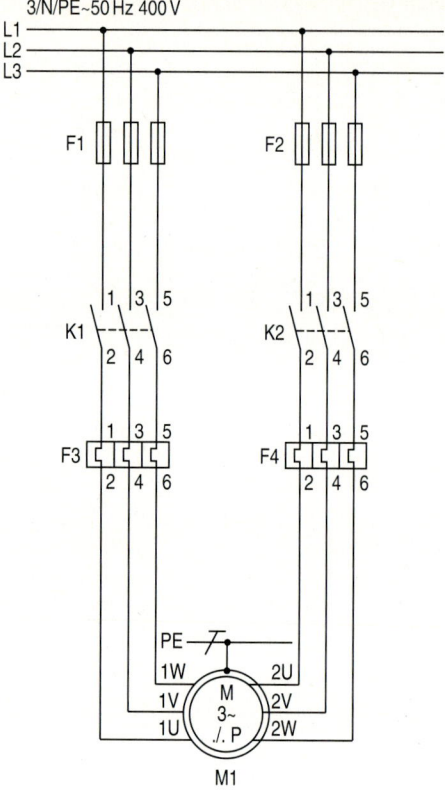

a) Main circuit

Figure 8.20
Circuit diagrams for a pole-changing three-phase induction motor with two separate windings for two speeds and one direction of rotation

Switching of pole-changing three-phase induction motors

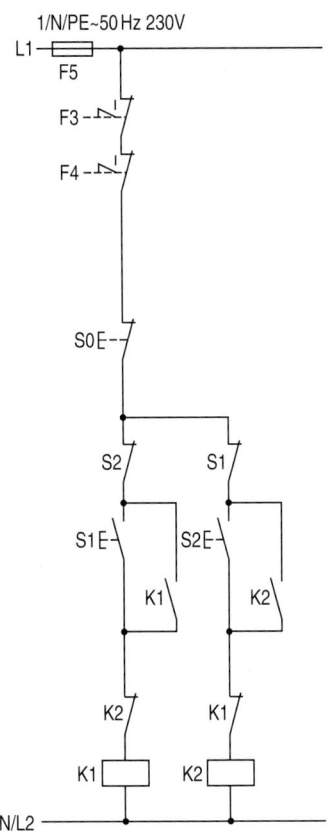

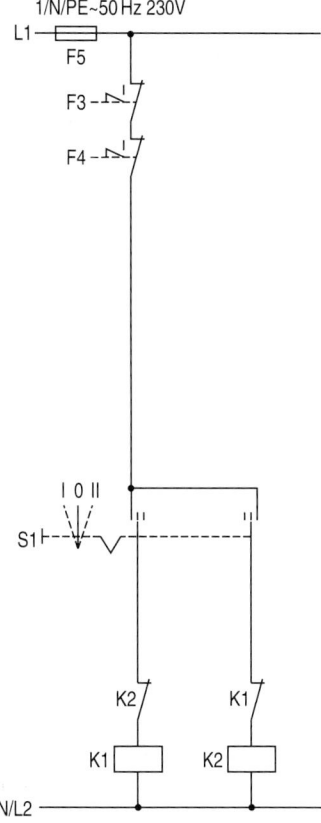

b) Auxiliary circuit for momentary-contact control

Required auxiliary contacts:

Contactors K1, K2 each with 1NO+1NC,
Momentary-contact switch S0 with 1NC,
Momentary-contact switches S1, S2 each with 1NO+1NC,
Overload relays F3 and F4 each with 1NC

c) Auxiliary circuit for maintained-contact control

Required auxiliary contacts:

Contactors K1, K2 each with 1NC,
Maintained-contact switch S1 with 3 switch positions,
Overload relays F3 and F4 each with 1NC

8 Fundamental circuit diagrams

8.2.5.4 Pole-changing three-phase induction motor with two separate windings, two speeds, two directions of rotation

Figure 8.21 illustrates the circuit diagrams for a pole-changing three-phase induction motor with two separate windings, two speeds, and two directions of rotation.

Pushbutton control/momentary contact (Fig. 8.21 b)

Switching on, e.g., for clockwise rotation: Contactor K1 is energized by the actuation of momentary-contact switch S1 and the closing of its NO contact. The sealing-in contact K1 closes. The motor starts up at low speed (for example).

Change-over from low to high speed: Contactor K1 is de-energized by the actuation of momentary-contact switch S2 and the opening of its NC contact. Contactor K2 is energized by the closing of the NO contact of pushbutton S2. This energizing command on K2 only becomes effective once the NC contact of K1 has closed. Sealing-in contact K2 closes and the motor runs at high speed.

Change-over from clockwise to anti-clockwise rotation at, e.g., low speed: Contactor K1 is de-energized by the actuation of momentary-contact switch S3 and the opening of its NC contact. Contactor K3 is energized by the closing of the NO contact of S3. This energizing command on K3 only becomes effective once the NC contact of K1 has closed. Sealing-in contact K3 closes and the motor is stopped (braked) and then re-started in anti-clockwise direction.

Change-over from clockwise to anti-clockwise rotation and simultaneous change-over from low to high speed is accomplished in a similar way by the actuation of momentary-contact switch S4.

Switching off: The contactor presently in circuit is de-energized by the actuation of momentary-contact switch S0 and the motor is switched off.

Switch control/maintained contact (Fig. 8.21 c)

Contactors K1, K2, K3 or K4 are energized, de-energized or changed over by the actuation of maintained-contact switch S1, as with pushbutton control.

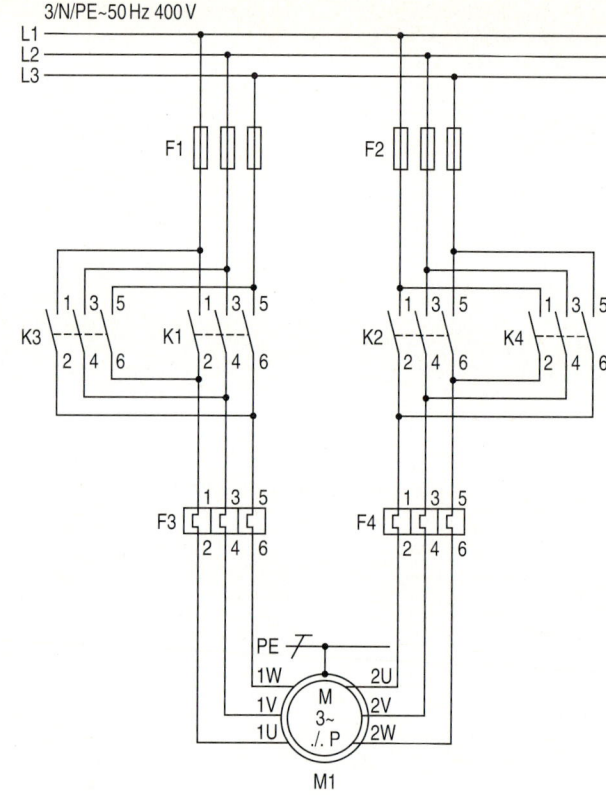

a) Main circuit

Figure 8.21
Circuit diagrams for a pole-changing three-phase induction motor with two separate windings, two speeds and two directions of rotation

Switching of pole-changing three-phase induction motors

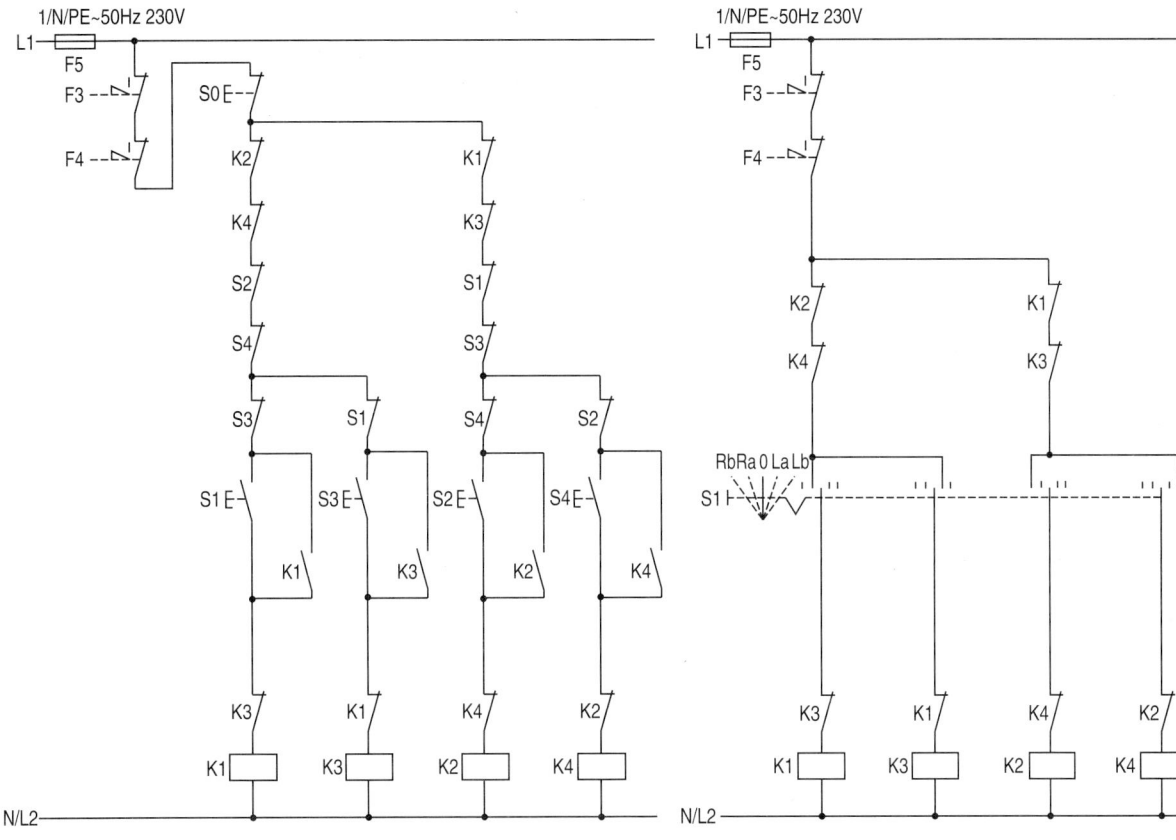

b) Auxiliary circuit for momentary-contact control

Required auxiliary contacts:

Contactors K1, K2, K3, K4 each with 1NO+2NC,
Momentary-contact switches S1, S2, S3, S4 each with 1NO+2NC,
Momentary-contact switch S0 with 1NC,
Overload relays F3 and F4 each with 1NC

c) Auxiliary circuit for maintained-contact control

Required auxiliary contacts:

Contactors K1, K2, K3, K4 each with 2NC,
Maintained-contact switch S1 with 5 switch positions,
Overload relays F3 and F4 each with 1NC

8.2.5.5 Pole-changing three-phase induction motor with <u>three</u> speeds, <u>one</u> direction of rotation, one winding in a Dahlander connection, and one separate winding for <u>low</u> speed

Figure 8.22 illustrates the circuit diagrams for a pole-changing three-phase induction motor with two separate windings, three speeds and one direction of rotation (one winding in a Dahlander connection and one separate winding for low speed).

Pushbutton control/momentary contact (Fig. 8.22b)

<u>Switching on</u>: Contactor K1 is energized by the actuation of momentary-contact switch S1 and the closing of its NO contact. The sealing-in contact K1 closes. The motor starts up at low speed.

<u>Change-over</u>: Contactor K1 is de-energized by the actuation of momentary-contact switch S2 and the opening of its NC contact. Contactor K2 is energized by the closing of the NO contact of S2. This energizing command on K2 only becomes effective once the NC contact of K1 has closed. The sealing-in contact K2 closes and the motor runs at the intermediate speed.

Contactor K3 is energized by the actuation of momentary-contact switch S3 and the closing of its NO contact. The energizing command on K3 only becomes effective once the NC contact of K2 has closed. The sealing-in contact and the NO contact of K3 close. Line contactor K4 is energized. The motor runs at high speed.

<u>Switching off</u>: All the contactors in circuit are de-energized by the actuation of momentary-contact switch S0 and the opening of its NC contact.

Switch control/maintained contact (Fig. 8.22c)

Contactors K1, K2, K3 and K4 are energized, de-energized or changed over by the actuation of maintained-contact switch S1, as with pushbutton control.

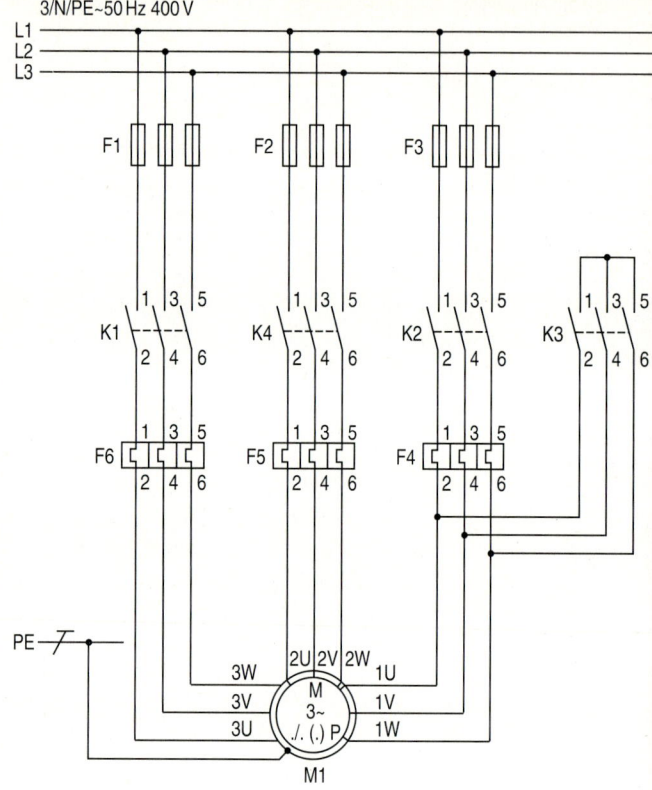

a) Main circuit

Figure 8.22
Circuit diagrams for a pole-changing three-phase induction motor with two separate windings, three speeds and one direction of rotation (one winding in a Dahlander connection and one separate winding for low speed)

Switching of pole-changing three-phase induction motors

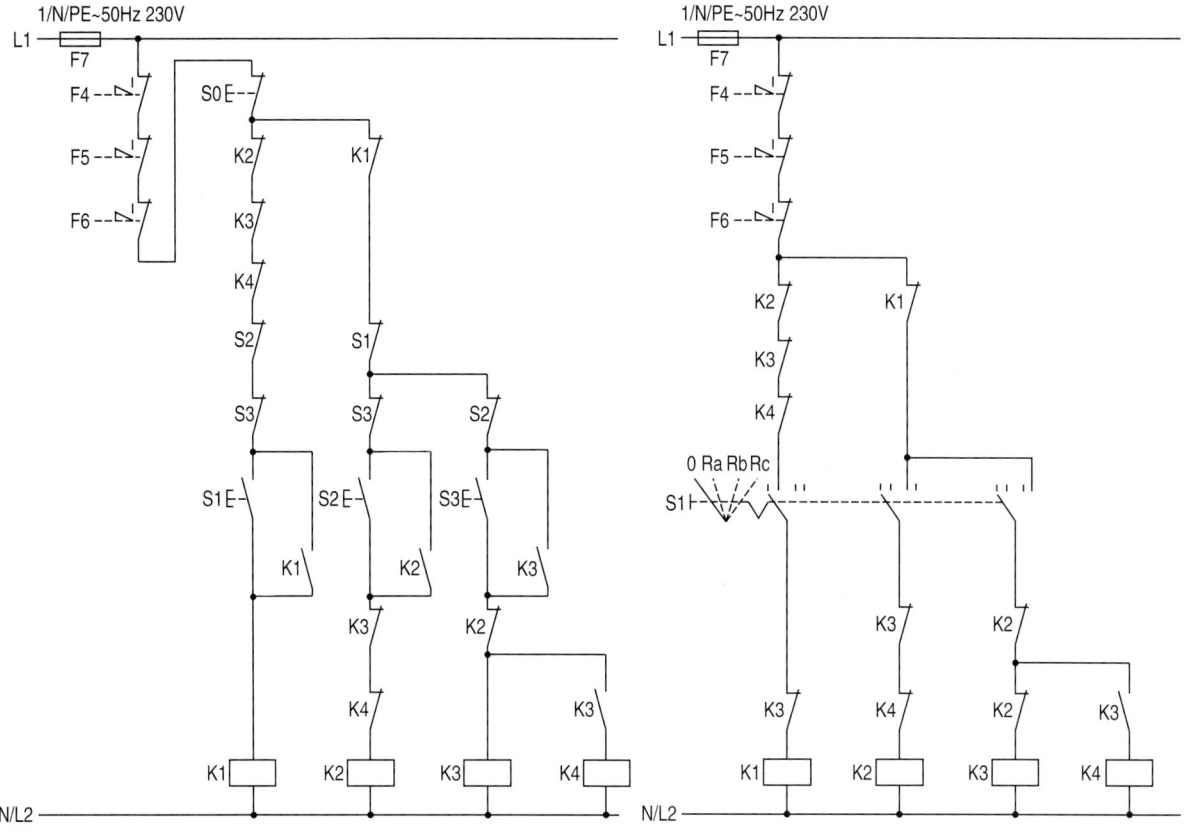

b) Auxiliary circuit for momentary-contact control

Required auxiliary contacts:

Contactor K1 with 1NO+1NC,
Contactor K2 with 1NO+2NC,
Contactor K3 with 2NO+2NC,
Contactor K4 with 2NC,
Momentary-contact switch S0 with 1NC,
Momentary-contact switch S1 with 1NO+1NC,
Momentary-contact switches S2, S3 each with 1NO+2NC,
Overload relay F4, F 5 and F6 each with 1NC

c) Auxiliary circuit for maintained-contact control

Required auxiliary contacts:

Contactor K1 with 1NC,
Contactors K2, K4 each with 2NC,
Contactor K3 with 1NO+2NC,
Maintained-contact switch S1 with 4 switch positions,
Overload relays F3, F4 and F5 each with 1NC

8.2.5.6 Pole-changing three-phase induction motor with three speeds, one direction of rotation, one winding in a Dahlander connection, one separate winding for the intermediate speed

Figure 8.23 illustrates the circuit diagrams for a pole-changing three-phase induction motor with two separate windings, three speeds and one direction of rotation (one winding in a Dahlander connection and one separate winding for the intermediate speed).

Pushbutton control/momentary contact (Fig. 8.23b)

Switching on: Contactor K1 is energized by the actuation of momentary-contact switch S1 and the closing of its NO contact. The sealing-in contact K1 closes. The motor starts up at low speed.

Change-over: Contactor K1 is de-energized by the actuation of momentary-contact switch S2 and the opening of its NC contact. Contactor K2 is energized by the closing of the NO contact of S2. This energizing command on K2 only becomes effective once the NC contact of K1 has closed. The sealing-in contact K2 closes and the motor runs at the intermediate speed.

Contactor K3 is energized by the actuation of momentary-contact switch S3 and the closing of its NO contact. The energizing command on K3 only becomes effective once the NC contact of K2 has closed. The sealing-in contact and the NO contact of K3 close. Line contactor K4 is energized. The motor runs at high speed.

Switching off: All the contactors in circuit are de-energized by the actuation of momentary-contact switch S0 and the opening of its NC contact.

Switch control/maintained contact (Fig. 8.23c)

Contactors K1, K2, K3 and K4 are energized, de-energized or changed over by the actuation of maintained-contact switch S1, as with pushbutton control.

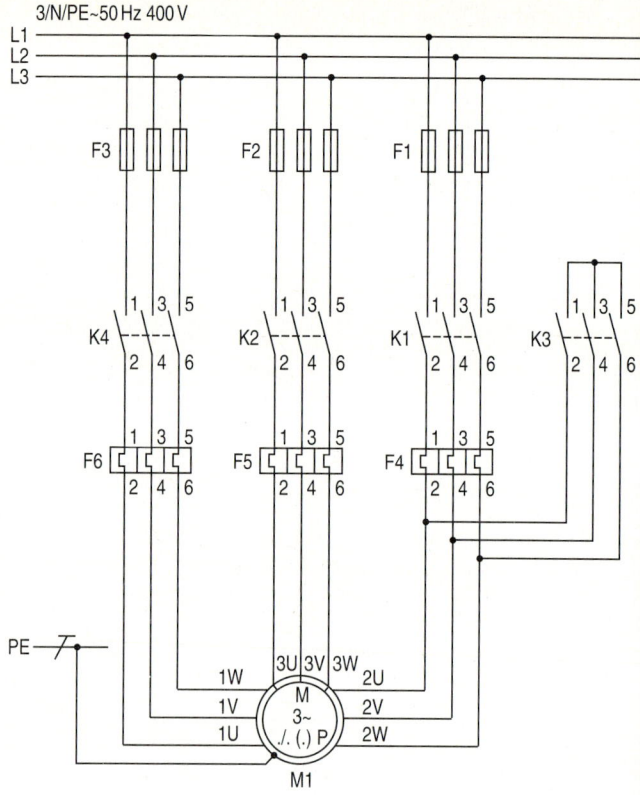

a) Main circuit

Figure 8.23
Circuit diagrams for a pole-changing three-phase induction motor with two separate windings, three speeds and one direction of rotations (one winding in a Dahlander connection and one separate winding for the intermediate speed)

Switching of pole-changing three-phase induction motors

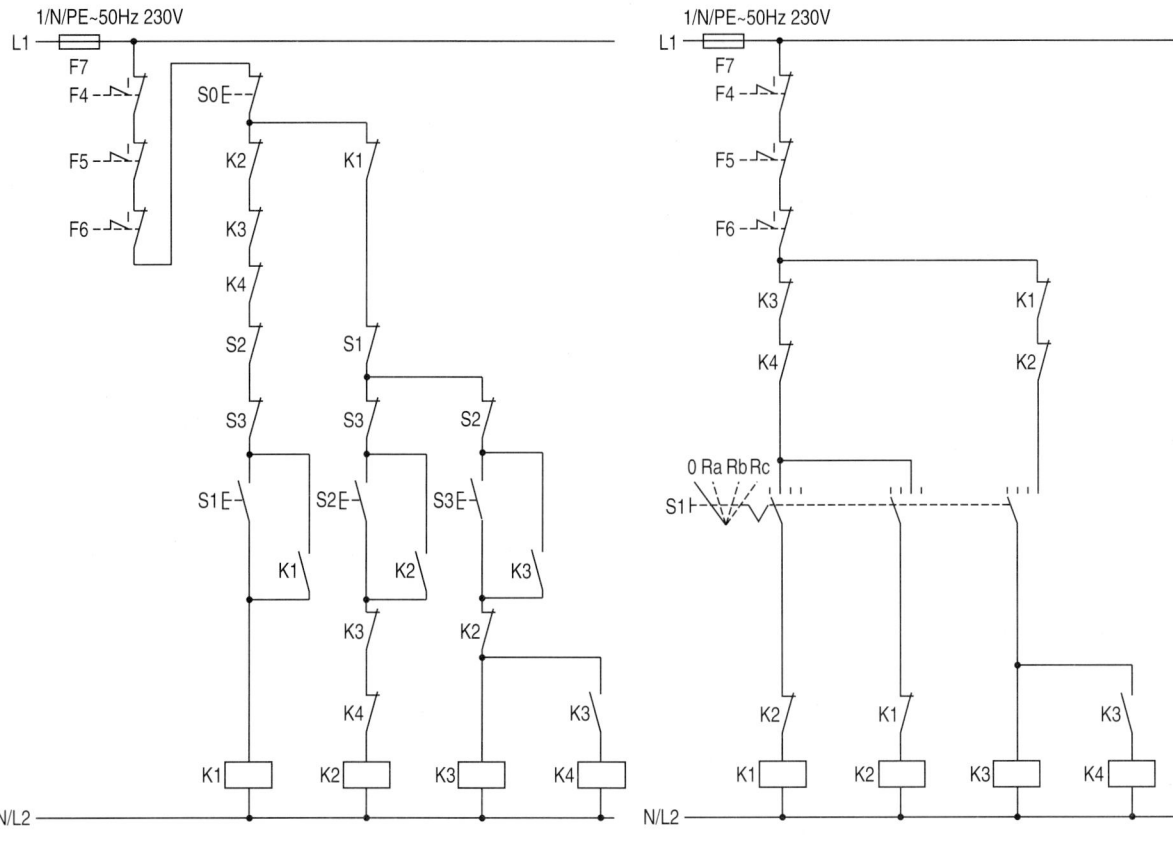

b) Auxiliary circuit for momentary-contact control

Required auxiliary contacts:

Contactor K1 with 1NO+1NC,
Contactor K2 with 1NO+2NC,
Contactor K3 with 2NO+2NC,
Contactor K4 with 2NC,
Momentary-contact switch S0 with 1NC,
Momentary-contact switch S1 with 1NO+1NC,
Momentary-contact switches S2, S3 each with 1NO+2NC,
Overload relays F4, F5 and F6 each with 1NC

c) Auxiliary circuit for maintained-contact control

Required auxiliary contacts:

Contactors K1, K2 each with 2NC,
Contactor K3 with 1NO+1NC,
Contactor K4 with 1NC,
Maintained-contact switch S1 with 4 switch positions,
Overload relays F4, F5 and F6 each with 1NC

8.2.5.7 Pole-changing three-phase induction motor with <u>three</u> speeds, <u>one</u> direction of rotation, one winding in a Dahlander connection, one separate winding for <u>high</u> speed

Figure 8.24 illustrates the circuit diagrams for a pole-changing three-phase induction motor with two separate windings, three speeds and one direction of rotation (one winding in a Dahlander connection and one separate winding for high speed).

Pushbutton control/momentary contact (Fig. 8.24b)

<u>Switching on</u>: Contactor K1 is energized by the actuation of momentary-contact switch S1 and the closing of its NO contact. The sealing-in contact K1 closes. The motor starts up at low speed.

<u>Change-over</u>: Contactor K1 is de-energized by the actuation of momentary-contact switch S3 and the opening of its NC contact. Contactor K2 is energized by the closing of the NO contact of S3. This energizing command on K2 only becomes effective once the NC contact of K1 has closed. The sealing-in contact and the NO contact of K2 close. The line contactor K4 is energized. The motor runs at the intermediate speed.

Contactor K3 is energized by the actuation of momentary-contact switch S4 and the closing of its NO contact. For the interlocking, the same applies as for contactor K2. The motor runs at high speed.

<u>Switching off</u>: All the contactors in circuit are de-energized by the actuation of momentary-contact switch S0 and the opening of its NC contact.

Switch control/maintained contact (Fig. 8.24c)

Contactors K1, K2, K3 and K4 are energized, de-energized or changed over by the actuation of maintained-contact switch S1, as with pushbutton control.

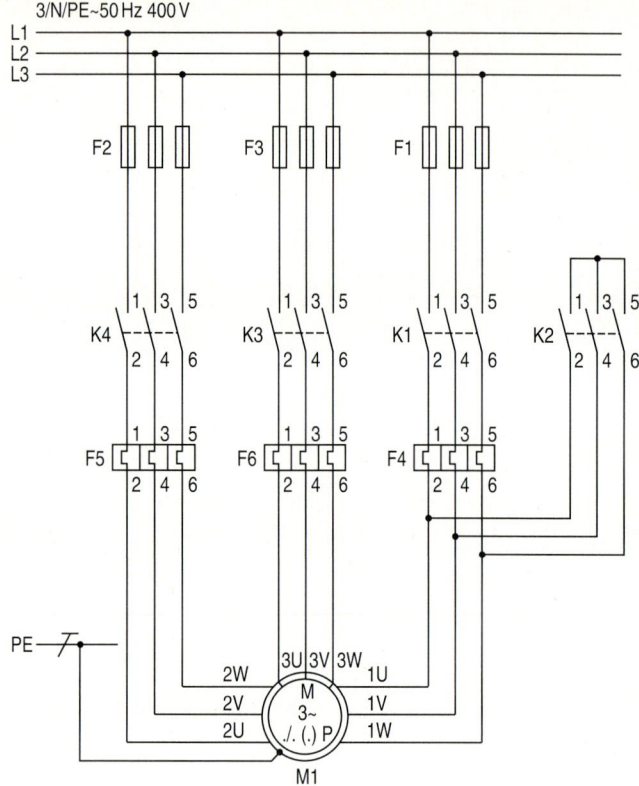

a) Main circuit

Figure 8.24
Circuit diagrams for a pole-changing three-phase induction motor with two separate windings, three speeds and one direction of rotation (one winding in a Dahlander connection, one separate winding for high speed)

Switching of pole-changing three-phase induction motors

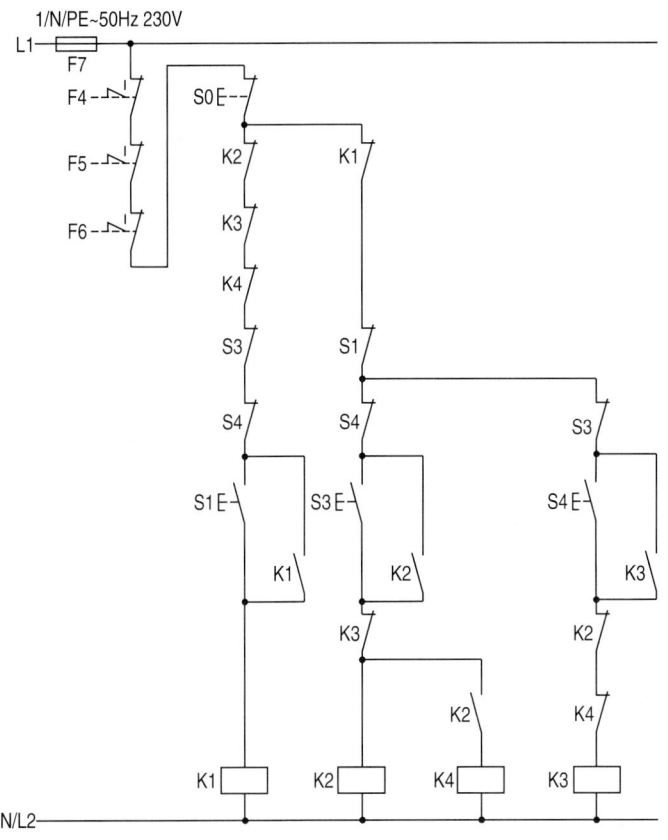
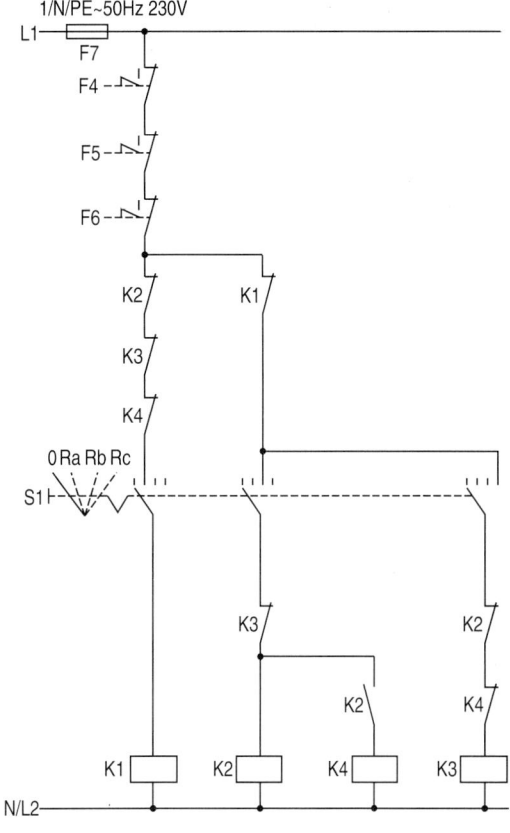

b) Auxiliary circuit for momentary-contact control

Required auxiliary contacts:

Contactor K1 with 1NO+1NC,
Contactor K2 with 2NO+2NC,
Contactor K3 with 1NO+2NC,
Contactor K4 with 2NC,
Momentary-contact switch S0 with 1NC,
Momentary-contact switch S1 with 1NO+1NC,
Momentary-contact switches S3, S4 each with
 1NO+2NC,
Overload relays F4, F5 and F6 each with 1NC

c) Auxiliary circuit for maintained-contact control

Required auxiliary contacts:

Contactor K1 with 1NC,
Contactor K2 with 1NO+2NC,
Contactors K3, K4 each with 2NC,
Maintained-contact switch S1 with 4 switch positions,
Overload relays F4, F5 and F6 each with 1NC

8.2.5.8 Pole-changing three-phase induction motor with <u>three</u> speeds, <u>two</u> directions of rotation, one winding in a Dahlander connection, one separate winding for <u>low</u> speed

Figure 8.25 illustrates the circuit diagrams for a pole-changing three-phase induction motor with two separate windings, three speeds and two directions of rotation (one winding in a Dahlander connection and one separate winding for low speed).

Pushbutton control/momentary contact (Fig. 8.25b)

Switching on, e.g., at low speed, clockwise rotation: Contactors K1 and K3 are energized by the actuation of momentary-contact switch S13 and the closing of its NO contacts. The energizing command on K3 only becomes effective once the NO contact of K1 has closed. The sealing-in contacts K1 and K3 close. The motor starts up at low speed in clockwise direction.

Change-over, e.g., to intermediate speed, anti-clockwise rotation: Contactor relay K24 is energized by the actuation of momentary-contact switch S24 and the closing of its NO contact. Contactors K1 and K3 are de-energized via the NC contact K24, and contactors K2 and K4 are energized via the NO contact K24. The energizing command on K2 only becomes effective once the NC contact of K1 has closed and the energizing command on K4 only becomes effective once the NO contact of K2 has closed. The motor runs at the intermediate speed in anti-clockwise direction.

Further switching-on and changing-over operations are described in Table 8.9.

Switching off: All the contactors in circuit are de-energized by the actuation of momentary-contact switch S0 and the opening of its NC contact. Thereby, the motor is disconnected from the supply and switched off.

Switch control/maintained contact (Fig. 8.25c)

Contactors K1, K2, K3 and K4 are energized, de-energized or changed over for clockwise and anti-clockwise rotation by the actuation of maintained-contact switch S1, as with pushbutton control.

Table 8.9
Pushbutton control and switching commands for the contactors in the control circuit of a pole-changing three-phase induction motor with three speeds and two directions of rotation according to Figure 8.25

Motor start-up	Actuation of momentary-contact switch [1]	Energizing command on contactor	De-energizing command on contactor
Clockwise rotation			
Low speed	S13	K1, K3	K2, K4, K5, K6
Intermediate speed	S14 (K14)	K1, K4	K2, K3, K5, K6
High speed	S16 (K16)	K1, K5, K6	K2, K3, K4
Counterclockwise rotation			
Low speed	S23	K2, K3	K1, K4, K5, K6
Intermediate speed	S24 (K24)	K2, K4	K1, K3, K5, K6
High speed	S26 (K26)	K2, K5, K6	K1, K3, K4

[1] Contactor relays for contact multiplication in brackets

Switching of pole-changing three-phase induction motors

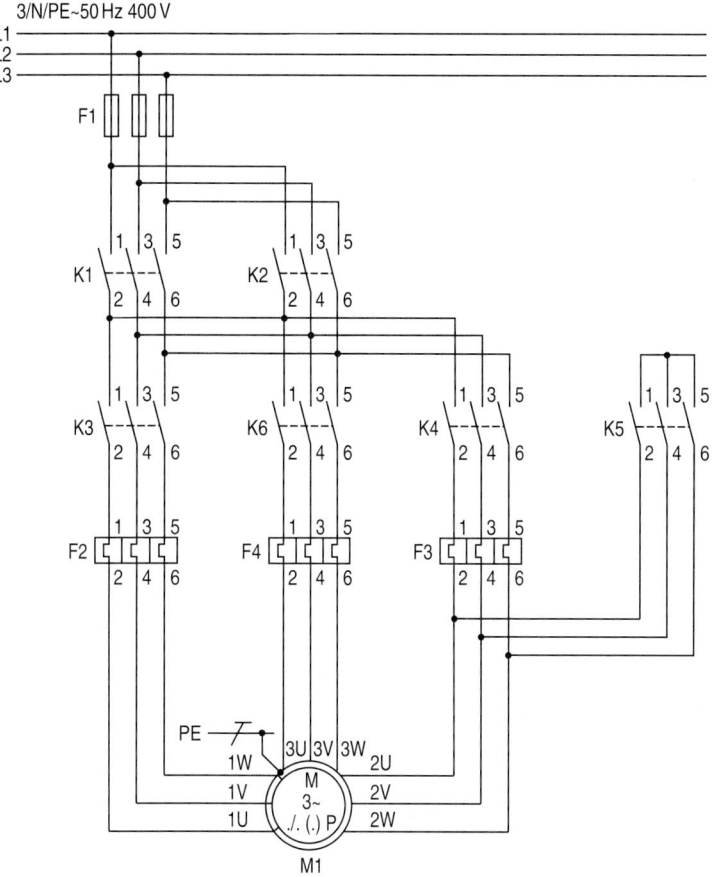

a) Main circuit

Figure 8.25
Circuit diagrams for a pole-changing three-phase induction motor with two separate windings, three speeds and two directions of rotation (one winding in a Dahlander connection and one separate winding for low speed)

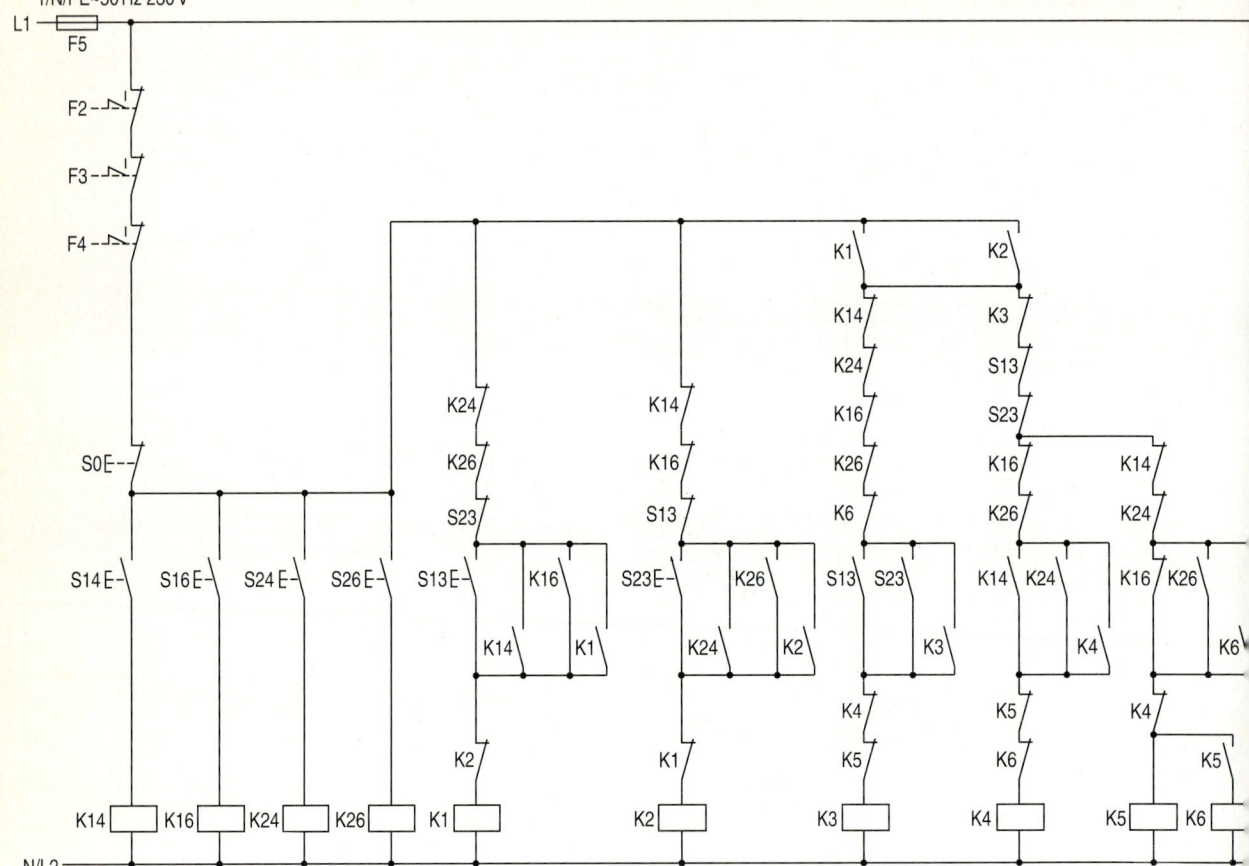

b) Auxiliary circuit for momentary-contact control

Required auxiliary contacts:

Contactors K1, K2 each with 2NO+1NC,
Contactor K3 with 1NO+1NC,
Contactors K4, K5, K6 each with 1NO+2NC,
Contactor relays K14, K16, K24, K26 each with 2NO+3NC,
Momentary-contact switch S0 with 1NC,
Momentary-contact switches S13, S23 each with 2NO+2NC,
Momentary-contact switches S14, S16, S24, S26 each with 1NO,
Overload relays F2, F3 and F4 each with 1NC

Switching of pole-changing three-phase induction motors

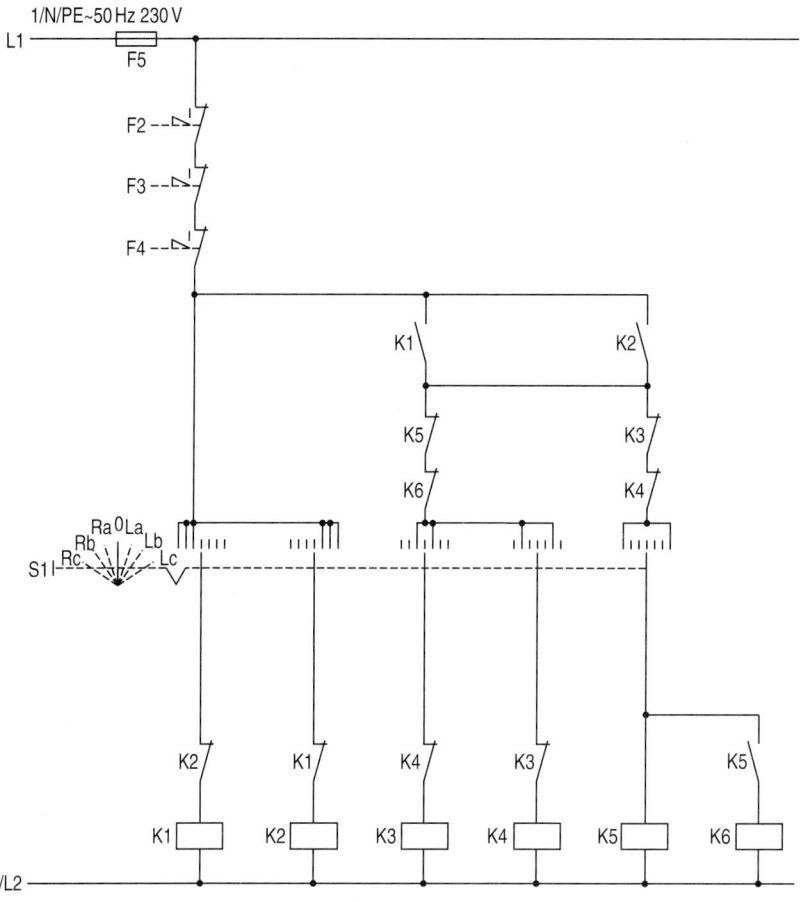

c) Auxiliary circuit for maintained-contact control

Required auxiliary contacts:

Contactors K1, K2, K5 each with 1NO+1NC,
Contactors K3, K4 each with 2NC,
Contactor K6 with 1NC,
Maintained-contact switch S1 with 7 switch positions,
Overload relays F2, F3 and F4 each with 1NC

8.2.5.9 Pole-changing three-phase induction motor with four speeds, one direction of rotation and two separate windings

Figure 8.26 illustrates the circuit diagrams for a pole-changing three-phase induction motor with two separate windings, four speeds and one direction of rotation (both windings in Dahlander connection).

Pushbutton control/momentary contact (Fig. 8.26b)

Switching on: Contactor K1 is energized by the actuation of momentary-contact switch S1 and the closing of its NO contact. The sealing-in contact K1 closes, and the motor starts up at speed 1.

Change-over: Contactor K1 is de-energized by the actuation of momentary-contact switch S2 and the opening of its NC contact. Contactor K2 is energized by the closing of the NO contact S2. This energizing command on K2 only becomes effective once the NC contact of K1 has closed. The sealing-in contact K2 closes, and the motor runs at speed 2.

Contactors K1 or K2 are de-energized by the actuation of momentary-contact switch S4 and the opening of its NC contact. Contactor K3 is energized by the closing of the NO contact S4. This energizing command on K3 only becomes effective once the NC contacts of the contactors K1, K2, K5, and K6 have closed. The sealing-in contact K3 and NO contact K3 close, the line contactor K4 is energized and the motor runs at speed 3.

Contactors K5 and K6 are energized by the actuation of momentary-contact switch S6 and the closing of its NO contact. The interlocking is the same as for energizing contactors K3 and K4. The motor runs at speed 4.

Switching off: All the contactors in circuit are de-energized by the actuation of momentary-contact switch S0. Thus the motor is disconnected from the supply.

Switch control/maintained contact (Fig. 8.26c)

Contactors K1 to K6 are energized, de-energized and changed over by the actuation of maintained-contact switch S1, as with pushbutton control.

Switching of pole-changing three-phase induction motors

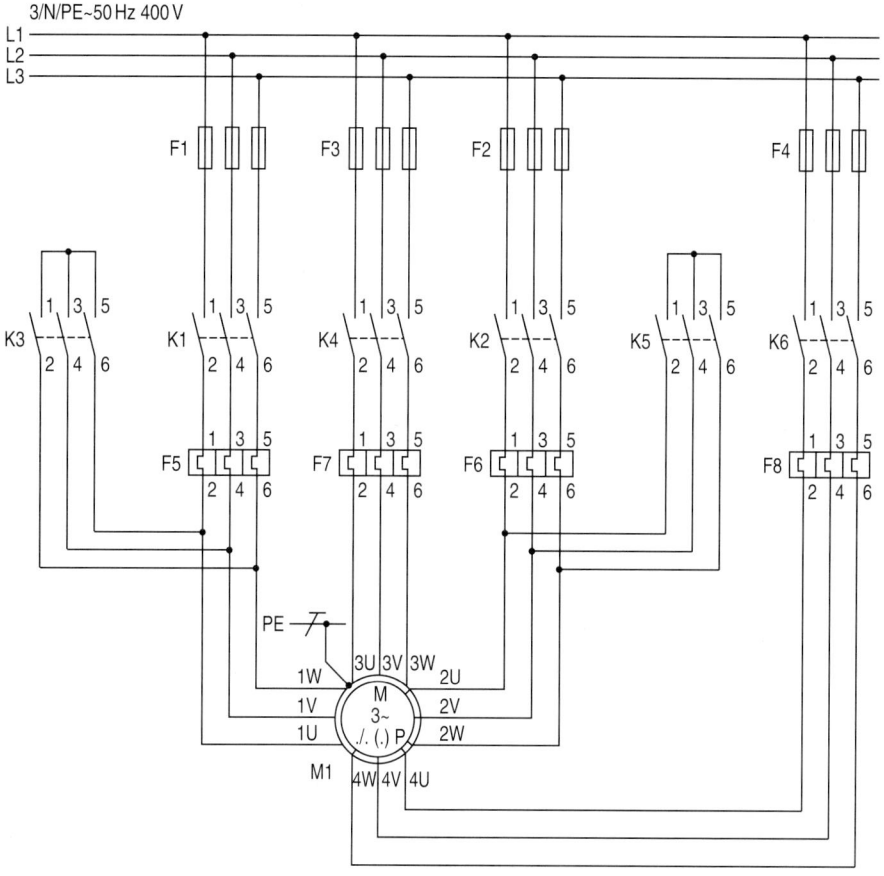

a) Main circuit

Figure 8.26
Circuit diagrams for a pole-changing three-phase induction motor with two separate windings, four speeds and one direction of rotation

▷

8 Fundamental circuit diagrams

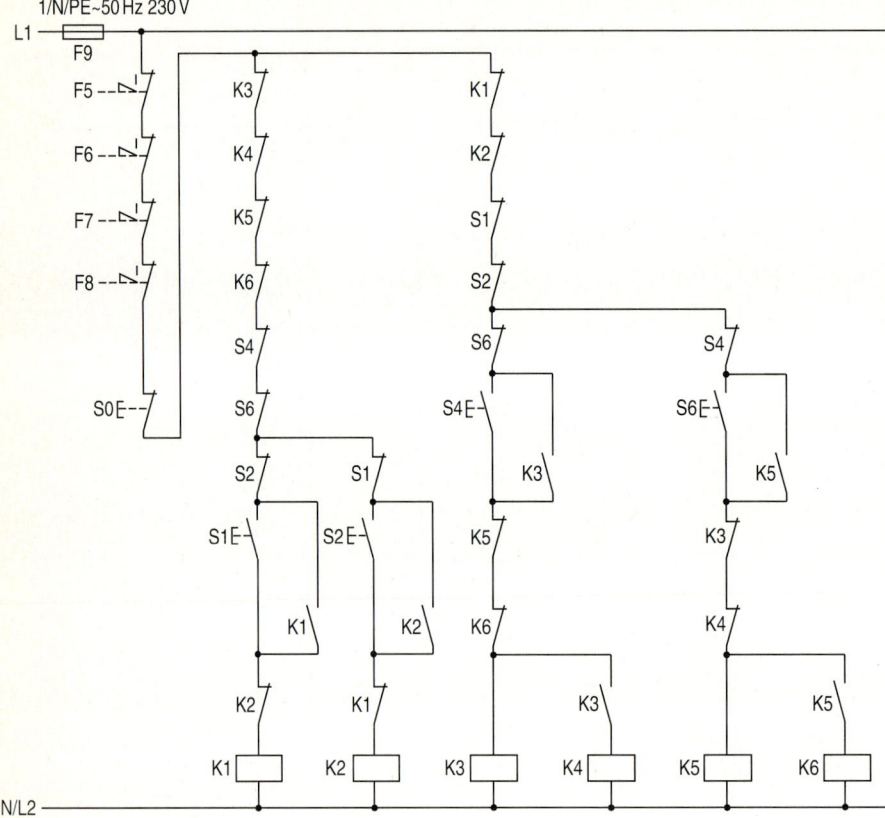

b) Auxiliary circuit for momentary-contact control

Required auxiliary contacts:

Contactors K1, K2 each with 1NO+1NC,
Contactors K3, K5 each with 2NO+2NC,
Contactors K4, K6 each with 2NC,
Momentary-contact switch S0 with 1NC,
Momentary-contact switches S1, S2, S4, S6 each with 1NO+2NC,
Overload relays F5, F6, F7 and F8 each with 1NC

Switching of pole-changing three-phase induction motors

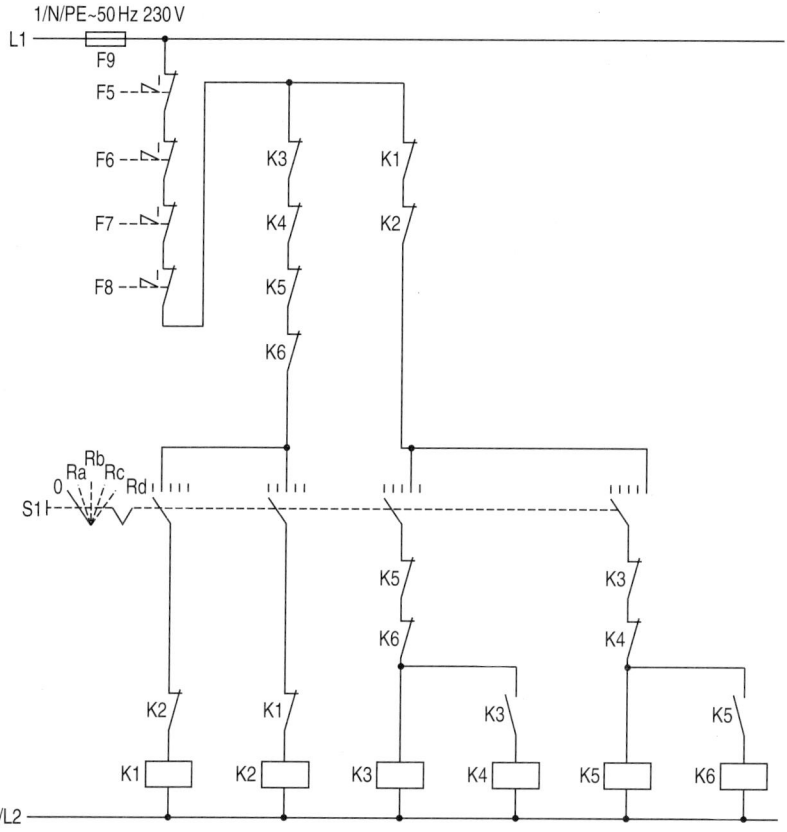

c) Auxiliary circuit for maintained-contact control

Required auxiliary contacts:

Contactors K1, K2, K4, K6 each with 2NC,
Contactors K3, K5 each with 1NO+2NC,
Maintained-contact switch S1 with 5 switch positions,
Overload relays F5, F6, F7 and F8 each with 1NC

8 Fundamental circuit diagrams

8.3 Starting of three-phase induction motors

In this section, circuit diagrams for the starting of three-phase induction motors using contactors are described. Circuit diagrams for soft starting and stopping of three-phase induction motors by means of SIKOSTART 3RW2 solid-state starters, are contained in the SIKOSTART handbook (Order No. E20001-P285-A484-7600). Also refer to Section 3.7.15 from page 230.

8.3.1 Star-delta starting of three-phase induction motors with star contactor, delta contactor and line contactor

Figure 8.27 illustrates the circuit diagrams for the star-delta starting of three-phase induction motors of various output power ratings. In order to provide motor protection in accordance with IEC 947-4-1 and DIN VDE 0660 part 104, the main circuits are to be configured for the respective output power ratings as illustrated in Fig. 8.27a to c.

To prevent short-circuits from occurring during the switch-over from star to delta during star-delta starting of three-phase induction motors, a star-delta time relay with change-over delay should be used. This ensures that the switching arc in the star contactor is completely quenched upon opening before the delta contactor closes. (for the auxiliary circuit, see Section 8.10.1, page 488, "Auxiliary circuits incorporating time relays".)

Since the overload relay is connected within the delta circuit, its setting range (rated current) must correspond to the load current in the motor windings themselves, i.e. $1/\sqrt{3}$ times the line current drawn by the delta connected motor.

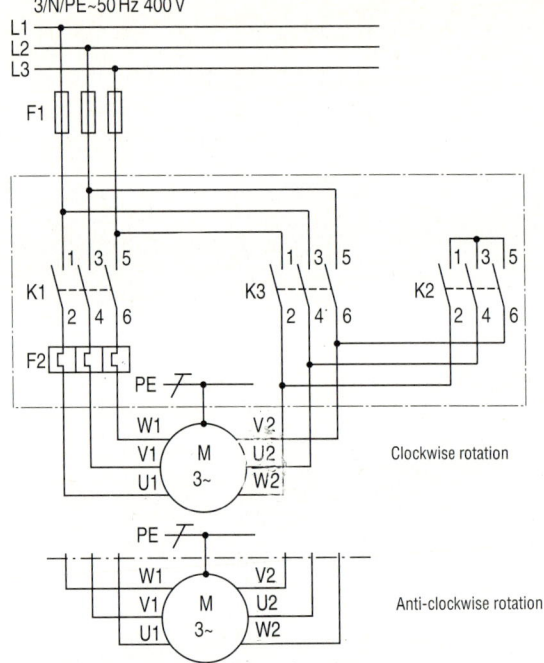

a) Main circuit for sizes 0 to 6, 3TE40 to 3TE50.
   Common feeder for line and delta contactors; *common* short-circuit protection (F1)

Figure 8.27
Circuit diagrams for star-delta starting of three-phase induction motors of various output power ratings using star, delta and line contactors

Table 8.10 Selection of the contactors for star-to-delta change-over times of up to 60 s

Contactor Size	Rated operational current I_e at 400 V A	Three-phase induction motor ratings at 50 Hz and					Contactors		
		230 V kW	400 V kW	500 V kW	690 V kW	1000 V kW	Line	Star	Delta
0	20	4,7	10	11,5	13,6	–	3TF40	3TF40	3TF40
1	30	7,5	15	18,5	19	–	3TF42	3TF42	3TF42
2	60	19	30	45	42	–	3TF44	3TF44	3TF44
3	66	19	33	44	55	–	3TF46	3TF46	3TF46
	90	26	45	60	75	–	3TF47	3TF47	3TF47
4	130	37	65	85	112	75	3TF48	3TF48	3TF48
6	160	45	80	105	135	95	3TF50	3TF50	3TF50
6	175	54	93	122	161	–	3TF51	3TF51	3TF51
8	250	75	135	175	230	156	3TF52	3TF52	3TF52
10	370	110	190	250	320	225	3TF54	3TF54	3TF54
12	500	160	280	350	480	480	3TF56	3TF56	3TF56
14	800	230	460	590	730	1000	3TF68	3TF68	3TF68

Star-delta starting of three-phase induction motors

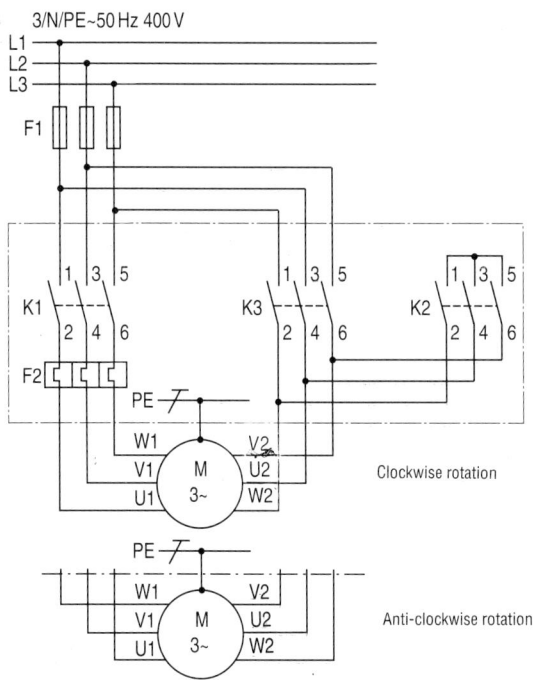

Clockwise rotation

Anti-clockwise rotation

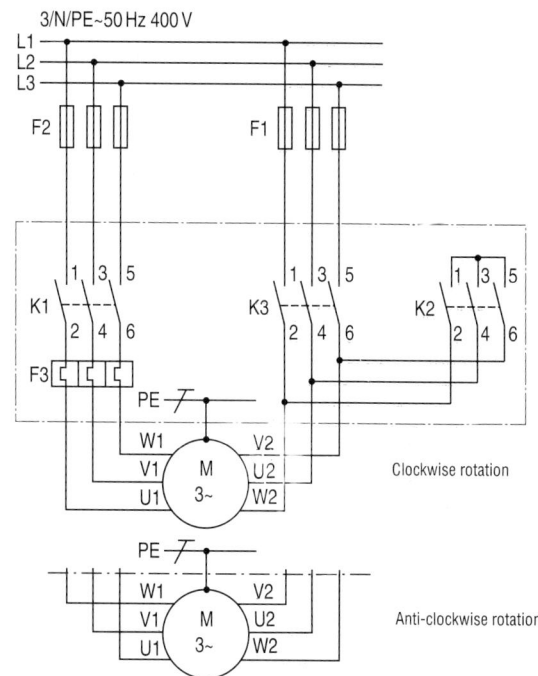

Clockwise rotation

Anti-clockwise rotation

b) Main circuit for sizes 8 to 14, up to 630 A, 3TE52 to 3TE68. *Separate feeders* for line and delta contactors; *common* short-circuit protection (F1, F2)

c) Main circuit for sizes 8 to 14, up to 630 A, 3TE52 to 3TE68. *Separate feeders* for line and delta contactors; *separate* short-circuit protection (F1, F2)

d) Auxiliary circuit for momentary-contact control of 3TE starters rated at 45 A to 630 A, with star-delta time relay 7PU62

The overload relay is therefor set to 0.58 times the operational full-load current of the motor, and provides overload protection in both the star and delta modes.

Rating of the contactors for run-up times of up to 10 s

The line contactor K1 must be rated for 58% of the motor output power, the star contactor K2 for 33% of the motor output power, and the delta contactor K3 for 58% of the motor output power.

Rating of the contactors for start-up times of up to 60 s

The power rating data of the contactors in a star-delta circuit for change-over times of up to 60 s are given in Table 8.10 and in the Siemens l.v. switchgear catalogue.

Pushbutton control/momentary contact (Fig. 8.27 d)

Switching on: Time relay K4 and star contactor K2 are energized by the actuation of momentary-contact switch S1 and the closing of its NO contact.

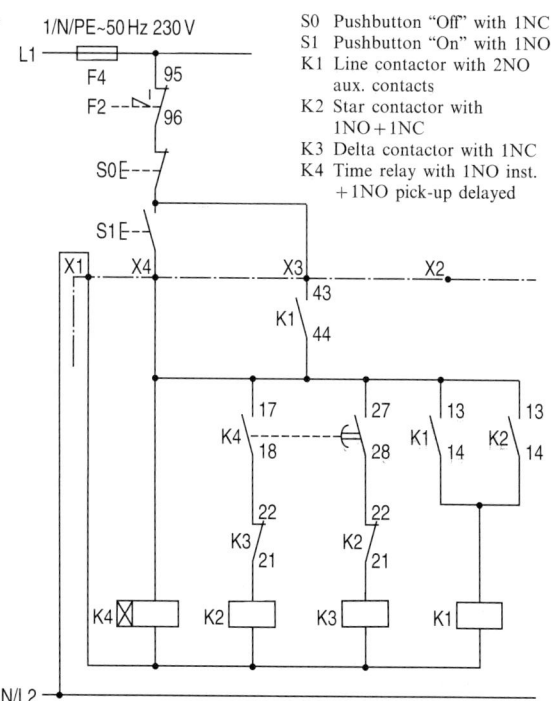

S0 Pushbutton "Off" with 1NC
S1 Pushbutton "On" with 1NO
K1 Line contactor with 2NO aux. contacts
K2 Star contactor with 1NO+1NC
K3 Delta contactor with 1NC
K4 Time relay with 1NO inst. +1NO pick-up delayed

In addition, contactor K2 is connected via the instantaneous NO contact 17/18 of the star-delta time relay K4. The motor is started in the star connection.

Change-over: After the set delay time of K4 has elapsed, the instantaneous NO contact K4 opens and star contactor K2 is de-energized. The delta contactor K3 is energized via the NC contact of K2 when the pick-up delayed NO contact K4 of the star-delta time relay closes (after the change-over delay of 50 ms). The motor continues running in the delta connection.

Switching off: The line contactor K1 is de-energized by the actuation of momentary-contact switch S0; the NO contact of K1 opens, the delta contactor K3 is de-energized and the motor is switched off.

The auxiliary contacts which have not been mentioned above are used for interlocking purposes between the star and delta contactors.

Switch control/maintained-contact control.

See Section 8.10.1, page 488.

8.3.2 Closed transition star-delta starting of three-phase induction motors

In conventional star-delta starting, a change-over delay of approximately 50 ms is usually observed when changing over from star to delta. For shorter change-over times, phase-to-phase short-circuits may occur. During the change-over delay the motor is, however, completely disconnected from the supply and, depending on the counter torque, its speed decreases to some extent (usually only slightly). When the delta step is now switched on, it may happen that the supply phase angle and the magnetic field in the motor are in opposition to each other, which will lead to transient phenomena causing high change-over current peaks and even mechanical shocks to the drive as a whole.

In closed-transition star-delta starting, the motor is delta-connected to the network via a transition contactor and resistors after the run-up in star. It is switched over to the normal delta step without disconnection from the supply after approximately 50 ms. Thus, no switching interval during which the motor is disconnected from the supply occurs, and the high transient change-over current peaks can be avoided.

Figure 8.28 illustrates the circuit diagrams for closed-transition star-delta starting.

Switch control/maintained contact (Fig. 8.28b)

Switching on: The star-delta time relay K7 is energized by the actuation of maintained-contact switch S1. As a consequence, contactor relay K5, star contactor K2 and line contactor K1 are energized. The motor starts up in the star connection.

Change-over: After the preset run-up time on K7 has elapsed, transition contactor K4 delta-connects the motor windings including the resistors R1, and causes the star contactor K2 to drop out. Subsequently, contactor relay K6 energizes the delta contactor K3 which connects the windings in normal direct delta configuration (bridging-out the resistors R1). The transition contactor K4 is de-energized by the opening of NC contact K3.

Switching off: By the opening of the maintained-contact switch S1, all the contactors in circuit are de-energized and the motor is disconnected from the network.

Rating of the contactors

The line contactor K1, delta contactor K3 and star contactor K2 must be rated for 58% of the rated operational current, and the transition contactor K4 must be rated for 26% of the rated operational current $\left(\triangleq \dfrac{1.5 \cdot I_e \cdot \sqrt{3}}{10}\right)$.

Time relay K7 is a pick-up-delayed (on-delayed) star-delta time relay, e.g. 7PU62.

The star contactor K2 has the task of breaking the star current as well as the current flowing via the transition resistors. A switching capacity P_{R1} which is higher than would normally be the case is required, since the current flowing in the transition resistors is approximately 1.5 times the rated operational current. Closed-transition star-delta starters are normally used at higher switching frequencies. For this reason too, the seemingly over-dimensioned star contactor of the closed-transition star-delta starter meets the requirements more readily than would be the case if the star contactor were rated as for the conventional star-delta connection.

Calculation of the transition resistors R_1:

$$R_1 = \dfrac{U_e}{1.5 \cdot I_e \cdot \sqrt{3}} \text{ (in } \Omega\text{)}$$

$$P_{R1} = \dfrac{U_e^2}{1200 \cdot R_1} \text{ (in W)} \quad \text{for max. 12 make-break operations/hour}$$

$$P_{R1} = \dfrac{U_e^2}{500 \cdot R_1} \text{ (in W)} \quad \text{for max. 30 make-break operations/hour}$$

Closed transition star-delta starting of three-phase induction motors

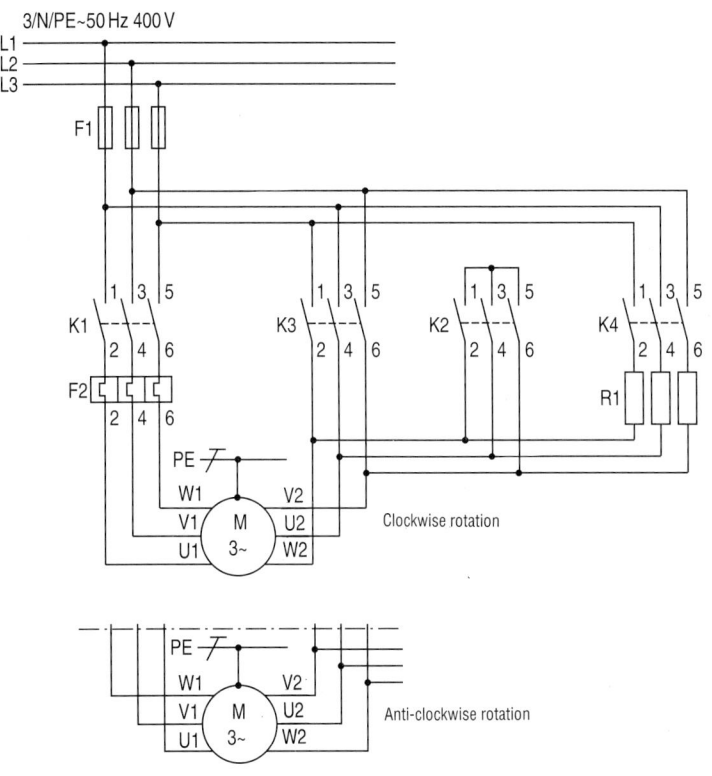

a) Main circuit

K1 Line contactor
K2 Star contactor
K3 Delta contactor
K4 Transition contactor
R1 Transition resistors

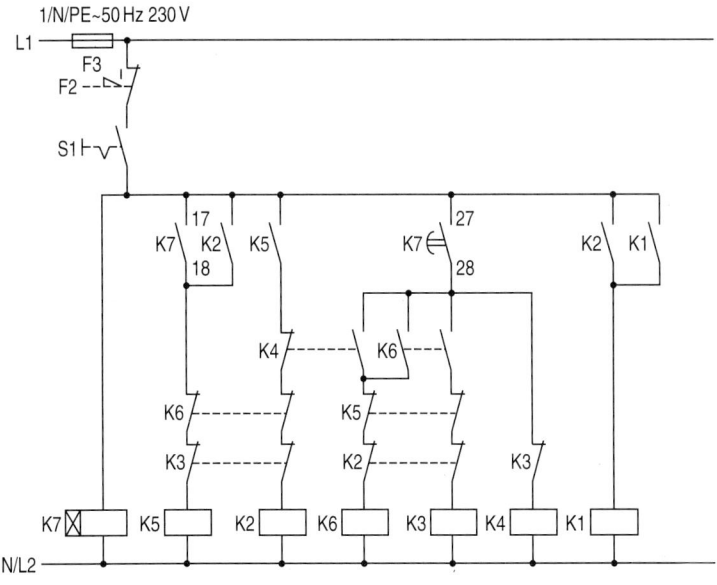

b) Auxiliary circuit for maintained-contact control with star-delta time relay 7PU62

Required auxiliary contacts:

Maintained-contact switch S1 with 1NO,
Contactor K1 with 1NO,
Contactor K2 with 2NO+2NC,
Contactor K3 with 3NC,
Contactor K4 with 1NO+1NC,
Contactor relay K5 with 1NO+2NC,
Contactor relay K6 with 2NO+2NC,
Time relay K7 with 1NO inst.+1NO pick-up delayed,
Overload relay F1 with 1NC

Figure 8.28
Circuit diagrams for closed-transition star-delta starting

8 Fundamental circuit diagrams

8.3.3 Four stage star-delta starting [1]

The four-stage star-delta connection is used for drives with such a high counter-torque that the normal star-delta connection would not be able to accelerate the drive to a sufficiently high change-over speed.

Figure 8.29 illustrates the torque and current characteristics during four-stage star-delta starting. Figure 8.30 illustrates the sub-division and arrangement of the windings in the three-phase induction motor for four-stage star-delta connection. Figure 8.31 illustrates the circuit diagrams of a four-stage star-delta starter.

Pushbutton control/momentary contact operation

Starting stage I: Contactor coil K3 is energized by the actuation of the momentary-contact switch S1 and the closing of its NO contact. Contactor K3 is sealed in by means of its own NO contact and the NC contact K4, and in turn energizes contactor K5 and time relay K6 via a further NO contact. Thus, the motor is connected to the network in the first starting stage configuration (K3 & K5).

Change-over to stage II: After the delay time on the pick-up-delayed time relay K6 has elapsed, its change-over contact de-energizes contactor K5 and energizes K1. Thus, the motor is connected to the network in the second starting stage configuration (K3 & K1).

Change-over step III: The closing of K1 causes the time relay K7 to become energized, and its change-over contact energizes K9 after its set delay time has elapsed. Contactor K3 is de-energized and K2 is energized after the time-delay on K9 has elapsed. A NO contact K2 energizes K4 and time relay K8. The motor is switched over to the third starting stage configuration (K4 & K2).

Change-over step IV: After the set delay time of K8 has elapsed, its change-over contact de-energizes K2 and energizes K3. The drive is now accelerated to its rated speed in this delta stage (K4 & K3).

Switching off: By the actuation of momentary-contact switch S0 and the opening of its NC contact, all the contactors in circuit are de-energized and the motor is disconnected from the supply.

The contactors are selected in accordance with the respective currents and starting times (refer to the Siemens low-voltage switchgear catalogue).

Owing to the different run-up currents of the motor in the respective stages, thermistor motor protection is recommended (also refer to Section 3.4.3.5, p. 138).

[1] also known as four step star-delta starting

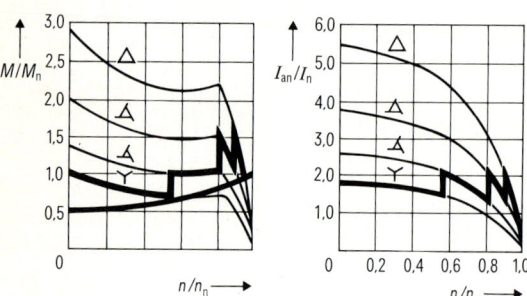

Figure 8.29
Torque and current characteristics during four-stage star-delta starting

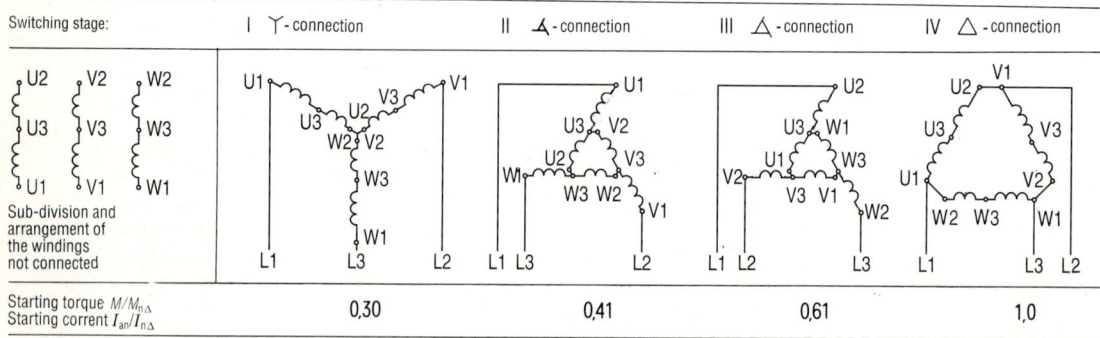

Figure 8.30
Illustration of the interconnection of the winding groups in a three-phase induction motor for special four-stage star-delta switching (for simplicity, the connections are shown as having been made within the motor itself). Change-over from the star connection (stage I) via intermediate stages II and III to the delta connection (stage IV)

Four stage star-delta starting of three-phase induction motors

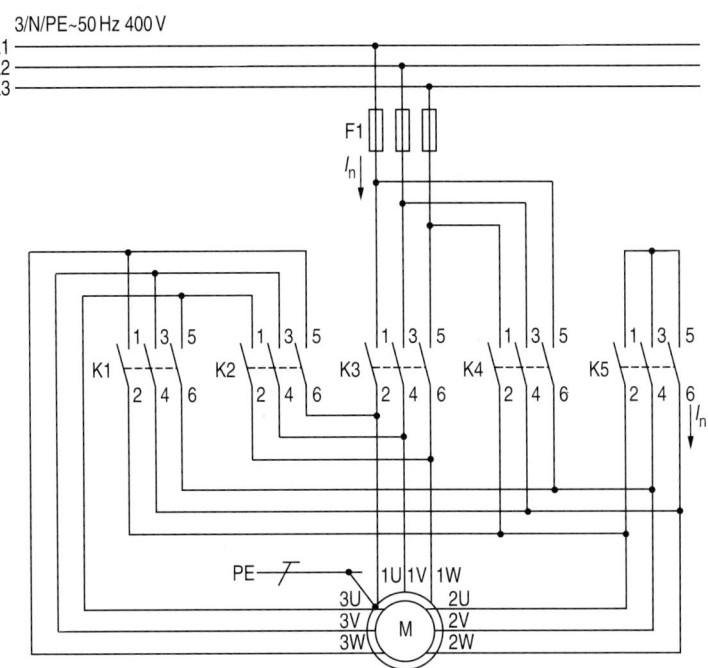

a) Main circuit

Stage 1: Contactors K3 and K5
Stage 2: Contactors K3 and K1
Stage 3: Contactors K4 and K2
Stage 4: Contactors K4 and K3

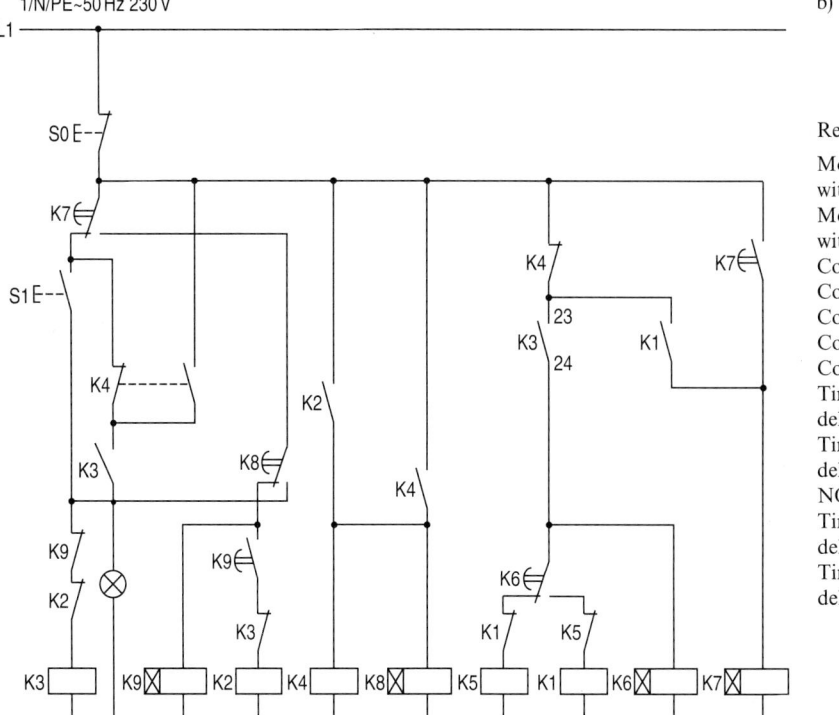

b) Auxiliary circuit for momentary-contact control and with time relays for setting the duration of each starting stage

Required auxiliary contacts:

Momentary-contact switch S0 with 1NC,
Momentary-contact switch S1 with 1NO,
Contactor K1 with 1NO+1NC,
Contactor K2 with 1NO+1NC,
Contactor K3 with 2NO+1NC,
Contactor K4 with 2NO+2NC,
Contactor K5 with 1NC,
Time relay K6 with 1 pick-up delayed CO,
Time relay K7 with 1 pick-up delayed CO+1 pick-up delayed NO,
Time relay K8 with 1 pick-up delayed CO,
Time relay K9 with 1 pick-up delayed NO+1 inst. NC

Figure 8.31 Circuit diagrams for the four-stage star-delta starting of a three-phase induction motor

8.3.4 Star-delta starting of three-phase induction motors in two directions of rotation

Figure 8.32 illustrates the circuit diagrams for the star-delta starting and reversing of a three-phase induction motor, using a star-delta time relay with change-over delay.

Pushbutton control/momentary contact (Fig. 8.32b)

Switching on (clockwise rotation of the motor): The star-delta time relay K5 is energized by the actuation of momentary-contact switch S1 and the closing of its NO contact. The star contactor K4 is energized via the instantaneous NO contact of the star-delta time relay K5. The line contactor K1 is energized by the closing of NO contact K4. Sealing-in contacts K1 and K4 close. The NC contacts S1 and K1 prevent line contactor K2 for anti-clockwise rotation from being energized in this mode of operation. The motor runs up in the star connection in clockwise direction of rotation.

Change-over to the delta step: After the set delay time t_1 on K5 has elapsed, the instantaneous NO contact of K5 opens and star contactor K4 is de-energized. The NC contact of star contactor K4 closes. After the change-over delay t_2 of approximately 50 ms has elapsed, the delayed NO contact of K5 closes. Thus, delta contactor K3 is energized, since the NO contact of line contactor K1 has already closed. The motor runs in the delta connection.

Switching on (anti-clockwise rotation of the motor): The star-delta time relay K5 is energized by the actuation of momentary-contact switch S2 and the closing of its NO contact. The star contactor K4 is energized by NO contact of the star-delta time relay K5. The line contactor K2 is energized by the closing of NO contact K4. Sealing-in contacts K2 and K4 close. The NC contacts S2 and K2 prevent line contactor K1 for clockwise rotation from being energized in this mode of operation. The motor runs up in the star connection in anti-clockwise direction of rotation.

Change-over to the delta step: as for clockwise rotation of the motor.

Switching off: Line contactor K1 (or K2) drops out by the actuation of the momentary-contact switch S0 and the opening of its NC contact. The delta contactor K3 and the motor are switched off by the opening of the NO contact K1 (or K2).

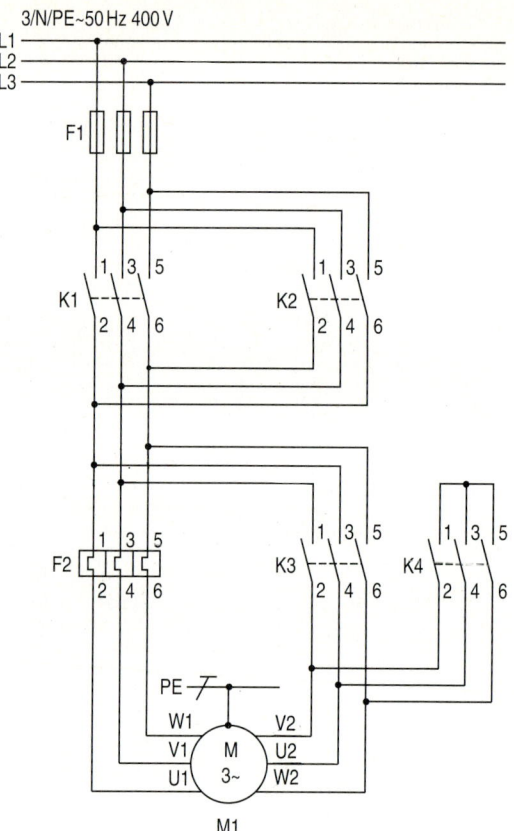

a) Main circuit

Figure 8.32
Circuit diagrams for the star-delta starting of a three-phase induction motor in two directions of rotation

Switch control/maintained contact (Fig. 8.32c)

According to its position, the maintained-contact switch S1 energizes and de-energizes the contactors for switching the motor on and off in clockwise or anti-clockwise direction, in the same way as with momentary-contact control.

Rating of the contactors

Line contactors K1 and K2 must be rated for 100% motor output power. The star contactor K4 must be rated for 33%, and the delta contactor K3 for 58% of the motor output power.

Star-delta starting of three-phase induction motors in two directions

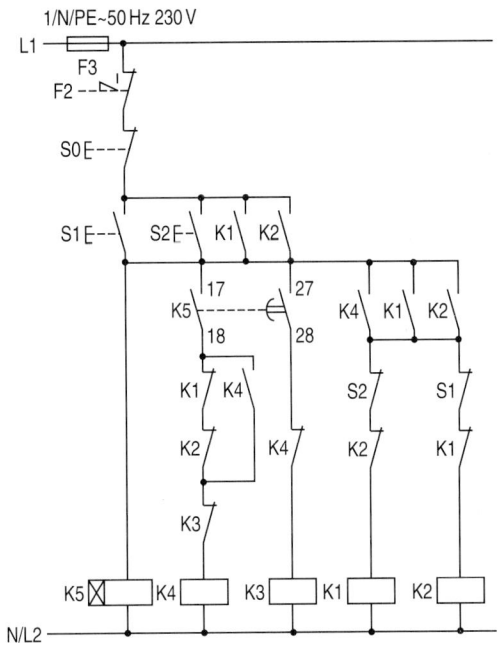

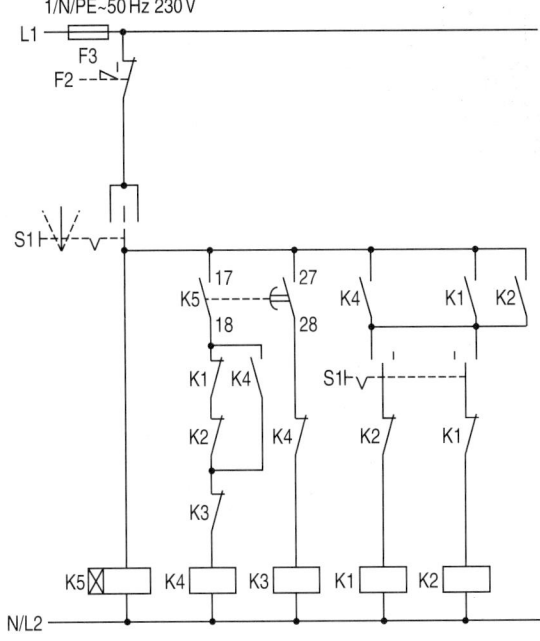

b) Auxiliary circuit for momentary-contact control

Required auxiliary contacts:

Contactor K1 with 2NO+2NC,
Contactor K2 with 2NO+2NC,
Contactor K3 with 1NC,
Contactor K4 with 2NO+1NC,
Time relay K5 with 1 inst. NO+1 pick-up delayed NO (with change-over delay),
Momentary-control switch S0 with 1NC,
Momentary-control switches S1, S2 each with 1NO+1NC

c) Auxiliary circuit for maintained-contact control

Required auxiliary contacts:

Contactor K1 with 1NO+2NC,
Contactor K2 with 1NO+2NC,
Contactor K3 with 1NC,
Contactor K4 with 1NO+1NC,
Time relay K5 with 1 inst. NO+1 pick-up delayed NO (with change-over delay)

8 Fundamental circuit diagrams

8.3.5 Star-delta starting of three-phase induction motors with power factor correction

Figure 8.33 illustrates the circuit diagrams for the star-delta starting of three-phase induction motors with power factor correction (the capacitor is switched separately from the motor winding), using a star-delta time relay with change-over delay.

Pushbutton control/momentary contact (Fig. 8.33b)

Switching on: The star-delta time relay K5 is energized by the actuation of momentary-contact switch S1 and the closing of its NC contact. The star contactor K2 is energized via the instantaneous NO contact K5. The NO contact K2 energizes line contactor K1 and capacitor contactor K4. Sealing-in contacts K2 and K1 close and the motor runs up in the star connection.

Change-over: After the set delay time t_1 on K5 has elapsed, the instantaneous NO contact K5 opens and star contactor K2 is de-energized. The NC contact of star contactor K2 closes. After the change-over delay t_2 of approximately 50 ms, the delayed NO contact of K5 closes and delta contactor K3 is energized as the NO contact of line contactor K1 is already closed. The motor runs in the delta connection.

Switching off: Line contactor K1 is de-energized by the actuation of momentary-contact switch S0 and the opening of its NC contact. The NO contact of K1 opens causing both delta contactor K3 and capacitor contactor K4 to be de-energized. Thus, the motor is disconnected from the supply and is switched off.

The contactor auxiliary contacts which have not been mentioned above are used for interlocking purposes between the star and delta contactors.

Switch control/maintained contact (Fig. 8.33c)

According to its position, maintained-contact switch S1 energizes and de-energizes the contactors in the same way as described for momentary-contact control.

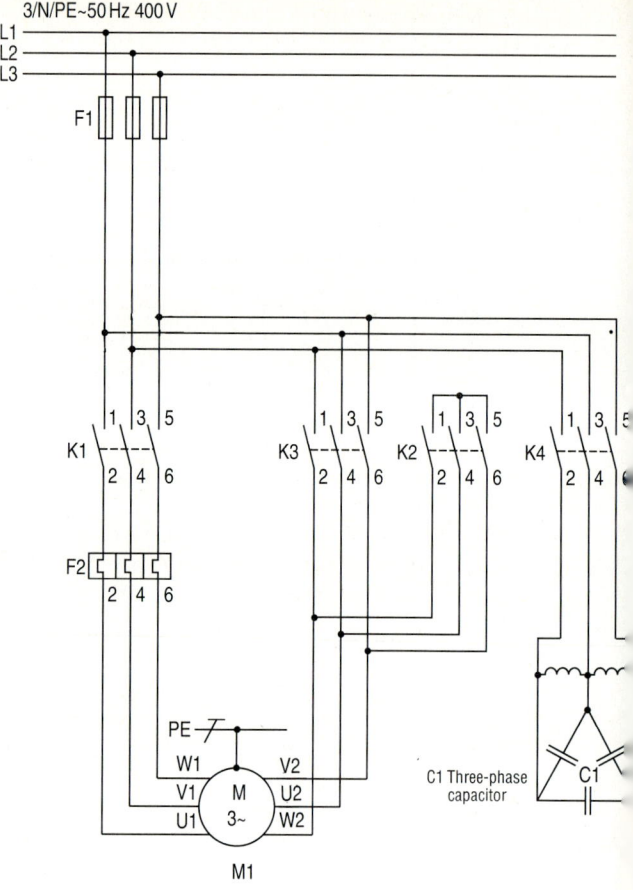

a) Main circuit

Figure 8.33
Circuit diagrams for star-delta starting of a three-phase induction motor with power factor correction

Star-delta starting of three-phase induction motors with power factor correction

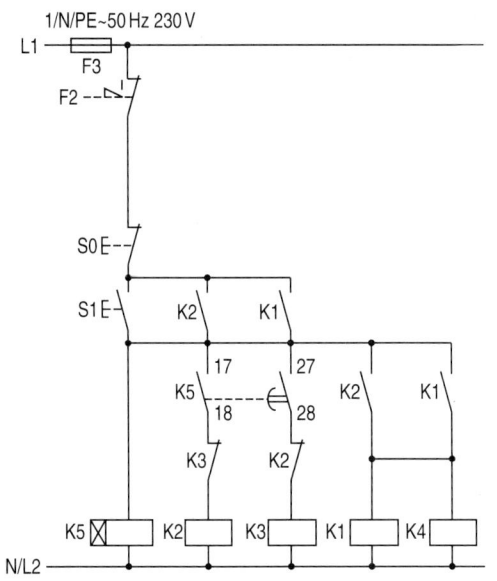

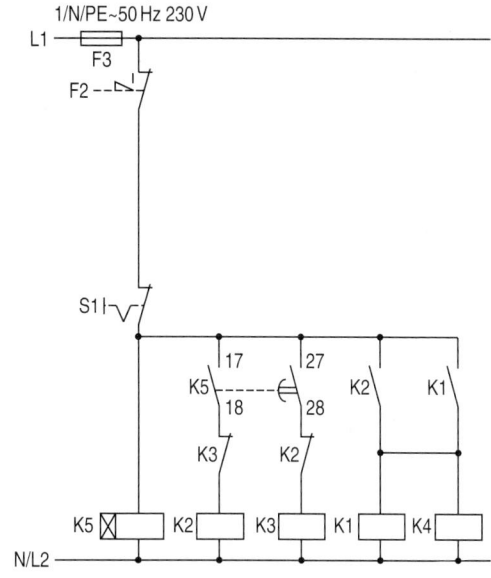

b) Auxiliary circuits for momentary-contact control

Required auxiliary contacts:

Contactor K1 with 2NO,
Contactor K2 with 2NO+1NC,
Contactor K3 with 1NC,
Momentary-contact switch S0 with 1NC,
Momentary-contact switch S1 with 1NO,
Time relay K5 with 1 inst. NO+1 pick-up delayed NO
(with change-over delay)

c) Auxiliary circuits for maintained-contact control

Required auxiliary contacts:

Contactor K1 with 1NO,
Contactor K2 with 1NO+1NC,
Contactor K3 eith 1NC,
Maintained-contact switch S1 with 1NO,
Time relay K5 with 1 inst. NO+1 pick-up delayed NO
(with change-over delay)

<u>Note on K5:</u>
There is no mechanical link between contacts 17/18 and 27/28

8.3.6 Automatic starting of three-phase squirrel-cage motors via a single-pole stator-resistor (KUSA connection) using a time relay

Figure 8.34b illustrates the circuit diagrams for the automatic starting of a three-phase squirrel-cage motor in a single-pole stator-resistance starting connection using a time relay.

Pushbutton control/momentary contact (Fig. 34b)

Switching on: Contactor K1 is energized by the actuation of momentary-contact switch S1 and the closing of its NO contact. Sealing-in contact K1 closes. Time relay K3 is energized by the closing of the NO contact of K1. The motor runs up at reduced speed.

After the set delay time has elapsed, contactor K2 is energized via the NO contact of time relay K3. Sealing-in contact K2 closes and the time relay K3 is de-energized via the NC contact of contactor K2. The NO contact K2 in the main circuit closes and bridges out resistor R1. The motor runs at rated speed.

Switching off: Contactors K1 and K2 are de-energized by the actuation of momentary-contact switch S0 and the opening of its NC contact. Thus, the motor is disconnected from the supply and is switched off.

Switch control/maintained contact (Fig. 8.34c)

According to its position, the maintained-contact switch S1 energizes and de-energizes the relevant contactors and switches the motor in the same way as described for momentary-contact control.

Rating of the contactors

Contactor K1 must be rated for 100% motor output power. Contactor K2 can be dimensioned for AC-1 duty. By the parallel connection of its three poles, the rated operational current of K2 can be increased by 2.4 times that stated in the catalogues for normal three-phase loading.

In order to calculate the series impedance for a particular starting torque M_{Ku}, the starting torque M_{an} and the starting current I_{an} of the motor under normal operating conditions must be known.

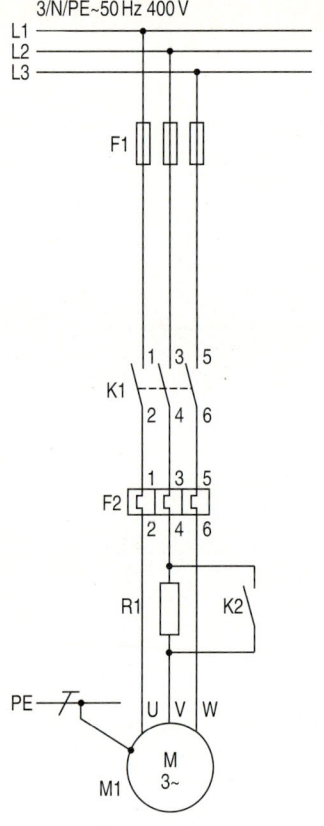

a) Main circuit

Figure 8.34
Circuit diagrams for the automatic starting of a three-phase squirrel-cage motor in single-pole stator-resistance starting connection (KUSA connection) using a time relay

A specific ratio $\alpha = \dfrac{I_{Ku}}{I_{an}}$ is associated with the ratio $K = \dfrac{M_{Ku}}{M_{an}}$, thus $I_{RV} = \alpha \cdot I_{an}$.

I_{an} Starting current
I_{Ku} Starting current with KUSA resistor (R1)
M_{an} Starting torque
M_{Ku} Starting torque with KUSA resistor (R1)

This ratio α may be found by using the diagram in Figure 3.103 on page 220 (Section 3.7.13).

Starting of three-phase squirrel-cage motors with the KUSA connection

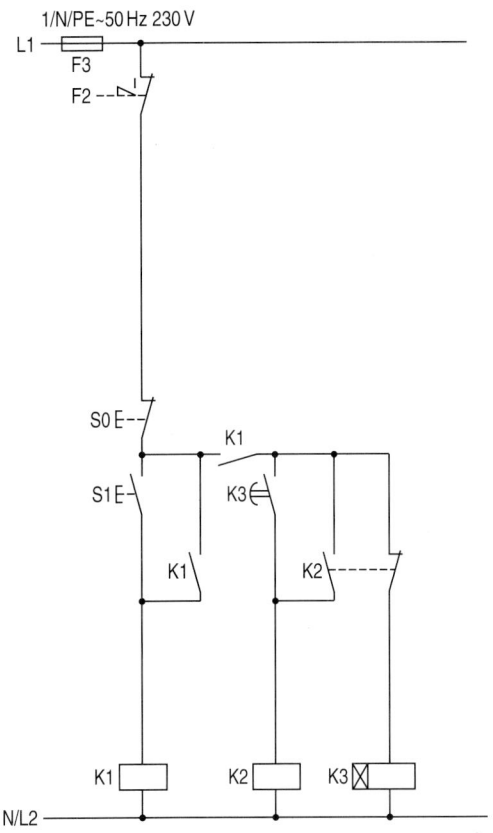
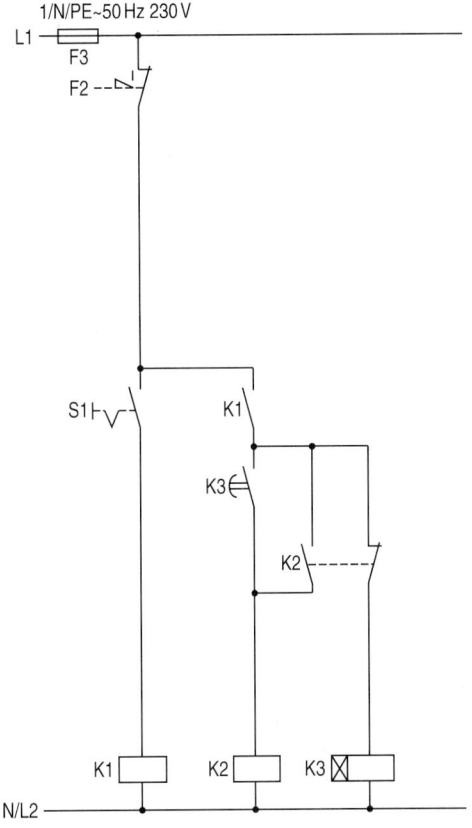

b) Auxiliary circuit for momentary-contact control

Required auxiliary contacts:

Contactor K1 with 2NO,
Contactor K2 with 1NO+1NC,
Time relay K3 with 1NO, pick-up delayed,
Momentary-contact switch S0 with 1NC,
Momentary-contact switch S1 with 1NO,
Overload relay F2 with 1NC

c) Auxiliary circuit for maintained-contact control

Required auxiliary contacts:

Contactor K1 with 1NO,
Contactor K2 with 1NO+1NC,
Time relay K3 with 1NO, pick-up delayed,
Maintained-contact switch S1 with 1NO,
Overload relay F2 with 1NC

8 Fundamental circuit diagrams

8.3.7 Automatic starting of three-phase induction motors via a three-pole stator-resistor using a time relay

Figure 8.35 illustrates the circuit diagrams for automatic starting of a three-phase induction motor via a three-pole resistor using a time relay (also known as 3-pole stator resistance starting).

Pushbutton control/momentary contact (Fig. 8.35b)

Switching on: Contactor K1 is energized by the actuation of momentary-contact switch S1 and the closing of its NO contact. Sealing-in contact K1 closes. Time relay K3 is energized by the closing of the NO contact K1. The motor runs up at reduced speed.

After the set delay time has elapsed, contactor K2 is energized via the NO contact of time relay K3. Sealing-in contact K2 closes, and the time relay K3 is de-energized via the NC contact of contactor K2. The NO contacts K2 in the main circuit close and bridge out resistor R1. The motor runs at rated speed.

Switching off: Contactors K1 and K2 are de-energized by the actuation of momentary-contact switch S0 and the opening of its NC contact. Thus the motor is disconnected from the supply and switched off.

Switch control/maintained contact (Fig. 8.35c)

Maintained-contact switch S1 energizes and de-energizes the motor in the same way as described for momentary-contact control.

Rating of the contactors

The contactor K1 must be rated for 100% motor output power. Contactor K2 can be selected for AC-1 duty (see Section 3.3.4.1, page 118).

For the calculation of a starting resistance for a particular starting torque M_{RV}, the starting torque M_{an} and the starting current I_{an} of the motor under normal operating conditions must be known.

By means of the ratio $K = \dfrac{M_{RV}}{M_{an}}$, the value for β can be taken from the diagram in Figure 3.104, see Section 3.7.13, page 221.

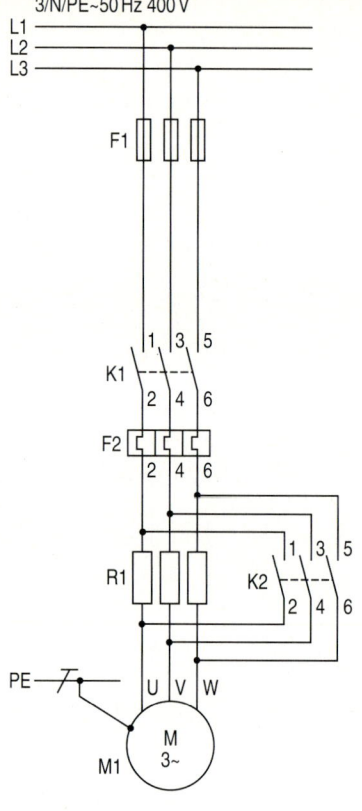

a) Main circuit

Figure 8.35
Circuit diagrams for the automatic starting of a three-phase induction motor via a three-pole stator resistor using a time relay

A specific ratio $\alpha = \dfrac{I_{RV}}{I_{an}}$ is associated with the ratio $K = \dfrac{M_{RV}}{M_{an}}$, thus $I_{RV} = \alpha \cdot I_{an}$.

I_{an} Starting current
I_{RV} Starting current with stator resistor (R1)
M_{an} Starting torque
M_{RV} Starting torque with stator resistor (R1)

This ratio α may be found by using the same diagram in Figure 3.104 on page 221 (Section 3.7.13).

Starting of three-phase induction motors via three-pole resistors

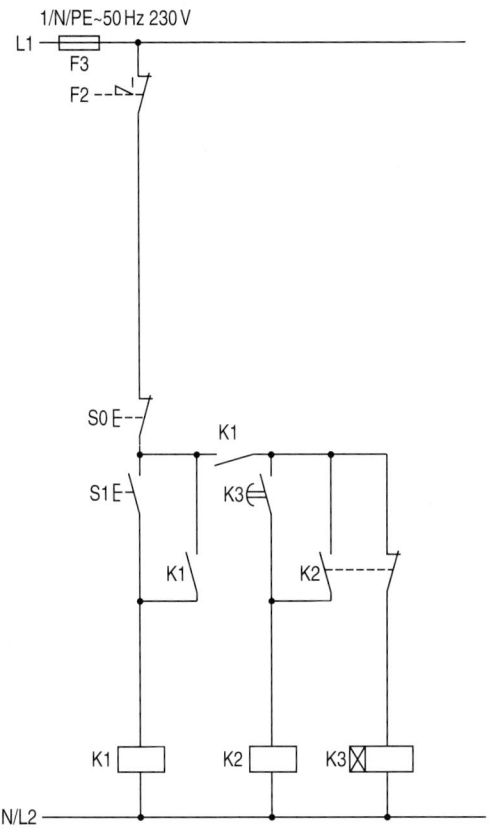

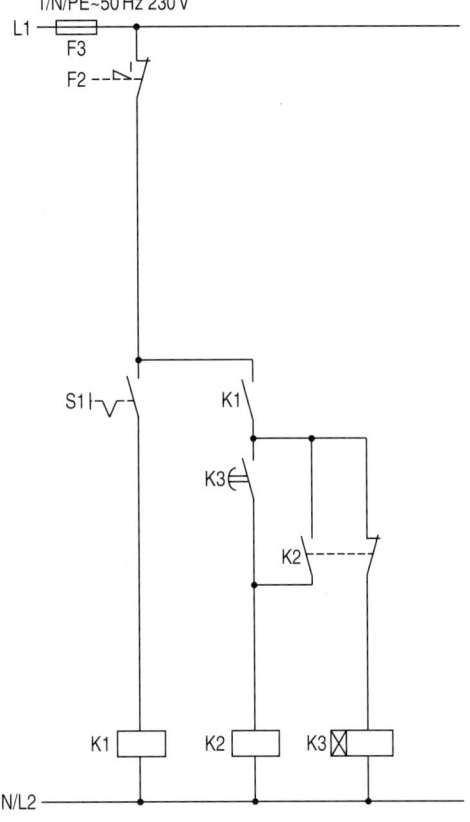

b) Auxiliary circuit for momentary-contact control

Required auxiliary contacts:

Contactor K1 with 2NO,
Contactor K2 with 1NO+1NC,
Time relay K3 with 1NO, pick-up delayed,
Momentary-contact switch S1 with 1NC,
Momentary-contact switch S1 with 1NO,
Overload relay F2 with 1NC

c) Auxiliary circuit for maintained-contact control

Required auxiliary contacts:

Contactor K1 with 1NO,
Contactor K2 with 1NO+1NC,
Time relay K3 with 1NO, pick-up delayed,
Maintained-contact switch S1 with 1NO,
Overload relay F2 with 1NC

8.3.8 Automatic starting of three-phase slip-ring motors

Figure 8.36 illustrates the circuit diagrams for automatic starting of a three-phase slip-ring motor (resistance steps in the rotor circuit). This circuit includes four time relays, and is also known as a rotor resistance starter.

Pushbutton control/momentary contact (Fig. 8.36b)

Switching on and starting the motor: Contactor K1 is energized by the actuation of momentary-contact switch S1 and the closing of its NO contact. Sealing-in contact K1 closes and time relay K2 is energized. The motor runs up with full resistance R1 in the rotor circuit.

After the set delay time has elapsed, rotor contactor K11 is energized via the NO contact of time relay K2. Time relay K2 is de-energized via the NC contact of rotor contactor K11 and sealing-in contact K11 closes. The NO contact of rotor contactor K11 energizes time relay K3, and the motor runs with remaining resistance R2 in the rotor circuit.

After the set delay time has elapsed, rotor contactor K12 is energized via the NO contact of time relay K3. Rotor contactor K11 and time relay K3 are de-energized via the NC contact of rotor contactor K12. The sealing-in contact K12 closes. The NO contact of rotor contactor K12 energizes time relay K4. The motor runs with remaining resistance R3 in the rotor circuit.

This process continues until rotor contactor K14 is energized and the complete rotor resistance is bridged out. The motor runs at rated speed.

Switching off: All contactors and time relays in circuit (after complete starting procedure stator contactor K1 and rotor contactor K14) are de-energized by the actuation of momentary-contact switch S0 and the opening of its NC contact. The motor is disconnected from the supply and switched off.

Switch control/maintained contact (Fig. 8.36c)

Maintained-contact switch S1 energizes and de-energizes the contactors and time relays in the same way as described for momentary-contact control.

For an automatic starter with six resistance steps, refer to the Siemens low-voltage switchgear catalogues and to Section 3.7.12, page 210.

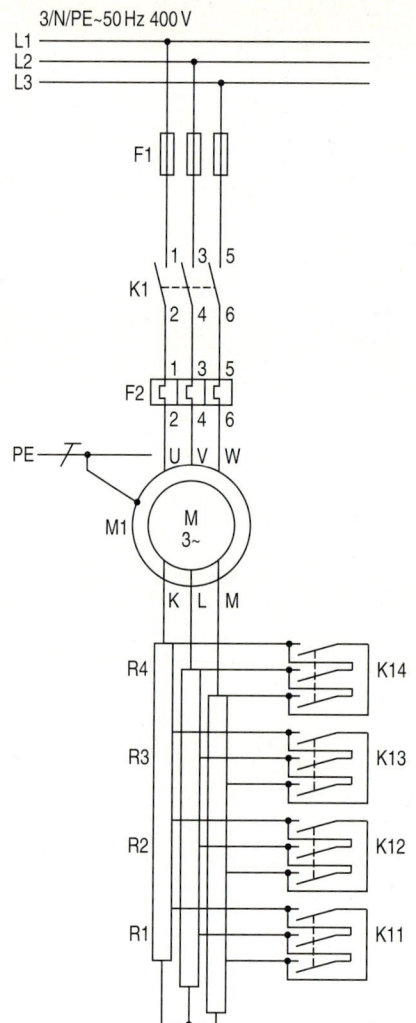

a) Main circuit

Figure 8.36
Circuit diagrams for the automatic starting of a three-phase slip-ring motor

Starting of three-phase slip-ring motors

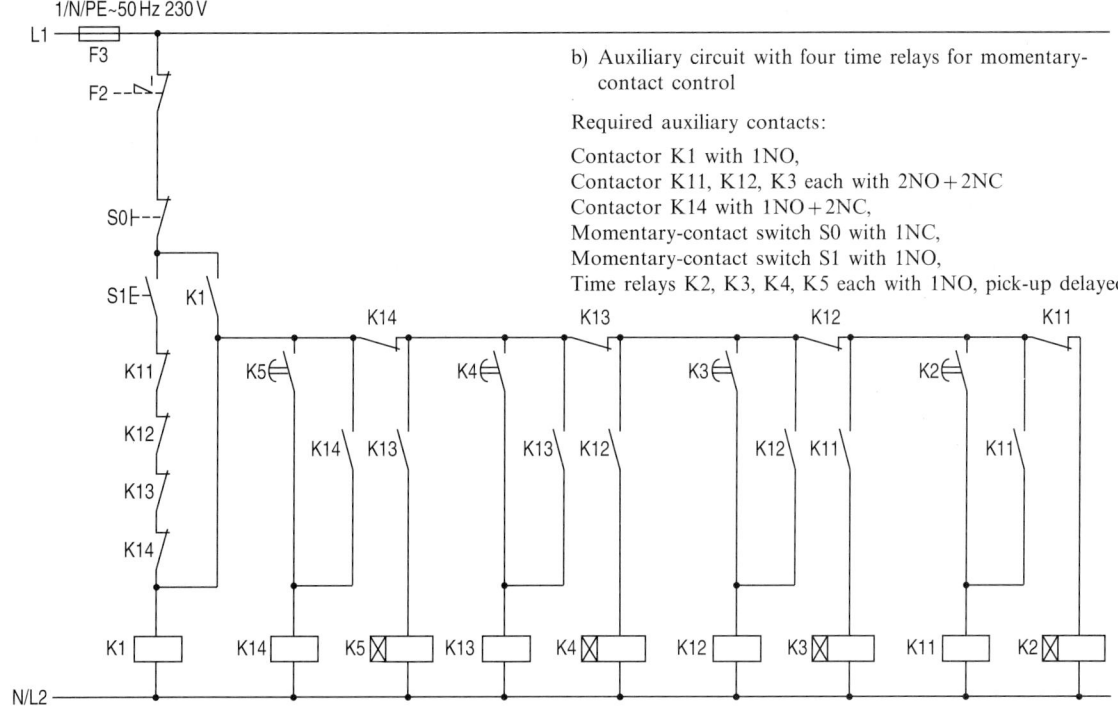

b) Auxiliary circuit with four time relays for momentary-contact control

Required auxiliary contacts:

Contactor K1 with 1NO,
Contactor K11, K12, K3 each with 2NO+2NC
Contactor K14 with 1NO+2NC,
Momentary-contact switch S0 with 1NC,
Momentary-contact switch S1 with 1NO,
Time relays K2, K3, K4, K5 each with 1NO, pick-up delayed

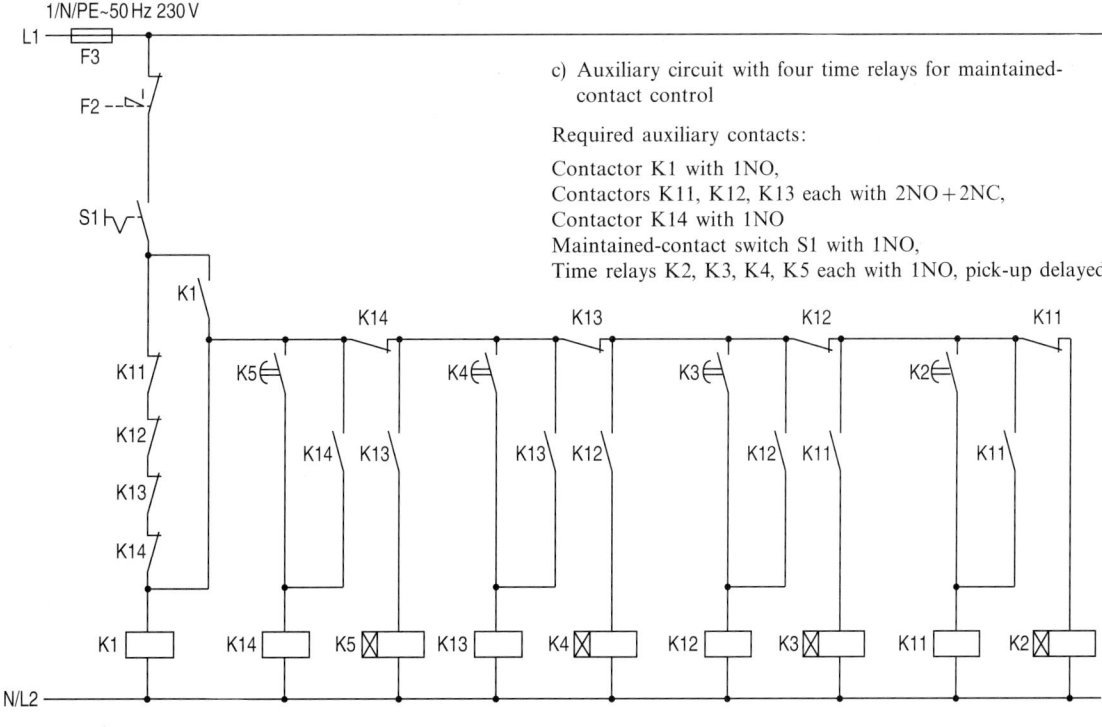

c) Auxiliary circuit with four time relays for maintained-contact control

Required auxiliary contacts:

Contactor K1 with 1NO,
Contactors K11, K12, K13 each with 2NO+2NC,
Contactor K14 with 1NO
Maintained-contact switch S1 with 1NO,
Time relays K2, K3, K4, K5 each with 1NO, pick-up delayed

8.3.9 Closed transition autotransformer starting of three-phase squirrel-cage motors (Korndörfer connection)

Starting transformers with three windings and contactors may be used for the automatic starting of motors under no-load or very low load conditions, if the network does not permit direct-on-line starting owing to the high starting current which would otherwise result (Fig. 8.37).

Pushbutton control/momentary contact (Fig. 8.37b)

Switching on: Contactor relay K5 and time relay K4 are energized by the actuation of momentary-contact switch S1 and the closing of its NO contact. Star contactor K1 is switched on and energizes transformer contactor K2 via its auxiliary contact. The motor runs up at reduced voltage and with low current. The motor torque is a function of the square of the voltage ($M \sim U^2$).

For the optimal adaptation of the motor torque to the starting conditions, the starting transformers usually offer three selectable tappings in each phase.

Change-over: After run-up at reduced voltage and expiring of the time set on K4, the transformer star contactor K1 opens. The transformer part-winding now acts as a reactor. At this instant, the motor is supplied with the line voltage reduced by the reactors and the motor speed remains constant. The motor contactor K3 is energized via the NC auxiliary contact of the star contactor K1 once it has been de-energized, and thus the motor is supplied with full line voltage. Transformer contactor K2 is de-energized by the opening of the NC contact of motor contactor K3. The change-over from supply via the transformer and the direct supply via the motor contactor, is carried out without disconnection of the motor (closed transition).

Switching off: Contactor relay K5 is de-energized by the actuation of momentary-contact switch S0 and the opening of its NC contact. The motor is switched off by the dropping out of contactor K1.

Example

A motor draws eight times its rated current (starting current) during direct on-line starting. The starting current is to be limited to four times the rated current (50%).

Required:
The tapping of the transformer winding.

The starting current is reduced by a square-law, and thus in this case the tapping is

$$\frac{I_{an1}}{I_{an}} = \sqrt{\frac{50}{100}} \approx 70\%.$$

The voltage at standstill is thus 70%, the static torque is 49%, the reduced line current is 49% of the d.o.l. starting current.

Rating of the contactors

Star contactor K1 and transformer contactor K2 must be rated for 49% of the motor output power.

The transformer contactor K2, however, must be selected to have a making capacity at least as high as the switch-on current of the starting transformer.

Motor contactor K3 must be rated for 100% of the motor output power.

Closed transition autotransformer starting of three-phase squirrel-cage motors

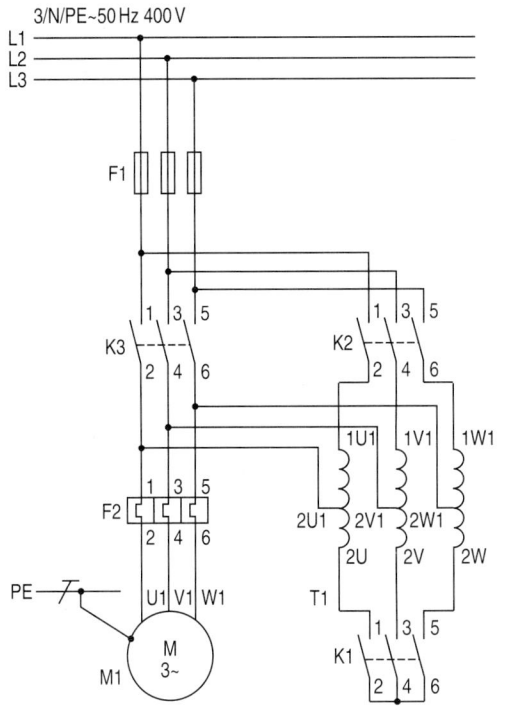

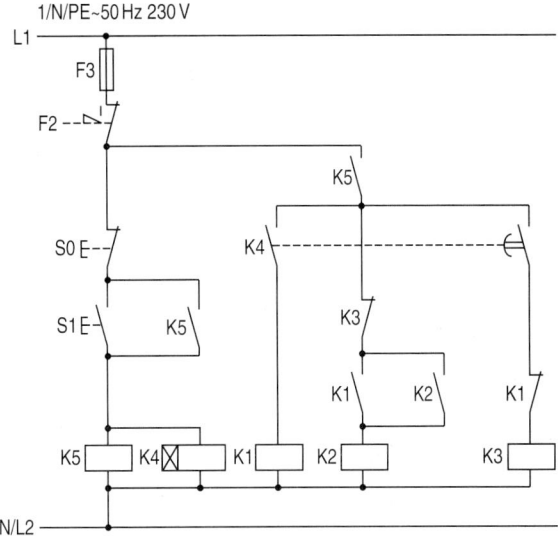

a) Main circuit

Figure 8.37
Circuit diagram for closed transition autotransformer starting of a three-phase induction (squirrel-cage) motor

b) Auxiliary circuit for momentary-contact control

Required auxiliary contacts:
Contactor K1 with 1NO+1 NC,
Contactor K2 with 1NO,
Contactor K3 with 1 NC,
Contactor relay K5 with 2NO,
Time relay K4 with 1NO, inst.+1NO pick-up delayed,
Momentary-contact switch S0 with 1NC,
Momentary-contact switch S1 with 1NO

8.4 Circuits with thermistor motor protection

8.4.1 Thermistor motor protection with positive temperature coefficient (PTC) temperature sensors

8.4.1.1 Thermistor motor protection for a pole-changing three-phase induction motor with <u>two</u> separate windings and <u>two</u> speeds

Figure 8.38 illustrates the circuit diagrams for thermistor motor protection with tripping unit 3UN2 and a pole-changing three-phase induction motor with two separate windings.

Switch control/maintained contact (Fig. 8.38b)

<u>Switching-on</u>: The energizing circuits of tripping units F4 and F5 for the individual windings are supplied with voltage via maintained-contact switch S3. NC contact 97–98 is opened and NO contact 95–96 is closed.

Contactor K1 is energized by the actuation of momentary-contact switch S1 and the closing of its NO contact. The sealing-in contact K1 closes. The motor runs up, for example, at low speed.

Contactor K1 is de-energized by the actuation of momentary-contact switch S2 and the opening of its NC contact. Contactor K2 is energized by the closing of NO contact S2. This energizing command on K2 becomes effective once the NC contact of K1 has closed. Sealing-in contact K2 closes, and the motor runs at high speed.

<u>Switching off via B1 or B2</u>: If the motor is overloaded, for example, during operation at low speed, and the nominal tripping temperature of the PTC temperature sensor B1 is reached, tripping unit F4 is tripped. NO contact 95–96 opens and NC contact 97–98 closes. Contactor K1 is de-energized and the motor is switched off. Indicator light H3 lights up.

In the event of a motor overload during operation at high speed, contactor K2 is de-energized and thus the motor is switched off via tripping unit F5. Indicator light H4 lights up.

<u>Switching off manually</u>: The contactor K1 or K2 is de-energized by the actuation of pushbutton S0 and the opening of its NC contact. The motor is disconnected from the supply and switched off.

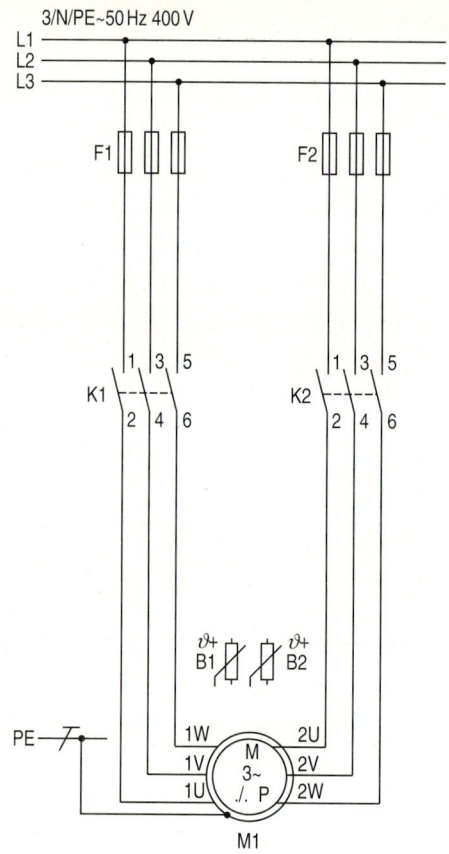

a) Main circuit

Figure 8.38
Circuit diagrams for thermistor motor protection in a pole-changing three-phase induction motor

Thermistor motor protection with PTC temperature sensors

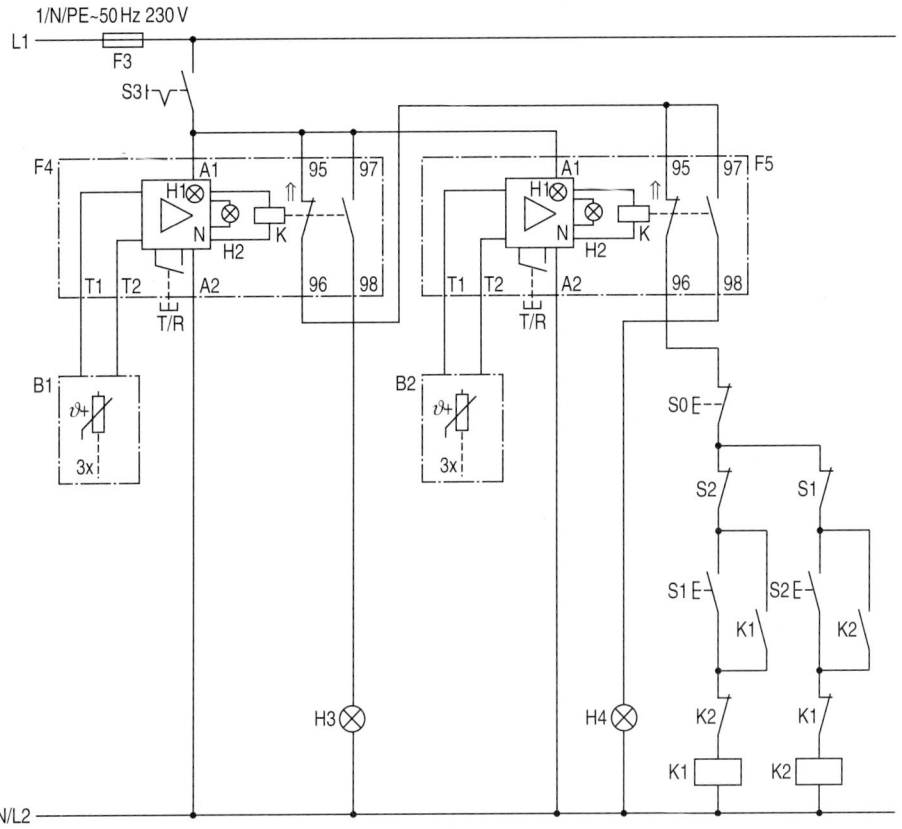

b) Auxiliary circuit

Required auxiliary contacts:

Contactors K1, K2 each with 1NO+1NC,
Maintained-contact switch S3 with 1NO,
Momentary contact switch S0 with 1NC,
Momentary contact switches S1, S2 each with 1NO+1NC,
Tripping units F4, F5 each with 1NO+1NC

T/R TEST/RESET button
⇑   NO and NC contacts are shown in an operated state (see also Table 8.4, page 402)

8.4.1.2 PTC thermistor motor protection for alarm and switch-off of a three-phase induction motor with six sensors via a circuit-breaker for motor protection equipped with overload and short-circuit releases

Figure 8.39 illustrates the circuit diagrams for thermistor motor protection with six sensors[1] and two 3UN2 tripping units for alarm and switch-off of the motor respectively. Switch-off is accomplished using a circuit-breaker which incorporates a current-dependent delayed overload release (a) and an instantaneous electro-magnetic short-circuit release (n), as well as an undervoltage release.

Switch control/maintained contact (Fig. 8.39b)

Switching on: Voltage is applied to the energizing circuits of tripping units F3 for alarm and F4 for shut-down respectively via maintained-contact switch S2. The NC contacts 97–98 open and the NO contacts 95–96 close. Voltage is applied to the undervoltage release F1 of circuit-breaker Q1 via maintained-contact switch S3. Circuit-breaker Q1 is closed manually, and the motor is switched on.

If the motor is overloaded and the nominal tripping temperature (TNF) of PTC sensor B1 for alarm is reached, tripping unit F3 responds. NC contact 97–98 in F3 closes, indicator light H3 for alarm lights up. Although the NO contact 95–96 of F3 opens, voltage is still applied to the undervoltage release via the auxiliary contact of the circuit-breaker Q1, so that the circuit-breaker does not trip in the event of an alarm.

Switching off via B2: If the motor attains the nominal tripping temperature (TNF) of PTC sensor B2 for shut-down, tripping unit F4 responds. NO contact 95–96 of F4 opens and undervoltage release F1 trips the circuit-breaker Q1 and switches off the motor. Indicator light H4 is illuminated by the closing of NC contact 97–98 in F4.

After such a tripping action, the circuit-breaker can only be reclosed again once *both* tripping units have been reset, since the u.v.r. is supplied via the NO contacts 95–96 of both F3 *and* F4.

Switching off manually: The motor can be switched off manually by the actuation of maintained-contact switch S3 or by switching off the circuit-breaker by means of its operating mechanism.

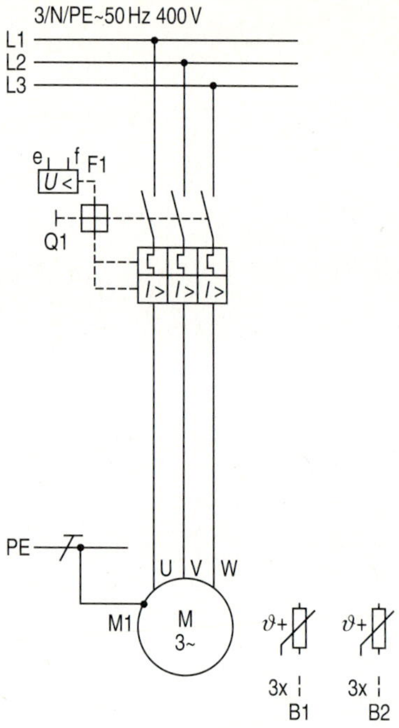

a) Main circuit

Figure 8.39
Circuit diagrams for thermistor motor protection incorporating alarm and switch-off of a motor having six sensors, via a circuit-breaker with a current-dependent delayed overload release (a) and an instantaneous electro-magnetic short-circuit release (n).

---

[1] 1 per phase for alarm, plus 1 per phase for switch-off temperature sensing respectively = 6 sensors.

Thermistor motor protection with PTC temperature sensors

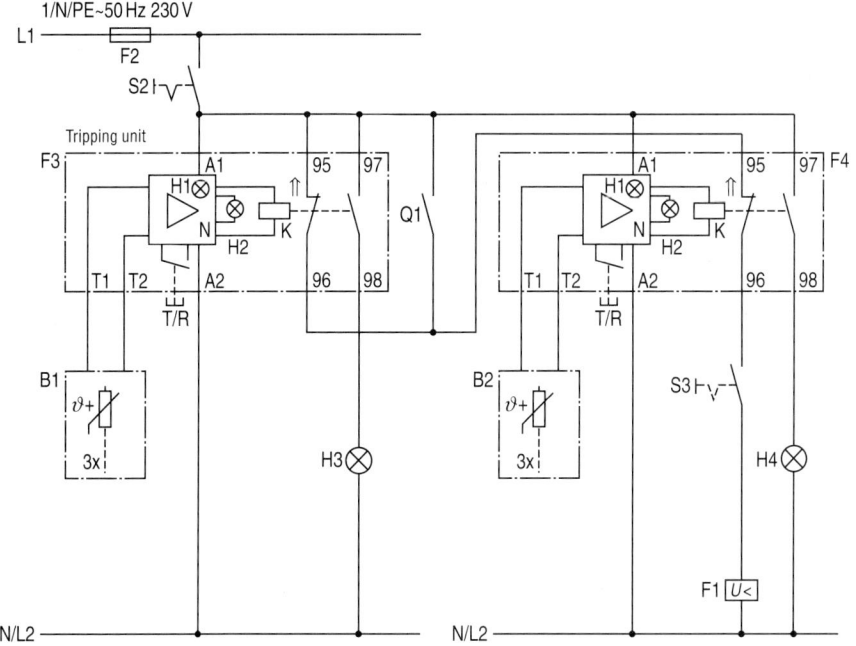

b) Auxiliary circuit

Required auxiliary contacts:

Circuit-breaker Q1 with 1NO,
Maintained-contact switch S2, S3 each with 1NO,
Tripping units F3, F4 each with 1NO+1NC

T/R TEST/RESET button
⇑   NO and NC contacts are shown in an operated state (see also Table 8.4, page 402)

8.4.1.3 Thermistor motor protection for the switching off of six three-phase induction motors via contactors

Figure 8.40 illustrates the circuit diagrams of a thermistor motor protection with tripping unit 3UN26 for the shut-down of six three-phase induction motors via contactors.

Up to six three-element (1 per phase) PTC sensor loops can be connected to the terminals nT1 and nT2 of the tripping unit 3UN26. The tripping unit is operated by any one of the PTC sensor loops B1 to B6. Terminals nT1/nT2 which are not used, must be bridged out.

Pushbutton control/momentary contact (Fig. 8.40b)

Switching on: Voltage is applied to the energizing circuits of the tripping unit F8 via maintained-contact switch S2. NO contact 95–96 is closed and NC contact 97–98 is open. Contactor relay K7 is energized by the actuation of momentary-contact switch S1 and the closing of its NO contact. The sealing-in contact and NO contacts K7 close. Contactors K1 to K6 are energized and motors M1 to M5 are switched on. If the simultaneous switching on of the six motors is not permissible, the automatic sequential switching according to Figure 8.16, Section 8.2.3, page 418 must be applied.

If one (or more) of the motors is (are) overloaded and the nominal tripping temperature of the corresponding PTC sensor is reached, tripping unit F8 responds. The NO contact 95–96 of F8 opens and de-energizes all contactors and thus all the motors are switched off. The NC contact 97–98 of F8 closes and indicator light H9 lights up. By means of the LED's H1 to H6 respectively, the sensor loop B1 to B6 which caused the tripping unit to operate, can be identified. The LED H7 indicates that the unit is operational and LED H8 indicates that the unit has tripped.

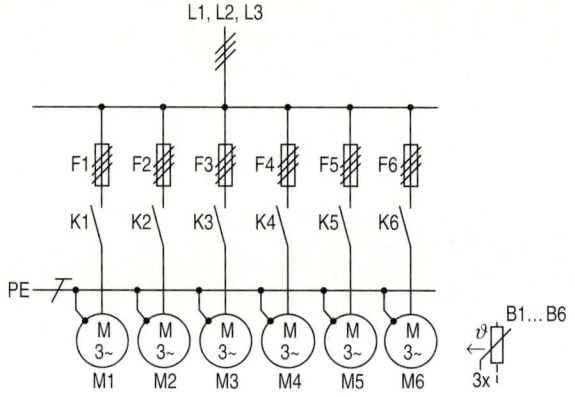

a) Main circuit (single-line representation)

Figure 8.40
Circuit diagrams of a thermistor motor protection for the switching off of six three-phase induction motors via contactors

The resetting of F7 by means of depressing the T/R button is only possible once the temperature sensor which caused the unit to trip has cooled down suficiently. By the bridging out of terminals Y1 and Y2 an automatic reset may be achieved. By the inclusion of a NO momentary-contact switch (pushbutton) between terminals Y1 and Y2, remote resetting may be accomplished.

Switching off: By the actuation of momentary-contact switch (pushbutton) S0 and the opening of its NC contact, the motor can be switched off manually.

Further circuit diagrams with thermistor motor protection units employing PTC temperature sensors are illustrated in the Siemens l.v. switchgear catalogue.

Thermistor motor protection with PTC temperature sensors

Thermistor motor protection with PTC temperature sensors

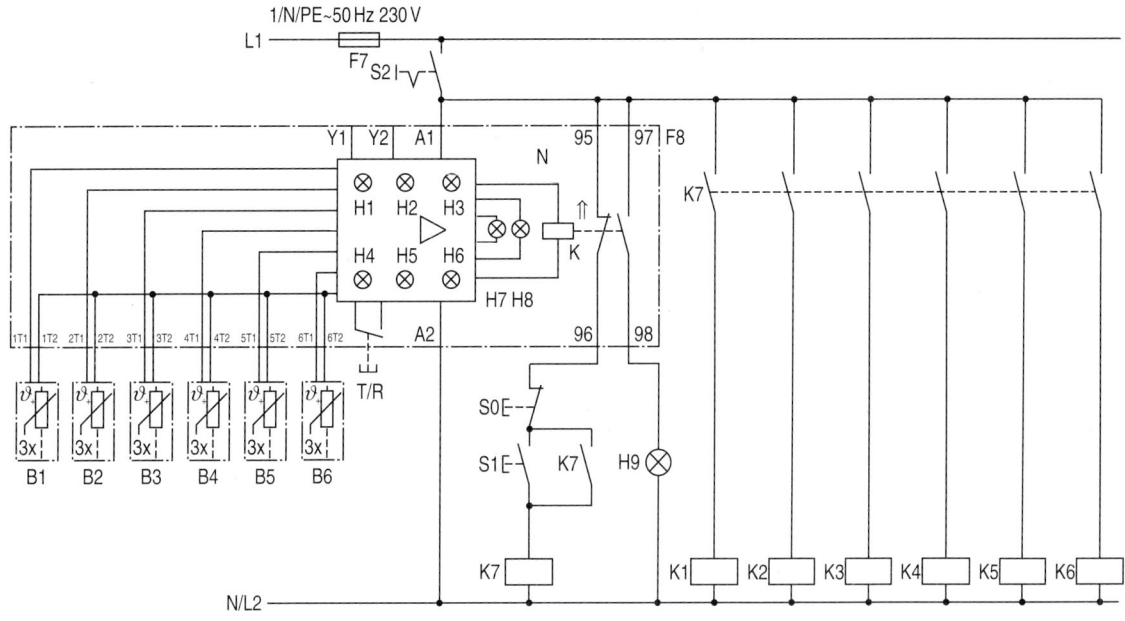

b) Auxiliary circuit for momentary-contact control

Required auxiliary contacts:

Maintained-contact switch S2 with 1NO,
Momentary-contact switch S1 with 1NC,
Momentary-contact switch S0 with 1NO,
Tripping unit F7 with 1NO+1NC

T/R TEST/RESET button
⇑   NO and NC contacts are shown in an operated state (see also Table 8.4, page 402)

8.4.2 Thermistor motor protection with negative temperature coefficient (NTC) temperature sensors

This Siemens thermistor motor protection comprises the tripping unit F2 shown in Figure 8.41 and NTC temperature sensors.

The protective device operates on the closed-circuit (quiescent-current) principle and is self-monitoring in the event of a short-circuit or an interruption of (or break in) the sensor circuit.

The tripping unit F2 is equipped with two operational ampliers each having one output relay. Each output relay has one change-over contact. Up to six sensors can be connected to the tripping unit.

The tripping unit F2 is delivered with the link B fitted. In this case the same NTC sensors can be used for "trip" and "alarm".

If the link B is removed, the two ampliers operate separately, each with three sensors.

Figure 8.41 illustrates the circuit diagram for alarm indication and shut-down of a three-phase induction motor.

Setting the response temperature RT

The value of resistance assigned to the desired response temperature RT can be obtained from the resistance table of the NTC sensors (Table 8.11). The appropriate resistor for a response temperature RT between 90 to 160 °C can be selected from the calibrating resistors which are delivered with the tripping unit.

The sensor circuits have to be routed as separate control circuits. Since the measuring circuit has a relatively high resistance, a line resistance of up to 50 Ω has hardly any effect on the operation of the protective device. As soon as one of the sensors attains the response temperature RT, the amplier adjusted to this temperature operates and the associated output relay drops out.

Further circuit diagrams for thermistor motor protection devices with NTC temperature sensors are illustrated in the Siemens l.v. switchgear catalogue.

Table 8.11
Response temperature RT of the NTC thermistors

RT °C	Thermistor resistance kΩ
60	7.16
70	4.93
80	3.46
90	2.47
100	1.80
110	1.33
120	1.00
130	0.76
140	0.59
150	0.46
160	0.36

Thermistor motor protection with NTC temperature sensors

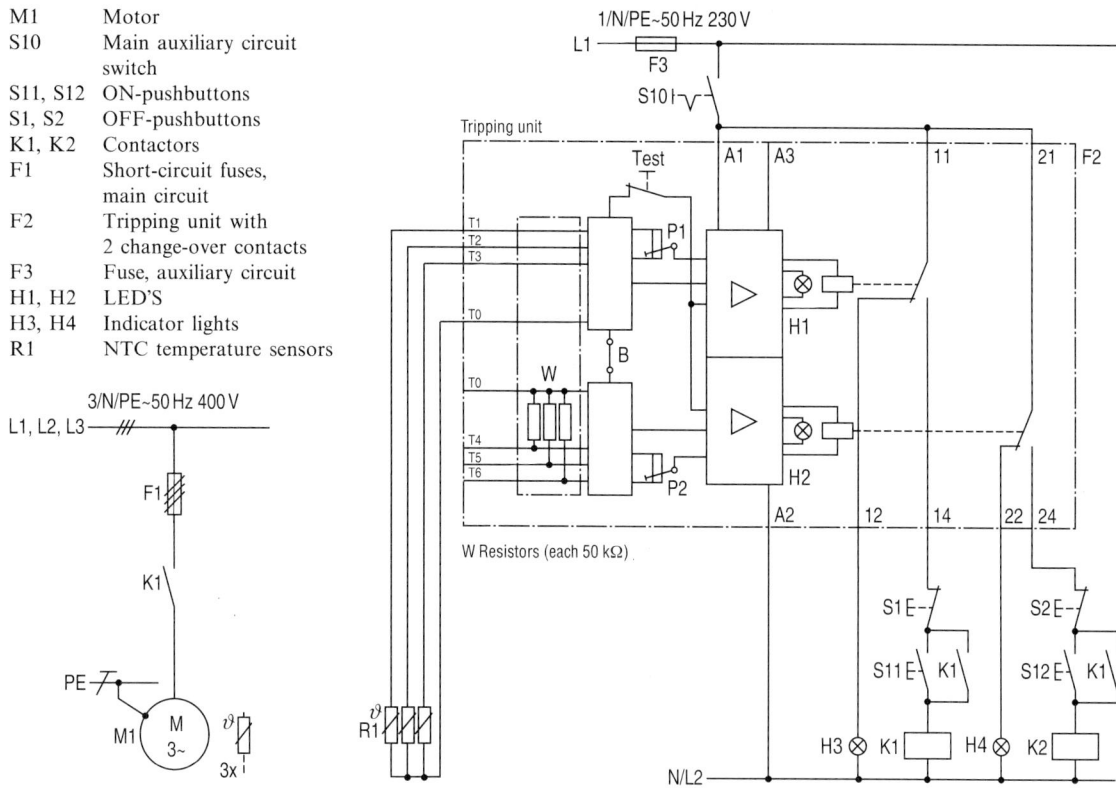

a) Main circuit

b) Auxiliary circuit for momentary-contact control

Required auxiliary contacts

Maintained-contact switch S10 with 1NO,
Momentary-contact switches S1, S2 each with 1NC,
Momentary-contact switches S11, S12 each with 1NO,
Contactors K1, K2 each with 1NO,

Figure 8.41
Thermistor motor protection with NTC temperature sensors for alarm and shut-down of a three-phase induction motor

8.5 Circuits with monitors

8.5.1 Circuits with speed monitors

8.5.1.1 Direct-on-line starting of three-phase induction motors. Stopping by reverse-current braking (plug braking)

Figure 8.42 illustrates the circuit diagrams for direct-on-line switching of three-phase induction motors, using plug or reverse-current braking.

Pushbutton control/momentary contact (Fig. 8.42b)

Switching on: Contactor K1 is energized by the actuation of momentary-contact switch S1 and the closing of its NO contact. The sealing-in contact K1 closes and the motor is switched on (d.o.l. starting). The contact of speed monitor F3 changes over once the motor has run up to speed.

Switching off: By the actuation of the OFF-pushbutton S0, contactor K1 is de-energized by the opening of its NC contact and contactor K2 is energized by the closing of its NO contact. The energizing command on K2 only becomes effective once the NC contact of K1 has closed. Sealing-in contact K2 closes and the motor speed is reduced by plug braking.

Once the motor reaches the low speed set on the speed monitor, the change-over contact on F3 returns to its initial position, contactor K2 is de-energized and the plug braking operation is completed.

Switch control/maintained contact (Fig. 8.42c)

Contactor K1 is energized by the closing of maintained-contact switch S1. The plug braking operation is initiated by the opening of the maintained-contact switch (the other procedures are the same as with momentary-contact control).

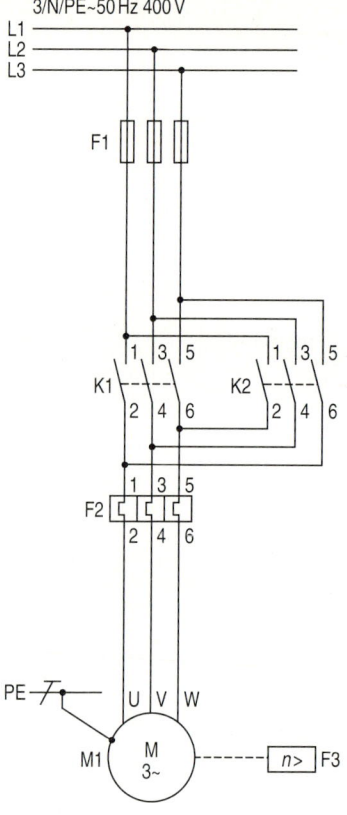

a) Main circuit

Figure 8.42
Circuit diagrams for direct switching on and off of a three-phase induction motor with plug braking

Circuits with speed monitors · Direct-on-line starting

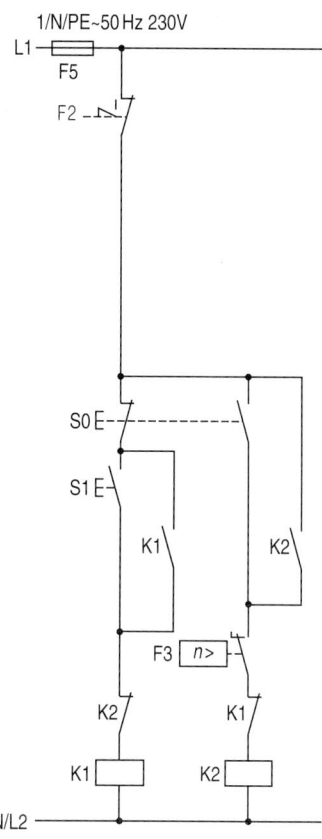

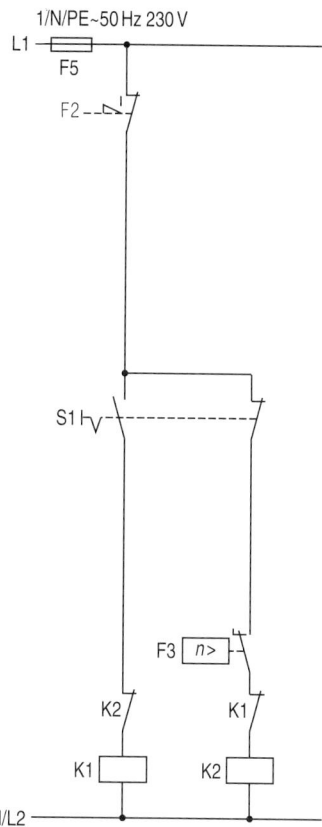

b) Auxiliary circuits for momentary-contact control

Required auxiliary contacts:

Contactors K1, K2 each with 1NO+1NC,
Momentary-contact switch S0 with 1NO+1NC,
Momentary-contact switch S1 with 1NO,
Speed monitor F3 with 1CO,
Overload relay F2 with 1NC

c) Auxiliary circuits for maintained-contact control

Required auxiliary contacts:

Contactors K1, K2 each with 1NC,
Maintained-contact switch S1 with 1NO+1NC,
Speed monitor F3 with 1CO,
Overload relay F2 with 1NC

8.5.1.2 Direct-on-line starting of three-phase induction motors. Stopping by reverse-current braking (plug braking). Circuit with contactor relay

This circuit with contactor relay is required if dangerous unexpected manual starting is to be prevented during the commissioning and setting-up of the drive.

Figure 8.43 illustrates the circuit diagrams for direct-on-line starting of three-phase induction motors and stopping by plug braking.

Pushbutton control/momentary contact (Fig. 8.43b)

Switching on: Contactor K1 is energized by the actuation of momentary-contact switch S1 and the closing of its NO contact. The sealing-in contact K1 closes and the motor is switched on (d.o.l. starting).

The contact of speed monitor F3 changes over once the preset speed has been attained. As the NO contact of contactor K1 is closed, contactor relay K3 is energized. The sealing-in contact and the NO contact of K3 close.

Switching off: By the actuation of the OFF-pushbutton S0, contactor K1 is de-energized via its NC contact and contactor K2 is energized via its NO contact. The energizing command on K2 only becomes effective once the NC contact of K1 has closed. The sealing-in contact of K2 closes. The motor speed is reduced by plug braking. At a specific low speed preset on the speed monitor, contact F3 changes over to its initial position. Contactor relay K3 is de-energized. The NO contact of K3 opens and de-energizes contactor K2. The plug braking operation is completed.

Switch control/maintained contact (Fig. 8.43c)

Contactor K1 is energized on the closing of maintained-contact switch S1. The braking operation is initiated on the opening of the maintained-contact switch (the other procedures are the same as with momentary-contact control).

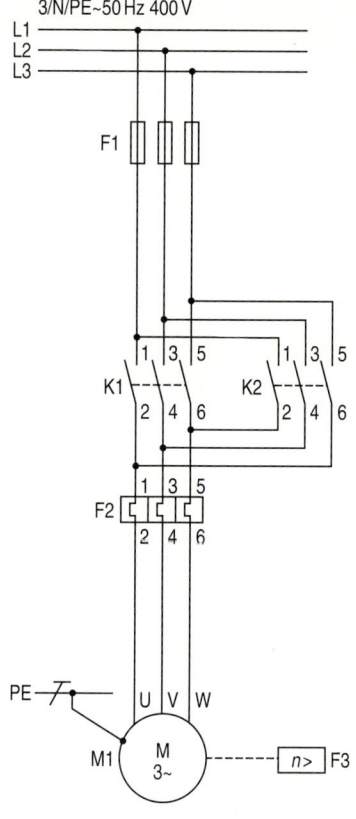

a) Main circuit

Figure 8.43
Direct-on-line starting of a three-phase induction motor. Stopping by plug braking. Circuit with contactor relay

Circuits with speed monitors · Direct-on-line starting

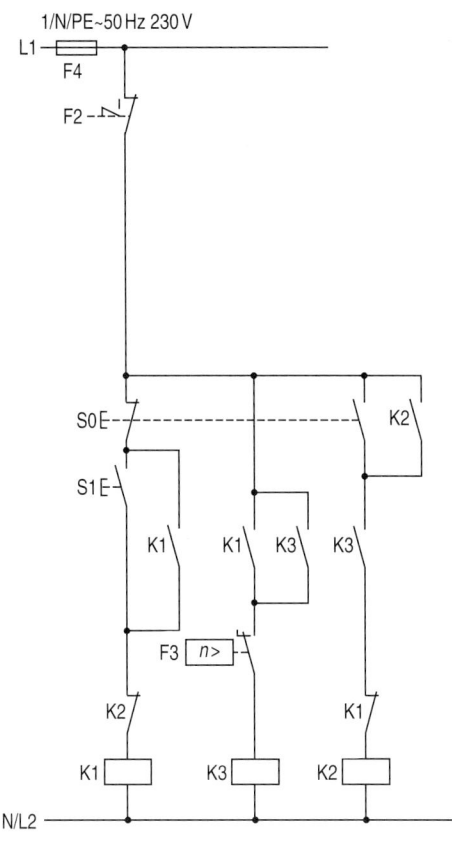

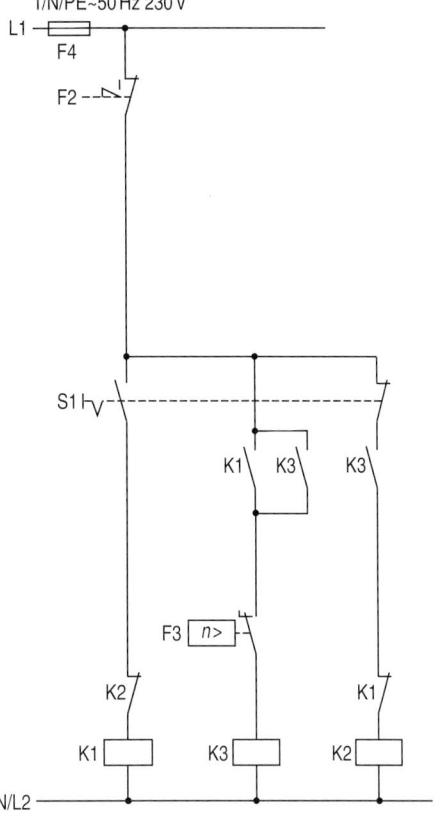

b) Auxiliary circuit for momentary-contact control

Required auxiliary contacts:

Contactor K1, with 2NO+1NC,
Contactos K2, with 1NO+1NC,
Contactor relay K3 with 2NO,
Momentary-contact switch S0 with 1NO+1NC,
Momentary-contact switch S1 with 1NO,
Speed monitor F3 with 1CO,
Overload relay F2 with 1NC

c) Auxiliary circuit for maintained-contact control

Required auxiliary contacts:

Contactor K1 with 1NO+1NC,
Contactor K2 with 1NC,
Contactor relay K3 with 2NO,
Maintained-contact switch S1 with 1NO+1NC,
Speed monitor F3 with 1CO,
Overload relay F2 with 1NC

8.5.1.3 Direct reversal of three-phase induction motors with reverse-current or plug braking in both directions of rotation

Figure 8.44 illustrates the circuit diagrams for the direct reversal of a three-phase induction motor with plug braking in both directions of rotation.

Pushbutton control/momentary contact (Fig. 8.44b)

Switching on: Contactor K1 is energized by the actuation of momentary-contact switch S1 and the closing of its NO contact. The sealing-in contact K1 closes. The motor runs up, for example, in clockwise direction. Contact F3 of the speed monitor for clockwise rotation changes over.

Reversing: Contactor K1 is de-energized by the actuation of momentary-contact switch S2 and the opening of its NC contact. Contactor K2 is energized by the closing of NO contact S2. The energizing command on K2 only becomes effective once the NC contact of contactor K1 has closed. Sealing-in contact K2 closes. The motor speed is reduced and then the motor runs up in anti-clockwise direction.

Switching off: Contactor K1 is de-energized via the NC contact S0 by the actuation of the OFF-pushbutton S0 (assuming that the contactor K1 was energized). Contactor K2 is energized by the NO contact of S0 via the change-over contact of speed monitor F3. The energizing command on contactor K2 only becomes effective once the NC contact of K1 has closed. The sealing-in contact and the NO contact of K2 close, contactor relay K3 is energized. Sealing-in contact K3 closes, NC contact K3 opens and sealing-in contact K2 becomes ineffective. The motor speed is reduced by plug braking.

At the low speed preset on the speed monitor, the contact of F3 changes over to its initial position and contactor K2 is de-energized. The NO contact of K2 opens and de-energizes contactor relay K3. The plug braking operation is completed.

If contactor K2 was energized, the process is carried out analogously.

Switch control/maintained contact (Fig. 8.44c)

Switching on: Contactor K1 is energized by the actuation of maintained-contact switch S1 which is, for example, in position R(clockwise). The motor starts up in clockwise direction. The change-over contact of speed monitor F3 for clockwise rotation changes over. The position 0 contact of S1 is open.

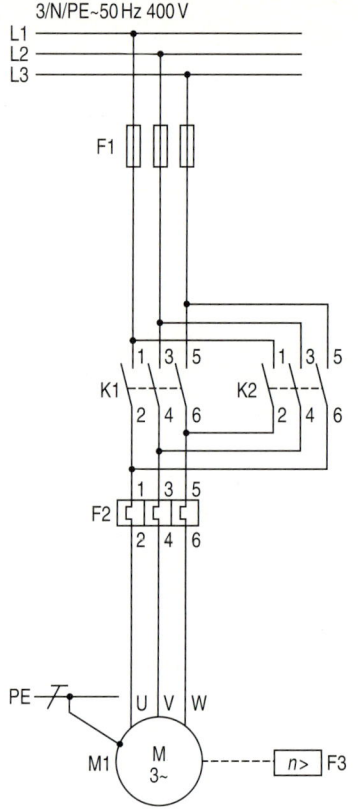

a) Main circuit

Figure 8.44
Circuit diagrams for direct reversal of a three-phase induction motor by plug braking in both directions of rotation

Reversing: Maintained-contact switch S1 has to be switched over from position R through position 0 to position L (anti-clockwise). Firstly, contact of position R opens (de-energizing contactor K1), then the contact of position 0 closes (energizing contactor K2 as the NC contact of K1 is closed and the change-over contact of F3 for clockwise direction has operated). The motor speed is reduced by plug braking. Afterwards, the contact of position L closes (after the contact of F3 for clockwise rotation has dropped back contactor K2 remains energized via the L position of S0), the motor starts up in anti-clockwise direction.

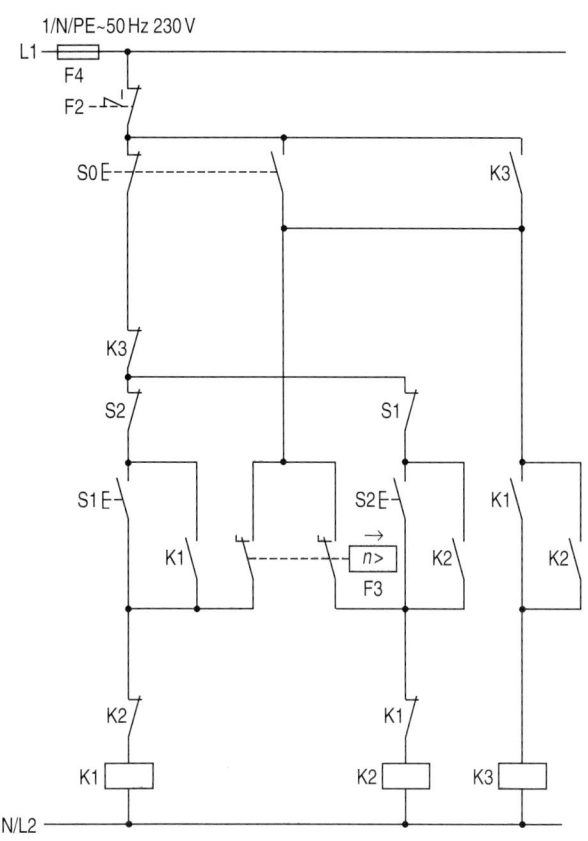
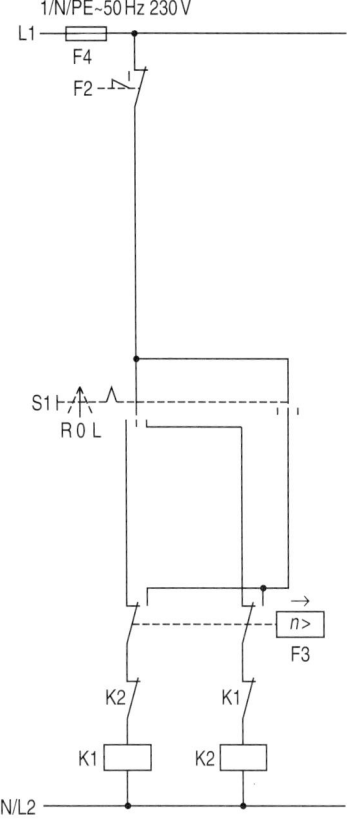

b) Auxiliary circuit for momentary-contact control

Required auxiliary contacts:

Contactors K1, K2 each with 2NO+1NC,
Contactor K3, with 1NO+1NC,
Momentary-contact switches S0, S1, S2 each with
  1NO+1NC,
Speed monitor F3 with 2CO,
Overload relay F2 with 1NC

c) Auxiliary circuit for maintained-contact control

Required auxiliary contacts:

Contactors K1, K2 each with 1NC,
Maintained-contact switch S1
  with 3 switch positions,
Speed monitor F3 with 2CO,
Overload relay F2 with 1NC

Switching off with, e.g., contactor K1 energized (motor is running in clockwise direction): Maintained-contact switch S1 has to be switched over from position R to position 0. The contact of position R opens, contactor K1 is de-energized. The contact of position 0 closes, contactor K2 is energized as soon as NC contact K1 has closed. The motor speed is reduced by plug braking. At the low speed preset on the speed monitor, the contact of F3 changes over to its initial position, and contactor K2 is de-energized. The plug braking operation is completed.

If contactor K2 was energized, the process is carried out analogously.

8.5.2 Circuits with conveyor belt monitors

A conveyor belt monitor sends a de-energizing command to the motor control gear if the conveyor belt jams, breaks, slips excessively, or if the drive speed drops below a certain value.

A conveyor belt monitor can also monitor the correct start-up of a conveyor belt system. In this function, the conveyor belt monitor permits the motor of the feed-in belt drive to start only after the preset speed of the feed-out belt drive has been attained (Fig. 8.45).

During the belt start-up, time relay K4 and contactor relay K2 bridge out the change-over contact of conveyor belt monitor F3 for the preset change-over time period.

Pushbutton control/momentary contact (Fig. 8.45b)

Switching on: Contactor relay K2 and time relay K4 are energized by the actuation of momentary-contact switch S1 and the closing of its NO contact. The sealing-in contact K2 closes. NO contact K2 and thus contactor relay K3 close. The sealing-in contact and NO contact K3 close. Contactor K1 switches on the motor. The change-over contact of F3 changes over. After the preset delay time has elapsed, the NC contact of time relay K4 de-energizes contactor relay K2.

Switching off: If the conveyor belt jams, the change-over contact of F3 changes over to its initial position. The open NC contact K1 prevents contactor relay K2 and time relay K4 from being energized again. Contactor relay K3 is de-energized and thus contactor K1 drops out and the motor is switched off.

The use of a drop-out delay unit on contactor relay K3 prevents the motor being switched off at temporary belt speed variations (see Section 8.1.5.1, page 412).

Switch control/maintained contact (Fig. 8.45c)

Switching on: Contactor relay K2 and time relay K5 are energized by the operation of maintained-contact switch S1. The sealing-in contact K2 closes. Contactors K3 and K4 are energized via NO contact K2. The sealing-in contact K4 closes. Contactor K1 is energized via NO contact K3. The motor is switched on. If the motor is switched off by conveyor belt monitor F3, it is prevented from restarting automatically, since the NC contact of contactor K4 is open.

(For further description of the circuit, see "Pushbutton control / momentary contact" above).

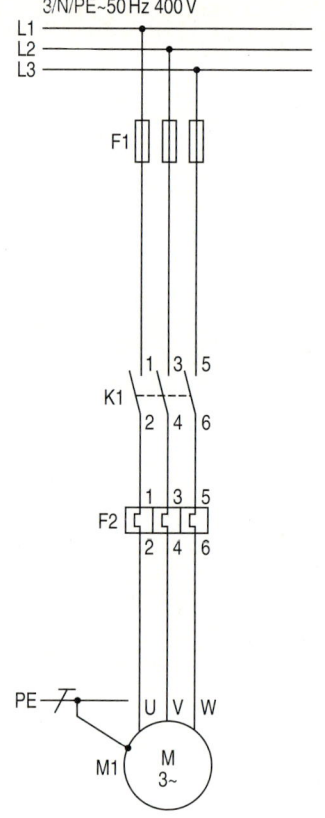

a) Main circuit

Figure 8.45
Circuit diagrams of a three-phase induction motor with a conveyor belt monitor

Circuits with conveyor belt monitors

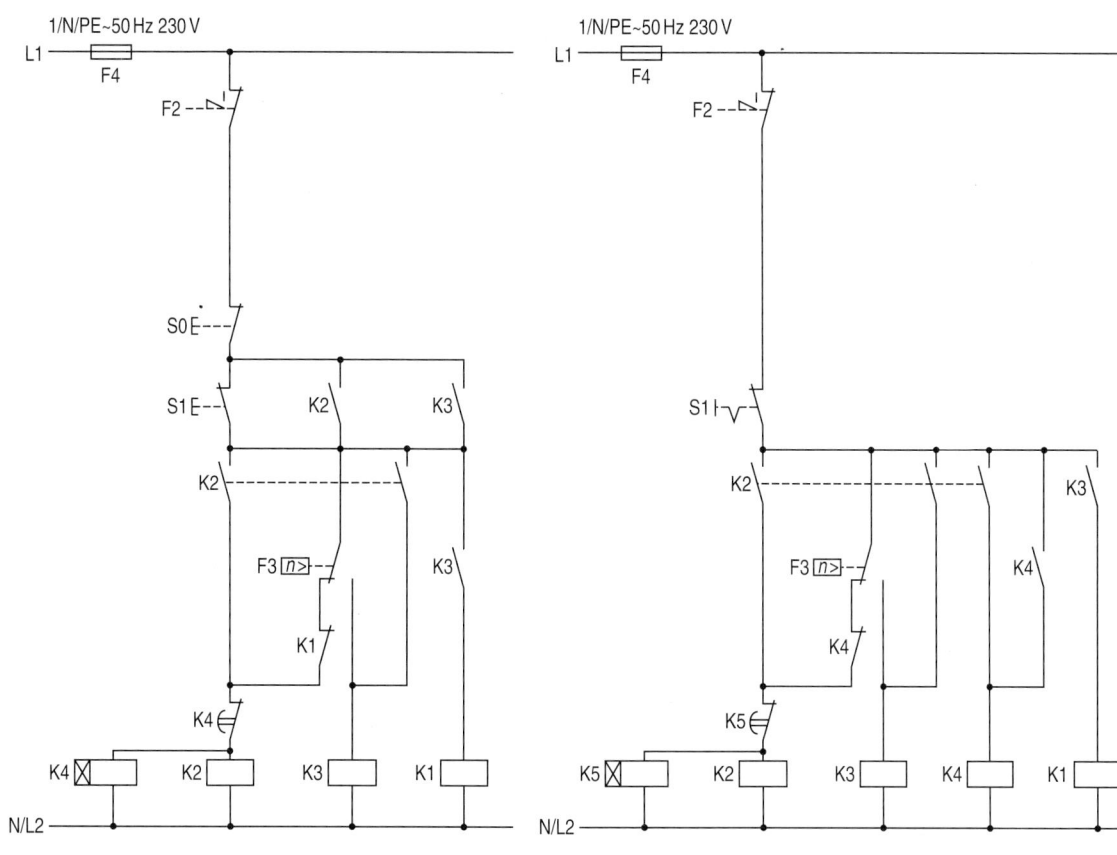

b) Auxiliary circuit for momentary-contact control

Required auxiliary contacts:

Contactor K1 with 1NC,
Contactor K2 with 3NO,
Contactor K3 with 2NO,
Momentary-contact switch S0 with 1NC,
Momentary-contact switch S1 with 1NO,
Time relay K4 with 1 NC, pick-up delayed,
Conveyor belt monitor F3 with 1CO,
Overload relay F2 with 1NC

c) Auxiliary circuit for maintained-contact control

Required auxiliary contacts:

Contactor K2 with 3NO,
Contactor K3 with 1NO,
Contactor K4 with 1NO+1NC,
Maintained-contact switch S1 with 1NO,
Time relay K5 with 1NC, pick-up delayed,
Conveyor belt monitor F3 with 1CO,
Overload relay F2 with 1NC

8 Fundamental circuit diagrams

8.5.3 Circuits of contactor control systems with pressure monitors

Figure 8.46 illustrates the use of a pressure monitor in the contactor control system for a three-phase induction motor.

Contactor K1 is energized and de-energized by pressure monitor F3 via its NC contact. Thus, the motor is switched on and off by the direct-on-line starting method.

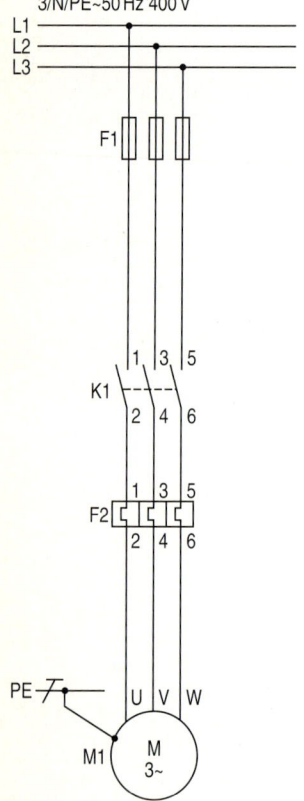

a) Main circuit

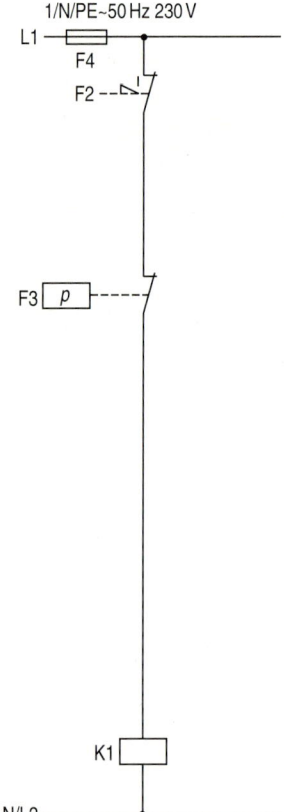

b) Auxiliary circuit

Figure 8.46
Circuit diagrams of a contactor control system with a pressure monitor

Required auxiliary contacts:
Pressure monitor F3 with 1NC,
Overload relay F2 with 1NC

8.6 Circuits with position switches

8.6.1 Reverser circuits with position switches (e.g. gate control)

Figure 8.47 illustrates the circuit diagrams for a reversing starter with position switches used to open and close a gate or door (gate control).

Pushbutton control/momentary contact (Fig. 8.47b)

Closing the gate: Contactor K1 is energized by the actuation of momentary-contact switch S1 and the closing of its NO contact. The sealing-in contact K1 closes. The motor runs up in clockwise direction and the gate is closed. When the gate is completely closed, position switch S3 operates. Contactor K1 is de-energized and the motor is switched off.

Opening the gate: Contactor K2 is energized by the actuation of momentary-contact switch S2 and the closing of its NO contact. The sealing-in contact K2 closes. The motor runs up in anti-clockwise direction and the gate is opened.

When the gate is completely open, position switch S4 operates. Contactor K2 is de-energized and the motor is switched off.

OFF-pushbutton S0 is provided to stop the opening and closing of the gate in intermediate positions.

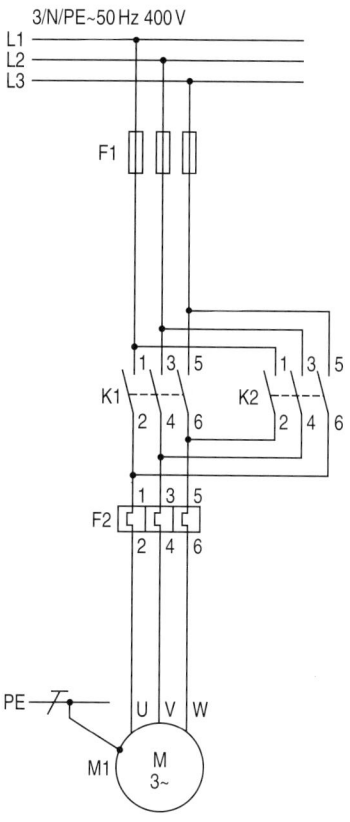

a) Main circuit

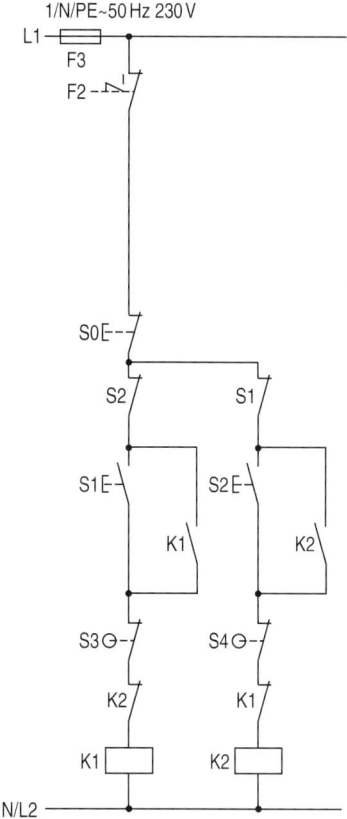

b) Auxiliary circuit

Figure 8.47
Circuit diagrams of a reversing starter with position switches for the opening and closing of a gate

Required auxiliary contacts:

Contactors K1, K2 each with 1NO+1NC,
Momentary-contact switch S0 with 1NC,
Momentary-contact switches S1, S2 each with 1NO+1NC,
Position switches S3, S4 each with 1NC,
Overload relay F2 with 1NC

8.6.2 Position switches with indicator lights

An indicator light H1 can be connected directly across (parallel to) the NO or NC contact of position switch S1, which energizes contactor coil K1 (Fig. 8.48a). When the contactor coil is de-energized, indicator light H1 is illuminated. H1 and K1 are connected in series. The lamp resistance, however, is so high that the contactor coil is not energized via the lamp itself. When the contactor coil K1 is energized, indicator light H1 is bridged out and switched off.

The 3SE3 position switches (see Siemens l.v. switchgear catalogue) can be delivered with integrated lamp holders, and thus the indicator light can easily be connected across the NO or NC contact of the position switch within the housing of the switch itself.

If contactor coil K1 is energized via the NC contact of position switch S1 and indicator light H1 via the NO contact of S1 (or vice versa), an indication is given when contactor coil K1 is de-energized (Fig. 8.48b).

If contactor coil K1 and indicator light H1 are energized via the same contact of position switch S1 (NO or NC contact), an indication is given when the contactor coil is energized (Fig. 8.48c).

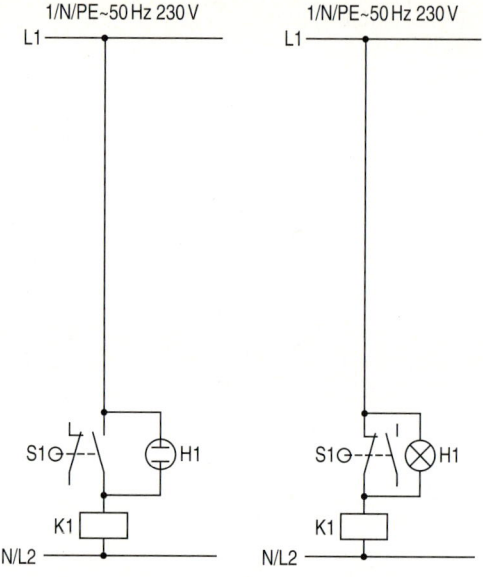

a) Indication with the contactor coil de-energized; indicator light connected across the contact of the position switch

Figure 8.48
Circuit diagrams for position switches with indicator lights

Position switches with indicator lights

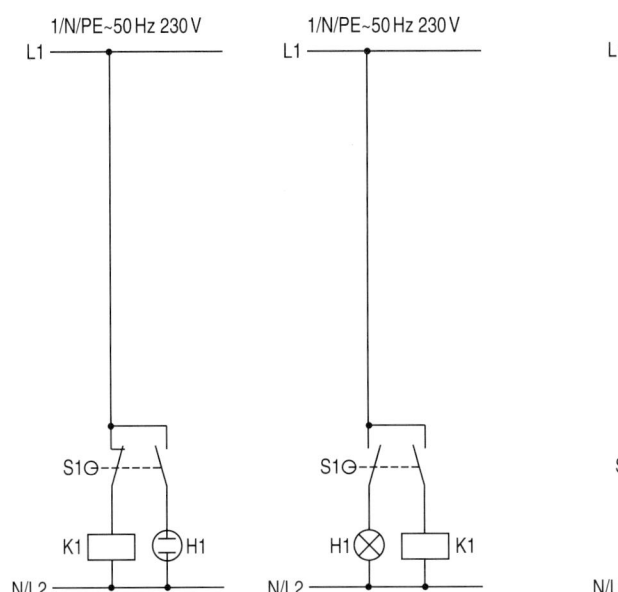
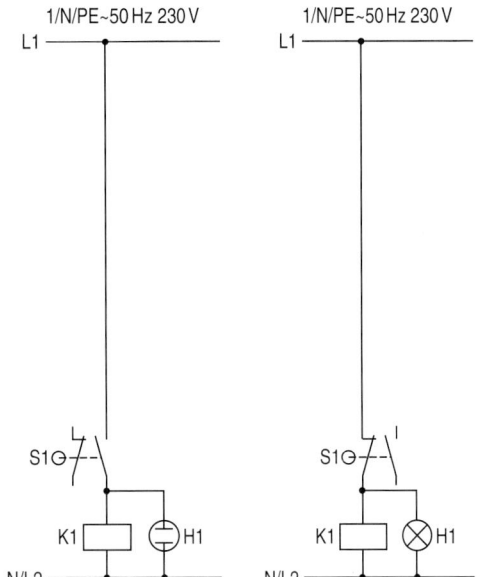

b) Indication with the contactor coil de-energized; indicator light connected in parallel with the contactor coil

c) Indication with the contactor coil energized; indicator light connected directly across the contactor coil

8.7 Terminal blocks

8.7.1 Circuits with isolating terminal blocks for current transformers

Isolating line-up terminals can be used in switchboards, control rooms and the like, for testing, isolating and switching of circuits without having to interrupt the operation.

Two terminal blocks can be connected in parallel by the insertion of a link enabling, for example, the current transformer to be shunted out or bridged on its secondary side. In this way measuring instruments can be replaced during operation (Fig. 8.49).

Testing:

(Current transformer without outgoing circuit, Figure 8.49a)

Test instrument P10G is connected to the sockets of terminal 2 (circuit 1, terminal strip X1) and the isolating link of terminal 2 is opened. The test instrument P10G is now connected in series with measuring instrument P1N.

Similarly, in circuit 2 (current transformer with outgoing circuit, Figure 8.49b), test instrument P10G is connected to the sockets of terminal 1 (circuit 2, terminal strip X2), and the isolating links of this terminal is opened. The test instrument is connected in series with measuring instrument P1N and P2N. Figures 8.49 c and d illustrate the corresponding test configurations with three current transformers without outgoing circuits and without or with a common star-point connection.

Replacing measuring instruments during operation:

Table 8.11 illustrates the switching methods for replacing measuring instruments in circuits 1 to 4 during operation.

Further examples and notes on terminal blocks are given in the Siemens low-voltage switchgear catalogue.

Table 8.12
Replacement of measuring instruments during operation using isolating line-up terminals, with reference to circuits 1 to 4, Figure 8.49

For shunting out the current transformers of				Replacement of measuring instrument
Circuit 1	Circuit 2	Circuit 3	Circuit 4	
Insert link on terminal strip X1 between terminals				
1 and 2	2 and 3	1 and 2	2 and 3	P1N
–	3 and 4 (1 and 2)[1]	3 and 4	4 and 5	P2N
–	–	5 and 6	6 and 7	P3N
–	1 and 2	–	–	all

[1] Terminal strip X2

Circuits with terminal blocks for current transformers

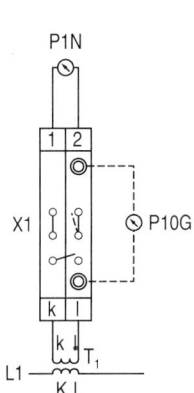

a) Circuit 1:
   one current transformer without outgoing circuit

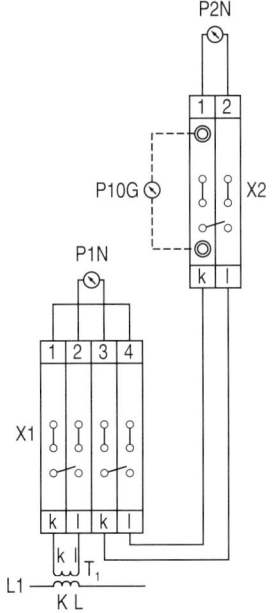

b) Circuit 2:
   one current transformer with outgoing circuit

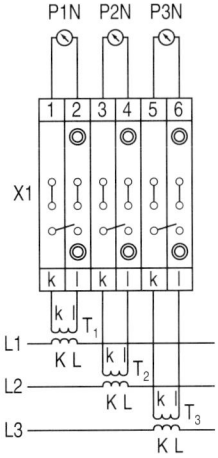

c) Circuit 3:
   three current transformers without outgoing circuit and *without* common star-point connection

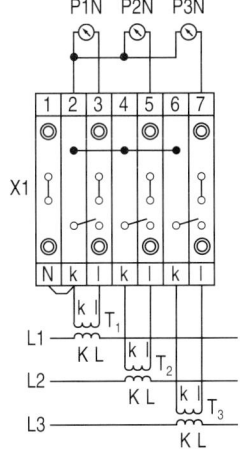

d) Circuit 4:
   three current transformers without outgoing circuit and *with* common star-point connection

Figure 8.49 Circuits with line-up terminals for current transformers

8.7.2 Circuit-breaker terminals for auxiliary circuits

Circuit-breaker terminals[1] serve as line protection in auxiliary circuits. Their main field of application is in the control wires leading to control devices, magnetic valves and other loads. They meet the market demands for a fuseless design. As the acquiring and stocking of fuses is not necessary, the downtime of an electrical installation after a shut-down due to a fault, is considerably reduced.

By mounting the circuit-breaker terminals directly onto the terminal strip (mounting rail), a direct positional correlation between the circuit-breaker and the protected auxiliary conductor, can be achieved.

The advantages are clear: In the event of a tripping operation, the indication is given directly at the auxiliary circuit concerned, thus trouble-shooting is significantly simplified. During commissioning, for example, or re-arrangement, the circuits can easily be connected and disconnected individually.

Where-ever applicable, the circuit-breaker terminals conform to the "Specifications for low voltage switchgear, miniature circuit-breakers" DIN VDE 0660 Part 1, as well as to IEC publication 947-2 and the "Specifications for terminal blocks for the connection of copper conductors" DIN VDE 0611 Part 1, 11.77 (Draft 12.85).

The circuit-breaker terminals with auxiliary contact are equipped with a NC contact which is open when the main circuit is closed and which closes upon tripping. Thus, a trip indication is facilitated. In addition, a feed-through connection for the return line of the protected circuit is provided (Fig. 8.50). All the screw-terminals are designed for the connection of two conductors, e.g. for the parallel connection of the auxiliary contacts to indicate a group alarm (Fig. 8.51).

A pushbutton-type toggle is provided for manual operation. In conjunction with the breaker latching mechanism, this button has the following functions:

Reset after a tripping operation by depressing the button,

Switching off by pulling out the button,

Trip indication by the popping up of the button,

Trip-free feature, i.e. tripping of the breaker even if the button is held in the depressed position.

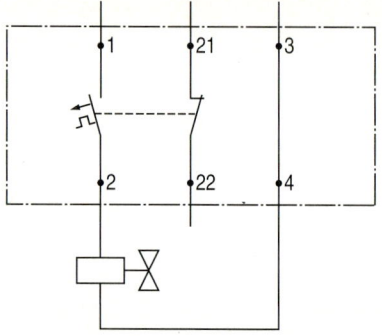

Figure 8.50
Use of the fitted feed-through connection for the return line from the load

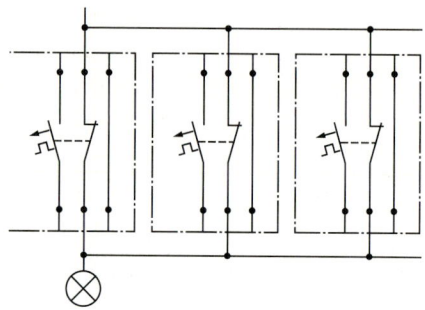

Figure 8.51
Parallel connection of the auxiliary contacts for group alarm indication

Selection

Inductive a.c. loads, such as contactor coils and magnetic valves, cause inrush current peaks of up to ten times their continuous current. This must be taken into consideration when selecting circuit-breaker terminals with an instantaneous short-circuit release to ensure that they do not trip as a result of the initial transient inrush current. The tripping characteristics of the circuit-breaker terminals with short-circuit release are illustrated in Figure 8.52. If necessary, it should also be checked that the connected conductors are of sufficient dimension for the selected circuit-breaker.

When using a circuit-breaker terminal incorporating an inverse time-delayed overload release, a device with a lower rated continuous current may be selected, since in this case, the short-circuit release trips at higher current values (Fig. 8.53).

---

[1] Circuit-breakers in line-up terminal form

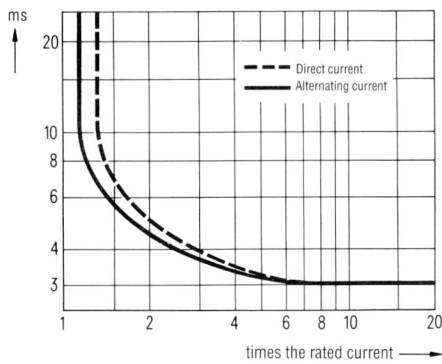

Figure 8.52
Tripping characteristics of the circuit-breaker terminals with short-circuit release (n-release)

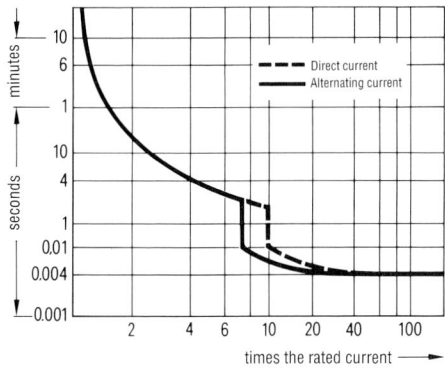

Figure 8.53
Tripping characteristics of the circuit-breaker terminals with combined overload and short-circuit release (an-release) at an ambient temperature of $+40\,°C$

Examples

Given: Magnetic valve AC 24 V, 50 Hz, 20 VA.

Required:

1. Circuit-breaker terminal with short-circuit release

   Selection according to the switch-on current I_u:
   Continuous current $I_u = 20\,\text{VA} : 24\,\text{V} = 0.83\,\text{A}$,
   Switch-on current = 10 times the continuous current = 8.3 A

   Selected:

   Circuit-breaker terminal 10 A
   Order number 8WA1011-1SF28 from Siemens l.v. switchgear catalogue.

2. Circuit-breaker terminal with overload and short-circuit release

   Selection according to the continuous current I_u:
   Continuous current $I_u = 20\,\text{VA} : 24\,\text{V} = 0.83\,\text{A}$,

   Selected:

   Circuit-breaker terminal 2 A
   Order number 8WA1011-2SF25 from Siemens l.v. switchgear catalogue.

   Check:

   The starting current is 10 times the continuous current = 8.3 A. This is 4.15 times the rated continuous current of 2 A. In this case, the short-circuit release does not trip upon switch-on.

8.8 Circuits with leakage-current (residual-current) protective devices

Figure 8.54 illustrates circuit protection using a circuit-breaker with a 3UL2 residual-current protective device. If there is no leakage fault current, "Ready" is indicated (H2) and undervoltage release F7 in the circuit-breaker is energized. Circuit-breaker Q1 can be switched on.

During operation, a leakage current or earth fault may occur in the monitored circuit and/or in the load, e.g. owing to faulty insulation. If the leakage current exceeds the value of the rated leakage (or residual) current $I_{\Delta n}$, the voltage induced in the secondary winding (Z1, Z2) of summation (core-balance) current transformer T1 causes the operation of the release F6 in tripping unit F5. The NC contact (15, 16) and the NO contact (27, 28) are operated mechanically by residual current release F6. This has the effect that undervoltage release F7 trips circuit-breaker Q1 and the indication "Earth-current fault" (H1) lights up. At the same time the "Ready" indication is extinguished. Circuit-breaker Q1 opens and the point of fault current is disconnected from the network.

As the residual current release F6 latches mechanically upon operation, the reset pushbutton R on the residual-current tripping device F6 must be actuated before circuit-breaker Q1 can be switched on again.

In order to carry out a function test, the secondary winding (Z1, Z2) of summation current transformer T1 is excited via the test winding (A1, A2) using test pushbutton T (S1). A test current flows corresponding to fault current $I_{\Delta n}$. The residual current release F6 is energized and causes the NC contact (15, 16) to open, thus interrupting the test circuit.

Figure 8.55 illustrates the monitoring of a melting furnace using a 3UL2 residual-current protective device. In the case of conductive melt material and a fireclay lining too thin or faulty due to ageing, current flows from the conducting material of the melt to the earthed outer wall.

If this earth-fault current exceeds the value of the rated leakage (or residual) current $I_{\Delta n}$ in the secondary winding (Z1, Z2) of summation current transformer T1, the horn is switched on via the NO contact (27, 28) of tripping unit F5. In addition, the NC contact (15, 16) is operated. Thus the function of the 3UL2 residual-current protection device can only be tested in the reset mode, i.e. after actuation of the reset pushbutton R.

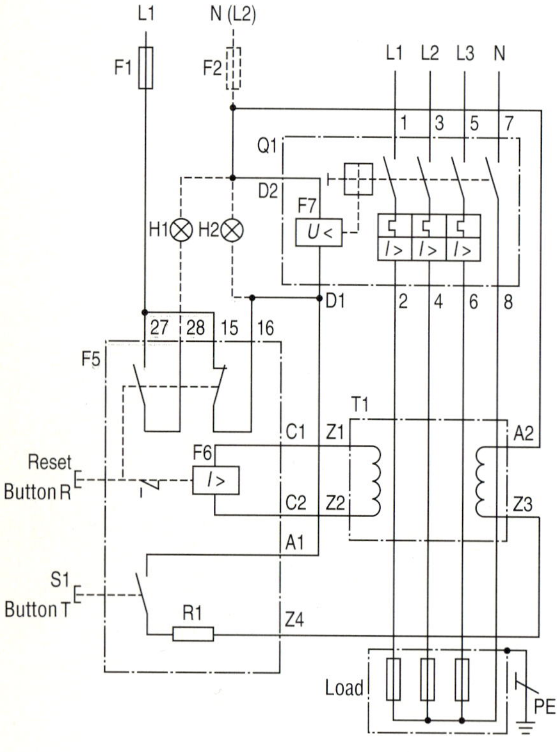

F1, F2 Control circuit fuses[1]
F5   Tripping relay
F6   Residual current release
F7   Undervoltage release
H1   Indication "Earth current fault"
H2   Indication "Ready"
Q1   Circuit-breaker
Reset Reset pushbutton R
R1   Test resistor
T1   Summation current transformer
S1   Test pushbutton T

[1] F2 may be omitted if the voltage of the test circuit is tapped between the phase and the neutral conductor, alternatively from the secondary of a control voltage transformer

Figure 8.54
Protection by disconnection using a circuit-breaker in conjunction with a 3UL2 residual-current protection device. Tapping of the voltage for the test circuit upstream of the circuit-breaker or from an external voltage source (interruption by auxiliary contact 15-16 is necessary)

Circuits with leakage-current protective devices · Interface units

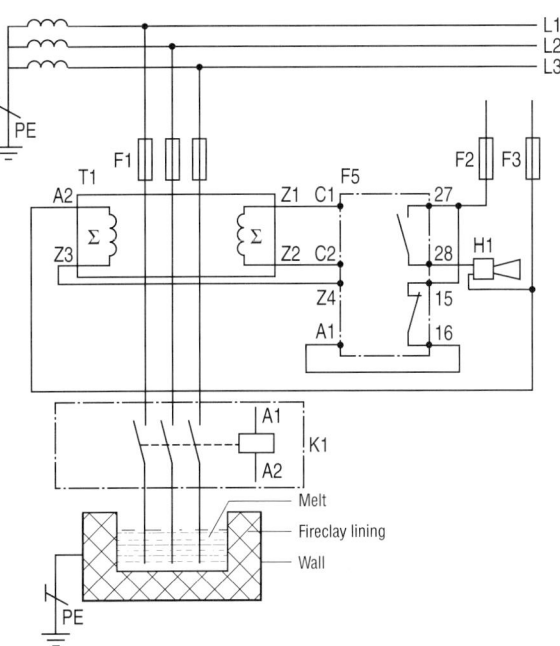

Figure 8.55
Example for the monitoring of "melting furnaces" using the 3UL2 residual-current protection device

units are marked in accordance with DIN EN 50005. Each interface unit is provided with a blank labelling plate on which its equipment code may be specified.

An LED indicates the operation of the interface unit. An internal bridge-connected rectifier enables the coil to be energized by a.c. or d.c. and serves to suppress overvoltage spikes upon de-energizing the coil.

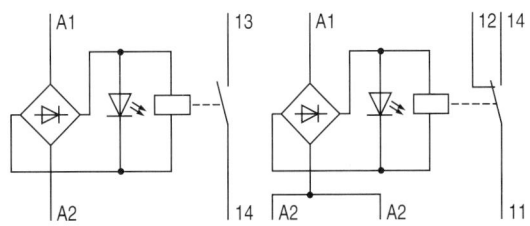

Figure 8.56
Unit wiring diagrams of terminal-type interface units

8.9 Interface units

Interface units simultaneously fulfill a number of functions:

▷ Interconnection of switchgear devices and installations with differing signal levels, especially the interconnection of electronic systems with electromechanical devices

▷ Galvanic separation of input (e.g. coil) and output (e.g. contacts) within the interface unit

▷ Arresting or trapping of overvoltages due to making-breaking processes or noise interference so that these are not transferred

▷ Amplification of weak signals. This allows, for example, the output loading of an electronic system to be kept low and permits a more compact configuration of the electronic system.

Terminal-type interface units (Fig. 8.56) are especially suitable for mounting directly on the line-up terminal rail with other terminals. The input and output interface units differ from each other by the location of their terminals (Fig. 8.57). The terminal-type interface

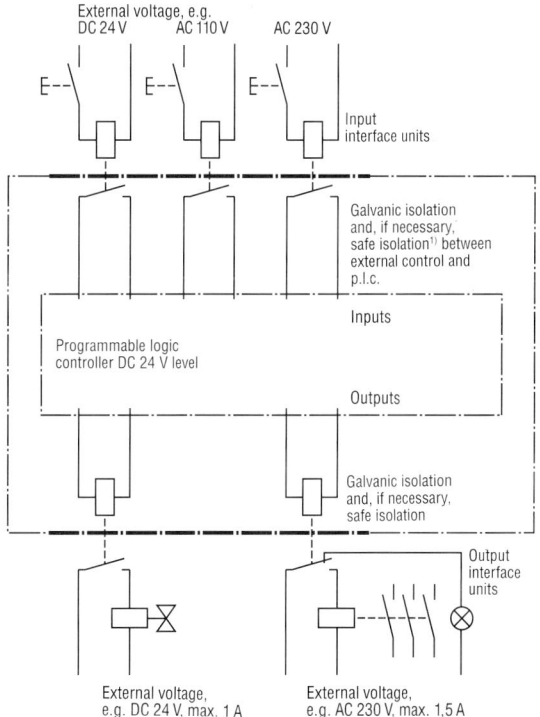

[1] Depending on the type of interface unit used

Figure 8.57 Input and output interface units

487

8 Fundamental circuit diagrams

8.10 Auxiliary circuits incorporating time relays

8.10.1 Star-delta starting of three-phase induction motors with star contactor, delta contactor, line contactor and time relay

Pushbutton control/momentary contact (Fig. 8.58a)

Switching on: The actuation of the momentary-contact switch S1 energizes star-delta time relay K4 and star contactor K2 via the instantaneous NO contact 17/18 of the time relay. Sealing-in contacts K1 and K2 close. The motor runs up in the star connection.

Change-over: After the preset change-over delay time on K4 has elapsed, its instantaneous NO contact K4 opens and the star contactor K2 is de-energized. When the delayed NO contact K4 of the star-delta time relay closes after the change-over delay of 50 ms, delta contactor K3 is energized via the NC contact of K2. The motor runs in the delta connection.

Switching off: Line contactor K1 is de-energized by the actuation of momentary-contact switch S0 and the opening of its NC contact; the NO contact of K1 opens. Delta contactor K3 is de-energized and the motor is switched off.

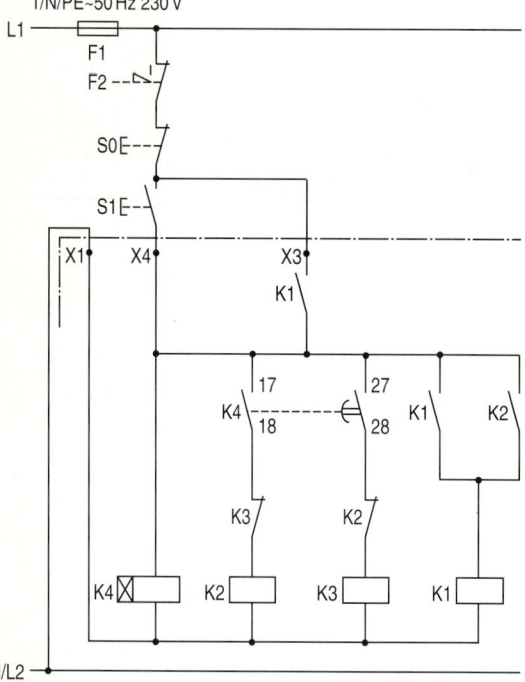

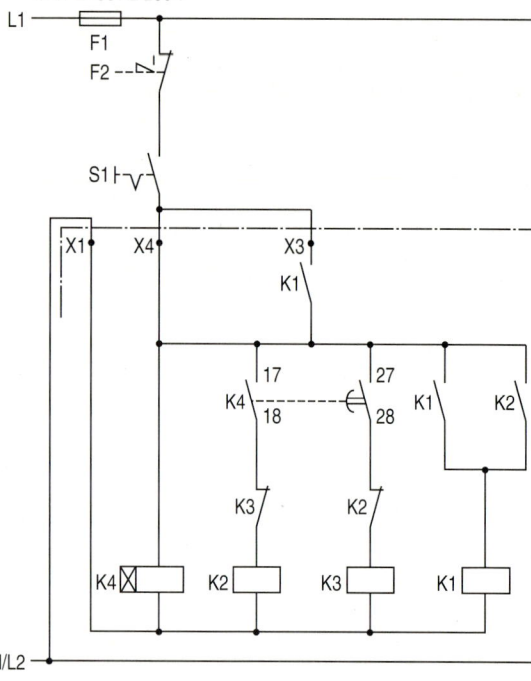

a) Auxiliary circuit for momentary-contact control

Required auxiliary contacts:

Momentary-contact switch S0 with 1NC,
Momentary-contact switch S1 with 1NO,
Time relay K4 with 1NO inst. + 1NO pick-up delayed,
Contactor K1 with 2NO,
Contactor K2 with 1NO + 1NC,
Contactor K3 with 1NC,
Overload relay F1 with 1NC

Figure 8.58
Auxiliary circuits for the star-delta starting of a three-phase induction motor with star contactor, delta contactor, line contactor and time relay

b) Auxiliary circuit for maintained-contact control: New start required after voltage recovery following temporary failure of control supply voltage

Required auxiliary contacts:

Maintained-contact switch S1 with 1NO,
Time relay K4 with 1NO inst. + 1NO pick-up delayed,
Contactor K1 with 2NO,
Contactor K2 with 1NO + 1NC,
Contactor K3 with 1NC,
Overload relay F1 with 1NC

The auxiliary contacts of the contactors which have not been mentioned above are used for electrical interlocking between the star and delta stages.

Switch control/maintained contact (Fig. 8.58b and 8.58c)

The contactors are energized and de-energized by the operation of the maintained-contact switch S1 in a similar way as is described for momentary-contact control above.

1/N/PE~50Hz 230V

c) Auxiliary circuit for maintained-contact control: Continued operation in delta after short-time (e.g. < 50 ms) failure of control supply voltage

Required auxiliary contacts:
Maintained-contact switch S1 with 1NO,
Time relay K4 with 1NO inst. + 1NO pick-up delayed,
Contactor K1 with 2NO,
Contactor K2 with 1NO + 1NC,
Contactor K3 with 1NC + 1NO,
Overload relay F1 with 1NC

Behaviour in the event of temporary failure of control supply voltage or voltage dip during operation in the delta connection

In the event of control voltage failure or even a short voltage dip, contactors K1 to K3 and time relay K4 may drop out. If the voltage returns within the minimum required recovery time of the star-delta time relay K4 (typically 50 ms), the delayed NO contact 27/28 of K4 may re-close immediately. If this happens, none of the contactors are, however, re-energized and the starter may remain "frozen" in this intermediate state. To restart the motor, the starter must first be switched off and then on again (e.g. by opening and reclosing S1 in the maintained contact conguration).

To achieve an automatic continued running of the motor (in delta) after a voltage dip or temporary voltage failure of a duration shorter than the minimum required recovery time of K4, the following changes to the standard control circuit are required (Fig. 8.58c):

▷ A bridge (or link) must be inserted between terminals 17 and 27 of the time relay, and

▷ an additional NO contact K3 is required for the switching of K1.

If the duration of the supply voltage failure is longer than the recovery time of the time relay, then this circuit will initiate an automatic restart from the star connection stage. If this is not wanted, then e.g. a circuit-breaker with a delayed undervoltage release may be fitted in the auxiliary supply circuit.

Figure 8.59 illustrates the function diagram of the star-delta time relay.

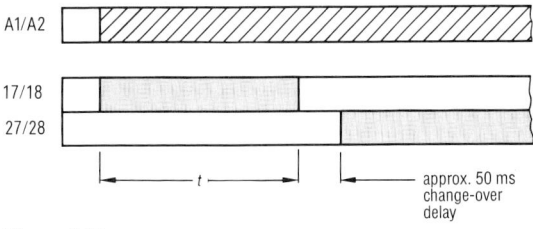

Figure 8.59
Function diagram of the star-delta time relay

8.10.2 Functions of the motor-driven time relay 7PR4140

Since the motor and the internal clutch relay of time relay 7PR4140 can be energized separately, the unit can be operated with or without the function "time addition". In this configuration, the selector button on the unit must be set to the position "without zero voltage latch". Figure 8.60 illustrates the function diagrams for the time relay with and without time addition.

Operation with time addition

When energizing the internal clutch relay (terminals B1/B2), the motor is engaged and the instantaneous NO contact (33/34) closes. The timing period starts upon the energizing of the motor (A1/A2). If the motor operation is interrupted, the timing period is also interrupted but continues upon voltage recovery. After the preset time has elapsed, both the delayed change-over contact (15/16 to 15/18) and the delayed NO contact (27/28) are operated. When the clutch relay is de-energized, the motor is disengaged and all contacts return to their initial position.

Operation without time addition

The internal clutch relay (terminals B1/B2) is energized with the motor already running („flying start") and the instantaneous NO contact (33/34) is operated. When the relay operation is interrupted, the motor is disengaged, the timing period is reset and the instantaneous NO contact opens. After voltage recovery, the timing period starts anew.

The repeat accuracy is improved in the time range of up to 6 s due to the "flying start". Based on the reference conditions (rated voltage, ambient temperature $+25\,°C$), the accuracy is $\pm 0.3\%$.

The 7PR4140 time relay can also be operated by energizing both motor and internal clutch relay simultaneously. The zero voltage latch may then be switched on or off as required using the selector button on the unit itself.

If the unit is operated with the button in position "with zero voltage latch", the internal clutch relay must not be de-energized separately during the timing period, since the delayed contacts will adapt an undefined state after the set time delay has expired!.

Drop-out delay (Off-delay) operation

The separate energizing of the motor and the internal clutch relay enables the use of the 7PR4140 in a drop-

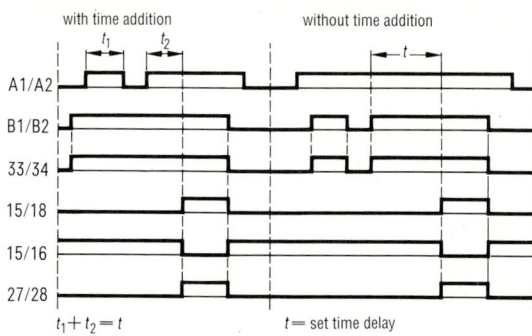

Figure 8.60
Time sequence function diagrams for the 7PR4140 time relay with and without time addition

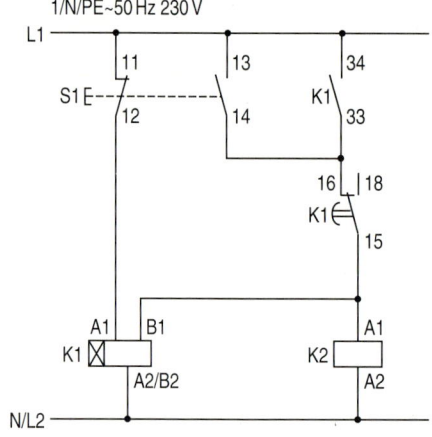

Figure 8.61
Circuit diagram of the 7PR4140 time relay in drop-out delay operation (OFF-delay)

out delay function in conjunction with an external clutch supply voltage during the timing period (Fig. 8.61). By the closing of NO contact S1, the internal clutch relay (terminals B1/B2) of time relay K1 is energized, instantaneous NO contact 33/34 closes and voltage is applied to load K2. Voltage is applied to the motor (terminals A1/A2) when the NO contact S1 opens and the NC contact S1 closes, and the timing period begins. At the end of the timing period, the delayed change-over contact changes over from 15/16 to 15/18 and thus disconnects the load.

8.11 Switching of an electrical heating system using a thermostat and contactor combination

Figure 8.62 illustrates the circuit diagrams for an electrical heating system controlled via a thermostat and contactor combination.

Switch control/maintained contact (Fig. 8.62b)

Switching on: The heating system is switched on manually using maintained-contact switch S1. The energizing command on contactor K1 only becomes effective once the NC contact of thermostat F3 has closed, thus energizing contactor relay K2 and closing NO contact K2. Thus, the heating elements R1 are switched on.

Switching off via F3: As soon as the room temperature has reached the preset value, the thermostat F3 operates. The NC contact of F3 opens and de-energizes contactor relay K2. The NO contact of K2 opens and de-energizes contactor K1. Heating elements R1 are switched off.

Switching off manually: The heating elements can be switched off manually by operation of the maintained-contact switch S1.

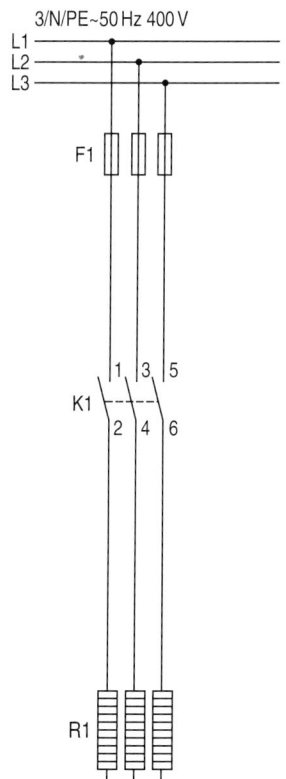

a) Main circuit

Figure 8.62
Circuit diagrams for an electrical heating system controlled via a thermostat and contactor combination

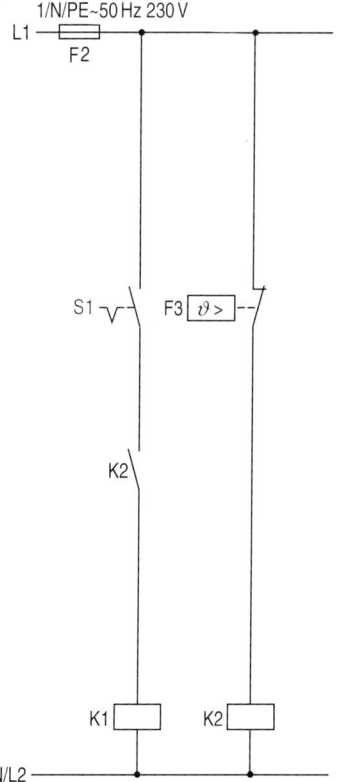

b) Auxiliary circuit for maintained-contact control

Required auxiliary contacts:
Maintained-contact switch S1 with 1NO,
Contactor K2 with 1NO,
Thermostat F3 with 1NC

8.12 Stand-by power supply installations

The need for autonomous stand-by power supply systems increases with the complexity of line-dependent systems and the severity of the consequences which may result from prolonged power failure. This is especially true in cases where high-tech units and systems are in operation, where information and communication are top priorities, where production, automation and administration take place and where even a small disturbance in the supply may be extremely detrimental.

The hierarchy and task distribution in stand-by power supply systems are structured as follows:

▷ The most sensitive loads are supplied without interruption via UPS systems (UPS = uninterruptable power supply). The period between power failure and start-up of the stand-by power supply system (diesel generator) is bridged by a battery.

▷ In addition to the UPS, the generator supplies power to all the critical systems necessary for smooth and safe operation during main supply failure (e.g. lighting systems, air conditioning, elevators, etc.).

The power supply of critical loads, however, can only be guaranteed if the UPS is an integrated component of the whole system. Uninterruptable power supply systems ensure that voltage fluctuations, temporary failures and power dips do not have any adverse effect on the system. This must be true regardless of whether the disturbances are erratic and last only a few milliseconds or whether they last for a few minutes.

The main components of a UPS are rectifiers, inverters, static bypass switches (SBS) and manually-operated bypass switches. Combined with a battery as back-up supply, this configuration provides an absolutely uninterruptable power supply.

Owing to the interaction and interdependence of the various system components, particular care must be taken in their selection to ensure optimal performance during operation in the stand-by power supply mode.

In the event of power failure, the stand-by power supply system is usually expected to take over the power supply automatically. Electrical starting equipment, consisting of starter, starter battery and generator, is required for the automatic operation of the diesel engine.

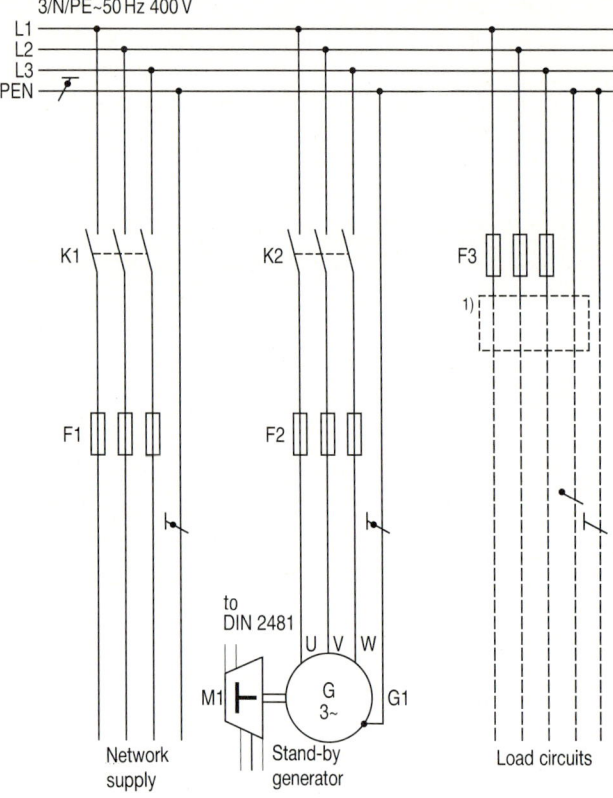

[1] Earth-leakage protection required! (DIN VDE 0100 Part 728)

a) Main circuit

Figure 8.63
Circuit diagrams for a three-pole change-over from network supply to stand-by supply using contactors

In automatic operation, the change-over (or load transfer) from network supply to generator supply is initiated by a voltage monitor which in turn is operated by a control unit connected to the battery supply.

In addition to automatic operation, manual and test operation is generally also provided.

Switching of stand-by power supply installations

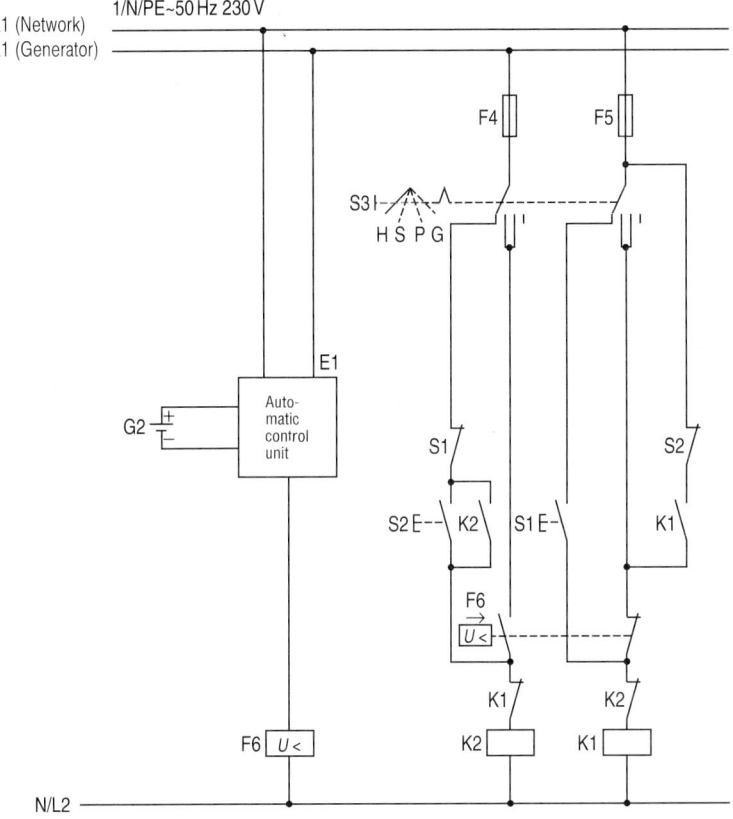

b) Auxiliary circuit

Required auxiliary contacts:

Contactors K1, K2 each with 1NO+1NC,
Pushbuttons S1, S2 each with 1NO+1NC,
Voltage monitor F6 with 1NO+1NC
Selector switch S3 with 4 switch positions

8.12.1 Three-pole change-over from network supply to stand-by supply using contactors (generator operation)

Figure 8.63 illustrates an economical solution for three-pole change-over (or transfer) from network supply to stand-by supply via contactors.

Operation (Fig. 8.63b)

Selector switch S3 in position "Manual" (H): The generator supply is switched on by the actuation of pushbutton S2. The network supply is switched on by the actuation of pushbutton S1.

Selector switch S3 in position "Automatic" (S): In the event of a power failure, the NC contact of voltage monitor F6 opens and the NO contact of F6 closes. Line contactor K1 drops out. NC contact K1 closes and energizes generator contactor K2. The automatic control unit starts the generator. Upon line voltage recovery, the NO contact of voltage monitor F6 opens. Generator contactor K2 drops out. The automatic control unit switches off the generator. The NC contact of voltage monitor F6 and the NC contact of generator contactor K2 close. Line contactor K1 is energized and the line supply is re-connected.

8 Fundamental circuit diagrams

8.12.2 Change-over from network supply to stand-by supply with four-pole disconnection of the distribution system via two three-pole contactors

Figure 8.64 illustrates the circuit diagrams for the change-over (or load transfer) from network supply to stand-by supply with four-pole disconnection of the distribution system using two three-pole contactors.

Operation (Fig. 8.64b)

Selector switch S3 in position "Manual" (H): The generator can be switched on via generator contactor K2 using pushbutton S2. The neutral contactor K3 and line contactor K1 are energized by the actuation of pushbutton S1.

Selector switch S3 in position "Automatic" (S): In the event of a power failure, the NC contact of voltage monitor F6 opens and the NO contact of F6 closes. The neutral contactor K3 and the line contactor K1 are de-energized via the NC contact of F6.

Generator contactor K2 is energized via the NO contact of F6. The energizing command on K2 only becomes effective once the NC contact of K1 has closed. K2 changes over to generator operation.

Upon voltage recovery, the NO contact of voltage monitor F6 opens. Contactor K2 and the generator are switched off. The NC contacts of F6 and K2 close, the neutral contactor K3 and line contactor K1 are energized.

If contactor K1 or K3 is not switched on or off as required, a fault is indicated by H1.

For further details on three-pole or four-pole disconnection of the distribution system with respect to stand-by power supply systems, see DIN VDE 0100, part 728.

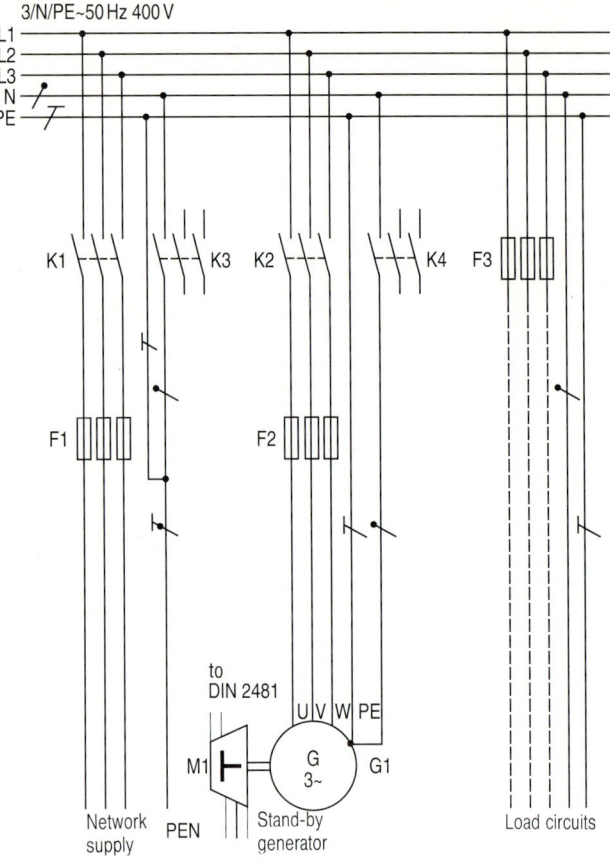

a) Main circuit

Figure 8.64
Circuit diagrams for the change-over from network supply to stand-by supply with four-pole disconnection of the distribution system using two three-pole contactors.

Switching of stand-by power supply installations

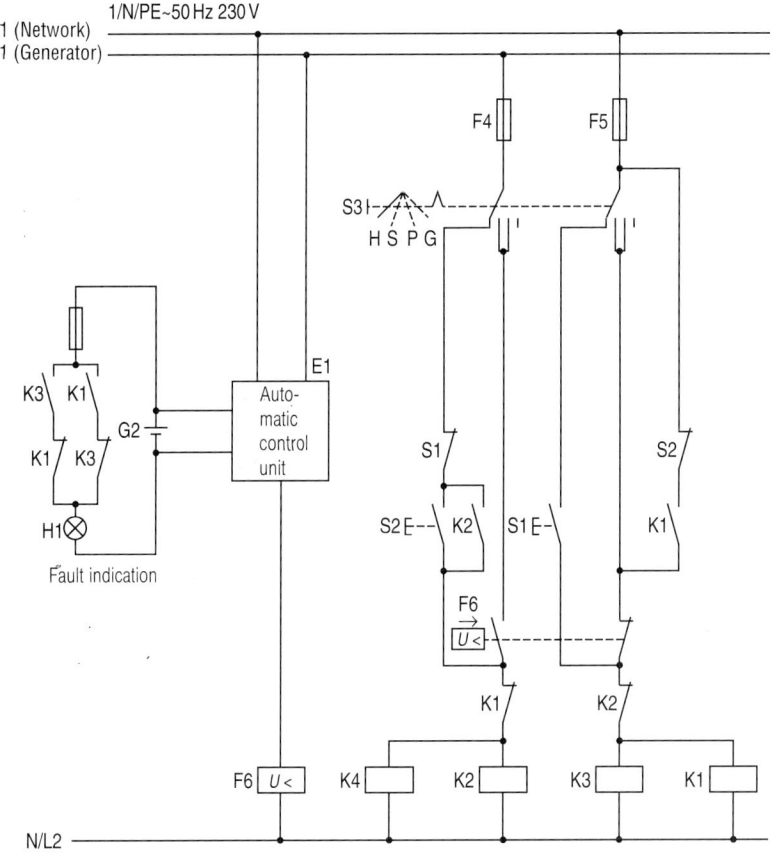

b) Auxiliary circuit

Required auxiliary contacts:

Contactor K1 with 1NO+1NC,
Contactor K2 with 1NO+1NC,
Contactor K3 with 2NO+1NC,
Pushbuttons S1, S2 each with 1NO+1NC
Voltage monitor F6 with 1NO+1NC
Selector switch S3 with 4 switch positions

8.13 Project planning and engineering aids

Unless otherwise indicated, the project planning and engineering aids listed below may be ordered via your local Siemens branch or agent. Most of the items are available in English – please enquire for the latest editions and translations.

Switchgear selection slide-rule

The switchgear selection slide-rule is available for the most commonly used motor feeder circuits in drive engineering. All switchgear items represented in the corresponding circuits can be read off in windows. The order numbers of the devices are also given in these windows, facilitating their correct identification. After setting to the motor rating on the slide rule, the components can be dimensioned and selected for the corresponding circuit and the given three phase operating voltage e.g. 400 V, 50 Hz.
Order number E20001-P285-Y167.

Adhesive switchgear labels

The time-consuming repetitive drawing of circuit diagrams can be avoided by using the adhesive labels which also guarantee uniform representation. Owing to the increasing use of CAD systems and computer workstations, however, the labels are steadily losing their popularity. Therefore, they are only still supplied occasionally.

Pre-printed order forms for command devices and control switches with customized switching and contact arrangements

Order forms for command devices and control switches are provided to facilitate the correspondence regarding quotations and orders and to avoid misunderstanding of the customers needs.
For further details, see Siemens low-voltage switchgear catalogue.

Design drawing and lay-out templates for distribution systems

Design drawing and lay-out templates in a scale of 1:10 are available to facilitate the selection and configuration of the 8HP and 8HS distribution systems, as well as for the control mounting and wiring systems 8LW and 8LX:

For distribution systems:
  8HP, Order no. E 20001-A285-Y156,
  8HS, Order no. E74/5056 and A19100-E74-E266,

for control mounting and wiring systems:
  8LV/8LW/8LX, Order no. A19100-E74-W102,

for SIKUS free-standing distribution boards:
  8GF, Order no. A19100-J21-Y226.

Templates and graph paper for grading diagrams of circuit-breakers

The tripping characteristics of the circuit-breakers 3VE, 3VU, 3VT, 3VN, 3WE and 3WN are available as templates in a standardized DIN scale. Used in combination with double-logarithmic graph paper, these templates provide a quick and simple method of sketching circuit-breaker tripping characteristics and grading diagrams.

Templates
  for 3WE, Order number A19100-E732-E153,
  for 3V., Order number A19100-E732-E344,
  for 3WN, Order number A19100-E732-W108.

Double logarithmic graph paper (A4 pad)
  Order number A19100-E732-E104.

Calculation sheets for the calculation of short-circuit currents in low-voltage networks to DIN VDE 0102 can be ordered under the number H30-E1056-N41.

Planning sheets for distribution systems:
Please refer to the Siemens catalogue "Produkte Systeme und Dienstleistungen – PSD" (German) for the relevant order numbers.

Computer programs (Software packages)

The following PC programs for the MS-DOS[1] operating system are available at a nominal fee as planning and selection aids:

"KUBS"
This software package assists in the calculation of single and three-phase short-circuit currents, current discrimination and back-up protection, as well as in the selection of Siemens circuit-breakers. The program is available in six languages (see also Section 2.1.5, page 93).
  Order number E86010-D1801-A107-A2.

"SIKOSTART 3RW22" (Handbook and software package) for the correct dimensioning and selection of the 3RW22 solid-state motor starters,
  Order number E20001-P285-A485-X-7600
            (including software)

"ELLE", Version 2.0
for the calculation of the electrical service life of a.c. contactors (multilingual),
  Order number E86010-D1802-A117-A2,

---
[1] MS-DOS is a registered trademark of the Microsoft Corporation

"ASIST"

for the selection of Siemens control circuit transformers, (multilingual),
  Order number E86010-D1802-A127-A1,

"PROKOM"

for the updating of commercial data for selected low-voltage switchgear products from the Siemens master data bank to customer systems.
  Order number for program disk E20002-D1803-A107-A2,
  Order number for PROKOM standard products data disk (pre-selection of most commonly used products) E20002-D1803-A607-A3

"PROCAT"

Complete data bank program containing the commercial data of the most commonly ordered low-voltage switchgear products. By means of this program, customers may e.g. select a range of products to produce their own price lists in ASCII format for importation into other application software programs or for printing out in standard format.
  Order number for program and data disks (multilingual) E20002-D1002-A107-A2

"AUSTER"

in off-line PC version (see below for description),
  Order number E20002-D1000-A107-A2 (basic-software)
  E20002-D1000-A307-A2 (low voltage product data)
  E20002-D1000-A387-A2 (installation material product data)

"SIPUK"

for the calculation and design of Siemens distribution systems incorporating domestic installation products (MS-DOS version 3.2 or higher),
  Order number 8GD9588 (SIPUK 8HP)
  8GD9396 (SIPUK-CAD, basic package)
  8GD9397 (for STAB and SIKUS distribution systems)

To be ordered from: Händler GmbH &Co KG
  P. O. Box 1341
  D-93413 Cham-Altenmarkt
  Germany

"SIEBKO"

for the design and engineering of power factor correction units,
  Order number E50001-V213-Y27-X-7600

Updating of commercial product data for computer applications using PROKOM

PROKOM provides the user with a software utility by means of which he can set up and maintain a data bank in an economical and trouble-free way. PROKOM supplies files which contain product-specific commercial data. This data may be used to facilitate project calculations, quotations and ordering procedures.

PROKOM can be run on any AT-compatible PC which is equipped with a hard disk and a floppy disk drive (minimum configuration). The operating system MS-DOS version 3.2x, 4.0 or MS-DOS 5.0 must be installed.

Initially, PROKOM is used to send a list of the relevant type/order numbers to Siemens (request file). To the data on this file, all the currently up-to-date commercial data (e.g. list price, description text, mass, packing unit, etc.) is added, and an answer file is created. Upon return of this answer file, the user can evaluate and select the data which is of importance to his application. The user's own disks are used for the data exchange and communication with Siemens. Typically, a 360 KB disk can store about 600 order numbers complete with all available commercial data; disks with a higher storage capacity can, of course, store more.

The sequence and the syntax of the order numbers in the answer file remain the same as in the request file. The commercial data of each product number can thus be copied directly into the user's own application data bank files. Furthermore, "master disks" are available which already contain the commercial data on selected low voltage switchgear in a ready-made answer file. If the data on other products is desired, it may be requested from Siemens (Department ASI 2, Werner von Siemensstr. 50, Erlangen, D-91052 Federal Republic of Germany) with the help of the PROKOM request file program as described above.

Interactive on-line procedures via main frame (BS2000)[1] computers

Product selection data using AUSTER on computer workstation systems

Depending on the experience of the user, the selection of task-specific products on the basis of their attributes and related characteristics is often a time-con-

---
[1] Operating system 2000 of the Siemens main frame computers

suming and difficult process. This is especially true if one considers the steadily increasing range of products and the increase in the rate of innovation.

As an alternative to paging through catalogues, the complete order numbers of products for a specific application can now be ascertained via an easy-to-understand interactive dialog, in which selections are made on the basis of the technical attributes and their related characteristics. The user interface has been harmonized for all product groups.

For a specified or selected order number, data pertaining to technical, commercial, text and graphic aspects, as well as digitized photographic representations (in the near future), can be called up to on the computer screen. This data may, naturally, also be printed out.

AUSTER (from the German *Auswahl Technischer Erzeugnisse*) is available as an on-line version for BS 2000 applications, and is also available as off-line versions for AT-compatible personal computers with the operating system MS-DOS, SINIX or UNIX-V [1]. For the off-line versions, users are supplied with disks containing the necessary software and the selected product groups.

Based on its five data libraries and its interactive selection dialog, AUSTER offers a uniform procedure for preparing quotations and orders, for the engineering and designing of electrical installations and for developing publications, catalogues and other product documentation.

Computer program NORIS, Part NA –
a specification search program for low-voltage switchgear and systems

Low-voltage switchgear and systems which are to be used worldwide, are not only subject to a number of national specifications, norms and standards, but also to many European and international stipulations.

The part "NA" of the NORIS (from the German "*Normungs-Informationssystem*") was created with this in mind, and offers a list of relevant and current specifications, norms and standards based on search criteria entered by the user.

– Research may be carried out, for example, in terms of specific products, countries or key words.

– Cross-references are provided in cases where the search criteria indicate relevance in further documentation.

– Information on planned changes, drafts and correlation with related standards is provided.

NORIS operates in dialog under the BS2000 operating system.

Computer program PROBAT for the engineering and design of 8PU low-voltage switchboards

PROBAT 8PU.011 (from the German "*Projektierungsbaustein mit Auswahltabellen*")

Functions:
Engineering and design of low-voltage switchboards in interactive dialog
Selection of equipment, calculation of the space requirements of auxiliary devices in the draw-out rack, calculation of the cubicle load, preparation of quotations, calculation, price calculations, order processing, placing of orders on suppliers, plant documentation.

Inputs:
The user is guided by means of an interactive dialog which prompts him for all the required input data and which carries out a number of plausibility evaluations and limit value checks.

Results:
Complete customer-orientated documentation for quotation purposes
Quotation text including a list of the designed draw-out units, cubicle lay-out, lists of components, specifications including material and installation costs, product lists (via DIVA interface), frontal view, circuit diagrams (via Malta/SICAD interface).

Advantages:
Quick and accurate preparation of quotations,
Processing of the orders from start to finish with a single computer-based procedure,
Exact calculation of costs,
No need for the entry of complete order numbers,
Simple copying of repeated draw-out units, cubicles and installation sections.

Operating mode:
BS 2000 [2] in interactive or batch mode.

---

[1] UNIX is a registered trademark of AT & T

[2] Operating system 2000 on Siemens main frame computers

9 Appendix

9.1 Fundamental equations, characteristic quantities and units of electricity

9.1.1 Fundamental equations of electrical engineering

$U = I \cdot R$ Ohm's law

$R = \dfrac{2 \cdot l}{\varkappa \cdot q}$ Resistance (forward and return line)

$R_T = \dfrac{u_{Rr}}{100} \cdot \dfrac{U^2}{P}$, $X_T = \dfrac{u_s}{100} \cdot \dfrac{U^2}{P}$ (Transformers)

$R_C \sim \dfrac{1}{F}$, $X_C = \dfrac{3 \cdot U^2}{P_c}$ (Capacitors)

$Z = \sqrt{R^2 + (X_L - X_c)^2}$, $Z = \dfrac{R}{\cos \varphi}$ (Impedance)

$u_{kr} = \sqrt{u_{Rr}^2 + u_x^2}$

$W = I^2 \cdot R \, t = \dfrac{U^2}{R} \cdot t = P \cdot t$ Joule heat

$P = U \cdot I$ Direct current power

$P = U \cdot I \cdot \cos \varphi$ Active (real) power of single-phase alternating current

$P = \sqrt{3} \cdot U \cdot I \cdot \cos \varphi$ Active (real) power of three-phase alternating current

$\eta = \dfrac{P_{out}}{P_{in}}$ Efficiency

U_e Rated operational voltage in V (in 2-wire systems across the two phases (conductors); in three-wire d.c. systems across the two outer lines; in three-phase systems across two lines (phase-to-phase))

I Current in a conductor in A
R Resistance in Ω
R_T Transformer resistance
X_T Transformer reactance
u_{kr} Impedance voltage in %
u_{Rr} Ohmic voltage drop in %
u_X Inductive voltage drop in %
W Electric work in Ws

P Active (real, effective) power in W
P_{out} Active (real, effective) power output in W
P_{in} Active (real, effective) power input in W
P_C Capacitor power
η Efficiency
$\varkappa$ Electrical conductivity in $\dfrac{S\,m}{mm^2} = \dfrac{m}{\Omega\,mm^2}$

(e.g., for copper ≈ 55, for aluminium ≈ 35, for silver ≈ 63, for steel wire ≈ 7 to 10, at 20 °C; see Table 9.14 on page 514).

$\cos \varphi$ Power factor
q Conductor cross-section in mm²
l Single length of the line section under consideration in m
t Time in s

Table 9.1 lists the equations for calculating the voltage drop Δu in conductors, as well as the equations for calculating the conductor cross-section.

Table 9.1
Equations for calculating the voltage drop Δu (from the beginning to the end of the line) and for calculating the conductor cross-section q

	Direct current and single-phase alternating current (non-inductive, $\cos \varphi = 1$)	Three phase current
Voltage drop Δu (in V)		
current given	$\Delta u = \dfrac{2l \cdot I}{\varkappa \cdot q}$	$\Delta u = \dfrac{\sqrt{3} \cdot l \cdot I \cdot \cos \varphi}{\varkappa \cdot q}$
power given	$\Delta u = \dfrac{2l \cdot P}{\varkappa \cdot q \cdot U}$	$\Delta u = \dfrac{l \cdot P}{\varkappa \cdot q \cdot U}$
Conductor cross-section q (in mm²)		
current given	$q = \dfrac{2l \cdot I}{\varkappa \cdot \Delta u}$	$q = \dfrac{\sqrt{3} \cdot l \cdot I \cdot \cos \varphi}{\varkappa \cdot \Delta u}$
power given	$q = \dfrac{2l \cdot P}{\varkappa \cdot \Delta u \cdot U}$	$q = \dfrac{l \cdot P}{\varkappa \cdot \Delta u \cdot U}$

Table 9.2
Equations for calculating the effective (active) electrical power of motors

	Motor output power (in W)
Direct current	$P_1 = U \cdot I \cdot \eta$
Single-phase alternating current	$P_1 = U \cdot I \cdot \cos \varphi \cdot \eta$
Three-phase current	$P_1 = \sqrt{3} \cdot U \cdot I \cdot \cos \varphi \cdot \eta$
Efficiency	$\eta = \dfrac{P_1}{P_2} \cdot (100\%)$

P_1 Mechanical output power at the motor shaft, as given on the rating plate
P_2 Electrical input power consumption

Table 9.2 lists the equations for calculating the effective (active) power of motors.

9.1.2 Characteristic quantities and units of electricity in accordance with DIN VDE and IEC

The relevant DIN VDE regulations and IEC publications specify equation symbols for the most important characteristic quantities of electricity. Complex time-dependent quantities are represented underlined in accordance with DIN 5483 part 3.

If different equation symbols are used in DIN VDE regulations and IEC publications, this is indicated in Table 9.3. It is recommended that the latest respective issues of the relevant DIN VDE regulations and or IEC publications be used.

The relevant DIN VDE regulations and IEC publications are (among others):

▷ DIN VDE 0660, regulations on low-voltage switchgear,
▷ DIN VDE 0532, regulations on transformers and reactors,
▷ DIN VDE 0102, 01.1990, calculation of short-circuit currents in three-phase networks,
▷ IEC publications 947-1 to 947-7 – Low-voltage switchgear and controlgear:
    947-1, 1988 "General rules",
    947-2, 01.89 "Circuit-breakers" (replaces IEC 157-1),
    947-3, 03.90 "Switches, disconnectors, switch-disconnectors, and fuse combination units" (replaces IEC 408),
    947-4, 05.90 "Contactors and motor starters" (replaces IEC 158 and IEC 292-1),
    947-5, 03.90 "Control-circuit devices and switching elements" (replaces IEC 337-1),
    947-6, 06.89 "Multiple function equipment",
    947-7, 1989 "Ancillary equipment" (e.g. terminal blocks for copper conductors).

In addition to the valid standards, drafts of DIN VDE 0660 Parts 107 and 109 in conformance with IEC 947- have been published (also refer to Section 1.2.2, page 17).

Information on nominal values, limiting values, rated values, and ratings is given in DIN 40200, 10.81, which is based on IEC 50(151).

The following definitions are given:

Nominal value
A suitable approximate quantity value used to designate or identify a component, device or equipment.

Limiting value
In a specification, the greatest or smallest admissible value of one of the quantities.

Rated value
A quantity value assigned, generally by the manufacturer, for a specified operating condition of a component, device or equipment.

Rating
The set of rated values and operating conditions.

Thus, the „nominal voltage" of an electrical network is a value, e.g. 10 kV, which serves to designate the network (see DIN 40002 "Rated voltages of 100 V to 380 kV"). A voltage which is approx. 20% higher than the nominal voltage – 12 kV in our example – is taken as the base for rating certain items of equipment in the network. Thus, this value is the "rated voltage". It was formerly called "upper rated voltage", but this contradicts the current standard.

In the past, little or no distinction was made between "nominal" and "rated" values.

---

[1] Since the publication and enforcement of the new DIN VDE specifications, which are based on IEC 947-., a distinction has been made between the *nominal values* of low-voltage switching devices ("Nennwerte"), and the *rated values* ("Bemessungswerte"). To a large extent, the term "Nennwert" has been regarded as synonymous with *rated value* and this explains apparent contradictions in the use of some suffixes as shown (e.g. I_n for rated current) ▷

Characteristic electrical quantities and units

Table 9.3
Some important characteristic electrical quantities, equation symbols and indices for low-voltage switchgear, three-phase transformers and other items of equipment in accordance with DIN VDE and IEC

Symbol	Characteristic electrical quantity	Symbol	Characteristic electrical quantity
U_i	Rated insulation voltage [1] to DIN VDE 0110/DIN VDE 0660	I_{er}	Rated rotor operational current (DIN VDE 0660, IEC 947-1)
U_{imp}	Rated impulse withstand voltage to IEC 947-	I_r	Setting current ("current setting") to DIN VDE 0660
U_e	Rated operational voltage [1]	I_B	Take-over current
U_c	Rated control circuit voltage (IEC 947-1) at which an operating mechanism or release is rated, e.g coil voltage to DIN VDE 0660 part 102	$I_{\Delta n}$	Rated leakage fault current [1] to DIN VDE 0664
U_s	Rated control supply voltage [1] (control voltage) to DIN VDE 0660 part 102, IEC 947-1	S_{nT}	Rated power of a transformer to DIN VDE 0532 (apparent power in kVA or MVA)
U	No-load voltage to IEC 947-2, -3, -5	c	Factor to determine the equivalent driving voltage of a short-circuit
U_r	Power-frequency recovery voltage (IEC 947-.)		
U_n	Rated network voltage (phase-to-phase voltage) to DIN VDE 0102, Rated voltage [1] (across the terminals) of a transformer winding, DIN VDE 0532	I_{kK}	Sustained short-circuit current at the terminals of a low-voltage generator under influence of the excitation apparatus
		R	Ohmic resistance
U_0	Transformer no-load voltage to DIN VDE 0532	R_{sG}	Ohmic resistance of a generator as relevant to the calculation of impulse short-circuit current; also taking into account the decreasing symmetrical short-circuit current
U_k	Short-circuit impedance voltage to DIN VDE 0532		
u_{kr}	Rated value of the impedance voltage in % to DIN VDE 0102, 01.90		
u_{Rr}	Ohmic voltage drop (transformers) to DIN VDE 0532 / DIN VDE 0102	S''_k	Initial symmetrical a.c. short-circuit power (simplified: apparent short-circuit power)
		$\ddot{u}_n$	Rated transformation ratio of the principal tapping of a transformer
I_n	Rated current [1] to IEC 947-.		
I_{th}	Eight-hour-current to DIN VDE 0660, conventional free-air thermal current to IEC 947- (defined as eight-hour-current) thermally equivalent short-time current (r.m.s. value) to DIN VDE 0103	X	Reactance, reactive impedance
		Z	Impedance (apparent resistance)
		$\varkappa$	Factor to determine the peak short-circuit current i_p
I_{the}	Conventional enclosed thermal current	\multicolumn{2}{l	}{Suffixes referring to symmetrical components (added as the first suffix)}
I_u	Rated uninterrupted current [1] to IEC 947-1		
I_e	Rated operational current [1]	1	Positive-sequence system
I_s	Selectivity (discrimination) limit current (DIN VDE 0660, IEC 947-1)	2	Negative-sequence system
		0	Zero-sequence system
I_{cm}	Rated short-circuit making capacity [1] to IEC 947-1	\multicolumn{2}{l	}{Suffixes referring to condition (added as the second suffix)}
I_{cn}	Rated short-circuit breaking capacity [1] to IEC 947-1	$''$	Superscript suffix: Initial condition (subtransient value)
I_{cs}	Rated service short-circuit breaking capacity [1] to IEC 947-1	k	Short-circuit If no particular index is given, a three-phase fault is assumed.
I_{cu}	Rated ultimate short-circuit breaking capacity [1] to IEC 947-1		
I_{cw}	Rated short-time current withstand capacity [1] to IEC 947-1	k2p	Phase-to-phase fault clear of earth (2-phase fault)
		k1p	Single-phase fault, earth fault
I_p	Test current (general) to DIN VDE 0660, prospective current to DIN VDE 0636	r	Rated value
		n	Nominal value [1] (DIN 40200, DIN 1304, parts 1 and 3)
I_a	Breaking current (r.m.s. value) to DIN VDE 0102		
i_p	Peak short-circuit current (maximum instantaneous value) to DIN VDE 0102	\multicolumn{2}{l	}{Suffixes referring to items of equipment (preferably upper case, added at the last suffix)}
I''_k	Initial symmetrical short-circuit current (r.m.s. value) to DIN VDE 0102	A, B (C)	Busbars in a low voltage network, switchboard or installation
I_k	Sustained (symmetrical) short-circuit current (r.m.s. value), DIN VDE 0102. Rated short-time withstand current [1] to DIN VDE 0660	E	Earth (ground)
		F	Fault location, point of short-circuit
		G	Generator
i_D	Let-through current of fuses and rapidly operating switching devices (maximum instantaneous value during the break time) to DIN VDE 0102	K	Terminal
		L	Line (overhead line or cable)
		M	Motor or group of induction motors
I_0	No-load current at the input side of a transformer (unloaded output side) to DIN VDE 0532	N	Star point of a three-phase system (neutral)
		Q	Connection point of a network (e.g. incomer)
I_z	Current carrying capacity (ampacity)	T	Transformer

501

9.1.3 Differences in the IEC 157-1 and IEC 947-2 publications

Table 9.4 shows a comparison of important terms and the differences between IEC 157-1 and IEC 947-2. The contents of various international standards were combined into a uniform, new harmonized IEC 947 standard which is intended initially only to apply as a European Norm (if applicable also worldwide). National regulations are drafted within this harmonized framework or they may adopt complete paragraphs and sections.

Table 9.4 Comparison of important terms as used in IEC 157-1 and IEC 947-2

IEC publication 157-1	IEC publication 947-2
–	Utilization category A: – not suitable for discrimination (selectivity) – no short-time current withstand capacity
–	Utilization category B: – circuit-breaker suitable for discrimination (selectivity) – short-time current withstand capacity $\geq 12 \cdot I_r$ for 50 ms Utilization categories must be stated on the rating plate.

Short-circuit categories
P-1: (switching sequence O-t-CO)
 	
    reduced operation after short-circuit interruption
and/or
P-2: (switching sequence O-t-CO-t-CO)
 	
Testing of the temperature rise on the connection terminals, in case of doubt <70 K, thereafter tripping in accordance with the characteristic ±20% tolerance at $2.5 \cdot I_r$
	Short-circuit switching capacity
	I_{cu}: rated ultimate short-circuit breaking capacity
	    (switching sequence O-t-CO),
	    reduced operation after short-circuit interruption
	and
	I_{cs}: rated service short-circuit breaking capacity
	    (switching sequence O-t-CO-t-CO),
	    temperature rise on connection terminals must be <80 K, there-after release at $1.45 \cdot I_r$ within 2h
	
	In the case of utilization category A:
	$I_{cs} = 25, 50, 75$ or 100% of I_{cu}
	
	In the case of utilization category B:
	$I_{cs} = 50, 75$ or 100% of I_{cu}

P-1 and/or P-2 must be stated on the rating plate	I_{cu} and I_{cs} must be stated on the rating plate
Type test on a single new and clean circuit-breaker	Type tests on several circuit-breakers
Thermal release: without temperature compensation, trip (release) at:	Thermal release: without temperature compensation, trip (release) at:
$I_r \leq 63$ A +20°C or 40°C 1.05–1.35 within 1h $I_r > 63$ A +20°C or 40°C 1.05–1.25 within 2h	$I_u \leq 63$ A +30°C±2°C 1.05–1.30 within 1h $I_u > 63$ A +30°C±2°C 1.05–1.30 within 2h
with temperature compensation, trip (release) at:	with temperature compensation, trip (release) at:

$I_r \leq 63$ A   −5°C   1.05–1.40
            +20°C   1.05–1.30 within 1h
            +40°C   1.00–1.30
	$I_u \leq 63$ A −5 to +40°C 1.05–1.30 within 1h

$I_r > 63$ A   −5°C   1.05–1.35
            +20°C   1.05–1.25 within 2h
            +40°C   1.00–1.24
	$I_u > 63$ A −5 to +40°C 1.05–1.30 within 2h

9.1.4 Equation symbols and SI units
International System of Units (SI)

The base units of the "International System of Units" (SI) and units derived from them are recommended by international associations of experts and form the basis for standardization. In Germany, for example, their use is compulsory in trade, commerce and official government business.

The SI base units are:

Base quantity	Base unit	
	Symbol	Name
Length	m	the meter
Mass	kg	the kilogram
Time	s	the second
Electric current	A	the Ampere
Thermodynamic temperature [1]	K	the Kelvin
Amount of substance	mol	the mole
Luminous intensity	cd	the candela

[1] A special name for the Kelvin in quoting Celsius temperatures is the degree Celsius; unit symbol: °C

In a universal system of units, one and only one unit is provided for each quantity.

A system of units is called coherent (self-consistent), if the units of the system are related solely by unit equations containing no numerical factor other than unity.

Decimal multiples and sub-multiples (Table 9.5) of SI units, created by the use of prefixes, are by definition not to be termed as being SI units.

Apart from SI units and their decimal multiples and sub-multiples, it is permissible to use legal units defined independently of the International System of Units, e.g., the minute, the hour, the day and the angular units degree (old degree) with the minutes and seconds as well as gons (new degree). Prefixes must not be used to form decimal multiples or sub-multiples of the time units minute, hour, day and year, or of the angular units degree, minute and second. The ratio of two equal SI units is represented by 1.

Table 9.7 [2] shows a selection of quantities and SI units in accordance with DIN 1301, part 1, 12.85, DIN 1304, part 1, 03.89, and part 3 "Equation symbols for electrical power supply". DIN 13304, 03.82 must also be observed in electronic data processing.

[2] Also refer to: Siemens AG, Technical Tables – Quantities, Equations, Definitions, Order no. A 19100-L532-K529

Table 9.5
Prefixes for decimal multiples and their designations

Powers of ten	Prefix	Prefix symbol
10^{18}	exa	E
10^{15}	peta	P
10^{12}	tera	T
10^{9}	giga	G
10^{6}	mega	M
10^{3}	kilo	k
10^{2}	hecto	h
10^{1}	deca	da
10^{-1}	deci	d
10^{-2}	centi	c
10^{-3}	milli	m
10^{-6}	micro	µ
10^{-9}	nano	n
10^{-12}	piko	p
10^{-15}	femto	f
10^{-18}	atto	a

Table 9.6 A selection of useful values

Symbol	Numerical value	Symbol	Numerical value
$\sqrt{2}$	1.414213	$\sqrt{\pi}$	1.772454
$\sqrt{3}$	1.732051	$\sqrt{2\pi}$	2.506628
$\sqrt{10}$	3.162278	$1/\sqrt{\pi}$	0.564190
π	3.141593	π^2	9.869604
3π	9.424778	$4\pi^2$	39.478418
4π	12.566371	$\pi^2/\sqrt{2}$	6.978864
$\pi/\sqrt{2}$	2.221441	$\pi^2/4$	2.467401
$\pi/1.8$	1.745329	$1/\pi^2$	0.101321
$\pi/2$	1.570796	$(2/\pi)^2$	0.405285
$\pi/3$	1.047198	e	2.718282
$\pi/4$	0.785398	e^2	7.389056
$1/\pi$	0.318310	$1/e$	0.367879
$1.8/\pi$	0.572958	ln 2	0.693147
$2/\pi$	0.636620	ln 10	2.302585
$2.5/\pi$	0.795775	log 2	0.301030
$3/\pi$	0.954930	M = log e	0.434294
$5/\pi$	1.591550	= 1/ln 10	

Table 9.7 Quantities and SI units

Quantity		SI unit		Relationship, conversion
Symbol	Meaning	Symbol	Name	
Quantities and units of space				
l	Length	m	meter	
w	Width	m	meter	
h	Height	m	meter	
d, δ	Thickness	m	meter	
a	Distance	m	meter	
d, D	Diameter	m	meter	
r	Radius	m	meter	
s	Distance (e.g. along a curve)	m	meter	
H	Altitude (height above sea level)	m	meter	
A	Area, general	m^2	square meter	
Q, q	Cross-section, cross-sectional area	m^2	square meter	
O	Surface area, cooling area	m^2	square meter	
V	Volume	m^3	cubic meter	
Quantities and units of time				
t	Time, time interval, duration	s	second	
τ	Time constant	s	second	
T	Period, period of oscillation	s	second	
f, ν	Frequency	Hz[1]	Hertz	1 Hz = 1/s[1]
ω	Angular frequency	1/s, s^{-1}	per second	$\omega = 2\pi f$, $\omega = 2\pi \nu$
n	Speed, frequency of rotation	1/min, min^{-1}	per minute	
v	Velocity	m/s	meters per second	
g	Acceleration due to gravity (local)	m/s^2	meters per second squared	
Quantities and units of mechanics				
m	Mass	kg	kilogram	
ρ	Density	kg/m^3	kilogram per cubic meter	
F	Force, applied force	N	Newton	1 N = 1 kg m/s^2
p	Pressure (force per unit area),	Pa	Pascal	1 Pa = 10^{-5} bar = 0.1 N/mm^2
	Pressure in the case of fluids	bar	bar	1 bar = 10^5 Pa
W	Energy, work	J	Joule	1 J = 1 Ws = 1 Nm
P	Active power, energy flow	W	Watt	1 W = 1 J/s = 1 Nm/s
S	Apparent power	VA	Volt-ampere	
Q	Reactive power	var	Var reactive Volt-ampere	
Φ	Heat flow	W	Watt	1.163 W = 1 kcal/h
Quantities and units of heat transmission				
T	Thermodynamic temperature	K	Kelvin	
ϑ	Celsius temperature	°C	degree Celsius (centigrade)	$\vartheta = T - T_0$ with $T_0 = 273.15$ K
$\Delta T, \Delta \vartheta$	Temperature difference, Temperature rise	K, permissible °C	Kelvin permissible degree Celsius	1 K = 1 °C
Q	Quantity of heat	J	Joule	1 J = 1 Ws = 1 Nm

[1] Hertz (Unit symbol: Hz) when quoting frequencies. It is the special term for the SI unit "reciprocal second" or "per second" (symbol: 1/s)

Table 9.7 Quantities and SI units *(continued)*

Quantity		SI unit		Relationship, conversion
Symbol	Meaning	Symbol	Name	
Quantities and units of electricity				
Q	Electric charge	C	Coulomb	$Q = n \cdot e$ (n integer, e charge on the electron $\approx 1.6 \cdot 10^{-19}$ C)
φ	Electric potential	V	Volt	$\varphi = W/Q$ (W Work)
U	Voltage, Potential difference	V	Volt	$U_{12} = \int_1^2 E \cdot ds = \varphi_1 - \varphi_2$
				$1\,\mathrm{V} = 1\,\mathrm{W/A} = 1\,\mathrm{A\Omega}$
$\overset{\circ}{U}$	Potential difference along a closed path	V	Volt	$\overset{\circ}{U} = \oint E \cdot ds$
E	Electric field strength	V/m	Volt per meter	$E = F/Q = dU/dx$ (F Force)
D	Electric flux density	C/m²	Coulomb per square meter	$D = \varepsilon \cdot E$
Ψ	Electric flux	C	Coulomb	$\Psi = \int_A D \cdot dA$
ε	Permittivity, Dielectric constant	F/m	Farad per meter	$\varepsilon = D/E$
ε_0	Electric field constant	F/m	Farad per meter	$\varepsilon_0 = \dfrac{1}{\mu_0 \cdot c_0^2} = 8.854188 \cdot 10^{-12}$
ε_r	Permittivity ratio, Relative dielectric constant	1	–	$\varepsilon_r = \varepsilon/\varepsilon_0$
C	Electric capacitance	F	Farad	$C = Q/U$
				$1\,\mathrm{F} = 1\,\mathrm{C/V}$
P	Electric polarization	C/m²	Coulomb per square meter	$P = D - \varepsilon_0 E$
P/ε_0	Electrification	V/m	Volt per meter	$P/\varepsilon_0 = (\varepsilon_r - 1)E$
χ_e	Electric susceptibility	1		$\chi_e = \dfrac{P}{\varepsilon_0 \cdot E} = \varepsilon_r - 1$
F_e	Forces in an electric field	N	Newton	$F_{e1} = \dfrac{Q_1 \cdot Q_2}{4\pi\varepsilon a^2}$
				$F_{e2} = Q \cdot E$
W_e	Energy in an electric field	J	Joule	$W_e = \tfrac{1}{2} D \cdot E \cdot V$
				$= \tfrac{1}{2}\varepsilon E^2 \cdot V = \tfrac{1}{2} Q \cdot U$
				$= \tfrac{1}{2} Q^2/C = \tfrac{1}{2} C \cdot U^2$
I	Electric current	A	Ampere	$I = Q/t$
				for direct current
				$i = dQ/dt$
				instantaneous value
S	Electric current density	A/m²	Amps (Ampere) per square meter	$S = I/q$
A	Specific electric loading	A/m	Ampere conductors per meter	$A = \Sigma I/l$
G	Electrical conductance	S	Siemens	$G = 1/R = 1/U = 1/\Omega$
B	Susceptance, reactive conductance	S	Siemens	$B_L = -1/\omega \cdot L$ (inductance)
				$B_C = \omega C$ (capacitance)

Table 9.7 Quantities and SI units *(continued)*

Quantity Symbol	Meaning	SI unit Symbol	Name	Relationship, conversion
Y	Admittance	S	Siemens	$Y=\sqrt{G^2+B^2}$, $\tan\varphi=\dfrac{G}{B}$
$\varkappa, \gamma$	Electrical conductivity	S/m	Siemens per meter	$\varkappa = Gl/q$ $1\,\text{S/m}=1/(\Omega\,\text{m})=1/\varrho$
R	Electrical resistance	Ω	Ohm	$R=1/G=U/I$ $1\,\Omega = 1\,\text{V/A} = 1/\text{s}$
X	Reactance, reactive impedance	Ω	Ohm	$X_L = \omega \cdot L$ $X_C = -1/\omega \cdot C$
Z	Impedance	Ω	Ohm	$Z=\sqrt{R^2+X^2}$
ϱ	Specific electrical resistance	$\Omega\,\text{m}$	Ohm meter	$\varrho = Rq/l$ $1\,\Omega\,\text{m} = 1\,\text{m/S} = 1\,\text{V/A}$
U	Electrical potential	V	Volt	$U = I\cdot R = I/G$
P	Electrical power	W	Watt	$P = I^2\cdot R = U\cdot I$
Q, P_q	Reactive power	var	Volt-ampere, reactive	$Q = I^2\cdot X = U\cdot I \sin\varphi$
S, P_s	Apparent power	VA	Voltampere	$S = \sqrt{P_w^2 + Q^2} = U\cdot I$

Quantities and units of magnetism

Symbol	Meaning	SI unit Symbol	Name	Relationship, conversion
Θ	Current linkage	A	Ampere	$\Theta = \int_A S_A\cdot \mathrm{d}A = \Phi H_s\,\mathrm{d}s$ $\Theta = I\cdot w$
V	Magnetic potential	A	Ampere	$V = \int_1^2 H\cdot \mathrm{d}s$
H	Magnetic field strength	A/m	Ampere per meter	$H = B/\mu$
Φ	Magnetic flux	Wb	Weber	$\Phi = V\Lambda = V/R_m$ $\Phi = \int_A B\cdot \mathrm{d}A$ $1\,\text{Wb} = 1\,\text{V\,s}$
B	Magnetc flux density, magnetic induction	T	Tesla	$B = \Phi/A = \mu_0\cdot H$ $1\,\text{T} = 1\,\text{Wb/m}^2$
ψ	Linkage flux, magnetic linkage	Wb	Weber	$\psi = \xi w\Phi$
Λ	Permeance	H	Henry	$\Lambda = \Phi/V$ $1\,\text{H} = 1\,\text{Wb/A}$
R_m	Reluctance, magnetic resistance	1/H	1 per Henry	$R_m = V/\Phi$ $1/\text{H} = 1\,\text{A/Wb}$
μ	Permeability	H/m	Henry per meter	$\mu = B/H$
μ_0	Magnetic constant	H/m	Henry per meter	$\mu_0 = 4\pi\cdot 10^{-7}$ $= 1.256637\cdot 10^{-6}$
μ_r	Relative permeability	1		$\mu_r = \mu/\mu_0$
J	Magnetic polarization	T	Tesla	$J = B - \mu_0\cdot H$
M	Magnetization	A/m	Ampere per meter	$M = J/\mu_0 = (\mu_r - 1)\cdot H$
χ_m	Magnetic susceptibility	1		$\chi_m = M/H = \mu_r - 1$

Table 9.7 Quantities and SI units *(continued)*

Quantity		SI unit		Relationship, conversion
Symbol	Meaning	Symbol	Name	
L	Inductance, inductivity	H	Henry	$L = \psi/i = \xi w^2 \Lambda$ $1\,\text{H} = 1\,\text{Wb/A}$
u_i	Instantaneous value of an induced voltage	V	Volt	$u_i = \dfrac{d\psi}{dt} = L\dfrac{di}{dt}$ $1\,\text{V} = 1\,\text{Wb/A}$
F_m	Forces in a magnetic field	N	Newton	$F_{m1} = i \cdot Bl$ $F_{m2} = \dfrac{\mu_0}{2\pi} \cdot \dfrac{i_1 i_2}{a} \cdot l$
m	Magnetic dipolar moment	Wb m	Weber meter	$m = M/H$
m'	Electromagnetic moment	A m²	Ampere square meter	$m' = M/B$
W_m	Energy in a magnetic field	J(Ws)	Joule (Watt second)	$W_m = \tfrac{1}{2} BHV = \tfrac{1}{2}\psi \cdot i$ $= \tfrac{1}{2}\psi^2/L = \tfrac{1}{2} L \cdot i^2$ (V Volume)

9.1.5 Conversion of international, British and American units

Table 9.8 illustrates how international, British and American units may be converted from one to the other. Detailed tables for the conversion of inches to millimeters are given in DIN 4890, 02.75 and DIN 4892, 02.75. Details on converting millimeters into inches are given in DIN 4893, 03.65.

Table 9.9 on page 510 gives a comparison between the British copper wire gauge G and the *American Wire Gauge* (AWG). Table 9.10 shows a comparison between standardized conductor cross-sections in conformance with British *Standard Wire Gauge* (SWG) and *American Wire Gauge* (AWG).

Table 9.8 Table for converting international, British and American units

Length

Unit			mm	cm	m	in	ft	yd
millimeter	1 mm	=	1	0.1	0.001	0.03937	0.0032808	0.0010936
centimeter	1 cm	=	10	1	0.01	0.3937	0.032808	0.010936
meter	1 m	=	1000	100	1	39.3701	3.28084	1.09361
inch	1 in	=	25.4	2.54	0.0254	1	0.08333	0.02777
foot	1 ft	=	304.8	30.48	0.3048	12	1	0.3333
yard	1 yd	=	914.4	91.44	0.9144	36	3	1
mile	1 mile	=			1609.344	63360	5280	1760
nautical mile	1 nmile	=			1852			

1 fathom = 2 yd = 1.8288 m;
1 rod = 1 pole = 1 perch = 5.5 yd = 5.0292 m;
1 link = 0.201168 m;
1 chain = 4 rods = 22 yd = 100 links = 20.1168 m;

1 furlong = 10 chains = 220 yd = 1000 links = 201.168 m;
1 mile = 8 furlongs = 80 chains = 1760 yd = 1609.344 m;
1 mil = 0.001 in = 0.0254 mm
1 hand = 10.16 cm

Area

Unit			mm²	cm²	m²	in²	ft²	yd²
square millimeter	1 mm²	=	1	0.01	0.000001	0.00155	0.0000107	0.00000119
square centimeter	1 cm²	=	100	1	0.0001	0.154999	0.0010763	0.00011959
square meter	1 m²	=	1000000	10000	1	1550	10.7639	1.19599
square inch	1 in²	=	645.16	6.4516	0.00064516	1		
square foot	1 ft²	=	92903	929.03	0.0929	144	1	0.1111
square yard	1 yd²	=	836130	8361.3	0.83613	1296	9	1

1 circular mil = $\frac{\pi}{4}\left(\frac{in}{1000}\right)^2 = 5.06707 \cdot 10^{-4}$ mm²;

1 square chain = 404.686 m²;

1 circular inch = 5.06707 cm²

1 rood = 1011.71 m²

Volume/Capacity

Unit			dm³ (l)	in³	ft³	yd³
cubic decimeter (1 Liter)	1 dm³	=	1	61.0237	$35.3147 \cdot 10^{-3}$	$1.3080 \cdot 10^{-3}$
cubic inch	1 in³	=	0.01638	1	$0.5787 \cdot 10^{-3}$	$21.4335 \cdot 10^{-6}$
cubic foot	1 ft³	=	28.3168	1728	1	$37.0370 \cdot 10^{-3}$
cubic yard	1 yd³	=	764.5549	46656	27	1

UK-Units:
1 bushel = 8 gallons = 36.3687 dm³ (l);
1 quart = 2 pints = 1.1365 dm³ (l);
1 gill = 5 fluid ounces = 142.065 cm³;
1 fluid drachm = 60 minims = 3.5516 cm³;
1 UK-gallon = 4 quarts = 4.5461 dm³ (l);
1 pint = 4 gills = 0.5683 dm³;
1 fluid ounce = 8 fluid drachms = 28.4131 cm³

US-Units:
1 US-Barrel = 42 US-gallons = 158.987 dm³ (l);
1 US-gallon = 0.8327 UK-gallon = 6.6614 pints
                = 3.7855 dm³ (l)
1 board foot = 2.35974 dm³
1 cord (cd) = 3.62456 m³

Table 9.8 Table for converting international, British and American units *(continued)*

Mass

Unit			kg	lb (av)[1]
kilogram	1 kg	=	1	2.2046
pound (av)[1]	1 lb (av)[1]	=	0.45359	1

[1] avoirdupois (used for commercial weight and for ordinary goods)

1 ton = 2240 lb (av); 1 shtn = 2000 lb (av); 1 cwt = 112 lb (av); 1 drachm = 60 grain (av) = 60/7000 lb (av) = 3.8879 g;
1 lb (av) = 16 oz (av); 1 grain (av) = 1/7000 lb (av) = 64.7989 mg; 1 oz (tr) = 480/7000 lb (av); 1 lb (tr) = 5760/7000 lb (av)
1 dram (av) = 1.7718 g; 1 stone = 6.35029 kg; 1 quarter = 12.70 kg
1 ounce (oz) = 437.5 grain (av) = 16 dram (av) = 28.3495 g;

Velocity

Unit			m/s	km/h	mile/h	kn
meters per second	1 m/s	=	1	3.6	2.2369	1.9438
kilometers per hour	1 km/h	=	0.277$\overline{7}$	1	0.6214	0.5400
miles per hour	1 mile/h	=	0.4470	1.6093	1	0.8690
knots	1 kn	=	0.514$\overline{4}$	1.852	1.1508	1

1 kn = 1 sm/h;
1 ft/s = 0.3048 m/s;

Standard gravitational acceleration:
$g_n = 9.80665$ m/s^2 = 32.17405 ft/s^2

Pressure (fluids and gases)

Unit			bar	at	Torr	mm WS	inch of mercury	foot of water
bar	1 bar	=	1	1.0197	750	10197	29.53	33.4553
technical atmosphere	1 at	=	0.980	1	735.559	104	28.959	32.8084
Torr	1 Torr	=	0.00133	0.001359	1	13.5951	–	–
head of water in mm	1 mm Water	=	98.0665	10^{-4}	73.5559·10^{-3}	1	–	–
inch of mercury	1 in of Hg.	=	0.03386	34.531·10^{-3}	–	–	1	1.1329
foot of water	1 in of w.	=	0.02989	30.48·10^{-3}	–	–	0.8827	1

1 Pa = 1 N/m^2 = 10^{-5} bar;
1 MPa = 1 N/mm^2 = 10 bar;
1 atm = 760 Torr = 1.01325 bar;

1 at = 1 kp/cm^2;
1 Torr ≈ 1 mm Hg;
1 mm WS = 1 kp/m^2.

Temperatures

degrees Celsius[2]	°C	−10	−5	0	+5	+10	+15	+20	+25	+30	+40	+50	+60	+70	+80	+90	+100	+110	+120
degrees Fahrenheit	°F	14	23	32	41	50	59	68	77	86	104	122	140	158	176	194	212	230	248

1 °C = 5/9 ϑ − 32 °F;
1 °F = 9/5 ϑ + 32 °C

[2] Centigrade

Table 9.8 Table for converting international, British and American units *(continued)*

Power

Unit			W	kcal/h	Btu/h	ft·lbf/s
Watt	1 W	=	1	0.8598	3.4121	0.7376
kilocalories per hour	1 kcal/h	=	1.163	1	3.9683	0.8578
British thermal unit/hour	1 Btu/h	=	0.2931	0.2520	1	0.2162
foot pound force/second	1 ft·lbf/s	=	1.3558	1.1658	4.6263	1

1 hp = 550 ft·lbf/s = 0.735 kW;
$1 \frac{Btu}{lbh} = \frac{5}{9} \frac{kcal}{kgh} = 2326 \frac{J}{kgh}$;

1 kcal/h = 1.163 W;
1 SKE[1] = 7000 kcal = 29.3076 MJ = 8.141 kWh.

[1] SKE = "Stein*kohlen*einheit" (unit of coal)

Force

Unit			N	kp	lbf	pdl
Newton	1 N	=	1	0.101972	0.22481	7.2230
kilopond	1 kp	=	9.80665	1	2.20462	70.9316
pound force	1 lbf	=	4.44822	0.45362	1	32.1740
poundal	1 pdl	=	0.13826	14.0981·10⁻³	31.0810·10⁻³	1

1 tonf (ton force) = 2240 lbf = 9.9640 kN

Table 9.9
Comparison between the British copper wire gauge G (in lb/mile) and the American Wire Gauge (AWG)

G[2]	lb/mile	4	6.5	10	20	40	50	100	160	220
D	mm	0.40	0.51	0.63	0.90	1.27	1.42	2.01	2.54	2.98
AWG[3]		26	24	22	19	16	–	12	10	–
D	mm	0.40	0.51	0.64	0.91	1.29		2.05	2.59	

[2] As a rule of thumb, the conductor diameter D can be converted into the British copper wire gauge G by $G = 25 \cdot \left(\frac{D}{mm}\right)^2 \approx \frac{G}{lb/mile}$
[3] An increase in the AWG number by 1 corresponds to a reduction in the wire diameter D to about 89.05%

Table 9.10
Comparison of standardized rated cross-sections q_n for solid and flexible conductors in accordance with the British Standard Wire Gauge (SWG) and the American Wire Gauge (AWG), alt. Mille Circular Mil (MCM) and kcmil (kilo circular mil)

Rated cross-section q_n			Wire-Gauge No.		Rated cross-section q_n			Wire Gauge No.	
solid mm²	flexible mm²	in²	SWG	AWG[1]	solid mm²	flexible mm²	in²	SWG	AWG[1]
0.196	–	0.000304	–	–	27.27	–	0.0424	4	–
0.2	**0.2**			(24)	33.62	–	0.0521	–	2
0.203	–	0.000314	25	–	**35**	**25**	0.0542	–	(2)
0.205	–	0.000317	–	24	38.60	–	0.0598	2	–
0.245	–	0.000380	24	–	42.41	–	0.0657	–	1
0.283	–	0.000438	–	–	45.60	–	0.0707	1	–
0.292	–	0.000452	23	–	**50**[2]	**35**	0.0728	–	(0)
0.324	–	0.000504	–	22	53.20	–	0.0824	0(1/0)	–
0.397	–	0.000616	22	–	53.51	–	0.0829	–	0(1/0)
0.5	**0.5**	0.000775	–	(20)	61.40	–	0.0952	00(2/0)	–
0.519	0.5	0.000802	21	20	67.44	–	0.1045	–	00(2/0)
0.657	–	0.001018	20	–	**70**	**50**	0.1085	–	(00)
0.75	**0.75**	0.001162	–	(18)	70.12	–	0.1087	000(3/0)	(00)
0.811	–	0.001257	19	–	81.08	–	0.1257	0000(4/0)	–
0.821	–	0.001272	–	18	85.03	–	0.132	–	000(3/0)
1.0	**1.0**	0.001550	–	–	**95**	**70**	0.147	–	(000)
1.168	–	0.001810	18	–	107.22	–	0.166	–	0000(4/0)
1.307	–	0.002026	–	16	**120**	**95**	0.186	–	–
1.5	**1.5**	0.002325	–	(16)					MCM[1]
1.589	–	0.002463	17	–	126.68	–	0.196	–	250
2.082	–	0.003228	16	14	(120)				
2.5	**2.5**	0.003875	–	(14)	**150**	**120**	0.233	–	(300)
2.627	–	0.004072	15	–	152.01	–	0.236	–	300
3.243	–	0.005027	14	–	177.35	–	0.275	–	350
3.309	–	0.005129	–	12	**185**	**150**	0.287	–	(400)
4	**2.5**	0.006200	–	(14)	202.68	–	0.314	–	400
4.289	–	0.006337	13	–	**240**	**185**	0.372	–	(500)
5.260	–	0.008152	–	10	253.35	–	0.393	–	500
5.481	–	0.008495	12	–	**300**	**240**	0.471	–	600
6	**4**	0.009300	–	(10)	(304.0)				
6.818	–	0.010568	11	–	354.71	–	0.55	–	700
8.302	–	0.012868	10	–	380.0	–	0.589	–	750
8.365	–	0.012967	–	8	**400**	**300**	0.628	–	800
10	**6**	0.01550	–	(8)	(405.36)				
10.507	–	0.01628	9	–	456.04	–	0.768	–	900
12.972	–	0.02011	8	–	**500**	**400**	0.785	–	1000
13.296	–	0.02061	–	6	(506.70)				
15.700	–	0.02433	7	–	633.38	–	0.982	–	1250
16	**10**	0.02480	–	(6)	760.05	–	1.178	–	1500
18.700	–	0.02895	6	–	886.7	–	1.374	–	1750
21.150	–	0.03278	–	4	1013.4	–	1.571	–	2000
22.773	–	0.0353	5	–					
25	**16**	0.0388	–	(4)					

[1] AWG American Wire Gauge, standard American wire gauge for cross-sections of up to 107.2 mm²
   MCM Mille Circular Mil (1 MCM ≙ 0.5067 mm²) or kcmil, standard American wire gauge for cross-sections greater than 126.7 mm².
[2] Actual cross-section 47 mm²

9.2 Enclosures for electrical equipment to American and Canadian standards

The IEC publication 529 provides a system for specifying the enclosures of electrical equipment on the basis of the IP degree of protection provided by the enclosure (see Section 7.4.7, page 360). IEC 529 indicates the degrees of protection afforded against the ingress of foreign bodies and water into the enclosure, but does not specify protection against mechanical damage of equipment, risk of explosions or conditions such as internal moisture (e.g. due to condensation), corrosive vapour, fungus or vermin.

American and Canadian standards for enclosures of electrical equipment, on the other hand, do test for environmental conditions such as corrosion, rust, icing, oil and coolants. Furthermore, they distinguish between enclosures intended for indoor or outdoor use, and between enclosures for hazardous and non-hazardous locations. For this reason, and also because tests and evaluations for other characteristics are different, the IP degrees of protection to IEC 529 cannot be equated exactly with e.g. NEMA (USA) or EEMAC (Canada) enclosure types.

In the United States of America, the following specifications apply:
▷ NEC NFPA 70 "National Electrical Code"
▷ UL 508 "Industrial Control Equipment"
▷ ANSI/NEMA 250 – 1985 "Enclosures for Electrical Equipment (1000 Volts Maximum)"

In Canada, the relevant specifications are:
▷ CSA-C22.1 "Canadian Electrical Code Part I"
▷ CSA-C22.2 No 94 "Special Purpose Enclosures 2, 3, 4 and 5"
▷ EEMAC E14-2 "Industrial Controls and Systems"

The specifications for the various enclosure types are virtually identical in the USA and Canada although the various methods of testing differ in some respects. For the purposes of illustration, the following comments are based on ANSI/NEMA 250 (Revision No. 2 – May 1988). This publication deals with the classification and description of the enclosures for electrical equipment (excluding rotating machinery). It provides the reader with information to enable correct specification of the enclosure appropriate to the application at hand. Furthermore, it describes the protective features the enclosures are expected to have, and specifies the tests applied to demonstrate compliance with the descriptions.

Table 9.11 shows a comparison of the specific applications and afforded protection of the various types of enclosures for indoor and outdoor non-hazardous locations. The NEMA 250 standard does not specify structural details except where these are essential to the identification of the enclosure type. Structural details pertaining specifically to the enclosure of e.g. a motor starter may, however, be contained in the corresponding NEMA product standard publication. Examples of such structural details may include specific hinge arrangements, external mounting flanges, etc.

In terms of NEMA 250, outdoor locations are those areas which are exposed to the weather. Indoor locations are protected from direct exposure to weather influences (also refer to Section 1.6.1, page 65).

Enclosures do not normally protect devices against conditions such as condensation, icing, corrosion or contamination which may occur inside the enclosure itself or may enter via cable entries, conduit or unsealed openings. For advice on these problems, please refer to Section 7.4.7 on page 360.

The enclosure Types 7, 8, 9 and 10 are intended for use in hazardous (classified) locations as defined in the National Electrical Code (risk of explosion). Type 7 and 10 enclosures are designed to contain an internal explosion and thereby prevent an external hazardous condition. Type 8 enclosures are intended to prevent combustion through the use of oil-immersed equipment. Type 9 enclosures are designed to prevent the ignition of combustible dust.

Table 9.11
Comparison between the specific applications and afforded protection of the various types of enclosures to NEMA 250 – 1985

Use and degrees of protection	Type of enclosure													
	1[1]	2[1]	3	3R[1]	3S	4	4X	5	6	6P	11	12	12K[2]	12
Indoor (I)	I	I				I	I	I	I	I	I	I	I	I
Outdoor (O)			O	O	O	O	O		O	O				
Incidental contact with enclosed equipment	+	+	+	+	+	+	+	+	+	+	+	+	+	+
Falling dirt	+	+	−	−	−	+	+	+	+	+	+	+	+	+
Circulating dust, lint, fibres and flyings[3]	−	−	−	−	−	+	+	−	+	+	−	+	+	+
Settling airborne dust, lint, fibres and flyings[3]	−	−	−	−	−	+	+	+	+	+	−	+	+	+
Windblown dust	−	−	+	−	+	+	+	−	+	+	−	−	−	−
Falling liquids and light splashing	−	+	−	−	−	+	+	−	+	+	+	+	+	+
Hosedown and splashing water	−	−	−	−	−	+	+	−	+	+	−	−	−	−
Rain, snow and sleet[4]	−	−	+	+	+	+	+	−	+	+	−	−	−	−
Sleet[5]	−	−	−	−	+	−	−	−	−	−	−	−	−	−
Oil and coolant seepage	−	−	−	−	−	−	−	−	−	−	−	+	+	+
Oil and coolant spraying and splashing	−	−	−	−	−	−	−	−	−	−	−	−	−	+
Corrosive agents	−	−	−	−	−	−	+	−	−	+	+	−	−	−
Occassional temporary submersion	−	−	−	−	−	−	−	−	+	+	−	−	−	−
Occassional prolonged submersion	−	−	−	−	−	−	−	−	−	+	−	−	−	−
Associated classification designation to IEC 529[6]	IP 10	IP 11	IP 54	IP 14	IP 54	IP 56	IP 56	IP 52	IP 67	IP 67	IP 11	IP 52	IP 52	IP 54

[1] These enclosures may be ventilated. Type 1 enclosure may not provide protection against small particles of falling dirt when ventilation is provided in the enclosure roof.
[2] Type 12K enclosures incorporate knock-outs. The degrees of protection against dust, falling dirt and dripping non-corrosive liquids do not apply at the knock-outs.
[3] These fibres and flyings are of non-hazardous (non-combustible) materials.
[4] External operating mechanisms (e.g. operating handles) need not operational if the enclosure is covered with ice.
[5] External operating mechanisms (e.g. operating handles) remain operational even if the enclosure is covered with ice.
[6] This line indicates a conversion *from* NEMA enclosure type numbers *to* approximate IEC enclosure classification designations. The NEMA types meet or exceed the test requirements for the associated IEC classification. Thus, the table should be viewed as a guideline only, and cannot be used to convert from IEC classifications to NEMA (or CSA) types of enclosures.

9.3 Climatic values, influence of temperature and thermal conduction

9.3.1 Climatic values

Standard atmosphere

Table 9.12 illustrates how the various values specified in DIN 5450 change relative to the altitude.

Table 9.12
Standard atmospheres in accordance with DIN 5450

Altitude m	Temperature °C	Pressure mbar	Density kg/m³	Saturation pressure mbar	Boiling point of Water °C
0	15	1013	1.226	17	100
500	11.8	955	1.168	13.7	98
1000	8.5	899	1.112	11	97
2000	2	795	1.007	7	93
4000	−11	616	0.819	2.4	87
8000	−37	356	0.525	0.13	74
15000	−56.5	120	0.194	−	51
30000	−56.5	11	0.02	−	15

Table 9.13
Absolute humidity of the air as a function of the temperature and the relative humidity

Temperature °C	Relative humidity							
	20	30	40	50	60	70	80	90	100%
	Absolute humidity in g/m³								
0	1.0	1.5	1.9	2.4	2.9	3.4	3.9	4.3	4.8
2	1.1	1.7	2.2	2.8	3.3	3.9	4.4	5.0	5.6
4	1.3	1.9	2.5	3.2	3.8	4.5	5.1	5.7	6.4
6	1.5	2.2	2.9	3.6	4.4	5.1	5.8	6.5	7.3
8	1.7	2.5	3.3	4.1	5.0	5.8	6.6	7.4	8.3
10	1.9	2.9	3.8	4.7	5.6	6.5	7.5	8.5	9.4
12	2.1	3.2	4.3	5.3	6.4	7.4	8.5	9.6	10.7
14	2.4	3.6	4.8	6.0	7.3	8.5	9.7	10.9	12.1
16	2.7	4.1	5.5	6.8	8.2	9.5	10.9	12.3	13.6
18	3.1	4.6	6.2	7.7	9.2	10.7	12.3	13.8	15.4
20	3.5	5.2	6.9	8.6	10.4	12.1	13.8	15.6	17.3
22	3.9	5.9	7.8	9.7	11.7	13.6	15.5	17.5	19.4
24	4.4	6.5	8.7	10.9	13.1	15.3	17.4	19.6	21.8
26	4.9	7.4	9.8	12.2	14.6	17.1	19.6	22.0	24.4
28	5.5	8.2	10.9	13.6	16.3	19.1	21.8	24.5	27.3
30	6.1	9.2	12.1	15.2	18.2	21.3	24.3	27.4	30.4
32	6.7	10.1	13.5	16.9	20.3	23.7	27.1	30.5	33.8
34	7.5	11.3	15.0	18.8	22.6	26.3	30.1	33.9	37.7
36	8.3	12.5	16.7	20.9	25.0	29.2	33.4	37.6	41.7
38	9.2	13.9	18.6	23.2	27.8	32.4	37.0	41.6	46.2
40	10.2	15.3	20.4	25.6	30.7	35.8	40.9	46.0	51.1

Relative and absolute humidity of the ambient air

Atmospheric humidity is usually measured with a hygrometer, which indicates the relative atmospheric humidity as a percentage of saturation ($=100\%$) at the prevailing room temperature. Table 9.13 illustrates the absolute humidity of the air in g/m^3 at a given temperature and a given relative humidity.

9.3.2 Effects of temperature and thermal conduction

Specific resistance

Table 9.14 shows the specific resistance ϱ with the associated temperature coefficient α and the specific conductivity of various metals.

Thermal conduction

The heat flow Φ through a plane wall of area A, thickness δ and heat conductivity λ is

$$\Phi = \frac{\lambda \cdot A \cdot \Delta \vartheta}{\delta}. \tag{9.1}$$

The heat flow through the wall of a straight pipe with a circular cross-section (outer diameter d_2, inner diameter d_1) and length l is

$$\Phi = \frac{2\pi \cdot \lambda \cdot l \cdot \Delta \vartheta}{\ln(d_2/d_1)}. \tag{9.2}$$

Table 9.14
Specific resistance ϱ with associated temperature coefficient α and conductivity $\varkappa$ at 20 °C

Material	ϱ_{20} $\Omega\,mm^2/m$	α_{20} K^{-1}	$\varkappa_{20}$ $m/\Omega\,mm^2$
Aluminium	0.029	$4.1 \cdot 10^{-3}$	34.7
Lead	0.21	$4 \cdot 10^{-3}$	4.8 to 10
Iron/Steel wire	0.10 to 0.15	$4.5 \cdot 10^{-3}$	6.7
Gold	0.023	$3.8 \cdot 10^{-3}$	43.5
Constantan	0.49 to 0.51	$\pm 0.03 \cdot 10^{-3}$	2.04 to 1.96
Copper	0.0182	$3.95 \cdot 10^{-3}$	55
Magnesium	0.043	$4.1 \cdot 10^{-3}$	23.2
Manganin	0.42	0.001 to $0.01 \cdot 10^{-3}$	2.4
Brass, CuZn 40	0.067	$2 \cdot 10^{-3} \leq 15$	
Nickelin	0.40 to 0.44	$0.1 \cdot 10^{-3}$	2.5 to 2.27
NiCr 80 20	1.09	$0.1 \cdot 10^{-3}$	0.92
Mercury	0.962	$0.99 \cdot 10^{-3}$	1.04
Silver	0.016	$4.1 \cdot 10^{-3}$	62.5
Zinc	0.06	$4.2 \cdot 10^{-3}$	16.7

Forced convection

The turbulent flow along a plane wall (l length of the wall along the flow, λ thermal conductivity of the flowing fluid) is

$$\mathrm{Nu} = \frac{\alpha \cdot l}{\lambda} = 0.037\,\mathrm{Re}^{0.8} \cdot \mathrm{Pr}^{0.4}. \qquad (9.3)$$

The turbulent flow through a pipe with a circular cross-section (d internal pipe diameter) is

$$\mathrm{Nu} = \frac{\alpha \cdot d}{\lambda} = 0.023\,\mathrm{Re}^{0.8} \cdot \mathrm{Pr}^{0.4}. \qquad (9.4)$$

From these equations, the heat transfer coefficient α can be calculated using the parameters:

Nusselt number: $\quad \mathrm{Nu} = \dfrac{\alpha \cdot l}{\ell}$

Reynolds number: $\quad \mathrm{Re} = \dfrac{v \cdot l}{\nu}$

Prandtl number: $\quad \mathrm{Pr} = \dfrac{v \cdot \varrho \cdot c_p}{\lambda}$

- l characteristic length
- u mean flow velocity
- ϱ fluid density
- ν kinematic viscosity of the fluid
- c_p specific thermal capacity of the fluid at constant pressure

Heat radiation

The heat flow during radiation exchange between two diffusely reflecting areas A_1 and A_2 with thermodynamic temperatures T_1 and T_2 and emittance values ε_1 and ε_2 respectively is:

$$\Phi = \Phi_{12} - \Phi_{21} = \varepsilon_{12} \cdot A_1 \cdot C_S \left[\left(\frac{T_1}{100}\right)^4 - \left(\frac{T_2}{100}\right)^4 \right], \qquad (9.5)$$

with the technical radiation constant σ of the standard black body

$$C_S = \sigma \cdot 10^{-8} = 5.67\,\mathrm{W\,m^{-2}\,K^{-4}}.$$

σ Stefan Boltzmann constant

The resulting degree of emission ε_{12} in the case of concentric spheres is:

$$\frac{1}{\varepsilon_{12}} = \frac{1}{\varepsilon_1} + \frac{A_1}{A_2}\left(\frac{1}{\varepsilon_2} - 1\right), \quad A_2 > A_1,$$

in the case of a relatively small body in a large room:

$$\varepsilon_{12} \approx \varepsilon_1, \quad A_2 \gg A_1,$$

in the case of parallel walls:

$$\frac{1}{\varepsilon_{12}} = \frac{1}{\varepsilon_1} + \frac{1}{\varepsilon_2} - 1, \quad A_2 \approx A_1.$$

In the case of ribbed surfaces, only the outer profile and not the total ribbed surface should be taken as the radiating area for calculation purposes.

Table 9.15 shows the thermal characteristics of some solid materials.

Table 9.16, overleaf, shows the emittance values for some industrial surfaces.

Table 9.15   Thermal characteristics of some solid materials

Material	ϑ_S °C	ϱ kg/dm³	λ W/(K·m)	c_W J/(kg·K)	α $10^{-6}\,\mathrm{K}^{-1}$
Aluminium	658	2.7	210	910	23.9
Concrete, air-dry	–	2.2–2.5	0.8–1.3	880	10–14
Iron (steel)	≈1350	7.7–7.9	40–60	460	11–13
Glass	≈1000[1]	2.2–3.9	≈0.8	500–840	7–10
Copper	1080	8.96	380	386	16.9
Brass (MS 67)	900	8.4–8.9	117	380	18.7
Silver, pure	961	10.5	408	237	19.3
Zinc	419	7.13	113	390	30

[1] Processing temperature

ϑ_S melting point ϱ density λ thermal conductivity c_w specific thermal capacity α coefficient of linear expansion

9 Appendix

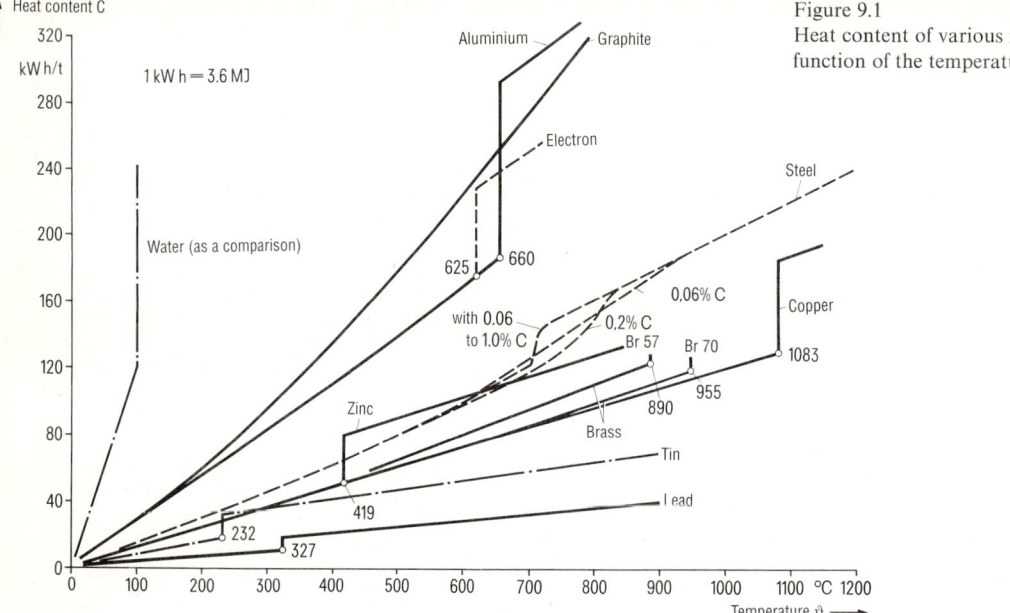

Figure 9.1
Heat content of various metals as a function of the temperature

Table 9.16 Emittance values for some industrial surfaces

Material	Surface	Temperature °C	Emissions ε
Black body	–	any	1.0
Aluminium	uncoated, polished	20	0.039
Copper	uncoated, polished	20	0.030
Silver	uncoated, polished	20	0.020
Iron	corroded	20	0.61–0.85
Copper	oxidized black	20	0.78
Zinc	oxidized grey	20	0.23–0.28
Aluminium bronze coating		20	0.20–0.40
Enamel lacquers		20	0.85–0.95
Black lacquer	matt	80	0.97
Red lead oxide coating	matt	100	0.93
Tar-paper (roofing)		20	0.91–0.93
Ice	smooth	0	0.97
Wood (beech/oak)		20	0.89–0.93
Brickwork (tiles, mortar, plaster)		20	0.93
Porcelain	glazed	20	0.93
Textiles (cotton, wool, silk)		20	0.78

Heat content of metals

Figure 9.1 illustrates the heat content of various metals as a function of the temperature.

The heat transfer coefficient through free convection and radiation in the case of a painted metal surface with a surface temperature of 40 to 50 °C is:

$\alpha = 11$ to 12 W/(K·m^2) with totally undisturbed convection and

$\alpha = 15$ W/(K·m^2) in normal ambient conditions.

9.4 Current carrying capacity and overcurrent protection of insulated wires, cables and busbars

The protection of cables, insulated wires and busbars is important not only with regard to excessive heat (danger of fire), but also with regard to service and reliability, i.e. protective devices should not trip unnecessarily. Although DIN VDE 0100 part 430 includes no specific requirements regarding discrimination, it should be one of the major considerations during the planning phase (see Section 3.4.6, page 163).

Overcurrent protection devices must be fitted in all earthed or unearthed wires but are not permitted in protective conductors and equipotential bonding conductors. Protection devices must effect disconnection of the overloaded wire only, and must leave the other conductors unaffected unless this could cause a dangerous condition.

If the power source supplies a current which does not exceed the current carrying capacity I_z of the wires, e.g. bell transformers, welding transformers and certain types of thermoelectric generators, overcurrent and short-circuit protection is provided even in the absence of protection devices.

Overcurrent protection devices may be omitted in neutral conductors of TT and TN systems (networks), if:

— the cross-section of the neutral conductor is equal to the cross-section of the outer (line) conductor,
— the cross-section of the neutral conductor is smaller than the cross-section of the outer (line) conductor,

but only if the following conditions are fulfilled at the same time:

— the neutral conductor is protected by the protection device of the outer (line) conductor in the event of a short-circuit and
— the maximum current flowing in the neutral conductor under normal operating conditions does not exceed its current carrying capacity I_z. This requires that the consumer loads are distributed as evenly as possible among the outer conductors (balanced loading).

In all other cases, overcurrent detection devices are required in the neutral conductor in order to precipitate disconnection of the outer conductors, although not necessarily of the neutral conductor itself.

In IT systems (networks), an overcurrent detection device must be fitted in the neutral conductor of each circuit unless:

— the particular neutral conductor is protected by a preceding device or
— this circuit is protected by a fault current protection device whose rated fault current does not exceed 0.15 times the current carrying capacity of the respective neutral conductor and which disconnects all conductors including the neutral conductor.

In the specifications, overcurrent protection is divided into the following categories:

— Coordination of overload and short-circuit protection devices in accordance with DIN VDE 0100 part 430, 11.91; supplement 5 to DIN VDE 0100, draft 3.88 and
— Current carrying capacity of insulated cables and wires with regard to installation conditions, bundling of cables, and conductor/insulating materials in accordance with
DIN VDE 0100 part 430, supplement 1, 11.91 [1],
DIN VDE 0298 part 4, 2.88 [1],
DIN VDE 0298 part 2, 11.79,

of busbars in accordance with DIN 43670 (aluminium) or alternatively DIN 43671 (copper) and of switchgear assemblies to DIN VDE 0660 part 500 A5 (also refer to Section 1.3.1.1, pages 28 and 29, Tables 1.11 and 1.12).

9.4.1 Coordination of protection devices

9.4.1.1 Overload protection

Overload currents in individual conductors are to be interrupted before insulation, termination points, connections and other joins, or surrounding materials are damaged by heat.

If the preceding protection device cannot provide reliable protection, then overload protection *must* be provided if the current carrying capacity is reduced in any way, e.g. by:

— a reduction of the conductor cross-section (parallel conductors of smaller cross-sections, e.g. 3×35 mm^2 instead of 1×95 mm^2, are not regarded as a reduction in this sense),
— a change in the type of installation/routing,
— a change in the type and or construction of the cables or wires.

The overload protection device may be positioned *anywhere* in the cable run, if there are neither branch circuits nor plug-and-socket devices along its length and if short-circuit protection is provided, or if the conductor is not longer than 3 m, is installed as short-circuit and earth fault proof, and it is not in the vicinity of inflammable materials.

Overload protection *may be omitted* (not applicable e.g. in areas subject to fire and explosion risks),

— if overload currents are unlikely to occur and there are neither branch circuits nor plug-and-socket devices along the length of the current path, e.g. connecting cables or busbars between

---
[1] Supersedes DIN VDE 0100 part 523

the power source (transformer, generator) and the switchboard,
- in auxiliary circuits,
- in public distribution networks, or
- if the preceding protection device can provide sufficient protection.

Note: In IT systems (networks) the above-mentioned simplifications are only applicable to a limited degree.

For safety reasons, *omission* of overload protection devices is recommended in:
- excitation field circuits,
- supply circuits to solenoid actuators,
- circuits which serve the purpose of safety only, for example, circuits in fire extinguishing systems.

An overload *indicating* device is recommended in these cases.

In the selection of overload protection devices, the following important conditions must be met, although they alone do not always guarantee complete protection:

Condition 1
$$I_b \leq I_n \leq I_z \tag{9.6a}$$

Condition 2
$$I_2 \leq 1.45 \cdot I_z \tag{9.6b}$$

I_b Operating current of the circuit

I_z Permissible current carrying capacity in accordance with
Supplement 1 of DIN VDE 0100 part 430, DIN VDE 0298 parts 4 and 2

I_n Rated current of the protection device,

I_2 Current which causes tripping under certain pre-defined conditions (conventional tripping current)

The rated current of the protection device or the set value may be equal to the value of the permitted current carrying capacity I_z in the following cases:
- miniature circuit-breakers with characteristics B, C where $I_2 \leq 1.45 \cdot I_n$ applies,
- circuit-breakers where $I_2 \leq 1.35 \cdot I_n$ applies,
- fuses with $I_2 \leq 1.45 \cdot I_n$,
- thermally-delayed overload relays where $I_2 \leq 1.25 \cdot I_n$ applies.

In the case of adjustable protection devices, I_n equals the set value.

Overload protection for conductors connected in parallel

If conductors connected in parallel have a common overload protection, the sum of the individual I_z values is taken as the current carrying capacity I_z if
- the electrical properties (conductor material, insulation and cross-section) as well as
- the installation conditions (bundling, length)

are similar and no branch circuits exist. Branch circuits are permitted within switchgear assemblies.

9.4.1.2 Short-circuit protection

Protection against short-circuits means that protective devices are provided which interrupt short-circuit currents before they cause harmful heat or mechanical damage to insulation, connections, termination points or other joins, or to the surroundings.

Short-circuit protection is guaranteed if the breaking capacity of the overcurrent protection device corresponds at least to the maximum prospective short-circuit current which can occur at the point of installation in the circuit. Should this condition not be met, DIN VDE 0100 part 430, 11.91 section 6 must be observed.

Short-circuit protection is required
- at the origin, or source, of each circuit and
- at a reduction in the short-circuit current carrying capacity, e.g. a reduction in the cross-section, at branch circuit nodes or if the type of insulation is changed.

The short-circuit protection *may* be displaced by up to 3 m along the length of the conductor to be protected, provided that the preceding piece of conductor with reduced short-circuit capacity meets three conditions:

1. its length may not exceed 3 m,
2. the risk that a short-circuit will occur is very small, and
3. the risk of fire damage and injury to persons is minimal.

The switching device(s) provided for short-circuit protection must be capable of interrupting the maximum prospective short-circuit current which can arise at its (their) point of installation. In some cases back-up protection, i.e. more than one switching device is required (see Section 3.4.6.2, page 172).

Overload and short-circuit protection

Example

A miniature circuit-breaker with a rated current of 16 A and a short-circuit breaking capacity of 6 kA is to be installed at a location where the prospective short-circuit current I_k'' is 10 kA. The circuit-breaker is protected by the series connection of a 63 A gL back-up fuse on its line side. (The appropriate back-up protection is given in the manufacturer's documentation for each individual case). This means, however, that there will no discrimination (selectivity) in the case of a short-circuit occurring immediately downstream from the circuit-breaker, i.e. both protection devices operate to interrupt the current. Only in the case of faults occurring at a certain distance from the point of installation will the conditions for discrimination be fulfilled, as the short-circuit current is then limited by line impedances to below the value necessary to rupture the fuse.

LV HRC fuses seldom present problems with regard to the required short-circuit interrupting capacity. In terms of specifications and standards, they must be capable of interrupting 50 kA. Typically, interrupting capacities of 80 to 100 kA are given in manufacturers' documentation.

Short-circuit protection *may be omitted*:
— if cables or wires connect the power source (transformer, generator) to the switchboard and the latter is equipped with a protection device,
— in circuits which would cause a danger if they were opened,
— in certain measuring circuits (e.g. current transformer circuits),
— in public distribution networks with underground cables or overhead lines,
— in circuits where the following conditions are met simultaneously:
    the risk of short-circuit is reduced to a minimum *and*
    the conductors are not in the vicinity of inflammable materials.

The condition to be met for short-circuit protection is:

$$q = \frac{\sqrt{t \cdot I_a}}{k} \qquad (9.7)$$

q Conductor cross-section in mm² [1]
I_a Effective (r.m.s.) value of the current in the case of a short-circuit measured in A; it corresponds to the current which causes the protection device to trip within the time t
t Switch-off time for protection against hazardous shock currents-max. 0.2 or 5 s, as applicable.

Note:
If the short-circuit current is determined from the loop impedance, then the corresponding switch-off time must be read from the fuse characteristic curves and inserted into the equation.

k Material coefficient
for PVC-insulated Cu conductors   115 As/mm²
for PVC-insulated Al conductors   76 As/mm²
for rubber-insulated Cu conductors 141 As/mm²

The equation (9.7) is only applicable up to a maximum switch-off time of 5 s. This means that short-circuits must be interrupted within 5 s.

Only "zero-impedance" faults must be considered (consideration of line impedance only)!

The period of 5 s need not be adhered to in the case of coordinated protection (refer to page 519)! Protection against hazardous shock currents "protection by interruption" must, however, be guaranteed within a switch-off time of 0.2 or 5 s, as applicable.

Both requirements illustrate that the length of cables and wires is an important factor in fulfilling the break conditions.

The following equation applies for three-phase loads and may be used to calculate the maximum permissible conductor length on the load side of a short-circuit protection device. It takes the conductor temperature during operation and short-circuit into account (also refer to the standard calculation sheet shown on page 45):

$$l_{perm} = \frac{\frac{U_0 \cdot 10^3}{I_{kreq.}''} - Z_v}{Z'} . \qquad (9.8)$$

l_{perm} Maximum permissible cable length (distance) on the load side of the protection device in m
U_0 Voltage to earth or earthed conductor in V
$I_{kreq.}''$ Required short-circuit current in A which will cause the short-circuit protection device to trip (rupture) within the period of time before the conductor heats up to an excessive degree and, under consideration of "Protection against hazardous shock currents", within 0.2 or 5 s (see Section 1.4.2.1, pages 49 and 50).

---
[1] Designated with S in the relevant standard

Z_v Loop impedance from the power source (transformer) to the protection device in mΩ

Z' Loop impedance for 1 m of cable/wire on the load side of the protection device in mΩ/m (see Section 1.4.2.1, page 42)

The following equation is valid for a.c. loads (also refer to the standard calculation sheet shown on page 45):

$$l = \frac{\varkappa \cdot q \cdot 10^{-3} \left(\dfrac{U_0 \cdot 10^3}{I''_{kreq.}} - Z_v \right)}{2}. \qquad (9.9)$$

$\varkappa$ Conductivity in mΩ/mm^2
q Cross-section in mm^2
2 Length/distance conversion factor, insert 3 in the case of half return lines.

Consideration of $I^2 \cdot t$ let-through values of short-circuit protection devices

In cases where the protection device interrupts the current extremely quickly in a.c. and three-phase circuits (<0.1 s), for example if the occurring short-circuit current is extremely large due to low impedances or in the case of current-limiting protective devices (e.g. fuses), the $I^2 \cdot t$ let-through value (either quoted by the manufacturer or to be obtained) should be considered since its value is determined to a large extent by the d.c. component in the short-circuit current (see page 146). This could cause problems in the use of protection devices which interrupt in ≤ 0.1 s, e.g. circuit-breakers and miniature circuit-breakers. The following equation applies:

$$I^2 \cdot t \leq k^2 \cdot q^2 \qquad (9.10)$$

I Effective value of the current in the case of an absolute short-circuit (let-through current) in A
t Break time in s
k Material coefficient (Table 9.16)
q Cross-section in mm^2

The $k^2 \cdot q^2$ values for insulated wires are listed in Table 9.17. The values should be compared to the $I^2 \cdot t$ let-through values of the protection devices. Tables 9.18a and b contain the $I^2 \cdot t$ values for miniature circuit-breakers and Table 9.19 those for fuses.

CENELEC is soon to issue the following supplementary clause for the short-circuit protection of conductors *connected in parallel*:

"A common protection device for short-circuit protection may be used in the feed line to the parallel

Table 9.17 $k^2 \cdot q^2$ values for insulated wires

Cu conductors mm^2	Insulation			
	PVC ($k=115$) A$^2 \cdot$s^2	Rubber ($k=141$) A$^2 \cdot$s^2	VPE, EPR ($k=143$) A$^2 \cdot$s^2	II K ($k=134$) A$^2 \cdot$s^2
0.5	3.306·10^3	4.97 ·10^3	5.112·10^3	4.489·10^3
0.75	7.439·10^3	11.183·10^3	11.502·10^3	10.100·10^3
1	13.225·10^3	19.881·10^3	20.449·10^3	17.956·10^3
1.5	29.756·10^3	44.732·10^3	46.010·10^3	40.401·10^3
2.5	82.856·10^3	124.3 ·10^3	127.8 ·10^3	112.225·10^3
4	211.6 ·10^3	318.1 ·10^3	327.2 ·10^3	287.3 ·10^3
6	476.1 ·10^3	715.7 ·10^3	736.2 ·10^3	646 ·10^3
10	1 322.5 ·10^3	1 988.1 ·10^3	2 044.9 ·10^3	1 795.6 ·10^3
16	3 385.6 ·10^3	5 089.5 ·10^3	5 234.9 ·10^3	4 596.7 ·10^3
25	8 265.6 ·10^3	12 426 ·10^3	12 781 ·10^3	11 223 ·10^3
35	16 200 ·10^3	24 354 ·10^3	25 050 ·10^3	21 996 ·10^3
50	33 063 ·10^3	49 703 ·10^3	51 123 ·10^3	44 890 ·10^3
70	64 803 ·10^3	97 417 ·10^3	100 200 ·10^3	87 984 ·10^3
95	119 356 ·10^3	179 426 ·10^3	184 552 ·10^3	162 053 ·10^3
120	190 440 ·10^3	286 286 ·10^3	258 566 ·10^3	258 566 ·10^3
150	297 563 ·10^3	447 323 ·10^3	460 103 ·10^3	404 010 ·10^3
185	452 626 ·10^3	680 427 ·10^3	699 867 ·10^3	614 544 ·10^3
240	761 760 ·10^3	1 145 146 ·10^3	1 177 862 ·10^3	1 034 266 ·10^3
300	1 190 250 ·10^3	1 789 290 ·10^3	1 840 410 ·10^3	1 616 040 ·10^3
400	2 116 000 ·10^3	3 180 960 ·10^3	3 271 840 ·10^3	2 872 960 ·10^3
500	3 306 250 ·10^3	4 970 250 ·10^3	5 112 250 ·10^3	4 489 000 ·10^3

Table 9.18 $I^2 \cdot t$ values for miniature circuit-breakers

a) Characteristic B, C and former characteristic L, in accordance with DIN VDE 0641 including alteration A4 (part 11)

Rated current in A	$I^2 \cdot t$ at $I_k'' = 3$ kA $A^2 \cdot s$	$I^2 \cdot t$ at $I_k'' = 6$ kA $A^2 \cdot s$	$I^2 \cdot t$ at $I_k'' = 10$ kA $A^2 \cdot s$
4 10 6 12 8 16	$15 \cdot 10^3$	$35 \cdot 10^3$	$70 \cdot 10^3$
20 25	$18 \cdot 10^3$	$45 \cdot 10^3$	$90 \cdot 10^3$
32–64	No data in DIN VDE		

b) Former characteristic G in accordance with DIN VDE 0660 part 101

Rated current in A	$I^2 \cdot t$ $A^2 \cdot s$	Rated current in A	$I^2 \cdot t$ $A^2 \cdot s$
0.5	260	10	$13 \cdot 10^3$
1	700	16	$13 \cdot 10^3$
1,6	2200	20	$14 \cdot 10^3$
2	$3 \cdot 10^3$	25	$15 \cdot 10^3$
3	$7 \cdot 10^3$	32	$15 \cdot 10^3$
4	$9 \cdot 10^3$	40	$26 \cdot 10^3$
6	$12 \cdot 10^3$	50	$30 \cdot 10^3$

Table 9.19 $I^2 \cdot t$ values for fuses to DIN VDE 0636

Rated current in A	$I^2 \cdot t$ $A^2 \cdot s$	Rated current in A	$I^2 \cdot t$ $A^2 \cdot s$
2	$0.0144 \cdot 10^3$	160	$185 \cdot 10^3$
4	$0.0576 \cdot 10^3$	200	$302 \cdot 10^3$
6	$0.16 \cdot 10^3$	224[1)]	$293 \cdot 10^3$
10	$0.64 \cdot 10^3$	250	$557 \cdot 10^3$
16	$1.21 \cdot 10^3$	300[1)]	$750 \cdot 10^3$
20	$2.5 \cdot 10^3$	315	$900 \cdot 10^3$
25	$4 \cdot 10^3$	355[1)]	$950 \cdot 10^3$
32	$5.75 \cdot 10^3$	400	$1600 \cdot 10^3$
35	$6.75 \cdot 10^3$	425[1)]	$1410 \cdot 10^3$
40	$9 \cdot 10^3$	500	$2700 \cdot 10^3$
50	$13.7 \cdot 10^3$	630	$5470 \cdot 10^3$
63	$21.2 \cdot 10^3$	800	$10000 \cdot 10^3$
80	$36 \cdot 10^3$	1000	$17400 \cdot 10^3$
100	$64 \cdot 10^3$	1250	$33100 \cdot 10^3$
125	$104 \cdot 10^3$	–	–

The corresponding values for circuit-breakers (e.g. moulded case circuit-breakers) must be obtained from the respective manufacturers!

[1)] Not standardized values; Supplied by the manufacturer

conductors if it reacts to the smallest possible short-circuit current on only one parallel line."

This supplement must already to be taken into consideration if a common protection device is used.

One may, however, still provide short-circuit protection devices in each individual conductor of a parallel run. However, it is then necessary to provide protection devices for the individual conductors at both ends of the cable run.

Coordinated overload and short-circuit protection

A common protection device is permitted for overload and short-circuit protection. Here, the conductor *length* is not relevant. Since "protection against hazardous shock current" must nevertheless be ensured, this possibility is permitted only

– where leakage current protection devices with $I_{\Delta n} \leq 30$ mA are used, or
– in IT-systems (networks) with protection by indication, e.g. by insulation monitoring devices and additional equipotential bonding.

9.4.2 Current carrying capacity

The specifications relevant to practical installation work and the corresponding requirements are described below:

DIN VDE 0100 part 430, supplement 1

This specification contains tables on current carrying capacities and the assignment of overload protection devices for frequently used types of installation and cable routing. It is mainly applicable for installations in buildings and an ambient conductor temperature of 25 °C.

DIN VDE 0298 part 4

This specification applies for all the usual types of installation of cables or wires in free air; mainly in residential and office buildings. Distinctions are made between:

permanently laid cables, flexible cables and cables with increased heat resistance.

The permissible current carrying capacity I_z depends on the material and the insulation of the conductor as well as on

– the ambient temperature (f_1) (DIN VDE 0298 part 4, Table 10),
– the bunching of cables, type of installation (f_2) (DIN VDE 0298 part 4, Tables 11 and 12),

9 Appendix

Table 9.20
Current carrying capacity of copper wires and cables for permanent installation and installation in free air; uninterrupted operation at an ambient temperature of **30 °C**; permissible operating temperature 70 °C (Summary of Tables 3 and 4 of DIN VDE 0298 part 4, 02.88) [1]

a) Types of installation A, B1, B2, and C

Insulating material	PVC							
Cable-type designation [2]	NYM, NHXMH, NBUY, NHYRUZY, NYIF, NYIFY, SIENOPYR, SIFLA, H07V-U, H07V-R, H07V-K							
Permissible operating temperature	70 °C							
Ambient temperature	30 °C [3]							
Number of cores under load	2	3	2	3	2	3	2	3
Type of installation	A		B1		B2		C	
	in cavity walls filled with insulating material		on or in solid walls or under the plaster in electric conduits or ducting				directly attached/laid, incl. on the floor	
	single-core insulated wires in conduit [4] [7]; multi-core cable in conduit [5]; multi-core cable in the wall		single-core insulated wires in conduit on the wall surface [5]; single-core insulated wires in ducting on the wall surface; single-core insulated wires, single-core sheathed cables, multi-core cables in conduit in the brickwork		multi-core cable in conduit on the wall or on the floor; multi-core cable in ducting on the wall or on the floor		multi-core cable fixed to the wall or on the floor [6]; single-core sheathed cables fixed to the wall or on the floor; multi-core cable, flat webbed cable in the wall or under the plaster	

Rated cross-section q_n	Current carrying capacity I_z and rated current I_n [8] of the overload protection device where $I_2 \leq 1.45 \cdot I_n$ in A															
mm²	I_z	I_n	I_z	I_n	I_z	I_n	I_z	I_n	I_z	I_n	I_z	I_n	I_z	I_n	I_z	I_n
1.5	15.5	13	13	13	17.5	16	15.5	13	15.5	13	14	13	19.5	16	17.5	16
2.5	19.5	16	18	16	24	20	21	20	21	20	19	16	26	25	24	20
4	26	25	24	20	32	32	28	25	28	25	26	25	35	35	32	32
6	34	32	31	25	41	40	36	35	37	35	33	32	46	40	41	40
10	46	40	42	40	57	50	50	50	50	50	46	40	63	63	57	50
16	61	63	56	56	76	80	68	63	68	63	61	50	85	80	76	63
25	80	80	73	63	101	100	89	80	90	80	77	63	112	100	96	80
35	99	80	89	80	125	125	111	100	110	100	95	80	138	125	119	100
50	119	100	108	100	151	125	134	125	–		–		–		–	
70	151	125	136	125	192	160	171	160	–		–		–		–	
95	182	160	164	160	232	200	207	200	–		–		–		–	
120	210	200	188	160	269	250	239	250	–		–		–		–	

See page 523 for footnotes

Table 9.20 (continued)

b) Type of installation E, in free air

Insulating material	PVC	
Cable-type designation[2]	NYY, NYCWY, NYKY, NYM, NYMZ, NYMT, NYRUY, NHYRUZY	
Permissible operating temperature	70 °C	
Ambient temperature	30 °C[3]	
Number of cores under load	2	3
Type of installation	E	E
	Installed in free air, in compliance with specified distances	
	≥0.3d	≥0.3d
Rated cross-section q_n mm²	Current carrying capacity I_z and rated current I_n[8] of the overload protection device where $I_2 \leq 1.45 \cdot I_n$ in A			

mm²	I_z	I_n	I_z	I_n
1,5	20	20	18,5	16
2,5	27	25	25	25
4	37	35	34	32
6	48	40	43	40
10	66	63	60	63
16	89	80	80	80
25	118	100	101	100
35	145	125	126	125
50	–	–	153	125
70	–	–	196	160
95	–	–	288	250
120	–	–	–	–

[1] Notes on planning and configuration from Heinhold, L.; Stubbe, R.: Kabel und Leitungen für Starkstrom, part 1 (1987) and part 2 (1987) Publ. Siemens AG. Berlin, Munich and DIN VDE 0298 part 2.
[2] List of type designations with indication on standards to which the cables and wires conform see DIN VDE 0298 part 1, 11.82 and part 3, 08.83 (draft 01.90 available).
[3] If the ambient temperature differs from this value, the current carrying capacity I_z can be converted using Table 9.21. This table corresponds to Table 10 of DIN VDE 0298 part 4.
[4] Also applies for single-core insulated wires in conduits laid in closed underfloor ducting.
[5] Also applies for single-core insulated wires in conduits laid in ventilated underfloor ducting.
[6] Also applies for multi-core cables in open or ventilated electrical ducting.
[7] Also applies for single-core insulated wires, single-core sheathed cables and multi-core cables in electrical underfloor ducting.
[8] In accordance with DIN VDE 0100 part 430.

— number of conductors (f_3) (DIN VDE 0298 part 4, Table 13).

Therefore the total conversion factor is:

$$f_{tot} = f_1 \cdot f_2 \cdot f_3 \qquad (9.11)$$

DIN VDE 0298 part 2

This specification applies for cables laid underground or installed in free air with U_0/U of 0.6/1 kV up to 18/30 kV. To some extent, it also applies for single conductors (NYM).

The current carrying capacity is dependent on the material and on the insulation of the conductor, as well as on the

— ambient temperature (f_{L1}, f_{E1}),
— bunching, type of installation (f_{L2}, f_{E2}),
— number of conductors (f_{L3}, f_{E3}),
— underground installation (f_{E4}).

This leads to the total conversion factors

underground: $f_{Utot} = f_{E1} \cdot f_{E2} \cdot f_{E3} \cdot f_{E4}$, (9.12)

in free air: $1 f_{Atot} = f_{L1} \cdot f_{L2} \cdot f_{L3}$. (9.13)

Table 9.21
Current carrying capacity I_z as percentages of the values in Table 9.20 for insulated wires and cables not laid underground and at ambient temperatures above 20 to 60 °C. For current carrying capacities of cables with mass-impregnated paper insulation and metal-sheathed cables, refer to DIN VDE 0255, 11.72

Ambient temperature	Current carrying capacity I_z as % of the values in Table 9.20	
	Rubber insulation,[1] permissible conductor temperature 60 °C	PVC insulation,[2,3] maximum permissible conductor temperature 70 °C
above 15 up to 20	115	112
above 20 up to 25	107	106
above 25 up to 30	100	100
above 30 up to 35	92	93
above 35 up to 40	82	87
above 40 up to 45	71	79
above 45 up to 50	58	71
above 50 up to 55	41	61
above 55 up to 60	–	50

[1] NR Natural rubber
    SR Synthetic rubber
[2] PVC Polyvinyl chloride
[3] Cross-linked polyethylene (XLPE) 90 °C and silicone rubber (SiR) 180 °C

9 Appendix

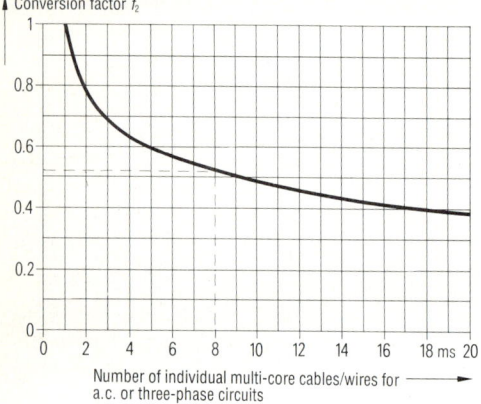

    ⊙ Symbol for a single or multi-core cable

Bunched arrangement directly on the wall or underfloor, in conduit or electric ducting mounted on or in the wall.

Figure 9.2
Conversion factors f_2 for bunching of cables, to be applied to the current carrying capacity values of installation types B1, B2 and C. For cable bunching in installation type E, refer to DIN VDE 0298 part 4, Table 11 or 13; for cable bunching in accordance with DIN VDE 0298 part 2, 11.79, see Tables 22 and 23

Table 9.20 shows the current carrying capacity for different types of installation at an ambient temperature of 30 °C. Table 9.21 gives the conversion factors for current carrying capacity as percentages of the values in Table 9.20 at ambient temperatures of 20 to 60 °C.

Figure 9.2 provides conversion factors f_2 related to the bunching (or bundling) of cables. These factors are to be applied to the current carrying capacity values of installation types B1, B2 and C.

The determination of the conductor cross-section and the maximum permissible length of the conductors, with regard to overload and short-circuit protection, is illustrated in the following example.

Example

Determination of the cross-section and the maximum permissible length of the conductors with regard to overload and short-circuit protection.

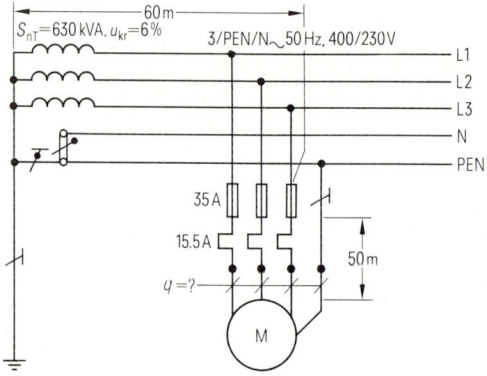

Overload protection

Given:
NYM, bunched with 8 further three-phase systems; ambient temperature 40 °C (intermediate ceiling); thermally-delayed overload relay set to 15.5 A.

To be found:
Required conductor cross-section q.

Solution:

f_1 = 0.87 from DIN VDE 0298 part 4, Table 10, column 3; at 40 °C

f_2 = 0.52 from Figure 9.2; with 8 systems

f_{tot} = 0.87 · 0.52 = 0.45

q in accordance with DIN VDE 0298 part 4, Table 3, type of installation C, column 9 (3 cores under load) for a calculated current I_{cal} of:

$$I_{cal} = \frac{I_n}{f_{tot}} = \frac{15.5 \text{ A}}{0.45} = \underline{34.4 \text{ A}};$$

$q = \underline{6 \text{ mm}^2 \text{ Cu}}$

For the same example without overload relay but with thermistor motor protection, where the fuse is responsible for protection in case of overload and short-circuit, the equation is:

$$I_{cal} = \frac{I_n \cdot I_2}{1.45 \cdot f_{tot}} = \frac{35 \text{ A} \cdot 1.6}{1.45 \cdot 0.45} = \underline{85.8 \text{ A}};$$

$q = \underline{25 \text{ mm}^2 \text{ Cu}}$

Short-circuit protection

Given:

Transformer with $S_{nT} = 630$ kVA;

$u_{kr} = 6\%$;

$Z_{trafo} = 15$ mΩ (corresponding value from Table 1.17, page 47);

Connection between transformer and distribution:
NYY $3 \times 95/50$ mm, $l = 60$ m;

$Z'_{cable95/50} = 0.681$ mΩ/m (corresponding value from Table 1.18b, page 48).

Motor branch circuit:
NYM-J 4×6 mm², $l = 50$ m;

$Z_{cable6} = 8.93$ mΩ (corresponding value from calculation sheet 1, page 45).

To be found:

Maximum permissible conductor length l_{perm}.

Solution:

$Z_v = Z_{trafo} + Z_{cable60m}$

$Z_{trafo} = 15$ mΩ

$Z_{cable95/50} = 0.681 \cdot 60 = 40.86$ mΩ

$Z_v = \underline{55.86}$ mΩ/m

$Z' = \underline{8.93\ m\Omega}$

The maximum permissible length l_{perm} downstream from the protective device can be calculated by means of equation (9.8):

$$l_{perm} = \frac{\dfrac{230 \cdot 10^3}{173} - 55.86}{8.93} \approx \underline{142\ m}.$$

Current carrying capacity of busbars

The current carrying capacity of busbars must comply with DIN 43670 or 43671, as already mentioned on page 515 above.

The tables in these regulations do not, however, apply directly to switchgear assemblies in terms of DIN VDE 0660 part 500. The applicable tables are currently in preparation.

9.4.3 Load ratings of insulated conductors at ambient temperatures of 30° to 70 °C and the assignment of cable protection fuses in accordance with USA and Canadian standards

Table 9.22 lists the permissible current carrying capacities (ampacities) of insulated conductors in accordance with USA and Canadian specifications for ambient temperatures up to 30 °C. Table 9.23 lists the corresponding values for ambient temperatures above 30° up to 70 °C.

The information in Tables 9.22 and 9.23 is based on the following specifications:

USA   National Electrical Code (NEC), 1993, Tables 310-16 and 310-17,

Canada   Canadian Electrical Code (CEC), part I, 1993, Tables 1–4, 13.

The following insulated conductors are normally used for the wiring of industrial switchgear and installations:

– for rated currents up to 100 A:
  Conductor type TW (flame-retardant, thermoplastic-insulated, moisture-resistant), temperature limit 60 °C,

– for rated currents above 100 A:
  Conductor type RH (rubber-insulated, heat-resistant) or type THW (thermoplastic-insulated, moisture-resistant, heat-resistant), limit temperature 75 °C.

In order to determine conductor cross-sections and fuse ratings at ambient temperatures above 30 °C, proceed as follows:

The permissible conductor rating (ampacity) for uninterrupted duty at the required ambient temperature may be obtained from Table 9.23 as a percentage of the values in Table 9.22. If a larger conductor cross-section is determined for temperatures above 30 °C, a fuse must be assigned which would correspond to the conductor cross-section which is suitable for 30 °C according to Table 9.22.

9 Appendix

Table 9.22
Permissible continuous current rating (ampacity) of insulated Cu or Al conductors at ambient temperatures up to **30 °C** (in accordance with USA (NEC) and Canadian regulations) with assigned line protection fuses

Wire size	Rated cross-section q_n (equivalent)	Rated cross-section acc. to ISO	Up to three insulated conductors installed in a raceway e.g. **conduit**, alternatively three-core cable						Single insulated conductor in **free air**					
			Cu conductor			Al conductor			Cu conductor			Al conductor		
			Permissible continuous current rating to [5)/6)]		Rated current of the fuse [4)]	Permissible continuous current rating to [5)/6)]		Rated current of the fuse [4)]	Permissible continuous current rating to [5)/6)]		Rated current of the fuse [4)]	Permissible continuous current rating to [5)/6)]		Rated current of the fuse [4)]
			Type	Type		Type	Type		Type	Type		Type	Type	
			TW [2)], UF	RH [3)], THW, RW75, TW75		TW [2)], UF	RH [3)], THW, RW75, TW75		TW [2)], UF	RH [3)], THW, RW75, TW75		TW [2)], UF	RH [3)], THW, RW75, TW75	
AWG [1)]	mm²	mm²	A	A	A	A	A	A	A	A	A	A	A	A
14	2.08	2.5	20/15	20/15	15	–	–	–	25/20	30/20	15	–	–	–
12	3.31	4	25/20	25/20	20	20/15	20/15	15	30/25	35/25	20	25	30/20	15
10	5.26	6	30	35/30	30	25	30/25	25	40	50/40	30	35	40/30	25
8	8.37	10	40	50/45	40/50[4)]	30	40/30	30/40	60/55	70/65	60/70[4)]	45	55/45	45/60[4)]
6	13.3	16	55	65	60/70[4)]	40	50	40/50[4)]	80	95	80/100[4)]	60	75	60/80[4)]
4	21.15	25	70	85	70/90[4)]	55	65	60/70[4)]	105	125	110/125[4)]	80	100	80/100[2)]
3	26.66	–	85/80	100	90/100[4)]	65	75	70/80[4)]	120	145	125/150[4)]	95	115	100/125[2)]
2	33.63	35	95/100	115	100/125[4)]	75	90	80/90[4)]	140	170	150/175[4)]	110	135	110/150[2)]
1	42.41	–	110	130	110/150[2)]	85	100	90/100[4)]	165	195	175/200[4)]	130	155	150/175[4)]
1/0	53.51	50	125	150	125/150[4)]	100	120	100/125[4)]	195	230	200/250[4)]	150	180	150/200[4)]
2/0	67.44	70	145	175	150/175[4)]	115	135	125/150[4)]	225	265	225/300[4)]	175	210	175/225[4)]
3/0	85.02	95	165	200	175/200[4)]	130	155	150/175[4)]	260	310	300/350[4)]	200	240	200/250[4)]
4/0	107.22	–	195	230	200/250[4)]	150/155	180	150/200[4)]	300	360	300/400[4)]	235/230	280	250/300[4)]
kcmil (MCM) [1)]														
250	126.8	120	215	255	225/300[4)]	170	205	175/225[4)]	340	405	350/450[4)]	265	315	300/350[4)]
300	152.01	150	240	285	250/300[4)]	190	230	200/250[4)]	375	445	400/450[4)]	290	350	300/350[4)]
350	177.35	185	260	310	300/350[4)]	210	250	225/250[4)]	420	505	450/600[4)]	330	395	350/400[4)]
400	202.68	–	280	335	300/350[4)]	225	270	225/300[4)]	455	545	500/600[4)]	355	425	400/450[4)]
500	253.35	240	320	380	350/400[4)]	260	310	300/350[4)]	515	620	600/700[4)]	405	485	450/500[4)]
600	304.0	300	355	420	400/450[4)]	285	340	300/350[4)]	575	690	600/700[4)]	455	540/545	500/600[4)]
700	354.71	–	385	460	400/500[4)]	310	375	350/400[4)]	630	755	700/800[4)]	500	595	500/600[4)]
750	380.0	–	400	475	400/500[4)]	320	385	350/400[4)]	655	785	700/800[4)]	515	620	600/700[4)]
800	405.36	400	410	490	450/500[4)]	330	395	350/400[4)]	680	815	700/1000[4)]	535	645	600/700[4)]
900	456.04	–	435	520	450/600[4)]	355	425	400/450[4)]	730	870	800/1000[4)]	580	700	600/700[4)]
1000	506.70	500	455	545	500/600[4)]	375	445	400/450[4)]	780	935	800/1000[4)]	625	750	700/800[4)]
1250	633.38	–	495	590	500/600[4)]	405	485	450/500[4)]	890	1065	1000/1200[4)]	710	855	700/1000[4)]
1500	760.05	–	520	625	600/700[4)]	435	520	450/600[4)]	980	1175	1000/1200[4)]	795	950	800/1000[4)]
1750	886.7	–	545	650	600/700[4)]	455	545	500/600[4)]	1070	1280	1200/1600[4)]	875	1050	800/1000[4)]
2000	1013.4	–	560	665	600/700[4)]	470	560	500/600[4)]	1155	1385	1200/1600[4)]	960	1150	1000/1200[4)]

[1)] AWG American Wire Gauge, standard American unit of measurement for conductor cross-sections of up to 107.2 mm², see also Table 9.10 on page 511
 kcmil = kilo circular mil (1 kcmil ≙ 0.5067 mm²), standard American unit of measurement for conductor cross-sections greater than 126.7 mm²
[2)] Conductor temperature 60 °C
[3)] Conductor temperature 75 °C
[4)] The first value applies to conductor types TW and UF, the second value to types RG and THW, RW75 and TW75
[5)] National Electrical Code 1993 (NPFA 70)
[6)] Canadian Electrical Code part I (CEC), 1993

Table 9.23
Permissible continuous current rating (ampacity) of insulated conductors expressed as % of the values in Table 9.22 at ambient temperatures between 20 and 70 °C

Ambient temperature	Permissible continuous current rating as % of the values in Table 9.22	
	Insulated conductors	
°C	Type TW, UF	Type RH, THW
above 20 up to 25	108	105
above 25 up to 30	100	100
above 30 up to 35	91	94
above 35 up to 40	82	88
above 40 up to 45	71	82
above 45 up to 50	58	75
above 50 up to 55	41	67
above 55 up to 60	–	58
above 60 up to 70	–	33

Example

Three insulated copper conductors of the type RH, installed in a conduit, are to be rated for an ambient temperature of 50 °C and a continuous current of 100 A.

In terms of Table 9.23, the permitted current carrying capacity is 75% of the values given in Table 9.22. Consequently, a conductor cross-section for 100 A : 0.75 = 133 A must be selected from Table 9.22. For a conductor of the type RH, the size AWG 1/0 is required.

The rated current of the fuse is found to be 100 A in accordance with the continuous current carrying capacity (corresponding to a conductor of the type RH for a continuous current carrying capacity of 100 A at 30 °C, i.e. corresponding to a conductor of the size AWG 3).

9.4.4 Thermal ratings of busbars and device terminals

When selecting the cross-sections of busbars, the following factors must be taken into consideration:

▷ the permissible temperatures of busbars and equipment during rated operation and under short-circuit conditions, as specified in the relevant DIN VDE regulations,

▷ any relevant additional information on connected items of equipment as supplied by the manufacturer,

▷ the permissible temperature rise limits of items of equipment in the vicinity,

▷ the permissible temperatures rise limits of insulating materials, e.g. the insulation of conductors connected directly to the busbars in question.

Permissible *temperature limits for rated operation* are:

▷ for busbar connections (joins)
 in accordance with DIN 43673, 02.82
  oxide-free and greased 120 °C,
  silver-coated or treated equivalently 160 °C,

▷ for supports and bushings
 in accordance with
 DIN VDE 0674 part 1, 12.84 85 °C,

▷ for insulating materials classified
 according to the DIN VDE draft 0301
 part 1, 08.83 (IEC publ. 85, 1984) >90 °C.

DIN VDE 0103, 04.88 recommends the following *temperature limits in the case of a short-circuit*:

▷ for bare conductors, solid or stranded,
 made of copper or aluminium 200 °C,

▷ for bare conductors, solid or stranded,
 made of steel 300 °C,

▷ the relevant temperatures for paper and/or plastic-insulated conductors, solid or stranded, may be found in the applicable equipment standards.

Permissible *temperature-rise limits*[1] on device terminals for external connections in accordance with IEC 947-1 are:

▷ for bare copper 60 K,

▷ for bare brass or for
 tin-plated (tinned) brass or copper 65 K,

▷ for copper or brass,
 silver- or nickel-plated 70 K.[2]

---

[1] The specified temperature rises may be exceeded by a maximum of 10 K for units with small dimensions as specified for the individual equipment.
[2] Temperature-rise limit if PVC cables are connected

9 Appendix

Copper busbars

Table 9.23 lists the current carrying capacity values for copper busbars at an ambient temperature of 35 °C and a busbar operating temperature of 65 °C.

At other ambient and busbar temperatures, the values for copper busbars given in Table 9.23 must be converted by *division* using factor k_2 from Figure 9.3.

Example

A flat, painted copper busbar for 250 A (AC 50 Hz) is required for a low-voltage switchboard. At an ambient temperature of 55 °C, the permissible busbar temperature should be 75 °C.

In this case factor $k_2 = 0.76$ from Figure 9.3. Consequently, the conductor cross-section must be selected for a current of 250 A : 0.76 ≈ 330 A.

Aluminium busbars

In consideration of the dynamic strength and the use of bolts with quality class 8 and thereby the current carrying capacity and thermal stresses, it is suggested that aluminium busbars of type E-Al Mg Si 05 F17 be used. The current carrying capacity is lower by more than a third and can be determined by using DIN 43670.

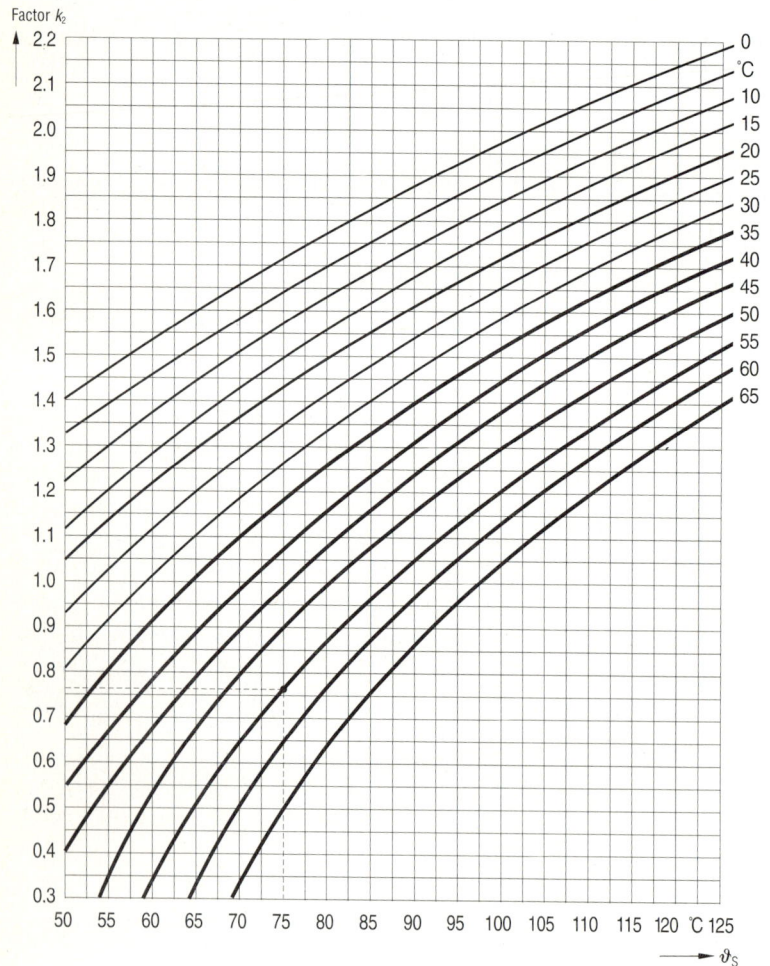

Figure 9.3
Factor k_2 for determining the conductor cross-section of copper busbars at various ambient temperatures ϑ_u from 0 to 65 °C and/or at busbar operating temperatures ϑ_s up to 125 °C

Table 9.24
Copper busbars in accordance with DIN 43671, 12.75 (excerpt) for an ambient temperature ϑ_u around the busbars of **35 °C** and a busbar operating temperature ϑ_s of 65 °C; material: E-CuF30

Width × thickness	Cross-section q	Width × thickness	Mass	Continuous current							
				Alternating current up to 60 Hz				Direct and alternating current up to $16^2/_3$ Hz			
				painted number of busbars		bare number of busbars		painted number of busbars		bare number of busbars	
				I	II	I	II	I	II	I	II
mm	mm²	inch	kg/m	A	A	A	A	A	A	A	A
20 × 5	99.1	0.75 × 0.20	0.882	319	560	274	500	320	562	274	502
30 × 5	149	1 × 0.25	1.33	447	760	379	672	448	766	380	676
40 × 5	199	1.25 × 0.25	1.77	573	952	482	836	576	966	484	848
50 × 5	249	1.5 × 0.25	2.22	697	1140	583	994	703	1170	588	1020
20 × 10	199	0.75 × 0.5	1.77	497	924	427	825	499	932	428	832
30 × 10	299	1 × 0.5	2.66	676	1200	573	1060	683	1230	579	1080
40 × 10	399	1.25 × 0.5	3.55	850	1470	715	1290	865	1530	728	1350
50 × 10	499	1.50 × 0.5	4.44	1020	1720	852	1510	1050	1830	875	1610
60 × 10	599	2 × 0.5	5.33	1180	1960	985	1720	1230	2130	1020	1870
80 × 10	799	2.5 × 0.5	7.11	1500	2410	1240	2110	1590	2730	1310	2380
100 × 10	999	3 × 0.5	8.89	1810	2850	1490	2480	1940	3310	1600	2890
120 × 10	1200	3.75 × 0.5	10.7	2110	3280	1740	2866	2300	3900	1890	3390
160 × 10	1600	5 × 0.5	14.2	2700	4130	2220	3590	3010	5060	2470	4400
200 × 10	2000	6 × 0.5	17.8	3290	4970	2690	4310	3720	6220	3040	5390

9.4.5 Resistance of copper and aluminium conductors

In order to determine 2-pole and 3-pole short-circuit currents in accordance with DIN VDE 0102, 01.90, the resistance per unit length of copper and aluminium conductors at a conductor temperature of 20 °C can be found in Table 9.25 and can be converted for a conductor temperature of 80 °C using equation (9.14). To determine the value of a single pole short-circuit current (fault to frame), the resistance values of PVC-insulated cables with copper conductors referred to 55 °C can be found in Tables 1.18a to 1.18d of Section 1.4.2, page 48.

Depending on the length and cross-section of the conductors between the transformer terminals and the point of fault, their resistance can significantly reduce the (3-pole) prospective short-circuit current from that which would occur directly at the transformer. The lower the rated output and the secondary voltage, and the higher the u_{kr} of the transformer is, the more effective this limiting will be. An example is given in Section 2.1.6 on page 97. The following applies as a rule of thumb:

In a 400 V network, a 10 mm² copper wire will limit a prospective short-circuit current of e.g. 50 kA occurring directly at the secondary-side transformer terminals, to 10 kA at the end of the cable after only 10 m; a 25 mm² copper wire will have the same effect with a length of 25 m between the transformer terminals and the point of fault (see Fig. 2.17 on page 97).

The values for the resistance R'_L of copper and aluminium conductors given in Table 9.25 can approximately be converted for conductor temperatures above 20 °C as shown below:

$$R_{\vartheta_2} = R_{20}(1 + \alpha_{20} \cdot \Delta\vartheta) \tag{9.14}$$

for copper: $\alpha_{20} = 0.00393$,
for aluminium: $\alpha_{20} = 0.00403$.

Table 9.25
The ohmic resistance R'_L per single unit length of copper and aluminium conductor at a conductor temperature of 20 °C and a frequency of 50 Hz

Copper		Aluminium[1]	
Rated cross-section q_n mm²	Ohmic resistance mΩ/m	Rated cross-section q_n mm²	Ohmic resistance mΩ/m
0.75	24.17	–	–
1	18.09	–	–
1.5	12.06	–	–
2.5	7.26	–	–
4	4.55	–	–
6	3.02	–	–
10	1.81	10	2.88
16	1.14	16	1.89
25	0.73	25	1.20
35	0.53	35	0.88
50	0.39	50	0.64
70	0.27	70	0.45
95	0.20	95	0.32
120	0.157	120	0.26
150	0.129	150	0.21
185	0.10	185	0.17
240	0.080	240	0.13
300	0.069	300	0.105

[1] The ohmic resistance of aluminium conductors is approximately 1.7 times higher than that of copper conductors.

Table 9.26
Inductive reactance X'_L per single unit length of copper and aluminium conductor at a conductor temperature of 20 °C and a frequency of 50 Hz. (For other frequencies, the inductive reactance has to be converted proportionally.)

Rated cross-section	Cable for 0.6/1 kV		
	N(A)KBA. 4 cores	N(A)KLEY. 3½ cores	N(A)YY. NYCY. N(A)YCWY. 4 cores
mm²	X'_L mΩ/m	X'_L mΩ/m	X'_L mΩ/m
1.5	–	–	0.115
2.5	–	–	0.110
4	–	–	0.107
6	–	–	0.100
10	0.103	–	0.094
16	0.099	–	0.090
25	0.094	–	0.086
35	0.092	–	0.083
50	0.090	0.071	0.083
70	0.087	0.069	0.082
95	0.086	0.068	0.082
120	0.086	0.068	0.080
150	0.085	0.067	0.080
185	0.085	0.067	0.080
240	0.084	0.066	0.079
300	0.084	–	0.079

Tables 9.25 and 9.26 give the resistance and the inductive reactance per unit of *single* length and for *a single* conductor respectively. The values given in Section 1.4 refer to "total distances" e.g. in an installation – i.e. to the outgoing and return line.

This means that at a temperature of slightly more than 80 °C, the resistance $R_{\vartheta 80}$ is about 25% higher than the values listed in Table 9.25.

The values for the inductive reactance X'_L of copper and aluminium conductors at 50 Hz listed in Table 9.26 can be converted proportionally for other frequencies, e.g. 60 Hz,

$X'_{60Hz} = 60/50 \cdot X'_L = 1.2 \cdot X'_L$.

9.5 Rated currents of three-phase induction motors

As an example, Table 9.27 lists the rated currents of four-pole three-phase induction motors for 50 Hz and 60 Hz networks. The rated currents may differ slightly from manufacturer to manufacturer. Table 9.28 gives the relationship between the number of poles and the motor speed at no-load and at rated load.

If a motor rated at 50 Hz is operated in a 60 Hz network, then the motor speed is increased by approximately 20% and the rated output by a factor of 1.15 respectively. The power factor (cos φ) remains virtually unchanged. For supply voltages other than those given, the currents may be converted accordingly.

Rated currents of three-phase induction motors

Table 9.27
Rated currents of four-pole three-phase induction motors in their basic version; degree of protection IP 54; for 50 Hz or 60 Hz networks (1500 or 1800 min^{-1})

Standardized motor power ratings		200 V [1]	208 V [1]	230 V [2]	230 V [1] (220–240 V)	400 V/ 380 V [2]	460 V [1] (440–480 V)	500 V [2]	575 V [1] (550–600 V)	690 V [2]
kW [3]	HP [1]	\multicolumn{9}{c}{Corresponding rated currents I_n in A}								
0.25	1/3	1.75	1.7	1.6	1.5	0.9	0.75	–	–	–
0.37	1/2	2.2	2.1	2.0	1.9	1.1	0.95	–	0.8	–
0.55	3/4	3.1	3.0	2.6	2.7	1.5	1.4	1.2	1.1	0.9
0.75	1	3.8	3.7	3.3	3.3	1.9	1.8	1.5	1.4	1.1
1.1	1.5	6.2	5.9	4.3	5.4	2.5	2.6	2.0	2.1	1.4
1.5	2	7.8	7.5	5.8	6.9	3.3	3.4	2.6	2.7	1.9
2.2	3	11	10.6	8.3	9.7	4.7	4.8	3.8	3.9	2.7
3	–	–	–	11.0	–	6.3	–	5.0	–	3.6
(3.7)	5	17.5	17	–	15.2	–	7.6	–	6.1	–
4	–	–	–	14	–	8	–	6.5	–	4.7
5.5	7.5	25.3	24	19	22	11	11	9.0	9.0	6.3
7.5	10	32.2	31	26	28	15	14	12	11	8.5
11	15	48.3	46	36	42	21	21	17	17	12
15	20	62.1	59.5	50	54	28	27	23	22	16
18.5	25	78	75	61	68	35	34	28.5	27	20
22	30	92	88	73	80	42	40	33	32	24
30	40	120	114	99	104	57	52	45	41	33
37	50	150	143	120	130	69	65	55	52	40
45	60	177	170	140	154	81	77	65	62	47
55	75	221	211	173	192	100	96	80	77	58
75	100	285	273	228	248	131	124	105	99	76
90	125	359	343	281	312	162	156	129	125	94
110	150	414	396	339	360	195	180	156	144	113
132	–	–	–	405	–	233	–	187	–	135
–	200	552	528	–	480	–	240	–	192	–
160	–	–	–	496	–	285	–	228	–	165
–	250	693	662	–	602	–	302	–	242	–
200	–	–	–	611	–	352	–	281	–	204
220	300	830	794	674	722	388	361	310	289	225
250	–	–	–	785	–	431	–	360	–	262
–	350	952	911	–	828	–	414	–	336	–
–	400	1097	1050	–	954	–	477	–	382	–
315	–	–	–	925	–	532	–	426	–	308
335	450	1230	1177	975	1070	561	535	448	428	325
355	–	–	–	–	–	608	–	486	–	352
375	500	1357	1298	1057	1180	637	590	509	472	369
400	–	–	–	1190	–	684	–	547	–	397
450	600	–	–	1338	1416	770	708	615	567	446
500	–	–	–	1470	–	846	–	676	–	490
530	700	–	–	1578	1652	907	826	726	661	526

[1] in accordance with UL 508 (19.09.77), Table 24.4 for 60 Hz
[2] in accordance with NF C63-110, annexe IV for 50 Hz (only up to 250 kW)
[3] in accordance with DIN 42973

Table 9.27 (continued)

Standardized motor power ratings		200 V[1]	208 V[1]	230 V[2]	230 V[1] (220–240 V)	400 V/ 380 V[2]	460 V[1] (440–480 V)	500 V[2]	575 V[1] (550–600 V)	690 V[2]
kW[3]	HP[1]	\multicolumn{9}{c}{Corresponding rated currents I_n in A}								
560	–	–	–	–	–	950	–	760	–	551
600	800	–	–	–	1888	1017	944	813	755	589
630	–	–	–	–	–	1064	–	851	–	617
670	900	–	–	–	2124	1140	1062	912	850	661
710	–	–	–	–	–	1216	–	973	–	705
750	1000	–	–	–	2360	1283	1180	1026	944	743
800	–	–	–	–	–	1378	–	1102	–	799
850	–	–	–	–	–	1463	–	1170	–	841
900	1200	–	–	–	2832	1549	1416	1239	1132	898
950	–	–	–	–	–	1634	–	1307	–	947
1000	–	–	–	–	–	1720	–	1376	–	997

[1] in accordance with UL 508 (19.09.77), Table 24.4 for 60 Hz
[2] in accordance with NF C63-110, annexe IV for 50 Hz (only up to 250 kW)
[3] in accordance with DIN 42973

Table 9.28 Relationship between the number of poles and the motor speed in 50 Hz and 60 Hz networks

Number of poles	Synchronous speed n[1]		Speed at rated load	
	50 Hz r.p.m.	60 Hz r.p.m.	50 Hz r.p.m.	60 Hz r.p.m.
2	3000	3600	2800–2950	3360–3540
4	1500	1800	1400–1470	1680–1765
6	1000	1200	900– 985	1080–1180
8	750	900	690– 735	830– 880
10	600	720	550– 585	660– 700

[1] approximate speed at no-load operation

9.6 Three-phase power transformers

Table 9.29 shows important characteristics of three-phase power transformers. The maximum permissible sustained short-circuit currents and short-circuit power of the networks are given in Table 9.30. Table 9.31 lists the rated currents and short-circuit currents of three-phase power transformers of 50 to 250 kVA.

Quick procedures to estimate the prospective short-circuit currents in radial low voltage networks

Rules of thumb:
Rated transformer current (9.15)

I_n [in A] $= K \cdot S_{nT}$ [in kVA]   at 400 V: $K = 1.45$,
at 525 V: $K = 1.1$,
at 690 V: $K = 0.85$

Initial symmetrical short-circuit current of transformer

$$I_k'' \approx I_k = \frac{I_n}{u_{kr}} \cdot 100 \qquad (9.16)$$

The two characteristic curves in Figure 9.4 show the range of the sustained short-circuit currents of typical power transformers as a function of the rated power and the rated value of the impedance voltage. It can be seen that the transformer short-circuit currents are usually not as high as is generally assumed. As typical examples, the short-circuit current of a 630 kVA transformer with $u_{kr} = 6\%$ and $u_{kr} = 4\%$ as well as that of a 2000 kVA transformer with $u_{kr} = 6\%$, are illustrated.

Table 9.29 Important characteristic data of three-phase power transformers

Version		Rated power S_{nT} kVA	Maximum equipment voltage U_m kV	Rated value of the impedance voltage u_{kr} %	Cooling (s. pg. 534)	DIN
Oil-filled transformers		50– 630	24	4	ONAN	42500 [1]
		1000– 2500	24	6	ONAN	42500 [1]
		2000–10000	123	6–10	ONAN	42504
		12500–80000 [3]	123	10–14	ONAN [2]	42508
with tap changer switch		up to 40 MVA [3]	123	6,6–12	s. DIN 42504 alt. 42508	42515
Dry-type transformers with insulation class	E, B, F	100– 630	1.1	4	AN	42524
	E, B, F	100– 630	12	4		
	E, B, F	250–1600	12	6		
	H	100– 630	1.1	4		
	H	250–1600	1.1	6		
	H	400– 630	12	4		
	H	400–1600	12	6		
Resin-encapsulated transformers		100– 630	12	4	AN	42523 [4]
		1000–2500	12	6		
		250–2500	24	6		

[1] EC standard, corresponds to HD 428 of the European Committee for Electrotechnical Standardization CENELEC
[2] Above 31.5 MVA ONAF (also OFAF or OFWF)
[3] Higher outputs are not standardized
[4] A *Normal* type (N) and a *Reduced* type (R) are planned, whereby the *R* type will have no-load losses reduced by around 23% and audio sound power levels reduced by 8 dB compared to the N type.

Table 9.30
Sustained short-circuit currents and their maximum permissible duration in accordance with DIN VDE 0532 part 5, 05.84

Typical values of the rated percentage value of the short-circuit voltage for transformers with two separate windings		Max. permissible short-circuit duration in German and Austrian networks [2]	Short-circuit power of the network which may be used as a basis if no other data is specified	
Rated output power S_{nT} [1] kVA	Rated impedance voltage u_{kr} %	s	Highest voltage for electrical equipment U_m kV	Short-circuit power S of the network MVA
up to 630	4.0	2	7.2; 12; 17.5; 24	500
631 to 1250	5.0	3	36	1000
1251 to 3150	6.25	4	52; 72.5	3000
3151 to 6300	7.15	5	100; 123	6000

$$I_k = \frac{U_{nT} \cdot 10^3}{(Z_r + Z_s)\sqrt{3}} \quad \text{where} \quad Z_s = \frac{U_s^2}{S} \quad \text{and} \quad Z_T = \frac{10 \cdot u_{kr} \cdot U_{nT}^2}{S_n}; \quad \text{if } Z_s = 0 \text{ then } I_k/I_n = \frac{100}{u_{kr}} \text{ [3]}$$

I_k Sustained short-circuit current in A
I_n Rated current of winding concerned in A
U_{nT} Rated voltage of winding concerned in kV
U_s Rated voltage of network in kV
Z_s Short-circuit impedance of network in Ω/phase [3]
Z_T Short-circuit impedance of transformer in Ω/phase

[1] In the case of single-phase transformers used to make up a three-phase bank, the rated power refers to the three-phase bank
[2] IEC publication 76-5 (1976) specifies a uniform short-circuit duration of 2 s for $I_k/I_n \leq 25$
[3] Z_s is only taken into account if $S_{nT} \leq 3150$ kVA with $Z_s/Z_T > 0.05$ or if $S_{nT} > 3150$ kVA

9 Appendix

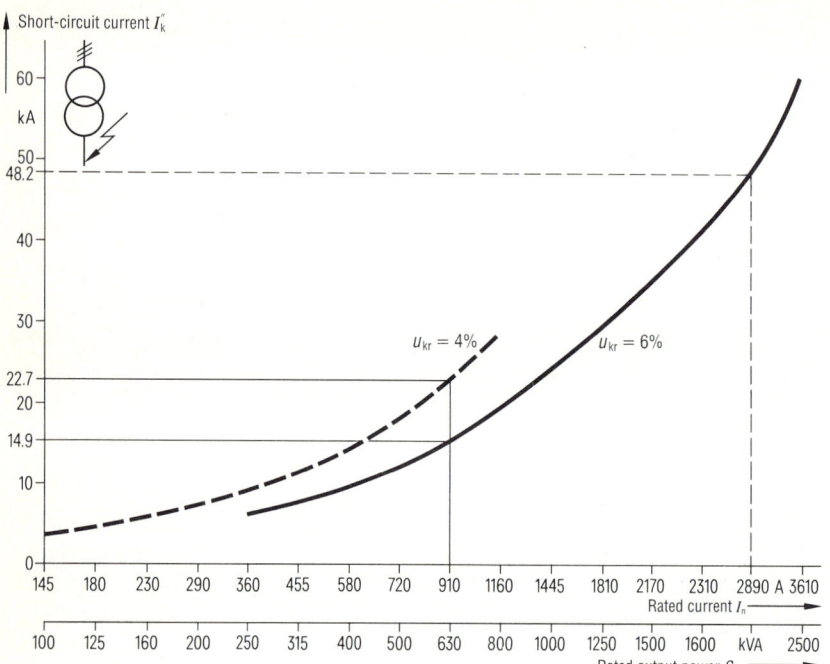

Figure 9.4
Prospective initial symmetrical a.c. short-circuit current I_k'' of transformers (400 V, 50 Hz), as a function of the rated transformer output power S_{nT} and the rated value of the impedance voltage u_{kr}

Code letters and their order of sequence to indicate the method of cooling in accordance with DIN VDE 0532 part 2, 01.89 [1]

Coolant

O	Mineral oil or synthetic cooling and insulating liquids with combustion points ≤ 300 °C
K	Synthetic cooling and insulating liquids with combustion points > 300 °C
L	Synthetic cooling and insulating liquids with non-measurable combustion points
G	Gas
W	Water
A	Air

Movement of the coolant

N	Natural circulation (self-cooling)
F	Forced movement (non-directed oil circulation)
D	Forced movement (directed oil circulation)

Sequence of the letter codes, e.g. ONAN:

For winding coolant		For external cooling	
1st letter	2nd letter	3rd letter	4th letter
Coolant	Circulation of coolant	Coolant	Circulation of coolant

[1] Corresponds to CENELEC-HD 398-2 including alteration 1

Permissible temperature rise in accordance with DIN VDE 0532 part 2, 01.89

Oil-filled transformers:

Thermal class A;
Temperature of winding 65 K with ON or OF cooling,
70 K with OD cooling;

Oil (at the top):
60 K for transformer with expansion tank or with hermetically sealed transformer,
55 K without expansion tank or without hermetic sealing of the transformer.

Dry-type transformers:

Thermal class		A	E	B	F	H	C
Winding	K	60	75	80	100	125	≥ 150

Coolant temperature:

– with air cooling
maximum temperature: 40 °C,
minimum temperature: −25 °C
(for short periods −30 °C),
average temperature: per day: max. 30 °C,
per year: max. 20 °C;

– with water cooling
inlet temperature max. 25 °C.

Table 9.31
Rated currents and initial symmetrical short-circuit currents of three-phase power transformers with 50 to 2500 kVA

Rated voltage U_r	400/230 V, 50 Hz			525 V, 50 Hz			690/400 V, 50 Hz		
Rated value of impedance voltage u_{kr}		4%[1]	6%[2]		4%[1]	6%[2]		4%[1]	6%[2]
Rated output power	Rated current I_n	Initial symmetrical short-circuit current I_k'' [3]		Rated current I_n	Initial symmetrical short-circuit current I_k'' [3]		Rated current I_n	Initial symmetrical short-circuit current I_k'' [3]	
kVA	A	A	A	A	A	A	A	A	A
50	72	1800	1200	55	1375	910	42	1040	690
100	144	3600	2400	110	2750	1830	84	2080	1390
160	230	5770	3850	176	4400	2930	133	3320	2230
200	288	7200	4810	220	5500	3660	167	4160	2780
250	360	9025	6015	275	6875	4580	209	5220	3480
315	455	11375	7580	346	8560	5770	262	6650	4360
400	578	14440	9630	440	11000	7330	335	8330	5580
500	722	18050	11030	550	13750	9160	418	10450	6960
630	910	22750	14860	693	17300	11550	525	13120	8760
800	1154	28850	19260	880	22000	14660	670	16750	11160
1000	1444	36100	24060	1100	27500	18330	836	20900	13930
1250	1805	45125	30080	1375	34375	22910	1046	26160	17430
1600	2310	57800	36530	1760	44000	29300	1330	33250	22170
2000	2887	–	48180	2200	55000	36660	1674	41850	27890
2500	3608	–	60150	2749	–	45800	2090	–	34840

[1] $u_{kr} = 4\%$, standard value to DIN 42503 for $S_{nT} = 50...630$ kVA
[2] $u_{kr} = 6\%$ standard value to DIN 42511 for $S_{nT} = 100...1600$ kVA
[3] I_k'' Prospective initial symmetrical short-circuit current of the transformer in the case of connection to a supply network with infinitely high short-circuit power

9.6.1 Graphic symbols and vector groups of three-phase power transformers

Since modern power supply is based on three-phase systems, power transformers are almost exclusively constructed as three-phase units. There is a number of possibilities for the electrical connection of the three higher-voltage and three lower-voltage windings respectively, the most important of which are summarized in DIN VDE 0532.

The representation of the windings with their complete terminal designations is in accordance with DIN 42402.

The numbers 1 and 2 preceding the letters U, V and W designate the higher-voltage (HV) and lower-voltage (LV) windings respectively. The numbers 1 and 2 following the letters designate the beginning and the end of a phase winding (these numbers are often omitted).

The vector group serves to indicate of connection of the HV and LV windings as well as the phase-angle relationship between the voltages (Table 9.32). The star connection, delta connection and zigzag connection provide three possibilities to interconnect the windings on each side of the transformer. These connections are designated by the letters Y, D and Z on the HV side, and by y, d and z on the LV side.

The letter N or n is added if the star point of a winding is brought out in a star connection or zigzag connection, for example Dyn5, Yzn, YNd5.

Table 9.32 illustrates the most commonly used vector groups of three-phase power transformers.

Table 9.32
Common vector groups for three-phase power transformers in accordance with DIN VDE 0532 part 4, 03.82

Numerical index	Vector group[1]	Vector diagram HV / LV	Connection diagram[2] HV / LV
0	Dd0		
	Yy0		
	Dz0		
5	Dy5		
	Yd5		
	Yz5		
6	Dd6		
	Yy6		
	Dz6		
11	Dy11		
	Yd11		
	Yz11		

[1] In case the neutral point is brought-out, N or n respectively is added at the end of the vector group designation.
[2] The same winding direction is assumed for all windings, i.e. seen three-dimensionally, the windings must be imagined as having been folded downwards in the connection diagrams. Brought-out neutral points are marked with 1N or 2N.

9.7 Tripping behaviour of line protection and switchgear protection devices

9.7.1 Time-current tripping characteristics of circuit-breakers, miniature circuit-breakers and overload relays

The time-current tripping characteristics (tripping curves) of

3VU, 3VF and 3VE circuit-breakers for motor protection,

3VN and 3VF circuit-breakers for starter combinations,

3VT and 3VF circuit-breakers for system or plant protection,

3VS circuit-breakers,

3WN and 3WS air and vacuum circuit-breakers (with solid-state overcurrent releases) for overload and short-circuit protection,

3WE air circuit-breakers (with electro-mechanical or solid-state overcurrent releases)

are given in the relevant Siemens catalogue.

Templates and double-logarithmic draft paper in a standardized DIN scale are available for circuit-breakers 3VE, 3VF, 3VT and 3VN (see also Section 8.13, "Project planning and engineering aids", page 496). For the tripping characteristics of 3UA, 3UB and 3UC overload relays, please refer to the appropriate Siemens catalogue.

Line protection (miniature) circuit-breakers with type B and C characteristics conform to DIN VDE 0641, A4, 11.88 and part 11, 08.92 and can also be looked up in various Siemens catalogues and brochures.

9.7.2 Pre-arcing time-current characteristics of fuses (operating classes gL/gG and aM)

In principle, a fuse link is a piece of wire with a reduced cross-section. Owing to its higher resistance, this „weak point" (invented by Edison) will heat up and melt earlier under the influence of excessive current than the rest of the wire of normal cross-sectional area.

The melting time (or "virtual pre-arcing time") is represented graphically in a so-called time-current characteristic as a function of the overcurrent flowing through the fuse element. The time-current characteristic lies between two asymptotes. The first (vertical), is the minimum melting current which only just causes the fuse element to melt through, and the second is a diagonal line of equal current heat values (I^2t line), based on higher short-circuit currents, and representing the constant melting heat I^2t of the element (also refer to Section 3.4.3.1, page 125).

The conventional non-tripping and tripping currents are of interest. In the case of gL or gG [1]) fuses and rated fuse currents of 63 to 160 A, they amount to 1.3 and 1.6·I_n respectively. If a current equal to the conventional non-tripping current is passed through the fuse, then the fuse may not rupture within the specified period of time, i.e. in this case, the fuse permits a continuous overload of 30%. During the testing of fuses, the load current is increased to the value of the conventional tripping current (here 1.6·I_n) after 2 hours. Now the fuse must react within 2 hours and interrupt the current flow.

In the case of large short-circuit currents, the fuse element will melt through and the resulting arc will be extinguished even before the peak value of the current has been reached. In this way, the fuse limits the short-circuit current to a value lower than the prospective short-circuit current which could otherwise occur at the particular point of installation.

The degree to which the short-circuit current is limited, depends to a great extent on two factors:

▷ high current density and short melting time t_s,

▷ a high arc voltage.

Owing to this current-limiting effect, industrial fuses are able to interrupt very high prospective short-circuit currents viz. up to 120 kA at 500 V/690 V AC. Low-voltage fuses are classified according to their use (function) and operating characteristics and according to their construction in terms of DIN VDE 0636 and IEC publication 269 (Table 9.33).

In addition, DIN VDE 0680, 01.83 which deals with protection devices and equipment for working on live

[1]) To replace the gL fuses in future

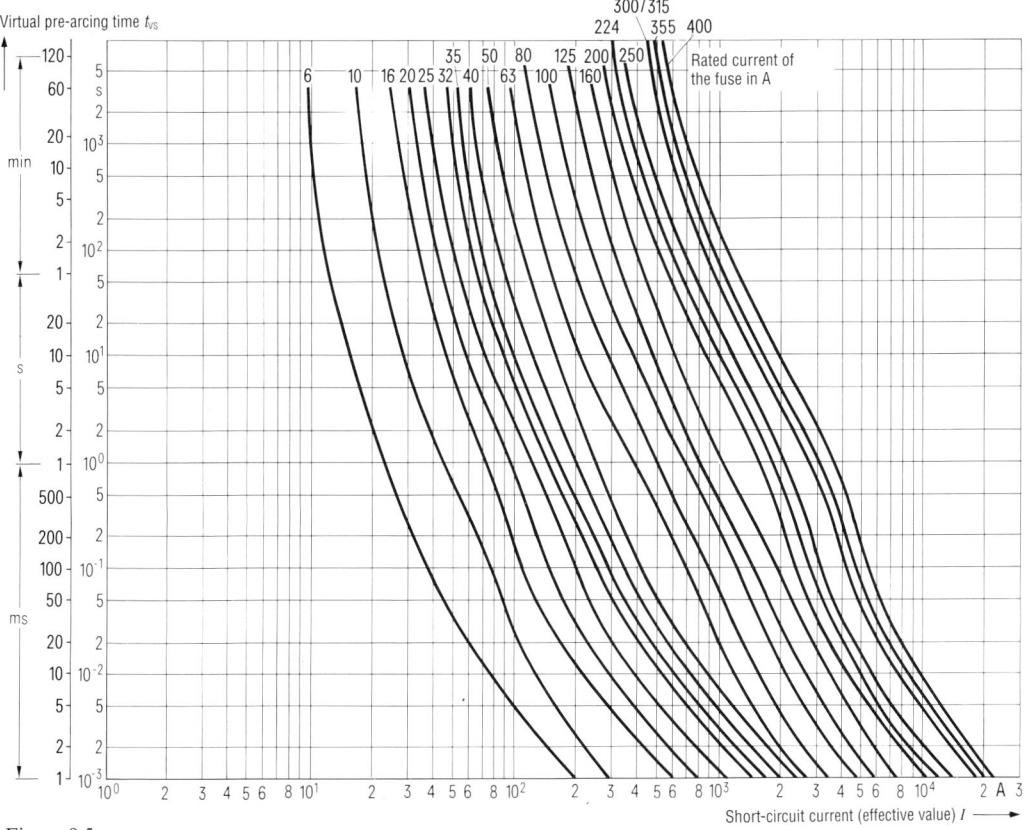

Figure 9.5
Melting (pre-arcing) time-current characteristics of LV HRC fuses, operation class gL, type 3NA/3NA5

Table 9.33
Function and operation classes of low-voltage high rupturing capacity (LV HRC) fuses

Function class			Operation class	
Designation	Continuous current up to	Rupturing current	Designation	Protection of
General purpose fuses (full range)				
g	I_n	$\geq I_{min}$	gL/gG gR gB	Cables and conductors Semiconductors Mining equipment
Accompanied fuses (partial range)				
a	I_n	$\geq 4 \cdot I_n$ $\geq 2.7 \cdot I_n$	aM aR	Switching devices Semiconductors

Of particular interest here are the fuses of operation classes gL or gG for cable and wire protection, as well as fuses of operating class aM for switchgear protection.

Figure 9.5 illustrates the time-current characteristic of gL-type LV HRC fuses for the full range protection of e.g. load cables. Figure 9.6 illustrates the time-current characteristic of switchgear protection fuses of operating class aM for accompanied switchgear protection (partial range protection). The fuse characteristics for 50 to 160 A can be extrapolated downwards for even higher short-circuit currents (up to approx. 10^{-4} s).

parts up to 1000 V, protective insulation and protection against electric shock, should also be taken into account.

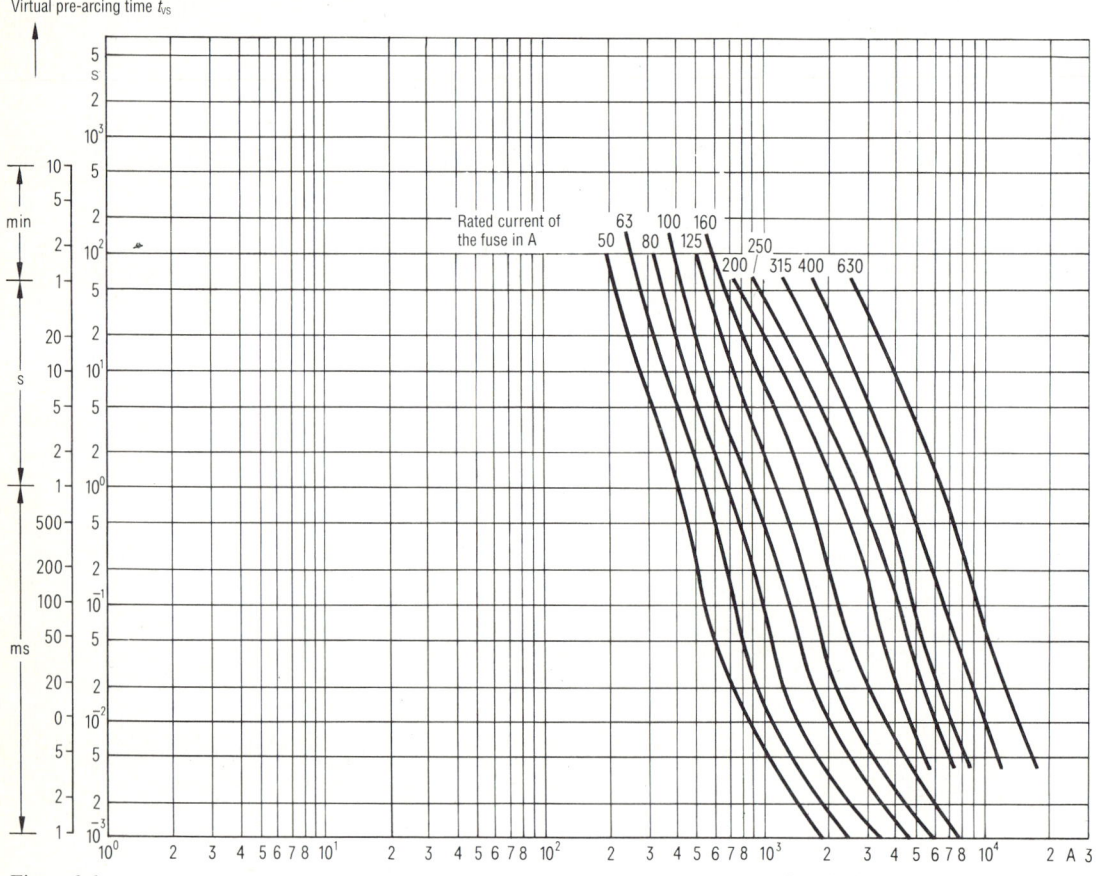

Figure 9.6
Melting (pre-arcing) time-current characteristics of switchgear protection fuses, operating class aM, types 3ND8 and 3ND1

Further definitions on operation and function classes are given in Section 9.12, pages 608 and 589 resp.

On comparing the two characteristics one can see that fuses of operating class aM do not offer overload protection. Therefore, an additional overload protection must be provided for the range of up to approximately 4 times the rated current.

9.7.3 Characteristic curves and tripping behaviour of circuit-breakers

Miniature circuit-breakers generally serve to protect cables and wires in the event of an overload or short-circuit. To ensure that the wire insulation does not suffer damage even in the case of fault, it is imperative that its temperature does not exceed a certain specified value. In the case of PVC insulation, this value is 70 °C continuously, or 160 °C for a maximum duration of 5 s (e.g in the event of a short-circuit). Also refer to Section 1.4.2.1 on page 51.

Normally, miniature circuit-breakers are equipped with two independent trip mechanisms for overcurrent protection. In the event of an overload, the bimetallic mechanism trips with a time-delayed function relative to the load current passing through the m.c.b. If, however, the load current exceeds a defined threshold value in the case of a short-circuit, the electromagnetic mechanism trips the circuit-breaker instantaneously. The time-current tripping characteristic and the tripping tolerance band of miniature circuit-breakers complying with DIN VDE 0641, are defined by the reference values I_1 to I_5 as illustrated in Figure 9.7. These values are referred to the characteristic values I_b and I_z of the line to be protected.

Until 1978 both the H and L characteristics were specified in the above-mentioned equipment specification. Since then, only the L characteristic is recognized as standard.

In addition, CEE publication 19, 1st edition, defines a G characteristic which is derived from the characteristics of moulded case circuit-breakers and circuit-breakers for motor protection as defined in IEC 947-2 and DIN VDE 0660 part 101 (Figure 9.8).

New characteristics B, C and D were defined internationally with the appearance of the new IEC 898/1987 publication for miniature circuit-breakers. By means of the draft 1/09.87 and alteration A4/11.88, DIN VDE 0641 was the first national specification to accept this new development and has adopted the B and C characteristics as standard. They have been included in DIN VDE 0641 part 11 since 08.92.

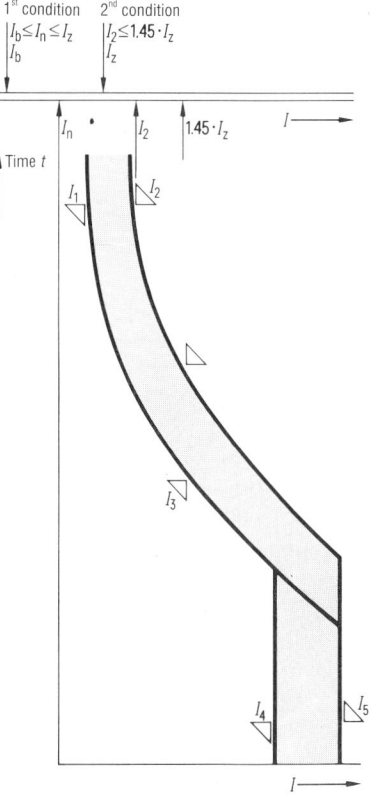

I_b Expected operating current, i.e. current during normal operation, as determined by the consumer load

I_z Permissible continuous load current for a conductor at which the continuous temperature rise limit of its insulation is not exceeded

$1.45 \cdot I_z$ Maximum permissible temporary overload current at which the brief exceeding of the continuous temperature rise limit of the insulation does not reduce its insulation properties to such an extent that it becomes unsafe.

I_n Rated current, i.e. the current rating of the circuit-breaker to which other rated quantities refer (setting value)

I_1 Conventional non-tripping test current, i.e. the current which is just not high enough to produce a tripping operation under specified conditions

I_2 Conventional tripping test current, i.e. the current which causes a tripping operation within one hour ($I_n \leq 63$ A) under specified conditions

I_3 Tolerance band limits

I_4 Hold-in current of the instantaneous electromagnetic overcurrent release (short-circuit release)

I_5 Tripping current of the instantaneous electromagnetic overcurrent release (short-circuit release)

Figure 9.7
Characteristic reference values for the conductor and the line protection device

9 Appendix

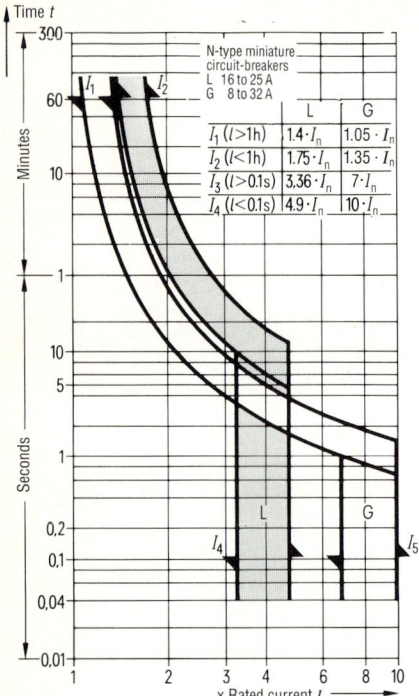

Figure 9.8
Previously valid: L and G characteristic curves in accordance with DIN VDE 0641, 6.78 or CEE 19/1

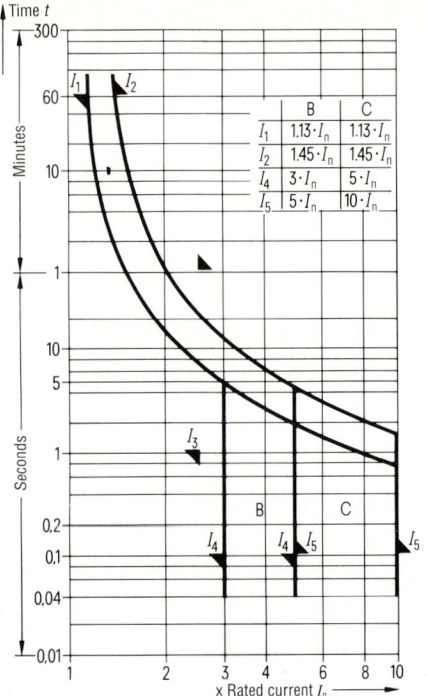

Figure 9.9
New: B and C characteristic curves in accordance with DIN VDE 0641 A 4, 11.88 and part 11, 08.92. (For direct current I_4/I_5 increased by a factor of 1.4 with respect to a.c. values)

The L and G characteristics (Fig. 9.8) differed in both their thermal and electromagnetic tripping behaviour. In terms of the new B and C curves, the tripping characteristics in the thermal overload range are identical. The new characteristics differ only in terms of the current/time response of their instantaneous electromagnetic releases (Fig. 9.9). In the overload range, the tolerance band of the L curve has been shifted towards the G characteristic while the tripping limits of electromagnetic instantaneous short-circuit releases have remained the same as before.

These new specifications for the tripping behaviour facilitate the correlation between conductor cross-

Table 9.34
Assignment of miniature circuit-breakers to conductor cross-sections.
Example: Flat webbed cable, multi-core cable installed on or in the wall, installation type C[1)] at an ambient temperature of **30 °C**

Rated cross-section q_n mm²	Protection by miniature circuit-breaker for				I_z (conductor, e.g. wire)	
	2 conductors under load		3 conductors under load		2 conductors under load	3 conductors under load
1.5	L 16	B/C 16	L 10	B/C 16	19.5 A	17.5 A
2.5	L 20	B/C 25	L 20	B/C 20	26 A	24 A
4	L 32	B/C 32	L 25	B/C 32	35 A	32 A
6	L 40	B/C 40	L 32	B/C 40	46 A	41 A
10	L 50	B/C 63	L 50	B/C 50	63 A	57 A
16	–	B/C 80	–	B/C 63	85 A	76 A

[1)] Installation type C in accordance with DIN VDE 0298, part 4, 2.88 and DIN VDE 0100 part 430, Supplement 1. The cables are installed in such a way that the distance between them and the wall surface is less than 0.3 times their outer diameter (refer to Section 9.4.2, page 521 et seq.)

sections and miniature circuit-breakers. The following two conditions are listed in DIN VDE 0100 part 430 (see equations (9.6a) and (9.6b)):

1st condition

$I_b \leq I_n \leq I_z$ (rule for rated current),

2nd condition

$I_2 \leq 1.45 \cdot I_z$ (rule for tripping current)

Since the second condition is automatically fulfilled by the new characteristics B and C ($I_z = I_n$ given), miniature circuit-breakers may nowadays be selected by means of the simplified relation $I_n \leq I_z$ only.

Consequently, the new assignment Table 9.34 can be derived for an ambient temperature of 30 °C, complying with DIN VDE 0100 part 430 supplement 1 (also refer to Table 9.20, page 522).

Table 9.34 serves the assignment of miniature circuit-breakers with B or C characteristics to conductor cross-sections at an ambient temperature of 30 °C.

Since the new characteristics do not differ from the old ones with regard to instantaneous magnetic tripping, the well-proven safety-relevant capabilities of miniature circuit-breakers (e.g. switching capacity, discrimination and back-up protection) still apply.

The Siemens miniature circuit-breakers of types 5SN... and 5SX... are available with the new characteristics B and C and display the relevant VDE mark of approval.

9.7.4 Current-limiting diagrams of fuses

The current-limiting effect is represented in the current-limiting diagram by the so-called let-through current characteristic curves. Current limiting starts at the value of short-circuit current I_k at which the let-through current line i_D intersects with the line representing the prospective (unlimited) peak short-circuit current i_p. The smaller the rated current of the fuse and the greater the short-circuit current I_k, the more pronounced the current-limiting effect will be (also refer to Section 3.4.3.1, page 125).

Figures 9.10 and 9.11 illustrate the let-through current characteristics of LV HRC fuses for 500 V and 660 V networks.

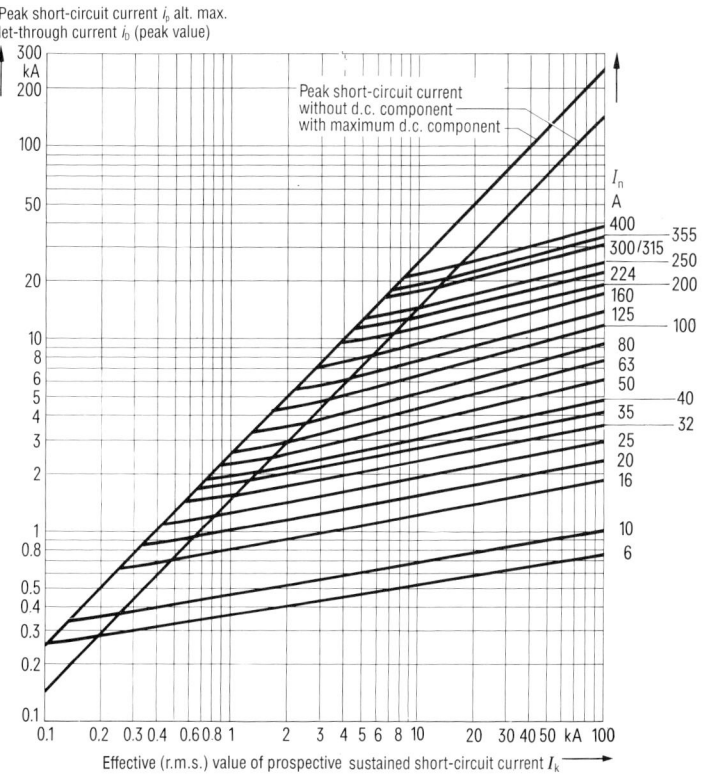

I_n Rated current of the fuse

Figure 9.10
Let-through current characteristics of LV HRC fuses, operation class gL, type 3NA3/3NA5 for AC 500 V, cos $\varphi = 0.1$ to 0.7

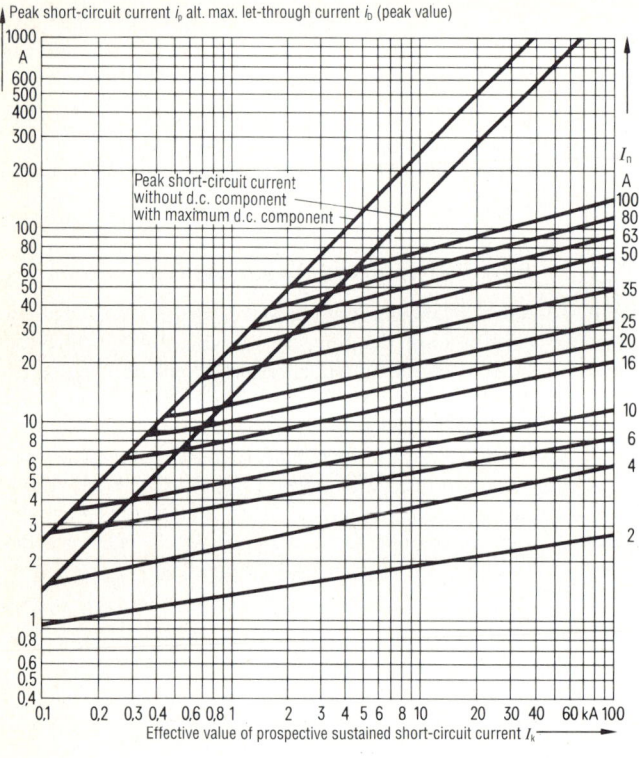

Figure 9.11
Let-through current characteristics of LV HRC fuses of operation class gL, type 3NA3/3NA5 for AC 660 V, cos $\varphi = 0.1$ to 0.7

I_n Rated current of the fuse

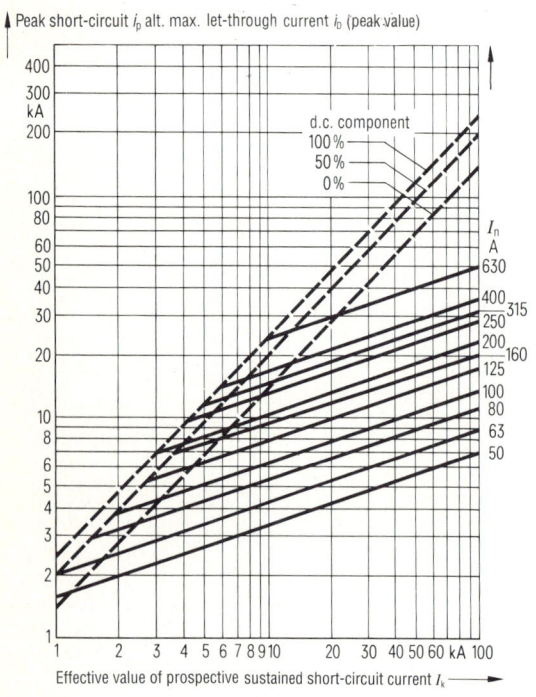

Figure 9.12
Let-through current characteristics of switchgear protection fuses of operation class aM, type 3ND8 (rated currents 50 to 160 A) for AC 500 V, and type 3ND1 (rated currents 200 to 630 A) for AC 690 V, cos $\varphi \leq 0.7$

I_n Rated current of the fuse

Figure 9.12 illustrates the let-through current characteristics of types 3ND8 and 3ND1 switchgear protection fuses with aM operating class.

Current-limiting diagrams (let-through current characteristics) for the 3VE, 3VF and 3VU circuit-breakers may be found in the relevant Siemens low-voltage switchgear catalogue.

9.7.5 Discrimination (selectivity) between fuses and circuit-breakers

According to DIN VDE 0636, the successful discrimination (selective rupturing) between fuses of operation class gL requires that their rated currents differ by the ratio of 1:1.6 at approximately 320 V.

For Siemens LV HRC, as well as for DIAZED and NEOZED fuses, the following ratios between the rated currents of consecutive fuses ensure discriminative rupturing:

at a rated operational a.c. voltage of:

≤400 V   1:1.25,
≤500/660 V   1:1.6.

The LV HRC fuse characteristics of operation class gL, type 3NA3/5 in accordance with DIN VDE 0636 will be harmonized with IEC specification 269 and will be included in the characteristics of operating class gG (gL/gG type).

Table 9.35 shows the discrimination between LV HRC fuses of different operation on classes.

For discrimination between LV HRC cable protection fuses 3NA and HV HRC fuses, please refer to the Siemens HG 12 catalogue.

For discrimination between circuit-breakers and between circuit-breakers and fuses, refer to Section 3.4.6.1 on page 164 et seq.

Tables 9.36 to 9.39 show the coordination between fuses and Siemens miniature circuit-breakers and the discrimination (selectivity) limits for I_{cn} = 3 kA, 6 kA or 10 kA. Since the new characteristics do not differ from the old ones with regard to instantaneous electromagnetic tripping, the safety-relevant capabilities of circuit-breakers (switching capacity, discrimination and back-up protection) remain unchanged.

To enable one to predict the degree of conductor protection which will be achieved in the event of a short-circuit as well as the discriminative behaviour with respect to the upstream fuse, m.c.b.'s with rated short-circuit breaking capacities I_{cn} = 3 kA, 6 kA or 10 kA are divided into three current limiting classes in terms of their discrimination capabilities. Here, different values apply for the tripping characteristics B and C.

a) Conductor (e.g. cable) protection

   In the event of a short-circuit, conductor protection is ensured up to a current where the let-through $I^2 \cdot t$ heat value of the circuit-breaker (Joule integral) is smaller than the max. permissible $I^2 \cdot t$ value for the conductor (e.g. which results in the max. temperature rise of the cable insulation).

b) Discrimination (selectivity)

   Discrimination between a circuit-breaker and an upstream fuse is ensured up to a current where the $I^2 \cdot t$ let-through value of the circuit-breaker is smaller than the virtual pre-arcing $I^2 \cdot t$ value of the fuse (see also Section 3.4.6.1, page 164).

Table 9.35
Discrimination between switchgear protection fuses of type 3ND1 and cable protection fuses of type 3NA

Downstream: Switchgear protection fuses type 3ND1, operation class aM	Upstream: Cable protection fuses type 3NA3 or 3NA5, operation class gL, at rated operational voltage:		
	400 V, AC	500 V, AC	660 V, AC
I_{n2} A	I_{n1} A	I_{n1} A	I_{n1} A
200	200	250	250
250	250	300/315	300/315
315	300/315	300/315	400
400	400	425	500
630	630	800	800

I_{n1} Rated current of upstream fuse link
I_{n2} Rated current of downstream fuse link

9 Appendix

Table 9.36
Discriminative coordination between fuses and miniature circuit-breakers with characteristics C 1.6 to C 10, at AC 250 V, cos $\varphi = 0.8$ to 1, (type 5SN with $I_{cn} = 6$ kA)

Upstream fuse, operation class gL, type			Discrimination (selectivity) limit values up to the following short-circuit currents:						
DIAZED A	LV HRC A	NEOZED A	C 1.6 A	C 2 A	C 3 A	C 4 A	C 6 A	C 8 A	C 10 A
10	–	–	1500	230	200	200	200	200	–
–	10	–	130	130	130	130	130	130	–
–	–	10	180	150	150	150	150	150	–
16	–	–	>6000	2000	350	350	350	320	320
–	16	–	>6000	800	350	320	320	270	270
–	–	16	>6000	2000	400	400	400	380	380
20	–	–	>6000	>6000	1400	700	600	580	580
–	20	–	>6000	>6000	450	400	400	400	390
–	–	20	>6000	>6000	850	600	550	520	510
25	–	–	>6000	>6000	5000	1900	1000	920	880
–	25	–	>6000	>6000	2200	1000	700	650	620
–	–	25	>6000	>6000	2600	1100	800	720	700

Table 9.37
Discriminative coordination between fuses and miniature circuit-breakers with characteristics B10 to B25, or C8 to C32, at AC 250 V, cos $\varphi = 0.8$ to 1, (types 5SN and 5SX with $I_{cn} = 6$ kA)

Upstream fuse, operation class gL, type			Discrimination (selectivity) limit values up to the following short-circuit currents:			
DIAZED A	LV HRC A	NEOZED A	B 10, C 8 A	B 16/20, C 10 A	B 25, C 16/20 A	C 25/32 A
16	–	–	320	–	–	–
–	16	–	270	–	–	–
–	–	16	380	–	–	–
20	–	–	580	550	–	–
–	20	–	400	390	–	–
–	–	20	520	510	–	–
25	–	–	920	880	–	–
–	25	–	650	620	–	–
–	–	25	720	700	–	–
35	–	–	1500	1400	1350	–
–	35	–	1300	1200	1150	–
–	–	35	1350	1300	1250	–
50	–	–	2650	2500	2300	2000
–	50	–	2300	2150	2050	1800
–	–	50	2250	2150	2050	1850
63	–	–	4300	4000	3600	3000
–	63	–	2900	2650	2500	2200
–	–	63	3100	2950	2800	2500
80	–	–	5350	5100	4700	4000
–	80	–	4300	4000	3700	3100
–	–	80	3900	3700	3500	3200
100	–	–	>6000	>6000	>6000	6000
–	100	–	>6000	>6000	5900	5200
–	–	100	>6000	>6000	6000	5300

Discrimination between fuses and miniature circuit-breakers

Table 9.38
Discriminative coordination between fuses and miniature circuit-breakers, at AC 250 V, cos $\varphi \approx 0.8$ to 1, (type 5SQ with $I_{cn} = 3$ kA)

Upstream fuse, operation class gL, type			Discrimination (selectivity) limit values up to the following short-circuit currents:				
DIAZED A	LV HRC A	NEOZED A	C 6 A	C 8 A	C 10 A	C 16 to 32 A	C 40 to 63 A
16 – –	– 16 –	– – 16	250 220 280	250 220 280	250 220 280	– – –	– – –
20 – –	– 20 –	– – 20	370 290 380	370 290 380	370 290 380	– – –	– – –
25 – –	– 25 –	– – 25	500 400 450	500 400 450	480 400 450	480 390 450	– – –
35 – –	– 35 –	– – 35	800 700 880	700 640 800	580 620 770	540 600 720	– – –
50 – –	– 50 –	– – 50	1450 1300 1500	1150 1050 1250	1100 1000 1200	950 900 1100	920 870 1050
63 – –	– 63 –	– – 63	2500 1800 2000	1750 1500 1650	1600 1400 1550	1350 1250 1350	1250 1200 1300
80 – –	– 80 –	– – 80	> 3000 2700 2800	2200 2000 2200	2000 1900 2000	1650 1650 1750	1500 1500 1600
100 – –	– 100 –	– – 100	> 3000 > 3000 > 3000	3000 > 3000 3000	2750 2900 2800	2250 2400 2300	2000 2150 2100

Back-up protection[1] for miniature circuit-breakers of type 5SQ is ensured up to a maximum of
— 50 kA with a 100 A fuse upstream,
— 6 kA with circuit-breaker upstream.

[1] also refer to Section 3.4.6.2, page 172

Table 9.39
Discriminative coordination between LV HRC fuses and AC/DC miniature circuit-breakers with B or C characteristics at $U_C = 220$ V, $\tau \leq 4$ ms

Upstream fuse, operation class gL, type LV HRC A	Discrimination (selectivity) limit values up to the following short-circuit currents:					
	B 6/C 8 A	B 10/C 10 A	B 16/20 A	B 25/C 16 A	C 20/25 A	C 32 A
16	460	420	–	–	–	–
20	840	740	–	–	–	–
25	2000	1600	1500	–	–	–
32	> 4500	3700	3300	2900	2600	–
35		> 4500	> 4500	> 4300	3800	3300
40				> 4500	> 4500	> 4500

9.8 Short-circuit currents

9.8.1 Limiting effect of conductors and cables on short-circuit currents

Excessive overloading often destroys insulation very quickly. This can lead to fault arcing and ultimately to a short-circuit.

A short-circuit is without doubt the severest and most unpleasant fault which can occur within an electrical installation. Short-circuits interrupt the power supply and thereby cause production losses in the case of manufacturing plants. This, in turn, often leads to considerable financial losses.

In most cases, short-circuits manifest themselves in the form of electrical arcs. These not only destroy equipment parts but also pose a danger to operating staff. In addition, strong short-circuit currents exert considerable mechanical forces on the system components in the circuit concerned and cause these to heat up rapidly to high temperatures. Whereas the maximum dynamic forces are directly related to the square of the peak value of the initial short-circuit current i_p, the thermal stressing is a function of the square of the effective value of the sustained short-circuit current I_k and its duration t, i.e. on the thermal value $I_k^2 \cdot t$.

The actual value of the short-circuit current is a function of all resistances along the current path from the generator or transformer to the point of the short-circuit. In a low-voltage network, the short-circuit current is determined by the short-circuit power of the supply, reduced by the power loss in wires and cables and other conductors in the current path. The conductor resistance between the transformer and the point of short-circuit can significantly reduce the value of the prospective short-circuit current from that which would occur directly across the terminals of the transformer.

The smaller the output power and the secondary voltage, and the higher the rated value of the transformer impedance voltage u_{kr} is, the more effective this limiting effect will be. It is likewise more effective, the smaller the conductor cross-section and the longer the current path between the transformer and the point of short-circuit.

These short-circuit currents only occur in the case of a full three-pole short-circuit through virtually zero

Table 9.40
Limiting of the short-circuit current by a length of copper conductor. The values are based on a three-pole short-circuit in a TN-C system, $I_k = 10$ kA at the infeed point; the rated voltage $U_r = 240$ or 420 V, 50 Hz

Conductor length	0.3 m		1.0 m		3.0 m		10 m		30 m		50 m	
Rated cross-section q_n mm²	Rated voltage U_r											
	240 V	420 V	240 V	420 V	240 V	420 V	240 V	420 V	240 V	420 V	240 V	420 V
	kA	kA	kA	kA	kA	kA	kA	kA	kA	kA	kA	kA
240	9.98	9.99	9.92	9.96	9.77	9.87	9.26	9.58	8.06	8.83	7.14	8.19
120	9.97	9.98	9.89	9.94	9.69	9.82	9.02	9.42	7.48	8.41	6.36	7.58
95	9.96	9.98	9.88	9.93	9.64	9.79	8.88	9.33	7.17	8.19	5.96	7.27
70	9.96	9.97	9.85	9.92	9.56	9.75	8.64	9.19	6.64	7.82	5.32	6.75
50	9.94	9.97	9.81	9.89	9.44	9.68	8.27	8.96	5.89	7.27	4.50	6.02
35	9.93	9.96	9.76	9.86	9.29	9.59	7.83	8.69	5.14	6.65	3.75	5.28
25	9.91	9.95	9.69	9.82	9.08	9.47	7.24	8.30	4.31	5.88	3.01	4.45
16	9.86	9.92	9.53	9.73	8.63	9.20	6.18	7.54	3.17	4.66	2.10	3.29
10	9.79	9.88	9.29	9.59	7.95	8.79	4.92	6.49	2.20	3.44	1.40	2.29
6	9.65	9.80	8.84	9.33	6.86	8.07	3.51	5.07	1.40	2.29	0.87	1.46
4	9.48	9.70	8.30	9.01	5.77	7.24	2.56	3.92	0.96	1.61	0.59	1.00
2.5	9.18	9.53	7.42	8.45	4.42	6.04	1.71	2.76	0.61	1.04	0.37	0.64
1.5	8.65	9.22	6.13	7.54	3.07	4.57	1.07	1.79	0.37	0.64	0.23	0.39

Table 9.41
Limiting of the short-circuit current by a length of copper conductor. The values are based on a three-pole short-circuit in a TN-C system, $I_k = 50$ kA at the infeed point; the rated voltage $U_r = 240$ V or 420 V, 50 Hz

Conductor length	0.3 m		1.0 m		3.0 m		10 m		30 m		50 m	
Rated cross-section q_n mm²	Rated voltage U_r											
	240 V kA	420 V kA	240 V kA	420 V kA	240 V kA	420 V kA	240 V kA	420 V kA	240 V kA	420 V kA	240 V kA	420 V kA
240	49.49	49.71	48.32	49.03	45.24	47.18	36.77	41.57	23.57	30.71	17.23	24.20
120	49.37	49.64	47.93	48.81	44.04	46.50	33.36	39.37	18.68	26.24	12.78	19.31
95	49.31	49.60	47.70	48.68	43.32	46.10	31.49	38.08	16.56	24.02	11.05	17.16
70	49.20	49.54	47.32	48.47	42.07	45.40	28.37	35.83	13.57	20.61	8.78	14.11
50	49.03	49.45	46.67	48.12	39.96	44.22	24.09	32.33	10.40	16.53	6.59	10.86
35	48.81	49.33	45.84	47.69	37.36	42.71	20.05	28.49	8.03	13.14	4.98	8.39
25	48.49	49.15	44.54	47.00	33.75	40.22	15.95	23.97	6.02	10.08	3.69	6.30
16	47.76	48.77	41.68	45.47	27.35	35.70	10.98	17.56	3.92	6.7	2.38	4.11
10	46.50	48.12	37.06	42.81	20.37	29.21	7.24	12.05	2.51	4.33	1.51	2.63
6	43.98	46.81	29.81	37.87	13.55	21.18	4.44	7.57	1.51	2.62	0.91	1.58
4	40.65	45.02	23.30	32.29	9.44	15.43	2.98	5.14	1.01	1.75	0.61	1.06
2.5	35.0	41.63	16.37	24.82	6.08	10.26	1.88	3.26	0.63	1.10	0.38	0.66
1.5	27.06	35.76	10.54	17.07	3.72	6.38	1.13	1.97	0.38	0.66	0.23	0.40

resistance (e.g. a bolted bridge across all three phases). In general, short-circuits flow through an arc in air which typically causes a reverse arc-voltage of about 70 V in the current path. This naturally also has a pronounced limiting effect on the short-circuit current. In addition, there are a number of ohmic and inductive resistances in the short-circuit current path which are very difficult to determine and which are generally not taken into consideration. These include contact resistances of terminals and contacts as well as the inductive influence (eddy currents) of e.g. structural steel elements located in the neighbourhood of the cables.

From practical experience, one knows the order of magnitude of these impedances and contact resistances. Approximately 10% allowance has been made for them in Tables 9.40 and 9.41. The conductors must be short-circuit proof up to the value of the prospective short-circuit current.

Tables 9.40 and 9.41 show the prospective short-circuit currents to be expected at the line end of copper conductors of different lengths and of different cross-sections at operational temperature. The values are based on a three-pole short-circuit with $I_k = 10$ kA or $I_k = 50$ kA at the infeed point in 230 or 400 V networks (TN-C systems).

Further information, also for other line voltages, may be found in the Siemens publication "Overload and short-circuit protection in low voltage installations", Order number E 20001-P285-326 (presently only in German, for English edition please enquire).

9.8.2 Dynamic forces created by short-circuit currents

The force which fixed main parallel conductors (e.g. busbars) exert on one another while a short-circuit current flows through them can be extremely high.

Figure 9.13 illustrates the force F per unit length as a function of the peak value of the initial asymmetrical short-circuit current i_p and the distance between the conductors a, in the case of a two-pole short-circuit.

Example

Peak of the initial asymmetric short-circuit current $i_p = 80$ kA;
Distance between the parallel conductors $a = 30$ cm;
Conductor length between two supporting insulators $l = 50$ cm.

A value of 43 N/cm is read from the diagram in Figure 9.13. This must be multiplied by the conductor length:

$F = 43$ N/cm $\cdot 50$ cm $= 2150$ N.

9 Appendix

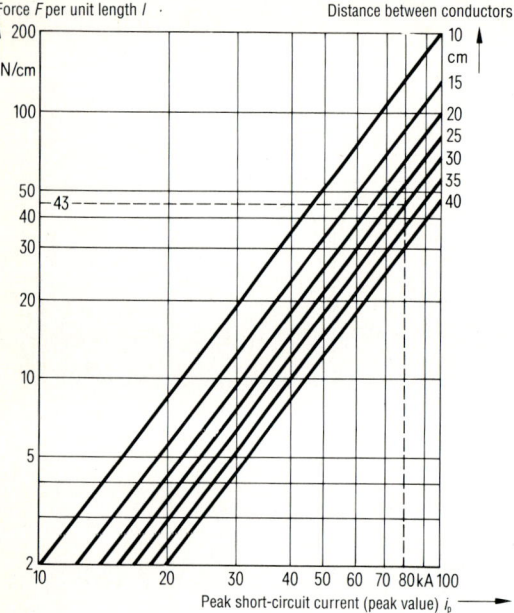

Figure 9.13
Force per unit length created by the initial asymmetric short-circuit current peak for various distances between the parallel conductors

The force applicable in the case of a *two-pole* short-circuit can be calculated using equation (2.9) (refer to Section 2.1.4, page 84):

$$F_{p2} = 0.2 \cdot i_{p2}^2 \cdot \frac{l}{a} = 0.2 \cdot (80)^2 \cdot \frac{50}{30} = 2133 \text{ N}.$$

For a *three-pole* short-circuit:

$$F_{p3} = 0.17 \cdot i_{p3}^2 \cdot \frac{l}{a} = 0.17 \cdot (80)^2 \cdot \frac{50}{30} = 1813 \text{ N}.$$

9.9 Number of switching operations of switching devices subjected to different operating periods per day

Table 9.42 provides a simple way to estimate the total number of switching operations a switching device has performed while in service over a period of several years. Various make-break operations/h can be taken as the basis for single-shift, two-shift or three-shift operation in a year with 250 working days (also refer to Section 5.5 on page 280).

Example

In the case of 100 make-break operations per hour and a daily operating period of 16 hours, the switching device will have performed 2 million switching operations after 5 years.

Table 9.42
Number of switching operations relative to the operating period and make-break operations per hour.
A make-break operation is a complete cycle consisting of a making and a breaking operation

Operating period		Make-break operations/h						
Years	Hours per day h	1	5	10	50	100	500	1000
		Total number of switching operations in millions						
3	8	0.006	0.030	0.060	0.300	0.6	3	6
	16	0.012	0.060	0.120	0.600	1.2	6	12
	24	0.018	0.090	0.180	0.900	1.8	9	18
5	8	0.010	0.050	0.100	0.500	1	5	10
	16	0.020	0.100	0.200	1.000	2	10	20
	24	0.030	0.150	0.300	1.500	3	15	30
10	8	0.020	0.100	0.200	1.000	2	10	20
	16	0.040	0.200	0.400	2.000	4	20	40
	24	0.060	0.300	0.600	3.000	6	30	60

9.10 International network voltages and frequencies

In the following, voltages of public power networks other than those of the Federal Republic of Germany (old federal states) are listed. The data is subject to change and the authors do not claim to have included all countries and possibilities. For network data in the Federal Republic of Germany, refer to Section 2, page 76.

If two network voltages lower than 1000 V are given, then the lower value applies primarily to lighting systems and small devices (e.g. radios) while the higher value applies to devices with higher outputs (e.g. motors) and industrial applications.

If two values are given and the higher voltage value is exactly double the lower one (e.g. 100/200 V), this indicates a single-phase three-conductor system with two lines and one neutral conductor.

If two values are given and the higher voltage value (line voltage = phase-to-phase voltage) is 1.73 times higher than the lower value (star voltage = phase voltage, e.g. 220/380 V), this indicates a three-phase four-conductor system with three line and one star point conductors. The voltages 220/380 V will gradually be changed to 230/400 V over the next few years (see Section 2.1.1 on page 76).

If only one value is indicated, e.g. 500 V, this indicates a three-phase three-conductor system with line conductors.

Western Europe

Austria
50 Hz   220/380 V,
        $3-5-6-10-20-25-27.5-30-35-45^{2)}-110-220-380$ kV

Belgium
50 Hz   220/380 – 127/220 – 220 V,
        $3-6-10-11-12-15-30-36-70-150-220-380$ kV

Denmark
50 Hz   220/380 V,
        $6-10-15-20-30-50-60-120-132-150-380$ kV

Finland
50 Hz $220/380-500^{1)}-660^{1)}$ V,
        $3-6-10-20-30-45-70-110-220-400$ kV

France
50 Hz $127/220-220/380-500^{1)}-380/660^{1)}-525/910^{1)}$ V,
        $3^{1)}-5.5^{1)}-6^{1)}-10-12.5^{1)}-13.8^{1)}-15-20-30-45-63-90-150-225-400$ kV

Area of former German Democratic Republic (new federal states)
50 Hz $220/380-660^{1)}$ V,
        $10-15-20-30-110-220-380$ kV

Great Britain
50 Hz   240/415 V,
        $3.3^{2)}-6.6-11-22^{2)}-33-66-132-275-400$ kV

Greece
50 Hz $220/380-127/220^{2)}$ V,
        $6.6^{2)}-15^{2)}-20-22^{2)}-150-400$ kV

Iceland
50 Hz $127/220^{2)}-220/380$ V,
        $6^{2)}-11-19-33-66-132-220-(380)$ kV

Ireland
50 Hz   220/380 V,
        $10-38-110-220-(400)$ kV

Italy
50 Hz $127/220-220/380$ V,
        $6-8.4-10-15-20-30-60-130-150-220-380$ kV

Luxembourg
50 Hz   220/380 V,
        $5^{1)}-15-20-65-150-220-(380)$ kV

Netherlands
50 Hz $220/380-660^{1)}$,
        $3-6-10-12,5-25-50-110-150-220-380$ kV

---

[1] Industry only
[2] No changes expected
(...) Undergoing change at the moment or changes planned

Northern Ireland
50 Hz   230/400-Belfast 220/380 V,
         6.6 − 11 − 33 − 110 − 275 kV

Norway
50 Hz   230 − 220/380[1] − 660 V,
         3.3[1] − 5[1] − 6.6 − 12 − 24 − 36 − 47 − 72.5 −
         110 − 132 − 220 − 275 − 380 kV

Portugal
50 Hz   220/380 V,
         3[1] − 6[1] − 10 − 15 − 30 − 60 − 150 − 220 −
         400 kV

Spain
50 Hz   220/380 V,
         6 − 10 − 13.2 − 20 − 30 − 45 − 50 − 110 −
         132 − 220 − 380 kV

Sweden
50 Hz   127/220 − 220 − 220/380 V,
         3.3[1] − 6 − 10 − 20 − 30 − 40 − 50 − 70 − 130 −
         220 − 400 kV

Switzerland
50 Hz   220/380 − 500 − 230/400 V,
         3[1] − 5[1] − 6 − 8 − 10 − 11 − 16 − 18 − 22 −
         30 − 40 − 50 − 60 − 75 − 110 − 132 − 150 −
         220 − 380 kV

| **Eastern Europe** |

Albania
50 Hz   220/380 V,
         6 − 10 − 16 − 35 − 110 − 220 − (420) kV

Bulgaria
50 Hz   220/380 V,
         6[2] − 10[2] − 20 − 35[2] − 110 − 220 − 400 kV

Area of former Czechoslovakia
50 Hz   220/380 V,
         3[2] − 6[2] − 10 − 20 − 30[2] − 35 − 60[2] − 110 −
         220 − 400 kV

---

[1] Industry only
[2] No changes expected
(…) Undergoing change at the moment or changes planned

Hungary
50 Hz   220/380 V,
         3[2] − 6[1] − 10 − 20 − 35 − 110 − 220 − 400 −
         750 kV

Poland
50 Hz   220/380 V,
         6 − 10 − 15 − 20 − 33[2] − 110 − 220 − 440 kV

Rumania
50 Hz   220/380 V,
         6 − 10 − 20 − 30 − 110 − 220 − 400 kV

Area of former Soviet Union
50 Hz   220/380 − 660[2] V,
         3 − 6 − 10 − 30 − 60 − 110 − 127[2] − 220 −
         380 − 500 − 775 − (800) kV

Area of former Yugoslavia
50 Hz   220/380 V,
         6 − 10 − 20 − 35 − 110 − 220 − 380 kV

| **Middle East** |

Afghanistan
50 Hz   220/380 V,
         6 − 15 − 20 − 33 − 110 kV

Bahrein
50 Hz   230/400 V,
         6.6[1] − 11 − 33 − 66 − (220) kV

Cyprus
50 Hz   240/415 V,
         11 − 66 − 132 kV

Iran
50 Hz   220/380 V,
         6 − 11 − 20 − 33 − 66 − 132 − 230 − 400 kV

Iraq
50 Hz   220/380 V,
         3.3[2] − 6.6[2] − 11 − 33 − 66[2] − 132 − 400 kV

Israel
50 Hz   230/400 V,
         22 − 33 − 160 − (220) kV

Jordan
50 Hz   220/380 V,
         6.6 − 11 − 33 − 66 − 132 − 230 − (400) kV

Kuwait
50 Hz   240/415 V,
         $3.3^{1)} - 6.6^{1)} - 11 - 33 - 132 - 300$ kV

Lebanon
50 Hz   110/190 − 220/380 V,
         $5.5^{1)} - 6^{1)} - 11 - 15 - (20) - 33 - 66 - 150 -$
         (230) kV

Oman
50 Hz   220/380 − 240/415 V,
         $6.6^{1)} - 11 - 33 - 132$ kV

Qatar
50 Hz   240/415 V,
         $6.6^{1)} - 11 - 33 - 66 - 132 - (220)$ kV

Saudi Arabia
60 Hz $127/220 - 220/380 - 480^{1)}$ V,
         (220/380 − 240/415 V, 50 Hz: only remnants)
         4.16 − 13.8 − (33) − 34.5 − 69 − 110 − 115 −
         132 − 230 − 380 kV

Syria
50 Hz $115/200 - 220 - 380 - 400^{1)}$ V,
         $6^{1)} - 6.6^{1)} - 11 - 20 - 66 - 230 - (400)$ kV

Turkey
50 Hz   220/380 V,
         (parts of Istanbul: 110/190 V)
         6.3 − 10 − 15 − 34.5 − 66 − 154 − 380 kV

United Arab Emirates
(Abu Dhabi; Ajman; Dubai; Fujairah; Ras al-Khaimah; Sharjah; Umm al-Gaiwain)
50 Hz   220/380 − 240/415 V,
         $6^{1)} - 11 - 33 - 132 - 220$ kV

Yemen (North)
50 Hz   220/380 V,
         11 − 15 − 33 − 132 kV

Yemen (South)
50 Hz   230/400 V,
         6.6 − 11 − 33 − (132) kV

Far East

Bangladesh
50 Hz   230/400 V,
         11 − 33 − 66 − 132 − 230 kV

Burma
50 Hz   230/400 V,
         3.3 − 6.6 − 11 − 33 − 66 − 132 − 230 kV

Cambodia
50 Hz   120/208 V − Phnom Penh: 220/380 V,
         20 − 110 kV

China
50 Hz   127/220 − 220/380 V (in the mining industry: 1140 V),
         3 − 6 − 10 − 35 − 60 − 110 − 220 − 330 − 380 −
         (420) − 500 kV

Hong Kong
50 Hz   200/346 V,
         3.3 − 6.6 − 11 − 33 − 66 − 132 − (345) −
         (400) kV

India
50 Hz   220/380 − 230/400 − 240/415 V,
         3.3 − 6.6 − 11 − 33 − 66 − 110 − 132 − 220 −
         400 (±500) kV

Indonesia
50 Hz $127/220 - 220/380 - 400^{1)}$ V,
         6 − 20 − 30 − 69 − 150 − (500) kV

Japan
50 Hz $100/200 - 400^{1)}$ V,
         Hokkaido, northern part of Honshu
60 Hz $110/220 - 440^{1)}$ V,
         southern part of Honshu, Shikoku, Kyushu
50 Hz   3.3 − 6.6 − 11 − 13.8 − 22 − 33 − 66 − 77 −
         110 − 138 − 154 − 187 − 220 − 275 − 500 kV

Korea (North)
60 Hz   220/380 V,
         3.3 − 6 − 10 − 66 − 110 − 220 − (400) kV

Korea (South)
60 Hz $100/200^{2)} - 220/380 - 440^{1)}$ V,
         $3.3^{2)} - 6.6^{2)} - 11^{2)} - 13.2 - 23 - 66^{2)} - 154 -$
         345 kV

---
1) Industry only
2) No changes expected
(...) Undergoing change at the moment or changes planned

9 Appendix

Malaysia
50 Hz 240/415 V,
 6.6 – 11 – 22 – 33 – 66 – 132 – 275 kV

Mongolia
50 Hz 220/380 V,
 6 – 10 – 110 – 220 kV

Pakistan
50 Hz 230/400 V,
 6.6 – 11 – 66 – 132 – 230 – 500 kV

Philippines
60 Hz 110/220 – 440 V,
 2.4 – 3.3 – 4.16 – 4.8[2] – 6.24[2] – 13.8 – 23 – 34.5 – 69 – 115 – 138 – 230 – (400) kV

Singapore
50 Hz 240/415 V,
 6.6 – 11 – 22 – 66 – 230 kV

Sri Lanka
50 Hz 230/400 V,
 3.3 – 11 – 33 – 66 – 132 kV

Taiwan
60 Hz 110/220 – 220/440 V,
 3.3 – 4[1] – 6.6 – 11 – 22 – 33 – 69 – 161 – 345 kV

Thailand
50 Hz 220/380 V,
 3.3 – 11 – 22 – 33 – 69 – 115 – 132 – 230 (500) kV

Vietnam
50 Hz 220/380 V,
 6.6 – 13.2 – 15 – 35 – 66 – 132 – 230 kV

North America

Canada
60 Hz 600 – 120/240 – 460 – 575 V,
 2.4 – 4.16 – 4.8 – 6.9 – 7.2 – 8.9 – 12.0 – 13.2 – 13.8 – 24 – 24.5 – 34.5 – 46 – 72 – 115 – 138 – 230 – 345 – 550 – 735 – (765) $\pm$ 250 – ($\pm$ 500) kV

[1] Industry only
[2] No changes expected
(...) Undergoing change at the moment or changes planned

USA
60 Hz 120/208 – 120/240 – 277/480 – 600[1] V,
 2.4 – 4.16 – 4.8 – 7.2 – 8.3 – 8.9 – 12.0 – 12.47 – 13.2 – 13.8 – 14.4 – 23 – 24 – 27.6 – 34.5 – 46 – 69 – 115 – 138 – 161 – 230 – 345 – 500 – 765 – $\pm$ 400 kV

Central America

Bahamas
60 Hz 115/200 – 120/208 V,
 2.4 – 7.2 – 11 – 33 – 66 kV

Barbados
50 Hz 110/190 – 120/208 V,
 3.3 – 11 – 24.9 kV

Belize
60 Hz 110/220 – 220/440 V,
 6.6 kV

Costa Rica
60 Hz 120/208[2] – 120/240 – 127/220[2] – 254/440[2] – 277/480[1] V,
 13.2[2] – 24.9 – 34.5 – 138 – 230 kV

Cuba
60 Hz 120/240 – 220/380 – 277/480[1] – 440[1] V,
 2.4 – 4.16 – 6.0 – 6.3 – 7.2 – 13.2 – 22 – 33 – 66 – 110 – 220 kV

Dominican Republic
60 Hz 120/208 – 120/240 – 480[1] V,
 4.16[2] – 34.5 – 69 – 138 kV

Guatemala
60 Hz 120/208 – 120/240 – 127/220 – 277/480[1] – 480[1] – 550[1] V,
 2.4 – 4.16 – 6.6 – 13.8 – 22 – 33[2] – 50 – 69 – 138 – 230 kV

Haiti
50 Hz 220/380 V (Jacmel),
60 Hz 110/220 V,
60 Hz 7.2 – 12.5 kV

Honduras
60 Hz 110/220 – 127/220 – 277/480 V,
 2.4 – 4.16 – 13.8 – 34.5 – 69 – 138 – (230) kV

Jamaica
50 Hz 110/220 – 440[1] V,
2.2[1] – 3.3[1] – 11[1] – 66 kV

Mexico
60 Hz 127/220 – 440[1] V,
2.4[1] – 4.16 – 6.6 – 13.8 – 23 – 34.5 – 69[2] – 85[2] – 115 – 230 – 400 kV

Nicaragua
60 Hz 110/220 – 120/240 – 127/220 – 220/440 – 254/440[1] V,
2.4 – 4.16 – 13.8 – 24.9 – 50 – 69 – 138 – 230 kV

Panama
60 Hz 120/208[1] – 120/240 – 254/440[1] – 277/480[1] V,
13.2 – 34.5 – 115 – 230 kV

Puerto Rico
60 Hz 120/208 – 480 V,
7.6 – 13.2 – 38 – 115 – 230 kV

El Salvador
60 Hz 110/220 – 120/208 – 127/220 – 220/440 – 240/480[1] – 254/440[1] V,
2.4 – 3.3 – 4.16 – 13.8 – 22 – 35 – 44 – 69 – 115 – 230 – (500) kV

Trinidad
60 Hz 110/220 – 120/240 – 230/400 V,
2.3 – 4 – 12 – 13 – 66 – 132 kV

South America

Argentina
50 Hz 220/380 V,
6.6 – 13.2 – 13.8 – 27 – 33 – 66 – 132 – 145 – 170 – 220 – 500 kV

Bolivia
60 Hz 220/380 – 480 V,
50 Hz 110/220 – 220/380 V (exceptions)
50/60 Hz 2.4 – 6.9 – 10 – 24 – 38 – 66 – 115 – 220 kV

Brazil
60 Hz 110/220 – 220/440 – 127/220 – 220/380 V,
2.4[2] – 4.16 – 6.6 – 11.5 – 13.8 – 24 – 33[2] – 44[2] – 69 – 88[2] – 138 – 230 – 345 – 460[2] – 500 – 800 kV

Chile
50 Hz 220/380 V,
3.3[1] – 5[1] – 6.0[1] – 6.9 – 12 – 13.2 – 23 – 44 – 66 – 88 – 110 – 154 – 220 – (500) kV

Colombia
60 Hz 110/220 – 150/260 – 440 V,
2.4 – 4.16 – 6.6 – 7.2 – 11.4 – 13.8 – 34.5 – 44 – 66 – 115 – 230 – 500 kV

Ecuador
60 Hz 120/208 – 127/220 V,
2.4[1] – 4.16[1] – 6.9[1] – 13.8 – 24 – 34.5 – 46 – 69 – 115 – 138 – 230 – (380) kV

Guyana
50 Hz 110/220 V (Georgetown),
60 Hz 110/220 – 240/480 V,
50/60 Hz 11.5 – 13.8 – 66 kV

Paraguay
50 Hz 220/380 – 220/440 V,
6 – 13.8 – 23 – 66 – 220 kV

Peru
60 Hz 220 – 220/380 – 440 V,
2.3 – 6.6 – 10 – 13.8 – 30 – 60 – 66 – 138 – 220 kV

Surinam
60 Hz 115/230 – 127/220 V,
6 – 12 – 33 kV

Uruguay
50 Hz 220 V,
6 – 15 – 30 – 60[2] – 110 – 150 – (500) kV

Venezuela
60 Hz 120/208 – 120/240 – 208/416 – 240/480 V,
2.4 – 4.16 – 4.8 – 6.6 – 8.3 – 12.4 – 13.8 – 24 – 34.5 – 69 – 115 – 138 – 230 – 400 – 800 kV

---

[1] Industry only
[2] No changes expected
(...) Undergoing change at the moment or changes planned

9 Appendix

Africa

Algeria
50 Hz 127/220 – 220/380 V,
       $5.5^{2)} – 10^{2)} – 15^{2)} – 20^{2)} – 30 – 60 – 90 – 150 – 220$ kV

Angola
50 Hz 220/380 V,
       6.6 – 30 – 60 – 150 kV

Benin
50 Hz 220/380 V,
       15 kV

Cameroon
50 Hz 127/220 – 220/380 V,
       3 – 6 – 10 – 15 – 20 – 30 – 45 – 90 kV

Congo
50 Hz 220/380 V,
       6 – 30 – 110 – 225 kV

Egypt
50 Hz 110/220 – 220/380 V,
       $3^{1)} – 6^{1)} – 10^{1)} – 11 – 33 – 66 – 132 – 220 – 500$ kV

Ethiopia
50 Hz 220/380 V,
       15 – 33 – 45 – 132 – 220 kV

Gabon
50 Hz 220/380 V,
       5.5 – 20 – 90 – (225) kV

Ghana
50 Hz 127/220 – 220/380 V,
       3.3 – 6.6 – 11 – 33 – 161 kV

Guinea
50 Hz 220/380 V,
       5.5 – 10 – 15 – 20 – 66 – 132 – (150) kV

Ivory Coast
50 Hz 220/380 V,
       5.5 – 15 – 33 – 90 – 225 kV

Kenya
50 Hz 240/415 V,
       3.3 – 11 – 33 – 66 – 132 – 275 kV

Liberia
60 Hz 120/208 – 120/240 V,
       7.2 – 12.5 – 24.5 – 69 kV

Libya
50 Hz $127/220^{2)} – 220/380$ V,
       6 – 11 – 33 – 66 – 150 – 220 kV

Madagascar
50 Hz 127/220 – 220/380 V,
       5 – 20 – 35 – 60 kV

Malawi
50 Hz 220/380 V,
       3.3 – 11 – 33 – 66 – (132) kV

Mali
50 Hz 220/380 V,
       5.5 – 15 – 30 kV

Mauritius
50 Hz 240/415 V,
       6.6 – 22 – 66 kV

Morocco
50 Hz $115/200 – 127/220 – 220/380 – 500^{1)}$ V,
       $5.5 – 6^{1)} – 6.6 – 20 – 30 – 22 – 60 – 72.5 – 150 – 225$ kV

Mozambique
50 Hz 220/380 V,
       $6.6 – 22 – 30 – 66 – 110 – 220 – \pm 533$ kV

Namibia
50 Hz 220/380 V,
       6.6 – 11 – 66 – 110 – 220 – 330 kV

Niger
50 Hz 220/380 V,
       5.5 – 15 (132) kV

Nigeria
50 Hz 240/415 V,
       3.3 – 6.6 – 11 – 33 – 66 – 132 – 330 kV

Rwanda
50 Hz 220/380 V,
       6 – 30 – 110 kV

Senegal
50 Hz 127/220 – 220/380 V,
       5.5 – 6.6 – 15 – 30 – 90 kV

[1)] Industry only
[2)] No changes expected
(...) Undergoing change at the moment or changes planned

Sierra Leone
50 Hz   220/380 V,
         3.3 – 11 – 33 kV

Somali Democratic Republic
50 Hz   220 – 220/440 V,
         3.3 – 15 – 33 kV

South Africa
50 Hz   220/380 – 500[1] – 550/950[1] V,
         3.3 – 6.6 – 11 – 22 – 33 – 44 – 66 – 88 – 132 –
         220 – 275 – 400 – ±533 – 800 kV

Sudan
50 Hz   240/415 V,
         6.6 – 11 – 33 – 66 – 110 – 220 kV

Swaziland
50 Hz   220/380 V,
         3.3 – 6.6 – 11 – 66 – 132 kV

Tanzania
50 Hz   230/400 V,
         3.3 – 11 – 33 – 66 – 132 – 220 kV

Togo
50 Hz   127/220 – 220/380 V,
         5.5 – 20 – 66 kV

Tunisia
50 Hz   115/200 – 220/380 V,
         10 – 15 – 22 – 30 – 90 – 150 – 220 kV

Uganda
50 Hz   240/415 V,
         11 – 33 – 66 – 132 kV

Zaïre
50 Hz   220/380 V,
         6.6 – 11 – 15 – 30 – 50 – 66 – 70 – 120 –
         132 – 220 – ±350 kV

Zambia
50 Hz   220/380 – 415 – 550[1] V,
         3.3 – 6.6[1] – 11 – 22 – 33 – 66 – 88 – 132 –
         220 – 330 kV

---

[1]   Industry only
[2]   No changes expected
(...) Undergoing change at the moment or changes planned

Zimbabwe
50 Hz   220/380 V,
         11 – 33 – 66 – 88 – 110 – 132 – 220 – 330 kV

Oceania

Australia
50 Hz   240/415 V, Western Australia: 254/440 V,
         3.6 – 7.2 – 11 – 22 – 33 – 36 – 44 – 66 – 88 –
         110 – 132 – 220 – 275 – 330 – 500 kV

New Zealand
50 Hz   230/400 V,
         3.3 – 6.6 – 11 – 22 – 33 – 50 – 66 – 110 –
         220 – ±250 – (±350) kV

9.11 EC guidelines for low voltage equipment

Low-voltage switchgear devices from Siemens are designed, constructed and tested for the world market. They are based on the applicable standards, DIN norms, DIN VDE specifications and IEC publications. In addition, they conform to the guidelines of 19.2.73 set out by the EEC council (73/23/EEC). These are known as the EC guidelines for low-voltage equipment, and were intended by the EC council to be of a binding nature on all members of the European Community and the European Free Trade Association (EFTA). The guidelines are presently being revised for the future European market.

To indicate that particular devices conform with these guidelines, the manufacturer may include a conformity declaration in his catalogues. He may also use the CE mark in descriptions of the devices, in publications, or may display the mark on the devices themselves.

Through the adoption of the EC council global concept, the guidelines have regained their importance. We therefore draw your attention to the fact that the conformity declaration for low-voltage switchgear, as described in the "Low-voltage guidelines", is included in the Siemens catalogues.

9.12 Glossary – Brief explanations of some technical terms

Not all of the technical terms explained in the following glossary are standardized internationally. Owing to differences in the English language and its usage in various countries, and particularly to the use of different technical terms to describe similar devices, quantities, processes, etc., the translator cannot accept responsibility for possible discrepancies between the explanations given here and locally accepted definitions. Wherever synonymous terms or phrases are known, or if the translator has become aware of different meanings for the listed terms, this has been indicated. Furthermore, it should be borne in mind, that the words and phrases are only explained in terms of their relevance to the field of low-voltage switchgear covered by this handbook.

The arrow → indicates that the term following is also explained elsewhere this section.

"a" contact

See → Normally open contact

a-release

The short designation for a → current-dependent delayed overload release (→ inverse-time release), trip mechanism or relay. Also see → Thermal overload release.

Abrasion

Wear or erosion of material caused by friction between two solid bodies rubbing against each other.
In the context of switchgear, abrasion leads to the loss of → contact material owing to the wiping or rubbing action of one contact surface against the other after initial touching. Abrasion between the moving parts in any switching device limits the → mechanical service life.

Accept switch

See → Discrepancy switch

Accuracy class (of a → current transformer)

In the case of a current transformer for measurement purposes, the accuracy class expresses the maximum permissible error in the secondary current at $1 \cdot I_n$ and $1.2 \cdot I_n$ (I_n is the rated primary current). For example, class 0.5 ($\pm 0.5\%$ error) is required for accurate billing purposes, where-as class 3 ($\pm 3\%$ error) is usually adaquate for operational measurement and indication.
Only a limited transfer error is permitted in the case of current transformers for protection purposes; even in the event of an overcurrent. At the rated → burden, the total error may be -5% (class 5P) or -10% (class 10P) respectively.

Across-the-line starter

See → Direct-on-line starter

Active power

The portion of → apparent power which is transformed into another non-electrical form (e.g. heat, mechanical power or energy). Also known as "effective power". Refer to the definition of → power.

Actuating angle

The angle through which the → actuator of e.g. a → position switch or the knob (→ actuating element) of a → control switch has to be rotated to operate the associated → switching element.

Actuating distance (of a → proximity switch)

See → Operating distances

Actuating element

a) of a → command device: The the part of the unit (also referred to as the → actuator) which is normally fitted into a hole in the control panel and which is touched (e.g. pushed, pulled, rotated or otherwise moved) by the operator in order to cause a change in the → switching state of the → switching element. Types of actuating element include pushbuttons, illuminated pushbuttons, knobs, twist levers, toggle switches, keys, etc.
b) of a → position switch: The device, cam, switching bar etc. which moves the → actuator head in such a way as to operate the → switching element of the position switch. Its leading angle, trailing angle, maximum operating speed and direction of operation are defined for each actuator type in the relevant Siemens catalogue.
c) of a → proximity switch: A body, also known as the → target, (e.g. metal plate in the case of an inductive → BERO) which causes operation of the proximity switch when it is brought into the → sensing zone. The shape, material and dimensions of this body can influence the → response characteristic of the proximity switch. Also refer to → Reduction factors.

Actuator (general use)

A device, part or body which moves, is moved, or produces movement (e.g. linear motor, servo motor, drive, → operating mechanism, → magnetic drive, fingertip, foot, etc.), with the purpose of causing a → response in a second device, part or piece of equipment. Also used as a verb, i.e. "to actuate", meaning e.g. "to operate", "switch on", "energize", etc.
In the field of automation, the word actuator is sometimes used as a general term for complete items of switchgear (including valves) which operate when a command signal is applied to their input (see → Actuator Sensor Interface).

Actuator head

The part on a → position switch (→ limit switch) which must be actuated or moved to produce a switching action in the switch itself.

Actuator of a command device

See → Actuating element

Actuator of a position switch

See → Actuating element, → Actuator head

Actuator Sensor Interface (ASI)

The Actuator Sensor Interface (ASI) is a field communication system based on internationally recognized specifications and standards for the master-slave interconnection of a controlling device and → low-voltage switch and/or controlgear (e.g. to IEC 947 or other standards). The Actuator Sensor Interface connects the controlling device or "master" (e.g. part of a → programmable controller) to → binary → actuators, such as hydraulic valves, → contactors, → operating mechanisms, and to → sensors such as → proximity switches, → position switches or even → pushbuttons (via the "slaves"). The master and slave units may be integrated into the actual controller and actuators/sensors respectively or may be self-contained units. ASI is classified as a simple, low-level interface system using, in particular, the cyclic → data exchange of short data values. The actual interconnection is achieved by a mechanically coded (by shape) unshielded 2-wire cable which transfers power and data from the master to the slaves, and data from the slaves to the master.

AF

Audio frequency. See → AF ripple control (Audio-frequency remote control)

AF ripple control

In this technology, audio-frequency (AF) impulses are superimposed on the network supply to switch remotely installed receiver relays on and off (also known as "audio-frequency remote control systems"). Typical applications include load shedding and the changing-over from one billing tariff to another. Special pracautions may be required in installations which incorporate → power factor correction equipment. See → Audio-frequency blocking circuits.

AF traps

See → Audio-frequency blocking circuits

Ageing (of fuse links)

This term is used to describe the deterioration of a → fuse link (i.e. the changing of its → time-current characteristic) over a period of time as a result of changes in the constituent materials. The resistance to ageing is an important measure of the fuse-link quality. A fuse must be able to withstand the effects of continuous normal current, cyclic overload and pulsed overload without deterioration to an unsafe state or to a state in which nuisance rupturing occurs. It must also interrupt the current safely in the event of an overcurrent or short-circuit - even after several years of apparently uneventful service. The choice of materials, the design of the → fuse element and the ongoing consistency of manufacturing quality are vital in determining the resistance to deterioration. Siemens fuse links are well known for their resistance to ageing effects.

Regular overcurrents which almost cause melting of the fuse element can have a negative effect on the ageing of a fuse element - particularly one of lower quality. For example, localized destruction of the fuse element can occur if the eutectic alloy formed between e.g. tin deposits on a silver element is sufficient to penetrate the element thickness (see → Metcalf effect).

Air circuit-breaker (ACB)

Strictly speaking, the term refers to all → circuit-breakers in which the → main contacts open and close in air at atmospheric pressure. Traditionally, however, an a.c.b. is assumed to be of higher current rating (e.g. >630 A), to be of the → zero-point quenching type, and to employ a mechanically strong metal supporting chassis. If the supporting chassis is of reinforced moulded insulating material, the circuit-breaker may be referred to as an ICCB (→ Insulated Case Circuit-Breaker). Air circuit-breakers are typically used in power distribution networks where → time-based discrimination is called for.

Air gap (between conductive parts)

See → Clearance

Air gap (in a magnet system)

This term is used to refer to the portion(s) of the → magnetic circuit of a → magnet drive in which the flux path passes from the fixed portion of the → magnet core to the moving → armature through air. The force by which the armature is pulled against the fixed portion of the magnet core depends on the magnetic flux density in the air gap(s). Analogously, the clearance between the stator and the rotor of a motor, is also known as the air gap.

Air-break contactor

See → Contactor

Alarm switch

→ Auxiliary switch of a → circuit-breaker which is actuated only when the circuit-breaker is tripped e.g. by its → overcurrent release, → shunt trip or → undervoltage release. It is not operated by the normal opening and closing of the circuit-breaker contacts by means of the → operating mechanism.

ALPHA

The "ALPHA Association for Testing and Certification of Low Voltage Switchgear" is an institution of the German electrical industry. It was founded as an extension to existing national and international standards authorities. It aims to strengthen the accountability of the manufacturer, and to achieve the correct degree of acceptance for inhouse testing, certification and conformity declarations. This is ensured by means of standard ALPHA test modules, uniform documentation of test results and professional application of valid specifications. Through participation in →LOVAG, it strives for reciprocal recognition of certification, a reduction in the number of approval procedures and the creation of a single standard certificate for the whole of Europe (ultimately for world-wide acceptance).

Alternating current

A current which reverses its direction of flow in a periodic manner with respect to time (i.e. at a specific frequency) owing to the application of →alternating voltage in an electric circuit. Only in the case of current through a pure ohmic resistance will the alternating current by "in phase" with the applied alternating voltage. If a.c. voltage is applied across an inductive load (motor, solenoid, transformer) the alternating current will lag the voltage (theoretically by up to 90° in the case of a pure inductance). In the case af capacitive loads, the current leads the voltage (refer to →Phase angle and →Power factor).

Alternating voltage

A potential difference (or voltage) which reverses polarity in a periodic manner with respect to time (i.e. at a specific frequency). A sinusoidal alternating voltage is produced when a conductor loop, e.g. the winding in an alternator, rotates in a magnetic field (see →Alternating current). Although, strictly speaking, the abbreviation "a.c." means "alternating *current*", it is also used with reference to alternating voltage (as in "an a.c. voltage").

Ambient (air) temperature

The temperature of the air surrounding the complete switching device or fuse, determined under prescribed conditions (Note: In the case of e.g. circuit-breakers or fuses inside a housing or casing, it is the temperature of the air outside the housing). Since the heat dissipation from an item of switchgear depends to a large extent on this ambient temperature, derating of switchgear may be necessary at high ambient temperatures. The →rated values quoted for a switching device are always based on a specific maximum ambient temperature.

American Wire Gauge (number)

A number assigned to a specific conductor or wire cross-sectional area. The wire cross-sectional area jumps by 26% for each successive gauge number in the geometric number series and the following interesting rules apply:

1. The thicker the wire, the higher the AWG number.
2. An AWG 10 copper wire has a diameter of approximately 0.1 inches (= 2.54 mm), a cross-sectional area of 10.000 circular mils (approx. 5 mm^2) and a resistance of 1.0 →per 1000 foot (= 304.8 m).
3. For an increase in dimensions by 3 AWG numbers (e.g. AWG 10 to AWG 7), the cross-sectional area and the mass (weight) are doubled. The d.c. resistance is halved. Conversely, for a decrease in the dimensions by 3 AWG numbers, the cross-section and weight are halved; the d.c. resistance is doubled.
4. For an increase in dimensions by 6 AWG numbers (e.g. AWG 10 to AWG 4), the diameter is doubled. Conversely, the diameter is halved for a decrease in dimensions by 6 AWG numbers (e.g. AWG 10 to AWG 16)

The AWG numbers cannot be converted directly to square millimeters. Also refer to →kcmil and to Tables 9.10 and 9.22 on Pages 511 and 526 resp.).

Ampacity

See →Current carrying capacity

Ampere turns

The strength of the magnetic field created by a current flowing through a coil depends on the sum of the currents in all the conductors lying next to each other (or turns of the coil). This sum is also known as the magnetomotive force (m.m.f.) and is expressed in ampere turns.

an-release

The combination of an →**a**-release and an →**n**-release.

AND-gate

A device or electronic component (integrated circuit) which combines two or more input signals according to the logic AND function of Boolean algebra to a single output signal. Symbol: A. The inverse of the AND function, is the NAND function (=*not* AND), with the symbol $\bar{A}$. Examples 1 A 1=1; 1 A 0=0; 1 $\bar{A}$ 0=1

Anti-pumping device

A device or circuit which prevents repetitive restarting (e.g. of a motorized →operating mechanism on a →circuit-breaker) after the initial ON-command, while the energizing signal is still applied.

Anticondensation heater

A device used to heat up the air inside cabinets and enclosures (e.g. switchboards) particularly during the periods in which the equipment inside is switched off and no heat losses are produced. It reduces the relative humidity inside the cabinet and prevents condensation.

Antivalent output (of a →proximity switch)

If a proximity switch has two current sourcing outputs (or electronic →switching elements) of which one has an →NO and the other an →NC function, then it is said to have antivalent outputs.

Apparent power

In a.c. systems, the apparent power is the product of voltage and current, regardless of the →power factor. The apparent power is a measurable quantity and is given in VA (see →Power).

Glossary

Applied voltage

The voltage which is present on the line terminals (usually designated 1/L1, 3/L2, 5/L3) of a switching device prior to it being operated or energized (i.e. prior to current flow through the device).

Approach angle

The angle between the direction of approach to a → position switch and the reference axis of the → actuator head.

Approach speed

The speed with which a → position switch (→ limit switch) is actuated mechanically (also known as the operating speed). The maximum permissible approach speed is dependent on the shape and the → approach angle of the operating cam. The higher the approach speed, the more acute the approach angle must be.

Approval

Permission to use an item of switchgear in a particular country or application on the basis of obligatory national specifications or standards. UL or CSA approvals, for example, are required for the North American market (USA, Canada) where additional marking and designation is also obligatory, i.e. a label with the approval symbol must be attached to the device (see Section 1.2, Page 17). Approvals are normally based on specifications other than the internationally recognized → IEC, → CENELEC, and → CEE.

Arc blow-out space

During switching operations of a → switching device and particularly during the interruption of highly inductive load currents or → short-circuit currents, → ionized gases produced by the → switching arc are expelled from the openings in the → arc chamber. To ensure that the concentration of this ionized gas does not reach a dangerous level, a minimum free space may be required above or in front of the device. This arc blow-out space is specified by the manufacturer (usually on the dimension drawings) and may depend on the presence of exposed live conductors (e.g. → busbars), conductive structures or insulating barriers close to the switching device. Arc chamber extensions (or arc traps) may be fitted to larger → air circuit-breakers to reduce this space and thereby reduce the effective space required in the switchboard. In the case of → vacuum circuit-breakers and → vacuum contactors the switching arc is contained within the vacuum chamber, no ionized gases are released, and consequently no arc blow-out space is required.

Arc chamber

A specially designed cover of ceramic or heat-resistant plastic material which encloses the → contacts of a → switching device to contain and extinguish the → switching arc and to protect against flashover between the individual → poles. The switching arc is driven/guided into the arc chamber where it is lengthened and subdivided into a number of smaller arcs by the → arc splitters. In addition, the switching arc is cooled intensively by the flue-shaped construction of the arc chamber walls and the vapours created by the switching arc are → deionized (the arc chamber is also known as the arc chute or deionization chamber). Slit-like openings permit expulsion of remaining ionized gases. In the case of larger → circuit-breakers, arc chamber extensions may be necessary to reduce the → arc blow-out space required above the device. In the case of a d.c. switching device, the arc chamber may contain a → blow-out coil and/or a permanent magnet to drive the switching arc into the arc chamber.

Arc chute

See → Arc chamber

Arc extinction

See → Arc quenching

Arc quenching

Whenever an electrical circuit is opened, typically by the moving apart of the → contacts in a → switching device, the current continues to flow via an → arc drawn between the separating conductors. This → switching arc must be extinguished or quenched before the current will cease to flow in the circuit. One can distinguish between several methods of arc quenching: The switching arc of *alternating current* is usually extinguished as the current passes through the natural zero of its waveform. It must, however, be prevented from restriking as the voltage builds up across the air gap formed by the separating contacts e.g. by sufficient distance, cooling and removal of ionized air. In a vacuum, the switching arc can be extinguished before the current passes through zero (→ chopping current) by the mere separation of the contacts. The switching arc of a *direct current* is extinguished only once the current in the circuit has been forced to virtually zero. This is achieved by extending the length of the arc, preferably via a number of → arc splitters and by intense cooling in the → arc chamber so that the reverse voltage built up along the length of the arc will increase to a value exceeding the driving voltage of the circuit (i.e. the driving voltage will be unable to sustain a current flow through the arc) Also refer to → Blow-out coil. In the case of → LV HRC fuses, an arc develops in place of the melted → fusible conductor. The arc is so strongly cooled by the quartz sand which is packed tightly around the → fusible conductor, that a high → arc voltage is achieved very rapidly and the current is forced to zero.

Arc splitters

Pieces of metal plate in the path of a → switching arc designed to increase the → arc voltage. An → arc chamber may contain several arc splitters so that the arc is forced to "jump" from one arc splitter (or splitter plate) to the next. The base or root of every arc, as it impinges upon or leaves an arc splitter, experiences a voltage drop which appears as cathode or anode drop according to the direction of current flow. In the case of a steady d.c. arc, this voltage is typically 10 to 20 V. As the current reverses (a.c.) this voltage

559

drop near the surface of the arc splitter plate is substantially higher. It is added to the voltage drop of the arc column itself.

Arc voltage

The highest instantaneous voltage value which occurs between the line and load terminals of a pole of a switching device during the → arcing time under specified conditions.

Arc, Arcing

Arcing is the phenomena whereby electric current flows through air or more precisely, through a column of ionized gas. (see → Ionized gases). While, in the case of an ohmic resistance the voltage across a length of conductor increases in proportion to the current flowing through it, the opposite occurs in an arc. Assuming a constant length, the voltage drop across an arc decreases with increasing current. This is explained by the fact that the conductivity of a column of ionized gas depends on its temperature; i.e. the greater the arc current, the higher its temperature and the greater the extent of ionization will be. Typically, in the field of low-voltage technology, the temperature in the interior of an arc can be between 10.000 to 15.000 °C.

Arcing occurs whenever an electrical circuit operating at a voltage exceeding 15 V is opened (see → Switching arc). Short-circuits are normally associated with (or result from) uncontrolled arcing between → conductive parts at differing potential (see → Flashover). They represent a fire hazard and can cause severe damage if they are not contained and/or switched off rapidly by the interruption of the short-circuit current (see → Short-circuit protection device)

Arc-quenching medium

The medium in which the → switching arc is extinguished or quenched. The following types of arc-quenching media are used for low voltage switching devices:
air,
sand (in → LV HRC fuses) and
vacuum.

Arcing contact (arcing horn, arc runner)

The part or extension of a → contact element, usually made of copper or iron, on which the → switching arc burns upon separation of the → contacts. It reduces erosion of the actual → contact material (e.g. expensive silver alloy) during switching of inductive loads (i.e. where extensive arcing is expected). The arcing contact may be part of the main contact itself or it may be a separate contact designed to open after and close before the main contact which it is intended to save from contact erosion.

Arcing time

... of a pole or a fuse: The period of time between the instant of the initiation of the → arc in a pole or a fuse and the instant of final → arc quenching or extinction in that pole or that fuse.

... of a multipole switching device: The period of time between the instant of the first initiation of an → arc (first pole to open) and the instant of final → arc quenching (last pole to break) or extinction in all poles. Also known as arc duration.

Armature (of a magnetic drive)

The moving part of a → magnetic drive (→ solenoid, → electromagnetically operated mechanism) as used e.g. in → contactors and → electromagnetic releases. In a.c. magnetic drives it is usually a moveable portion of the → laminated iron core (typically U or E-shaped), where-as solid → clapper-type, or hinged armatures are used mainly in d.c. applications. When the coil is energized, the armature is attracted to or drawn into to the fixed magnet core. In the case of contactors, the → moving contacts are connected to the armature.

ASI

See → Actuator Sensor Interface

Askarel

A generic term for a group of nonflammable synthetic chlorinated hydrocarbons. See → PCB

Assured operating distance s_a (of a → proximity switch)

The distance of a given → target from the → sensing face at which correct operation of the proximity switch is assured under specified conditions. In terms of IEC-947-5-2, it must be between 0% and 81% of the → rated operating distance s_n for → inductive proximity switches and between 0% and 72% of s_n for → capacitive proximity switches.

ASTA

This abbreviation stands for the independent **A**ssociation of **S**hort-Circuit **T**esting **A**uthorities. Today, the association is known as ASTA Certification Services. It is based in England and carries out testing and certification of short-circuit protection devices and their combinations with other items of low-voltage switchgear. Also refer to → LOVAG.

Asymmetric short-circuit current

→ Short-circuit current waveform which is displaced asymmetrically to the zero line immediately after occurrence of the fault. It comprises the symmetrical short-circuit current and the → direct current component. The d.c. component decays to zero and the short-circuit current becomes symmetrical about the zero line.

Asymmetry (in %)

Maximum deviation of the current in one phase from the arithmetic mean value of all three phases.

ATSE

See → **A**utomatic **t**ransfer **s**witching **e**quipment

Audio-frequency blocking circuit

If power capacitors or → power factor correction units are installed in networks which incorporate audio-frequency re-

Glossary

mote control systems (→ AF ripple control), then high frequency signals can cause large audio-frequency currents to flow in the network, since the impedance of the capacitive load is inversely proportional to frequency. This not only leads to additional loading of the audio-frequency signal generator, but may also cause the voltage of the switching signal to drop below the value necessary to operate the AF receiver relays in the vicinity of the capacitors. A remedy for this problem, is the fitting of → audio-frequency blocking circuits to the capacitive load. These are almost invariably parallel resonance blocking circuits which are connected into the supply line upstream of the capacitors. Since they have a relatively low impedance at the system frequency and a high impedance at the audio frequency, → filter circuits of this type are also known as "low-pass filters" or "audio-frequency traps" (AF traps).

Audio-frequency remote control

See → AF ripple control

AUSTER (in German: Auswahl Technischer Erzeugnisse)

Siemens trademark for a computer programme which facilitates the selection of technical products and which offers complete technical and commercial product data on each item (central databank of products).

Automatic (rapid) recharging

After each power failure, the back-up battery may be recharged automatically with an increased voltage (high-rate or boost charging). After a specific period of time, the charger is switched back to the trickle-charging, or floating voltage again.

Automatic power factor correction

By means of a → power factor controller, the instantaneous value of the phase angle between voltage and current in a supply conductor may be monitored continuously and capacitive load switched into or from the circuit as required to maintain a value of → cos φ as close to unity as is practicable. Thus, the → p.f.c. unit responds automatically to changes in the load and optimal use of the supplied electrical energy, transformers and connecting cables may be achieved.

Automatic reset feature (on an → overload relay)

By means of the automatic reset feature on a → bimetal overload relay, the device can be made to return to its initial → switching state automatically after a → tripping operation and after a cooling down period. This feature is not recommended for → motor starters employing → maintained-contact control (also known as two-wire control), as sudden unexpected restarting of the motor may result if the → control circuit were not opened by some other means during the reset (or cooling-down) period (also refer to → Manual reset feature) It may be of interest to note, that → eutectic alloy overload relays generally do not offer this feature.

Automatic transfer switching equipment (ATSE)

Self-acting apparatus (or → switchgear assembly) comprising the → transfer switch(es) and other necessary devices for the monitoring of supply circuits and for transferring one or more load circuits from one source of power supply to another. A distinction is made in IEC 947-6-1 between Class PC and Class CB automatic transfer switching equipment:
Class PC: ATSE capable of making and withstanding, but is not intended for breaking short-circuit currents.
Class CB: ATSE equipped with → overcurrent releases and capable of making and breaking short-circuit currents.
Furthermore, IEC 947-6-1 specifies that the → operating mechanism of ATSE must be equipped with → electrical interlocking and → mechanical interlocking to prevent simultaneous connection of both the normal and the alternative supplies.

Autotransformer

A transformer in which at least two windings have a part in common (i.e. share the same current path). For example, the secondary winding may be regarded as a tap off the primary winding.

Autotransformer motor starter

A form of → reduced voltage starter for squirrel-cage motors in which the → terminal voltage is lowered by means of an autotransformer during the starting period.

Auxiliary circuit

The term encompasses all the conducting parts of a switchgear assembly or a switching device other than the → main circuit which serve control, measurement, indication, regulation, interlocking, data processing purposes, etc. (→ control circuit).
Note:
Auxiliary circuits serve other functions, too. They may be part of the → control circuit of other switching devices.

Auxiliary contact

A contact included in an → auxiliary circuit and mechanically operated by the → switching device in or on which it is mounted. Depending on its function, the auxiliary contact is either a → NC (normally closed) contact, a → NO (normally open) contact, a → changeover contact or a → fleeting contact. Also see → Control contact, → Auxiliary contact block and → Auxiliary switch
Note:
An auxiliary contact fulfills additional functions such as indication, interlocking, etc.; in this case it can a be part of the control circuit of other switching devices.

Auxiliary contact block

An → auxiliary switch designed for retrofitting onto a switching device (e.g. snap-on auxiliary contact blocks for → contactors)

Auxiliary contactor

See → Contactor relay

Auxiliary release

An electromagnetic mechanism, usually supplied as an optional accessory module or built into a → circuit-breaker or other mechanically latched switching device and used to → trip or delatch the breaker or switching device by means of an electrical signal from an → auxiliary circuit. Examples include → shunt releases and → undervoltage releases.

Auxiliary supply voltage

See → Control supply voltage

Auxiliary switch

A switch or switch module containing one or more → control and/or auxiliary contacts mechanically operated by a switching device. Auxiliary switches are either incorporated in the basic device (→ contactor relay), or may be retrofitted (→ auxiliary contact block)

Availability delay t_v (of a → proximity switch)

see → Time delay before availability

AWG

→ American Wire Gauge

Axial approach (to a → proximity switch)

This is the direction of approach of the → target (→ actuating element) to the → sensing face of a proximity switch when its centre moves on a path which coincides with the → reference axis Z of the proximity switch.

azn-release

The combination of an → a-release, a → z-release and an → n-release.

azng-release

The combination of an → a-release, a → z-release, an → n-release and a → g-release.

"b" contact

See → Normally closed contact

Back-up protection

The → short-circuit protection of an overload or → short-circuit protection device in the range of → prospective short-circuit current greater than its own → short-circuit breaking capacity by the simultaneous operation of a second protection device connected in series on the incoming supply side. The devices must be carefully matched so that they operate at the same time in the event of a high → short-circuit current in such a way that the breaking capacity of the combination is sufficiently high. In the event of a lower short-circuit current, the downstream unit may interrupt the current alone while the upstream unit does not respond. Back-up protection can be achieved by means of → circuit-breakers or → fuses. In the case of two circuit-breakers in a back-up combination, one also speaks of a → "cascade connection".

Barrier, Flash barrier

Walls or sheets of insulating material mounted between the poles or terminals of a switching device. They are used to increase the → clearance and → creepage distances between poles, and to increase the degree of protection against accidental contact with live parts (also refer to → Partitioning)

Baud rate

See → Data transfer

BERO

BERO is the trade name for Siemens → proximity switches (it is the abbreviation for **B**erührungslose **E**ndtaster mit **r**ückgekoppeltem **O**szillator; transl. "proximity limit switch with feedback oscillator"). They are available in inductive, capacitive, ultrasonic and photoelectric versions.

Bimetal relay

This term is commonly used to refer to thermally delayed → overload relays (or inverse time-delayed overload relays) which operate on the bimetallic principle. In bimetal relays, each bimetallic strip (or element) is heated directly or indirectly by the phase load current. A special → current transformer may be used to reduce actual current which passes through the overload relay in order to limit the heat losses, or to alter the → tripping characteristic of the overload relay (e.g. for → heavy-duty starting). Bimetal relays usually offer such features as → phase-loss sensitivity, → temperature compensation and → manual reset.

Bimetal release

An → overload release or → trip of a → circuit-breaker which operates on the bimetallic principle (see → Bimetal relay, → Bimetallic strip)

Bimetallic strip or element

A bimetallic strip or element comprises two thin strips of dissimilar metals fused together along their length. The two metals each have a different coefficient of expansion, which causes the element to bend as its temperature increases. This bending movement is used to → trip an → overload relay or → overload release mechanically when the end of the bimetallic element operates the tripping latch at a preset point. The bimetallic strip (or element) is heated directly or indirectly by the phase load current passing through the main poles of the overload tripping device. The degree of bending movement resulting from the temperature or alternatively the load current determines the → tripping characteristic of the release or relay.

Binary

Binary means "consisting of two units". The two values are usually designated as 0 or L (low) and 1 or H (high). The advantage of binary representation is that the two values are easily translated into electrical or electronic signals, i.e. by ON or "voltage applied" and OFF or "no voltage applied". Most electromechanical →switching devices such as →contactors and →circuit-breakers are either open or closed (ON or OFF), and are therefore termed "binary devices".

Bit

Abbreviation for "binary digit". A bit is the smallest information unit in digital technology. Each bit represents a binary decision (→ binary), e.g. H (high) or L (low).

Blind zone (of a → proximity switch)

The space between the →sensing face and the →minimum operating distance of an →ultrasonic or →photoelectric proximity switch, where no object (→ target) can be detected reliably.

Blocking circuits

see → Filter circuits

Blocking factor a_f

The blocking factor a_f of an audio-frequency blocking circuit is the ratio between the audio-frequency impedance Z_f of a capacitor blocked by this circuit, and the reactance X_{Cf_1} of the capacitor at sinusoidal voltage with the fundamental frequency f_1 (50 Hz).

$$a_f = \frac{Z_f}{X_{Cf_1}}.$$

Blow-out coil

This is a special coil with a number of turns incorporated in each main current path (or pole) of a d.c. switching device to improve the → arc-quenching characteristics during opening of the →contacts. The magnetic field created by the current flowing through the blow-out coil is used to force the →switching arc into the →arc chamber where it is cooled and ultimately quenched to interrupt the current flow. In cases where the magnetic field from the blow-out coil would be too weak to force the arc away from the contacts and into the →arc chamber (e.g. if the current to be interrupted is low), permanent magnets fitted inside the switching chamber of the switching device ensure the same effect. In the case of a.c. switching devices, a similar effect is achieved by a single curve or loop in the current path, but here the need for arc blow-out is not as great, since the current passes through zero every half cycle. The technique is also known as "magnetic blow-out".

Blow-out space

See → Arc blow-out space

Bonding

The permanent joining (e.g. welding, soldering) of metallic parts to form an electrically conductive path of sufficient conductivity and →current-carrying capacity for the application at hand. In cases where the required electrical conductivity is not assured by the join between metallic parts (e.g. between metal →conduit and a metal enclosure), a bonding jumper (i.e. extra conductor or wire) may be required.

Bounce duration

The time duration from the instant when →contacts first touch during closing until full and stable contact is established (→ contact bounce). The bounce duration is an important factor determining the →electrical service life and operational reliability of a switching device.

Bounce-free switching operation

In the case of mechanical switching devices and contacts, →contact bounce cannot be eliminated altogether at reasonable cost. If, however, the →bounce duration is limited to such an extent (e.g. to a few milliseconds) that it has no effect on the electrical function of the →switching device, one speaks of "practically bounce-free operation". True bounce-free switching can be achieved electronically.

Bowden cable

A steel cable or wire guided within a sleeve (often of spiralled material), and used to apply pulling or pushing force on an →actuator located in an inaccessible position. Bowden cables are used, for example, to operate the reset button on an →overload relay not situated directly under an operator panel and in →mechanical interlocks of →air circuit-breakers

Box terminal

A type of →terminal or point of connection (e.g. on an electrical →switching device) in which the wire is clamped within a frame or box by the tightening of a screw or bolt. The terminal may or may not incorporate a washer or pressure plate to distribute the clamping force of the screw on the wire.

Box-type assembly or construction

A switchboard construction which comprises several modular boxes or enclosures bolted together. This type of construction is commonly used for →totally insulated switchboards, distribution boards and control cubicles.

Break function (of a → proximity switch)

The output of a proximity switch is said to have a break function if load current is caused:
not to flow when the presence of a →target is sensed, and
to flow when the target is not detected.
It corresponds to the →normally open contact function (NO)

Break time

The time duration between the beginning of the → opening time of a mechanical switching device (or → pre-arcing time in the case of a fuse) and the end of the → arcing time.

Breakaway pulse

See → Start impulse

Breakdown voltage

This is the value at which the voltage between two electrodes breaks down and destroys the insulating material between them (see → Flash-over)

Breaking capacity

The value of prospective → breaking current, at specified cos φ and test voltage, which a → switching device or a → fuse is capable of breaking reliably and safely under prescribed conditions of use and behaviour. In a.c. systems the current is expressed as the symmetrical r.m.s. value of the a.c. component.
In the case of *short-circuit* breaking capacity, the prescribed conditions include a short-circuit at the load-end terminals of the switching device. In IEC 947-2, a distinction is made between the → ultimate short-circuit breaking capacity and the → service short-circuit breaking capacity. These differ mainly in terms of the prescribed → test sequence and the condition of the → circuit-breaker after performance of the test.

Breaking current

The current which flows in a pole of a switching device or a fuse at the instant of initiation of the arc during the → opening or → breaking operation. In a.c. systems the current is expressed as the symmetrical r.m.s. value of the a.c. component. The *prospective* breaking current is the prospective current evaluated at the instant of the initiation of the breaking process - usually the instant of initiation of the arc.

Breaking operation

In the case of → fuses, this term refers to the melting through (or rupturing) of the → fuse element as a result of the temperature caused by current flowing through the → fuse link and the consequent opening of the current path or circuit. For switching devices, refer to → Opening operation.

Build-up time t_s (of a → proximity switch)

The time delay between the instant an → inductive proximity switch is de-actuated by a → target leaving the → sensing zone, and moment its output state changes (i.e. the time required for the damped electromagnetic field to be re-established).

Bunched

Cables are said to be bunched if two or more are contained within the same → raceway or, if not enclosed, are not separated from each other by a specified distance.

Burden

This term is used to mean the total resistance of the secondary circuit of a → current transformer. The burden includes e.g. the resistance of all the connecting leads in the meter circuit and the internal resistance of the measuring instrument itself.

Burst

This term refers to one of the standardized forms of → electromagnetic interference used in the → EMC testing of low-voltage switch and controlgear devices containing electronic components (e.g. → proximity switches). In terms of IEC 801-4, for example, it consists of repeated "bursts" of low energy switching transient (with defined rise time and frequency) coupled e.g. into the → control supply voltage of the device. The severity levels are defined in terms of the test voltage (typically > 1 kV but appropriate to the particular device or part being tested) and the frequency at which the bursts are repeated. The ability of a device to withstand burst testing at a specific severity level, is referred to as its "fast transient withstandability". Also refer to → Surge.

Bus (data bus)

The connecting lead (multi-core or twisted pair) between individual functional units, protection devices, logic controllers, remote control stations, etc. with communication capabilities (e.g. → SIMOCODE and various → SIMATIC systems). The → data transfer for the purposes of e.g. control, signalling, measurement, and parameterization takes place via the bus. Each bus (or data bus) system is characterized by its particular protocol, data transfer rate (Baud rate), etc. A number of standard bus systems, suitable for various levels and applications in the automation hierachy, have found acceptance in the international industrial markets. These include, for example, → ASI, SINEC L2 and the → PROFIBUS.

Bus coupler

A → circuit-breaker (usually an → air circuit-breaker) which serves to join (or couple) two → busbar sections.
Application example: Each busbar section may have its own → incomer circuit-breaker and associated transformer. If the power supply on one busbar section should fail or be switched off (e.g. transformer failure, or maintenance) the bus coupler is closed so that both busbar sections are supplied from the one incomer (also see → Mechanical interlock)

Busbar

A conductor, usually one or more copper bars of low impedance with rectangular cross-section, to which several electrical circuits are connected and from which they derive power (voltage and current). Typically, the busbars are arranged in a busbar system comprising a conductor for each phase or line (and N, PE or PEN if applicable) and are mounted horizontally across the switchboard sections. The load circuits are connected to the busbars by means of cables bolted directly to the copperwork via cable lugs, via branch busbar systems or dropper busbars. Alternatively, certain switchgear devices (e.g. →circuit-breakers and →fuse switch disconnectors) may be mounted directly on the busbars and derive their electrical connection via special busbar clamps.

Busbar trunking

A special type of switchgear assembly which consists of an insulated busbar system mounted inside modular sections of ducting. Tap-off units are used to connect electrical circuits to the busbars.

Bushing

An insulating component which enables an exposed conductor to pass through a metallic partition, cover or wall. It includes the means of attachment to the partition or cover.

Byte

A unit of data or information consisting of a sequence of 8 →bits.

Cabinet

An →enclosure for a →switchboard, →distribution board or →panelboard designed either for surface or flush mounting and provided with a frame or trim in which a swinging door or doors is/are hung. Cabinets are also referred to as →cubicles, particularly when two or more are joined together to form a switchboard →section.

Cable ducting

A manufactured, →raceway of metal or insulating material (other than →cable trunking or →conduit) intended for the holding and protection of cables which are drawn in after installation of the ducting.

Cable entry

A flange, base, roof or cover plate containing openings for cables and insulated wires entering or leaving a switchboard or cubicle.

Note:
A cable entry may also incorporate the →cable glands (gland plate).

Cable gland

A fitting which secures and protects an insulated electric cable as it passes through a →cable entry into the →housing of a →switching device (e.g. →position switch), a switchboard or other piece of electrical equipment. Cable glands also ensure that the →degree of protection is not reduced by the entry of the cable. They often consist of two threaded ring elements which are slipped over the cable and which clamp the cable by means of a compression ferrule when they are screwed together (compression glands). Cable glands may be of insulating material (e.g. PVC) or metal and do not necesarily provide additional insulation between the conductor and the partition, wall or cover through which it passes (also refer to →Bushing). Cable glands may or may not provide a means of clamping cable armouring to provide an earth connection.

Cable tray

A shelf or "ladder rack" used to support power cables. It is usually attached to a wall or suspended from the ceiling.

Cable trunking

A manufactured, →raceway, usually of insulating material intended for the holding and protection of cables. It is normally of rectangular cross-section and has one side which is removable or hinged. Also see →Cable trunking, →Conduit).

Cam switch

A →control switch which employs the rotation of a cam shaft to operate a number of →switching elements simultaneously.

Cam-operated switch (cam switch)

A (usually) hand-operated rotary-action switching device with a number of switching elements (containing →NO or NC contacts) for →main circuits and/or →auxiliary circuits. The moving contacts of the switching elements are operated by means of a cam shaft which is rotated to pre-defined latching positions. The shapes of the cam disks on the shaft and the latch positions (i.e. the "switching program"), as well as the number of switching elements may be specified. Cam switches are generally used for manual switching sequences, stepped starters, crane controllers, etc.

Capacitive proximity switch (→BERO)

A →proximity switch which produces an electric field from its →sensing face (sensing zone). It operates when the steady-state condition of this field is disturbed, e.g. by the introduction of conducting or non-conducting →target (change in dielectric constant) into the sensing zone. The steady-state condition, and therefor the sensitivity of the capacitive proximity switch to various target materials can usually be set up by means of an adjustment potentiometer on the switch body. Capacitive proximity switches employ completely encapsulated semiconductor (i.e. electronic or solid state) switching elements and are therefor not subject to mechanical wear. They are further characterized by virtually unlimited →service life, high →operating frequency, resistance to environmental influences and insensitivity to vibrations.

Capacitor bank

Power factor correction units (→p.f.c. unit) are designed to compensate inductive reactive power in an installation by adding the correct amount of capacitive reactive load to the circuit. Particularly in the case of →centralized power factor correction in which several loads are compensated, the amount of capacitive load to be added (or removed) varies as sections of the installation are switched on and off. To facilitate the selective switching of parts of the total installed capacitive load, it is split into a number of groups or banks. Each bank contains one or more three-phase →power capacitors connected in parallel which are switched in or out of circuit by a common command signal from the →power factor controller. It is common for the individual capacitors of a capacitor bank to be equipped each with its own →capacitor contactor. These are switched on in quick succession following the command from the power factor controller so that the sudden current surges may be reduced in amplitude and the →electrical service life of the switchgear extended. The kvar ratings of the banks relative to one another, and particularly the rating of the smallest bank, determine the degree to which the power factor can be held constant at varying load.

Capacitor contactor

This term is used to refer to →contactors which are particularly suitable for the switching of capacitive loads. The 3TK capacitor contactor, for example, incorporates precharge resistors which are switched into the supply current path by means of →early-make contacts prior to the closing of the →main contacts. In this way, the →electrical service life of the contactor is increased, and the need to overdimension a normal contactor for capacitor switching is avoided.

Cartridge fuse link

See → Fuse link

Cascade connection

The series connection of two (or more) →circuit-breakers with similar → opening times in order that their combination will have a higher short-circuit →breaking capacity than that of the single circuit-breaker closest to the point of fault. In the case of low level short-circuit and overload currents the downstream circuit-breaker opens alone to interrupt the current flow. In the event of a higher short-circuit current, e.g. one which exceeds the breaking capacity of this unit, both circuit-breakers trip simultaneously and the series connection of the two switching arcs has an additional limiting effect on the current so that one may speak of an increased breaking capacity. Naturally, the upstream circuit-breaker must have a sufficiently high breaking capacity to interrupt the short-circuit current in the event of a fault occurring between the two units (see → Back-up protection)

Cast-resin current transformer

The cores and the windings of cast-resin →current transformers are encapsulated in synthetic resin compound. This renders them corrosion-proof and suitable for use in tropical climates. The high dielectric strength and high resistance to tracking of the resin permits favourable c.t. dimensions. Cast-resin current transformers are shock-proof and impact-resistant.

Cast-resin transformer

See →GEAFOL

Castell interlock

See → Mechanical interlock

CBR

This abbreviation is used to refer to circuit breakers which provide →residual-current protection.

c.d.f.

See → Cyclic duration factor

CEE

International commission which publishes "Regulations for the assessment of electrotechnical equipment" (**C**ommission de l'**é**quipement **é**lectrique). The commission has undertaken the conversion of national specifications and regulations into editions which may be applied internationally. These editions are intended to assist manufacturers of electrotechnical equipment in the world-wide marketing of their products.

CENELEC

European committee for electrotechnical standardization (**C**omité **E**uropéen de **N**ormalisation **É**lectrotechnique). CENELEC has the task of developing international standards from the existing national specifications and regulations of its member countries on the basis of the IEC publications (→ IEC) for the benefit of the European electrical industry.

Central processing unit (CPU)

The main module or unit of a → programmable controller This module normally contains the microprocessor, the program and the main memory of the controller. From the CPU the complete system comprising inputs, outputs, indicators, etc. is monitored, controlled and supplied with the necessary data.

Centralized power factor correction

Centralized → power factor correction of a plant or distribution system is usually achieved by means of →automatic power factor correction units, which switch power capacitors into or from the load circuit as dictated by the actual load conditions. They are usually directly associated with, and installed centrally at, a particular switchboard, distribution board or incoming supply.

CFS

Combination **f**use-**s**witch unit.
See → Fuse combination unit.

Change-over auxiliary contact (Abbreviation: CO)

A contact included in an →auxiliary circuit and mechanically operated by a switching device. It has one moving contact (common terminal) and two separate fixed contacts. Depending on the →switching state of the switching device (either opened or closed), the moving contact closes the auxiliary circuit between itself and one of the fixed contacts, i.e. it has a separate closed position for the closed and open state respectively of the switching device.

Choke

This term is used to refer to an inductive reactance (e.g. several turns of wire around a core of magnetic material) which serves as a blocking device or filter. The impedance of a coil is directly proportional to the frequency of alternating current passing through it ($X_L = \omega L = 2\pi f_L$). Also refer to → discharge choke, → choked capacitor and → audio-frequency blocking circuit

Choked capacitor

If → power factor correction capacitors are installed in a network which incorporates a significant number of loads which produce → harmonics, then measures must be taken to avoid → resonance effects at the → harmonic frequencies. Appropriately dimensioned → chokes or inductors (reactor coils) may be connected in series with the capacitors to form series oscillating circuits of which the natural (or → resonance frequency) is lower than the frequency of the first expected harmonic (also known as a "detuned resonance circuit"). In this way, the power factor correction unit as a whole will appear to be inductive at all possible harmonic frequencies and resonance is no longer possible. Power factor correction capacitors with series-connected chokes are referred to as choked capacitors.

Chopping current

The value of the current during a breaking operation at which the → switching arc is extinguished (chopped) prior to the natural current zero (only relevant in vacuum switchgear). Also refer to → Chopping effect

Chopping effect (current chopping)

Abrupt ending of the current flow (including arc quenching) prior to natural (power-frequency) current zero during operational switching off of alternating current. This effect or phenomenon is found particularly in vacuum switching technology (see → Vacuum circuit-breaker) in which the absence of ionized air does not allow the current to flow via a switching arc until the natural current zero is reached. This abrupt ending or "chopping" of the current can give rise to extremely high voltage spikes in inductive load circuits. The value of the current at the instant prior to chopping is known as the chopping current I_{ab} and may be 2…5 A. These values are typical for all switching techniques which produce a chopping effect (including SF_6 switchgear technology).

Circuit diagram

A circuit diagram shows the electrical connections between the various parts of a network, an installation, a plant, a group of devices or parts of an electrical device and indicates the influence of one part on the other. The circuit diagram (or connection diagram) is used for the design and configuration of a system, to indicate the external line connections and during trouble-shooting or fault-finding.

Circuit-breaker

A mechanical switching device incorporating a → latching mechanism, capable of → making, carrying and → breaking currents under normal circuit conditions and also of making, carrying for a specified time and breaking currents under specified abnormal circuit conditions such as those of → short-circuit

Circuit-breakers can normally be classified in terms of their → current-limiting capabilities, their application (e.g. motor protection) and their construction. Circuit-breakers may be of the fixed-mounted, → plug-in or → withdrawable types. More detailed descriptions of the various versions are found elsewhere in this section. Examples include the:
→ vacuum circuit-breaker, → air circuit-breaker (ACB), → insulated case circuit-breaker (ICCB), → moulded-case circuit-breaker (MCCB), → miniature circuit-breaker (MCB), etc.

Note:
Circuit-breakers are normally not designed for a particularly high number of switching operations. Particular models are, however, also available for frequent switching operations → circuit-breakers for motor protection.

Circuit-breaker for distribution protection

These are → circuit-breakers specially designed for the protection of cables and insulated wires to subdistribution systems (also known as "distribution circuit-breakers" or "circuit-breakers for system protection"). They usually incorporate a non-adjustable → current-dependent overload release set to the rated current (preferred values) of the cable or wire to be protected. The → instantaneous short-circuit release is often of the adjustable type, since the fault level or prospective short-circuit current of the point at which the circuit-breaker is to be installed may vary considerably from one application to the next. Alternatively it is fixed; typically at $10 \times I_n$.

Circuit-breaker for mesh-connected systems

A three-pole power → circuit-breaker combined with a → network master relay (or reverse power relay) and a special → shunt release (shunt release with energy-storage capacitor unit, or "fc-release") which operates positively at between 10 and 110% of the rated operational voltage. The network master relay monitors the direction of power flow at its location in the installation. If the power flows from the transformer to the load (forward power), the relay does not respond. If the direction of power flow is reversed owing to a fault in the transformer or in the high voltage supply network, i.e. if the power flows towards the transformer (reverse power) - a possible fault condition in a meshed network with several infeeds - the relay operates and trips the circuit-breaker instantaneously via the shunt release and capacitor unit. These circuit-breakers are also referred to as "network circuit-breakers".

Circuit-breaker for motor protection

→ Circuit-breakers for the protection of motors against overload and short-circuit currents. They incorporate both current-dependent → overload releases and → instantaneous short-circuit release (or trip) mechanisms. The overload releases have a → tripping characteristic specifically designed in terms of the thermal behaviour of a motor and its windings, and are usually adjustable. The instantaneous short-circuit release must be insensitive to the initial magnetizing → inrush current experienced upon motor switch-on, and are typically set to $12 \times I_n$. These units are also known as "motor protection circuit-breakers".

Circuit-breaker for starter assemblies

In a contactor/overload relay combination (typical motor starter circuit), the overload relay is responsible for the overload protection of the motor. Short-circuit protection for the starter and connecting cables can be provided by a special type of → circuit-breaker for starter combinations. These circuit-breakers are equipped with a pre-set or adjustable → instantaneous short-circuit release only (→ n-release).

Circuit-breaker terminal

a) This term can be used to mean the point at which the incoming (circuit-breaker line terminals) or outgoing (circuit-breaker load terminals) wires are connected to the circuit-breaker. They may, for example, be of the → box terminal or → stub terminal types.
b) The term also refers to a special type of rail-mounting terminal block which incorporates a circuit-breaker for auxiliary circuit protection.

Circuit-breaker with mechanical lock-out

To prevent the inadvertent reclosing of a → circuit-breaker e.g. after it has tripped and before the fault has been repaired, a mechanical lock-out device may be fitted. The lock-out must prevent the contacts from closing, even if a closing command is given. Before the circuit-breaker can be reclosed, the lock-out must first be reset.

Clapper-type armature (e.g. in a contactor

See → Armature

Class of insulating material

Insulating parts and winding insulation are divided into insulation classes according to the materials used. A specific maximum temperature and/or → temperature-rise limit is allocated to each insulation class in the product standards. For example, in IEC 947-4-1 (→ contactors and → motor starters) the following temperature-rise limits are specified for insulated coils in air (→ ambient temperature -5 to $+40$ °C).

Class of insulating material	Temperature-rise limit (K)
A	85
E	100
B	110
F	135
H	160

CLASS (trip class)

The trip class of a → current-dependent overload relay (including thermal and electronic overload relays and releases) specifies the time limits of operation under given load conditions starting from the cold state. The trip class number (e.g. class 10, 20, 30) coincides with the maximum permissible → tripping time in seconds if the relay is loaded from the cold state with a balanced three-pole current of 7.2 times the current setting (IEC 947-4-1). The tripping classes 20 and 30 are used e.g. to protect motors used under → heavy-duty starting conditions.

Clearance (in air)

The shortest distance between any two conductive parts along a string stretched the shortest way between them. The clearance is a decisive factor in determining the insulation level of a → switching device or → switchgear assembly. The clearance is also known as the "air gap".

Clearance between open contacts (contact gap)

The total → clearance between the → contacts (or any conductive parts connected to them) of a pole of a mechanical switching device in the open position.

Clearance between poles

The → clearance between any conductive parts of adjacent poles in a switching device.

Clearance to earth

The → clearance between any conductive parts and parts which are earthed or intended to be earthed.

Climates

In terms of technical norms and standards, the term "climate" is used to describe the physical and chemical states of the atmosphere, including typical local weather factors such as temperature and humidity.

Closed position

The closed position of (the → contacts of) a mechanical switching device is the position in which the predetermined continuity of the → main circuit of the device is secured.

Closed transition star-delta starter

A special → star-delta starter by means of which the motor is delta-connected to the network via a transition → contactor and resistors after the run-up in star. It is switched over to the normal delta step without disconnection from the supply after approximately 50 ms. Thus, no switching interval during which the motor is disconnected from the supply occurs, and the high transient change-over current peaks can be avoided. Closed transition starting is used if the counter-torque from the driven machine is such that an appreciable drop in motor speed would occur during normal switch-over from the star to the delta configuration.

Glossary

Closing delay

See → Make time

Closing duration

The time taken for a → moving contact to move from its position of rest in the open state to the point of first contact (time of contact travel)

Closing operation

The closing operation of a mechanical switching device is the → operation by which the device is brought from the → open position to the → closed position by means of its → operating mechanism

Closing time

The period of time between the initiation of the → closing operation of a switching device and the instant when the contacts touch in all poles. Note, that the "total closing time" includes the → bounce duration - see Fig. 9.14 on page 637.

CO

See → Change-over auxiliary contact

Coercive force

The attracting force which remains between the → poles of an → electromagnet system (→ solenoid system) after de-energizing of its coil as a result of the → residual magnetism (or remanence) if no appropriate counter-measures are taken. The value of this force is determined to a large extent by the material and design of the magnet poles. It is used in → remanence contactors.

Coil voltage tolerance

See → Operating voltage tolerance and → Limits of operation.

Coincidence factor

See → Demand factor

Colour coding

The colour coding for → command and signalling devices of industrial equipment is internationally standardized in accordance with DIN VDE 0199 and 0113 as well as IEC 73. Red - yellow - green - white and blue are regarded as the five basic colours. The colours red, yellow and green have similar meanings as they have in traffic lights: Red – Danger/Stop, Yellow – Caution/Attention, Green – Go, Start. In terms of the standards, only the actual operating states of machines, as opposed to the switching states of contacts and circuits, may be displayed on operator panels. The coil terminals of Siemens → contactors and → contactor relays are colour coded in terms of IEC 309-1 to facilitate the identification of the → rated control circuit voltage.

Combination fuse-switch (CFS)

See → Fuse-combination unit

Command device

This term is used to refer to a → control switch used in the → control circuit of one or more → switching devices in associated → auxiliary or main circuits. It is used to produce the → command signal(s) to operate a machine or other piece of electrical equipment. Examples of command devices include → pushbuttons, toggle switches, key switches, etc.

Command signal

A command signal (electrical) may be a pulse or the continuous application of → control supply voltage to the input terminals of the → operating mechanism in a → switching device or piece of electrical equipment (e.g. to the coil terminals of a → contactor). It may also be the interruption of the control voltage, e.g. the "off" command. The command signal should be of a duration longer than the → minimum command time.

Comparative tracking index (CTI)

A measure of the resistance to tracking of a particular insulating material. A CTI of 600, for example, means that a test sample of the material can withstand a maximum applied voltage of 600 V without tracking if 50 drops of a test solution are applied to its surface. Each → material group is assigned a range of CTI values.

Compartment

A switchboard or switchboard → cubicle may comprise a number of compartments or enclosed structural units, each containing a → switchgear assembly (e.g. motor → starter, distribution circuit or → withdrawable unit). Each compartment usually has its own door to provide access for maintenance, repair or adjustment purposes. By means of a compartmentalized switchboard design, faults may be contained within a predefined space and prevented from spreading to other parts.

Compensation of reactive power

See → Power factor correction

Conditional short-circuit current I_q

This value is determined and stated by the manufacturer of e.g. a → switching device (see → rated value). It indicates the value of the prospective short-circuit current that the piece of equipment, protected by a specified → short-circuit protection device (→ back-up protection), can withstand satisfactorily under the test conditions specified by the applicable product standard. For example, a manufacturer may choose the value of I_q at which type 2 coordination applies for a particular → motor starter with short-circuit protection by → HRC fuses (see → Coordination with short-circuit protection devices)

Conducting path

See → Current path

Conductive part

A part which is capable of carrying current even if this is not its intended operational purpose.

Conductor

General:
The term conductor can be used to apply to any piece of conductive material used to transfer electrical energy. In the context of power distribution and control, it is understood to mean the metal part(s) of an electric wire, cable, line or busbar. Conductors may consist of a single wire, of several stranded or braided wires or of a solid section of metal bar.
Copper is the preferred conductor material as it has a high electrical conductivity (i.e. low resistance). In the case of larger conductor cross-sections, aluminium is frequently used because of its lower weight and the comparative ease with which it is bent, drilled, etc.

Conductor designations:
Phase conductors, line conductors, or outer conductors of a three-phase system (L1, L2, L3).
These are conductors which connect power sources to loads, but which do not emanate from the center or star point.
Neutral conductor (N)
A conductor which is connected to the center or star point and conducts electrical energy.
Protective (earth) conductor (PE)
A conductor which is used with certain → protective measures to interconnect → conductive parts which do not belong directly to the operating circuit or to connect them to other non-associated conductive parts as well as to → earth
PEN conductor
A conductor which combines the functions of a neutral and a protective conductor.

Auxiliary conductor
A wire or e.g. a multi-core cables used in an → auxiliary circuit.

Conductor cross-section

The total cross-sectional area of round and flat → conductors expressed in mm^2 (or → AWG). It determines the impedance of the conductor and is therefor a decisive factor in determining the thermal and short-circuit characteristics and ratings of the conductor. In the case of direct current, the current flow density through a conductor is uniformly distributed across its cross-section. Alternating current, on the other hand is not uniformly distributed but with increase in frequency tends to flow more densely towards the outer perimeter of the conductor. At extremely high frequency, virtually no current flows through the centre of the conductor. This phenomenon is known as the skin effect.

Conduit

A → raceway or pipe-like enclosure of circular cross-section primarily for the holding and protection of wires and cables in an electrical installation. The cables may be drawn in or replaced, but cannot be inserted laterally (also refer to → Cable ducting and → Cable trunking). Several different types of conduit are available, depending mainly on the material from which it is manufactured. The conduit systems usually include a host of associated fittings, couplings, connectors, supports, etc. The dimensions (e.g. diameter unit lengths) are standardized. The use of a particular type and/or size may be restricted to specific applications and environments and is governed by the relevant installation regulations (e.g. National Electrical Code). Examples of conduit types include:

- rigid metal conduit,
- rigid non-metallic conduit (e.g. PVC conduit),
- intermediate metal conduit (a thinner-walled rigid metal conduit suitable for a wide range of application, also known as IMC,),
- flexible metal conduit.

Connected load

See → Effective installed load

Consumer installation

A term for all items of electrical equipment connected downstream from a → local service box, or in cases where a local service box is not required, downstream from the output terminals of the last distribution level before the consumer equipment.

Contact bounce

This phenomena occurs during the closing of → contacts (in some cases during opening too, but this is rarely of interest). It is caused by the elasticity of the spring mounted → moving contact elements, the hardness off the → contact material and the dynamics of the contact system as a whole. As the moving contacts strike their fixed counterparts, they bounce back and restrike several times at a high frequency before coming to rest. Depending on the nature and magnitude of the → prospective making current, which in turn depends on the load being switched, each separation of the contacts draws a → switching arc between the surfaces of the fixed and moving contacts. These arcs can cause severe → contact erosion, and even → contact welding if the making current exceeds the specified values of the switching device.

Contact bridge

This type of moving → contact element forms a link between two fixed contact elements to close the electrical circuit. In the → closed position, each end of the contact bridge is pressed onto its respective fixed contact. The current flows from one fixed contact element, through the contact bridge to the other fixed contact element (i.e. via two sets of contacts). During the → opening operation, the bridge is lifted from the fixed contacts, and the current path is interrupted at each end (→ double break contacts). To increase the → contact reliability, the contact bridge of → auxiliary contacts may consist of two independently supported contact elements, each capable of the same full → rated current carrying capacity and → switching capacity. This configuration is known as a "double moving contact bridge" and is employed in virtually all Siemens low-voltage auxiliary switches and contacts.

Contact element

The parts of a → switching device, fixed and movable, conducting and insulating which are directly associated with the closing and opening of one single conducting path of a circuit (excluding the → operating mechanism).
Each contact element consists of one or more "moving" and "fixed" parts to which the actual contacts (→ contact tips) themselves are bonded in such a way, that they touch when the contact element closes. In the closed position, the contact elements (including supports, springs, links, etc.) transfer the force from the operating mechanism to the contacts to produce the necessary contact pressure between their surfaces.

Note:
The word "contact" (e.g. as in "spare contacts") often includes the fixed or moving contact "arm" (contact carrier), since the contact material is usually permanently bonded to it and cannot be replaced separately (exception: some types of → arcing contacts)

Contact erosion

Loss of → contact material caused during the switching of electrical current. The thickness of the contact material is reduced through evaporation and splashing of the material under the intense heating influence of the → switching arc. Furthermore, contact erosion is caused by → abrasion between the contact surfaces.
Important factors which influence contact erosion are the type and nature of the contact material, the current density, the phase angle of the current during contact separation, the velocity with which the contacts move together and apart, as well as the heat dissipation and thermal capacity of the → contact elements.

Contact failure

Contact failure has been defined as the condition or state of → auxiliary contacts, which leads to malfunction of a → control circuit owing to excessive → contact resistance and the resulting → voltage drop when the contacts close. It is often caused by the presence of contamination on the → contact surfaces and the switching of low voltages and/or currents (see → Contact reliability).

Contact failure rate H_F

The contact failure rate H_F is defined as the number of → contact failures occurring within a specified number of switching operations. Hereby, a large quantity of → current paths are tested so that the published results represent statistic mean values. The average failure rate H_F can be calculated by dividing the sum of all failures F by the number of tested current paths N multiplied by the number of mechanical → make-break operations M:

$$H_F = \frac{F}{N \cdot M}.$$

Contact gap

The distance between open (or separated) → contacts. See → Clearance between open contacts.

Contact life (contact service life)

The number of → make-break operations a set of → contacts can perform under specified electrical and/or mechanical load conditions. For example, the contact life is expired when the volume or thickness of the → contact material of the → contact tips is reduced by → contact erosion to such an extent that the contact tip carrier material is exposed and the risk of → contact welding becomes high. Contacts are also subject to mechanical wear (see → wiping contacts) so that there is a limit to the contact life even if the contact erosion due to → switching arcs is negligible. If, for example, the load conditions for a → contactor are known, the expected service life may be predicted using service life diagrams or the corresponding computer programs (→ ELLE). The validity of these predictions are, of course, strongly subject to the switching of short- circuit or unexpected overload currents (e.g. in the case of circuit-breakers). Also refer to → Contact surface.

Contact material

To a large extent, the properties of the contact material (material of the contact tips bonded to the moving and fixed → contact elements) determine the quality, reliability and → service life of the switching device. The following characteristics are required for contact material:

a) high electrical and thermal conductivity,
b) low tendency for the formation of contaminating surface layers,
c) low burning tendency,
d) high melting point and resistance to welding,
e) correct hardness, strength and rigidity (balance between → contact pressure, closing speed, bounce, etc.
f) ecologically friendly (i.e. non-toxic in manufacture and disposal)
g) easy to work and favourably priced

There is, indeed, no ideal material which satisfies all these different requirements, even approximately! Modern materials for low-voltage switchgear include silver, silver alloys and a number of sintered materials. Although it has a high resistance to welding and has other favourable properties, silver-cadmium oxide has become unpopular over the last few years owing to its toxic nature. This contact material is no longer used in the manufacture of Siemens contactors.

Contact movement delay

See → Movement delay

Contact piece

See → Contact(s)

Contact potential

See → Touch voltage

Contact pressure

The force (per unit area) by which two conductive elements (e.g. → contacts) are pressed together in order that electric current will pass from the one to the other. The higher the contact pressure, the higher the conductivity between the elements will be, since more non-conductive foreign layers (oxidation) on the contact surfaces will be penetrated. The → contact resistance will be lower, or more accurately, the contact voltage-drop will be lower and heat losses will be reduced. The contact pressure directly affects the → making capacity, the → heat losses and the → electrical service life of a → switching device.

Contact reliability

Contact reliability of e.g. an → auxiliary contact is defined as the probability of a contact condition or state, which leads to malfunction of a → control circuit (see → Contact failure). Thereby, the contact condition is described by the permissible → contact resistance at a specified current value. It is strongly influenced by the presence of contamination or oxide films between the → contact surfaces. A number of techniques are employed to enhance the contact reliability of mechanical → switching devices. These include:
- high → contact pressure,
- specific shapes of the → contacts (e.g. knife edge)
- wiping or self-cleaning movement (→ wiping contacts)
- single instead of → double-break contacts (single contact resistance)
- parallel connection of current paths (e.g. → double moving contact bridges)
- encapsulation of contacts in a sealed → switching chamber.

Contact resistance

The electrical resistance across a set of closed → contacts. It is dependent on a number of different physical factors. It is influenced, for example, by the shape of the contacts and the nature of the → contact surfaces, the → contact material, the → contact pressure, as well as by oxide layers and contamination between the contact surfaces. The contact resistance is not constant, but changes significantly with each switching operation. In practice, the contact resistance is determined by measuring the voltage drop across the closed contacts at realistic values of voltage and current.

Note:
In the case of power switching devices (industrial switchgear), the measurement of contact resistance by means of ohm-meters, multimeters, and the like (high impedance, low voltage) should be avoided as the results tend to be misleading and usually meaningless. Also refer to → Contact reliability.

Contact surface

The areas on the fixed and moving → contacts which are brought together during the → closing operation are known as the contact surfaces. Depending on the → contact material, the contact surfaces of → main contacts may become rough, pitted and/or discoloured during switching operations. Generally speaking, modern contacts as used in low-voltage a.c. switching devices should not be cleaned, filed or dressed. It is advisable to check the maintenance instructions or to consult the supplier before contact surfaces are cleaned or contact sets are replaced. The condition of the contact surfaces and the volume of → contact material remaining on the → contact elements (main power contacts) after a period of operation can give a good indication of the remaining → contact life.

The contact surfaces of → auxiliary contacts and other contacts designed to switch low current and voltage may have knife edges or point shapes to puncture possible contamination layers each time the contacts close, thereby enhancing the → contact reliability. If these contacts are used at higher voltages and/or currents, these edges may be burnt away by the → switching arc, which itself has a cleaning effect on the contact surfaces. By a suitable mechanical arrangement, contact surfaces may be wiped against each other during each → closing and opening operation. This self-cleaning effect is also designed to increase the → contact reliability, but may lead to unwanted → contact erosion.

Contact tip

See → Contact(s)

Contact wear

See → Contact erosion

Contact welding

Solid bonding between → contacts caused by the solidification of molten → contact material between closed contacts. The melting of contact material surfaces is caused by the intense heat generated through the presence of a → switching arc between opening contacts. Contact welding may be caused by: - impermissibly high currents (e.g. → short-circuit current, high → starting current) which cause such high electrodynamic forces that the contacts fly apart and reclose before the switching or tripping mechanism of the switching device has operated to keep the contacts apart, - insufficient → contact pressure, - contact chatter (opening and closing commands in rapid and uncontrolled succession).

Contact(s)

a) The condition which is the result of two conductive elements (e.g. contacts, copper wires) touching each other (or the action which causes it), so that electrical current may flow from the one to the other. Contact usually refers

an intentional action (as opposed to accidental touching) and is generally understood to be of low impedance (ref. "good contact").
b) The term is also used to refer to a → contact element, or more precisely to the mass of → contact material bonded to the contact elements in a switching device (also called "contact tip", "contact piece").
c) A "contact" (e.g. as in an → auxiliary contact) is also used to mean a complete set of → contact elements, contacts and associated terminals. Also refer to → Auxiliary contact block, → Main contacts

Contactor

A switching device with only one position of rest, i.e. only one position when not energized (in general, a contactor does not have a mechanical → latch mechanism). It is operated otherwise than by hand, and is capable of making, carrying and breaking currents under normal circuit conditions, including operational overload conditions. Contactors are usually designed for a high → operating frequency. A distinction is made between contactors for switching motors or other load circuits and → contactor relays for control purposes:
Contactors for the switching of motors and three-phase loads usually have three main poles (one set of normally open contacts per phase) and are selected (rated) in terms of the load current or power, the → utilization category and the → operational duty. Single, double and four pole versions, as well as contactors specially designed for the switching of direct current, as well as a number of special purpose contactors are also common.
Contactor relays are predominantly used in auxiliary circuits

Comment:
Contactors are capable of making and breaking short-circuit currents provided that they are selected (rated) accordingly. In general, they are not suitable for isolation (disconnection) purposes. A contactor of which the main contacts are closed in the position of rest is termed "rupteur" in French. In English there is no equivalent short term, and these contactors are referred to as e.g. contactors with normally closed main contacts.

Contactor relay

A → contactor which is designed for switching operations in → auxiliary circuits (e.g. controlling, signalling, → interlocking). It usually contains a number of → auxiliary contacts. Typically, these contacts are rated at the AC-15 → utilization category (previously AC-11) for the switching of electromagnetic loads and are designed for high → contact reliability even at low voltages and currents. Contactor relays are subject to the same generic specifications as contactors, and usually have a rated → operational current I_e/AC-1 in excess of 10 A. Contactor relays are also known as auxiliary contactors or power relays.

Comment: The use of relatively inexpensive cradle-type relays (often plug-in devices with transparent plastic covers) as a direct replacement of contactor relays, should be considered carefully with due regard to the application and the possible consequences of contact failure, welding, etc.

Contactor safety combinations

If → contactor relays are used in circuits intended to ensure the safety of operating personnel, e.g. in circuits of emergency-stop arrangements (→ safety circuits), then in terms of DIN VDE 0113 part 1/EN 60204 part 1 Paragraph 5.7.1 redundancy (or back up) through the use of more than one contactor relay per function must be provided. The → auxiliary contacts (more specifically, the → control contacts) of the contactor relays must be so interconnected as to ensure that in the case of a fault in any one of them (e.g. failure to operate or → contact welding), the safety circuit must remain operational and reliable (at least until the start of the next → operating cycle, at which time it should indicate the failure by not operating at all, i.e. remaining in a safe state). The contactor relays must fulfill the conditions for → positively driven contacts. These circuits are also known as "back-up combination" units. Also refer to → Fail-safe operation.

Continuous (operation) duty (S1)

For a general explanation of the various duty types of motors, see → operational duty. In terms of switching devices, continuous duty is also known as → uninterrupted duty and is a rated duty type in terms of the relevant specifications. Please refer to → Uninterrupted duty for more information.

Continuous duty with intermittent periodic loading (S6)

In the case of motors (DIN VDE 0530), this type of duty is understood to mean a sequence of identical → load cycles, each consisting of a time period at constant specified load (rated output power) and a time period at → no-load operation (idle). There are no intervals during which the current is zero. The cycle time is generally too short for thermal equilibrium to be reached within a single load cycle.
A further illustrative example: The primary side of a transformer is continuously connected to the supply, while the secondary is loaded cyclically with the specified permissible load. The cycle time is too short for the permissible → temperature rise limit to be exceeded, and the subsequent period during which the secondary is not loaded, is not long enough for cooling down to thermal equilibrium (also refer to → Continuous duty with *short-time loading*).
Continuous duty with intermittent periodic loading is not a rated duty in terms of the specifications for switching devices (see → Intermittent periodic duty).

Continuous duty with short-time loading

This type of duty is understood to mean a sequence of identical → load cycles, each consisting of a time period at high load (e.g. operational overload) and a time period at no-load operation (idle). There are no intervals during which the current is zero. The → on-load period is too short for the permissible temperature rise to be exceeded, while the → no-load period is long enough for the temperature to drop to thermal equilibrium.

A further illustrative example: The primary side of a transformer is continuously connected to the supply, while the secondary is loaded cyclically with the specified permissible load. The cycle time is too short for the permissible → temperature rise limit to be exceeded, and the subsequent period during which the secondary is not loaded, is long enough for cooling down to thermal equilibrium (also refer to → continuous duty with *intermittent periodic loading*).

Continuous duty with short time loading is not defined as a rated duty in terms of the DIN VDE specifications for switching devices or motors (see → Intermittent periodic duty).

Control circuit

This term encompasses all the conductive parts (other than the → main circuit) of a switching device which are included in a circuit used for the closing operation or opening operation, or both, of the device. It is part of the → auxiliary circuit. Typical uses include:
- signal shaping and signal input,
- signal processing, including transformation, storage, interlocking and amplification,
- signal output and the control of actuators and signal transmitters.

Control circuit voltage

This is, for example, the voltage across the opened → NO auxiliary contacts in the → control circuit of a → switching device. It may differ from the → control supply voltage owing to the presence of transformers, rectifiers, resistors, etc.

Control contact

A contact included in a → control circuit of a mechanical → switching device and mechanically operated by this device. Also see → Auxiliary contact, and → Auxiliary switch.

Note:
The control contact of e.g. a → contactor is an integral part of the contactor or becomes this once it has been attached by means of the → auxiliary contact block in which it is contained.

Control relay

See → Contactor relay

Control supply voltage

The voltage at which generally → auxiliary circuits, and more precisely, the → control circuits of → switching devices are operated (also known as the "auxiliary supply voltage" or "operating voltage") or to which they must be connected for correct operation. For control circuits which are fed via a → control transformer, AC 230 V, 50 Hz is the preferred value for the control supply voltage. It is quoted as the → rated value U_s for items of switchgear and other electrical equipment. The correct operating conditions of an item of switchgear are based on a value of control supply voltage not less than 85% of its rated value, with the highest value of control circuit current flowing, and not more than 110% of its rated value (IEC 947-1)

Note:
A distinction is made between the → control circuit voltage e.g. as may be measured directly across the terminals of a contactor coil in the control circuit of a switching device or piece of equipment, and the control supply voltage as applied to the input terminals of the control circuit as a whole. The two voltages may differ owing to the presence of transformers, rectifiers, resistors, etc.

Control switch

a) A mechanical (usually manually operated) → switching device used in → main circuits for the direct control of items of equipment. Examples of control switches for main circuits include → motor switches, operating sequence step switches, on/off switches for heating equipment, etc.

b) A mechanical (usually manually operated) → switching device used in → control circuits and → auxiliary circuits and which serves the purpose of controlling the operation of → switchgear or → controlgear, including signalling, → electrical interlocking, etc. Examples of these control switches include pushbuttons, twist knobs, key switches, etc.

Control transformer

A transformer with electrically separated primary and secondary windings for the power supply of → auxiliary circuits. A control transformer is characterized by a low → voltage drop under inductive load. In terms of commonly applied regulations for electrical equipment in industrial machines, the use of a control transformer is recommended in auxiliary circuits with more than 5 electromagnetic loads (e.g. → contactors).

Controlgear

A general term for → switching devices and their combination with associated control, measuring, protective and regulating equipment. It includes assemblies of these devices and equipment with the corresponding interconnections, accessories, enclosures and supporting structures. In principle, controlgear is intended for the control of all equipment which consumes electric energy.

Conventional enclosed thermal current I_{the}

The conventional *enclosed* thermal current (I_{the}) of a → switching device, (e.g. → circuit-breaker) is the maximum test current to be used for temperature-rise tests of the device when mounted in a specified → enclosure. The switching device must be capable of conducting this test current without the temperature rise limits given in the relevant specifications being exceeded. It must be at least equal to the maximum rated → operational current of the enclosed equipment in → eight- hour duty. It is not a → rated value and does not have to be marked on the switching device.

Glossary

Conventional free-air thermal current I_{th}

The conventional *free air* thermal current (I_{th}) of a → switching device, (e.g. → circuit-breaker) is the maximum test current to be used for temperature-rise tests of the unenclosed device in free air, as stated by the manufacturer. The switching device must be capable of conducting this test current without the temperature rise limits given in the relevant specifications being exceeded. It must be at least equal to the maximum rated → operational current of the unenclosed equipment in → eight-hour duty. It is not a → rated value and does not have to be marked on the switching device.

Note 1:
Free air means air in normal rooms, as far as possible free of draught and heat radiation.

Note 2:
An unenclosed switching device is understood be a device which is normally delivered from the factory without an enclosure, or a device with an integral enclosure which is, however, not normally intended to be the only protective cover for the device once it is installed.

Conveyor belt monitor

A → pilot switch used for the switching and monitoring of conveyor belts. A conveyor belt monitor issues a stop or warning command to the → controlgear if the belt slips, jams or breaks, or if its speed varies from the setpoint value by an amount greater than that specified.

Cooling time-constant T_K

See → Temperature-rise time-constant T_A

Coordination of insulation

The correct relationship between the insulation characteristics of electrical devices and equipment, the expected overvoltages and the characteristics of overvoltage protection devices on the one hand, and the expected → micro-environment as well as measures taken against → pollution on the other hand.

Coordination (of overcurrent protection equipment)

The selection and arrangement of two or more → overcurrent protection devices in series in such a way as to ensure → discrimination (in the overload range) and/or back-up protection under short-circuit conditions (see → Cascade connection).

Coordination with short-circuit protection devices

The manufacturer of a → switching device (or piece of electrical equipment) is obliged to indicate the type and/or the characteristics of the → short-circuit protection device to be used with or in the device or equipment, as well as the maximum → prospective short-circuit current for which the device (including the short-circuit protection device) is suitable at the stated operational voltage. A distinction is made between various types of coordination based principally on the degree of damage which may result during the interruption of a specified short-circuit current by the short-circuit protection device.
As an example, for → contactors and → motor starters, IEC 947-4-1 distinguishes between two types of coordination (to be verified in mandatory testing at the appropriate → prospective current "r" and the rated → conditional short-circuit current I_q):

Type 1 coordination specifies that, under short-circuit conditions, the contactor or starter may not cause danger to persons or neighbouring parts of the installation. After the short-circuit condition, the equipment *does not have to be suitable for further service* without repair or replacement of parts (including e.g. the → overload relay of a motor starter, the → contacts or the contactor itself, etc.).

Type 2 coordination specifies that, under short-circuit conditions, the contactor or starter may not cause danger to persons or neighbouring parts of the installation and *must be suitable for further service* after the short-circuit condition. No replacement of parts is permissible (e.g. the overload relay of a motor starter may not be damaged, and it must not be necessary to replace the → contacts of the contactor). → Contact welding may occur, but the contacts must be easily separable (e.g. by means of a screwdriver).

Note:
The use of short-circuit protection which is not in accordance with the recommendations made by the manufacturer, may invalidate the coordination.

Core-balance current transformer

This is a special form of → summation current transformer. In the → window-type, all the active phase → conductors including the neutral conductor (if present) pass through the same ring band core. In a fault-free installation, the geometric sum of the currents flowing to and from the load equals zero. In this case the flux induced in the core is balanced and no current flow is induced in the single secondary winding on the same core. If, however, a leakage current flows e.g. to → earth downstream from the unit, the sum of the primary currents will no longer be equal to zero (i.e. a → residual current exists). This differential current causes a magnetic flux to circulate within the transformer core which in turn induces a current in the secondary winding. This secondary current is used to operate a tripping device or a → residual current release

cos φ

The cosine of the angular phase displacement (φ) between the voltage and alternating current waveforms in a → conductor. In a rotating vectorial representation, the current can be shown to comprise one vector (or portion) which is perfectly in phase with the voltage waveform and one vector which is displaced by 90° (leading or lagging) from the voltage. The active power (kW) flowing through a conductor is only that portion produced by the current vector which is in phase with the voltage. The ratio of the active to the → apparent power therefor corresponds to the cosine of the angle between voltage and current and is also known as the → power factor. It is dependent on the nature of the load (resistive, inductive or capacitive) and may be altered (corrected) be means of → power factor correction.

Counter cells

Counter cells consist of a string of series-connected silicon diodes and are connected between a battery and the consumer load during charging phases and when the battery is fully charged. The voltage drop across this chain of diodes is independent of the load current. As the voltage of the battery drops during discharge, these counter cells are automatically bridged out (in one or two stages) so that the consumer voltage remains constant within a narrow tolerance band.

CPU

See → Central processing unit

Creep action

If the → operation of a → switching device (e.g. opening or closing of its → contacts) is dependent on the velocity of the → actuator or → actuating element, then it is said to possess creep-action. The term is commonly used in reference to the operation of → position switches. Also refer to → Snap action.

Creepage current

The current which flows across the surface of an insulating material between two live parts as a result of contamination (also refer to → creepage distance, and → pollution degree). This phenomenon is also referred to as → tracking.

Creepage distance

The shortest distance along the surface of insulating material between two → conductive parts (→ insulation group, → rated insulation voltage).
Note:
A joint between two insulating elements is regarded as part of the total insulator surface.

Critical breaking current

A value of the → breaking current which is lower than the rated short-circuit breaking capacity, but at which the energy of the → switching arc is significantly higher than it would be at the rated short-circuit breaking capacity.

Cross-section

See → Conductor cross-section

c.t.

See → Current transformer

Cubicle (switchboard cubicle)

→ Switchboards of the enclosed construction type normally comprise a single → cabinet or cubicle closed on all sides, or a number of cubicles bolted together side-by-side. Each cubicle is self-supporting (i.e. has its own chassis frame) and has its own door (or a number of door - each opening an individual → compartment). Typically, a set of → busbars is supported in such a way that it passes horizontally through the adjoining cubicles (sometimes in its own busbar compartment or chamber), and a group of up to four cubicles (known as a → section) normally constitutes a transport unit

Cubicle-type construction (of a switchboard)

An enclosed type of → switchgear assembly which is made up of one or more (several) → cubicles joined together side-by-side.

Current carrying capacity

The maximum current a → conductor or a → current path in a → switching device can carry under specified conditions without itself or its associated insulation (if applicable) becoming damaged. The current carrying capacity of a switching device is dependent on (among other factors) its physical size, the → enclosure and the ambient temperature. The current carrying capacity of a conductor is also known as its "ampacity".

Current limiting

In practice, the → prospective short-circuit current never flows, but is reduced by the circuit impedances encountered between the source of current and the point of fault. This phenomenon is generally known as current limiting. Factors contributing to the current-limiting effect include:
– the impedance of transformers in the short-circuit → current path,
– the impedance of the connecting wires, cables and → conductors in general, as a function of their → cross-sections, lengths, arrangements and the materials they are made from,
– voltage drops across → terminals and other points at which conductors are joined together, and
– the presence of → switching arcs in the current path.
Particularly → circuit-breakers and → fuses are known for their current-limiting effects, and can contribute significantly to reducing thermal and dynamic stresses during fault conditions. Also refer to → current-limiting circuit-breakers, → let-through current and → inherently short-circuit proof.

Current path (general)

The route or path the current (flow of free electrons) will take from one point to another through conductive material (e.g. wire, → busbar, electrical circuit) if a potential difference (voltage) exists between the two points. The current path may be influenced by the relative resistance of parallel-connected conductors, or by the presence of a magnetic field. It may be interrupted by the presence of an insulator (e.g. air gap or clearance of sufficient length).

Glossary

Current path (in a switching device)

The set of conductive parts in a → switching device associated exclusively with one electrically separated pole of the circuit to be switched. Distinctions are made between the main current path, the control current path and the auxiliary current path (for indication and interlocking purposes). The current path includes the line and load-side terminals, the conductive → contact elements, the → contacts themselves as well as any interconnecting conductors within the switching device (also refer to → Pole (of a switching device))

Current transformer

Current transformers convert larger currents to values which are easier to measure and/or use in monitoring or protection devices. Owing to the characteristics of the current transfer technique, they also protect measuring instruments from short-circuit currents and overvoltages. Current transformers are commonly referred to as "c.t.'s" and are classified in terms of their transformation ratio, rated → burden, → accuracy class, → rated overcurrent factor and purpose (i.e. measurement or protection). Current transformers for measurement purposes (connection to measuring instruments) generally have a rated secondary current of 1 Amp or 5 Amp, whereby the 1 Amp versions produce lower losses in connecting wires and are therefor more popular. Special forms of c.t.'s include → summation current transformers, and → core-balance current transformers.

Current-based discrimination

This term refers to the technique of discriminative protection (→ discrimination) which uses the different response times of series connected → switching devices (e.g. → fuses, → circuit-breakers) at various load currents. For example, series connected fuses tend to behave in a discriminative manner if their respective → melting-time characteristic curves are sufficiently far apart. In general, this is the case if the ratio between the rated currents of the successive fuses is greater than, or equal to 1:1.6. For example, conditions for discrimination are met when a 100 A and a 160 A fuse are connected in series

In the case of two series-connected circuit-breakers one must distinguish between discrimination in the → overload and the → short- circuit current ranges respectively. In the overload current range, current-based discrimination between the two breakers is ensured if the → tripping curves of their respective → overload releases are sufficiently far apart (also refer to → Take-over current).

In the short-circuit current range, in which the → instantaneous short- circuit releases of both circuit-breakers operate, current-based discrimination can only be achieved up to the short-circuit current value at which the → tripping current of the upstream breaker exceeds the peak → let-through current of the downstream unit (see → Discrimination limit). For comparison, also refer to → Time-based discrimination.

Current-dependent delayed overload relay (or release)

See → Overload relay or release

Current-limiting circuit-breaker

The term is used to refer to → circuit-breakers and → miniature circuit-breakers which have a pronounced → current-limiting effect on the short-circuit current upon opening. The current-limiting effect in circuit-breakers is usually achieved by means of one or both of the following techniques:
- by the rapid opening of the → contacts at the onset of a short-circuit (i.e. before the natural peak value of the first half wave is reached) and the introduction of a high impedance → switching arc into the → current path. In these circuit-breakers the contact opening is usually achieved by dynamic forces produced by the short-circuit current itself; the relatively long → mechanical delay of the → latch mechanism is saved. The high impedance arc is achieved by the highly developed arc-quenching techniques employed. The current is interrupted (including → arc quenching) before it passes through the natural zero. Also refer to → Zero-point quenching.
- by the value of the inherent impedance of the circuit-breaker. Also refer to → Inherently short-circuit proof.

Current-limiting circuit-breakers are normally → moulded case circuit-breakers, and are particularly used to reduce the damaging thermal and dynamic effects produced by the flow of strong short-circuit currents (the "survival" of their own moulded casings and construction may depend on this!). Owing to their instantaneous rapid operation under short-circuit conditions, they cannot provide any useful degree of → time-based discrimination between distribution levels (also refer to → Current-based discrimination). Circuit-breakers of the current-limiting type are also used in → cascade connection arrangements.

Current-time characteristic

See → Time-current characteristic

Cut-out box

An → enclosure designed for surface mounting and having swinging doors fitted directly to its walls (see → Cabinet).

Cycle duration

The period of the → duty cycle, alternatively the time taken to complete each repetitively recurring → load cycle (also refer to → Operational duty)

Cyclic duration factor (c.d.f.)

The ratio of the → on-load time (→ load duration) including → start-up time and → plug-braking time, to the total period of the → load cycle (→ duty cycle), usually expressed as a percentage. It is also known as the → on-load factor and is decisive in determining the thermal loading of a → switching device. Also see → Intermittent periodic duty.

Dahlander connection

The preferred winding configuration for pole-changing motors with one winding (each winding comprises two coil groups per phase) and two speeds of operation in the ratio 1:2. This connection has the advantage that the whole winding is used in both speeds and only six motor terminals are required.

Damping material

A material which has an influence on the → response characteristic of a → proximity switch.

Dark-ON function (of a → photoelectric proximity switch)

The dark-ON function means that the relevant output is switched ON (i.e. closed or conducting) when light is *not* detected by the → receiver. This corresponds to the → normally closed contact function (NC). The opposite is known as the → light-ON function.

Data bus

See → Bus

Data exchange

See → Data transfer

Data transfer

The transfer of electronic signals (i.e. data or information) between systems and system parts with communication capabilities. One can distinguish between two types of data transfer:

1. Serial data transfer.
The signals are transferred one after the other along a single line or twisted pair (e.g. → ASI, → PROFIBUS, SINEC L2 → bus).
Advantage: Only one line is required for covering long distances.

2. Parallel data transfer.
The signals are transferred in parallel at the same time along several lines (multi-core cables).
Advantage: Signals which are critical with respect to time can be transferred very rapidly.
The rate at which data is transferred in code elements per second, is known as the Baud rate.
Data transfer is understood to mean the flow of data in one direction. In most cases, data is transferred in both directions, and one speaks of "data exchange".

DC magnet system

See → Magnet system

DC or d.c.

See → Direct current

Dead front

A → switching device, switchboard or piece of electrical equipment is said to have a "dead front" if no live parts are exposed to a person on the operating side.

Decoupling

The various functional units of an electrical or electronic system may be rendered independent of one another by means of a number of decoupling techniques. Electronic signals may be decoupled by means of opto-electronic coupling elements (opto-couplers) or diodes.

Definite-time-delay overcurrent release

An → overcurrent release which operates with a delay independent of the overcurrent intensity. The delay may be adjustable. Also see → Short-time delayed short-circuit release.

Degree of pollution

See → Pollution degree

Degrees of protection

Various standards and specifications stipulate definitions and test procedures to classify enclosures and equipment housings in terms of the protection they provide against the ingress of foreign bodies/substances and, in some cases, in terms of the protection they provide against contact with live parts and moving parts. Characteristic designations or digits/numerals are assigned to the defined degrees of protection.
In Europe, the most commonly applied standard for degrees of protection is the IEC Publication 529. Other IEC publications use this document as a basis, although specific deviations may be noted. For example, IEC 947-1 makes an additional distinction between protection of persons and protection of equipment inside the housing against ingress of solid foreign bodies. The designation to indicate the degree of protection consists of the letters IP followed by two numerals:
The first characteristic numeral indicates conformity with the specifications for the protection of *persons* against contact with live parts and against contact with moving parts as well as the specifications for the protection of the *equipment* inside the enclosure against ingress of *solid foreign bodies*
The second characteristic numeral indicates conformity with the specifications for protection of the *equipment* inside the enclosure against harmful ingress of *water*.
A short-form description of the IP degrees of protection is given on page 63.

Deionization

A physical process whereby the current carriers (ions) in → ionized gases (plasmas) are neutralized (recombined). During the switching of electric current, → switching arcs are created which ionize the vapours and air in the vicinity of the → contacts and thereby cause a breakdown in the dielectric strength of → clearances within and around the → switching device. This potentially dangerous condition is limited to a minimum time duration by the action of the → arc chamber (Deionization chamber). In the arc chamber, the arc is split (by → arc splitters) into several smaller sections and energy is withdrawn (through cooling) until it is quenched. Through the flue action of the arc chamber construction, remaining ionized gases are expelled from the switching device.

Deionization chamber

See → Arc chamber

Demand factor

The ratio between the power consumption of all loads which operate simultaneously and the total rated power of all loads in the system or installation (i.e. the relationship between the actual power consumption and the theoretical maximum if all load were connected at the same time). The demand factor is one of the parameters which may be used, for example, to determine the ratings of the main fuses or → incomer → circuit-breaker in electrical distribution systems. It is also known as the "coincidence factor".

Dependent manual operation

See → Operation (of a switching device)

Dependent power operation

See → Operation (of a switching device)

Dependently delayed overcurrent relay, or release

→ Overcurrent relays or releases which operate with a time delay. The extent of the time delay is dependent on a specific parameter, usually the increase in the value of → overload current.

Note:

A relay or release of this kind can be designed in such a way that the time delay approaches a specific minimum value at high overload currents.

Also refer to → Current-dependent delayed overload release (→ inverse-time release, → thermal overload relay or release, → bimetal relay)

Destruction characteristic (or curve)

This term refers to a curve entered e.g. on the → time-current tripping characteristic of an → overload relay or release. For values of load current and duration which fall below this curve, the relay will trip before it is destroyed or its characteristic is permanently damaged (e.g. by the → let-through current of the → back-up protection). It is also known as the "thermal stress limit".

Detuned resonance circuits

See → Choked capacitor

DIAZED

The trade name for a range of Siemens → fuses and accessories. A DIAZED, or "D" fuse consists of a → fuse base, a protective insulating cover, an adapter sleeve, the → fuse link and a screw cap. DIAZED wire and cable protection fuses are available in 5 standardized sizes from 2 to 200 A

rated current and with a rated voltage of AC 750 V/DC 750 V. The adapter sleeve has an annular recess with a diameter corresponding to that of the appropriate fuse link. The fuse links are marked according to their → operation class or → time-current characteristic (also see → Flink and → Träg).

Dielectric strength

This is defined as the quotient of the → breakdown voltage and the distance apart of two electrodes between which the voltage is applied, whereby the concept of dielectric strength is only strictly valid for a uniform field between the electrodes. Note, that dielectric strength is not a property specific to the material (or "dielectric") between the electrodes or its thickness only. It also depends, for example, on the wave form of the test voltage, the rate of increase in the voltage, the ambient temperature, the air pressure, the relative humidity, etc.

Differential current

See → Residual current

Differential travel H (of a → proximity switch)

The distance between the → response point A and the → reset point B. It is usually quoted as a percentage of the → effective operating distance s_r and is also known as the "hysteresis" of the switch. Refer to → Response characteristic A_k

Diffuse photoelectric proximity switch

This term is used to refer to a type **D** photoelectric proximity switch using **d**iffuse reflective sensing in which the light beam travels from an → emitter-receiver unit to the → target (object) from which it is either reflected back to the receiver part, or is diffused. If the intensity of the light returning to the emitter-receiver unit is sufficient, then the output is switched. The → sensing range depends on the size and colour of the target as well as on its surface texture and is usually quoted as the value obtainable using a → standard target. The sensitivity of → Opto-BERO units may be adjusted by means of a built-in potentiometer. Guideline correction factors are given in the Siemens catalogue to indicate the sensing ranges which may be achieved using various target surfaces/materials.

Digital

The representation of a quantity in the form of characters or digits. The functional variation of a changing analog quantity is represented in pre-determined steps which are assigned definite numerical values. For example, a digital input may be a series of numbers (a value) set up on a thumb-wheel switch. In → data transfer systems, all digital quantities are represented by one or more → bits of information.

Digital display

The optical output of a measuring instrument which indicates the measured result as a sequence of digits or numbers (as opposed to e.g. a scale and needle type of display). The most common forms of digital displays utilize a series of "seven-segment" display elements. These derive their name from the fact that all the numbers from 0 to 9 can be made up from seven individual straight lines (when all are on or present, the figure 8 is depicted).

Direct current

Electrical current which continuously flows in the same direction, and which has a constant value. It is created by the application of a constant potential difference (→ direct voltage) in an electric circuit (after decay of initial transient effects). Abbreviation: DC or d.c.

Direct current component (d.c. component)

If a short-circuit occurs exactly as the → alternating voltage is passing through zero, the current experiences a phase displacement of about 90° because of the almost purely inductive reactance of the → current path (owing mostly to the leakage reactance of generator(s) and transformer(s)). At the instant of short-circuit, therefor, the current would have to "jump" to its maximum value. It cannot attain this value instantaneously, and the necessary equalization is provided by the so-called d.c. component, the initial value of which can be as high as 100% of the peak value of the symmetrical alternating short-circuit current at the instant of short-circuit, and which decays after a few cycles as a function of the circuit power factor. The short-circuit current thus consists of two components, namely the a.c. component symmetrical about the zero line and the d.c. component displaced to one side of the zero line. Summation of the d.c. and a.c. components at every instant gives the actual curve of the short-circuit current.

Note:
In conventional low-voltage networks, the d.c. component is usually about 50% of the peak value of symmetrical short-circuit current.

Direct starter

See → Direct-on-line motor starter

Direct voltage

A potential difference (or voltage) which is constant with respect to time. Direct voltage sources are batteries, direct voltage generators or stabilized → rectifier circuits. (see → direct current). Although, strictly speaking, the abbreviation "d.c." means "direct *current*", it is also used with reference to direct voltage (as in "a d.c. voltage").

Direct-on-line motor starter (d.o.l. starter)

In this type of → motor starter, the full operational voltage of the motor (typically a squirrel-cage motor) is applied to its terminals upon switch-on. The most common examples of d.o.l. starters include:
a → switchgear assembly comprising → contactor, → overload relay the associated → command devices,
a → motor protection circuit-breaker, and
a motor → control switch.
Direct-on-line, or d.o.l. starters are also known as "full voltage starters", "across-the-line (XL) starters" or simply "line starters".

Direct opening action

See → Positive opening operation

Discharge choke

A device used for the rapid discharging of → power factor correction capacitors.
Capacitors in automatic power factor correction units must be discharged sufficiently after being disconnected from the network and before they are switched in again (typically within 10 seconds), to avoid excessive stressing of the system components e.g. the → capacitor contactors. This is often achieved by connecting a resistance ("rapid discharge resistor") across the terminals of the capacitor after it has been disconnected, using the → auxiliary contacts of the associated → switching device.
A perhaps a more important specification states that a means of discharging the capacitor (typically to max. 50 V within 1 minute) must be *directly connected* to the terminals of the capacitor (slow discharge resistor).
Discharge chokes (reactance made up of several windings around a core of magnetic material) can be connected directly across the capacitor terminals to perform both functions, since their impedance is high at 50 Hz but progressively lower at the decaying transient discharge frequency ($X_L = \omega L = 2\pi \cdot f_L$).

Disconnected position

Usually, this is the designation for the physical position of a → withdrawable unit in its guide frame in which predefined → isolation distances have been established in the main and auxiliary circuits, although the withdrawable unit remains mechanically attached to the → switchgear assembly (e.g. behind a closed door).

Note:
The isolating distances may also be established by the operation of a suitable → switching device, i.e. the withdrawable unit may not have to be moved.

Disconnector

A mechanical → switching device which, in the open position, complies with the specific requirements for the → isolation function, and which provides a reliable switching state indication to confirm this open condition in all of its current paths or poles. Disconnectors are also known as isolators.
A distinction is made between "on-load" and "off-load" disconnectors, whereby the term "switch disconnector" is understood to be a mechanical → switch capable of making, carrying and breaking currents under normal as well as certain abnormal circuit conditions (e.g. overload and short-circuit). → Knife switches and load transfer switches are typical examples of switch disconnectors. The isolation function,

which is largely but not exclusively based on the isolating distance, may also be achieved by means of →fuses and →fuse bases. Special forms of switch disconnectors which also provide short-circuit protection by incorporating fuse bases and fuses, i.e. so-called →fuse-combination units include →fuse switch disconnectors and →switch disconnectors with fuses.

Discrepancy switch

A →control switch used to operate a remotely located →switching device by →momentary contact operation. An integrated lamp or acoustic signalling device indicates (acknowledges) that the desired operation has taken place. If there is no acoustic or optical signal when the discrepancy switch is actuated, then this indicates a "discrepancy" between the operating command and the actual state of the switching device being controlled, or alternatively, that the operating command has not been "accepted". These control switches are also known as "accept switches".)

Discrimination (Discriminative protection)

Discriminative protection means that only those protective →switching devices (e.g. →circuit-breakers, or →fuses) which are located closest to the point of fault will operate, and this in the quickest manner possible, so that no widespread shutdown results and no loads or distribution levels are unnecessarily disconnected from the supply.
One can distinguish between two techniques employed to achieve discriminative protection, viz. →current-based discrimination and →time-based discrimination.

Discrimination limit

If e.g. two →moulded case circuit-breakers are connected in series, then the conditions for →current-based discrimination can be fulfilled up to a value of short-circuit current known as the discrimination limit. The discrimination limit applies to the combination of the particular devices in question, and is usually stated by the manufacturer in special selection tables for combinations of →short-circuit protection devices (→discrimination tables). At short-circuit currents exceeding the discrimination limit, the peak →let-through current of the downstream unit exceeds the →tripping current of the upstream breaker, and both circuit-breakers trip.

Discrimination tables

By means of these tables, pairs of →short-circuit protection devices (e.g. →circuit-breakers, →HRC fuses) which fulfil the conditions for →current-based discrimination with respect to one another, may be selected. The tables state the maximum value of short-circuit current at which the downstream unit will trip and clear the fault before the upstream unit reacts (i.e. →discrimination limit)

Disruptive discharge

This term relates to phenomena associated with the failure of insulation under the influence of electric stress (see →arcing, →flashover and →sparkover). The discharge completely bridges the insulation in question, reducing the voltage between the electrodes to zero or nearly zero.

Distribution board

An assembly (see →switchgear assembly), or type of →switchboard containing switching and protective devices (e.g. →fuses and →circuit-breakers) primarily concerned with the distribution of electrical energy. It usually contains more than one outgoing →feeder circuit, and is fed from one or more →incomers. It is equipped with the necessary →terminals for the →neutral and →protective conductors.

Distribution circuit-breakers

See →Circuit-breakers for distribution protection

d.o.l. starter

See →Direct-on-line starter

Door interlock

See →Operating mechanism with detachable coupling, and →Mechanical interlook

Double insulation

Insulation comprising both basic inculation and supplementary insulation (also see → Total insulation).

Double moving contact bridge

The →contact bridge of →auxiliary contacts may consist of two independently supported contact elements, each capable of the same →current carrying capacity and →switching capacity. This configuration is known as a "double moving contact bridge" and is employed in virtually all Siemens low-voltage auxiliary switches and contacts to increase the →contact reliability.

Double-break contacts

In a set of double-break →contacts, the →current path is interrupted at two points during an →opening operation. The typical form of double-break contacts consists of a moving →contact bridge between two fixed contacts - one at each end of the bridge. As the bridge is lifted from the fixed contacts, the current path from one fixed contact to the other is interrupted at each end of the bridge. During opening, a →switching arc is drawn at each break point. The two arc voltages are in series and add to oppose the driving voltage, thereby extinguishing the arc and forcing the current to zero. Double-break contacts have the advantage that the switching arc is split into two smaller arcs (lower energy per arc, longer →service life, space saving), and are used extensively in →contactors and →auxiliary switches (also refer to → Multiple-breaking)

Double-throw switch

A →change-over switch with two distinctly separate rest positions of its →operating mechanism and corresponding sets of → NO contacts per → pole for each of these positions. Double-throw switches may also have a centre position in which none of the contacts are operated. Typical examples are → control switches such as toggle switches, two-position selector switches, load transfer switches etc. Double-throw switches for use in → auxiliary circuits are classified as → maintained-contact command devices.

Note:
The term is also used to refer to → change-over ⟋ switching devices in general.

Draw-out unit (Draw-out circuit-breaker)

See → Withdrawable unit and → Withdrawable circuit-breaker

Drive

The term drive or group of drives is used to refer to those parts of a machine which cause movement of some kind. It usually includes mechanical parts, electric motors and actuators as well as electric control and starting equipment.

Drop-out delay

The time period between the specified instant of initiation of the release or → opening operation of a → switching device and the moment the contacts of its → NC auxiliary contact touch (alternatively: the moment the contacts of its → NO auxiliary contacts separate). In the case of a drop-out delayed → time relay, the instant of initiation of the opening operation corresponds to the instant the preset time period starts expiring. The drop-out delay of auxiliary contacts is also known as the "OFF delay" or the "opening delay". An intentional drop-out delay may be achieved by means of an → energy storage device.

Drop-out delay unit

See → Energy storage device

Drop-out time

The period of time from the opening of the coil energizing circuit to the instant at which the initial or rest position of the → magnet drive (→ solenoid, → electromagnetically operated mechanism) is attained. For example, in the case of a → contactor, it is the time between deenergizing of the coil and the instant the magnet cores reach the fully opened position. Also known as "release time"

Drop-out voltage

Voltage, usually expressed as a percentage of the → rated control supply voltage (→ rated voltage), at which the pulling force curve of a → magnetic drive (→ solenoid, → electromagnetically operated mechanism) falls below the curve representing the force required. For example, a → contactor will release (or drop out) from the operated (closed) position to the rest position if the voltage on its coil is reduced to the drop-out voltage.

Dry-type power capacitors

→ Power capacitors which do not contain mineral oil or other liquid for insulation or cooling purposes (older designs of liquid-filled capacitors may still contain a toxic liquid known as → PCB or Askarel). Dry-type power capacitors may be resin-encapsulated or contain → SF_6 gas. See → MKK dry-type power capacitors.

Dry-type transformer

A transformer whose core and windings are not immersed in a cooling and insulating liquid (see → GEAFOL).

Duration factor

See → Cyclic duration factor

Duty cycle

See → Load cycle. Also refer to → Duty types

Duty types

The following duty types are defined:
1. → Continuous operation duty (S1) (→ Uninterrupted duty)
   Operation under constant load conditions, e.g. at rated power, of sufficient duration for thermal equilibrium of the motor to be attained.
2. → Short-time operation duty (S2)
   Operation at constant load for a period of time which is not long enough for thermal equilibrium to be attained, and followed by an interval of sufficient duration to allow the machine temperature to be within 2K of the coolant temperature.
3. → Intermittent periodic duty (S3 to S6)

This refers to operation which consists of a sequence of identical → load cycles (on/off periods).
The operational duty types S7 to S9 are described in Section 2.2.3 on Page 99.

Dynamic strength

Two parallel conductors either attract or repel each other, depending on whether the currents in them flow in the same or in opposite directions. Assuming that the same current flows in both conductors, the forces of repulsion and attraction are roughly proportional to the square of the current peak value and to the distance between the conductors. During short-circuit conditions, the oscillating mechanical stresses due to these dynamic effects may be extremely severe and continue until the short-circuit current is interrupted. Items of switchgear, supporting structures and the conductors themselves must therefor have sufficient dynamic strength to withstand the mechanical forces due to short-circuit currents which may arise at their point of installation.

Early-make NO auxiliary contact

An → auxiliary contact on a → switching device (→ normally open contact) which closes ahead of the → main contacts (and other NO auxiliary contacts). Physically, this may be achieved by extending the height of the → contact tips in the NO → switching element (hence the term "extended contacts"). An early-make NO contact may also be termed "leading contact". Also refer to → Make-before-break.

Glossary

Earth

This term refers to the conductive mass of earth which is considered to be at an electrical potential of zero. A → conductive part is said to be "earthed" if it is solidly connected to earth by means of a suitable conductor (see → earthing). In terms of DIN VDE 0100 part 200 both the location and the material are termed earth. The word → ground has the same meaning electrically speaking, and is used predominantly in areas influenced by American terminology.

Earth fault

A non-intentional conductive connection between a phase conductor or an operationally insulated neutral conductor (→ conductor) and → earth or earthed parts. It may be caused, for example, by → tracking, failure of insulation, or the presence of excessive → ionized gases between conductive parts at differing potential. It is a particular form of short-circuit (see → Residual current).

Earth terminal

A stud or → terminal to which the earthing → conductor is connected. Note, that in the case of totally insulated equipment, no protective conductor may be connected to earth terminals on switchgear, equipment or conducting parts installed inside the totally insulating housing. The looping through of protective conductors to equipment further downstream is, however, permitted.

Earth-electrode potential

The potential difference between an earthing system and the reference → earth.

Earth-fault current

The earth-fault current is that current which flows as the consequence of an earth fault (see → Residual current)

Earth-fault proof

Electrical equipment or → current paths may be considered to be earth-fault proof under specified operating conditions if appropriate measures are taken to ensure that the risk of an → earth fault is negligible (see → Short-circuit proof).

Earth-fault protection

See → Residual-current protection

Earth-leakage circuit-breaker (e.l.c.b.)

A type of → circuit-breaker (also known in Germany as an "FI" circuit-breaker) which is tripped if the → earth-leakage current (see → residual current) exceeds a specific limit value. It is extensively used in domestic applications with the purpose of protecting human life from dangerous → fault voltages created by the flow of leakage current (→ touch voltage) to earth, as well as from earth-leakage current which actually flows through the body in the case of accidental contact with a live part (typically units which respond at a residual current → 10 mA are used). These devices are also termed → CBR or → differential current protection devices.

Earth-leakage current

This term is generally used to refer to → residual current flowing from live → conductive parts to earth in the *absence of* a specific insulation fault (also see → Creepage current)

Earthing

The conductive connection between the → conductive parts to be earthed and the earth electrode via an earthing system. The earth electrodes which are ultimately embedded in the mass of earth, may either be designed as an earth strip, earthing rod or earth plate. The earthing is said to be "open" if overvoltage protection features, e.g. protective spark gaps, are included in the earthing conductor.

Economy connection

This term is used to refer to a technique whereby an essentially a.c. → magnet system (i.e. one with a laminated magnet core) in a → contactor is operated by means of a d.c. → control supply voltage. To produce the necessary pull-in force with a direct current, an a.c. magnet system needs to be over-excited. Once the magnet system has closed, however, the ampere turns in the coil may be reduced considerably by, for example, the series connection of an "economy resistor". In the unenergized state, the economy resistor is usually bridged out by a → late-break NC auxiliary contact which opens once the magnet system has enough energy to complete the → closing operation. An alternative method, employs two separate coil windings, i.e. a "pull-in" and a "hold-in" winding, to achieve the economy operation. It should be noted, that the → pick-up power of the magnet system may be considerably higher than the → hold-in power. This may have to be taken into account when a → control transformer is selected, or if the contactor is to be switched by an electronic switching component.

Economy resistor

See → Economy connection

Eddy-current losses

Heat losses caused by the flow of induced current in → conductive parts (e.g. metal supporting structures, → magnet cores) situated in a changing magnetic field. The changing magnetic field may be caused by the flow of alternating current in a nearby → conductor or coil. A number of techniques are employed to reduce or contain these losses: Magnet cores used in a.c. systems (e.g. in motor armatures, transformers, → magnet drives) consist of stacked thin plates or laminates so that the current paths of induced eddy currents are reduced in length to the thickness of the laminations.

If the three conductors of a three phase current are to pass through a metal panel or structure, then they should be fed through the same hole. Thereby, the vectorial sum of the magnetic fields resulting from the current in each phase will be small (or nearly zero in the case of balanced three-phase) and the induced circulating current in the panel will be small.

EEMAC sizes

See → NEMA sizes/EEMAC sizes (for → contactors and → motor starters)

Effective installed load (Total load, Connected load)

Maximum power consumption of all the electric → loads in one power system.

Effective operating distance s_r (of a → proximity switch)

The → operating distance of a particular proximity switch, as determined under specified temperature, voltage and mounting conditions. In terms of IEC-5-2, it is measured at 23 °C ± 5 °C and at rated → control supply voltage (see → rated value). For → inductive and → capacitive proximity switches it must be between 90% and 110% of the → rated operating distance s_n. In the case of → ultrasonic proximity switches it may be any distance between the → maximum and → minimum operating distances.

Effective power

See → Active power and → Power

Effective value (of a sinusoidal waveform)

See → Root-mean-square value

Efficiency

The ratio between output and input power. Owing to losses such as resistance losses (heat) or friction, the efficiency of a motor, for example, is always <1 or <100%.

Eight-hour duty

Eight-hour duty is an operating mode used e.g. for the rating of → contactors in which the main contacts (→ main contact elements) are kept closed and a constant load current is passed through the contactor until thermal equilibrium is attained, but not longer than eight hours without interruption.

e.l.c.b.

See → Earth-leakage circuit-breaker and → Residual current protection.

Electric shock

A pathophysiological effect resulting from an electric current passing through the body of a human or an animal.

Electrical installation

An assembly of associated and interdependent electrical equipment (including all items for the purposes of generation, conversion, transmission, distribution or utilization of electrical energy - e.g. motors, transformers, measuring instruments, protective devices, wiring luminaires, etc.) supplied from a common source (or common sources) to fulfil a specific purpose.

Electrical interlock

An → interlock which operates electrically. It is typically a reciprocal arrangement of → NC auxiliary contacts in the → control circuits of the respective → switching devices. For example, if two → contactors are to be electrically interlocked, then a NC contact of each contactor may be inserted in series with the → operating mechanism (coil) of the other. In this way, each contactor can only be energized once the NC contact of the other unit has closed (i.e. once the other contactor has dropped out). An additional → mechanical interlock is sometimes used if strong vibrations or mechanical shocks can cause both contactors to close simultaneously, or if the risk of → contact welding and the consequent failure of the electrical interlocking is unacceptably high.

Electrical isolation

See → Isolation

Electrical service life

The → service life of the → contacts in a → switching device under specified operating and load conditions (also see → Contact life).

Electromagnet

A piece of magnetic material or a structure (e.g. stack of laminations, → magnet core) which develops a north and south magnet pole (magnetic field) as a result of current flowing in a coil around it. It concentrates or bundles the lines of magnetic force passing through it and thereby increases the flux density. The moment the current is switched off, the magnetic effect is lost (with exception of the so-called → residual magnetism). Also refer to → Permanent magnet. The electromagnetic principle is used extensively in → electromagnetic operating mechanisms.

Electromagnetic compatibility (EMC)

The electromagnetic compatibility, or EMC, of an item of → switchgear or other piece of electrical equipment describes its ability to operate reliably (i.e. without change in its characteristics) when subjected to specified levels of → electromagnetic interference (EMI). EMC requirements and test levels for industrial-process measurement and control equipment are laid down in the various sections of the → IEC Publication 801. The electromagnetic compatibility of, for example, a → proximity switch may be expressed in terms of the maximum levels of electromagnetic interference at which the operating characteristics are maintained, whereby a distinction is made between electrostatic discharge (ESD), → burst and → surge levels.

The electromagnetic compatibility also describes the ability of a piece of equipment to operate without itself creating excessive electromagnetic noise or disturbances. Also refer to → Interference suppression.

Electromagnetic interference (EMI)

Changing electric and magnetic fields resulting from switching operations in high current circuits or → magnet systems cause electromagnetic disturbances in the vicinity of → switching devices. As a result of inductive or capacitive coupling effects, these can adversely influence the operation of other items of equipment, especially those which operate at low currents and voltages such as (high impedance) electronic devices. Depending on the → electromagnetic compatibility of these devices, → auxiliary and/or control circuits and conductors may need to be → screened or located remotely from e.g. high current cables, welding machines, or transformers. Electromagnetic interference is also known as electromagnetic → noise.

Electromagnetic latching

See → Remanence contactor

Electromagnetic operating mechanism

A common form of → operating mechanism providing → independent power operation and → remote control to the → switching device of which it is part, e.g. a → contactor. In principle, it consists of an → electromagnet or coil wound around the yoke or limbs of a → magnet core which has a moving part known as the → armature. If a magnetic flux passes through the → magnetic circuit owing to a current flowing through the coil (→ ampere turns), then the armature is attracted to the rest of the magnet core to close the → air gap. This movement and force is used in the mechanism to operate the switching device. In addition to their widespread application in contactors and relays, electromagnetic operating mechanisms are used extensively to operate → circuit-breakers, valves, as well as a host of → actuators. Electromagnetic operating mechanisms are also referred to as "solenoids" and "magnet drives"

Electromagnetic overcurrent relay or release

An → overcurrent relay or release which is operated by a generated magnetic field. The field is generated by the current being sensed - either directly, or via a → current transformer. → Instantaneous short-circuit releases and → short-time delay short-circuit releases are typical examples.

Note:
A relay or release of this kind may operate with a delay.

Electronic motor starter

See → Semiconductor motor controller/starter and → SIKOSTART

Electrostatic discharge (ESD) withstandability

The ability of a device (e.g. proximity switch) to operate under the influence of electrostatic discharges (direct contact). Severity levels (test voltages) are specified in IEC 801-2. For example, → BERO proximity switches satisfy the conditions for level 3 (8 kV). See → Electromagnetic compatibility.

ELLE

A Siemens computer program for the selection of three-phase a.c. → contactors in terms of the load current and → utilization category as well as the → electrical service life or alternatively the number of sets → contacts required (German: **el**ektrische **L**ebensdauer = **ELLE**).

Embeddable proximity switch

A → proximity switch is said to be "embeddable" if any → damping material can be placed around its → sensing face plane without it effecting the → response characteristic. Refer to → Mounting (of a → proximity switch). The term "flush-mounting" is also common. For comparison, also refer to → Non-embeddable proximity switch.

EMC

See → Electromagnetic compatibility

Emergency-Off apparatus

In terms of DIN VDE 0100 part 460, the Emergency-Off apparatus serves to disconnect the supply voltage (→ Emergency-Off switching device).

Emergency-Off switching device

In terms of DIN VDE 0113 part 1, an Emergency-Off switch must be able to shut down or stop the operation of a system, parts of a system or a machine in the event of danger in such a way, that persons or equipment are protected. The → actuator or → operating mechanism of an Emergency-Off switch must be clearly visible to the operator from his workplace and must be easily accessible. The actuator or operating handle must be red and it must be mounted on a yellow background for contrast and recognition purposes. Also refer to → Main switch.

EMI

See → Electromagnetic interference

Emitter (of a → photoelectric proximity switch)

The light source, lens arrangement and necessary circuitry which produce the light beam to be detected by the → receiver. In → diffuse and → reflex sensors the emitter and receiver are combined into a single unit.

Emitter-receiver unit (of a → photoelectric proximity switch)

A unit which comprises both the → emitter and → receiver of a photoelectric proximity switch in one housing.

Encapsulation

See → Housing

Enclosed construction

The constructional design of a switchboard or → switchgear assembly which is enclosed on all sides - with the possible exception of the mounting surface - in such a way that at least the → degree of protection IP 2X is achieved.

Enclosure

The term enclosure is generally understood to refer to an *additional* cover, shrouding, cabinet, or box over or around an item of electrical equipment or a → switchgear assembly. It provides protection against direct touching of live parts from all directions and is generally expected to provide a → degree of protection in excess of IP2X (also refer to → Housing, → Cabinet and → Cut-out box).

Endurance

See → Service life

Energizing command

Used to refer to a → command signal which causes → operation of a → switching device by activating its → operating mechanism.

Energy storage device

A unit in which the energy required to operate e.g. a → shunt release, an → undervoltage release or a → contactor is stored in a capacitor circuit.
Used in conjunction with a shunt release, it may be employed to trip a circuit-breaker in the event of auxiliary power loss or to achieve an extremely fast reaction time (→ fc-release).
These units are also used to prevent the drop-out of a contactor or the operation of an undervoltage release during a short interruption of their normal power supply (e.g. power supply voltage dip)

ENT

Electrical **n**on-metallic **t**ubing typically used in America as a → raceway to hold wires in household distribution systems. It may be concealed in walls, floors and ceilings (provided the building does not exceed three floors above grade - see NEC).

Environmental influences

Climatic and other influences on the operation or reliability of a piece of electrical equipment owing to or resulting from its location. Examples include humidity, high or low temperatures, pollution, etc.

ESD

Electro**s**tatic **D**ischarge. See → Electromagnetic compatibility

Eutectic alloy overload relay

Eutectic alloy → overload relays utilize a spring-loaded mechanism that is held in the rigid loaded position by a solid eutectic alloy substance. Eutectic materials have the characteristic that they change abruptly from a solid to a liquid state at a specific temperature. If enough heat is generated by the load current passing through the overload relay, the alloy changes from solid to liquid, releasing the spring-loaded mechanism ("paddle wheel") and thereby tripping the overload relay. The → tripping current and the → CLASS of eutectic alloy overload relays may be altered by the changing of the → heater elements. This feature allows one overload relay to serve a very wide range of motor full load currents. Eutectic alloy overload relays offer → manual reset only.

EVU

Abbreviation for the German "Elektrizitäts-Versorgungsunternehmen" which means electricity supply company.

Explosion group

In terms of the specifications for → explosion protection, electrical equipment is divided into two groups viz.:
Group I - electrical equipment for mining, and
Group II - electrical equipment for all other → hazardous locations.
Electrical apparatus of Group II may be classified further in terms of the so-called *explosion group* and → temperature class of the actual → potentially explosive atmosphere which may be encountered at its point of application. This classification may, for example, indicate the strength of → flameproof enclosures and/or the maximum permissible current and voltage values for → intrinsically safe circuits.

Explosion proof

See → Explosion protection

Explosion protection

Conformity to the specifications of explosion protection is a precondition for the use of electrical equipment in areas deemed to be → hazardous locations in terms of e.g. DIN VDE 0170/0171. The basic principles of explosion protection are the same worldwide:
The simultaneous presence of flammable substances (including gas, vapour, mist and dust) in potentially dangerous quantities, air (i.e. oxygen) and possible sources of ignition, must be prevented. In areas and applications in which the occurrence of potentially explosive mixtures of flammable substances and air cannot be avoided by the application of so-called "primary explosion protection", special constructional and installation specifications apply to the electrical equipment. These are intended to prevent or avoid all possible sources of ignition. In DIN VDE 0170/0171 part 1, DIN EN 50014 various *types of protection* are defined for the manufacture of explosion protected equipment. These include (examples):

– *flameproof enclosure "d"*, in which parts which could ignite explosive atmospheres are installed in an enclosure (the enclosure is designed to withstand the pressure developed by the internal explosion and prevents the transmission of the explosion to the surrounding explosive atmosphere),
– *increased safety "e"*, in which measures are taken to prevent the occurrence of excessive temperatures, → switching arcs or sparks in the interior and on external parts of

equipment which does not itself produce any of these effects under normal operating conditions,
- *pressurized apparatus "p"*, in which ingress of the surrounding explosive atmosphere is prevented by an over-pressure of non-explosive gas inside the enclosure,
- *intrinsic safety "i"*, in which the electrical apparatus is made up of intrinsically safe circuits (i.e. in which no spark or thermal effect is capable of causing ignition as determined under specified normal and fault test conditions),
- *oil immersion "o"*,
- *powder filling "q"*, and
- *moulding "m"* or → encapsulation

Equipment which complies with the specifications for one of these types of protection is also said to be "explosion proof".

Explosive atmosphere

See → Potentially explosive atmosphere

Exposed conductive part

Accessible → conductive parts of items of electrical equipment which are normally not live but which can become live in the event of a malfunction.

Extra-low voltage

Defined as a potential difference which does not exceed AC 50 V/DC 120 V between → conductors or between conductors and → earth or, in three-phase networks, between conductors and the → neutral conductor.

Extraneous light limit (of a → photoelectric proximity switch)

Although most photoelectric proximity switches use modulated light (see → pulsed light) to prevent interference from other light sources (e.g. Opto-BERO), there is nevertheless an upper limit for the intensity of the external, or ambient light. It is normally specified by the manufacturer for sunlight (unmodulated light) and halogen lamps (light modulated or pulsing at twice the mains frequency).

f-release

The short designation for a → shunt release.

Factory-built assembly (FBA)

FBA's *were previously* defined as combinations of → switching devices and other elements of low-voltage → switch and controlgear together with all the necessary interconnections including → busbars and other wiring - assembled, wired and tested by the manufacturer (DIN VDE 0660 part 5). Today, in terms of DIN VDE 0660 part 500, these combinations are known as → type-tested or partially type tested assemblies (TTA and PTTA respectively).

Fail-safe operation

The operation of a piece of electrical equipment, a control circuit or an item of switchgear is said to be "fail safe" if the failure of any one of its component parts does not cause a potentially dangerous situation to arise. Circuits which specifically serve the purposes of operator safety, should have fail-safe operation. Also refer to → Contactor safety combinations

Note:
It is extremely difficult to achieve fail-safe operation with electronic circuitry, as conditions of intermittent and/or semi-failure of components (e.g. damaged or overheated semiconductor) may have different effects on the operation of the device as a whole.

Fast transient withstandability

See → Burst

Fault

An unintentional circuit condition in which the current flows through an abnormal → current path. This may be caused by insulation failure or the bridging of insulation. Faults normally develop into → overcurrents or even → short-circuits if the fault current is not interrupted in time.

Fault voltage

A potential difference which exists between → exposed conductive parts which do not belong to the operational circuit or between them and → earth in the event of a → fault or insulation failure. It may or may not be caused by the flow of an → earth-leakage current through a current path of finite resistance. If a person comes into contact with such faulty equipment or systems, and is e.g. standing on a conductive floor, a part of the fault voltage, called the → touch voltage, can affect him and possibly cause → electric shock.

Fault-voltage protection circuit-breaker

A special type of → circuit-breaker (also known in Germany as an "FU" circuit-breaker) which is tripped if the potential difference between an → exposed conductive part and → earth exceeds a specific limit value (→ touch voltage).

FBA

→ Factory-built assembly. Also refer to → Type-tested assemblies and → Partially type-tested assemblies

fc-release

Short designation for a → shunt release with an energy-storage capacitor unit (see → Energy storage device). The energy stored in the capacitor circuit is sufficient to operate the shunt trip, so that → tripping of the circuit-breaker is ensured even if the → control supply voltage drops to zero e.g. owing to a short-circuit (see → Circuit-breaker for mesh-connected systems).

Feeder

Used to refer to a branch circuit which transfers (distributes) electrical energy from a source or distribution level to a → load. Also refer to → Radial network.

FELV

See → Functional extra-low voltage

Fibre-optic amplifier

A length of → fibre-optic conductor specially designed for use with → diffuse and → reflex photoelectric proximity switches. It may be fitted with special end-pieces to facilitate mounting.

Fibre-optic conductor

A fibre-optic conductor consists of a core with a high refractive index (bundle of glass fibres or one or more plastic fibres respectively) and an outer cladding with a low refractive index. The optical fibre is used to guide light from one place to another, even around bends and curves.
In the field of low-voltage switch and controlgear, fibre-optic conductors are used in conjunction with → diffuse and → reflex photoelectric proximity switches. They are fitted in front of the → emitter and → receiver and act as an extended "eye" of the switch (→ fibre-optic amplifier). Since they are highly flexible and no electrical potential is transferred along their length, they facilitate sensing in inaccessible, problematic or even → hazardous locations.
Most types of plastic fibre-optic conductors may be cut to length as required. Owing to their stronger attenuation at lower frequencies, plastic fibre-optic conductors are not recommended for use with → infrared light.

Field bus

A → bus primarily designed for → data transfer between remotely located → switching devices (field devices) with communication facilities (e.g. → SIMOCODE) and process automation equipment (e.g. → programmable controllers) in industrial networks. Example: → PROFIBUS.

Filament lamp

See → Incandescent lamp

Filter circuits

In industrial power distribution systems, filter circuits are used to remove (or block) currents and/or voltages, at frequencies other than that of the supply, from the network. Filter circuits (also known as "blocking circuits") use the relationships between frequency and impedance as found in inductive and capacitive reactances. In the case of a → choke or inductive reactor coil, the reactance increases in direct proportion to the frequency ($X_L = \omega L = 2\pi f L$). The reactance of a capacitor increases with the reciprocal of the frequency ($X_C = 1/\omega C = 1/2\pi f C$). One can distinguish between parallel and series LC filter circuits in which the capacitance C and the inductance L are connected in parallel or series respectively. At the → resonance frequency, the vectorial sum of the inductive and capacitive reactances is zero. Typical applications for filter circuits include → choked capacitors and → tuned resonance circuits e.g. in the field of → power factor correction, as well as → audio-frequency blocking circuits used in conjunction with → AF ripple control systems.

Flameproof enclosure "d"

A → type of protection as defined in the general requirements DIN EN 50014 and the specific standard DIN EN 50018. See → Explosion protection

Flange plate

A part of the enclosure of a → switchgear assembly which covers an opening to the outside. It is usually fixed in place by means of screws. Once the switchgear assembly is put into operation, the flange plate is normally not removed again.

Note:
The flange plate may be equipped with openings or knock-outs for cable entry, and may be fitted with suitable cable glands or grommets.

Flashover

The term flashover is understood to mean the striking of an → arc (i.e. a → disruptive discharge) between two live → conductive parts at differing potential over the surface of a dielectric (e.g. insulator) in a gaseous or liquid medium (usually in air). Also see → breakdown voltage and → sparkover).
In the field of low-voltage technology, flashovers are usually initiated by conductive contamination, pollution or precipitation causing → tracking and → creepage currents.

Fleeting contact element

See → Pulse contact element

flink

A German term meaning "quick". It is used to designate the "quick response" → time-current characteristic of a → fuse link. It is particularly used in conjunction with → DIAZED fuses. Fuse links with a quick response characteristic are "faster" than those with the → operation class gL to VDE 0636.

Flip-flop (also RS-flip-flop)

A logic device or electronic component (integrated circuit) which requires a "set" command on the S input for the output to switch high. The output remains high even after the S input has been removed, and is only switched low again (reset) once a high signal has been applied to its R input. In terms of low-voltage switchgear, a latching → contactor (or → contactor relay) with separate main and reset → magnet drives operates in a similar way.

Flush-mounted proximity switch

See → Embeddable proximity switch

Four-stage star-delta starter

The four-stage star-delta connection (only possible with motors having split windings and 9 terminals – see page 448) is used for drives with such a high counter-torque that the normal star-delta connection would not be able to accelerate the drive to a sufficiently high change-over speed (see → Star-delta starter)

Fourier analysis

See → Harmonics

Free zone (of a → proximity switch)

The space around a proximity switch which must be kept free of any material (→ damping material) which could affect its → response characteristic. It includes the → sensing zone (also known as "damping zone" particularly in the case of → inductive proximity switches) and is determined by the dimensions r, c, w, and g.

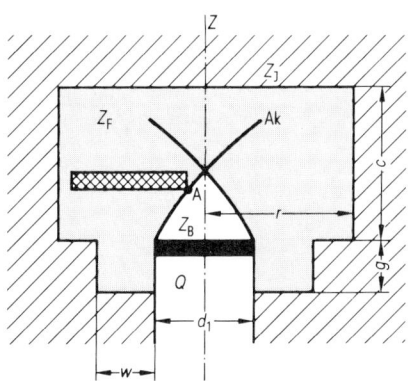

Z_j Inactive zone
Ak Response characteristic
A Response point
Z_F Free zone
Z_B Damping (operating) zone
Q Sensing face
Z Reference axis
c Partial height of the transition zone
g Partial height
r Radius of the damping zone
d_1 Diameter of the proximity switch
w Mounting distance

Freewheel diode

A diode connected across the terminals of an inductive load (e.g. across the coil of a → contactor → magnet drive) in a d.c. circuit (observe correct polarity!). Refer to → Surge suppressor.

Front time T_1, unipolar

The rise-time of a positive or negative → overvoltage spike, measured from zero to its maximum value (peak).

Front time T_2, bipolar

The rise time measured from the negative to the positive peak value of an → overvoltage transient oscillation or spike.

FU circuit-breaker

See → Fault-voltage protection circuit-breaker

Full voltage starter

See → Direct-on-line starter

Function class

In terms of DIN VDE 0636, low-voltage high rupturing-capacity → fuse links are classified in terms of their function and operating characteristics (also see → Operation class). The function class indicates the range of current the fuse link is designed to interrupt (also known as the "utilization class").

Function class g:

General purpose fuse links, which can carry currents of up to at least their → rated current continuously, and which interrupt currents from the lowest → pre-arcing current up to their rated → rupturing capacity. The **gL** fuse links, for cable and busbar protection, fall into this group.

Function class a:

Accompanied or partial range fuse links, which can carry currents of up to at least their → rated current continuously and which interrupt currents from a specific multiple of their rated current up to their rated rupturing capacity. The **aM** fuse links for switchgear protection fall into this group. Their rupturing current begins at approximately 4 times the rated current and they therefore serve only for short-circuit protection. The **aR** fuse links for semiconductor protection have a particularly "quich" → time-current characteristic from about $2.7 \cdot I_n$.

Functional extra-low voltage

This is a potential difference defined in terms of the specifications for → protective measures in which circuits are operated at up to AC 50 V or DC 120 V rated voltage. The functional extra-low voltage (FELV), however, does not meet all the demands required for → safety extra-low voltage and is therefore subject to additional protective specifications. A distinction is made between functional extra-low voltages with and without → safety separation.

Functional unit

Used to refer to a part of a → switchgear assembly comprising all electrical and mechanical components necessary to perform one function.

Fundamental

See → Harmonics

Fuse

A protective device that opens the circuit in which it is inserted by breaking the current when one or more of its specifically designed and proportioned components (i.e. the → fuse-element) melts (or "fuses") as a result of the current exceeding a given value for a sufficient period of time (see → Time-current characteristic)

9 Appendix

Note:
Strictly speaking, a fuse is understood to comprise all the elements which constitute the complete protective apparatus. The term is, however, commonly used to mean the → fuse link only (also see → Fuse base)

Fuse base

The fixed-mounted part of a → fuse which contains the terminals and the necessary parts to hold the → fuse links in place. They may be single-pole or three-pole in design, and are dimensioned for specific fuse link types and sizes. They may be coded (by shape) to accept only a fuse link with a specific → rated current.

Fuse element (or fusible element)

A part of the → fuse link designed to melt if the current passing through it exceeds a specific value for a definite period of time. In → LV HRC fuse links, it typically consists of one or more specially shaped copper or silver conductors, so designed as to achieve the desired → time-current characteristic. Several constrictions may be used to ensure rapid multiple rupturing of a short-circuit current. The fuse element melts through very rapidly or vaporizes at these constrictions under the influence of short-circuit current. At lower fault currents, the temperature at these reductions in the cross-section may not increase fast enough for the → fuse link to operate in time and provide adequate protection. In such cases, a technique which employs the → Metcalf Effect may be used to reduce the melting point of the fuse element. The fuse element in an HRC fuse link is usually embedded in quartz sand of a particular quality and grain size. Not only does the sand serve to cool the arc which is created when the fuse element ruptures, it also increases the arc voltage rapidly by splitting the arc into several small sections (i.e. between the grains).

Fuse link

The part of a → fuse intended to be replaced after the fuse has operated. It contains the → fuse element(s) and is equipped with blades or terminals for the electrical connection to the → fuse base. Fuse links used in low-voltage installations usually comprise one or several parallel-connected fuse elements enclosed in a cartridge which is filled with an arc-extinguishing medium (e.g. quartz sand) - hence the term "cartridge fuse link". If the fuse operates, the whole cartridge is discarded and replaced with a new one. → LV HRC fuse links are classified in terms of their → function class and their → operation class.

Note:
Strictly speaking, a distinction is made between a fuse link and the fuse as a whole. The word fuse is understood to mean all the elements which constitute the complete protective apparatus. The term is, however, commonly used to mean the fuse link only.

Fuse switch disconnector

This form of → fuse-combination unit utilizes the actual blades of the → HRC fuses as its moving → contacts (also refer to → Switch disconnector with fuses). It consists of a shrouded and partitioned base element, which incorporates lyre-shaped fixed contacts and → arc chambers for each → pole, and a separate fuse carrier tray. The fuses are plugged into the lyre-shaped fixed contacts when the carrier is inserted or hinged into the base. Rapid withdrawal or opening of the hinged fuse carrier tray is usually ensured by a snap-action feature so that the fuses may be withdrawn safely under load conditions. The units are often used in distribution systems and may be mounted directly on busbar systems. In-line fuse switch disconnectors are specifically designed for busbar mounting. In these units the fuses are arranged one-below-the-other, and may be inserted or withdrawn individually (single-pole switching) or switched simultaneously in a single operation (three-pole switching).

Fuse-combination unit

An assembly comprising a mechanical → switching device and one or more → fuse bases with → fuses, usually intended for use as a → switch disconnector offering short-circuit protection. Such a combination unit, may be a single factory-built device which incorporates the fuses on or within one housing, or it may be an assembly of discreet devices combined according to manufacturer specifications. Units of this nature are also referred to as "combination fuse-switch units" or "CFS units". Typical examples include → fuse switch disconnectors and → switch disconnectors with fuses.

g-release

Short designation for a → residual-current release (also known as an earth-fault release or ground-fault release).

GEAFOL

This is a Siemens trade name for a range of cast-resin power transformers. The GEAFOL cast-resin transformer is a self-cooling → dry-type transformer. Its insulation consists of a mixture of epoxy resin and quartz powder which is classified as an environmentally acceptable material and which renders the windings maintenance-free, moisture-proof and suitable for use in tropical environments, virtually non-inflammable and self-extinguishing. Even under the influence of fault arcs, no toxic gases are produced.

Glow lamp

A lamp which produces light by means electrical discharge effects in a rarefied gas. Its advantages include low heat loss, long → service life and a high resistance to mechanical vibrations and/or shocks.

Grading

The current path from the source of electrical energy to the load or point of fault usually passes through a number of series-connected → fuses and/or → switching devices with → overcurrent releases or relays (e.g. → circuit-breakers). As a general rule, the operating currents and/or response times of the fuses and overcurrent protection devices are selected in such a way that they become smaller the closer they are located to the load. This is known as grading. One may distinguish between → time grading and response current grading. → Discrimination in distribution systems can be achieved by the correct grading of protection apparatus.

Ground; Grounded

See → Earth.
The term ground is also used to mean a conducting connection, whether intentional or accidental, between an electrical circuit and the earth.

Ground-fault protection

See → Residual current protection

Group power factor correction

In the case of group → power factor correction an → automatic power factor correction unit or a fixed-value power capacitor (capacitor bank) is assigned and connected to a specific part of the installation (or group of loads) only. The load group may consist of specific motors or, e.g. a number of fluorescent lamps which are connected to the network via a common → contactor or other → switching device.

Hand reset

See → Manual reset feature

Harmonic content

This term is used to indicate the degree by which a particular waveform (e.g. load current) differs from that of a sine wave (e.g. owing to the presence of non-linear loads). It can be expressed as the ratio between the amplitude of the → fundamental and the sum of all the → harmonics. In this form, it does not indicate which harmonics are present. For any meaningful analysis of a waveform, the actual → harmonic numbers and their respective amplitudes in relation to that of the fundamental should be known.

Harmonic current

Current which flows at a → harmonic frequency.

Harmonic frequency

The frequency of a → harmonic waveform. It is usually expressed as a multiple of the → fundamental frequency known as the harmonic number v. In industrial networks, where → harmonic currents are mainly produced by → rectifier circuits, the harmonic numbers can be derived from the → pulse number p of the rectifier circuit in question by the equation $v = p \cdot k \pm 1$, where k is an integer (i.e. 1, 2, 3, etc.).

Harmonic number

See → Harmonic frequency

Harmonics

Any non-sinusoidal waveform can be expressed as the sum of a number of sinusoidal waves; one at the fundamental frequency plus a series of so-called harmonics at integer multiples of this frequency. The mathematical process by which a waveform is broken down into its "fundamental" and its constituent sinusoidal elements each having a particular frequency and amplitude, is known as "Fourier analysis".

The → alternating voltage of industrial power supply systems produces a purely sinusoidal current waveform in linear power circuits. If non-linear elements in the network distort this waveform, then the harmonic currents which result, cause a corresponding distortion of the voltage waveform across the network impedances.

In industrial networks, the presence of → rectifier load circuits is one of the main causes of → harmonic currents. Particularly in networks which incorporate capacitors (e.g. in → power factor correction units), currents flowing at → harmonic frequencies can pose a real danger since they may lead to → resonance effects.

Hazardous location

These are locations where → potentially explosive atmospheres can occur as a result of the specific local and/or operational conditions. Hazardous locations are classified into *zones* according to the probability of a potentially explosive atmosphere occurring. Examples of these include:

– *Zone 0* covers areas in which an explosive gas atmosphere is present continuously or for long periods of time,
– *Zone 1* covers areas in which an explosive gas atmosphere can be expected to be present occasionally,
– *Zone 2* covers areas in which an explosive gas atmosphere can only be expected to be present very occasionally and if it does occur, then only for short periods of time.

Only → explosion protected equipment may be used in Zones 0 an 1, whereby equipment destined for use in Zone 0 areas must be specifically certified for this purpose.

Heat exchanger

A device used to extract heat from e.g. a switchboard, → resistance starter or even large motor. It comprises two separate coolant circuits, and the heat passes through a membrane or dividing wall from one circuit to the other. In the case of switchboards, heat exchangers are used to extract heat losses from the interior of the enclosure while maintaining a high → degree of protection (e.g. IP 54).

Heat losses (of a switching device)

The heat (Watts) produced by a → switching device when conducting and switching its → rated load at a given → operating frequency. This heat is due to resistive losses in the copperwork of its current paths, losses produced at all → contact points (see → Contact resistance), as well as → eddy-current losses in the steel framework of the device itself.

Heat value

See → $I^2 \cdot t$ value

Heater element (of an → overload relay)

In → thermal overload relays the heat caused by the load current flowing through the three resistive heater elements

in the main current paths of the 3-pole device is utilized to produce its → tripping characteristic.
In the case of → bimetal overload relays, these heater elements are often wound around the → bimetallic strips themselves. The → tripping current is adjusted by moving the point at which the bimetallic strips cause a switch-over or delatching of the switch mechanism.
The heater elements of → eutectic alloy overload relays may be exchanged or retrofitted. The tripping characteristic (e.g. → CLASS) and the tripping current of the relay are determined by the heater elements which are fitted.

Heating unit (in a switchboard)
See → Anticondensation heater

Heavy-duty starting (of e.g. squirrel-cage motors)
If a motor requires more than approximately 10 to 15 seconds during → direct-on-line starting to reach its rated speed owing to the particular run-up torque characteristic of the driven machine, it is said to operate under heavy-duty starting conditions. Typical examples of driven machines which present a higher load torque during start-up than during rated operation are rolling mills and or centrifuges. Special → overload relays (heavy-duty relays, electronic overload relays) with the appropriate trip class settings (see → CLASS) or → thermistor motor protection devices are required to protect motors which start under heavy-duty starting conditions.

Hold current (of a → proximity switch)
See → Minimum operational current I_m

Hold-in power (also holding power)
The power consumption of a → magnet drive (coil) in the operated state. In the case of a → contactor, for example, it corresponds to the power consumption of the coil circuit during the operated state (i.e. after closing). It is expressed in VA for a.c., and in Watts for d.c. magnetic drives. Particularly in the case of contactors employing the → economy connection, it may be substantially lower than the → pick-up power. It is also known as the "seal-in power".

Hold-in winding
See → Economy connection

Housing or encapsulation
An cover or shrouding which is part of an item of electrical equipment and which serves to protect it from particular external environmental influences. It provides protection against direct touching of live parts from specific directions. The word housing is also sometimes used to refer to an *additional* cover, shrouding, cabinet, → enclosure or box which provides protection against direct touching of live parts from all directions and which is generally expected to provide a → degree of protection in excess of IP2X. The word encapsulation is also used to refer to the technique

whereby and item is completely encased in protective epoxy resin, e.g. → current transformers.

HRC fuse
High-rupturing-capacity → fuse. See → LV HRC fuse

Hysteresis H (of a → proximity switch)
See → Differential travel

$I^2 \cdot t$ characteristic (of a circuit-breaker)
A curve giving the maximum → Joule integral values (→ $I^2 \cdot t$ values) of the → circuit-breaker → let-through current related to the → break times as a function of the → prospective current (r.m.s. symmetrical in the case of alternating current) at a specific voltage. It is usually plotted up to the value of prospective short-circuit current corresponding to the rated → short-circuit breaking capacity. Note, that this curve alone cannot be used to determine the → coordination between → switching devices and → short-circuit protection devices

$I^2 \cdot t$ grading
A mixture of → time and current-based discrimination may be achieved by basing the time delays of the electronic → short-time delayed short-circuit releases in successive series-connected → air circuit-breakers on the $I^2 \cdot t$ value of the fault current (→ Joule integral). This has the advantage, that the time delay is shortened automatically for higher values of short-circuit current. In principle, this approximates the behaviour of → fuses under short-circuit conditions (see → Pre-arcing $I^2 \cdot t$ value and → Total $I^2 \cdot t$ value).

$I^2 \cdot t$ value
The thermal value (or heat value) usually of a high magnitude current flowing for a short duration (e.g. short-circuit current or → let-through current). Also refer to → $I^2 \cdot t$ characteristic (of a circuit-breaker) as well as → Pre-arcing $I^2 \cdot t$ value and → Total $I^2 \cdot t$ value of → fuses.

ICCB
→ Insulated Case Circuit-Breaker

IEC
(International Electrotechnical Commission)
Within the framework of international harmonization, the commission aims to produce standards and publications which combine the rules, requirements and specifications of the different national regulations, and thereby contribute to the removal of trade barriers.

IMC
Intermediate metal conduit. See → Conduit.

Impedance voltage u_{kr}
Voltage (expressed as a percentage of the rated input voltage), which has to be applied at the transformer input termi-

nals to produce the rated output current (I_2) in the secondary winding, if this is short-circuited at the output terminals. The reference temperature is 20° C. It is the vectorial sum of the → resistance voltage and the → reactance voltage.

Impulse voltage withstandability

See → Surge

Impulse withstand voltage

See → Rated impulse withstand voltage

Incandescent lamp

A lamp which produces light through the glowing of an intensely heated filament wire (also known as a filament lamp). To prevent the wire from burning, it is suspended in a vacuum. Larger incandescent lamps are filled with a neutral gas which reduces the rate at which the filament wire evaporates or "ages". Particular characteristics in comparison with → glow lamps for use in indicator lights include a higher colour fidelity and a stronger light output.

Inching

Repeated energizing of a motor (or → solenoid) for short periods (i.e. shorter than the → start-up time of the motor) to achieve small movements in the driven machine. The → switching device used must be capable of making and breaking the motor → starting current, i.e. a multiple of the rated operational current of the motor (see → Utilization category). Inching is also referred to as "jogging".

Incomer

Used to refer to the source of energy and/or the associated switchgear in a network or distribution level (e.g. incomer → circuit-breaker or incomer transformer).

Increased safety "e"

A → type of protection as defined in the general requirements DIN EN 50014 and the specific standard DIN EN 50019. See → Explosion protection

Independent manual operation

See → Operation (of a switching device)

Independent power operation

See → Operation (of a switching device)

Individual power factor correction

Individual → power factor correction means that the capacitors are connected directly to the individual loads. The capacitors and loads are switched on and off by a common command signal, and in some cases by the same → switching device. This form of power factor correction is favoured if a poor → power factor of the installation as a whole is due to the power factor of one or more large single loads.

Inductive proximity switch (→ BERO)

A → proximity switch which operates when energy is drawn from an oscillating electromagnetic field, which it produces in a specific zone in front of its → sensing face. When energy is drawn from the field, e.g. by a designated object or → target, the oscillations are damped and hence the zone from which the switch may be operated is known as the damping zone and the material of the target as → damping material (see illustration on page 588). The target of an inductive proximity switch must be metal (conductive). Inductive proximity switches employ completely encapsulated semiconductor (i.e. electronic or solid state) switching elements and are therefor not subject to mechanical wear. They are further characterized by virtually unlimited → service life, high → operating frequency, resistance to environmental influences and insensitivity to vibrations.

Infrared light (IR light)

This refers to electromagnetic radiation with a wavelength greater than that of normal visible light (380 to 780 nm). In industrial applications, for example in → photoelectric proximity switches), infrared light with a wavelength of approx. 780 to 1500 nm is used.

Inherent delay

See → Movement delay

Inherently short-circuit proof

In general, items of equipment or → current paths are termed inherently short-circuit proof if, under the specified operating conditions, the risk of a → short-circuit is made negligible through the application of suitable measures. (also refer to → Earth-fault proof).

Some → circuit-breakers may be inherently short-circuit proof at a particular → operational voltage by virtue of their internal resistance. If the resistance of the → current path through the circuit-breaker (e.g. through its → a-release and/or → n-release units and the → contact elements) is such that the maximum short-circuit current which can flow at the particular system voltage is less than the → breaking capacity of the unit, then it is short-circuit proof at this voltage and setting range of its → overcurrent release. In other words, the circuit-breaker may be installed at a point in the network where the → prospective short-circuit current may exceed its breaking capacity. In the event of a fault, only the current determined by the internal resistance of the breaker can pass through it.

Initial symmetrical short-circuit current I_k''

Effective value (r.m.s. value) of the short-circuit current at the instant of short-circuit. In the case of a short-circuit close to generator terminals, the value decreases to the continuous steady-state short-circuit current I_k. In the case of a short-circuit remote from the generator terminals, the value is equal to the continuous steady-state short-circuit current and remains nearly constant for the duration of the fault.

Inrush current

The transient current peak which flows the instant a →magnetic circuit (e.g. transformer, motor, →operating mechanism, etc.) is connected to the supply. It is caused by the initial build-up of the magnetic field (the so-called "rush effect"). The amplitude of the inrush current not only depends on the inductivity of the load circuit, but also on the instant in the supply voltage waveform at which the →contacts of the →switching device close. The inrush current has its greatest value, if the load is connected the instant the a.c. supply voltage waveform passes through zero. Typical values for the inrush current of a squirrel-cage motor are given below (applicable at rated voltage).

Switching on: $i_{max} = \sqrt{2} \cdot I_{an} \cdot (1.8 \text{ to } 2.0)$,

Star-delta changeover: $i_{max} = \sqrt{2} \cdot I_{an} \cdot (2.1 \text{ to } 3.7)$,

Reversing: $i_{max} = \sqrt{2} \cdot I_{an} \cdot (2.7 \text{ to } 5.0)$,

where I_{an} is the →starting current of the motor. In the case of transformers, the inrush current (also known simply as "rush current") depends on the transformer model, version, winding arrangement, application, power rating, etc. Its r.m.s. value may be as high as 15 to 30 times the primary rated current. The inrush current decays sharply after a few periods of the a.c. supply and is already considerably lower after 20 ms.

Instantaneous short-circuit release

An →overcurrent release intended to provide short-circuit protection by operating the instant the load current exceeds a predetermined set value and thereby →tripping the associated →switching device (usually the →circuit-breaker in which it is incorporated). The predetermined set value may be adjustable or fixed. The instantaneous short-circuit release is also referred to as the "n-release".

Instrument transformer

An item of electrical equipment which transforms a primary electrical quantity, i.e. current or voltage, into a secondary proportional quantity of the same kind for use in measuring instruments, meters, protective relays and similar devices. The magnetic core of an instrument transformer becomes saturated at relatively low values of →overcurrent to protect the connected devices from overload.

Insulated Case Circuit-Breaker (I.C.C.B.)

Insulated case circuit-breakers combine the high →breaking capacities of →moulded case circuit-breakers with the high →short-time withstand current capabilities of →air circuit-breakers. Their encapsulating housing, which also forms a major part of the supporting and bearing structure, is of a reinforced moulded insulating material.

Insulation class

See →Class of insulating material

Insulation coordination

See →Coordination of insulation

Insulation group

Until recently, electrical equipment was divided into 4 insulation groups depending on the application and the degree to which the →dielectric strength of insulating materials used was effected by dust or the humidity of the surrounding air (DIN VDE 0110). In terms of DIN VDE 0660, the insulation group C was specified for →low-voltage switchgear. Nowadays, it is commonly accepted that the use of electrical equipment depends primarily on the →pollution degree and the voltage or overvoltages which are likely to occur (see →Overvoltage category). See VDE 0110 parts 1 and 2. Minimum →creepage distances on items of →low-voltage switchgear, for example, are specified in terms of the →rated insulation voltage (or →working voltage), the →pollution degree, and the →material group of the insulation in question.

Insulation monitoring

A technique used to measure the →insulation resistance of an electrical system (in which neither a phase conductor nor a star point is directly earthed) continuously. A reduction in the insulation resistance (e.g. owing to insulation deterioration or failure) may be indicated, or may be used to shut down the relevant section of the plant or equipment.

Insulation rating

See →Rated insulation voltage

Insulation resistance

The smallest resistance value measured between parts which are insulated from each other, or between these parts and →earth.

Interface

A common point between two (or more) circuits or systems at which electrical signals, data or power is transferred from one to the other(s). This transfer may take place electronically, e.g. between a →switching device and a data →bus (see →Data transfer) or electromechanically (see →Interface units).

Interface units

This term is used to refer to →contactors, →contactor relays and p.c. board mounted →relays (e.g. contained in a terminal block-type housing) intended for use with and direct connection to →programmable controllers and other electronic devices. They have a particularly low coil power consumption and a →coil voltage tolerance matched to the technical data of solid-state systems, a high →contact reliability at low voltage and current, and they usually incorporate some form of →switching overvoltage suppression on the operating coil (see →Interference suppression). They are also known as "interface modules" or simply as "interfaces".

Interference

See → Electromagnetic interference, → Noise and → Electromagnetic compatibility

Interference suppression

Interference suppression is a measure to prevent or reduce high-frequency electromagnetic oscillations in electrical equipment and systems which may cause radio interference (DIN VDE 0857 part 11, 07.92; EN 55011, 1991. DIN VDE 0871, 06.78 is valid until 31.12.95). Also refer to → Electromagnetic compatibility.

Interlock (Interlocking device)

An interlock between two or more → switching devices (or between a switching device and some moving or moveable part of a piece of equipment or installation, e.g. a door or cover) is understood to mean some mechanism, device or circuit arrangement which renders the operation of each device dependent upon the → switching state of the other (or upon the position of the movable part). Typically, in the field of low-voltage → switchgear, one distinguishes between → electrical interlocks and → mechanical interlocks whereby they are often used in conjunction with one another and may or may not introduce a delay between the operation of the devices.

Intermittent periodic duty (or intermittent duty)

A rated → duty cycle with repetitive → on-load periods (during which the main contacts of the → switching device remain closed) and → off-load periods (during which the main contacts are open). The on-load periods have a definite relation to the off-load periods, both periods being too short to allow the switching device to reach thermal equilibrium. Intermittent duty is characterized by the value of the current, the duration of the current flow and by the on-load factor. The on-load factor (also known as the "cyclic duration factor") is the ratio of the on-load duration to the entire period of the duty cycle. The on-load factor (cyclic duration factor, abbr. c.d.f.) is usually expressed as a percentage. Standardized values are 15%, 25%, 40% and 60%. A number of standard intermittent duty classes are defined in terms of the number of → operating cycles (make/break operations) a switching device can carry out per hour, e.g. class 1 for 1 operating cycle, class 3 for 3 operating cycles, class 12 for 12 operating cycles, etc.

Example
A switching device intended for the intermittent duty "100 A load current every 20 minutes for a duration of 8 minutes" may be designated as follows:
Intermittent duty 100 A, class 3, 40%

Interposing current transformer

Interposing current transformers serve to change the secondary current of a main → current transformer to render it suitable for the particular measuring instrument. In addition they effectively reduce the overcurrent factor of main transformers (see → Rated overcurrent factor) and thus protect measuring devices from overload. Interposing current transformers are usually of the → pin-wound type. Their power consumption should be as low as possible to prevent excessive loading of the main current transformer.

Intrinsic safety "i"

A → type of protection as defined in the general requirements DIN EN 50014 and the specific standard DIN EN 50020. See → Explosion protection

Inverse-time delayed overload relay (or release)

See → Overload relay or release

Ionized gases

At the extremely high temperature developed in the gas column of an electric arc (see → Arcing), the molecules of the gas and vaporized matter (or plasma) move so rapidly (kinetic energy) that electrons are "struck" from them or they are split. These molecules are known as positive and negative ions or "current carriers". Depending on the density of the ions, the gas has low → dielectric strength properties and, unless special precautions are taken, it can lead to → flashover between other live → conductive parts in the vicinity of the original arc. In the case of circuit-breakers and large contactors, for example, a sufficient volume of space and distance to conductive parts must be allowed for deionization of expelled gases (see → Blow-out space).

IP Degree of protection

See → Degrees of protection

Isolation (isolating function)

A procedure or function which disconnects (see → Disconnector) all or specific parts of an electrical installation from the supply by separating the installation from every source of electrical energy for reasons of safety.

Isolation (of a conductive part)

The separation electrically conducting parts at different potentials (voltages) by means of insulating material or → clearance in air in such a way, that no current can flow between them.

Isolator

See → Disconnector

IT system (network)

See → Network types and systems

Jogging

See → Inching

Joule integral

The integral of the square of the current over a given time period:

$$I^2 \cdot t = \int_{t_0}^{t_2} i^2 \cdot dt.$$

It is used to calculate the calorific value of the heat produced e.g. by a current flow of high magnitude and short duration (→ short-circuit current).

Joy stick

A → control switch with an → actuator consisting of a stick or pin which protrudes from the mounting panel. The switch is operated by the angular displacement of this stick, whereby the direction of displacement determines which of the → switching elements is operated. Also refer to → Master controller.

kcmil (MCM)

(**ki**lo **c**ircular **mil**) **M**ille **C**ircular **M**il: A unit of conductor or wire cross-sectional area. 1 circular mil corresponds to the cross-sectional area of a wire with 1 mil diameter (1 mil = 0.001 inch = 0.0254 mm) It is used especially in America for wires thicker than approx. 126.8 mm^2 ($\hat{=}$ 250 kcmil). For thinner wires, an AWG number is assigned to each cross-section (AWG = → American Wire Gauge). Also see Tables 9.10 and 9.22 on Page 511 and 526).

Key interlock

See → Mechanical interlock

Kick start

See → Start impulse

Knife switch

The simplest form of → disconnector in which a the → making operation is achieved by forcing, or wedging, a hinged blade (moving → contact) between the two "wings" of the lyre-shaped fixed contact. Knife switches are normally used in → off-load applications to achieve clearly visible → isolation or to transfer a load from one supply to another (→ load transfer switch) If fitted with an appropriate → arc chamber, a knife switch may also be suitable for on-load switching operations.

KOALA

The Siemens KOALA system provides programmable interconnection and communication between items of switchgear. KOALA stands for "**Ko**mmunikation für einfachen **An**lagenaufbau" which means "communication for simple electrical installation design".

KUSA starter

KUSA is the abbreviation for the German "**Ku**rzschluß-läufer-**Sa**nftanlauf", which means soft-start for short-circuited rotors (squirrel-cage motors). Please refer to → Stator resistance starter and → Soft starter.

Lagging NC auxiliary contact

See → Late-break NC contact

Laminated core

See → Magnet core, → Eddy-current losses

Latch

The latch is that part of the → operating mechanism of e.g. a → circuit-breaker (or other → switching device with internally stored energy for the → opening operation) which prevents its opening until it receives the corresponding signal (mechanical or electromagnetic) to release the opening energy. Also see → Mechanical latching (of a contactor).

Latched contactor

A → contactor which incorporates → mechanical latching. The latching and unlatching may be effected mechanically, magnetically, electrically, pneumatically, etc.

Note:
Owing to the latching mechanism, the latched contactor has a second "rest position". Strictly speaking, it is therefor not a contactor in terms of the definition. Since the latched contactor is more closely related to a contactor than to any other type of → switching device with regard to its application and design, one may presume that it meets all the requirements which are relevant for contactors.

Latching mechanism

See → Latch

Latching pushbutton

See → Pushbutton

Late-break NC auxiliary contact

An → auxiliary contact on a → switching device (→ normally closed contact) which opens after of the → main contacts (and other NC auxiliary contacts). Physically, this may be achieved by extending the height of the → contact tips in the NC → switching element (hence the term "extended contacts"). A late-break NC contact may also be termed "lagging contact". Also refer to → Make-before-break.

Lateral approach (to a → proximity switch)

The approach of the → target along a path perpendicular to the → reference axis of the proximity switch.

LC circuit

See → Filter circuits

Leading NO auxiliary contact

See → Early-make NO contact

LED

A **l**ight-**e**mitting **d**iode which serves the indication of a signal state. The LED lights up ($\otimes$), e.g. if a load is applied.

Let-through current $\hat{i}_D$

The highest instantaneous value (peak value) of the current during the breaking operation (i.e. within the → break time) of a →switching device or →fuse link. If the let-through current is lower than the →prospective short-circuit current, then one speaks of →current-limiting (e.g. current-limiting fuse, current-limiting →circuit-breaker). The let-through current and the break time at the specific voltage determine the →let-through $I^2 \cdot t$ value which is decisive for the thermal stressing of all the equipment in the current path.

Let-through $I^2 \cdot t$ value

The heat value of the →let-through current.
In the case of →fuses, a comparison between the let-through $I^2 \cdot t$ value of an upstream →fuse link with the →pre-arcing $I^2 \cdot t$ value of a downstream counterpart (of lower →rated current), may be used to determine whether they will operate in a discriminative manner in the event of a short-circuit (see →Discrimination). For series-connected →circuit-breakers and other electromechanical devices, the let-through $I^2 \cdot t$ value is decisive in determining the thermal stressing. A comparison of maximum and minimum $I^2 \cdot t$ values alone is however not sufficient to determine the degree of →coordination or discrimination which may be achieved.

Light-ON function (of a →photoelectric proximity switch)

The light-ON function means that the relevant output is switched ON (i.e. closed or conducting) when light *is* detected by the →receiver. This corresponds to the →normally open contact function (NO) The opposite is known as the →dark-ON function.

Limit switch

See →Position switch

Limits of operation

Maximum and minimum values of e.g. →control supply voltage and →operating time under given operating conditions are specified in the corresponding product standards for electrical or pneumatic →operating mechanisms, →undervoltage releases, →shunt releases, →magnet drives, etc.
Examples (from IEC 947-1 unless otherwise stated):
- An undervoltage release shall operate to trip the associated →switching device (even on a slowly falling voltage) within the range between 70% and 35% of its rated control supply voltage U_s. It shall *prevent* closing of the associated device if the supply voltage is below 35% of U_s, and *permit* closing at voltages $\geq 85\%$ of U_s. The upper limit of the supply voltage is usually accepted as being 110% of U_s (Also refer to →No-volt release).
- A shunt release shall cause tripping of the associated switching device under all normal operating conditions if the control supply voltage measured during the tripping operation remains at between 70% and 110% of the rated value (at rated frequency in the case of a.c.)
- A contactor (IEC 947-4-1) shall close reliably at any value of control supply voltage between 85% and 110% of the rated value (→operating voltage tolerance). The limits at which contactors shall drop out and open fully are 75% to 20% of U_s for a.c. operation, and 75% to 10% of U_s for d.c. operation.

Line starter

See →Direct-on-line starter

Linear loads

When sinusoidal voltage is applied to a linear load, it draws a sinusoidal current. There may be a phase shift (lagging or leading) between the driving voltage and the drawn current.

Live parts

Conductive parts of electrical equipment which are live (carry voltage) under normal operating conditions. They include the →neutral conductor (N) and any conductive parts connected to it unless the neutral conductor is also the protective earth (→PEN conductor).

Load

Any consumer of electrical energy or, more precisely, any device, machine or piece of equipment which converts electrical energy into another, non-electrical form (e.g. heat, mechanical movement, light). Particularly in a.c. engineering, loads are divided into three categories:
Ohmic loads, which do not cause a significant phase shift between the driving voltage and the drawn current (e.g. heating, →incandescent lamps).
Inductive loads, which cause the current to lag behind the driving voltage (e.g. motors, coils, →electromagnets).
Capacitive loads, which cause the current to lead the driving voltage (e.g. capacitors).

Load cycle

If a load is increased and decreased repetitively (including being switched off i.e. the off-load state, being braked or reversed), then each distinct complete and recurrent operation (i.e. including high and low loading or the off-state) is known as a load cycle (or "duty cycle" see →Operational duty)

Load duration

The period of time for which a load current flows (see →Intermittent periodic duty)

Load factor

See →On-load factor

Load shedding

A technique whereby less important →loads are disconnected from the supply in time to prevent overloading of the supply network or before a specified →maximum demand value is exceeded.

Load switch

A →switching device for switching items of equipment and system parts on and off under normal operating conditions. Its →switching capacity is usually of an order of magnitude equal to its →rated operational current and is specified by the manufacturer.

Load transfer switch

See → Transfer switch

Local bus

A → bus primarily designed for → data exchange between → switching devices with communication facilities (e.g. → SIMOCODE, → KOALA) and process automation equipment (e.g. → programmable controllers) in a closely defined (local) protected environment (e.g. within a switchboard, control room or building).

Local control

The control or → operation of a → switching device (or other piece of equipment) from a point on or adjacent to the device itself (including on the device housing, cover or enclosure). Also see → Remote control).

Local service box

A point of power supply (distribution board, panel, → circuit-breaker or fuse base) in an electricity distribution network from which a consumer may draw electrical energy provided that certain regional connection specifications (refer to → TAB) have been fulfilled (e.g. a meter cabinet with an approved electricity meter has been installed by a recognized electrical contractor for billing purposes). See → Consumer installation

Locked-rotor current

The effective value of the maximum current drawn in by a three-phase induction motor at standstill when → rated operational voltage at → rated operational frequency is applied to its terminals (i.e. upon starting and prior to turning of the rotor, or under locked rotor conditions). It is a calculated value and disregards transient currents (see → Inrush current)

LOVAG

The **L**ow **V**oltage **A**greement **G**roup strives to achieve reciprocal acceptance of testing and certification documents for low-voltage switchgear between the various national associations in Europe. Its initial members include ASEFA (France), → ASTA (England) and → ALPHA (Germany).

Low voltage

Designation of the voltage range (rated voltages) up to 1000 V a.c. or 1500 V d.c.

Low-voltage switchgear and controlgear assembly

This term is used to refer to a combination of one or more items of → low-voltage switchgear (and/or controlgear) with the necessary associated equipment for control, measurement, signalling, protection and/or adjustment purposes, etc. together with all the internal electrical and mechanical connections as well as the structural parts (e.g. chassis, mounting hardware, cladding) which has been assembled or put together entirely by the manufacturer (or within his scope of responsibility). Previously, the term → factory-built assembly was used.

The requirements for low voltage switchgear and controlgear assemblies are specified in DIN VDE 0660 part 500 and IEC 439-1 standards, whereby a distinction is made between → type-tested assemblies (TTA) and → partially type-tested (PTTA) assemblies.

TTA: Switchgear assemblies which do not deviate significantly from the original type or system which has been → type-tested and shown to comply with the requirements of the standard.

PTTA: Switchgear assemblies which contain type-tested and/or non type-tested elements and for which compliance with the requirements of the relevant standard has been demonstrated.

The difference between TTA's and PTTA's lies essentially in the nature and the extent of tests to be carried out. For example, the determination of the heat rise may also be done by calculation instead of by actual measurement.

Low-pass filters

See → Audio-frequency blocking circuits

LV HRC fuse

Low-**v**oltage **h**igh-**r**upturing-**c**apacity → fuse. It consists of a → fuse base (single or triple pole) and the cartridge-type → fuse link. Typically, the → rupturing capacity of an LV HRC fuse link is in excess of 100 kA. Standardized sizes (e.g. size 00, 1, 2, etc.) and dimensions (e.g. as specified in DIN 43620) ensure the interchangeability between fuse links from different manufacturers. The basic system is designed for use (e.g. replacing of a fuse link) by skilled or suitably qualified personnel.

In the case of HRC fuses to DIN 43620, the fuse links are usually equipped with blade terminals which are plugged into lyre-shaped contacts on the fuse base. Special fuse pullers are used to withdraw the fuse links. These fuse pullers clip onto gripping lugs or tags on the fuse-link body. Depending on the fuse-link version, these lugs may be insulated or live.

In the case of HRC fuses to the BS 88 standard, the → cartridge fuse links of lower current rating may be plugged into corresponding fuse bases by means of terminal blades in a similar way to that described above for DIN fuse links. For higher rated currents, they are equipped with centre or offset tags which are bolted to the terminals of the fuse bases.

LV HRC fuse links are classified in terms of their → function class and their → operation class. LV HRC fuse links are also commonly used in conjunction with → fuse-combination units.

Lyre contact

A term used to refer to a fixed → contact which has a "pinched U" shape (form of a lyre) into which the blade-like → moving contact is inserted during the → closing operation. Since the currents flowing in each of the two parallel tines (sides or cheeks) of the lyre contact (in the closed position)

are always in the same direction, they attract each other electrodynamically. The higher the current, the greater the →contact pressure on the moving contact blade between the tines will be. Lyre contacts are used extensively on →fuse bases for →LV HRC →fuses as well as on →fuse switch disconnectors and →switch disconnectors with fuses. The terminal blades of the HRC fuses (DIN standard) themselves are inserted into the lyre contacts.

Magnet armature

See →Armature

Magnet core

The parts of a →magnetic circuit (e.g. in a transformer or a →magnetic drive) which pass through the inside of the coil(s), and in which the magnetic flux is concentrated. In magnet drives used in →contactors, for example, the magnet core usually consists of a stationary or fixed part and a movable →armature. One can distinguish between the magnet cores of →electromagnetic operating mechanisms for a.c.[1]) and d.c. excitation:

The magnet core of an a.c. magnet drive usually has a yoke with two or three limbs (U or E shaped core respectively) and is made up of a number of plates or laminations to reduce →eddy-current losses (→laminated core). →Shading rings are embedded in the →polefaces of the core to ensure that the attracting force between the limbs and the armature does not disappear every half cycle as the a.c. current in the coil passes through zero.

The magnet core of a d.c. magnet drive is usually made of a solid piece of soft iron (soft-iron core), and the armature may be in the form of a solid hinged metal plate or so-called "clap anchor".

Magnet drive

Commonly used to refer to the →electromagnetic operating mechanism of a →contactor or →moulded case circuit-breaker.

Magnet system

This term is often used when a distinction is made between →electromagnetic operating mechanisms specifically designed for a.c. or d.c. →control supply voltage respectively. An a.c. magnet system is characterized by a →laminated core, where-as d.c. magnet systems tend to use →soft iron cores for the flux path (also refer to →Economy connection). The term "magnet system" is also commonly used to refer to all the parts comprising the electromagnetic operating mechanism (→magnet drive) of e.g. a →contactor, →electromagnetic overcurrent release, etc.

Magnetic blowout

See →Blowout coil

---

[1]) Note that an a.c. magnet drive may be operated with direct current too. Refer to →Economy connection.

Magnetic circuit

Magnetic lines of force (making up a magnetic field) are created by a current flowing in a →conductor or a conductor coil. Although the magnetic field does not actually flow, the number of lines of force are referred to as the magnetic flux and the lines themselves form a closed path (from the north to the south →pole) known as the magnetic circuit. This flux path is determined by the permeability of the materials between the north and south pole of the magnetic field. It usually passes through a number of different substances, generally air (→air gap) and iron (→magnet core).

Magnetic field withstandability (of a →proximity switch)

The ability of the proximity switch to operate in a strong magnetic field. Special versions of →BERO proximity switches are offered for use in (or near to) welding machines. They are, for example, able to operate reliably at a distance of approx. 25 mm (1 inch) from a welding current of 21 kA (r.m.s.).

Magnetically latching contactor

See →Remanence contactor

Main circuit of a switching device

All the conductive parts of a switching device included in the circuit which it is designed to close or open (also refer to →Power circuit).

Main contact

A →contact in the →main circuit of a mechanical →switching device. In the closed position, it carries the current of the main circuit.

Main switch (Main →switching device)

Every industrial machine which is subject to DIN VDE 0113 part 1 must be equipped with a main switch which disconnects (→isolation) all the electrical equipment from the network for the duration of cleaning, maintenance and repair work as well as for longer shutdown periods.

Usually, it is a specified hand-operated switch intended to provide a means of protection in the event of electrical and/or mechanical dangers. The main switch may also serve as the →Emergency-Off switching device.

Main switches must fulfil the following specifications:

1. The hand-operated mechanism (→operating mechanism) must be easily accessible from the outside of the →enclosure.
2. It may have only one definite "ON" and "OFF" position respectively, each with assigned stops.
3. The two positions must be marked "O" and "I".
4. It must be lockable in the "OFF" position.
5. The incoming supply terminals must be suitably shrouded to prevent accidental touching of live parts.
6. The →switching capacity must correspond to AC-23 in the case of →motor switches and AC-22 for general →load switches (refer to →Utilization category and IEC 947-1).

7. A positive indication of the →switching state (of the →main contacts) is obligatory.

Maintained-contact command device

A →control switch or →command device with at least two positions of rest, and without stored energy (spring) return. Examples include ON/OFF switches (→single-throw switches), →double-throw switches, rotary selection switches (excluding spring return positions), latching →pushbuttons, etc.

Maintained-contact control

This term is used to mean the operation or control of a →switching device, →switchgear assembly (e.g. starter) or any piece of electrical equipment by *maintained* →command signals as opposed to short pulse-type application/interruption of the →control supply voltage. Typically, this is achieved by means of one or more →maintained-contact command devices of which the →contacts remain in the operated condition after the actuating force has been removed, e.g. →single-throw switches. This method of operation is also referred to loosely as "ON/OFF switch control"

Make function (of a →proximity switch)

The output of a proximity switch is said to have a make function if load current is caused:
to flow when the presence of a →target is sensed, and
not to flow when the target is not detected.
It corresponds to the →normally open contact function (NO)

Make time

This is the period of time between the initiation of the →closing operation and the instant the contacts of the first pole to close touch. (i.e. the instant when current begins to flow in the main circuit). The make time is also known as the "closing delay" and is the sum of the time taken for the operating mechanism to start moving (inherent mechanical delay, or →movement delay) and the time taken for the travel of the moving contacts.

Make-before-break

If, in a pair of →auxiliary contacts (i.e. 1NO+1NC, both operated by the same actuating force), the →normally open contact closes before the →normally closed contact has opened, one refers to the pair as "make-before-break" contacts. Physically, this may be achieved by extending the height of the →contact tips in both the NO and NC →switching elements (hence the term "extended contacts"), and thus creating an →early make NO and a →late break NC contact combination. Make-before-break contacts are also referred to as "overlapping contacts".

Make-break operation

A single →closing operation followed by an →opening operation of a →switching device. Also refer to →Operating cycle

Making capacity

This is the value of the →prospective making current which the switching device is capable of switching on under prescribed conditions of use and behaviour. The making capacity for a.c. contactors is stated as the symmetrical r.m.s. value of the a.c. component of the making current. For circuit-breakers, the making capacity is expressed as the maximum instantaneous value of the prospective current at the input terminals at the prescribed voltage. In the case of the *short-circuit* making capacity, the prescribed conditions include a short-circuit at the load-end terminals of the switching device.

Making operation

See →Closing operation

Malfunction

See →Operational reliability

Manual reset feature

A feature by which e.g. on overload protection device (→overload relay, →circuit-breaker) must first be reset after a →tripping operation before it can be put back into operation.
Example: After an →overload relay in a →motor starter has operated and caused the current flow to be interrupted, its →bimetallic strips cool down and return to their initial position. In the →automatic reset mode, the →auxiliary switch also returns to its initial →switching state. In the case of →maintained contact control, the starter would cause automatic restarting of the motor — a potentially dangerous situation. If the manual reset mode is selected on the overload relay, then the motor can only be restarted once the relay has been reset. Manual resetting is only possible once the bimetallic strips have cooled down sufficiently (or a delay time has expired in the case of electronic overload relays). Manual resetting may also be achieved by →remote control if a reset unit is fitted to the overload relay.
The manual reset may also be termed →reclosing lock-out.

Master controller (or master control switch)

A term commonly used to refer to a complex →control switch (usually a →cam switch with a →joy stick →actuator) which operates contactor combinations in a particular sequence, typically in crane control applications.

Material group

Insulating materials are grouped in terms of their →comparative tracking index. The material groups, in turn, are used to specify minimum →creepage distances.

Maximum demand

Particularly in industrial networks, the supply of electricity is not only billed in terms of the actual consumption units (e.g. kWh), but also in terms of the so-called maximum demand drawn during the billing period. It is usually determined as the maximum value of apparent power in kVA or MVA drawn over a fixed integral period of e.g. 30 minutes at any time during the billing period, and is based on the fact that conductors, transformers and other switchgear between the power source and the consumer must be dimensioned to deliver this power (even if only once per month!). Methods of reducing the maximum demand, and thereby saving electricity costs, include selectively controlled → load shedding and → power factor correction (a low → power factor means that the → apparent power kVA is significantly larger than the useful → active power!)

Maximum operating distance (of a → proximity switch)

The upper limit of the → sensing range of an → ultrasonic or → photoelectric proximity switch.

Maximum prospective peak current (in an a.c. circuit)

The → prospective peak current if the → contacts which close to initiate the current flow do so at the instant which leads to the highest possible value.

Note:
In the case of a multipole → switching device (e.g. 3-pole → circuit-breaker) in a multiphase circuit, the maximum prospective peak current refers to the current in one pole only.

MCB

See → **M**iniature **c**ircuit-**b**reaker

MCC

See → **M**otor **C**ontrol **C**entre

MCC handle mechanism

See → Operating mechanism, → Operating mechanism with detachable coupling. Also refer to → Mechanical interlock.

MCCB

See → **M**oulded **C**ase **C**ircuit-**B**reaker

MCM

Mille **C**ircular **M**il.
See → kcmil

MCP

Motor **C**ircuit **P**rotector. This term is used to refer to a → moulded case circuit-breaker which incorporates an → instantaneous short-circuit release only. Overload protection for the motor is provided, for example, by an → overload relay in the → motor starter.

Mechanical delay

See → Movement delay

Mechanical interlock

An → interlock which operates mechanically (i.e. by non-electrical means). It may take the form of a rocker assembly, a set of levers, a latch mechanism, → Bowden cables, key and lock arrangements, etc. Typical examples/applications include:
1. The interlocking device between two → contactors in a → reversing starter or between the star and delta contactors in an automatic → star-delta starter. Each contactor in the interlocked pair is prevented from closing if the other is already closed, thereby preventing a short-circuit upon load change-over from one contactor to the other. Note, that the mechanical interlock alone may not *delay* the successive operation of the contactors. If it is ensured electrically (i.e. with an → electrical interlock and/or timing device) that the first contactor has opened (including extinction of the → switching arc!) before the → contacts of the second contactor touch, then a mechanical interlock may not be necessary. Mechanical interlocks between contactors also prevent simultaneous closing of the two contactors as a result of strong vibration or mechanical shocks.
2. The interlocking device or mechanism between two or more circuit-breakers which determines that only a specific number (e.g. one or two) breakers may be closed at the same time. For example, in the case of two feeder circuit-breakers from two different transformers joined on their load side by a → bus coupler, a special mechanical interlocking arrangement can ensure that either both feeder breakers are closed and the coupler is open, or that only one feeder breaker is closed while the coupler is closed. This arrangement is often achieved by means of locks attached to the circuit-breakers which permit closing only once the key has been inserted, turned and held captive in the lock. Since only two keys are available for the three circuit-breakers, only two can be closed at any one time. This arrangement is also known as "Castell interlocking", whereby the word "Castell" is a trade name of a particular range of high security locks and keys and "Castell interlocking" could therefore be used to refer to any lock and key type of interlocking arrangement. Bowden cables or pushrods are also used to achieve this form of interlocking.
3. The door interlock on the → operating mechanism handle of a → switching device. Such an interlock may be used, for example, to prevent opening of a cabinet door while the disconnector behind the door is in the closed or "ON" position (also known as an "MCC handle mechanism").

Mechanical latching (of a contactor)

Some → contactors and → contactor relays incorporate a mechanism (or permit the fitting of an accessory block or module) which prevents the → contacts from returning to their position of rest after the energizing signal has been removed. Thus, the → contacts remain in the operated state even if the → control supply voltage fails. The mechanical latching module incorporates a delatching magnet system, and drop-out of the contactor is achieved by a voltage pulse on the coil terminals of this electromagnet. This technique is not only used to save power in the control circuit in cases where contactors are to remain energized for long periods of time, but is also used if a specific switching state of the contacts in a control circuit should be maintained during voltage interruptions. A similar effect may be achieved using → remanence contactors.

Mechanical service life

The → service life of a → switching device under → off-load switching conditions.

Melting time

See → Pre-arcing time

Meshed network

In a meshed network, power is distributed by a grid of power lines. In general, the power is fed in at several points. If one line fails, all loads are supplied via the remaining branches (switching over is not necessary). For comparison, also refer to → Radial network and → Ring network).

Metcalf Effect (or "M" effect)

This effect is used to reduce the melting point of a → fuse element at specific locations along its length. If a deposit of metal (e.g. tin) that will alloy with the element (e.g. silver) under the influence of temperature is placed at convenient points on the element surface, the eutectic alloy which forms will melt at a considerably lower temperature than that which would be necessary to melt the base material of the element alone. In this way, more rapid operation of the → fuse link may be achieved at low fault currents without significantly affecting its short-circuit rupturing performance.

Micro-environment

This term is used to define the quality of → clearance and → creepage distances. It refers to the ambient conditions immediately around the distances in question, and includes factors such as climatic conditions, electromagnetic field strength, local temperature, → pollution, etc.

Miniature circuit-breaker (MCB; m.c.b.)

Miniature → circuit-breakers typically have current ratings below 50 A and are predominantly used in domestic-type distribution boards to protect wiring in wiring systems and installations against → overload and → short-circuits. M.c.b.'s with the B-characteristic (see → Tripping characteristic) are designed for line protection. Their rating should correspond to the cross-section of the → conductor being protected in accordance with VDE 0100 part 430. M.c.b.'s with the C-characteristic may also be used for line protection, but are generally more suited to the protection of equipment - e.g. small transformers, motors and groups of → incandescent lamps. The two tripping characteristics B and C to DIN IEC 898/1987 simplify the coordination between protection devices in the overload range.
Siemens miniature circuit-breakers form part of the well-known → N-System range of products.

Minimum command time (or duration)

The minimum period of time for which the rated signal (→ command signal) must be applied to the terminals of the electrical → operating mechanism of a switching device (e.g. to the coil terminals of a → contactor or the input terminals of a → time relay) before it will cause the device to operate (e.g. open or close its → contacts) reliably. If the command signal is not applied for at least the minimum command time specified for the switching device, the operating mechanism should remain in (or return to) its initial unenergized state.

Minimum operating distance (of a → proximity switch)

The lower limit of the → sensing range of an → ultrasonic or → photoelectric proximity switch. It corresponds to the distance between the → sensing face and the end of the → blind zone.

Minimum operational current I_m (of a → proximity switch)

The lowest current which is necessary to maintain ON-state conduction of the electronic → switching element

MINIZED

A Siemens trade name for a range of → N-type → switch-fuses which utilize → NEOZED → fuse links. A mechanical locking arrangement prevents closing of the switch unless all the fuse links have been screwed or clipped securely in position.

MKK dry-type power capacitor

The abbreviation MKK refers to a particular design of → power capacitors. Siemens three-phase MKK capacitors consist of a roll of low-loss plastic foil dielectric. The metal is vapour-deposited directly onto the plastic foil which serves as both the dielectric and the electrode carrier. The three phases are wound successively on the same roll and are separated by an insulating layer. The roll is mounted in a compact metal canister filled with → SF_6 gas (i.e. they are → dry-type power capacitors). The connections to the terminals are accomplished via fuse wires which break if the pressure inside the capacitor should become excessively high. High voltage spikes in the network may cause flashover between the electrodes by puncturing the dielectric substrate. This would ultimately lead to overloading. MKK capacitors can "heal" these punctures through the localized evaporation of the metal layer around point of fault. They are therefor classified as "self-healing capacitors".

M = metallisierte (metalized)
K = Kunststoffolie (plastic foil)
K = kompakt (compact)

MKV power capacitor

The abbreviation MKV refers to a particular design of → power capacitors. Siemens three-phase MKV capacitors consist of rolls of metalized paper. Each layer of paper is separated by a low-loss plastic (polypropylene) foil dielectric.

The rolls are impregnated with a mineral oil and are mounted in a metal canister. The connections to the terminals are accomplished via fuse wires which break if the pressure inside the capacitor should become excessively high. High voltage spikes in the network may cause flash-over between the metalized paper electrodes by puncturing the dielectric. This would ultimately lead to overloading. MKV capacitors can "heal" these punctures through the localized evaporation of the metal layer around point of fault. They are therefor classified as "self-healing capacitors".

M = metallisiertes Papier (paper with a metal-vapour deposit)
K = Kunststoffdielektrikum (plastic dielectric)
V = verlustarm (low losses)

Modular width

This term is used to refer to the width of the narrowest standardized element in a system of equipment, housings or switchgear designed to be mounted side-by-side. For example, the modular width of → m.c.b.'s and → fuse bases belonging to the Siemens → N-System is 18 mm. Thus, a single-pole m.c.b. has the width 18 mm, the double-pole a width of 36 mm, etc. D01 and D02 → NEOZED fuse bases have a width of 27 mm, i.e. 1.5 modular widths. The modular width is also referred to as "spacing unit".

Modulated light beam

See → Pulsed light beam (of a → photoelectric proximity switch)

Momentary-contact command device

A → command device which gives a command or signal (e.g. → closes) only for as long as it is → actuated. The moment the actuating force is removed, the device returns to its initial → switching state (see → Pushbutton).

Momentary-contact control

This term is used to mean the operation or control of a → switching device, → switchgear assembly (e.g. starter) or any piece of electrical equipment by → command signals of *short duration* (typically by means of one or more → pushbuttons). Circuits with → contactors and/or → contactor relays employing momentary contact control (and → sealing-in contacts) generally have the characteristic, that a conscious renewed "start" or "on" command signal needs to be given after an interruption of the → control supply voltage. This can be of advantage in terms of operational safety.

Motor circuit-breaker

See → Circuit-breaker for motor protection

Motor Control Centre (MCC)

The term is used to refer to a special design of low-voltage switchboards containing several withdrawable → motor starter assemblies. Each motor starter (or feeder circuit) is contained in its own compartment and is equipped with an → operating mechanism with a detachable coupling (→ MCC handle mechanism).

Motor drive

See → Operating mechanism

Motor protection circuit-breaker

See → Circuit-breaker for motor protection

Motor starter

The combination of all the switchgear necessary to start (see → Starting) and stop a motor and to provide it with suitable overload protection. This could be, for example, a combination of one or more → contactors, → overload relays, → command devices and, if necessary, indicator lights and instruments, all wired together and fitted in an enclosure. Another example would be a → circuit-breaker for motor protection equipped with an → a-release and an → n-release.
A.c. starter types include → direct-on-line starters, → reduced voltage starters, → star-delta starters, → solid-state starters, → auto-transformer starters, as well as → resistance starters for slip-ring motors. (d.c. starters are not within the scope of this book).

Motor switch

A → control switch used in the → starting and control of induction motors. It must have a → switching capacity, which corresponds to the → starting current of the motor (see → Utilization category).

Moulded case circuit-breaker (MCCB)

A → circuit-breaker of which the internal mechanism (→ operating mechanism, → switching elements, → arc chamber) is enclosed and mechanically supported by a compact → housing of moulded insulating material. Since the supporting parts are not normally able to withstand the strong electrodynamic forces of severe short-circuit currents for long, these circuit-breakers either have a relatively low short-circuit → breaking capacity in relation to their → operational current (→ zero-point quenching types), or they open extremely quickly in the event of a short-circuit and are thus of the current-limiting type (see → Current-limiting circuit-breaker).

Mounting (of an → inductive or capacitive proximity switch)

Refer to → Embeddable and → Non-embeddable → BERO proximity switches

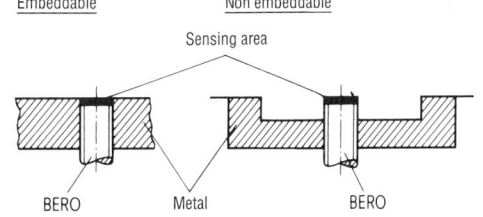

Mounting position

Normally, the permissible angular deviation from the perpendicular to a vertical mounting surface is given in the catalogues. During installation of a →switching device, this permissible mounting position must be taken into account. Larger deviations can, for example, cause a reduction in the →rated operating current, the permissible tolerance of the →control supply voltage for electromagnetic →operating mechanisms, the →switching capacity, the →electrical and mechanical service life and/or a change in tripping characteristics.

Movement delay (contact movement delay)

The time period between the specified instant of initiation of the →closing operation or the →opening operation of a switching device and the moment its moving contacts begin to move. It is determined by factors such as the →minimum command duration, a possible preset time delay (electronic or mechanical) as well as the dynamics of the contact system and operating mechanism themselves (e.g. inertia, friction, etc.). It is also known as the "time to contact movement" or the "inherent delay".

MSP

This abbreviation stands for "motor starter protector" and refers to a →moulded case circuit-breaker which is specifically designed for the starting and protection of motors (typically below 63 A). It incorporates an adjustable →a-release and a fixed →n-release. See →Circuit-breaker for starter assemblies and →Circuit-breaker for motor protection

Multiple breaking

The series connection of →contacts in one or more →switching devices to achieve a higher →breaking capacity. It relies on the (virtually) simultaneous opening of the individual contacts so that the →arc voltages of the →switching arcs are added to oppose the driving voltage and force the current to zero (or prevent restriking after a current zero).
Multiple breaking may, for example, be used to increase the d.c. switching capacity of an a.c. switching device at a particular voltage. It is also used in the so-called →cascade connection by means of which the short-circuit breaking capacity of two →MCCB's in series may be increased provided that both open at the same time.
The most common form of multiple breaking is that of →double-break contacts.

N System

A wide range of Siemens low-voltage switchgear (particularly →miniature circuit-breakers and →fuses) including cabinets and accessories specifically suited to space-saving installations; primarily in the field of domestic distribution. The flush-mounting distribution boards of the system require a recess only 65 mm deep, where-as the depth of the surface mounting cabinet distribution board is only 70 mm. The equipment is designed for simple snap-on mounting on 35 mm DIN mounting rails, and all items have a standard mounting depth. The equipment widths are always a multiple of the standard →modular width 18 mm.

n-release

Short designation for an →instantaneous short-circuit release.

N-type

Term used to refer to switchgear belonging to the Siemens →N-System range of equipment.

NAMUR (→proximity switches)

Abbreviation for **N**orme**na**usschuß für **M**eß- **u**nd **R**egelungstechnik (Standards committee for measurements and control). →BERO inductive proximity switches to NAMUR and DIN 19234 are 2 wire d.c. units which incorporate only the oscillator to produce the electromagnetic sensing field. Owing to their low-impedance input, they are particularly insensitive to inductive and capacitive coupling effect. In conjunction with special series-connected switching modules, these BERO's may be used in intrinsically safe circuits.

Natural frequency

See →Resonance

NBS

New **B**ritish **S**tandard. This is an outdated, non-metric wire gauge which expresses the conductor diameter in inches.

NC

See →Normally closed contact

NEC

National **E**lectrical **C**ode of the USA. It is sponsored by the National Fire Protection Association (NFPA) and although it is intended to be purely advisory, it is used in law and for regulatory purposes.

NEMA sizes/EEMAC sizes
(for →contactors and →motor starters)

Both the NEMA standard of the USA (National Electrical Manufacturers' Association) and the EEMAC standard (Electrical and Electronic Manufacturers' Association of Canada) define a system of contactor and motor starter sizes based on *standardized load ratings*. In contrast, the IEC philosophy is based on a set of →utilization categories or *standardized test criteria* to define the application and match the starter to the actual load and switching duty required. Traditionally, NEMA (or EEMAC) starters are also characterized by replaceable →heater elements on the →overload relay, replaceable main →contacts (even for low ratings) and standardized mounting dimensions (backplates).

NEOZED

The trade name for a range of Siemens →fuses and accessories. A NEOZED, or "D0" fuse consists of a →fuse base, a protective insulating cover, an adapter sleeve, the → fuse link and a screw cap. NEOZED wire and cable protection fuses are available in 3 standardized sizes (viz. D01, D02, D03) from 2 to 100 A rated current and with a rated voltage of AC 415 V/DC 250 V. The widths of the NEOZED fuses are multiples of the 18 mm → modular widths of the → N-type m.c.b.'s and other items, and they have a particularly low profile making them suitable for use in shallow distribution boards. The adapter sleeve has an annular recess with a diameter corresponding to that of the appropriate fuse link. NEOZED fuses may be used in conjunction with the N-type → MINIZED fuse-switch.

Network circuit-breakers

See → Circuit-breakers for mesh-connected systems

Network master relay (release)

These units are used in conjunction with → circuit-breakers for mesh-connected systems. They provide rapid and selective isolation of damaged high-voltage feeder cables. The relay senses the reversal in the direction of the energy flow if a short-circuit in the high voltage feeder cable causes high currents to flow from the → meshed network through the associated transformers to the point of fault, and causes instantaneous tripping of the associated circuit-breaker by means of a → shunt trip and → energy storage device.

Network types and systems

An electrical network is classified in terms of its → operational voltage, its type and the system employed. As far as the voltage is concerned, one can distinguish between:
– Low-voltage networks - up to 1000 V,
– Medium-voltage networks - above 1 and up to 60 kV,
– High-voltage networks - above 60 and up to 380 kV,
– Extra high-voltage networks - above 380 kV.

A distinction is made between two network types viz.:
– Open networks (e.g. → radial network) and
– Closed networks (e.g. → ring network, → meshed network).

An electrical system consists of a single source of electrical energy and an installation (see → Electrical installation). The type of system is determined by the relationship between the source (and/or exposed → conductive parts of the installation) and → earth. The following systems are common in the field of low-voltage engineering:

TN system
In TN systems, one or more points of the energy source are connected directly to earth (star point or phase conductor). The exposed conductive parts of the installation are connected to that point by means of protective → conductors (PE or PEN conductor). The following versions of TN systems are defined:

TN-S system
Neutral conductor (N) and protective conductor (PE) are separated from each other throughout the system.

TN-C system (network)
Neutral and protective functions are combined in a single conductor throughout the system (PEN).

TN-C-S systems (network)
Neutral and protective functions are combined in a single conductor, the PEN conductor, in part of the system. In the rest of the system, neutral and protective conductors are separate.

TT system
In a TT system, one point of the source is earthed directly. The exposed conductive parts of the installation are connected to separate earth electrodes.

IT system
In an IT system, there is no direct connection between live parts and earth. The exposed conductive parts of the installation are earthed.

Neutral conductor (N)

A conductor which is connected to the neutral point (or star point) of a system and contributes to the transmission of electrical energy. It may or may not be combined with the → protective conductor (also refer to → Network types and systems)

NFPA

National Fire Protection Association (USA)

NO

See → Normally open contact

No-load condition

The operating state of a motor when it is connected to its rated mains supply, but the total mechanical load is made up of its own frictional losses only (if the term is used to refer to the complete drive, i.e. including the driven machine, then the frictional losses include those of the unloaded driven machine). Under no-load conditions, the motor draws its "no-load current" which is mainly made up of the magnetizing current and the small load current required to overcome friction (also refer to → On-load condition and to → Off-load condition).

No-load current

The input current of an unloaded transformer (or motor) at → rated operational voltage and frequency.

No-load power

The power input to a transformer when it is connected to its → rated operational voltage at rated network frequency on the primary side, and the output (secondary) is not loaded (i.e. open circuited). Analogously, it is the power input to a motor (at rated voltage and frequency) when no (mechanical) power is drawn from its shaft.

No-load supply current I_o (of a → proximity switch)

The current drawn by a 3 or 4 wire proximity switch from its supply when all the loads in its output circuits are disconnected.

No-load voltage

The output voltage of an unloaded transformer at rated input voltage and rated frequency.

No-volt release

A special form of → undervoltage release. Their operating voltage is between 35% and 10% of the rated → control supply voltage. Also refer to → Limits of operation).

Noise

Electrically speaking, this term refers to random, undesirable voltage and/or electromagnetic disturbances (usually of higher frequency and amplitude) in the vicinity of electrical equipment or superimposed on the voltage/current waveform in the circuit conductors. In is also known as "electrical or electromagnetic interference" (see → Electromagnetic compatibility).

Nominal value

A suitable approximate quantity value used to designate or identify a component, device or equipment.

Non-damping material

A material which has a negligible influence on the → response characteristic of a → proximity switch.

Non-embeddable proximity switch

A → proximity switch is said to be "non-embeddable" if it requires a → free zone around its → sensing face in order to maintain its → response characteristic. Refer to → mounting (of a → proximity switch). The term "non-flush-mounting" is also common. For comparison, also refer to → Embeddable proximity switch.

Non-flush-mounted proximity switch

See → Non-embeddable proximity switch

Non-linear load

Non-linear → loads draw a non-sinusoidal waveform of current when a sinusoidal line voltage is applied to them. Examples include gas discharge lamps, rectifiers, thyristor drives, inverters, etc. This distortion of the current waveform produces → harmonics in the network.

Normally closed contact (NC)

A → control or → auxiliary contact which is closed when the main → contacts of the → switching device are open, and which is opened when the main contacts are closed. Alternatively, it may be defined as a → contact element which opens the → current path when e.g. the → control switch or → contactor is → actuated. It is also known as a "break" or "b" contact or contact element.

Note:
In special contactors in which the main contacts are closed in the rest position and open when the → magnet drive is energized, the main contacts are also referred to as normally closed.

Normally open contact (NO)

A → control or → auxiliary contact which is open when the main → contacts of the → switching device are open, and which is closed when the main contacts are closed. Alternatively, it may be defined as a → contact element which closes the → current path when e.g. the → control switch or → contactor is → actuated. It is also known as a "make" or "a" contact or contact element.

NPN output (of a → proximity switch)

This indicates a specific → type of output stage (semiconductor circuit) used in the electronic → switching element(s) of 3 or 4 wire d.c. proximity switches. If the load must be connected between L+ of the → control supply voltage and the output (i.e. the proximity switch "sinks" the load current when the output is ON or conducting), then the output is known as an "N" or NPN output (from npn transistor).

NTC thermistor (temperature detector)

Thermistor → temperature detectors are used extensively in the protection of motors and transformers against excessive temperature rise (see → Thermistor motor protection). They are embedded directly between the phase windings (in the overhang of the stator winding in the case of a motor); normally during manufacture. An NTC thermistor has a **n**egative **t**emperature **c**oefficient. In contrast to → PTC thermistor detectors, these devices exhibit gradual decreasing resistance (i.e. no abrupt changes) with increasing temperature. The *system* operating temperature (known as the "TFS") is set on the tripping unit (e.g. 3UP7). A separate warning temperature may also be set.

NTC thermistors are preferred in cases where the thermal characteristics of the motor or transformer are not known prior to manufacture (special production), and the temperature limits for warning and/or shut-down need to be set up after installation of the machine.

ODP (motor)

Open **D**rip **P**roof; i.e. an indication of the protection provided by a motor housing (approx. equal to IP 22). Motors with this → degree of protection are also known as "DP motors" (drip proof).

OFF delay

→ Drop-out delay

Off-load condition

A motor is said to be in the off-load condition, when its main terminals are not connected to the supply, i.e. when it draws zero load current (also refer to → On-load condition and to → No-load condition).

Off-load disconnector (isolator)

See → Disconnector

OFF-state current i_r (of a → proximity switch)

See → Residual current i_r

On-load condition

The operating state of a motor when the mechanical output power at its shaft is greater than zero, i.e. when the driven machine draws power from the motor. The on-load condition may correspond to the full-load condition (when the motor draws its rated full load current), or even to an → overload condition. Analogously, the on-load condition for a → switching device, is the switching state in which the → contacts are closed, and a finite current passes through them. Note, however, that the motor also draws a finite current under → no-load conditions (also see → off-load condition).

On-load disconnector (isolator)

See → Disconnector

On-load factor (or load factor)

The ratio of the → on-load time (→ load duration) including → start-up time and → plug-braking time, to the total period of the → load cycle (→ duty cycle), usually expressed as a percentage. It is also known as the → cyclic duration factor and is decisive in determining the thermal loading of a → switching device. Also see → Intermittent periodic duty)

Open position

The open position of (the → contacts of) a mechanical switching device is the position in which the predetermined dielectric withstand voltage requirements are satisfied between the separated contacts in the → main circuit of the device.

Open-type construction (or assembly)

The constructional design of a switchboard, → switchgear assembly (→ TTA or → PTTA) or chassis in or on which the items of electrical equipment are mounted and arranged in such a way that live → conductive parts are directly exposed and accessible.

Opening delay

See → Drop-out delay

Opening operation

The opening operation of a mechanical switching device is the → operation by which the device is brought from the → closed position to the → open position by means of its → operating mechanism or a → tripping operation.

Opening time

The opening time of a mechanical switching device is the interval of time between the specified instant of initiation of the → opening operation (e.g. de-energizing of a coil, application of voltage to a release) and the instant when the arcing contacts have separated in all poles.

Operating current (of an overcurrent relay or release)

The value of current at and above which the relay or release will operate. It is also referred to as the → tripping current.

Operating cycle

A sequence of one or more → switching states of a → switching device and the return to the initial position of rest. Normally, the operating cycle of a switching device is understood to mean e.g. a single → closing operation followed by an → opening operation i.e. a single "make-break" or "on-off" operation. If applicable, however, an operating cycle may include intermediate positions as well.

Note:

If the operating cycle includes several → make-break operations or intermediate positions, one also speaks of an "operating sequence" or "sequence of operations".

Operating distances (of an → inductive proximity switch)

A distance at which the → target approaching the → sensing face of the proximity switch along the → reference axis causes the output signal to change. Since the actual operating distance may be effected by factors such as manufacturing tolerances, material of the → target, temperature, etc. a number of more specific distances have been defined. Also refer to:

→ Rated operating distance s_n

→ Sensing range s_d

→ Minimum operating distance

→ Maximum operating distance

→ Effective operating distance s_r

→ Usable operating distance s_u

→ Assured operating distance s_a

Aktive Fläche = Sensing face

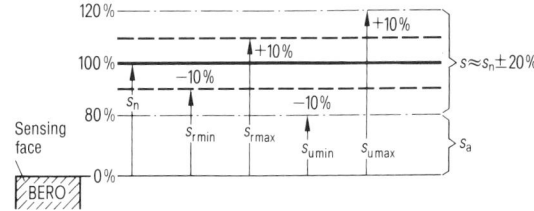

Operating frequency (of a switching device)

The number of → make-break operations per unit time (e.g. in 1 hour) that a → switching device can perform under given load and switching conditions without damage or a significant reduction in its → service life.

9 Appendix

Operating frequency f (of a → proximity switch)

Also known as the frequency of operating cycles, this parameter gives the maximum possible rate (in operating cycles per second) at which the proximity switch can change from the unoperated to the operated condition and back again, while still supplying clearly definable signals to the load circuitry (see DIN EN 50010).

Operating mechanism

Parts of an item of switchgear which cause or transfer movement to the → contact elements. The term is used to refer to mechanical levers, latches and shafts inside the switch (internal operating mechanism) as well as to parts which protrude from the switch or are retrofitted to facilitate operation in a particular application. A number of different operating mechanisms (also known as "operators") are common in the field of low-voltage switchgear. These include: (also refer to → Operation as well as to Section 5.3, Page 277)

a) Hand-operated mechanisms (e.g. toggle, rotary handle mechanism, spade-handle mechanism, MCC handle mechanism)
b) Foot-operated mechanisms,
c) Electromagnetically-operated mechanisms (e.g. solenoid operator on a circuit-breaker, → magnetic drive in a contactor)
d) Motor-operated mechanisms (e.g. motor drive on an air-circuit-breaker)
e) Spring-operated mechanisms (e.g. stored energy operator – spring either charged by hand or motor)
f) Pneumatically operated mechanisms.

Operating mechanism auxiliary switch

An → auxiliary switch mounted directly on the → operating mechanism of e.g. a circuit-breaker. It is directly actuated by the operating mechanism and is independent of the position of the circuit-breaker contacts. In the case of circuit-breakers, this auxiliary switch is usually of the → early-make type and is used to switch the → auxiliary supply voltage to the → undervoltage release. This ensures, that the undervoltage release is disconnected from the supply when the circuit-breaker is in the OFF position and that it is only operated once the operating mechanism is specifically moved towards the ON-position.

Operating mechanism with detachable coupling

A hand-operated switch mechanism in which the operating handle is connected via a coupling. Application: the switch is mounted on a base plate inside the → cubicle or enclosure; a shaft connects the → operating mechanism to the actuating handle mounted outside on the door. A coupling on the shaft permits the door to be opened. By means of an → interlock mechanism on the coupling, the door may be prevented from being opened while the switch is in the ON-position. (known as a "door interlocking handle mechanism" or an "MCC handle mechanism")

Operating speed (of a position switch)

See → Approach speed

Operating time

For example, in the case of a → time relay, this is the period of time between the start command and the operation of the output contacts (also known as run time).
In the case of an → overload relay, it is the time taken for the unit to → trip at a specific value of → overcurrent (which exceeds the setting value by a specified amount) from the cold state.

Operating voltage

See → Control supply voltage

Operating voltage tolerance

The range within which the → operating voltage may deviate from the → rated control supply voltage without affecting the operational reliability of the switching device. In the case of contactors and relays, this is also referred to as the "coil voltage tolerance". The maximum and minimum values of the range are specified in the relevant product standards (e.g. IEC 947-4-1) by the → limits of operation.

Operation (of a switching device)

The transfer of the → moving contact(s) from one position to an adjacent position, i.e. a change in the → switching state of the device.
A distinction is made between a number of types of operation:

1.1 Dependent manual operation refers to operation by means of directly applied manual force (energy) whereby the speed and force of the operation are *dependent* upon the action of the operator (person).
1.2 Independent manual operation refers to operation by means of applied manual force whereby the operating energy is applied, stored and released in one continuous movement so that the speed and force of the operation are *independent* of the action of the operator (e.g. → snap action)
2.1 Dependent power operation is by means of energy other than manual in which the completion of the operation is *dependent* on the continuity of the power supply.
2.2 Independent power operation is a form of stored energy operation in which the energy originates from an external power source (manual or non-manual) and is released in one continuous movement, such that the speed and force of the operation are *independent* of the action of the operator.
3 Stored-energy operation refers to operation by means of energy *stored* in the → operating mechanism at some stage prior to the actual operation itself. This form of operation may be classified in terms of the energy storage method (e.g. spring), the origin of the energy (e.g. electric motor) and/or the way in which the energy is released (e.g. manually).

Furthermore, one can distinguish between → closing operations and → opening operations. Also refer to → Positive opening and → Positively driven operations.

Operation class

In terms of DIN VDE 0636 (IEC 269), low-voltage high rupturing-capacity → fuse links are classified in terms of their function and operating characteristics (also see → Function class). The operation class of an → LV HRC fuse link is indicated by letters, whereby the first letter is used to identify the function class and the subsequent letters to identify the object to be protected. The following list includes a number of standardized identifying letters:

G   general purpose, i.e. cables and insulated wires, (previously L)
M   switching devices,
R   semiconductors,
B   mining equipment,
Tr   transformers.

These result in the following operation classes:

gG/gL   wide-range general-purpose protection for cables, → conductors and wires,
aM   partial range accompanied protection for → switching devices
aR   partial range accompanied protection for semiconductors
gR   wide range general-purpose protection for semiconductors
gB   wide range general-purpose protection for mining installations
gTr   wide-range general-purpose protection for transformers.

Further operation classes include:
träg   or "time delay" and
flink   or "quick response"

which are both offered as original → DIAZED fuse links for cable and wiring circuit protection.

Operational current

See → Rated operational current

Operational duty

Information on the operation of motors (or other loads) under specified conditions (and the consequent operation of the → switching devices. Also refer to → Duty types.

Operational reliability

This term is sometimes used to refer to the quality of an electrical control system, circuit or → switching device, with specific reference to its ability to operate continuously without malfunction.
Typically, malfunction occurs if the → control circuit voltage for an → electromagnetic operating mechanism of a switching device (e.g. a → contactor, → time relay, magnetically operated valve) is too low, so that the required operation cannot be executed when the corresponding → command signal is given.
In order to achieve a high degree of operational reliability, negative influences must be considered and reduced to an absolute minimum during the planning, engineering and construction of the control system. Examples of negative influences on the operational reliability include:
– unstable or unreliable → control supply voltage,
– too many series-connected → contacts in the → control circuit,
– → auxiliary and/or control circuit → conductors which are too long and/or which have insufficient → cross-section (→ voltage drop)
– excessive → contact resistance or low → contact reliability (especially important in the case of → command signals of low voltage and current, e.g. in conjunction with electronic systems).

Operational switching

The → operation of a switching device in such a way that the limits and tolerances specified in the → rated data, e.g. maximum permissible → operating frequency, → rated operational power, etc. are not exceeded.

Operational voltage

See → Rated operational voltage

Operator

This term is generally used to refer to the *person* who e.g. causes the → operation of the → switching device or who works the machine or piece of equipment in question. It is, however, also sometimes used to refer to the → operating mechanism of e.g. a → circuit-breaker.

Optical proximity switch

See → Photoelectric proximity switch

Opto-BERO

The trade name for Siemens → photoelectric proximity switches.

OR-gate

A device or electronic component (integrated circuit) which combines two or more input signals according to the logic OR function of Boolean algebra to a single output signal. Symbol: O. The inverse of the OR function, is the NOR function ($=not$ OR), with the symbol $\bar{\text{N}}$.
Examples: $1 \text{O} 1 = 1$; $1 \text{O} 0 = 1$; $1 \bar{\text{O}} 0 = 0$

Outgoing feeder

Functional unit which is normally used for the supply of electrical energy to one or more outgoing circuits.

Overcurrent

Any current which exceeds the → rated current. It includes all → overload and → short-circuit currents. Overcurrent usually poses a potential danger to insulation, conductors, electrical machines and equipment by excessive heat generation.

Overcurrent relay or release

A →relay or →release which operates instantaneously or with delayed action, if the load current exceeds a specified, value (see →Overload relay or release, →Instantaneous short-circuit release and →Short-time delay short-circuit release).

Overlapping auxiliary contacts

See →Make-before-break.

Overload

Operating conditions which cause →overcurrents to flow for a significant period of time in an otherwise electrically healthy and trouble-free circuit (e.g. through excessive friction in a driven machine). An overload condition usually causes damage of some kind (e.g. shortening the →service life of insulation) and may even lead to a →short-circuit if it is allowed to persist for too long. On the other hand, machines are usually designed to tolerate so-called "operational overloads" without damage.

Overload current

The →overcurrent which flows as a result of an →overload condition. Depending on its value and duration, it may or may not lead to the →tripping of overload protection devices (e.g. →overload relay).

Overload discrimination

This term is used to refer to discriminative protection in the →overload current range (see →Discrimination and →Current-based discrimination).

Overload relay or release

An →overcurrent relay or release intended for the protection against persisting →overload conditions. Its operation is determined by the value of load current it senses and its →time-current characteristic. Since the →tripping time is inversely proportional to the load current, overload relays and releases are often referred to as "current dependent" or "inverse-time" delayed overload relays and releases. They are also known as "a-releases".
The particular behaviour of the relay or release with respect to the load current and its duration may be achieved electronically, by employing the bimetallic principle (→bimetal relays) or by utilizing the melting characteristic of a eutectic material (→eutectic alloy overload relay).

Overvoltage category

A conventional number based on limiting the values of prospective →transient overvoltages occurring in a circuit and depending on the means employed to influence the overvoltages.

Overvoltage factor k

The ratio of the peak values of →overvoltage U_o and →rated operational voltage U_e to the star point.

$$k = \frac{U_o}{\frac{U_e}{\sqrt{3}\sqrt{2}}}$$

Overvoltage spike

See →Switching overvoltage

Overvoltage value

The peak value of a high frequency →transient overvoltage (e.g. →recovery voltage). It is higher than the →rated operational voltage U_e

$(U_o > U_e)$.

Panelboard

In terms of NEC (USA) a panelboard is a single panel (or group of panel units which may be combined into a single panel) on which switching and protective devices (e.g. →contactors, →overload relays) including connectors (e.g. solid →busbar stubs, if applicable) are mounted. It is specifically designed to be fitted into a →cabinet or →cut-out box from the front. The panel may also contain switches for the control of heat, light or power circuits. Also refer to →Switchboard.
A panelboard may also be understood to mean a complete →distribution board containing mainly →moulded case and →miniature circuit-breakers, primarily concerned with the distribution of electrical energy to several feeder or load circuits (UK).

Partial load optimization

A technique employed by →solid-state motor starters (e.g. →SIKOSTART) whereby the motor →terminal voltage is reduced under partial-load and →no-load conditions in order to improve the →power factor. During operation under partial-load and no-load conditions, the current of a three-phase induction motor does not drop to the same extent as the output power. This means that the power factor of the motor is reduced. The SIKOSTART unit recognizes the drop in power factor and improves it by reducing the motor supply voltage accordingly. Thereby, the electrical power consumption is also reduced. Should the load change (change in output power demand), the control unit immediately restores the full motor supply voltage. A drop in speed does not occur.
The extent of energy saving depends on the rated output power of the motor, the number of poles, the motor design and the nature of the load (→on-load factor). For smaller motors (e.g. ≤ 5 kW), a saving of up to 10% motor rated power can be achieved. In the case of large motors (> 110 kW), the saving drops to below 2% of the motor output rating.

Partially type-tested assembly (PTTA)

A → low-voltage switchgear and controlgear assembly in terms of DIN VDE 0660 part 500 (IEC 931-1) which contains type-tested and/or non type-tested elements and for which compliance with the respective requirements of the standard is demonstrated by testing and/or calculation. An element may, for example, comprise a single low-voltage → switching device with the associated electrical connections or an → enclosure.

Partitioning

Partitions are used to compartmentalize a switchboard or to separate functional units or even switching poles from one another. Partitioning is required to prevent any damage (e.g. a → fault arc) from spreading from its point of origin, especially if it involves danger to persons or the environment (also refer to → Barriers/flash barriers).

PCB

Printed **c**ircuit **b**oard – used extensively for the mounting and interconnection of electronic components

This is also the abbreviation for **P**oly-**C**hlorinated **B**iphenyls, an impregnation liquid previously used in (for example) transformers and → power capacitors. To a large extent, its use has been outlawed, or forbidden in most countries owing particularly to the highly toxic substances (e.g. polychlorinated dibenzofuranes) which are produced at high temperatures (fire or explosions involving PCB pose extremely serious risks and health hazards!). Askarel is the one of the general names for PCB used as insulation liquied in transformers and capacitors.

Peak withstand current

Maximum current value (peak value) which a circuit or → switching device can withstand in the closed position under specified conditions of use and behaviour (also see → Short-time withstand current).

PEN conductor

A → conductor which combines the functions of a → protective conductor (PE) and a → neutral conductor (N). Also refer to → Network types and systems.

Performance test

An obligatory test or testing procedure designed to prove compliance with the performance requirements as outlined in a product standard or specification (see → Testing of low-voltage switchgear and controlgear).

Permanent magnet

A magnet made of material with high → coercive force which retains the original magnetization at almost the same level throughout its service life. Permanent magnets are used in → contactors for → magnetic blowout of the → switching arc.

P.f.c.

See → **P**ower **f**actor **c**orrection

P.f.c. unit

A **p**ower **f**actor **c**orrection unit consists of a number of → capacitor banks, a → power factor controller as well as the necessary → capacitor contactors and other switchgear to achieve → automatic power factor correction of the electricity supply.

Pg-Thread

Standard (German) thread on steel → conduit piping (Pg = **P**anzerrohr**g**ewinde = conduit thread).

Phase angle

The angular displacement between two waveforms e.g. alternating current and voltage respectively (also refer to → Power factor).

Phase control

In e.g. battery chargers, variable d.c. power supplies, → solid-state motor starters (→ soft starters), etc. the half-waves of the sinusoidal voltage waveform are cut off at specific points in time or angular displacement (known as the firing angle) before the waveform passes through the zero line, by the triggering (firing, switching on) of thyristors or triacs in the current path. By varying the point on the half-wave at which the thyristors or triacs are triggered (i.e. by varying the firing angle) the effective value of the voltage may be reduced from the r.m.s. value of the full sinusoidal waveform. The general term for this process is phase control. In the case of thyristors, only one half-wave can be cut. By means of triacs, phase control can be carried out in both the positive and negative half-waves.

Phase-loss sensitivity

A property of certain → overload relays which ensures the relay will trip in such a way if one phase of the three-phase supply is interrupted (→ single-phasing), that the protection of the motor is not jeopardized. The → tripping characteristic becomes "quicker" if the current ceases to flow in one of the phases. This feature is also known as "phase-failure sensitivity" or "single-phasing protection" and is offered by most → bimetal and electronic overload relays.

Photoelectric proximity switch (→ Opto-BERO)

A → proximity switch which senses objects which either reflect, block or diffuse a beam of visible or invisible light. Photoelectric proximity switches are divided into three groups:
Type **T**, → emitter and → receiver - Through-beam sensing in which the light beam travels from an emitter to a separate receiver. The presence of a → target is recognized when the light beam is blocked or scattered between emitter and receiver.
Type **R**, emitter-receiver and reflector - Retroreflective sensing in which the light beam travels from an emitter-receiver

unit to a reflector and back to the same emitter-receiver unit. The presence of a target is recognized when the light beam is blocked or scattered on its path via the reflector between emitter and receiver.

Type **D**, emitter-receiver and object - **D**iffuse reflective sensing in which the light beam travels from an emitter-receiver unit to the target (object) from which it is either reflected back to the receiver part, or is diffused. The Siemens photoelectric proximity switches are known as Opto-BERO switches and employ completely encapsulated semiconductor (i.e. electronic or solid state) switching elements.

Pick-up current (pull-in current)

This is the initial current surge drawn by a magnet coil of a → magnetic drive immediately after it has been energized. It is also referred to as the → inrush current although this term is preferred to mean short transient magnetizing current peaks which result whenever a magnetic load is switched on.

Pick-up delay

The time period between the specified instant of initiation of the → closing operation of a switching device and the moment the → contacts of its → NO auxiliary contact(s) touch (alternatively: the moment the contacts of its NC auxiliary contacts separate). In the case of a pick-up delayed → time relay, the instant of initiation of the → closing operation corresponds to the instant the preset time period starts expiring. The pick-up delay of auxiliary contacts is also known as the "ON delay". (→ make time)

Pick-up power

The power input a → magnetic drive requires to initiate sufficient movement of the → armature for it to travel from its position of rest to the final operated position. It is expressed in VA for a.c., and in Watts for d.c. magnetic drives. Particularly in the case of contactors employing the → economy connection, it may be substantially higher than the → hold-in power.

Pick-up voltage (pull-in voltage)

The minimum voltage, usually expressed as a percentage of the → rated control supply voltage of an item of switchgear, at which its desired operation will be initiated and completed reliably. In the case of contactors, for example, the pick-up voltage is that minimum voltage which, when applied to the coil terminals will cause the contactor to energize ("pull-in") and seal-in with sufficient contact pressure for rated operation. The pick-up voltage must therefor be at least equal to the → seal-in voltage.

Pilot switch

A mechanical, but non-manual, → control switch for use in → control circuits and → auxiliary circuits which is → actuated in response to specified conditions of an actuating quantity such as temperature, pressure, speed, liquid level, elapsed time, etc. Examples of pilot switches include pressure monitors, speed monitors, thermostatic switches, etc.

Pin-wound current transformer

The primary of a pin-wound → current transformer comprises more than one turn of wire which is threaded through the hole (or window) of the toroidal core. In contrast, the primary "winding" of current transformers with higher current ratings consists of the → conductor, wire or → busbar which passes through the window of the current transformer once only. Up to an operating current of approximately 200 A, → window-type current transformers may be used as pin-wound transformers. At higher currents this is no longer practicable, as only conductors with cross-sections of up to 35 mm^2 are thin and flexible enough to be threaded and wound in this way. By means pin winding, different transformation ratios may be achieved for a particular current transformer (i.e. by varying the number of turns). Pin-wound current transformers are also known as "wound-primary current transformers".

PLC

See → Programmable controller

Plug-in circuit-breaker

In addition to its → main contacts, a plug-in → circuit-breaker is equipped with a plug and socket arrangement which allows the unit to be removed from the circuit and its mounting base. A mechanical tripping pin ensures that the circuit-breaker (usually an → MCCB type) cannot be removed from the base socket while its main contacts are closed. If an attempt is made to remove the closed breaker, it will trip before the plug and socket contacts separate.

Plugging or plug braking

The rapid stopping or reversing of a three-phase induction motor by the reversal (swapping) of two phases while the motor is running (alternatively, reversal of the current in the armature winding of a d.c. motor). In the case of plug braking, a speed monitor may be used to sense standstill of the motor so that it may be disconnected from the supply before starting up in the opposite direction of rotation commences.

PNP output (of a → proximity switch)

This indicates a specific → type of output stage (semiconductor circuit) used in the electronic → switching element(s) of 3 or 4 wire d.c. proximity switches. If the load must be connected between the output and L- of the → control supply voltage (i.e. the proximity switch "sources" the load current when the output is ON or conducting), then the output is known as a "P" or PNP output (from pnp transistor).

Pole (of a → switching device)

All the parts of a switching device (e.g. → circuit-breaker) associated exclusively with one electrically separated → current path of the → main circuit through the device. The term does not include parts which serve the common fixing and simultaneous operation of all the poles.

Note:
A switching device with only one pole is termed single-pole. Devices with several poles are termed multi-pole (two-pole, three-pole, etc.) if the poles are linked or can be linked in such a way that they operate together.

Pole (of a magnet, magnet core or coil)

The magnetic lines of force (magnetic flux), created by a permanent magnet or an → electromagnet are said to be orientated from the north pole to the south pole (see → magnetic circuit).

Pole number (of a motor)

Each coil in the stator of e.g. a squirrel-cage induction motor creates a magnetic pole as current flows through its windings. The successive energizing of the coils located around the circumference of the stator leads to a rotating magnetic field which causes rotation of the rotor. The synchronous speed n_s of the motor may be determined by a knowledge of the number of poles p and the frequency f of the a.c. supply voltage.

$$n_s = \frac{120 \cdot f}{p}$$

For example, in a 50 Hz system a 2-pole squirrel-cage motor has a synchronous speed of 3000 r.p.m. The most common no. of poles are: 2, 4, 6 and 8.

Poleface

The surfaces at which the concentrated magnetic flux passes from the → magnet core of an → electromagnet to air (e.g. in the → air gap) is known as the poleface.

Pollution

In the field of → low-voltage switch and controlgear, pollution refers to any condition or even presence of foreign matter (solid, liquid or gas) which may jeopardize the → dielectric strength or surface resistivity of insulating parts, air gaps or clearances. It includes, for example, the presence of excessive → ionized gases, hygroscopic dust, salt, etc.(see → pollution degree).

Pollution degree

The pollution degree of a piece of equipment refers to the environmental conditions for which the equipment is intended. It is an assigned numeral based on the amount of conductive or hygroscopic dust, ionized gas or salt in the operational environment (usually the → micro-environment) of switch and controlgear (→ pollution). It is also based on the relative humidity and on its frequency of occurrence as causes of hygroscopic absorption or condensation leading to deterioration of the → dielectric strength and/or surface resistivity of insulating materials.
Unless otherwise stated (e.g. in a specific product standard), pollution degree 3 (IEC 947-1) is used as a basis for equipment intended for use in industry. This states, that conductive pollution (or dry non-conductive pollution which becomes conductive as a result of condensation) may occur.

Position switch

An electro-mechanical → pilot switch of which the → actuator is operated by some external moving part (e.g. on a machine, door, or mechanism) when this part reaches a specific position. Particularly when used to signal the final or end position of a mechanism, position switches are also referred to as "limit switches".

Positive drive (positive action)

This term refers to the link or connection between an → actuator and a → contact element in a → switching device which is such that the force applied to the actuator is *directly* transmitted to the contact element.

Positive opening

An → opening operation which guarantees that all the → main contacts of the → switching device are in the open position when the → actuator is in the position corresponding to the open condition of the device. If the actuator of a device with positive opening feature cannot be brought into this "off" or "open" position (e.g. owing to solidly welded main contacts), then this should be clearly indicated e.g. by breakage of the operating handle or actuator if excessive force is applied.

Positive opening operation

An → opening operation of a mechanical → switching device which is designed to ensure that (under specified conditions) all the → main contacts are in the open position when the → actuator is in the open position. The positive opening operation will in future be known as "direct opening action" (IEC 947-5-1).

Positively driven contacts

→ Contacts of a mechanical → switching device which are designed and arranged in such a way that → normally open (NO) and → normally closed (NC) contacts can never be closed at the same time (see → positively driven operation). Thereby, it must be guaranteed that even under fault conditions (e.g. → contact welding of either the NO or NC contact), an effective → contact gap of at least 0.5 mm exists if the NO and NC contacts are considered to be in series. This property must be retained throughout the entire → service life of the device. Also refer to the (German) safety regulations for controls of power operated presses in the metal industry (ZH1/457, edition 2, 1978) and the specifications of the Swiss institute for accident insurance (SUVA).

Positively driven operation

This refers to the → operation of a → switching device which is designed to guarantee (under specified conditions) that the → auxiliary contacts are always in the respective → switching state corresponding to the → open or closed position of the → main contacts. If, for example, one or more main contacts of a →contactor with positively driven operation should weld, then the →NO auxiliary contacts may not open (i.e. they may not indicate that the main contacts have separated!)

Potential-to-earth

The potential difference (voltage) between a →conductive part and →earth (or another conductive part connected to earth). If current flows along/through conductive parts to earth (e.g. along a walkway railing or through equipment housings), a voltage drop is caused e.g. at each join or change in material. Such voltages may pose a serious danger to personnel. (see →Residual-current protection).

Potentially dangerous part

Potentially dangerous parts are accessible →conductive parts which can be touched easily by the operator; they are not live under normal operating conditions but may become live in the event of a fault. Also see →Exposed conductive part.

Note:
Typical parts of this type may include, e.g., enclosures and actuators.

Potentially explosive atmosphere

This is a mixture consisting of air (oxygen) and flammable gas, vapour, mist or dust under atmospheric conditions in which combustion continues to spread rapidly (i.e. which explodes) from the point of ignition after ignition has taken place (see →Explosion protection)

Potentiometer

An ohmic resistor of which the resistance is adjustable between a maximum and minimum value without intermediate steps.

Power

1. Physical
The work performed by a unit of force per unit of time, measured in watts.
2. Electrical
The electrical energy drawn from the network by a →load per unit of time.
In a.c. networks, a distinction is made between the following load-dependent types of power:
– Apparent power S, expressed in VA, kVA or MVA
   The product of the absolute r.m.s. values of driving voltage and current. It does not take the angular phase displacement φ between the current and voltage waveforms into account, and thus consists of the vectorial sum of the effective and reactive powers.
   The apparent power can be calculated using the equation $S = U \cdot I$.
– Effective (or active) power P, expressed in W (watt), kW or MW
   The amount of power which is transformed into another non-electrical form (e.g. heat, mechanical power or energy). Only the components of the voltage and current waveforms which are perfectly in phase (i.e. zero phase displacement) can produce effective power.
   The effective power component of the apparent power can be calculated using the equation $P = U \cdot I \cdot \cos \varphi$.
– Reactive power Q, expressed in var, kvar or Mvar
   The amount of power which cannot be transformed into another form. It is produced by the components of the voltage and current waveforms which have an angular displacement of 90° from each other. Reactive power is caused by inductive loads (current lags the voltage by 90°) and capacitive loads (current leads the voltage by 90°).
   The reactive power component of the apparent power can be calculated using the equation $Q = U \cdot I \cdot \sin \varphi$.

The geometrical interdependence of the three types of power can be represented in a vector power diagram.
The relationship between the apparent power and the effective power to is determined by the →power factor which is represented by an angular function of the cosine of the angle phi ($\cos \varphi$).

In three-phase networks, the total power consumed in all three phases is normally quoted. Since the network voltage is given as the phase-to-phase voltage, the total power in a balanced system (assuming the same current in each phase) may be calculated using the equations:

$S = 3 \cdot U \cdot I$ (VA)
$P = 3 \cdot U \cdot I \cdot \cos \varphi$ (W)
$Q = 3 \cdot U \cdot I \cdot \sin \varphi$ (var)

Power capacitor

Power capacitors are mainly used in →power factor correction applications and →filter circuits. They are mostly three-phase units, and their power rating is quoted in kvar (see →Power). Power capacitors often incorporate an internal →fuse system which disconnects the capacitor from the supply if there is an excessive increase in pressure inside unit. Modern power capacitors have a very stable value of capacitive reactance throughout their long →service life, and exhibit extremely low watt losses. Their low inductive reactance causes high transient inrush currents upon switch-on, especially when capacitors in a →capacitor bank are energized in succession (short, low-impedance connections between fully charged and discharged capacitors). The use of special →capacitor contactors is recommended. Also refer to →MKK and →MKV capacitors.

Power circuit

A circuit with items of equipment which generate, transform, distribute, switch and consume electrical energy (also referred to as the main circuit).

Power factor

The ratio of →active power (e.g. in kW) to →apparent power (in kVA) drawn by a load. It corresponds to the cosine of the angular displacement between the driving voltage and the alternating current, and is therefor also known as the → cos φ of the load. Most loads in industrial applications are inductive (e.g. transformers, motors) and the current lags the driving voltage. Capacitive load may be added to improve the power factor of a load or installation to optimize the sizing of e.g. transformers and conductors or to reduce the kVA → maximum demand (see → Power factor correction).

Power factor controller

A monitoring and controlling device used in automatic → p.f.c. units. It measures the instantaneous value of the → power factor in one or more → conductors (usually in one of the three phases) by comparing the voltage and current waveforms, and controls a number of → capacitor contactors which switch → capacitor banks into or from the circuit to maintain the power factor within a given range close to unity. Typically, the target power factor and the tolerance band may be set up on the controller. The tolerance band (i.e. the permitted variance of the power factor before a capacitor bank is switched on or off) is determined by the kvar rating of the first (smallest) capacitor bank in relation to the instantaneous power being drawn from the supply. A time delay on the output stage of the power factor controller ensures that the capacitor banks are not switched on and off too rapidly.

Power factor correction

A method or technique employed to increase the economical efficiency of an electrical installation. Reactive power drawn by an inductive consumer load is compensated by means of reactive power drawn by an added capacitive load so that the → power factor of the resultant total load current is as close to unity as is practicable. Since the value of this resultant current will be smaller than that drawn by the uncompensated load, savings can be achieved in cable cross-sections, transformer sizes, etc. (see → Maximum demand). Depending on the application, one can distinguish between → individual power factor correction, → group power factor correction and → centralized power factor correction (→ automatic power factor correction). Typically, power → capacitor banks are used for the capacitive load.

Power relay

See → Contactor relay

Power-frequency recovery voltage

The → recovery voltage after the transients have subsided (see → Transient recovery voltage).

Note:
The value of the power-frequency recovery voltage may be stated as a percentage of the → rated operational voltage. This also applies for d.c. voltage (power frequency is zero).

Pre-arcing current

The maximum value of current which can flow through a → fuse for a given time without the → fuse element melting through.

Pre-arcing $I^2 \cdot t$ value

The value of heat required for the → fuse element of a → fuse link to start to melt, i.e. to start the interruption process. It is dependent on the fuse type (→ operation class) and on its → rated current. It is expressed in A^2s, whereby the value of 1 ms is assumed for the duration (also see → Total $I^2 \cdot t$ value)

Pre-arcing time

The time interval between the onset of a current value which would ultimately lead to melting of the fuse element (e.g. → overload or → short-circuit current) and the initiation of the → switching arc (also known as "melting time")

Pre-charging resistors

Pre-charging resistors are used in → power factor correction units to limit the inrush current peaks associated with the switch-on of low-loss → power capacitors (particularly in → capacitor banks). Such limitation not only relieves the network of excessive disturbances and → noise, but serves to prolong the → electrical service life of the → switching devices. Refer to → Capacitor contactor.

Prepared conductor

A → conductor which has been cut and which has been fitted with e.g. a cable lug, eyelet or end sleeve, etc. (see → unprepared conductor).

Note:
Conductors may also be prepared by soldering the individual strands together. This practice is, however, not recommended in cases where the current flow can cause cyclic heating and cooling of the terminal (cf. different rates of expansion and contraction) as it may lead to progressive deterioration of the electrical contact and ultimate failure of the termination.

Pressurized apparatus "p"

A → type of protection as defined in the general requirements DIN EN 50014 and the specific standard DIN EN 50016. See → Explosion protection

PROFIBUS (process **fi**eld **bus)**

A → field bus for → data transfer between → switching devices with communication facilities (e.g. → SIMOCODE) and process automation equipment (e.g. → programmable controllers) in industrial networks.

Programmable controller

An electronic control unit in which the control sequences are stored in the form of a software program. A typical programmable controller comprises the → central processing unit (or CPU) which usually contains microprocessor circuits and the memory module with the program, as well as a number of input and output units or stages. Basically speaking, the program performs a sequence of logic combinations between input and output stages - hence the term "programmable logic controller" or "PLC". The program is typically stored on EPROM or EEPROM modules after being written or modified on a programming unit (also known as a "programmer"). SIMATIC is the generic trade name for Siemens programmable controllers.

Prospective current

The *prospective* current of a circuit at the point of installation of e.g. a → switching device or → fuse is that current which *would* flow if the → current path in each → pole of the device or fuse were replaced by a conductor of negligible impedance. It is of particular significance when considering → short-circuit currents and protection devices with → current-limiting characteristics as it indicates the value of short-circuit current which would flow if the devices were not installed. The method used to evaluate and to express the prospective current may differ from one product standard to the next.

Prospective current "r"

This refers to a short-circuit test current specified in the IEC product standard for → contactors and → motor starters (IEC 947-4). It is assigned a specific value for each range of → rated operational current I_e, e.g. 3 kA for starters with rated operational currents greater than 16 A and smaller than or equal to 63 A (I_e values stated at → utilization category AC-3).

Prospective fault current

The maximum value of steady-state → overcurrent at a given point in a circuit or electrical installation which would result from a → fault of negligible impedance (i.e. a → short-circuit) between the live → conductors (or between a conductor and → earth) at that point.

Prospective making current

The → prospective current in a single → pole of a → switching device when initiated under specified conditions (e.g. at the instant in the alternating current waveform which leads to the highest value of → prospective peak current)

Prospective peak current

The peak (instantaneous) value of a → prospective current during the transient period following initiation. It is determined by the impedance of the circuit only, i.e. it is assumed that the → contacts which close to initiate the current are able to provide an instantaneous transition from infinite to zero impedance (see → Maximum prospective peak current).

Prospective short-circuit current

See → Prospective current

Prospective symmetrical current (in an a.c. circuit)

The → prospective current if the (ideal) → contacts which close to initiate the current flow do so at such an instant that no transient phenomena are created. Alternatively, in short-circuit considerations, it is taken to be the amplitude of the prospective current which is symmetrical about the line representing the decaying transient (d.c. component). The prospective short-circuit symmetrical current is also referred to as the "steady-state" short-circuit current, and is expressed as an r.m.s. value.

Note:
In the case of a multipole → switching device (e.g. 3-pole → circuit-breaker) the condition of non-transient period can only apply for one pole at a time.

Protection against direct contact

See → Protective measures

Protection against indirect contact

See → Protective measures

Protection class

See → Degrees of protection

Protective conductor (PE)

This → conductor is used in a number of → protective measures against → electric shock and is intended for connecting together of any of the following parts:

– exposed → conductive parts of the installation itself,
– conductive elements which are not part of the installation, but, which could introduce an extraneous potential,
– the main → earth terminal,
– earth electrode(s),
– the earthed point of the source (or artificial neutral).

Also refer to → Network types and systems.

Protective cover

A part, screen, cowling, barrier, etc. which prevents direct access to live parts from all the usual directions of approach. It may also protect personnel and/or equipment from the electric arcs of switching devices and the like.

Protective measures

In the broadest sense, protective measures include all the precautions taken, the techniques applied and the specifications fulfilled to ensure that persons and animals are protected from harm caused by electricity. Protective measures are generally divided into three main groups:
Protection against electric shock,
Protection against overcurrent, and
Isolation and switching.
Depending on the site conditions and the degree of risk, specific measures for protection against electric shock may have to be applied. These include
– protection against direct contact (e.g. by means of insulation, → barriers or → enclosures, etc.)
– protection against indirect contact (e.g. by means of equipotential bonding, electrical separation, etc.), and
– protection against both direct and indirect contact with live parts (e.g. by means of → safety extra-low voltage SELV).

Glossary

Protective separation

See → Safety separation

Proximity switch

A → position switch which is actuated (→ operated) without mechanical contact. The Siemens range of proximity switches are known by the trade name → BERO Several types of proximity switches with differing methods of operation and → sensing ranges, each suited to its own field of applications, are offered. These include:
→ inductive proximity switches,
→ capacitive proximity switches,
→ ultrasonic proximity switches, (→ Sonar-BERO) and
→ photoelectric proximity switches (→ Opto-BERO).
Proximity switches are further classified in terms their behaviour when surrounded by various materials (see → Embeddable and → Non-embeddable), their construction and form (e.g. cylindrical, threaded), the function and → type of output (e.g. → NO and → PNP resp.), as well as their method of connection (e.g. plug-in, attached wires, terminals). Refer to IEC 947-5-2.
For ease of selection, BERO proximity switches are also categorized in terms of the special requirements of various conditions under which they are designed to operate (e.g. in conjunction with → programmable controllers, under extremely harsh environmental conditions, in → hazardous locations, etc.)

PTB certificate

Physikalisch-**T**echnische **B**undesanstalt - A federal institute for technological physics.
A PTB certificate is obligatory for motor protection devices (e.g. → motor starters, → overload relays, → thermistor motor protection units), if they are to be used for the protection of → explosion-proof motors in → hazardous locations.
In conjunction with an attached → tripping characteristic, the PTB certificate confirms that the motor protection device is of such a design and rating, that under the specified operating conditions of the switchgear and in the defined hazardous location, the risk of excessive surface temperatures on the motor can be excluded.

PTC thermistor (temperature detector)

Thermistor → temperature detectors are used extensively in the protection of motors and transformers against excessive temperature rise (see → Thermistor motor protection). They are embedded directly between the phase windings (in the overhang of the stator winding in the case of a motor); normally during manufacture. The most common type of temperature detector used in this way is the PTC thermistor which has a **p**ositive **t**emperature **c**oefficient. Its resistance increases sharply in the range of its rated operating temperature (known as the "TNF"). This abrupt change in the resistance is sensed by the connected tripping unit (e.g. 3UN21). PTC thermistors are primarily used in series production motors and transformers where the permissible temperature limits and thermal characteristics are known prior to manufacture.

PTTA

See → Partially type-tested assembly

Pull-in power

See → Pick-up power

Pull-in winding

See → Economy connection

Pull-through voltage

See → Seal-in voltage

Pulse (fleeting) contact element

An → auxiliary contact element which opens or closes a circuit only momentarily during the → operation of the switching device (i.e. during the transition of the → actuator from one position to another.

Pulse number

The pulse number of a → rectifier circuit indicates the number of times rectification takes place within one period of the → fundamental waveform. For example, in the case of a full three-phase bridge rectifier, the pulse number is 6.

Pulsed light beam (of a → photoelectric proximity switch)

In → Opto-BERO proximity switches the light beam used is switched on and off in a duty cycle of about 1:25 and a frequency of 5 to 10 kHz (i.e. it is modulated) by the → emitter. In → diffuse and → reflex photoelectric proximity switches the → receiver is only active during the light pulse and is disabled during the pulse gap. In this way, the units are rendered largely insensitive to extraneous light.

Pump control

This refers to a special form of → soft stopping by means of a → solid-state motor starter.
The inertia of pump drives (e.g. three-phase induction squirrel-cage motor with paddle-type pump) tends to be low in comparison to the force exerted by the head of liquid being pumped. As a result, the pump stops abruptly or may be driven backwards the moment the motor is switched off until the non-return valves are closed. This may produce a shock wave in the piping system (also known as a "water hammer") which can be prevented by the controlled gradual reduction of the motor → terminal voltage. Typically, the stopping time needs to be longer than the period of the standing wave for the length of pipe involved. Depending on the load/speed curve of the complete pump drive, however, it may be necessary to employ a non-linear control of the terminal voltage. For this reason, a simple linear soft stopping of the motor may not be suitable.

Puncture

The term puncture is used to refer to a → disruptive discharge through a solid dielectric.

Pushbutton

A →command device with stored energy (spring) return. A pushbutton is operated by a force exerted on its →actuator by a part of the human body (usually a fingertip or palm of the hand), but both actuator and →contacts return to their initial position once the actuating force is removed.

Note:
Special types of pushbutton include versions with →latch mechanisms (latching pushbutton) in which the actuator must be twisted, pulled, pushed a second time or unlocked with a key before the initial position is regained. These often have mushroom-shaped actuators, and are commonly used in emergency stop applications.

Pushbutton control

See → Momentary contact control

r-release

The short designation for an →undervoltage release

Raceway

An enclosed channel designed expressly for holding wires, cables, →busbars. Examples of raceways include →conduit, →cable ducting, →cable trunking →busbar trunking, electrical non-metallic tubing (ENT), flexible metallic tubing, etc.

Rack-out circuit-breaker

See →Withdrawable circuit-breaker

Radial network

The single-line →circuit diagram of a radial network resembles a tree with several branches. Power is supplied from a single source at the "trunk" (→incomer) and each →load is attached to a "branch" (→feeder). Radial networks are classified as "open networks". Also refer to →Network types and systems, →Ring networks and →Mesh networks.

Radiated electromagnetic field withstandability

The ability of a device (e.g. →proximity switch) to operate in a radiated electromagnetic field (e.g. from radio transmitters). Severity levels for the field strength (in V/m) in a frequency band ranging from 27 MHz to 500 MHz are specified in IEC 801-3. See →Electromagnetic compatibility. Also refer to →Magnetic field withstandability.

Radio interference

Radio interference (audible clicking or crackling noise from radio receivers) may be caused by the switching off of inductive circuits, e.g. →contactor coil circuits. This effect may be remedied by the use of an RC-element across the inductive load or the →contact gap. Also refer to →Surge suppressor.

Radio interference level

The radio interference level is a frequency-dependent limit for →radio interference (continuous noise and clicking). It is used in conjunction with the interference voltage, interference power and interference field strength (DIN VDE 0875 part 3, 12.88).

RAL

Short form for the German "**A**usschuß für **L**ieferbedingungen und Gütesicherung beim Deutschen Normen Ausschuß".
RAL is the central board of the German industry for material quality assurance. Leading industrial bodies as well as the appropriate federal ministries are represented in the advisory committees. One of the special tasks of RAL is the harmonization and rationalization colours in a standard register.

Ramp time

It is common practice in → solid-state motor starters (e.g. →SIKOSTART) to increase (or decrease) the motor →terminal voltage linearly from an initial to a final value to obtain a gentle starting (or stopping) characteristic. The time taken for the voltage to change from the initial to the final value is known as the ramp time.

Rapid discharge resistor

See → Discharge choke

Rapid multiple restriking

Immediately after → contact separation (→ opening operation) to interrupt an inductive current, the → switching arc is quenched and restruck at a high frequency (100 kHz range). This effect may be reduced by the fitting of → surge suppressors, preferably across the load itself.

Rated control voltage U_c

See → Control circuit voltage and → Control supply voltage

Rated frequency (of an item of switchgear)

The frequency of the supply voltage in the main circuit for which the item is designed and on which other characteristic values (e.g. the → rated operational current) are based. Several rated frequencies may be assigned e.g. to a → switching device, or it may be rated for use in a.c. and d.c. applications.

Rated impulse withstand voltage U_{imp}
(of an item of switchgear)

The peak value of a voltage pulse having a prescribed shape and polarity which the item can endure without failure under specified test conditions. The values of →clearances are based on this value.

Rated insulation voltage U_i (of an item of switchgear)

The value of voltage on which the → dielectric strength test voltage and → creepage distances are referred.

Note:

If no rated insulation voltage is specified for a device, the highest → rated operational voltage must be taken as the rated insulation voltage.

Rated operating distance s_n (of a → proximity switch)

A conventional quantity used to designate the → operating distances. It does not take manufacturing tolerances, material of the → target or variations resulting from external influences such as voltage and temperature into account (see → Rated value)

Rated operational current I_e (of an item of switchgear)

The value of current for normal operation as stated by the manufacturer. It takes the → rated operational voltage, the → Rated frequency, the rated → operational duty, the → utilization category and (if relevant) the type of specified → enclosure into account.

Note:

In the case of → switching devices used predominantly in motor control applications, the rated operational current may be replaced or supplemented by the rated operational power (maximum rated power output of the motor at rated operational voltage with assumed values for → efficiency and → power factor).

Rated operational voltage U_e (of an item of switchgear)

The value of voltage which, in combination with the → rated operational current, determines the application of an item of switchgear. The relevant tests and → utilization categories are based on this value. For → switching devices destined for use in three-phase networks, it is stated as the voltage between phases. The maximum rated operational voltage may not exceed the → rated insulation voltage.

Rated output power S_{nT} (P_s) of a transformer

The apparent output → power (in VA or kVA) is the product of rated output voltage and rated output current (for three-phase transformers it is 3 times this value) for which the transformer is designed under the applicable reference conditions.

Rated output voltage U_{2n}

The voltage across the secondary → terminals of a transformer (the phase-phase voltage in the case of three-phase transformers) for rated input voltage, rated frequency, rated output current, $\cos \varphi = 1$ and rated temperature.

Rated overcurrent factor (of a current transformer)

The overcurrent factor indicates that value of primary current, in multiples of the rated value, which will cause the core of a measurement → current transformer to become saturated, or alternatively the minimum value at which the core of a c.t. for protection purposes may not saturate. It is, however, dependent on the connected → burden.

Rated power of a current transformer

Apparent power in VA at rated secondary current and rated → burden.

Rated short-circuit making capacity I_{cm} (of an item of switchgear)

The value of short-circuit → making capacity assigned by the manufacturer for the → rated operational voltage, at → rated frequency and at a specified → power factor (or time constant in the case of d.c.). It is expressed as the maximum prospective peak current under prescribed test conditions. In the case of → circuit-breakers, IEC 947-2 specifies a fixed relationship between the rated short-circuit making capacity and the rated → ultimate short-circuit breaking capacity.

Rated uninterrupted current I_u (of an item of switchgear)

The value of current, as stated by the manufacturer, which the item can carry indefinitely (→ uninterrupted duty) if the main → contacts remain closed (i.e. if it does not perform switching operations).

Rated value

A quantity value assigned, generally by the manufacturer, for a specified operating condition of a component, device or equipment.

Rating

A set of → rated values and operating conditions as applied to e.g. a → conductor, a → switching device or a → fuse.

Note:

The term is often used rather loosely in conjunction the a particular quantity, as in e.g. "current rating" and "short-circuit rating". The context in which it is used does not always define which rated value is meant. For example, the current rating of a fuse may be understood to mean its rated operational current under normal operating conditions, where-as the rated current of a contactor needs closer definition in terms of the → duty cycle, the → utilization category, the system voltage, etc.

RC circuit

The series arrangement of a resistor and a capacitor usually intended for parallel connection across an inductive load. The capacitor is selected to produce a resonance circuit with the inductive load at a specific frequency. The resistor serves to damp resonance oscillations. The arrangement is typically used to suppress → switching overvoltages which are produced when an inductive load is disconnected from the supply.

RC element

The series connection of a resistor and capacitor (→ RC circuit) specifically designed to suppress → switching overvoltages which are produced when the → magnet drive of a → contactor is switched off (see → Surge suppression).

9 Appendix

rc-release

Short designation for an → undervoltage release equipped with a capacitor → drop-out delay unit (see → Energy storage device) to prevent short dips or interruptions in the supply from causing the release to operate.

Reactance voltage

The voltage, being the product of reactance and current, which is responsible for the → reactive drop.
Of a transformer: The reactance voltage at rated load in quadrature with the secondary current (90° out of phase), usually expressed as a percentage of the no-load secondary voltage.

Reactive drop (or reactive rise)

A decrease (or increase) of secondary voltage (of a transformer) under any given output conditions and (where applicable) with unchanged input or primary voltage, owing to the internal reactance. It is usually expressed as a precentage of the no-load secondary voltage.

Reactive power

The → power drawn by a reactive → load (e.g. a → power capacitor draws leading reactive power, an inductor or motor draws predominantly lagging inductive power). For closer definition, see → Power

Receiver (of a → photoelectric proximity switch)

The detector (or → sensor), lens arrangement and necessary circuitry to detect the presence of the light beam from the → emitter. In → diffuse and → reflex sensors the emitter and receiver are combined into a single unit.

Reclosing lock-out

A unit which prevents reclosing of a circuit after an → overload relay (release) or short-circuit protection device has operated until the lockout is released intentionally. Also refer to → Short-circuit lock-out and → Manual reset feature

Recovery time

The recovery time may be defined or described in a number of different ways, depending on the type of → switching device in question.

For electronic devices, the following usually applies:
The recovery time is the shortest period which must elapse after the → control supply voltage has been disconnected (or fails) and before the device may be re-energized without the risk of an intermediate operating state or a spurious switch-on signal at the output (see, for example, → Spurious switch-on signal suppression). This definition does not apply to electronic → overload relays, in which a long recovery time may be desirable to prevent a motor load from being switched on immediately after an overload tripping operation.

For thermal overload relays and releases:
The recovery time is the rest period required for the → bimetallic strips to cool down after a tripping operation until the relay or release can be reset and is again ready for operation.

For → operating mechanisms:
The rest period required for all moving parts to decelerate sufficiently so that the next ON command can be processed reliably (e.g. deceleration time of centrifugal mechanisms).

Recovery voltage

The voltage which is built up across the separated (opened) → contacts of a → switching device or across the → terminals of a → fuse after the interruption of the current flow. Also see → Transient recovery voltage)

Rectifier

A device which converts alternating current into direct current using electrical "valves" (e.g. diodes, thyristors). The valves allow the current to flow in one direction only; the reverse direction is blocked.

Reduced-voltage starter

See → Soft starter

Reduction factor (of an → inductive proximity switch)

The factor by which the → usable operating distance of an inductive proximity switch must be multiplied if the → target is made of material (e.g. copper, brass or aluminium) other than that specified for the → standard target .

Reed contact

A set of fine → contact elements mounted in a gas-filled tube. The contact is operated by a magnetic field.

Reference axis (of a → proximity switch)

An axis perpendicular to the → sensing face of a proximity switch and passing through its centre. In the case of → photoelectric proximity switches the reference axis is defined as follows:
for type T, it is perpendicular to the centre of the → emitter;
for types R and D, it is located midway between the optical axes of the emitter and → receiver sections respectively of the emitter-receiver unit.

Reflex photoelectric proximity switch

This term is used to refer to a type **R** photoelectric proximity switch using **r**etroreflective sensing in which the light beam travels from an emitter-receiver unit to a reflector and back to the same emitter-receiver unit. The presence of a → target is recognized when the light beam is blocked or scattered on its path between → emitter and → receiver.

Regulation

The difference between the → rated output voltage or the → nominal output voltage and the → no-load voltage U_0

of a transformer. See →Voltage rise U_A (V); u_A (%) and →Voltage drop U_R (V); u_R (%); U (of a transformer)

Relay (electrical)

A device designed to produce sudden, predefined changes in one or several output circuits when certain conditions are fulfilled in the electrical input circuits or stages controlling the device. Typical examples of relays include:
→contactor relays - the auxiliary contacts change state if suitable voltage is applied to the coil terminals,
→time relays - the output contacts change state once the preset time has elapsed,
→overload relays - the contacts change state if the current flowing in the main current paths exceeds the minimum tripping value and flows for a certain period of time, etc.

Comment: Generally speaking, one should make a distinction between relays which are used in electronic applications where the atmosphere is relatively clean and the load conditions are less demanding, and relays which are intended for use under harsh industrial conditions (i.e. →contactor relays) involving vibration, dirt, electromagnetic interference (noise), and the switching of large contactors, machinery, etc. Although the demanded level of →contact reliability and →service life may be similar in both cases, the consequence of failure may be very different indeed and a careful analysis of the application at hand is recommended.

Release

A mechanism, module or device mechanically connected to a switching device and which releases the holding means, or →latch, and permits the opening (or closing) of the switching device. It is usually used to cause a →tripping operation. A release can have instantaneous, time-delayed or current-dependent operation. Various types of releases are described in this section. They include →undervoltage releases, →shunt releases, →overload releases, etc.

Note:
The word "trip" is sometimes used instead of "release" (e.g. shunt trip, short-circuit trip, etc.). As far as possible, we have used the term "release" to refer to the device, mechanism or module itself, and the word "trip" to refer to the result of its operation. Also refer to →Tripping operation.

Release time

See →Drop-out time

Release-free mechanism (or operation)

See →Trip-free operation

Remanence

See →Residual magnetism

Remanence contactor

A →contactor (or more commonly, a →contactor relay) which employs the →coercive force of its →magnet system to remain in the operated state after the →control circuit

voltage has been removed from its coil terminals (i.e. after it has been de-energized). A voltage pulse of reversed polarity is used to release the magnet system and cause the contactor to drop out. These devices (also known as "magnetically latching contactors") are found particularly in older generation stepped sequence control circuits, and are replaced nowadays with →mechanically latching contactors (no direct replacement, wiring changes required!)

Remote control

The control or →operation of a →switching device (or other piece of equipment) from a point located some distance from the device (also see →Local control). Remote control usually involves a run of auxiliary and/or control →conductors between the device and the point of control, although the use of radio or infra-red signals is also common.

Removed position

Position of a removable part of e.g. a switchboard (e.g. →withdrawable unit), when it is outside the switchgear assembly (→TTA/→PTTA) and is mechanically and electrically disconnected from the assembly.

Repeat accuracy

The repeat accuracy of a →time relay describes its ability to reproduce the set time delay consistently even after several operations.
The repeat accuracy of a →position switch describes its ability to close (or open) its contacts at a consistent displacement of the →actuator (mechanical operating consistency).

Repeat accuracy R (of a →proximity switch)

The variation of the →effective operating distance s_r under specified conditions.

Reset

A device (e.g. pushbutton, toggle) or position of an →actuator used to reengauge the →operating mechanism of a protective →switching device (e.g. →overload relay or →circuit-breaker), or to initialize an electronic evaluating circuit, after a →tripping operation so that the device may be operational again.

Reset point B (of a →proximity switch)

A reset point is any position of the →target (as it leaves the →sensing zone) at which the output signal of the proximity switch change →switching state (returns to the unoperated state). The reference point is the lower trailing edge of the target (or →actuating element). Also see →Response point.

Residual current/Residual-current protection

In terms of Kirchhoff's first law, the sum of all currents flowing into a circuit junction is zero. This means, that if the (vectorial) sum of the currents flowing in the 3 phases and neutral (if applicable) of a three-phase system is not

zero, then a finite non-zero current (i.e. the residual current, expressed as an r.m.s. value) must be flowing elsewhere - usually to earth. A distinction is made between:
a) earth-fault current, which refers to residual current flowing from live → conductive parts to earth *owing to* a specific insulation fault, and
b) earth-leakage current which refers to residual current flowing from live → conductive parts to earth in the *absence of* specific insulation failure (see → Creepage current).

The presence of excessive residual current (also known as "differential" current) indicates a fault in the system and can develop into a short-circuit condition and/or endanger personnel as a result of hazardous increases in the → potential-to-earth of accessible parts. Special protective devices employing → core-balance current transformers are frequently used in conjunction with (or as an integral part of) → circuit-breakers (see → CBR) to provide residual current protection. Residual current devices may also provide additional protection against fire and other hazards which can develop as a result of an → earth fault of lasting nature (e.g. current through wet wood) which, owing to its magnitude, is not detected by the overload or short-circuit protection apparatus (known as "earth fault protection").
Residual current devices which are operated by a residual current not exceeding 30 mA are also used as a means for additional protection against direct contact in the case of failure of the relevant → protective measure (see → Earth-leakage circuit-breaker)

Residual current i_r (of a → proximity switch)

The current which flows through the load circuit of the proximity switch in the OFF-state. It is also known as the "OFF-state current". In a 2-wire proximity switch it is that current which flows to maintain the function of the switch. It should not be high enough to effect the operation (switching off) of the load (e.g. input stage of a → PLC or an electronic → time relay).

Residual current release

A → release which operates to trip a → circuit-breaker if the → residual current exceeds a specific value, whereby this value may be adjustable. The release may operate instantaneously or may incorporate a time delay to avoid nuisance tripping e.g. in the event of a self-clearing fault. Residual current releases are also termed "g-releases". Also refer to → earth-leakage circuit-breaker and → CBR.

Residual magnetism

The magnetization persisting in a material when the external magnetizing force (e.g. from a permanent magnet or produced by a current flowing through a coil) is removed, is referred to as residual magnetism or remanence (refer to → Coercive force, → Remanence contactor)

Residual ripple σ

The rectification of alternating current results in a pulsing d.c. waveform. The pulses may need to be "smoothed out" by means of smoothing capacitors or other means. The degree of smoothing demanded by various electronic devices may differ, as may the consequences of using rectified a.c. with a residual ripple (or "ripple content") which is too high. For most → BERO proximity switches, for example, it is specified that the residual peak-to-peak ripple must not exceed 10% of the rated → operational voltage, i.e. $\sigma \leq 0.1 \cdot U_n$. For particular BERO's, it is specified that the operational voltage may not exceed 30 V or 65 V, respectively *including* the residual ripple.

Residual voltage (of a → proximity switch)

The voltage measured across the load when the output of the proximity switch is in the OFF state.

Resistance drop

A decrease of no-load voltage under any given output conditions and (where applicable) with unchanged input or primary voltage, as a result of the internal resistance. In the case of transformers, it is usually stated as a percentage of the no-load secondary voltage.

Resistance starter

This is the generic term for → motor starters which employ an external resistance in either the stator (→ stator resistance starter) or the rotor (→ slip-ring motor starter) circuits.

Resistance to mechanical shock

The ability of a → switching device to withstand defined mechanical shocks without malfunction. This data is of particular importance if devices are to be mounted on moving cranes, elevators, vehicles, etc.

Resistance to vibration

The resistance to vibration of a → switching device is stated as a multiple of the gravitational acceleration g at which the switching device still operates safely and reliably. Together with the value g, the direction of the force, the function - according to which the acceleration is effected - and the amplitude of the vibration must be stated.

Resistance voltage

The voltage, being the product of resistance and current, which is responsible for the → resistance drop.

Of a transformer: The resistance voltage at rated load and at a specified temperature (20 °C) in phase with the secondary current. It is usually expressed as a percentage of the no-load secondary voltage.

Resonance

Resonance is a phenomena which occurs when the vectorial sum of the inductive and reactive impedance in an a.c. circuit is (or approaches) zero. Since inductive reactance increases in direct proportion to the frequency ($X_L = \omega L = 2\pi f L$) and capacitive reactance depends on the reciprocal of the frequency ($X_C = 1/\omega C = 1/2\pi f C$), resonance for a given combination of inductance and capacitance will occur at a particular frequency value known as the "resonance frequency" or the "natural frequency" of oscillation. If the waveform of the current flowing in an electrical circuit (or alternatively the voltage drop caused by it) approaches the resonance frequency, then the effective impedance is reduced to the ohmic resistance of the current path.

The large currents which flow can damage circuit components and/or cause the premature operation of overload protection devices. Since resonance only occurs if inductive and capacitive loads are present, and usually only at frequencies which are far higher than the frequency of the a.c. power supply, unwanted resonance effects in industrial applications are rare. Capacitive loads may be present if e.g. → power factor correction or fluorescent lighting loads are installed. Problems caused by resonance in industrial networks are invariably due to the presence of → harmonics.

Resonance effects are used in → filter circuits to remove unwanted harmonic currents from a network or to "block" audio frequency signals (see → audio-frequency blocking circuits)

Resonance frequency

See → Resonance

Response (of an item of switchgear)

The transition from one switching state to another or the beginning of a process at the end of which the transition to another state takes place. An item of switchgear responds as soon as a specific value is attained (→ response value). Current, voltage, heat, distance of → actuator travel and time are examples of response values.

Response characteristic Ak (of a → proximity switch)

The line on which all → response points A of an → inductive or → capacitive proximity switch lie. The standard characteristic of a particular proximity switch is measured under laboratory conditions using a → standard target. The device-related characteristic values shown in the diagram below are all found by means of this curve. The → reference axis Z of the → BERO coincides with the y-axis of the response characteristic.

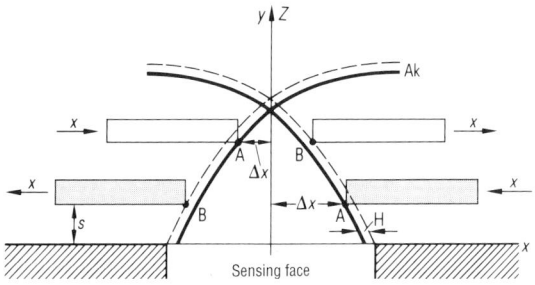

Z Reference axis
y Distance to the sensing face
A Response point
B Reset point
H Switching hysteresis
Ak Response characteristic
x Direction of travel
Δx Sensing radius
s Operating distance

Response delay

The period of time between the command or energizing signal and the → response of an item of switchgear.

Response point A (of a → proximity switch)

A response point is any position of the → target (as it approaches or enters the → sensing zone) at which the output signal of the proximity switch change → switching state. The reference point is the lower leading edge of the target (or → actuating element). Also see → Reset point.

Response time

– of a → proximity switch: The time delay between the instant a → target enters (or leaves) the → sensing zone of the proximity switch and the moment the → switching state of its output changes. In the case of → inductive proximity switches one distinguishes between the response time t_A, which refers to the time delay upon actuation (i.e. damping of the electromagnetic field), and the → build-up time t_s which is the time required for the field to be re-established after the target is removed. The response time is measured at an → operating distance of $s = 0.5 \cdot s_n$.

– general: The response time of an item of switchgear is the time between the command or energizing/de-energizing signal and the end of the switching process (e.g. contacts closed/open). See → Pick-up delay.

Response value

The specific value of e.g., current, temperature, voltage, distance which will cause the → response of an item of switchgear.

Reverse power relay

See → Network master relay

Reverse-current release (only with d.c.)

A → release which is designed to open a → circuit-breaker instantaneously or with a delay if a current flowing in the reverse direction exceeds a specified (preset) value.

Reversing (direction of rotation of a three-phase induction motor)

The direction of rotation of a three-phase induction motor may be reversed by swapping two of the three supply phases (i.e. by reversing the phase sequence).

Reversing contactor

This is a unit which essentially comprises two → contactors in a single casing. It has two → mechanically interlocked

→ magnet drives - one for each direction of rotation, and the main → poles are interconnected in such a way that depending on which magnet drive is operated, the phase sequence on the load side terminals (motor supply terminals) of the unit is either clockwise or anticlockwise. The term is also used to refer to two → interlocked contactors mounted on a common base plate - see → Reversing starter.

Reversing starter

A → motor starter which incorporates the facility for → reversing the direction of rotation of a three-phase induction motor. The simplest and most common form of a reversing starter (→ direct-on-line starting), is the assembly of two → contactors on a common base plate with a single → overload relay and associated → command devices. The two contactors are → electrically interlocked via their respective → control contacts, and a → mechanical interlock ensures that they cannot be closed at the same time - even in the event of a → control circuit failure.

Ring final network

See → Ring network

Ring network

In a ring network several transformers feed into a closed ring main system to which all → loads are connected (by so-called spur connections). If the ring is opened at any point, the loads are supplied from one of the two sides as if they were operating in a → radial network. A "ring final circuit" is a special form of a ring circuit in which a loads are connected via spurs to final circuits arranged in the form of a ring. Also refer to → Network types and systems, → Radial networks and → Mesh networks.

Ripple content

See → Residual ripple

Ripple control

See → Audio frequency remote control

RMS (or r.m.s value)

See → Root-mean-square value

Roller lever

A type of → actuator head on a → position switch. It consists of an arm or lever which is displaced by the → actuating element to → operate the switch. A wheel or roller is attached to the end of the lever to reduce friction to a minimum.

Root-mean-square value (r.m.s. value)

Also known as the "effective value", it corresponds to the peak instantaneous value of a sinusoidal waveform divided by the square root of two. Currents and voltages of a.c. systems are usually given as r.m.s. values since their momentary values depend on the corresponding phase position.

Measuring instruments (e.g moving-coil instruments) always indicate the r.m.s. value.
The r.m.s. value of an alternating current is equivalent to that value of direct current which would produce the same amount of heat (power); hence the term "effective value".

Rotary handle mechanism

See → Operating mechanism, → Operating mechanism with detachable coupling

Rotor class (KL)

See → Torque class

Routine test

See → Testing of low-voltage switchgear and control gear

Run-up current

See → Starting current

Run-up of a motor

See → Start-up ...

Run-up time (starting time)

See → Start-up time

Rupturing (of a fuse link)

When the → fuse element of a → fuse link melts through as a result of the high temperature caused by a current (overcurrent) flowing through it, it is said to "rupture". Refer to → Breaking operation.

Rupturing capacity

An alternative term for the → breaking capacity of a fuse.

Rush current, Rush effect

See → Inrush current

Safe locked-rotor time

See → t_E-time

Safety circuits

These are electrical circuits which specifically serve for the protection and safety of persons and equipment through the application of special measures and precautions.

Safety extra-low voltage

This is a potential difference defined in terms of the specifications for → protective measures in which circuits which are not → earthed are connected to a rated voltage of up to AC 50 V or DC 120 V. The power supply derived from circuits of a higher voltage must be safely separated from these SELV circuits (also see → Safety separation).

Glossary

Safety separation

The safe separation of circuits with → safety extra-low voltage (SELV) or → functional extra-low voltage (FELV) from other circuits, to ensure protection from dangerous shock currents. Generally speaking, the conditions for safety separation (isolation) between circuits are fulfilled if any single fault or damage (including breakage, cracking) does not lead to a voltage transfer from one circuit to another. This is of particular importance if e.g. safety extra-low voltages and operational voltages of up to, say, AC 1000 V are present in circuits of the same device or piece of equipment.

Sampling test

See → Testing of low-voltage switchgear and control gear

SCPD

See → Short-circuit protective device

Screening

Electrically or magnetically conductive enclosure, sheath or shrouding which serves to protect electronic devices from the effects of radiation or electrical noise interference.

Seal-in power

See → Hold-in power

Seal-in voltage

The voltage normally expressed as a percentage of the → rated control supply voltage of a → magnet drive. At this voltage, the force which is produced by the → operating mechanism after the initial closing (touching) of the → main contacts ensures sufficient → contact pressure. It is also known as the pull-through voltage.

Sealing-in contact

A particular → normally open → auxiliary contact of a → contactor connected into the → control circuit of the same contactor. It is connected in parallel to the NO → momentary contact of the ON → command device. Once the contactor has been energized (→ operated), the auxiliary contact bridges out the command device so that the contactor remains operated even after the command device has been released. Sealing-in contacts are also known as "self-holding contacts".

Section (switchboard section)

In switchboards of the enclosed construction type, a section is understood to mean a number of cubicles bolted together side-by-side and constituting a functional unit. All the cubicles in a section are usually supplied from the same set of busbars (busbar section).

Selectivity

See → Discrimination

Self-healing power capacitor

See → MKV power capacitor and → MKK dry-type power capacitor.

Self-holding contacts

See → Sealing-in contact

SELV

See → Safety extra-low voltage

Semiconductor contactor

In these → contactors, the power switching is carried out by means of semiconductor elements (e.g. thyristors). Each contactor consists of an electronic control circuit, the power semiconductors and a heat sink element. They are particularly suitable for applications which require a high number of switching operations. They are silent in operation, particularly robust and do not produce → ionized gases or other pollutants.

Semiconductor motor controller/starter

This term is generally used to refer to → motor starters which operate electronically. Examples include → semiconductor contactors and → soft starters. See → SIKOSTART.

Sensing face (of a → proximity switch)

The sensing face of a proximity switch is the area or surface on the unit which must be approached by the → target to initiate a switching operation. The sensing face is highlighted in the selection tables and is usually colour-coded on the device itself. By definition, the sensing face of:
an → inductive proximity switch is that surface through which the electromagnetic field emerges;
a → capacitive proximity switch is that surface through which the electric field emerges;
an → ultrasonic proximity switch is that surface where ultrasound is transmitted and received.

Sensing radius x (of a → proximity switch)

Distance between → target (→ actuating element) and the → reference axis Z of the proximity switch at any → response point A.

Sensing range s_d (of a → proximity switch)

The range within which the → operating distance may be adjusted. In the case of → ultrasonic proximity switches, the sensing range extends from the end of the → blind zone to the → maximum operating distance. → Sonar-BERO switches permit the initial and final values of the → effective operating distance to be adjusted.

Sensing zone (of a → proximity switch)

A volume of space in front of the → sensing face in which a → target will cause the proximity switch to operate. The boundary of the sensing zone is made up of all the → response points. Its length is determined by the → sensing range; its width by the values of →sensing radius applicable at each →operating distance.

Sensor

That part of measurement or detection apparatus (e.g. →thermistor motor protection, →proximity switch) which converts the physical quantity (e.g. temperature, light, pressure) to an electrical signal. Sometimes, e.g. in the field of automation, the word "sensor" is used to refer to the complete device (see → Actuator Sensor Interface (ASI)).

Service factor (of a motor)

If a service factor is indicated on the rating plate of a motor, it means that the motor may be continuously overloaded. The maximum permissible output power of the motor is calculated as follows:

$P_{perm.} = P_n \cdot$ service factor

The NEMA and EEMAC standards, for example, specify a service factor of 1.0 for → TEFC motors (IP 44) and a service factor of 1.15 for → ODP motors. The service factor is also known as the "overload factor" and may have to be taken into account when setting the tripping current on overload protection apparatus (see Section 3.4.2.2, page 124).

Service life

The duration for which an item of electrical equipment operates reliably (or is expected to operate reliably) under normal specified operating conditions. The service life of a →switching device is usually expressed as the number of →make-break operations (or →operating cycles) the device will or has performed and is also known as the "endurance". A distinction is made between "mechanical service life" and "electrical service life" (also see →contact life).
The end of the *electrical* service life is reached when the →contact material of the →contacts (→contact tips) is used up (see →contact erosion) to such an extent that reliable making and breaking operation is no longer ensured (e.g. the danger of →contact welding between the exposed contact elements is increased.)
The *mechanical* service life of a switching device is stated by the manufacturer as the number of make-break operations the switching device can be expected to carry out *without electrical load* before repair or replacement of parts may become necessary. It can be negatively influenced, for example, by harsh environments (e.g. abrasive dust; →abrasion), incorrect maintenance procedures, the wrong →control supply voltage, impermissible →mounting position, etc.

Service short-circuit breaking capacity
(of a → circuit-breaker)

A →breaking capacity defined in IEC 947-2 for which the prescribed test conditions include the capability of the circuit-breaker to carry its rated current continuously after a short-circuit interruption. The → test sequence is O–t–CO–t–CO. IEC 947-2 specifies a standard ratio (determined by the applicable →utilization category) between the rated service short-circuit breaking capacity I_{cs} and the rated → ultimate short-circuit breaking capacity I_{cu}. In fact, the rated service short- circuit breaking capacity may be expressed as a percentage of I_{cu} (e.g. $I_{cs} = 25\% \ I_{cu}$).

Set of contacts

a) This term is commonly used to refer to a complete mechanism comprising all the →contact elements, contacts, terminals and other parts which make up one main or auxiliary →pole of a switching device.
b) The term is also used to refer to the collection of spare parts (usually comprising the moving contact elements and the corresponding number of fixed contact elements) required to replace the contacts of a switching device (either for one pole, or for all the main contacts).

SF$_6$ gas

Sulphur hexafloride SF$_6$ is an excellent insulating medium. It is used extensively in medium and high-voltage →switching devices and installations to reduce the physical sizes (minimum →clearances) of the apparatus and to assist in the quenching of →switching arcs (→arc quenching). It is also used in low-voltage → dry-type power capacitors (Also refer to → MKK power capacitors).

Shading ring

A closed short-circuiting metal ring which is inserted in the →polefaces of the → laminated core employed in an a.c. →electromagnetic operating mechanism.
If no shading ring were present (or if it is broken), the flux in the → magnetic circuit would pass through zero as the alternating current in the coil passes through zero. Hence, the holding force on the → armature would also pass through zero and a pulsing actuation would result. The changing magnetic flux, however, induces a circulating current in the shading ring which in turn produces a lagging magnetic field. The force produced by this field ensures a non-zero force of attraction on the armature when the alternating current in the coil passes through zero. In this way, sufficient →contact pressure is retained throughout the entire a.c. cycle.

Shielding winding

A copper foil between the primary and secondary windings of a transformer to suppress →interference (see → Screening).

Short-circuit

The accidental (or intentional) connection of two or more points or →conductive parts in an electrical installation which are normally at different voltages through a relatively low resistance or impedance. Normally, a short-circuit is a fault condition and can have severe consequences if adequate protection is not provided.

Short-circuit current

An → overcurrent which flows as a result of a → short-circuit caused by a → fault or incorrect connection in an electric circuit.

Short-circuit lock-out

A mechanical latching arrangement in a → trip-free switching device (→ circuit-breaker) which prevents reclosing of the circuit after a short-circuit interruption. It is used to prevent the circuit-breaker from being reclosed onto a still remaining short-circuit. Once the fault has been repaired, the lock-out must first be reset before the breaker can be closed. Also refer to → Reclosing lock-out and → Manual reset feature

Short-circuit protective device (SCPD)

A device (e.g. → circuit-breaker or → fuse) which is designed to protect a circuit or parts of the circuit components against the effects of short-circuit currents by interrupting them and/or by limiting their value (see → Current limiting).

Short-circuit release

An → overcurrent release intended to provide short-circuit protection. Also refer to → Instantaneous short-circuit release and → Short-time delay short-circuit release.

Short-circuit strength

The withstand capability of a closed → switching device (including its accessories, e.g. → releases), or of a complete switchboard, against the electrodynamic (→ dynamic strength) and thermal stresses caused by the flow of → short-circuit current.

Short-circuiting ring (of a magnet core)

See → Shading ring

Short-time (operation) duty (S2)

For a general explanation of the various duty types of motors, see → operational duty. In terms of switching devices, short-time duty is also known as → temporary duty and is a rated duty type in terms of the relevant specifications. Please refer to → Temporary duty for more information.

Short-time current

The maximum current which can flow through a → switching device in the → closed position for a specified short period of time under given conditions.

Note:
The $I^2 \cdot t$ value during the current flow and the electromechanical effects due to the peak current value must also be taken into consideration.

Short-time delay short-circuit release

An → overcurrent release intended to provide discriminative short-circuit protection (see → Time-based discrimination) by operating after a delay period if the load current continues to exceed a predetermined set value and thereby → tripping the associated → switching device (usually the → circuit-breaker in which it is incorporated). The predetermined set value of load current and time delay is usually adjustable. The instantaneous short-circuit release is also referred to as the "z-release".

Short-time grading control

A technique whereby series-connected → circuit-breakers are interconnected with a data link to achieve discriminative protection (→ discrimination) in the event of a short-circuit. By means of the data link, and a process of mutual exclusion, the circuit-breaker immediately upstream from the point of fault can be identified and tripped with the shortest possible delay. The Siemens short-time grading control employed in 3WN air circuit-breakers is known as ZSS.

Short-time withstand current

The maximum permissible r.m.s. value of current a circuit or → switching device can carry for a specified short period of time, e.g. 0.5 to 1 s (1 s current) in the closed position under specified conditions of use and behaviour (also see → Peak withstand current).

Short-circuit proof

See → Inherently short-circuit proof

Shunt release

An → auxiliary release for the remote → tripping or → interlocking of a → mechanically latched switching device, e.g. a circuit-breaker. It is energized by a source of voltage which may be derived from the → main circuit It is also known as a shunt trip, or voltage trip and is often supplied as an accessory (possibly for retrofitting) to circuit-breakers.

Shunt trip

See → Shunt release

Shutter

An automatic-acting set of → barriers which covers the live parts (e.g. dropper → busbars) which remain behind in the switchboard when a → withdrawable unit or → withdrawable circuit-breaker is removed or racked out to the disconnected position.

SIGUT terminals

This is a Siemens trade name for → terminals (for up to approx. 63A and all auxiliary connections) incorporated in low-voltage switchgear. Among other advantages, the SIGUT terminals offer the following time-saving characteristics:

- Terminal screws are delivered in the opened state, so that the → unprepared (or → prepared) wires need simply be inserted and the terminal screw tightened.
- Terminal screws are captive, and cannot fall out even if loosened beyond the extent of their thread.

- Terminal screws and other conductive parts are protected from accidental contact (usually by the contour of the switchgear housing.
- Screwdriver guides facilitate the use of power-driven tools by guiding the screwdriver blades to the head of the terminal screw.
- Terminal openings are funnel shaped to guide the wire into the clamping unit.
- Definite end-stops ensure that the wire ends are inserted to the correct depth.

SIKOSTART

A Siemens → semiconductor motor controller/starter used for soft starting and → soft stopping of three-phase induction motors (→ soft starter). The 3RW22 SIKOSTART unit employs a micro-processor based digital phase-angle control and full-bridge connected power thyristors. The control unit (micro-processor) provides a linear increase of the motor → terminal voltage during the set → ramp time. Thereby, an almost linear increase in the motor current can be achieved. A → start impulse may be used to overcome initial friction in the drive. It is also possible to limit the motor current to a specified maximum value during starting. Soft stopping is achieved by a gradual controlled reduction of the motor terminal voltage. The 3RW22 SIKOSTART units also offer the features → pump control, and → partial load optimization.

SIKUFEST

The trade name for Siemens → motor starters comprising prewired, coordinated combinations of 3TF → contactors and 3VU1 → circuit-breakers (see → Coordination with short-circuit protection devices) on a common mounting bracket.

SIMATIC

The trade name for Siemens → programmable controllers.

SIMICONT

This a trade name for a uniform range or "family" of compact → contactors, → contactor relays, → overload relays, → time relays and associated accessories. The name SIMICONT is derived from **Si**emens **Mi**ni**cont**rol.

SIMOCODE

The **Si**emens **Mo**tor Protection and **C**ontrol **De**vice with communication capabilities offers intelligent → overload protection, an extensive range of programmable control functions as well as several diagnostic and monitoring features for → motor starter circuits. The → data exchange between SIMOCODE units and a process controller (e.g. → programmable controller) is accomplished via a serial → bus system (e.g. SINEC L2 or → PROFIBUS).

Single phasing

The abnormal operating condition of a three-phase induction motor (squirrel-cage motor) if one phase of the power supply is interrupted (e.g. broken wire, operation of a → fuse). If this condition is maintained for too long (especially under load conditions) severe thermal and mechanical damage to the motor may result. To increase the operational safety, it is recommended that overload protection devices incorporating → phase-loss sensitivity be used.

Single phasing protection

See → Phase-loss sensitivity

Single-break contacts

A → contact system in which the current is interrupted at one point in the → current path only (see also → Double-break contacts)

Single-throw switch

A → control switch with two distinctly separate rest positions of its → operating mechanism The contacts in each → pole of the switch are either all closed or all open depending on the position of the operating mechanism. Typical examples are toggle switches, → circuit-breakers and load disconnectors. Single-throw switches for use in → auxiliary circuits are classified as → maintained-contact command devices.

Note:
The term is also used to refer to → normally open (NO) or → normally closed (NC) → switching devices in general.

Site altitude

At altitudes of 2000 m above sea level, heat dissipation is lower owing to the lower air density. This applies to electronic items already at altitudes of 1000 m above sea level. For this reason, a derating of switchgear may be required at these altitudes. Furthermore, the dielectric strength may also be reduced.

SIVACON

The Siemens trade name for a range of versatile → type-tested low-voltage switchboards with → busbar ratings of up to 6300 A. The switchboard → cubicles are offered in both fixed and → withdrawable-type design, and are manufactured exclusively from standardized, series-produced structural parts and modules. The withdrawable design permits the → withdrawable units (e.g. → motor starter assemblies, or → air circuit-breakers) to be moved from the operational to the test and disconnected positions without the need for the compartment door to be opened (see → Motor Control Centre).

Skin effect

See → Conductor cross-section

Slip-ring motor starter

A slip-ring → motor starter is a → switching device or switchgear assembly (see → TTA) comprising a variable resistance (e.g. resistor banks, moveable electrodes in conductive liquid, etc.), and a controlling element (e.g. contactor stages with time relays and control circuit, → starter switch, etc.). It is used to accelerate the motor from zero speed (standstill) to its operating condition by reducing the resistance in the rotor circuit of the motor. During the starting sequence and during operation, the current and the torque may be limited to predefined desired values by regulating the rotor circuit resistance. To reduce the space required for the variable resistance, it may be force-cooled (e.g. by a fan, or circulating coolant) or it may be immersed in oil (oil-cooled starter).

Slow discharge resistor

See → Discharge choke

Smooth starting and stopping

See → Soft starting and → Soft stopping

Snap action

If the → operation of a → switching device (e.g. opening or closing of its → contacts) is substantially independent of the velocity of the → actuator or → actuating element, then it is said to possess snap-action. The term is commonly used in reference to the operation of → position switches. Also refer to → Creep action.

Soft iron core

See → Magnet core

Soft starter

This term is used to refer to a → motor starter which reduces the starting torque (locked rotor torque, breakaway torque) of the motor in order to reduce shock on the driven machine upon switch-on. In the case of squirrel-cage motors, the starting torque is usually reduced by the initial application of a supply voltage which is lower than the rated operational voltage of the motor (the starting torque is proportional the square of the applied voltage). This → terminal voltage may then be increased again once the motor is running. The classical methods by which the supply voltage may be reduced include → star-delta starting, → stator resistance starting and → autotransformer starting. The use of → semiconductor motor controllers and starters (e.g. → SIKOSTART) which employ switched thyristor circuits to control the terminal voltage of squirrel-cage motors, is increasing in popularity. Also refer to → Soft stopping.
In the case of slip-ring motors, soft starting is achieved by controlling the resistance in the rotor circuit (→ slip-ring starter).
Soft starting (or smooth starting) may also be achieved by mechanical means (e.g. fluid couplings, clutch mechanisms, etc.).

Soft stopping (of a squirrel-cage motors)

Soft stopping of squirrel-cage motors is achieved by reducing the → terminal voltage on the motor in accordance with a specific function relative to time (e.g. linear decrease or down ramp with a specified → ramp time). This is particularly useful if the inertia of the drive is small and a sudden stopping of the driven machine is undesirable (e.g. conveyer belts, escalators). A special form of soft stopping is the so-called → pump control by means of which shocks in pipe systems (e.g. "water hammer") upon switch-off of the pump motor can be avoided.

Solenoid

See → Electromagnetic operating mechanism

Solid-state motor starter

See → Semiconductor motor controller/starter

Solid-state overcurrent relay or release

Modern → circuit-breakers often incorporate electronic → overcurrent release modules. The current is generally sensed by means of → current transformers. Solid-state → overload relays offer advantages such as a wide setting range, warning of an eminent → tripping operation, tripping → CLASS selection, etc.

Sonar-BERO

The trade name for the Siemens range of → ultrasonic proximity switches.

SONPROG

This is a software package (MS DOS compatible) offered by Siemens for the direct monitoring, programming and adaptation of → Sonar-BERO proximity switches by the customer himself. By means of this program and an interface module, parameters such as the initial and final values of the → effective operating distance, the → hysteresis, the → blind zone, the switching function, the initial and final values of the analog output characteristic, average values, etc. may be monitored and/or adjusted.

Spacing unit

See → Modular width

Sparkover

The term sparkover is understood to mean the striking of an → arc (i.e. → disruptive discharge) between two live → conductive parts at differing potential in a gaseous or liquid medium (usually in air). Also see → Breakdown voltage and → Flashover.
Since the → dielectric strength property of clean, dry air (at 20°C and 50 Hz) is about 45 kV/mm, sparkovers occurring in the field of low-voltage technology are usually precipitated by the presence of → ionized gases of sufficient density between the conductive parts.

Spike suppressor

See → Surge suppressor

Splitter plates

See → Arc splitters

Spurious switch-on signal suppression
(in a → proximity switch)

A brief output signal may be generated by e.g. an → inductive proximity switch immediately after switch-on of its → control supply voltage even if no → target is present in the damping zone (see sketch illustrating → free zone). This transient phenomena occurs during the generation of the electromagnetic field. For this reason, the output signals of most → BERO's are suppressed during the initial switch-on time t_v (see → Time delay before availability).

Standard target (of a → proximity switch)

A specified object used for making comparative measurements of the → operating distances and → sensing ranges. The → response characteristics published for a particular proximity switch are based on the dimensions and material of the standard target. The following standard targets are defined in IEC 947-5-2.

For → inductive and → capacitive proximity switches:
A square shape of 1 mm thick carbon steel with a rolled finish. The length and width of the square must be equal to the diameter of the circle inscribed on the → sensing face of the proximity switch, or three times the → rated operating distance s_n (whichever is greater). In the case of capacitive proximity switches, the standard target is earthed.

For → ultrasonic proximity switches: A square shape of 1 mm thick, made of metal, and having a rolled finish. The dimensions of the square are specified to match the particular sensing range of the device (minimum sensing surface at maximum final value of → effective operating distance – refer to the relevant Siemens l.v. switchgear catalogue)

For → photoelectric proximity switches: For types R and T, the standard target is either supplied or specified by the manufacturer. For type D, it is a square of white paper having 90% reflectivity. The square is 100 x 100 mm for operating distances up to 400 mm. For longer operating distances, it may be 200 x 200 mm.

Star-delta starter

A → motor starter (generally for squirrel-cage motors) which causes the motor windings to be connected in the star configuration upon switch-on, and to change over to the delta configuration after a period of time. Star-delta starters are typically used in cases where a high starting current must be prevented in order to avoid voltage dips in the network. During the starting stage (star), the motor → starting current is limited to one third of the → d.o.l. value. The starting torque is also reduced to about a third of the value which would be produced during direct starting in the delta connection. After the motor has attained approx. 90% of its rated speed, the starter changes over to the operating stage (delta). In the case of automatic star-delta starters (typically a → switchgear assembly comprising 3 → contactors, an → overload relay and associated → controlgear), the change-over time is usually determined by the setting of a → time relay. In manual star-delta starters (e.g. → control switch), the change-over time is determined by the operator.

Special versions of star-delta starters include starters for closed transition from the star to the delta connection (→ closed transition star-delta starter) and → four-stage star-delta starters.

Star-delta starters are also known as "Y-D starters" and "wye-delta starters".

Start impulse

This term is used in connection with → semiconductor motor starters (e.g. → SIKOSTART). It refers to a feature by which a short voltage pulse is applied to the terminals of the motor before the normal starting procedure commences. This is designed (or set) to produce just sufficient torque to overcome the initial static friction in the drive. After application of the start impulse, the motor terminal voltage is increased from the initial set value according to the selected ramp function. It is also known as the "breakaway pulse", "impulse start" or "kick start" feature.

Start-up time (starting time)

The start-up time is the period of time taken from switching on of a motor (→ starting) until its shaft reaches the final operating speed.

Starter

See → Motor starter

Starter switch (for a slip-ring motor)

A → switching device (typically either manually operated or with a servo-motor → operating mechanism) used to alter the resistance in the rotor circuit of a slip-ring motor (by switching resistance banks in or out of the circuit) with the purpose of starting it or controlling its speed of rotation (also see → Slip-ring motor starter).

Starting

This term is principally applied to the switching on of electric motors, but is also used as general term for the switching on of any electrical equipment, machine or process. Usually, it involves at least one intermediate step or procedure before the final operating state of the motor or machine is achieved. For example, by the bridging out of resistors in the intermediate steps (e.g. → resistance starter), or by changing the interconnection of motor windings (e.g. → star-delta starter), or by using → semiconductor motor starters (e.g. → SIKOSTART), certain torque behaviour can be achieved and/or the → starting current may be reduced.

Starting (Start-up or Run-up) of a motor

Initialization of a rotating movement from standstill up to rated speed. This is achieved by the application of a suitable voltage (either the → rated operational voltage or an initially reduced value) to the motor terminals (see → Starter)

A precondition for successful motor start-up is that the torque of the drive motor is higher than the load torque of the driven machine at all speeds.

Starting current of a motor (Run-up current)

During direct-on-line starting of an induction motor, the supply current drawn by the motor may rise initially to anything from 6 to 13 times the rated operating current (this factor varies from one motor version to another and may differ from manufacturer to manufacturer). Once the shaft is turning and the speed increases, the starting current drops again to the value of the → operating current (→ rated full load current) as soon as the → operating speed (rated speed) is attained.

Various → starting methods are designed to reduce the starting current to lower levels (see → Motor starter)

Note:
Unless otherwise stated, the starting current given for a squirrel-cage motor usually refers to the initial maximum r.m.s. value of the supply current during → direct-on-line starting in the delta configuration. It is often stated as a multiple of the rated full load current and does does not include the initial transient peak, or → inrush current.

Stator resistance starter

This term applies to → motor starters which employ an external resistance in stator circuit of a three-phase induction squirrel-cage motor to reduce the voltage on the terminals during → start up. The resultant reduction in the starting current causes in marked drop in the available motor torque, since although the current is directly proportional to the applied voltage, the torque is reduced by the square of the voltage on the motor terminals. If a resistance is introduced into only one phase of the stator supply, the starting torque of the motor is also reduced, but the current is only reduced in this line (→ KUSA starter circuit). Also refer to → Soft starter.

Stored-energy operation

Operation by means of a stored-energy drive or → operating mechanism. Also see → Operation (of a switching device)

Striker armature

Part of an → instantaneous short-circuit release in a → current-limiting circuit-breaker.
In the event of a → short-circuit the → armature of the → electromagnetic release is caused to strike the moving contact element directly, and thereby force it from the fixed one at great speed, even before the tripping of the release can cause a delatching of the → switch mechanism. Thus, this striker armature increases the speed with which the contacts separate and reduces the → total break time in the event of high short-circuit currents (see → Time-related terms). The greater the rate-of-rise of short-circuit current, the more pronounced this effect will be.

Stub terminal

A terminal or point of connection on an electrical device in the form of a short busbar length (or stub) protruding from the device. Incoming and/or outgoing busbars are bolted directly to the stub terminals, or wires with cable lugs may be used. A special form of stub terminals are the so-called "rear-connecting stubs". These are usually round conductor sections which protrude from the back of a → moulded case circuit-breaker (line and/or load side) so that termination takes place behind the circuit-breaker mounting plate.

Summation current transformer

A → current transformer with more than one primary winding on a common core. Summation current transformers are generally used to measure the sum of the currents (being at the same frequency and at the same phase displacement, but possibly at differing power factors) in two or more incoming or outgoing feeder circuits. If the currents in all three phases (i.e. Ll, L2 and L3 respectively) of two feeder circuits are to be summated, then three → interposing summation c.t.'s and six main c.t.'s will be required. Current differences may also be measured by means of the reversed connection of the main current transformer secondary conductors. The → core-balance current transformer is a particular form of summation current transformer in which the difference between current flow to a load and returning from it is measured (see → Residual current protection)

Supply connection regulations

Regulations, guidelines and stipulations on the correct installation of electrical power systems and equipment with rated voltages up to 1000 V. These must be adhered to if power systems and equipment with rated voltages up to 1000 V are to be connected to the public low-voltage network or grid. In Germany, they are drawn up by the → VDE and DIN organizations . They are published and administered by the local or regional power supply authorities (→ **EVU** in German: = **E**lektrizitäts**v**ersorgungs**u**nternehmen). In Germany these rules are known as → TAB.

Surge

This term refers to one of the standardized forms of → electromagnetic interference used in the → EMC testing of low-voltage switch and controlgear devices containing electronic components (e.g. → proximity switches). In terms of IEC 801-5, for example, it consists of an impulse with 1.2 µs rise time and 50 µs half-wave duration. The severity levels are defined in terms of the test voltage (typically >1 kV) which is superimposed e.g. onto the maximum value of the → control supply voltage during testing. The ability of a device to withstand burst testing at a specific severity level, is referred to as its "impulse voltage withstandability". Also refer to → Burst.

Surge arrestor

A device designed to protect electrical equipment from high transient overvoltages (e.g. caused by lightning strikes on power transmission lines) and to limit the duration and frequency of the resulting currents.

Surge suppressor

A device or circuit designed, for example, to reduce the effects of → switching overvoltages created during the opening of inductive circuits (e.g. → contactor → control circuits). Usually, surge suppressors are included in the range of accessories offered for a particular contactor type and size or may be incorporated inside the device itself. Typically, they are mounted outside on the body of the contactor, and are connected in parallel with the coil terminals. Various techniques for the suppression of switching overvoltages are employed:

→ Freewheel diode: This method can only be used with d.c. → magnet drives. When the circuit is opened, the diode acts to short out the coil for all overvoltages having the opposite polarity to that of the driving voltage. Since it tends to extend the → drop-out time of the contactor, and may even cause two-stage drop-out in larger units, it may not always be a suitable method of surge suppression.

→ RC-element: This method (used in a.c. circuits) reduces the amplitude and frequency (rise-time) of the transient oscillations. It is particularly suitable for the protection of du/dt sensitive output stages of electronic controllers (e.g. triacs). They may be less suitable for output stages with a low-pass characteristic.

→ Varistor: In this method, the switching overvoltages are "chopped off" at a certain amplitude. The varistor circuit is well suited to electronic outputs with a low-pass characteristic, less so for the protection of du/dt sensitive output stages.

Combinations of the RC and varistor method are also offered.

Surge suppressors are also known as "spike" or "overvoltage" suppressors.

Surplus detection indication

(of a → photoelectric proximity switch)

An indication of the excess radiation power falling on the light incidence surface of the → receiver. The surplus light may decrease over a period time, e.g. as a result of dirt on the lenses and reflectors, deterioration in the reflection factor of the → target or even ageing of the → emitter diode, to such an extent that reliable operation may then no longer be guaranteed. Surplus detection indication may be by way of an LED which remains on while, say, 80% of the surplus light range is available or by means of a separate output at which the signal strength may be monitored externally.

SUVA

Schweizer Unfallversicherungsanstalt (Transl. Swiss institute for accident insurance). Refer to → Positively driven contacts.

SWG

British Standard Wire Gauge for conductor cross-sections of up to 81.08 mm² = SWG 0000. See also Table 9.10 on Page 511.

Switch (mechanical)

A → switching device capable of making, carrying and breaking currents under normal as well as under certain abnormal circuit conditions. Such conditions may include those of overload and short-circuit within specified limits of time and/or intensity.

Switch control, ON/OFF switch control

See → Maintained-contact operation

Switch disconnector

A → switch which, in the open position, satisfies the isolation requirements of a → disconnector

Switch disconnector with fuses

This form of → fuse-combination unit comprises a three or four pole → fuse base and separate → contacts to achieve the → isolation function (also refer to → Fuse switch disconnector). The contacts are usually operated by means of a rotary-drive → operating mechanism. In conjunction with special → door interlocking handle mechanisms, these units are extensively used in → MCC applications. Versions for plugging onto → busbars facilitate the construction of → withdrawable switchgear assemblies.

Switch mechanism

See → Operating mechanism

Switch-fuse

A switch in which one or more → poles incorporate a → fuse.

Switchboard

A switchboard is generally considered to be a large single panel, frame or assembly of panels on which various items of → switchgear, → busbars and instruments are mounted. Most modern switchboards are of the enclosed type and are either of a fixed-mounted or → withdrawable design. The term "switchboard" is therefor commonly used to refer to one or more → cubicles or → cabinets containing assemblies of switchgear and distribution equipment. Standard → type-tested modules are offered to facilitate extensions and save engineering costs (see → SIVACON). The term switchboard does not (normally) apply to groups of local switches in final circuits. Also refer to → distribution boards and → panelboards.

Switchgear

A general term for → switching devices and their combination with associated control, measuring, protective and regulating equipment. It includes assemblies of these devices and equipment with the corresponding interconnections, accessories, enclosures and supporting structures. In principle, switchgear is intended for use in the field of generation, transmission, distribution and conversion of electric energy.

Switchgear assembly

See → Low-voltage switchgear and controlgear assembly

Switching arc

A phenomenon which always occurs if an electrical circuit operating at a voltage which exceeds 15 V is opened. At the instant prior to physical separation of e.g. the → contacts in a → switching device upon opening (i.e. at zero → contact pressure), the current flows through a thin "bridge" of material. As this bridge is lengthened, the current density through it increases and the material of the bridge heats up so strongly that it melts and finally vaporizes. The current continues to flow through an → arc as the contacts move further apart and the bridge is broken. The arc intensity depends on various factors such as the nature and value of the current being interrupted, the → power factor and the instant of → contact separation with respect to the current waveform. Temperatures between 10.000 and 15.000 °C can be measured within the arc itself. Owing to this high temperature, → contact material is melted and even vaporized. Since vaporized contact material is displaced by the switching arc, the → electrical service life of the switching device is directly influenced by the switching arc and the way in which it is quenched within the → arc chamber.

Switching capacity

The maximum current which a → switching device is capable of switching on and off. If the device is able to switch on more current than it can switch off (or vice versa), then a distinction is made between its → making capacity and → breaking capacity.

Switching chamber

The enclosed area in a → switching device in which the → switching elements of each → pole are housed, and in which the switching operations take place. If necessary, it may be extended by an → arc chamber.

Switching device

A device designed to make and/or break the current in one or several electric circuits, whereby one can distinguish between switching devices intended primarily for use in → main circuits and in → auxiliary circuits. Also refer to → Making operation and → Breaking operation

Note:

Although → fuses do not contain any → contact elements as such, they may also be regarded as switching devices since they are designed to break → overload and/or short-circuit currents.

Switching element

A unit, component or mechanism designed to open and close an electric circuit. The term is widely used to refer to a single → contact element, an output relay of an electronic device or any semiconductor circuit in which the conductivity is controlled by an internal or external → command signal to switch current on and off.

Switching frequency (of a → switching device)

See → Operating frequency

Switching overvoltage

Overvoltage caused by the sudden interruption of current flowing in an inductive circuit. The opening of → contactor coil circuits (i.e. switching off of contactors) is one of the most common sources of switching overvoltage in industrial circuits. Owing to the high resonance impedance of the open-circuited coil, the amplitude of the oscillations are typically in a range of up to 10 kV, where-as the rate-of-rise of the overvoltage spikes may be as high as 1 kV/μs. Rapid restriking of the current initially occurs between the opening contacts, until the dielectric → recovery voltage exceeds the decaying amplitude of the voltage oscillations. The → rapid multiple restriking period lasting approx. 250 μs is followed by a damped oscillation with peak amplitude of approx. 3,5 kV. The rapid restriking can cause severe contact erosion. Furthermore, the steep wave flanks can introduce → radio interference or disturbances in neighbouring systems owing to capacitive coupling effects. An appropriate suppression circuit for this source of "switching noise" may be necessary. See → Surge suppressor.

Switching state

This term is used to refer to the position of the → operating mechanism of a → switching device (e.g. → actuated or not, energized or de-energized) and/or to the corresponding position of its → contacts (i.e. either open or closed).

System (Network)

See → Network types and systems

TAB (German: **T**echnische **A**nschlußbedingungen)

→ Supply connection regulations

Take-over current

The current coordinate of the point of intersection between the → time-current characteristics of two → overcurrent protection devices connected in series.

Note:

If, the upstream protection device (e.g. → circuit-breaker or → fuse) is provided for short-circuit current protection, and the prospective short-circuit current exceeds the breaking capacity of the downstream device (→ back-up protection), a distinction must be made between:
Take-over current I_B: The maximum overcurrent value above which the back-up protection device will definitely operate. (The *take-over current* I_B is also defined as the tripping current limit of the back-up switching device), and the Discrimination limit: The maximum value of the overcurrent for → current-based discrimination, i.e. at which only the downstream unit opens.

Target (of a → proximity switch)

A specified object used to operate a proximity switch (see → Standard target)

TEFC (motor)

Totally **E**nclosed **F**an **C**ooled; i.e. an indication of the protection of a motor. It corresponds approximately to IP 44 → degree of protection.

t_E-time

The so-called "safe locked-rotor time" t_E of a particular a.c. motor, is the time taken for the winding temperature to rise from its final operational value to a fixed maximum value (determined by the corresponding → temperature class of the motor) if the rotor is blocked. It is of particular significance when overload protection is specified for motors destined for use in → hazardous locations. For the → type of protection increased safety "e", the tripping time of overload protection apparatus from the cold state must be less than the t_E-time of the motor.

Temperature class

In terms of the specifications for → explosion protection, electrical equipment is divided into two groups viz.:
Group I - electrical equipment for mining, and
Group II - electrical equipment for all other → hazardous locations.
Electrical apparatus of Group II may be classified further in terms of the so-called *temperature class* and → explosion group of the actual → potentially explosive atmosphere which may be encountered at its point of application. The temperature at which an atmosphere will ignite owing to e.g. the hot surface of a piece of electrical equipment is dependent on the actual type of gases or vapours present. Temperature classes indicate the maximum surface temperature on a piece of electrical equipment (e.g. T1 up to 450 °C, T2 up to 300 °C, T3 up to 200 °C, etc.) which must always be lower than the ignition temperature of the gas or vapour mixture in which it is installed.

Temperature compensation

The → operating time of → bimetal overload relays and releases is influenced not only by the load current but also by the → ambient temperature. The effect of the ambient temperature may be removed by means of an additional → bimetal strip which is heated by the surrounding air only (i.e. not by the current). In this way, the relay retains its → tripping characteristic even if its ambient temperature varies.

Temperature detector (or sensor)

Temperature detectors may be installed directly between the phase windings in a motor or transformer. In this way, direct → overload protection may be achieved. In industrial low-voltage applications, small thermistors in conjunction with tripping units are used predominantly (see → Thermistor motor protection), although → bimetal switches and other special temperature-sensitive materials implanted in the windings or winding slots are also found. Also refer to → PTC thermistor and → NTC thermistor

Temperature-rise limit

In general, this refers to the maximum temperature which the individual parts of a → switching device are capable of enduring permanently without being damaged. It is the sum of the → ambient temperature and the permissible temperature-rise of the device itself. The functional safety of the device may be jeopardized if the temperature-rise limit is exceeded.
In terms of product standards, the temperature-rise limit of specific parts (e.g. terminals, coils, and accessible parts) may be assigned specific maximum values under given operating conditions (e.g. rated conditions).
For example, IEC 947-1 specifies the following temperature-rise limits for terminals on switching devices:

Terminal material	Temperature-rise limit (K)
Bare copper	60
Bare brass	65
Tin-plated copper or brass	65
Silver or nickel-plated copper or brass	70

The use of connected conductors significantly smaller than those specified for a particular terminal could result in higher terminal and internal part temperatures. Such conductors should therefor not be used without the manufacturer's consent.
Temperature-rise limits for accessible parts and operating means of switching devices are also specified – at much lower values. For example, IEC 947-1 specifies 25 K for non-metallic → actuators and 30 K for metallic parts intended to be touched but not hand-held.

Temperature-rise time-constant T_A, cooling time-constant T_K

The time-constant of a value which increases or decreases exponentially with respect to time and approaches its final value asymptotically, is the value on the time scale of a curve representing its successive values which corresponds to the point at which the initial tangent of the curve intersects with the asymptote of the final value.

Since the temperature of a body (e.g. motor) increases or decreases in this way (assuming constant cooling), the rate at which the temperature will change and approach a limit (e.g. maximum) value can be derived from the characteristic thermal time-constant of the body.

Temporary duty

The → duty cycle in which the → main contacts of a → switching device remain closed and conduct a specified load current for periods insufficient for the device to reach thermal equilibrium. After each → on-load period follows an → off-load period of sufficient duration to restore equality of temperature with the cooling medium. *Comment*: In practice, the maximum permissible loading of a switching device controlling a motor in → short-time duty may be limited by the → breaking capacity of the switching device (see Section 3.7.8, page 201).

Terminal

Connection parts comprising one or more clamping units or mechanisms on an item of switchgear which serve the exterior connection (including linking, branching) of cables, wires and/or solid →conductors. The clamping unit(s) must not only provide an enduring low-impedance electrical connection through maintained high →contact pressure, but must also ensure that the conductors are fixed securely against pull-out or mechanical movement. The wires and/or conductors may be →prepared (i.e. fitted with a cable lug, eyelet, etc.) or →unprepared (i.e. merely cut and stripped of insulation), depending on the design of the terminal (see →SIGUT terminals). The manufacturer's instructions regarding tightening torques, minimum and maximum conductor cross-sections, wire preparation, etc. must be followed to ensure correct termination.

Terminal block

An insulating housing containing one or more mutually insulated →terminals or terminal assemblies (two or more interconnected terminals). The insulating housing is designed to be fitted to a mounting rail or other support.

Terminal voltage

This term is used to refer to the actual (instantaneous) value of the voltage applied to the →terminals of a →switching device, or other item of electrical equipment (e.g. to the main terminals of a squirrel-cage motor)

Test sequence

The test sequence used for determining the short-circuit →breaking capacity of a →circuit-breaker is expressed by the symbols:
O for a →breaking (opening) operation
CO for a →making (closing) operation followed immediately by a breaking operation (i.e. closing onto an existing short-circuit)
t for a time period between successive short-circuit interruptions to permit cooling and resetting of the circuit-breaker.

Testing of low-voltage switchgear and control gear

A number of test sequences and test procedures are outlined in the general rules for low-voltage →switchgear and →controlgear (IEC 947-1) as well as in the standards and specifications for particular items of electrical equipment. These tests are designed to prove compliance with the constructional and performance requirements of the standards (e.g. →performance test) and may be categorized as follows:
- type tests which are performed on representative samples of each particular piece of equipment,
- routine tests which are performed on each individual piece of equipment manufactured to the particular relevant standard,
- sampling tests which are performed if called for in the relevant product standard.

The IEC regulations permit testing by the manufacturer, at his works or at any suitable laboratory of his choice. Where appropriate, special tests may also be performed. These are subject to the specifications in the relevant product standards and to agreement between the manufacturer and the user of the equipment.

Thermal overload relay or release

An inverse-time delayed →overload relay or release of which the function (including the delay) is based on the thermal effect of the current which flows through its main →current path. Also refer to →Bimetal relay and →Eutectic alloy overload relay.

Thermal short-circuit rating

See →Short-circuit strength

Thermal stress limit (of an overload relay)

See →Destruction characteristic

Thermal time-constant

See →Temperature-rise time-constant T_A

Thermal value

See →$I^2 \cdot t$ value

Thermistor motor protection

Thermistor →temperature detectors are used extensively in the protection of motors (and transformers) against excessive temperature rise. →PTC or →NTC thermistors embedded in the phase windings are connected to tripping units which sense the change in resistance brought about by the (excessive) rise in temperature inside the motor or transformer.

Through-beam photoelectric proximity switch

This term is used to refer to a type **T** photoelectric proximity switch using **t**hrough-beam sensing in which the light beam travels from an →emitter to a separate →receiver. The presence of a →target is recognized when the light beam is blocked or scattered between emitter and receiver.

Time delay before availability t_v (of a →proximity switch)

Also known as the "availability delay", this is the time period between switching on of the →control supply voltage and the instant the proximity switch becomes ready to operate correctly. During this time, the output element should not give any spurious or false signals, i.e. it should have →spurious switch-on signal suppression.

Time grading

In order to achieve →time-based discrimination between series-connected →circuit-breakers, the →short-time delayed overcurrent releases (z-releases) fitted to the respective circuit-breakers are set to operate with increasing delay at every distribution level closer to the source of power supply. This technique is also known as time grading (refer to →Short-time grading control)

Time relay

A → switching device (see → Relay) which introduces a preset time delay into the operating sequence of an electrical → control circuit. Time relays may be classified in terms of the following typical functions (or their combinations):
- On-delay (or pick-up delay) - the output → switching element(s) change → switching state after application of a switching command (or energizing of the relay) and expiry of the preset time delay,
- Off-delay (or drop-out delay) - the output switching element(s) change switching state after removal of a switching command (or de-energizing of the relay) and expiry of the preset time delay,
- Making and/or breaking pulse contact (or fleeting contact) – the output switching element(s) change switching state for a short preset time period upon application or removal respectively of a switching command,
- Flashing/pulsing - the output switching element(s) continually change switching state while the relay is energized. The ON:OFF ratio (or respective periods) may be separately adjustable or fixed.

The delay period may be achieved electromechanically (motor and clockwork mechanism), pneumatically, or electronically. Electronic time relays may be of the digital type (in which the time delay is determined by the "counting" of internal clock pulses) or of the analogue type (in which the time delay is determined by the charging or discharging of a capacitor). Time relays are also known as "timers", although this term is also used for devices which measure a time period (also see → Time switch).

Time switch

A → switching device incorporating a continuously running clock or equivalent mechanism (or electronic circuit) which may be set to make and/or break a circuit at a predetermined time or times (also refer to → Time relay).

Time to contact movement

See → Movement delay

Time-based discrimination

This term refers to the technique of discriminative protection (→ discrimination) which employs specific time delays between the operation of successive series-connected → circuit-breakers. The time delays, which increase with every distribution level closer to the source of power supply (→ time grading), may be achieved by means of → short-time delayed overcurrent releases (z-releases) fitted to the respective circuit-breakers. Since each circuit-breaker must withstand the short-circuit current for the delay time set on its z-release, they are usually → air circuit-breakers and employ the → zero-point quenching technique.

By means of → solid-state overcurrent releases, the heat value of the short-circuit current (→ I^2t-value) passing through each circuit-breaker may be used to determine the time delay of its z-release. Alternatively, a data link between the circuit-breakers may be used to minimize the delay time (see → Short-time grading control)

For comparison, also refer to → Current-based discrimination.

Time-constant ϱ

When the rate at which a function is decreasing can be expressed as that function divided by a constant, then the constant is known as the time-constant. It is equal to the time taken to fall to 0.368 (1/e) of the initial value. In the case of rising temperatures or currents the decreasing function is the difference between the instantaneous and the final values (see → Temperature-rise time-constant T_A, cooling time-constant T_K). In the case of d.c. test circuits (e.g. for determining d.c. → breaking capacity), the rated time-constant is taken to be 15 ms, unless otherwise stated.

Time-current characteristic

A curve of time (e.g. virtual pre-arcing time of a fuse, → tripping time of an overload relay) plotted as a function of → prospective current under specified conditions of operation. For example, the tripping time of a → bimetal relay at a given value of overload current (usually expressed in multiples of the set tripping current) from a point in time at which the relay was in the cold state, may be read from the time-current (tripping) characteristic of the particular overload relay. Similarly, the time it takes for a → fuse element to melt through at e.g. a particular value of → prospective short-circuit current may be found by means of the time-current characteristic of the fuse. A comparison of the time-current characteristics of particular items of protective switchgear, gives an indication of their discriminative operation (→ discrimination) with respect to one another.

Time-current characteristics are also referred to as "tripping curves". Also refer to → Tripping current.

Time-related terms

Figure 9.14 illustrates the definintions of the various terms used to describe times and durations during switching operations in → switching devices and → auxiliary switches.

TN, TN-C, TS-S system (network)

See → Network types and systems

Toggle, Toggle lever

See → Operating mechanism

Torque class (KL)

Siemens created this torque-classification system because the slot form of a squirrel-cage rotor does not generally provide enough information on the torque characteristic of the motor. A motor of torque class 16 (KL16 or "rotor class 16") for instance, can run up against a load counter torque of 160% of the rated motor torque.

Total break time

The period of time from the beginning of the → opening operation (e.g. de-energizing command in the case of a

Glossary

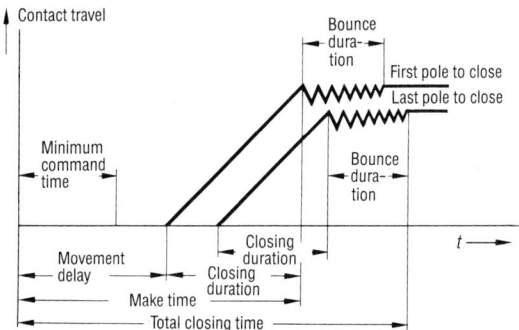

a) Closing operation of a switching device

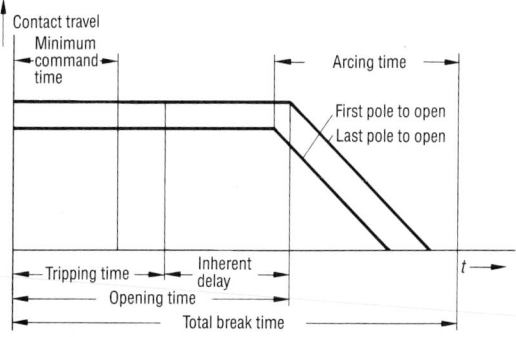

b) Opening operation of a switching device

→ contactor) to the end of the → switching arc duration (see → Arc extinction).

Total closing time

The period of time from the beginning of the energizing command to the end of the bounce duration (→ contact bounce) of the last pole to close (see → Time-related terms).

Total $I^2 \cdot t$ value (to rupture a fuse element)

The value of heat required for the → fuse element of a → fuse link to melt through, i.e. to start the interruption process irretrievably. It is expressed in $A^2 s$, and is dependant on the → rated current of the fuse link, on its → operation class and on the supply voltage. (also see → Pre-arcing $I^2 \cdot t$ value and → Let-through $I^2 \cdot t$ value)

Total insulation

This term is used for → protective measures taken against direct contact with live parts by means of protective insulation. It is achieved by means of
– insulation in addition to the basic insulation or
– reinforcement of the basic insulation
in such a way that dangerous leakage currents are prevented from flowing even if the basic or primary insulation fails. It is also referred to as "double insulation".

Total load

See → Effective installed load

Touch voltage

That part of the → earth-electrode potential which can be spanned by a person (e.g. from hand to foot or from hand to hand). Internationally, a voltage > AC 50 V or > DC 120 V is regarded as a dangerous touch voltage although these limits may be too high in particular cases (in agriculture operating areas or rooms used for medical purposes). Also refer to → Fault voltage. The touch voltage is also known as the "contact potential".

Tracking

The gradual formation of conducting paths on the surface of solid insulating material owing to effects of electric stress and electrolytic contamination (see → Creepage current).

◁ c) Switching operation of auxiliary switches
Change-over without interruption, e.g. from NC to NO, may be achieved by either a late-break NC or an early-make NO.

a Closed position of the normally closed contact (NC)
b Open position of the normally closed contact (NC)
c Closed position of the normally open contact (NO)
b Open position of the normally open contact (NO)

Figure 9.14
Definitions related to time during the switching operations of switching devices and auxiliary switches

träg

A German term meaning "slow or tardy". It is used to designate the "time delayed" or "slow response" → time-current characteristic of a → fuse link. It is used particularly in conjunction with → DIAZED fuses. See → Operation class

Trailing angle

The angle between the direction of departure from a → position switch → actuator (e.g. roller lever) and the actuating surface of the operating cam or bar.

Transfer switch

A → switching device specially designed to transfer one or more load circuits from one source of power supply to another. Load transfer switches are used particularly in cases where the continued power supply to the load must be guaranteed in the event of a power failure in the usual supply network. Also refer to → Automatic transfer switching equipment (ATSE).

Transient overvoltage

A voltage spike or series of decaying voltage oscillations superimposed on the system voltage. One can distinguish between → switching overvoltages (owing to specific switching operations e.g. → transient recovery voltages or faults), lightning overvoltages (owing to a specific lightning discharge) and functional overvoltages (deliberately imposed).

Transient recovery voltage (TRV)

The → recovery voltage during the period when it is of transient character.
Note:
Depending on the characteristics of the circuit and the → switching device (e.g. → circuit-breaker), the transient recovery voltage may be of an oscillatory or non-oscillatory nature (or a combination of these). It also includes the displacement of the neutral point in multi-phase systems.
If not otherwise stated, the transient recovery voltage in three-phase systems applies to the voltage of the first pole to be quenched (first pole to open), as this voltage tends to be higher than the voltage in the other two poles.

Transition contactor

A contactor used to switch in a resistance during the changeover from the star to the delta configuration in a → closed transition star-delta starter.

Transport unit

A → section or part of a → switchgear assembly or switchboard which can be transported without further dismantling. Its size is governed by the weight of the installed equipment, the construction of the switchboard, the transport route, etc.

Trickle charging

The technique whereby a battery is continuously supplied with a small amount of charging current - just sufficient to balance out natural losses and to maintain it at full charge.

Trip class

See → CLASS

Trip-free operation

If a switch, → circuit-breaker, → overload relay, release or operating mechanism in general trips without the need of an external energy supply (e.g. if it incorporates a → latch and charged opening springs), and the → tripping operation cannot be halted once it has started and cannot be blocked by holding the → actuator, toggle or tripped indicator in the ON position, it is said to have "trip-free" or "release-free operation". A trip-free switching device has been defined as one in which the → moving contacts return to and remain in the open position when the → tripping operation is initiated after the initiation of the → closing operation, even if the → closing command is retained.

Trip-free switching device

See → Trip-free operation

Tripped position

The position to which the → operating mechanism or → toggle of a → moulded-case circuit-breaker snaps after being → tripped. Before the circuit-breaker can be switched on (→ closed) again, the breaker must be → reset by moving the operating mechanism to the OFF-position.

Tripping characteristic

The tripping characteristic (characteristic curve) is the graphical representation of the relationship between some active quantity, e.g. current (→ tripping current), and the → tripping time. From the → time-current characteristic of an → overload relay, for example, the time required for the relay to → trip (or operate) when it is loaded at a constant value of overcurrent from the cold state may be determined.

Tripping current, earth fault (g)

The minimum value of → earth-fault current which causes an → earth-fault release or relay to operate. Typically, the earth-fault release is incorporated in a circuit-breaker (→ g-release, → CBR), and its operation will cause the circuit-breaker to trip and open its main contacts instantaneously. The value of this tripping current (or "operating current") may be adjustable. Also refer to → Residual-current protection.

Tripping current, instantaneous (n)

The minimum value of short-circuit current at and above which an → instantaneous short-circuit release will be caused to operate (also known as the "operating current"). Typically, the instantaneous electromagnetic release is incorporated in a circuit-breaker (→ n-release), and its operation will cause the circuit-breaker to trip and open its main contacts instantaneously. Depending on the version of the n-release, its tripping current may be adjustable.

Tripping current, overload (a)

The value of continuous current at and above which a → current-dependent delayed overload release (or relay) will operate within a specified time period (→ inverse-time delayed release). The → a-release may be incorporated in a circuit-breaker, and depending on its version, its tripping current (or "operating current") may be adjustable (see → Tripping characteristic).

Tripping current, short-time delayed (z)

The minimum value of extreme overload or short-circuit current which causes the operation of a → short-time delayed overcurrent release. Typically, the short-time delayed overcurrent release is incorporated in a circuit-breaker (→ z-release), and its operation will cause the circuit-breaker to trip after a pre-set time delay, provided that the minimum operating current is maintained or exceeded during the delay period. Normally, both the tripping current and the time-delay are adjustable.

Tripping curves

See → Time-current characteristic

Tripping operation

An → opening operation of a mechanical → switching device initiated by the → operation of a → relay, → release or tripping unit.

Note:
In the context of automatic or remote "switching off" of circuit-breakers, the terms "trip" and "release" are often used rather loosely or in different parts of the English-speaking world to mean the same thing! (e.g. shunt trip, shunt release). As far as possible, we have used the word "release" to mean the module or device which is used to initiate the "tripping" operation or to "release" the latch of a switching mechanism.

Tripping time

The period of time between the start of the tripping command and the moment when the → tripping operation becomes irrevocable.

TT system (network)

See → Network types and systems

TTA

See → Type-tested assembly

Tuned resonance circuit

This type of → filter circuit is used, for example, to remove selected → harmonic currents from the network. It comprises one or more series connections of chokes and capacitors, but in contrast to the → choked capacitor technique, the LC combinations are specifically tuned to → resonate at the frequencies of selected harmonics. At these frequencies, the impedance of the filter circuit is virtually zero. Since, as a first approximation, the harmonic currents are proportional to the → reactive power drawn by the load, tuned resonance circuits are also used to achieve → power factor correction.

Type 1 (Type 2) coordination

See → Coordination with short-circuit protection devices

Type of output (of a → proximity switch)

Proximity switches are classified in terms of their "type of output" stage. IEC 947-5-2 makes the following distinctions:
P for → PNP output (3 or 4 terminal/wire d.c.),
N for → NPN output (3 or 4 terminal/wire d.c.),
D for 2 terminal/wire d.c.,
F for 2 terminal/wire a.c.,
U for 2 terminal/wire a.c./d.c. and,
S for other.

Type of protection

In DIN VDE 0170/0171 part 1, DIN EN 50014 various types of protection are defined for the manufacture of explosion protected equipment. See → Explosion protection

Type test

See → Testing of low-voltage switchgear and control gear

Type-tested assembly (TTA)

A → low-voltage switchgear and controlgear assembly which is constructed and type-tested in accordance with DIN VDE 0660 part 500 (IEC 931-1), or which does not deviate in any significant way from an original type or system already shown to comply with the requirements of the standard by means of the specified → type-testing procedure. Switchgear assemblies may contain electromechanical and/or electronic items of equipment (also refer to → Partially type-tested switchgear assembly PTTA).

Ultimate short-circuit breaking capacity
(of a → circuit-breaker)

A → breaking capacity defined in IEC 947-2 for which the prescribed test conditions do not include the capability of the circuit-breaker to carry its rated current continuously after a short-circuit interruption. The rated ultimate short-circuit breaking capacity I_{cu} is expressed as the value of the prospective breaking current in kA (r.m.s. value of the a.c. component). The → test sequence is O–t–CO.

Ultrasonic proximity switch (→ Sonar-BERO)

A → proximity switch which operates by virtue of ultrasound waves transmitted and received within a sensing zone. The heart of an ultrasonic proximity switch is a piezoceramic transducer which can transmit and receive high frequency sound pulses. If the transmitted ultrasonic pulses are reflected from an object (→ target), the echoes which are produced impinge on the transducer and are reconverted to electric signals. The process of alternate pulse transmission and echo receiving is repeated continuously. By measurement of the time between transmitted and received pulses, an ultrasonic proximity switch is able to measure the distance from the → sensing face to the target (provided the target is within the → sensing range). The Siemens ultrasonic proximity switches are known as Sonar-BERO switches and employ completely encapsulated semiconductor (i.e. electronic or solid state) switching elements.

Undervoltage relay or release

A → relay or → release which enables a mechanical → switching device to open or close (instantaneously or with a time delay), if the voltage at the terminals of the release drops below a specific value. Undervoltage releases are typically used to prevent automatic restarting of machines after an unintentional supply voltage interruption. In addition, they are used for interlocking purposes and remote tripping of → circuit-breakers. It may be interesting to note, that a wire break in the → control circuit will lead to the operation of an undervoltage release and the consequent → tripping of the associated circuit-breaker (this fail-to-safe arrangement may offer advantages when compared to the use of a → shunt release for remote tripping).

Uninterrupted duty

Uninterrupted duty is a rated → duty type in terms of IEC 947-1. It is understood to mean an operational duty without any → off-load period in which the → main contacts of the → switching device remain closed while carrying a load current of constant value without interruption for periods longer than eight hours (for weeks, months or even years).

Note:
Distinction must be made to the → eight-hour-duty, as progressive increase in temperature may occur as a result of oxidation and/or dirt deposits on the → contacts. Uninterrupted duty has to be accounted for either by a derating factor or by special design features (e.g. silver contacts).

Unprepared conductor

A → conductor which has been cut and of which the insulation has removed for insertion into a → terminal (see → Prepared conductor).

Note:
If the strands of the conductor have merely been twisted together, or even if it has been specially shaped for the introduction into the clamping unit of the terminal, it is still regarded as unprepared.

Usable operating distance s_u (of a → proximity switch)

The → operating distance of a particular proximity switch, as determined under specified conditions. In terms of IEC 947-5-2, it is measured over a range of ambient tempera-

ture as specified by the manufacturer and at a → control supply voltage of 85% and 110% of the → rated value. For → inductive and → ultrasonic proximity switches it must be between 90% and 110% of the → effective operating distance s_r. In the case of → capacitive proximity switches it may be between 80% and 120% of the effective operating distance.

Utilization category

The utilization category defines the intended application of the → switching device. Although the actual utilization categories differ for each type of switching device (as defined e.g. in the various product standards or sections of IEC 947), they are all characterized by one or more of the following parameters:
- Values of → making and → breaking currents expressed as multiples of the → rated operational current,
- Values of the voltage as multiples of the → rated operational voltage,
- Values of the → power factor (or the time constant in the case of d.c. applications),
- Short-circuit performance
- → Discrimination capabilities, etc.

Examples of typical applications defined by utilization categories:

AC-3 starting and switching off during running of a squirrel-cage motor by means of a → contactor (IEC 947-4)

AC-14 control of a small a.c. electromagnetic load by means of a control device (IEC 947-5)

DC-3 starting, plugging, inching and dynamic braking of a d.c. shunt-wound motor by means of a → contactor (IEC 947-4)

A protection of an a.c. or d.c. circuit by means of a circuit-breaker with no rated short-time withstand current (IEC 947-2)

B protection of an a.c. or d.c. circuit by means of a circuit-breaker with a rated short-time withstand current, i.e. with capabilities for → time-based discrimination (IEC 947-2)

Utilization class

See → Function class

UVR

See → Undervoltage relay or release

Vacuum circuit-breaker

A → circuit-breaker in which the → main contacts open and close in a highly evacuated chamber (or "bottle"). Since virtually no → switching arc is drawn when the contacts part, vacuum circuit-breakers offer a particularly long → electrical service life, even if large short-circuit currents are interrupted regularly.

Glossary

Vacuum contactor

A → contactor in which the → main contacts open and close in a highly evacuated chamber (or "bottle"). Since even the small → switching arcs are contained within the vacuum bottle, these contactors offer a particularly clean and environmentally friendly alternative to classical contactors in which the main contacts operate in air at atmospheric pressure.

VAr controller

See → Power factor controller

Varistor

A non-linear resistor (semiconductor) of which the resistance value (ohms) decreases sharply when the voltage across its terminals exceeds a specific value. It may be used to suppress → switching overvoltages which are produced when the → magnet drive of a → contactor is switched off (see → Surge suppression).

VBG 4

A specification for the protection of personnel working with or on electrical systems and items of equipment. The abbreviation VBG is derived from the German "**U**nfall**v**erhütungsvorschrift der **B**erufs**g**enossenschaft" (→ "trade union accident prevention specification"). The applicable DIN VDE standards are used as a basis for implementation of the specifications.

VDE (**V**erband **D**eutscher **E**lektrotechniker)

This association of German electrical engineers and technicians was founded in 1893. As a non-profit-making association in the field of technology and science, its tasks include the support of scientific research and development, training of its members, the publication of legally recognized electrotechnical safety regulations (VDE regulations) and the testing of electrical devices and equipment (VDE test mark). Furthermore, it represents electrotechnical interests on a national and international level.

Virtual chopping

High frequency superimposed current zeros occurring in the → switching arc of the last pole to clear during the → opening operation of a 3-pole → switching device. Virtual chopping is caused by → rapid multiple restriking in the first pole to clear.

Voltage drop U_d (across a → proximity switch)

The voltage measured across the active output of the proximity switch under specified conditions while → operational current flows.

Voltage drop U_R (V); u_R (%); ΔU

The difference between the → no-load voltage U_0 and rated output voltage U_{2n} of a transformer at rated load, referred to an ambient temperature of 20 °C.
The voltage drop in → conductors and transmission lines depends on their length, cross-section and material. In the case of long distances, the prospective voltage drop along a conductor must be considered. The higher the current, the higher the voltage drop will be ($V = I \cdot R$).

Voltage rise U_A (V); u_A (%)

The difference between the → rated output voltage and the → no-load voltage U_0 of a transformer, usually stated as a % u_A of the rated output voltage at → rated frequency and → $\cos \varphi = 1$ (→ regulation). The voltage rise values given in selection tables, are based on transformers at operational temperature.

Voltage trip

See → Shunt release

Water hammer

See → Pump control

Window-type current transformer

In window-type → current transformers the → conductor e.g. → busbar or cable, passes through the unit (ring or doughnut-shaped) and the value of current flowing in the conductor is transformed to the secondary winding without the need for electrical connections. The current transformer merely contains the secondary winding (toroidal) and the terminals, where-as the conductor itself forms the primary winding.

Wiping contacts

To achieve a self-cleaning effect and thereby an increase in the → contact reliability, → contact elements are often designed in such a way, that the → contact surfaces rub against each other each time they are brought together (sometimes also during separation). This can e.g. be achieved by a sliding action, or by a cushioning movement (suspension) of the fixed contact element. Adverse effects of these designs is the tendency for increased → contact erosion and the possible production of → contact material dust in the switching chamber.

Withdrawable circuit-breaker

In addition to its → main contacts, a withdrawable → circuit-breaker is equipped with a set of isolating contacts (see → Isolation) that permit the circuit-breaker to be moved into a position which disconnects it from the → main circuit. It is mounted within a guide frame and can usually be moved to a number of predefined positions, for example:
1. Connected or operational position – main and auxiliary circuits of the breaker connected to their respective networks
2. Test position – main circuit disconnected, auxiliary circuit connected
3. Disconnected position – main and auxiliary circuits of the breaker disconnected (and isolated) from their respective networks

4. Maintenance position – as for the disconnected position, but the circuit-breaker is drawn from the switchboard to a position (e.g. on supporting rails) in which it may be inspected easily. When in the maintenance position, the breaker may also be removed from the switchboard.

Withdrawable unit

This refers to a → switchgear assembly which may easily be withdrawn from the switchboard (see → Motor Control Centre) as a complete unit. It is equipped with a set of isolating contacts (see → Isolation) that allows the unit to be moved into a position which disconnects it from the → main circuit (e.g. from dropper → busbars)

Working voltage

This is the maximum r.m.s. value of a.c. voltage (or highest d.c. voltage) which may exists across any insulating material (e.g. body of a → switching device between terminals) at rated supply voltage under open circuit or normal operating conditions. Transient voltages (e.g. spikes) are not taken into account.

Worst case

The simultaneous occurrence of specific influencing quantities (e.g. ageing, component tolerances, temperature, voltage) is termed the "worst case". It may be used as a basis for design and the selection of components. → BERO proximity switches, for example, are designed for worst case conditions.

Wound-primary current transformer

See → Pin-wound current transformer

Wye-delta starter

See → Star-delta starter

XL starter

An "across-the-line starter". See → Direct-on-line motor starter.

Y-D starter

See → Star-delta starter

z-release

Short designation for a → short-time delayed overcurrent release. See → Time-based discrimination

Zero-point quenching

Whenever an electrical circuit is opened, typically by the moving apart of the → contacts in a → switching device, the current continues to flow via an → arc drawn between the separating conductors. The technique whereby the switching arc of *alternating current* is extinguished as the current passes through the natural zero of its waveform is known as "zero-point quenching" (see → Arc quenching). This is of particular significance in the field of short-circuit protection, since it means that almost the full → prospective short-circuit current (limited only by the additional voltage drop across the arc) flows for up to half a cycle (10 ms in the case of a 50 Hz driving voltage).

ZH

Central body for accident prevention and work safety in Cologne, Germany.

Zone

See → Hazardous location

ZSS

This is the abbreviation for the German "**Z**eitverkürzte **S**elektivitäts-**S**teuerung". See → Short-time grading control

9.13 Addresses of important specification, standard and testing bodies

Standards of the Federal Republic of Germany

DIN-Standards:

Beuth Verlag GmbH
Postfach 1145
Burggrafenstraße 6
D-10787 Berlin

Beuth Publishers offer publications from most of the standards organisations in the world.

VDE-Regulations (DIN VDE):

VDE Verlag GmbH
Bismarckstraße 33
D-12169 Berlin

Information concerning technical regulations:

Deutsches Informationszentrum für
technische Regeln (DITR) im DIN
Postfach 1107
Burggrafenstraße 6
D-10787 Berlin

East European Testing Offices

EZU:

Elektrotechnický Zkubšební Ústav
Pod lisem 129
CR-17102 Prag-Troja

MEEI:

Hungarian Institut for Testing
Electrical Equipment
Váci ut. 48/A–B
H-1132 Budapest

SEP:

Stowarzyszenie Elektryków Polskich
Biuro Badawcze d/s Jakości
ul. Pożaryskiego 28
P-04-703 Warszawa

West European Testing Bodies

N.V. KEMA:

Nederlandse Vereniging tot Keuring van Electrotechniche Materialien
NL-6812 AR Arnheim
The Netherlands
Utrechtsweg 310

ALPHA:

Gesellschaft zur Prüfung und Zertifizierung
von Niederspannungsgeräten e.V.
Stresemannallee 19
D-60596 Frankfurt/M

9 Appendix

Internatioal Standards, National Standards

The international standards of ISO (International Standardization Organization) and IEC (International Electrotechnical Commission) are listed in the catalogues of these organizations.

IEC:

Central Office of the IEC (in Switzerland)
Postfach Nr. 131
3 rue de Varembé
CH-1211 Geneva 20

Office in Germany:

VDE-Verlag GmbH
Bismarckstrasse 33
D-12169 Berlin

AS – Australian Standards:

Standards Association of Australia
Standards House
80-86 Arthur Street
North Sydney, N.S.W. 2060

BS – British Standards:

BSI-Sales Department
Linford Wood
Milton Keynes MK 146 LE

CSA – Canadian Standards:

Canadian Standard Association
178 Rexdale Blvd.
Rexdale (Toronto), Ontario
Canada M9W 1R3

NEMA - National Electrical Manufacturers Association:

NEMA
2101 L Street, N.W.
Washington, D.C. 200037

UL Regulations:

Underwriters Laboratories Inc.
Publications Stock
333 Pfingsten Road
Northbrook IL 60062

Regulations of marine classification bodies:

ABS Americas:

American Bureau of Shipping
ABS Plaza
16855 North Chase Drive Houston, TX 77060

BV:

Bureau Veritas
17bis, Place de reflects La Défense 2
F-92400 Courbevoie

DNV:

Det Norske Veritas
Classification A/S
Postbox 300
N-1322 Hövik

GL:

Germanischer Lloyd
Vorsetzen 32
D-20459 Hamburg

LRS:

Lloyd's Register of Shiping
71 Fenchurch Street
London EC3M 4BS

PRS:

Polski Rejestr Statków
ul. Marynarki Polskiej 59
P.O. Box 445
PL-80-958 Gdańsk 1
Poland

RINa:

Registro Italiano Navale
Via Corsica, 12
P.O. Box 1195
I-16128 Genoa
Italy

MRS:

Maritim Register of Shipping
Leningrad, 191041
8, Dvortsovaya nab.

ZC:

China Classification Society
Hamburg Office
Breitestrasse 159 1/F
D-22767 Hamburg
Germany

Index

Entries marked with a * are also contained in Section 9.12 Glossary – Brief explanations of some technical terms. Refer to page 556 et seq.

Abbreviations 501
Abnormal operating conditions 65, 72
Abrasion* 556
ABS (American Bureau of Shipping) 17
AC-1 to AC-4 operation 119, 197, 198
ACB* 557
Accessibility of switchgear 336
Accept switch* 556
"a" contact* 556
Accompanied fuses 128, 538
Accuracy class* 390, 395, 556
Accuracy of tripping times 134
A.c. magnet system 192
Across-the-line starter* 556
A.c. semiconductor motor controllers and starters 223
A.c. switchgear in d.c. networks 192
Active power* 363, 556
Actuation speed 269
Actuating
– angle* 556
– distance* 556
– element* 269, 556
Actuator* 556
Actuator
– travel 270
– head* 556
Actuator Sensor Interface (ASI)* 557
Adhesive labels (switchgear symbols) 496
AF* 557
– blocking circuit 378
– operation 324
– ripple control* 377, 557
– trap* 378, 557
Ageing (of fuse links)* 127, 557
Air circuit-breaker (ACB)* 557
Air circulation 353
Air
– gap* 557
– pollution 69

– temperature 358
Air-break contactor* 557
Air-conditioning 351
Alarm switch* 557
ALPHA* 557
Alternating
– current* 558
– voltage* 558
Altitude 65
Aluminium busbars 528
Ambient conditions 64, 308
Ambient temperature* 65, 558
Ambient temperature compensation 135
American Bureau of Shipping 17
American Wire Gauge (number)* 511, 558
Ammonia 69, 72
aM fuses 128, 538
Ampacity* (see current carrying capacity) 558
Ampere turns* 558
Analog output 288
AN operation 324
an-release* 558
AND-gate* 408, 558
Angular roller lever 268
ANSI 16, 22
Anti condensation heaters* 357, 363, 558
Anti-pumping device* 558
Antivalent output* 558
Apparent power* 363, 558
Applied voltage* 559
Approach
– angle* 559
– speed* 559
Approvals* 21, 559
Arc blow-out
– chamber* 559
– chute* 559
– extinction* 559
– gap 178
– quenching* 559
– space* 275, 559

– splitters* 559
– voltage* 110, 560
Arc-quenching medium* 560
Arcing* 560
– contact* 560
– time* 560, 637
Area (units of area) 504
a-release* 122, 556
Armature* 560
Arrangement of actuators 278
AS (Australian Standard) 16, 22
ASI* 560
ASIST 497
Askarel* 560
Assured operating distance* 560
ASTA* 560
Asymmetric short-circuit current* 560
Asymmetrical loading 134, 137
Asymmetry (in %)* 560
Atmospheric conditions 65
ATSE* 560
Audio-frequency
– blocking circuit* 378, 560
– remote control* 377, 561
AUSTER* 497, 561
Automatic
– change-over (load transfer) 492
– power factor correction* 561
– (rapid) recharging* 382, 561
– reset feature* 132, 561
– restarting 489
– sequential starting 418
– shutters 318
– stator resistance starter 221
– transfer switching equipment (ATSE)* 561
Automation 297
Autotransformer* 561
– motor starter* 113, 460, 561
Autotransformers (control) 266
Auxiliary
– circuit* 561
– contact 295, 561
– contact block* 561

645

Index

Auxiliary contactor* 561
– release* 562
– supply voltage* 562
– switch* 562
Availability delay* 562
Average air temperature 358
AWG* 558
Axial approach* 562
azn-release* 562
azng-release* 562

Back-of-hand
– protection 276
– safe area 36
Back-up protection* 141, 173, 562
Ballast units 206, 209
Barrier, Flash barrier* 562
Battery charger 382
Baud rate* 562
"b" contact* 562
BCD (binary coded decimal) 288
Belt monitor 476
BERO* 282, 562
Bimetal
– relay* 131, 562
– release* 562
Bimetallic strip or element* 131, 562
Binary* 563
Bipolar wavefront time 176
Bit* 563
Blind zone* 563
Block diagram 410
Blocking
– circuit* 563
– factor* 379, 563
– range 288
Blow-out
– coil* 563
– space* 563
Boolean algebra (logic) 408
Bonding* 563
Boost charging (of batteries) 382
Bounce duration* 563, 637
Bounce-free switching operation* 563
Bowden cable* 563
Box terminal* 277, 563
Box-type assembly or construction* 304, 563
Break function* 563
Break time* 564
Breakaway pulse* 564

Breakdown voltage* 564
Breaking
– capacity* 564
– capacity of a fuse 126
– current* 564
– operation* 564
Breather gland 334, 362
Bridging time 153
BS (British Standard) 16, 22
BS2000 497
Build-up time* 564
Bunched* 564
Burden* 564
Burst* 564
Bus (data bus)* 564
Bus coupler* 171, 309, 564
Busbar* 310, 565
– adapter 273
– compartment 338
– material 310
– sectionalizing 309
– trunking* 302, 330, 565
Busbar trunking system 8PL 330
Bushing* 565
BV (Bureau Veritas) 17
Bypass contactor 228
Byte* 565

Cabinet* 565
Cable
– current transformer 388
– ducting* 565
– entry* 319, 565
– gland* 565
– resistance 48
– termination 320
– tray* 565
– trunking* 565
Calculation sheet 45, 92
CAD system 411, 497
Cam switch* 565
Cam-operated switch (cam switch)* 565
Campsite distribution board 342
Capacitance of control lines 243
Capacitive
– proximity switch* 283, 565
– shunt load 247
– stray currents 248
Capacitor 163, 195
– bank* 195, 366, 566
– contactor* 196, 452, 566
– switching of ... 114, 196

Capacity (units of capacity) 504
Captive screws 275
Carbon dioxide 72
Cartridge fuse link* 566
Cascade connection* 566
Cascaded circuit-breakers 141
Cast-resin
– current transformer* 388, 566
– power transformer* 323, 566
Castell interlock* 566
CBR* 566
c.d.f.* 99, 566
CEE* 15, 566
CEI 16
CEMA 16
CEN 15
CENELEC* 15, 18, 566
Central processing unit (CPU)* 566
Centralized power factor correction* 365, 566
CFS unit* (see combination fuse switch) 119, 569
Change-over auxiliary contact (Abbr. CO)* 566
Change-over
– current peak 446
– time* 637
– time (delay) 444, 489
– transient (in star-delta starters) 446
Characteristic tripping curves (of circuit-breakers) 539
Charging voltage 383
Chattering 238
Chlorine 69
Choke* 163, 374, 567
– ballast 206
– coils 196
Choked capacitor* 374, 567
Chopping current* 175, 567
Chopping effect (current chopping)* 175, 567
Circuit diagram* 410, 567
Circuit-breaker* 128, 567
– for distribution protection* 567
– for mesh-connected systems* 567
– for motor protection* 567
– for starter assemblies* 568
– terminal* 484, 568
– with mechanical lock-out* 568
Clapper-type armature, contactor* 568
CLASS (trip class)* 137, 568

Class of insulation material* 568
Clearance (in air)* 39, 56, 316, 568
– between open contacts (contact gap)* 568
– between poles* 568
– to earth* 568
Climate
– class 67
– group 67
– proof 65
Climates* 568
Climatic
– conditions 361, 65
– values 514
Closed position* 568
Closed transition
– autotransformer starter 460
– star-delta starter* 112, 446, 568
Closing
– delay* 568
– duration* 568, 637
– operation* 569
– time* 569
CO (change-over)* 569
Code letters 399
Code numbers 399
Coefficient of linear expansion 515
Coercive force* 569
Coil voltage at various system frequencies 191
Coil voltage tolerance* 279, 569
Coincidence factor* 569
Colour coding* 569
Colours
– of indicator lights 62
– of pushbuttons 62
Combination fuse-switch* 119, 569
Command
– device* 569
– signal* 569
Common star point 483
Comparative tracking index (CTI)* 569
Compartment* 569
Compensation of reactive power* 569
Computer programs 496
COM SIKOSTART 232
Condensation 334, 351
Condensors (see capacitors) 163, 195

Conditional short-circuit current* 105, 569
Conducting path* 569
Conductive part* 569
Conductor* 569
– capacitance 243
– cross-section* 570
Conduit* 236, 570
Connected load* 570, 583
Connection chamber 338
Constant torque d.c. drive 376
Constants 503
Construction specifications 26
Consumer installation* 570
Contact
– bounce* 570
– bridge* 570
– element* 570
– erosion* 118, 571
– failure rate* 294, 571
– gap* 571
– life (contact service life)* 189, 192, 197, 280, 571
– material* 571
– movement delay* 571
– piece* 571
– potential* 571
– pressure* 572
– reliability* 233, 293, 572
– resistance* 294, 572
– surface* 572
– travel* 637
– tip* 572
– voltage drop 293
– wear* 572
– welding* 238, 572
Contact(s)* 572
Contactor* 573
– relay* 573
– safety combinations* 254, 414, 573
Continuous operation 98
Continuous (operation) duty (S1)* 259, 573
– with intermittent periodic loading (S6)* 260, 573
– with short-time loading* 573
Control
– circuit* 574
– circuit supply voltage* 574
– contact* 574
– relay* 574
– switch* 574

– transformer* 257, 260, 574
– voltage loss 137
Controlgear* 574
Convection 351
Conventional enclosed thermal current* 574
– free-air thermal current* 106, 574
Conversion factors 508
Conveyor belt monitor* 159, 476, 575
Coolants 534
Cooling 324
– fans 353
– of switchboards 351
– time-constant TK* 575, 634
– times of contactors 201
Coordination
– of insulation* 575
– of overcurrent protection equipment* 575
– of protection devices 517
– with short-circuit protection devices* 575
Coordination Type "1", Type "2" 141, 639
Copper busbars 528
Core-balance current transformer* 183, 575
Corrosion 68, 363
Cos φ* 363, 575
Counter cells* 383, 575
CPU* 576
Crane duty 111
Creep action* 576
Creep-action contacts 270
Creepage current* 576
Creepage distance* 39, 56, 316, 576
Critical breaking current* 576
Critical length of control lines (conductors) 244
Cross-coupling 309
Cross-section* 576
– of PE/PEN conductors 310
CSA 16, 22, 24
c.t.* 385, 576
Cubicle (switchboard cubicle)* 576
– construction 303
– dimensions 314
Current
– break-off 175
– carrying capacity* 28, 159, 516, 522, 576
– chopping 175

647

Current limiting* 107, 126, 576
- limiting circuit-breakers 108
- limiting diagrams of fuses 541
- path (general)* 576
- path (in a switching device)* 576
- sharing (see parallel connection...) 185
- transformer* 385, 482, 577
- transformer for p.f.c. controllers 389
Current-based discrimination* (selectivity) 165, 577
Current-dependent delayed overload relay (or release)* 577
Current-limiting circuit-breaker* 577
Current-time characteristic* 577
Cut-off current 126
Cut-out box* 577
Cycle duration* 577
Cyclic duration factor (c.d.f.)* 99, 577

Dahlander connection* 113, 204, 422, 577
Damping material* 578
Dark-ON function* 578
Data
- bus* 578
- exchange* 578
- transfer* 578
DC or d.c.* 578
- braking 228
- component 80
- drives 376
- economy circuit 414
- magnet system* 192, 578
- power supplies 264
- switching capacity 193
Dead front* 578
Decontaminability 69
Decoupling* 578
Definite-time-delay overcurrent release* 578
Degree
- of damage 141
- of discrimination 127
- of pollution* 578
- of protection 63, 307, 360, 578
- of selectivity 127
Deionization* 578
- chamber* 578
Demand factor* 579

DEMKO (Denmark) 16, 21
Density 515
Dependent manual operation* 579
Dependent power operation* 579
Dependently delayed overcurrent relay, or release* 579
Derating factor 135, 188
Destruction characteristic (or curve)* 579
Detuned resonance circuit* 374, 579
DIAZED* 579
Dielectric strength* 579
Diesel generator 492
Differential
- current protection 183
- current* 579
- travel* 579
- tripping bar 135
Diffuse photoelectric proximity switch* 579
Diffuse-reflective photoelectric proximity switch 286
Digital* 579
- display* 579
- output 288
DIN* 580
- VDE specifications 15, 18
Diode chain 383
Direct
- contact 33
- current component (d.c. component)* 580
- current* 580
- opening action* 580
- point-to-point wiring 331
- reversing of three-phase induction motors 474
- starter* 580
- voltage* 580
Direct-on-line motor starter (d.o.l. starter)* 111, 223, 415, 580
Direction of operation 278
Discharge choke* 580
Disconnected position* 580
Disconnection 106
Disconnector* 119, 580
Discrepancy switch* 581
Discrimination (Discriminative protection, Selectivity)* 127, 163, 581
- and undervoltage protection 171

- between fuses and circuit-breakers 168, 543
- between series connected circuit-breakers 165
- between series connected fuses 164
- current-based 165, 577
- grading diagram 166
- in meshed networks 174
- in radial networks 164
- overload 610
- parallel incoming feeders 170
- short-time grading control 167
- time-based 166
- with leakage-current circuit-breakers 184
- limit* 166, 581
- tables* 172, 544, 581
Disposal of switchgear 74
Disruptive discharge* 581
Distribution
- board* 328, 581
- circuit-breakers* 581
- circuits with or without fuses 148
- systems 306
Domestic distribution boards 340
DNV 17
D.o.l. starter* 223, 581
Door
- coupling (interlocking) 277
- fastenings / locks 318
- interlock* 581
- interlocking (MCC) 315
Double
- insulation* 581
- moving contact bridge* 271, 295, 581
- star connection 204
Double-break contacts* 295, 581
Double-throw switch* 581
Double-wound motor 426
Draw-out circuit-breaker* 582
Draw-out unit 304
Drawing templates 496
Drip proof (see ODP*) 606
Drive* 582
Drop-out
- delay* 234, 582
- delay time relay circuit 490
- delay unit* 414, 582
- time 253, 582
- voltage* 279, 582
Dry-type power capacitors* 582

Index

Dry-type transformer* 582
Duration factor* 582
Dust filters 356
Duty cycle* 582
Duty types* 582
Dynamic
- effects 84
- forces (by short-circuit current) 547
- strength* 582

Early-make NO auxiliary contact* 582
Earth* 582
- fault* 582
- fault with two-pole switching 194
- terminal* 582
Earth-electrode potential* 583
Earth-fault
- current* 583
- proof* 583
- protection* 183, 583
Earth-leakage
- circuit-breaker (e.l.c.b.)* 583
- current* 183, 583
- device (see leakage-current device) 486
- protection 183, 583
Earthing* 583
- of current transformers 385
- switch 323
Earthquake 332
EC guidelines 555
Economy
- circuit 414
- connection* 583
- resistor* 583
- switching 192
ED 202
Eddy-current losses* 187, 583
EEMAC 16
- sizes / NEMA sizes* 583, 604
Effective
- installed load (Total load, Connected load)* 583
- operating distance* 583
- power* 584
- value (of a sinusoidal waveform)* 584
Efficiency* 584
Eight-hour duty* 584
Elastic door fastenings 318
E.l.c.b.* 183, 582

Electric shock* 584
Electrical
- installation* 584
- interlock* 237, 584
- isolation* 584
- operating mechanism 191
- service life* 118, 584
- shock 33
Electromagnet* 584
Electromagnetic
- compatibility (EMC)* 75, 332, 584
- interference (EMI)* 584
- latching* 584
- operating mechanism* 584
- overcurrent relay or release* 585
Electronic
- ballast units 115, 206, 209
- compatibility 291
- motor starter (see SIKOSTART)* 223, 585
- overload relay 136
Electrostatic discharge (ESD) withstandability* 585
ELLE* 119, 201, 280, 497, 585
Embeddable proximity switch* 585
EMC* 332, 585
Emergency power supply 492
Emergency-Off
- apparatus* 58, 585
- switching device* 585
EMI* 585
Emittance values 516
Emitter* 585
Emitter-receiver unit* 585
Enable circuit 255
Encapsulation* 585
Enclosed construction* 585
Enclosure* 585
- for electrical equipment 512
- material 307
- types 361
Endurance* 585
Energized* 585
Energizing command* 585
Energy saving at partial load 228
Energy storage device* 585
ENT* 586
Environmental acceptability 74
Environmental influences* 66, 586
Epoxy resin (transformer) 323
Equal potential 236
Equation symbols 501

Equations 499
Equipping specifications 31
Equivalent circuit diagram 410
Erection specifications 32
ESD* 586
Eutectic 131
Eutectic alloy overload relay* 586
EVU* 586
Explosion
- group* 586
- proof* 586
- proof motors 156
- protection* 586
Explosion-proof current transformers 388
Explosive atmosphere* 586
Exposed conductive part* 586
Extensions (to switchgear assemblies and switchboards) 38, 319
Extra-low voltage* 586
Extraneous light limit* 587
Extreme heavy duty starting 153

Factory wiring 23
Factory-built assembly (FBA)* 587
Fail-safe operation* 587
Failure of the supply voltage 489
Fan starting
- with SIKOSTART 223
- with slip-ring starter 217
Fast transient withstandability* 587
Fault* 587
- arcs 318, 320, 322
- voltage* 587
Fault-arc testing 318
Fault-voltage protection circuit-breaker* 587
FBA* 587
fc-release* 162, 587
Feeder* 587
FELV* 587
Fibre-optic amplifier* 287, 587
Fibre-optic conductor* 587
Field bus* 588
Field wiring 23
Filament lamp* 588
Filter circuits* 163, 374, 588
Finger-touch protection 276
First pole to close / open* 637
Fixed-mounted design 313
Flameproof enclosure "d"* 588
Flange plate* 588

649

Index

Flash barrier (see barrier)* 562
Flashover* 588
Fleeting contact element* 588
Flink* 588
Flip-flop* 588
Floor-standing distribution board 344
Flourescent lamps 115, 206
Flush-mounted proximity switch* 588
Flutter command signal 237, 412
Flying start (time relay) 490
Force (units of force) 506
Forced
– air cooling 324
– convection 515
– opening 270
Fork lever 268
Four speed control 440
Four-pole change-over (load transfer) 494
Four-pole switchgear 186
Four-stage star-delta starter* 448, 588
Fourier analysis* 588
Free zone* 588
Freewheel diode* 192, 253, 589
f-release* 587
Frequency 187, 190
Frequency meter 396
Fritting 233
Front time, bipolar* 589
Front time, unipolar* 589
FU circuit-breaker* 589
Full range fuses 538
Full-load starting 111, 217
Full voltage starter* 589
Fully withdrawable design 314
Function class of a fuse* 128, 538, 589
Function number 399
Functional extra-low voltage (FELV)* 38, 41, 589
Functional unit* 589
Fundamental* 589
Fuse* 125, 589
– base* 589
– element (or fusible element)* 589
– function class* 128, 538, 589
– general purpose 128, 538
– high rupturing capacity (HRC*) 592
– link* 590

– operation class* 128, 538, 608
– partial range 128, 538
– switch disconnector* 590
– switch unit 119
Fuse-combination units* 590
Fuseless distribution 149

Galvanic separation (of input and output) 487
Galvanic separation of contacts 271
GEAFOL transformers* 320, 590
General purpose fuses 128, 538
Generator 492
Generic specifications 31
GL 17
gL/gG fuses 128, 538
Glow lamp* 590
Grading* 590
Grading diagram 166
Graetz connection 373
Graphic symbols 400
g-release* 590
Ground* 590
Ground humidity 351
Ground-fault protection* 590
Group power factor correction* 365, 590

Half-load starting (slip-ring starter) 111, 217
Hand reset* 136, 590
Harmonic
– content* 590
– current* 187, 191, 591
– filter circuits 374
– frequency* 373, 591
– number* 591
– suppression 375
Harmonics* 163, 373, 591
Harmonizing Documents (HD) 15
Hazardous location* 591
Heat content 516
Heat
– exchanger* 353, 591
– extraction 353
– extraction fan 324
– losses* 591
– radiation 515
– transfer coefficient 351
– value* 591
Heater element* 591
Heating control 491

Heating unit* 357, 591
Heavy Duty 234
Heavy-duty starting* 111, 152, 591
High pressure vapour lamp 116, 207
High rupturing capacity fuse (HRC*) 592
High-voltage
– infeed 320
– motor 156
– module 320
– vacuum contactor 177
Hold current* 591
Hold-in power (also holding power)* 592
Hold-in winding* 592
Holding magnet 183
Holmgreen connection (current transformers) 388
Housing or encapsulation* 592
HQI (lamps) 116, 207
HQL (lamps) 116, 207
HRC fuse* 592
Humidity 68
HV module 320
Hydrogen sulphide 69
Hysteresis 592
Hysteresis losses 187

$I^2 \cdot t$
– characteristic (of a circuit-breaker)* 592
– grading* 592
– let-through value 520
– value* 127, 592
ICCB* 592
Identification of busbars 310
IEC* / IEC publications 15, 18, 592
IMC* 592
Impedance voltage 592
Impulse voltage withstandability* 592
Impulse withstand voltage* 592
Incandescent lamp* 115, 592
Inching* 113, 592
Inching duty 199
Incomer* 308, 593
Incomer position 310
Inconsistant loading 102
Increased safety 156
Increased safety "e"* 593
Independent manual operation* 593
Independent power operation* 593

650

Indicator lights   263
Indirect contact   40
Individual power factor correction*   365, 593
Indoor climate   65, 361
Indoor installation*   593
Induced vibration   73, 332
Induction motors (rated currents of ...)   531
Inductive proximity switch*   283, 593
Infeeders (Incomers)   308
Infrared light (IR light)*   593
Inherent delay*   593, 637
Inherently short-circuit proof*   593
Initial symmetrical short-circuit current*   593
Inrush current*   210, 593
Inside temperature   351
Inspection of contacts   280
Installed power   358
Instantaneous electromagnetic over-current relay   139
Instantaneous short-circuit release*   139, 593
Instrument transformer*   594
Insulated Case Circuit-Breaker (ICCB)*   594
Insulated distribution board system 8HP   329
Insulation
– coordination*   594
– class*   594
– group*   594
– monitoring*   594
– rating*   594
– resistance*   594
Interface*   594
– unit*   292, 487, 594
– contactors and relays   234, 292, 487, 594
Interference*   594
– suppression*   594
Interlock (Interlocking device)*   594
Interlocking   237
– electrical*   237, 584
– mechanical*   601
Intermittent periodic duty S3 (or intermittent duty)*   98, 201, 260, 594
Internal resistance   108
International network voltages and frequencies   549

International Standards Organization   15
Interposing current transformer*   386, 595
Intrinsic safety "i"*   595
Inverse-time delayed overload relay (or release)*   595
Ionized gases*   595
IP Degree of protection*   63, 595
IP2X   307
IS   16
Isle width   55, 334
ISO   15
Isolating terminal   482
Isolation
– isolating function*   595
– of a conductive part*   595
– see disconnection   106
– transformer   264
Isolator*   595
IT system (network)*   40, 595

JIS   16
Jogging*   113, 595
Jogging duty   199
Joule integral*   595
Joule value   127
Joy stick*   595

kcmil (MCM)*   595
KEMA   16, 23
Key interlock*   595
Kick start   226, 595
Knife switch*   595
KOALA*   596
Korndörfer starter   460
KUBS   93, 496
KUSA circuit (starter)*   112, 219, 454, 596

Lagging NC auxiliary contact*   596
Laminated core*   596
Lamps   115, 206
Last pole to close / open*   637
Latch*   596
Latched contactor*   596
Latching mechanism*   596
Latching pushbutton*   596
Late-break NC auxiliary contact*   596
Lateral approach*   596
LC – circuit*   596
– oscillating circuit   374
– tuned circuit   163

Lead batteries   383
Lead-lag circuit   115
Leading NO auxiliary contact*   596
Leakage-current device   486
LED*   596
Length (units of length)   504
Let-through
– current*   126, 596
– I^2 value*   596
Light-ON function*   597
Limit switch*   266, 597
Limiting
– of short-circuit current   97
– by length of conductor   546
– value   500
Limits of operation*   597
Line
– capacitance   243
– protection circuit-breakers   130
– starter*   597
Line-end coil   177
Line-up terminals   482
Linear
– distribution   302
– load*   597
– speed curve   224
Listed equipment   23
Live parts*   597
Lloyds Register of Shipping   17
Load*   597
– cycle*   597
– duration*   597
– factor*   597
– shedding*   597
– switch*   119, 597
– transfer (switch-over)   416, 492
– transfer switch*   597
Load-carrying capacity   192
Load-centre substation   320, 326
Local
– bus*   597
– control*   597
– service box*   598
Locked-rotor current*   598
Logic functions   408
Long control lines (conductors)   242
Loop impedance   30, 44
LOVAG*   598
Low voltage*   598
– switchgear and controlgear assembly*   598
– control transformer   257
– vacuum contactor   182

651

Low-pass filters* 598
Luminaires 206
LV HRC fuse* 598
Lyre contact* 598

Magnet
– armature* 598
– core* 599
– drive* 599
– system* 599
Magnetic
– blowout* 599
– circuit* 599
– field withstandability* 599
– operating mechanism 191, 194, 195
– operator 279
Magnetically latching contactor* 599
Main circuit
– contact* 599
– of a switching device* 599
– switch* 58, 599
Maintained-contact command device* 599
Maintained-contact control* 600
Maintenance 279
Maintenance switch 60
Make function* 600
Make time* 600, 637
Make-before-break* 600
Make-break operation* 548, 600
Making capacity* 600
Making operation* 600
Malfunction* 600
Manual reset feature* 132, 600
Manual/automatic reset 136
Manufacturers' Certificate 35
Maritime Register of Shipping 17
Mark A, Mark B detectors, tripping units 125
Marking of busbars 310
Mass (units of mass) 504
Master controller* 600
Material group* 600
Maximum
– demand* 369, 600
– operating distance* 601
– power loss 358
– prospective peak current* 601
MCB or m.c.b.* 130, 601
MCC* 315, 340, 601
– door interlock 315

– handle mechanism* 601
MCCB* 601
MCP (Motor Circuit Protector) 143
Measurement core current transformers 395
Measuring instruments 396
Measuring instrument connections 482
Mechanical
– delay* 601
– interlock* 601
– latching* 234, 601
– reclosing lock-out 129
– service life* 118, 601
Melting point 515
Melting time* 601
Mesh-connected systems* 160, 601
Meshed
– network* 160, 601
– release 130
– shunt trip 162
Metal-halide lamps 116, 207
Metcalf Effect (or "M" effect)* 602
Meter cubicles / cabinets 348
Method of cooling 534
Micro-environment* 602
Miniature circuit-breaker (MCB)* 130, 602
Minimum
– command time (or duration)* 602, 637
– current and voltage 293
– operating distance* 602
– operational current* 602
– short-circuit current 49
MINIZED* 602
Mixed operation 199
MKK dry-type power capacitor* 602
MKV power capacitor* 602
Modification of switchgear assemblies 38
Modular assemblies 314
Modular width* 343, 602
Modulated light beam* 603
Momentary-contact command device* 603
Momentary-contact control* 603
Motor
– circuit-breaker* 603
– Control Centre (MCC)* 315, 340, 603

– drive* 603
– power ratings 531
– protection circuit-breaker* 603
– starter* 603
– switch* 603
– switching 107
– switching capacity 188
Motor-driven time relay 490
Motorized operating mechanism 192, 279
Motorized time relay (see motor-driven time relay) 490
Moulded case circuit-breaker (MCCB)* 603
Mounting
– and wiring system 331
– of an inductive or capacitive proximity switch* 603
– position* 274, 603
Movement delay (contact movement delay)* 603, 637
Moving armature 109
Moving-coil instrument 396
Moving-iron instrument 396
MRS (Maritime Register of Shipping) 17
MSP* 604
Multiple breaking* 604
Multiple restriking 176, 178
Multiplexer 288

NAMUR (proximity switches)* 283, 604
National standards 16
Natural air cooling 324
Natural frequency* 604
NAV (lamps) 116, 207
NBN 16
NBS* 604
NC* 604
NEC 22
Negative temperature coefficient thermistors (NTC) 139, 468
NEMA 16, 22
NEMA sizes / EEMAC sizes* 604
NEMKO (Norway) 16, 21
NEN 16
NEOZED* 604
Network
– circuit-breaker* 604
– data 76
– frequency 187
– master relay (release)* 162, 605

Network types and systems* 605
– voltage and frequency 549
Neutral conductor (N)* 43, 605
NFPA 32
Nickel cadmium battery (NiCd) 383
Nitric oxide 69
NO* 605
No-load
– condition* 605
– current* 605
– power* 605
– reactive power 368
– supply current 605
– voltage* 605
No-volt release* 605
Node point fuse 174
Noise* 605
Nominal
– frequency 76
– value* 500, 606
– voltage 76
Non-damping material* 606
Non-embeddable proximity switch* 606
Non-linear load* 606
NORIS 498
Normal conditions 65
Normal Duty 234
Normally closed contact (NC)* 606
Normally open contact (NO)* 606
Norms 15
NPN output* 606
n-release* 122, 604
N System* 604
NTC thermistor (temperature detector)* 139, 468, 606
N-type* 604
Number of starts (slip-ring starter) 214
Number of switching operations 548
Nusselt number 515

Obligatory approvals 21
ODP* (open drip proof motor) 606
Off-delay* 637, 606
Off-delay unit 412
Off-load
– condition* 606
– disconnector (isolator)* 606
– period 202
– switching 107

OFF-state current* 606
Ohms law 499
Oil-cooled starter (slip-ring starter) 211
On delay* 637
One second current 129
On-load
– condition* 606
– disconnector (isolator)* 119, 607
– factor (or load factor)* 202, 607
– period 202
– switching 107
Open circuit shunt trip (see shunt release) 130, 627
Open position* 607
Open-air climate 65, 361
Open-air installation 334
Open-circuited current transformer 393
Open-frame switchboard 302
Open-type construction (or assembly)* 607
Opening
– delay* 607
– operation* 607
– time* 607
Operating
– conditions 64, 234, 237
– current* 607
– cycle* 607
– cycle time 203
– distance (of a BERO)* 284, 607
– frequency* 607
– mechanism auxiliary switch* 608
– mechanism with detachable coupling* 608
– mechanism* 607
– speed* 608
– time* 608
– voltage* 78, 279, 608
– voltage tolerance* 292, 608
Operation of a switching device* 608
Operation class* 128, 538, 608
Operational
– current* 106, 609
– duty* 98, 609
– reliability* 609
– switching* 609
– voltage* 609
Operator* 609
Optical proximity switch* 609

Opto-BERO* 286, 609
OR-gate* 408, 609
Outdoor
– climate 65, 361
– cubicle 350
– installation* 609
Outgoing feeder* 309, 609
Output power of a current transformer 391
ÖVE 16
Overburdened current transformer 392
Overcurrent* 122, 609
– factor (of current a transformer) 391, 395
– protection 516
– relay or release* 610
Overlap time* 637
Overlapping auxiliary contact* 610
Overload* 610
– causes of 150
– current* 610
– discrimination* 610
– protection 517
– protection (of insulated cables) 28
– relay or release* 129, 131, 610
– relay for heavy duty starting 152
– starting (slip-ring starter) 217
Overtravel absorber 269
Overtravel plunger 268
Overview diagram 410
Overvoltage
– category* 610
– factor* 175, 610
– limiter 178, 182
– protection 175
– spike* 175, 178, 182, 610
– suppression 249, 292
– transients 174, 182
– value* 610
Ozone 69

PAM connection 204
Panel construction 303
Panelboard* 610
Parallel
– connected transformers 161
– connection of current paths 185
– feeders 169
– resonance circuits 378

Index

Partial
- choking 378
- load optimization* 228, 610
- range fuses 128, 538
Partially type-tested assembly
  (PTTA)* 26, 611
Partitioning* 611
Pb battery 383
PCB* 611
P.c. interface 232
Peak withstand current* 129, 611
PEHLA 318, 320, 322
PEN conductor* 43, 611
Performance test* 611
Permanent magnet* 611
Permissible
- length of control lines 243
- load characteristic curve 154
- loading of contactors 201
- temperature limit 122
- voltage drop (across contacs) 293
P.f.c.* 611
Pg adapters 236
Pg thread* 611
Phase angle* 363, 611
Phase angle control* 611
Phase-loss sensitivity* 611
Photoelectric proximity switch*
  286, 611
Pick-up
- current* 611
- delay* 612
- power* 612
- voltage* 612
Piezo-ceramic transducer 287
Pilot switch* 612
Pin-wound current transformer*
  612
Plastic 72
PLC* 612
Plug braking (see reverse-current
  braking) 257, 470, 612
Plug-in circuit-breaker* 612
Plug-in systems 273
Plugging or plug braking* 257, 612
Plunger 268
PNP output* 612
Point distribution 299
Point-to-point wiring 331
Pole* 612
Pole number (of a motor)* 612
Pole-changing induction motors
  204, 113, 152

Poleface* 613
Pollution* 613
Pollution degree* 613
Position
- of incomers 311
- switch* 266, 479, 613
- switch with safety function 271
Positive
- drive (positive action)* 613
- opening operation* 613
- opening* 613
- temperature coefficient thermistors 125, 138, 462
Positively driven contacts* 254, 613
Positively driven operation* 613
Potential-to-earth* 613
Potentially dangerous part* 614
Potentially explosive atmosphere*
  614
Potentiometer* 614
Power* 614
- capacitor* 614
- circuit* 614
- consumption of contactor coils
  245, 261
- consumption of measuring instruments 396
- factor* 614
- factor controller* 614
- factor correction* 363, 615
- factor correction of motors 150,
  367, 452
- factor correction of transformers
  368
- losses 351
- losses of fuses 127
- relay* 615
- transformer 532
- units of power 506
Power-frequency recovery voltage*
  615
Prandtl number 515
Pre-arcing
- current* 126, 615
- $I^2 \cdot t$ value* 615
- time* 615
Pre-charging resistor* 115, 196, 615
Pre-striking 176
Preferred values
- nominal voltage 77
- rated voltage 104
Prefixes 503
Prepared conductor* 615

Pressure
- monitor 159, 478
- relief 326
- units of pressure 504
Pressurized apparatus "p"* 615
PROBAT 498
PROCAT 497
PROFIBUS (process field bus)*
  615
Programmable controller* 615
Project planning 333, 496
PROKOM 497
Prospective
- current* 615, 616
- fault current* 616
- making current* 616
- peak current* 616
- short-circuit current* 79, 616
- symmetrical current* 616
Protection
- against corrosion 68, 363
- against direct contact* 33, 336,
  616
- against earth-leakage and earth-
  fault currents 183
- against electrical shock 34
- against excessive temperature
  rise 123
- against fault arcs 325
- against inadvertent reclosure 133
- against indirect contact* 40, 336,
  616
- against overcurrent 516
- against overload 122, 517
- against overvoltage 175
- against short-circuit 123, 518
- against undervoltage 171
- class* 616
- core current transformers 389,
  395
- devices 122
- devices (coordination of ...) 517
- of capacitors 163
- of conductors and cables 159
- of three-phase induction motors
  150
- of transformers 160
Protective
- conductor (PE)* 43, 616
- cover* 616
- insulation 330
- measures* 42, 307, 616
- separation* 616

654

Index

Proximity switch* 616
- capacitive 283
- inductive 283
- magnetic 268
- photoelectric 286
- ultrasonic 287
PRS 17
PTB 157, 388
PTB certificate* 617
PTC thermistor (temperature detector)* 125, 138, 462, 617
PTTA* 26, 30, 617
Pull-in
- power* 617
- voltage 279
- winding* 617
Pull-through voltage* 617
Pulse (fleeting) contact element* 617
Pulse number* 373, 617
Pulsed light beam* 617
Pulsing d.c. leakage current 184
Pump control* 227, 617
Puncture* 617
Pushbutton* 617
Pushbutton control* 618
PVC cables 49, 519, 520

Quality assurance systems 17

Raceway* 618
Rack-out circuit-breaker* 618
Radial distribution 299
Radial network* 160, 618
Radiated electromagnetic field withstandability* 618
Radiation 72
Radio interference* 74, 618
Radio interference level* 618
RAL* 618
Ramp time* 618
Rapid
- discharge resistor* 618
- multiple restriking* 618
- recharging (of batteries) 382
- restriking 249
Rate-of-rise 175
Rated
- conditional short-circuit current 105
- control voltage* 618
- conventional free-air thermal current 106
- current 106

- current of induction motors 531
- current of three-phase power transformers 535
- frequency 76
- frequency of an item of switchgear* 618
- impulse withstand voltage* 618
- insulation voltage* 619
- operating distance (of a proximity switch)* 619
- operational current* 619
- operational voltage* 619
- output power of a transformer* 619
- output voltage* 619
- overcurrent factor* 619
- peak withstand current 129
- power of a current transformer* 619
- service short-circuit breaking capacity 105
- short-circuit making capacity* 105, 619
- short-circuit switching capacity (circuit-breaker) 128
- short-time withstand current 105, 129
- ultimate short-circuit breaking capacity 105
- uninterrupted current* 106, 619
- value* 500, 619
- voltage 76
Rating* 500, 619
RC circuit* 619
RC element* 178, 182, 249, 619
rc-release* 620
Reactance voltage 85, 620
Reactive
- drop* 620
- power compensation 365
- Reactive power* 620
Reactor 374
Receiver* 620
Reclosing lock-out* 129, 132, 620
Recognized equipment 23
Recording-type instruments 396
Recovery time* 132, 136, 489, 620
Recovery voltage* 620
Rectifier* 263, 620
Rectifier circuits 373
Reduced-voltage starter* 620
Reduction factor (of an inductive proximity switch)* 620

Redundancy 254
Reed contact* 620
Reference axis (of a proximity switch)* 620
Reflection of travelling wave 177, 179
Reflex photoelectric proximity switch* 620
Refrigeration unit 356
Registration 23
Regulation* 239, 621
Relative humidity 362
Relay (electrical)* 621
Release* 621
Release time* 621
Release-free mechanism (or operation)* 621
Remanence* 621
Remanence contactor* 621
Remote control* 621
Remote reset 133
Removed position* 621
Repair switch 60
Repeat accuracy* 621
Replacement of contacts 280
RER (Iceland) 16, 21
Reset point (of a proximity switch)* 621
Reset* 621
Residual
- current* 622
- current of a proximity switch 622
- current protection* 622
- current protective device 486
- current release* 622
- magnetism* 622
- ripple* 622
- voltage of a proximity switch* 622
Resistance
- drop* 622
- of copper and aluminium conductors 529
- of detector circuit 138
- starter* 622
- to mechanical shock* 622
- to vibration* 622
- voltage* 622
Resistive shunt load 247
Resonance* 163, 622
- effects 374
- frequency* 623

655

Response
- characteristic of a proximity switch* 623
- delay* 623
- of an item of switchgear* 623
- point (of a proximity switch)* 623
- time* 623
- value* 623

Retroreflective photoelectric proximity switch 286
Reverse power relay* 623
Reverse-current braking 257, 470
Reverse-current release* 623
Reversing* 623
- circuit 113
- contactor* 623
- direction of rotation 420
- of a 3-phase motor 156
- starter* 223, 420, 623

Reynolds number 515
RINA 17
Ring final network* 624
Ring network* 624
Ring-main feed 321
Ripple content* 624
Ripple control* 377, 624
Rising mains 330
RMS or r.m.s value* 624
Rod actuator 268
Roller crank 268
Roller lever* 268, 624
Roller plunger 268
Root-mean-square value (r.m.s. value)* 624
Rotary handle mechanism* 277, 624
Rotor
- class* 624
- critical motor 143, 155
- resistance starter 210, 458
- standstill voltage 212

Routine test* 624
r-release* 618
RS 17
Run-up
- current* 624
- of a motor* 624
- time (starting time)* 624

Rupturing of a fuse link* 624
Rupturing capacity* 126, 624
Rush
- current* 624

- effect* 624
- factor 210

S1 duty 98, 259
S2 duty 98, 260
S3 duty 99, 260
S4 duty 100
S5 duty 100
S6 duty 100, 260
S7 duty 101
S8 duty 101
S9 duty 102
SABS 16
Safe locked-rotor time* 624
Safety
- circuit* 254, 624
- extra-low voltage (SELV)* 33, 38, 41, 624
- function 271
- isolation transformer 265
- separation* 38, 39, 624

Same polarity 236
Sampling test* 624
SASO 16
Saturation c.t. relay 153
SBS (static bypass switch) 492
SCPD (short-circuit protection device)* 625
Screening* 625
Screwdriver guides 275
Sea air 72
Seal-in power* 625
Seal-in voltage* 625
Sealing-in contact* 625
Section (switchboard section)* 303, 625
Selection slide-rule 496
Selective protection (see discrimination) 163
Selectivity* (see discrimination) 127, 625
Self-healing power capacitor* 625
Self-holding contacts* 625
SELV* 33, 625
Semi-withdrawable design 314
Semiconductor
- contactor* 625
- fuse 231
- motor controller / starter 223, 625

SEMKO (Sweden) 16, 21
SEN 16

Sensing
- face of a proximity switch* 625
- radius* 625
- range of a proximity switch* 288, 625
- zone of a proximity switch* 625

Sensor* 625
Separate actuator 268
Sequence number 399
Sequential starting 231, 418
Series connection of poles 186
Service
- factor 124, 625
- life* 117, 189, 626
- life of indicator lights 263
- short-circuit breaking capacity* 105, 626
- switch 60

Set of contacts* 626
SETI (Finland) 16, 21
SEV (Switzerland) 16, 21
SF6 gas* 626
Shading ring* 626
Sheet-steel enclosed distribution boards 8HS 328
Sheet-steel enclosure 328
Shielding winding* 626
Short-circuit* 626
- breaking category 105
- calculation 85
- calculation sheet 92
- current* 79, 626
- current of transformers 312, 535
- lock-out* 626
- making capacity 105
- proof* 109, 627
- protection 518
- protection of control transformers 237
- protection of insulated cables 28
- protection in auxiliary circuits 236
- protective device (SCPD)* 626
- release* 129, 626
- strength* 335, 626
- switching capacity (circuit-breaker) 128
- withstand capability 306
- ring of a magnet core* 627

Short-time
- current* 627
- delay short-circuit release* 627
- duty 199

Index

Short time grading control* 167, 627
– operation 98
– operation duty (S2)* 260, 627
– power rating of control transformers 262
– withstand current* 105, 129, 627
Shower discharge 249
Shunt load 247
Shunt release* 130, 627
Shunt trip* 130, 627
Shutters* 318, 627
SI units 503
SICONTROL 297
SIEBKO 369, 497
Signal evaluation unit 287
Signal processing systems 282
SIGUT terminals* 275, 627
SIKOSTART* 223, 627
SIKUFEST* 627
SIKUS 342, 344
SIMATIC* 297, 628
SIMICONT* 628
SIMOCODE* 628
Single-break contacts* 628
Single-line diagram 410
Single-phasing* 628
– protection* 628
– connection 135
Single-pole loading 134, 137
Single-throw switch* 628
SIPRO 342, 348
SIPUK 8HP; SIPUK-CAD 341, 497
Site altitude* 628
SIVACON* 299, 628
Six-pulse rectification 373
Skin effect* 187, 628
Slide-rule 496
Slip-ring motor 111
Slip-ring motor starter* 210, 458, 628
Slow discharge resistor* 628
Smallest short-circuit current 96
Smooth starting and stopping* 628
Snap action* 279, 628
Snap-action contacts 269
Snap-mounting 273
Sodium chloride 72
Soft
– iron core* 628
– starter (see SIKOSTART) 223, 629

– stopping* 227, 629
Software 496
Solenoid* 629
Solid-state motor starter* 223, 629
Solid-state overcurrent relay or release* 629
Sonar-BERO* 287, 629
Sonar sensor 287
SONPROG* 291, 629
Sound cone 288
Space requirements (for switchgear) 337
Spacing unit* 629
Sparkover* 629
Specific resistance 93, 514
Specific thermal capacity 515
Specifications 15
Speed changing of induction motors 422
Speed monitor 159, 470
Spike
– suppression 178, 182
– suppressor* 629
– voltage 175, 249
Split batteries 384
Split-core current transformer 388
Splitter plates* 629
Spring rod actuator 268
Spring-loaded door locks 318
Spur feed 321
Spurious switch-on signal suppression* 629
Square wave voltage 194
Squirrel-cage motor 111
STAB 342, 344
Stand-by power supply 492
Standard
– atmosphere 514
– modular width 343
– motor power ratings 531
– switchboard 308
– target* 629
– wire guage 511
Standards 15
Standby batteries 382
Standstill monitor 159
Star-delta
– reversing starter 450
– starter* 111, 223, 444, 630
– starter with power factor correction 452
– time relay 444, 489
Start-control switch 219

Starter* 630
Starter characteristic value (slip-ring starter) 214
Start impulse* 226, 630
Start-up time (starting time)* 630
Starter switch* 630
Starting
– current* 111, 630
– energy (slip-ring starter) 214
– frequency (slip-ring starter) 214
– inrush current 111
– load factor (slip-ring starter) 213
– of a motor* (Start-up or Run-up) 630
– of a slip-ring motor 210, 458
– of a three-phase induction motor 444
– resistors (slip-ring starter) 211
– time (slip-ring starter) 213
Static bypass switch 492
Stator resistance starter* 210, 454, 630
Stator-critical motor 143
Stefan Boltzmann constant 515
Storage 72
Stored-energy operation* 279, 631
Stray current 245
Striker armature* 631
Stub terminal* 631
Substations 8FA 320
Subtransient short-circuit current 80
Sulphur dioxide 69, 72
Summation current transformer* 183, 386, 486, 631
Sunlight radiation 362
Supply connection regulations* 631
Supply voltage failure 489
Suppression
– of harmonics 375
– of overvoltages 249
– of radio interference 74
– of switching spikes 178, 182
Surge* 631
– arrestor* 631
– suppressor* 631
– voltage 176
– wave 177
Surplus detection indication* 287, 632
Survey diagram 410
SUVA* 254, 632
SWG* 632

657

Switch
- control, ON/OFF switch control* 632
- disconnector* 632
- disconnector with fuses* 632
- mechanism* 632
Switchboards 8PU 313
Switch-fuse* 632
Switch-on rush 210
Switch-over (also see change-over) 416
Switchboard* 306, 632
- section 303
- standardized designs 308
Switchgear* 632
- assembly* 632
- combination 139, 147
Switching
- arc* 632
- arc gases 275
- capacity* 131, 189, 194, 633
- capacity class 131
- chamber* 633
- cycle 105
- device* 633
- element* 633
- frequency* 117, 155, 633
- overvoltage* 249, 633
- spikes 175, 249
- state* 633
- transients 175, 249
Switching of
- capacitors 114, 196
- heating equipment 115
- high-voltage motors 114
- lamps 115, 206
- low voltage / current 292
- low-voltage motors 111
- three-phase induction motors 415
- transformers 117, 210
Symbols 400
System (Network)* 633
System frequency 187
System frequency > 50 Hz 188

Tab connectors 276
TAB* 145, 633
Take-over current* 633
Tap-off box 302
Target* 633
TE-time* 633
Technical terms 556

Temperature
- class* 634
- compensation* 135, 634
- detector (or sensor)* 634
- inside switchboards 351
- limit 122, 527
- rise 122, 337
- units of temperature 504
Temperature-rise
- characteristic 99
- limit* 634
time-constant* 634
Templates 496
Temporary duty* 634
Terminal* 634
- block* 482, 634
- designations 397
- markings 397
- voltage* 634
Terminal-type circuit-breaker
- see circuit-breaker terminals 484, 568
Termination 275
Termination chamber 338
Termites 74
Test
- button (overload relay) 134
- current 123, 124
- finger 36, 276
- reset button (overload relay) 136
- sequence* 634
Testing of low-voltage switchgear and control gear* 635
Thermal
- conductivity 515
- current 106
- derating 338
- detector 125, 138
- equilibrium 99
- overload relay or release* 635
- overloading of motors 150
- ratings of busbars and device terminals 527
- short-circuit rating* 635
- stress limit* 635
- time-constant* 635
- value* 635
Thermistor 125, 138
Thermistor motor protection* 138, 143, 158, 462, 635
Thermistor protection in control transformers 237
Thermostat control 491

Thread-through current transformers 388
Three poles in parallel 185
Three speed control 430
Three-phase capacitors 195
Three-phase transformers 210, 532, 535
Three-pole change-over (load transfer) 492
Three-pole stator resistance starter 220, 456
Through-beam photoelectric proximity switch* 286, 635
Thyristor malfunction 227
Time
- addition 490
- constant* 80, 636
- delay before availability (of a proximity switch)* 635
- grading* 635
- relay* 635
- switch* 635
- to contact movement* 635
Time-based discrimination (selectivity)* 166, 636
Time-current characteristic* 136, 636
- of a fuse 125, 536
- of an overload relay or release 134
Time-related terms in switching devices* 636
TN, TN-C, TS-S system (network)* 41, 636
TNF 125, 138
Toggle, Toggle lever* 636
Torque class* 229, 636
Torque-speed characteristic 224
Total break time* 636
Total closing time* 636
Total $I^2 \cdot t$ value* 636
Total insulation* 53, 330, 337, 636
Total load* 583, 636
Totally enclosed fan cooled (see TEFC*) 633
Touch voltage* 636
Tracking* 636
Träg* 636
Trailing angle* 269, 637
Transducer 287
Transducing sensor 282
Transfer (switch-over) 416
Transfer switch* 637

Index

Transformer 160, 210
– cubicles 324
– for indicator lights 263
– for rectifier operation 263
– load-centre substation 320, 326
– types 260
Transient
– frequency 175
– overvoltage* 637
– phenomena 79
– recovery voltage (TRV)* 637
– switching current 175
Transition contactor* 446, 638
Transition resistor 446
Transport 72
Transport unit* 638
Travelling wave 177
Trickle charging (of batteries)* 382, 638
Trip class* 638
Trip-free operation* 638
Trip-free switching device* 638
Tripped position* 638
Tripping
– characteristic* 134, 136, 638
– current 130
– current, earth leakage (elcb) 183
– current, earth fault (g)* 638
– current, instantaneous (n)* 638
– current, overload (a)* 638
– current, short-time delayed (z)* 638
– curves* 638
– operation* 638
– response of releases and relays 189
– time* 123, 134, 637, 639
– unit 138
TRV (see transient recovery voltage*) 637
TT system (network)* 40, 639
TTA* 26, 30, 299, 639
Tuned RC circuit 179
Tuned resonance circuit* 375, 639
Two poles in parallel 185
Two range current transformer 390
Two speed control 422
Two-stage drop-out 192
Type 1 / Type 2 coordination* 141, 639
Type 1, Type 2 overload relay (release) 123, 129
Type A PTC thermistor 138

Type
– of enclosure 513
– of output (of a proximity switch)* 639
– of protection* 639
– test* 639
Type-tested assembly (TTA)* 26, 299, 639

UL 16, 22, 24
Ultimate short-circuit breaking capacity* 105, 639
Ultrasonic proximity switch* 287, 639
Underburdened current transformer 391
Undervoltage protection 171
Undervoltage relay or release* 130, 639
Uninterruptable power supply (UPS) 492
Uninterrupted current 106
Uninterrupted duty* 640
Unipolar wavefront time 175
Units of electricity 499
Unprepared conductor* 640
UPS (uninterruptable power supply) 492
Usable operating distance (of a proximity switch)* 640
USSR 17
UTE 16
Utilization category* 119, 197, 234, 640
Utilization class of a fuse* 640
UVR* 640
UVV 34

Vacuum circuit-breaker* 182, 640
Vacuum contactor* 114, 177, 182, 640
VAr controller* 640
Variable speed drive 224
Varistor* 178, 253, 640
VBG* 34, 37, 640
VDE (Verband Deutscher Elektrotechniker)* 640
VDEW 379
Vector groups of transformers 536
Vectors 372
Velocity (units of velocity) 504
Ventilation 353
Vibration 73, 332

V.I.K: (Vereinigung Industrielle Kraftwirtschaft) 158
Virtual current chopping* 176, 640
Visual inspection 280
Voltage
– arrestor 178
– dip 237, 489
– drop* 238, 242, 641
– drop across contacs 293
– drop across a proximity switch* 641
– limiting 227
– ramp 226
– rise* 226, 641
– rise owing to capacitive load 372
– trip* 641
Voltage-dependent resistors 253
Volts per winding 178
Volume (units of volume) 504

Wall box 343
Water hammer* 641
Wavefront time 175
Welding of contacts 238
Wide range current transformer 390
Window-type current transformer* 385, 641
Wiping contacts* 294, 641
Withdrawable
– circuit-breaker* 318, 641
– construction 304
– design 274, 314
– unit* 641
Working voltage* 641
Worst case* 641
Wound-primary current transformer* 385, 641
Wye-delta starter* 641

XL starter* 641

Y-D starter* 642

ZC 17
Zero-point quenching* 107, 642
Zero-voltage latch (on motor-driven time relay) 490
ZH* 254, 642
Zinc oxide varistor 178
Zone* 642
z-release* 122, 642
ZSS* 167, 642

659